PRODUCTION AND NEUTRALIZATION OF NEGATIVE IONS AND BEAMS
FOURTH INTERNATIONAL SYMPOSIUM

AMERICAN INSTITUTE OF PHYSICS
CONFERENCE PROCEEDINGS NO. 158
NEW YORK 1987

PARTICLES AND FIELDS SERIES 35

SERIES EDITOR: RITA G. LERNER

PRODUCTION AND NEUTRALIZATION OF NEGATIVE IONS AND BEAMS

FOURTH INTERNATIONAL SYMPOSIUM

BROOKHAVEN, NY 1986

EDITOR:
JAMES G. ALESSI
BROOKHAVEN NATIONAL LABORATORY

Authorization to photocopy items for internal or personal use, beyond the free copying permitted under the 1978 US Copyright Law (see statement below), is granted by the American Institute of Physics for users registered with the Copyright Clearance Center (CCC) Transactional Reporting Service, provided that the base fee of $3.00 per copy is paid directly to CCC, 27 Congress St., Salem, MA 01970. For those organizations that have been granted a photocopy license by CCC, a separate system of payment has been arranged. The fee code for users of the Transactional Reporting Service is: 0094-243X/87 $3.00.

Copyright 1987 American Institute of Physics

Individual readers of this volume and non-profit libraries, acting for them, are permitted to make fair use of the material in it, such as copying an article for use in teaching or research. Permission is granted to quote from this volume in scientific work with the customary acknowledgment of the source. To reprint a figure, table or other excerpt requires the consent of one of the original authors and notification to AIP. Republication or systematic or multiple reproduction of any material in this volume is permitted only under license from AIP. Address inquiries to Series Editor, AIP Conference Proceedings, AIP, 335 E. 45th St., New York, NY 10017.

L.C. Catalog Card No. 87-71695
ISBN 0-88318-358-7
DOE CONF-8610115

Printed in the United States of America

CONTENTS

FUNDAMENTAL PROCESSES

Review of Progress in the Theory of Volume Production 2
    J. R. Hiskes

Electron Energy Distributions in Magnetic Multicusp Hydrogen Discharges   16
    W. F. Bailey and R. G. Jones

Electron-Ion Collision Processes Relevant to H^- Ion Sources 26
    J. B. A. Mitchell

Electron Excitation of $H_2(v'')$ Levels to Yield Vibrationally Excited H_2
Molecules 34
    J. R. Hiskes

The Role of H_3^+ Ions in $H_2(v \geqslant 6)$ Production in Negative Hydrogen Ion Sources   39
    J. B. A. Mitchell and W. G. Graham

Modeling of Electron Energy Distribution Functions and Vibrational
Distributions in Volume H^- Ion Sources 48
    J. Bretagne, M. Capitelli, and C. Gorse

Mutual Neutralization—Three Body Effects 59
    J. M. Wadehra

Present Status of the Measured Dissociative Attachment Cross Sections 69
    S. K. Srivastava

Electron Stimulated Desorption Negative Ions from Condensed Molecules 83
    R. Azria and L. Sanche

Energy and Angular Distribution of Electrons Detached in H^-–He Collisions
(50 eV–2 keV) 91
    F. Penent, J. P. Grouard, R. I. Hall, and J. L. Montmagnon

Computer Simulations of Particle-Surface Dynamics 97
    A. M. Karo, J. R. Hiskes, and T. M. DeBoni

Diffusion and Free Flow Through a Magnetic Filter 106
    R. G. Jones and W. F. Bailey

Comments on H^- Volume Production in Cs-Seeded Ion Sources 113
    J. R. Peterson

SOURCE DIAGNOSTICS

Diagnostic Techniques for Negative Ion Sources 120
    M. Bacal

Exploration of a Hydrogen Discharge Using Resonant Multiphoton Ionization   133
    J. H. M. Bonnie, P. J. Eenshuistra, and H. J. Hopman

Visible and V.U.V. Emission Measurements in a Tandem Multicusp Ion Source   145
    W. G. Graham and M. B. Hopkins

Spectroscopic Measurements of Atomic Species in Volume Sources 154
    M. P. S. Nightingale and M. J. Forrest

Spatially and Temporally Resolved EEDF Measurements in a Hydrogen
Discharge 166
    M. B. Hopkins and W. G. Graham

Spectroscopic Investigation of H^- and D^- Ion Source Plasmas 181
    H. V. Smith, Jr., P. Allison, and R. Keller

Discharge Characteristics of a Plasma Generator for SITEX and VITEX Ion Sources 194
    C. C. Tsai, W. K. Dagenhart, W. L. Stirling, G. C. Barber,
    H. H. Haselton, P. M. Ryan, D. E. Schechter, J. H. Whealton,
    and J. J. Donaghy

H^- SOURCES

A Model for H^- Volume Production Ion Sources 208
    T. S. Green, A. J. T. Holmes, and M. P. S. Nightingale

A Numerical Model of the Mirror Electron Cyclotron Resonance (MECR) Source 219
    G. Hellblom

Analysis and Interpretation of a High Density Tandem Negative Ion Source   231
    J. R. Hiskes, A. F. Lietzke, and C. Hauck

Volume Production of H^- Ions at Ecole Polytechnique. A Method for Extracting Volume Produced Negative Ions 246
    M. Bacal, J. Bruneteau, P. Devynck and F. Hillion

H^- Ion Source Scaling Studies at LBL 259
    A. F. Lietzke and C. A. Hauck

Operation of a Magnetically Filtered Multicusp Volume Source 271
    R. R. Stevens, Jr., R. L. York, K. N. Leung,
    and K. W. Ehlers

Characteristics of a Small Multicusp H^- Source 282
    K. W. Ehlers, K. N. Leung, R. V. Pyle, and W. B. Kunkel

Extraction of Negative Ion in Reflex Type Sheet Plasma Negative Ion Source .. 289
    T. Kuroda, K. Sakurai, Y. Oka, and O. Kaneko

Production and Formation of Intense H^- Beams 298
    R. McAdams, A. J. T. Holmes, M. P. S. Nightingale,
    L. M. Lea, M. D. Hinton, A. F. Newman, and T. S. Green

A High Current Volume H^- Ion Source with Multi-Aperture Extractor 309
    Y. Okumura, H. Horiike, T. Inoue, T. Kurashima, S. Matsuda,
    Y. Ohara, and S. Tanaka

Volume Produced H^-, D^- Ion Source for Proton Accelerator and Thermo-Nuclear Fusion Research by Sheet Plasma 319
    J. Uramoto

Production of Negative Ions by Double Charge Exchange and Their Acceleration 334
    N. N. Semashko, V. V. Kuznetsov, A. I. Krylov, and P. S. Firsov

Extraction of H^- Ions from a DC Cusp Source 346
    D. H. Yuan, R. Baartman, K. R. Kendall, M. McDonald,
    D. R. Mosscrop, and P. W. Schmor

Operation of a Dudnikov Type Penning Source with LaB_6 Cathodes 356
    K. N. Leung, G. J. DeVries, K. W. Ehlers, L. T. Jackson,
    J. W. Stearns, M. D. Williams, M. G. McHarg, D. P. Ball,
    W. T. Lewis, and P. W. Allison

Accelerated Beam Experiments with the ORNL SITEX and VITEX H^-/D^- Sources 366
    W. K. Dagenhart, C. C. Tsai, W. L. Stirling, P. M. Ryan,
    D. E. Schechter, J. H. Whealton, and J. J. Donaghy

The Cusp H^- Ion Source at KEK 378
    Y. Mori, A. Takagi, K. Ikegami, and S. Fukumoto

A Surface Conversion Source of H^- with Hot Walls and Variable Converter Temperature 384
    O. F. Hagena and P. R. W. Henkes

Surface–Plasma Source of H^- Ions 395
    G. E. Derevyankin and V. G. Dudnikov

High Brightness Ion Source: The Penning Ringatron 404
    J. H. Whealton, W. L. Stirling, P. M. Ryan, M. A. Akerman,
    W. R. Becraft, W. L. Gardner, H. H. Haselton, K. E. Rothe,
    M. A. Bell, R. J. Raridon, D. E. Schechter, L. A. Berry,
    P. L. Goranson, J. T. Greer, and W. K. Dagenhart

A Circular Aperture Magnetron for Injection into an RFQ 419
    J. G. Alessi

Operation of the Fermilab H^- Magnetron Source 425
    C. W. Schmidt and C. D. Curtis

H^- PRODUCTION AND BEAM FORMATION

Surface Production of Negative Hydrogen Ions by Hydrogen and Cesium Ion Bombardment 432
    M. Seidl, W. E. Carr, J. L. Lopes, S.T. Melnychuk,
    and G. S. Tompa

Relative Importance of Reflection and Desorption in Surface H^- Sources (Discussion) 447

Production of Intense H^-–Ion Beams in High-Power Pulsed Diodes 453
    H. J. Doucet

Some Comments on Emittance of H^- Ion Beams 465
    P. Allison

Review of Computer Modeling of Negative Ion Beam Formation 482
    J. H. Whealton

Transverse-Field Focussing Beam Transport Experiment 507
    J. W. Kwan, G. D. Ackerman, O. A. Anderson, C. F. Chan,
    W. S. Cooper, L. Soroka, and W. F. Steele

PANEL: TRANSPORT OF NEGATIVE ION BEAMS

Panel Session: Transport of Negative Ion Beams 520
    P. Allison, A. J. T. Holmes, A. F. Lietzke,
    J. H. Whealton, and L. Wright

PRODUCTION OF OTHER NEGATIVE IONS

Fundamental Processes in Low Energy Collisions of Alkali Anions and Atoms 536
    R. L. Champion, L. D. Doverspike, D. M. Scott, and Y. Wang

Dissociative Attachment to Lithium Dimers 547
    J. M. Wadehra
Negative Ion Formation in Lithium Atom Collisions 555
    H. H. Michels and J. A. Montgomery, Jr.
Density Scaling of an Optically Pumped Lithium Negative Ion Source 567
    M. W. McGeoch and R. E. Schlier
Production of a Cooled, Polarized Atomic Hydrogen Beam 573
    P. A. Schmelzbach, W. Gruebler, D. Singy, and Z. W. Zhang
Polarized H^- Source Development at BNL 585
    A. Kponou, J. G. Alessi, A. Hershcovitch, and T.O. Niinikoski
Optical Pumping of Na for Use in Polarized H^- Ion Sources 593
    L. W. Anderson, D. R. Swenson, and D. Tupa
Recent Progress on the Optically Pumped Polarized H^- Ion Source at KEK   605
    Y. Mori, A. Takagi, K. Ikegami, S. Fukumoto, A. Ueno,
    C. D. P. Levy, and P. W. Schmor
Status of the TRIUMF Optically Pumped Polarized H^- Ion Source 610
    M. Law, C. D. P. Levy, M. McDonald, P. W. Schmor, and J. Uegaki

NEUTRALIZATION

Neutralizer Options for High Energy H^- Beams 618
    J. H. Fink
A High-Charge-State Plasma Neutralizer for an Energetic H^- Beam 631
    A. S. Schlachter, K. N. Leung, J. W. Stearns, and R. E. Olson
Measurement of Plasma Production and Neutralization in Gas Neutralizers   643
    D. Maor, M. Meron, B. M. Johnson, K. W. Jones, A. Agagu,
    and B.-L. Hu
Characteristics of an RF Plasma Neutralizer 651
    J. R. Trow and K. G. Moses
Oscillations in Photodetachment Cross Sections for Ions in Magnetic Fields   663
    O. H. Crawford
A Novel High-Resolution Experimental Approach for Studying Laser
Photodetachment in a Magnetic Field 673
    H. F. Krause
Determination of Neutral Beam Direction from Radiation Emitted by
Photodetached H^- 682
    A. Hershcovitch

SYSTEMS AND APPLICATIONS

Japanese Negative Ion Beam Program 694
    T. Kuroda
Strategic Defense Initiative Organization Neutral Particle Beam Program
Overview 701
    B. Strickland
Magnetic Fusion Interest in Negative Ions 712
    H. S. Staten

Neutral-Beam Current Drive in Tokamaks 717
    R. S. Devoto
Proposed Neutral-Beam Diagnostics for Fast Confined Alpha Particles in a
Burning Plasma 727
    A. S. Schlachter and W. S. Cooper
Concluding Remarks 739
    K. W. Ehlers

APPENDICES

Appendix 1: List of Participants 744
Appendix 2: List of Authors 751
Appendix 3: Symposium Program 753

Foreword

It was an honor for me to get the opportunity to be Chairman and Proceedings Editor for the Fourth International Symposium on the Production and Neutralization of Negative Ions and Beams. Since 1977, this series has tracked the development of negative ion sources (primarily H^-), and the Proceedings have nicely documented the steady progress in the field. While there was concern over the future of this Symposium as funding for the development of intense negative ion beams for fusion decreased, activity in the field has remained high due to the interest in neutral particle beams for the SDI program. There were 93 participants from 11 countries, making this meeting larger than any of the previous three Symposia. In all, there were nine half-day sessions, one of which was a poster session. Topics covered at the meeting included fundamental processes in negative ion production (H^-, Li^-, and polarized H^-); source diagnostics; sources for fusion, SDI, and accelerator applications; neutralizers; and finally, overviews of the fusion and SDI applications of neutral particle beams. The shift this time in the emphasis of papers on fundamental processes from surface production to volume production was quite apparent.

I would first like to thank the sponsors of the Symposium—the U.S. Army Strategic Defense Command, the Air Force Office of Scientific Research, and Brookhaven National Laboratory. The continuing support of this Symposium by the AGS Department at Brookhaven is greatly appreciated. In the past, this Symposium had been organized by Theo Sluyters and Krsto Prelec, and their experience and advice were invaluable to me in trying to maintain the high standards they had set for this series. The Program Committee provided many helpful comments and suggestions. The experience of the Local Organizing Committee also made my job much easier. I would like to thank Ady Hershcovitch for his help as Co-chairman of the Symposium. Marion Heimerle was excellent as Symposium Secretary, and would typically have a job completed before I even knew that it had to be done. I would also like to thank Ron Clipperton, who acted as Symposium Coordinator, and among other things, arranged the highlight of the week—the banquet cruise to the Statue of Liberty. Proceedings Secretary Lisa Chimento deftly handled the difficult job of transcribing all the taped discussions from the week, which are included here after each paper. Her work in preparing the manuscript for publication was superb. Finally, I would like to thank the participants for once again making it a truly enjoyable and productive Symposium.

J. G. Alessi
Brookhaven National Laboratory

Local Organizing Committee

J. G. Alessi (Chairman)
R. L. Clipperton
A. I. Hershcovitch (Co-Chairman)
V. Kovarik
A. Kponou
K. Prelec
Th. Sluyters

International Program Committee

J. G. Alessi	Brookhaven National Laboratory Upton, NY
P. W. Allison	Los Alamos National Laboratory Los Alamos, NM
M. Bacal	Ecole Polytechnique, Palaiseau, France
W. S. Cooper	Lawrence Berkeley Laboratory Berkeley, CA
D. H. Crandall	U.S. Department of Energy Washington, DC
A. J. Holmes	Culham Laboratory Abingdon, England
K. N. Leung	Lawrence Berkeley Laboratory Berkeley, CA
K. Prelec	Brookhaven National Laboratory Upton, NY

Symposium Secretary

Mrs. Marion V. Heimerle

Proceedings Secretary

Mrs. Lisa Chimento

FUNDAMENTAL PROCESSES

REVIEW OF PROGRESS IN THE THEORY OF VOLUME PRODUCTION*

J. R. Hiskes
Lawrence Livermore National Laboratory, University of California
Livermore, CA 94550

I. INTRODUCTION

With the demonstration[1] of large current densities extracted from hydrogen-discharge-type negative ion sources there has been a new emphasis directed toward the further development of these volume-type sources. Along with this emphasis has been a rapid increase in our understanding of the underlying atomic processes that occur in hydrogen-negative-ion discharges, together with a rapid evolution of the geometric configuration of these ion sources. An account of the development of the atomic processes in negative hydrogen discharges has been given in a recent review.[2] Here we shall emphasize these atomic developments as they bear on the tandem high-density ion-source configuration.

II. DISCUSSION

Figure 1 shows a schematic of a tandem negative ion generator in which vibrationally excited molecules are generated in chamber one and dissociative attachment to these molecules to form negative ions occurs in chamber two. The magnetic filter supports a

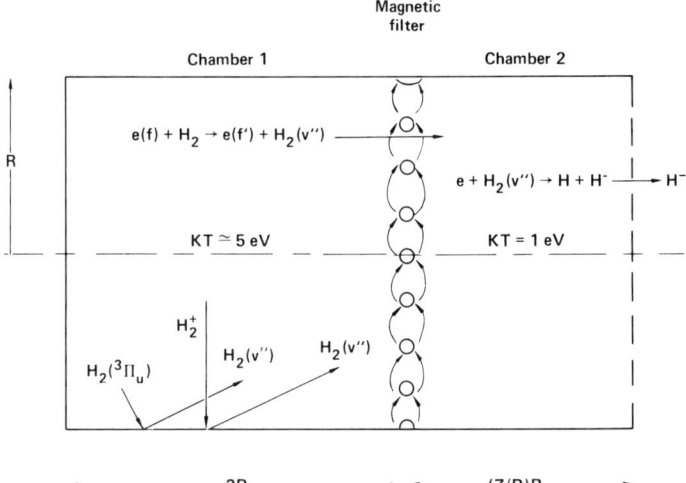

Figure 1. Schematic of a tandem negative ion generator.

---

*Work performed under the auspices of the U.S. Department of Energy by the Lawrence Livermore National Laboratory under contract number W-7405-ENG-48.

© American Institute of Physics 1987

relatively large concentration of fast electrons, E > 25 eV, in
chamber one while maintaining a low temperature, $kT \simeq 1$ eV, thermal
electron bath in the second chamber. In chamber one, three sources
of vibrationally excited molecules, $H_2(v")$, are shown: Auger
relaxation of electronically excited $H_2(^3\Pi_u)$ molecules while
undergoing wall collisions,[3,4] Auger neutralization of H_2^+ molecular
ions during wall collisions,[4,5] and excitation by fast electrons
through the excited electronic singlet states to provide residual
vibrational excitation.[6-9] In different regimes each of these
sources of vibrational excitation can predominate. At low electron
densities, $n < 10^{11}$ cm^{-3}, and for discharge potentials below 30-40
eV the $^3\Pi_u$ relaxation will dominate.[3] For discharges rich in H_2^+
ions and operating at moderately low pressures the H_2 Auger
neutralization can dominate.[10] But for most high density operation,
$n > 10^{12}$ cm^{-3}, the singlet excitations will represent the principal
source of vibrational excitation and it is these that we shall
mainly be concerned with here.

As this system is optimized to provide the maximum extracted
current density, the position along the axis of the second chamber,
z/R, of the current density maximum shifts to smaller values of z/R
as the first chamber and second chamber electron densities, n(1) and
n(2), increase.[10] These increases in electron density increase the
atomic number density and hence attenuate the vibrational
excitation; the negative ions are also directly attenuated by
associative detachment.[11,12] The optimum current density along the
second chamber, as indicated in Fig. 2, is obtained at an axial
position that is approximately equal to a mean free path beyond the
filter against these collision processes. As the electron density
in the two chambers is increased so as to enhance the negative ion
yield, considerable emphasis must be placed on achieving the
shortest possible length for the second chamber.

Leung et al.[13] have observed a variation of extraction current
with second chamber length but at intermediate electron densities,
$n \simeq 10^{11}$ cm^{-3}. These densities are too small to generate a short
mean free path for $H_2(v")$ or H^- attenuation, and there is apparently
an additional cause for second chamber attenuation. This may be
related to a rapid falloff of underlying electron density in moving
through the filter into the second chamber. To achieve the maximum
negative ion current density, the requisite underlying electron
density must be maintained as far as the extraction plane, i.e., up
to the point of optimum negative ion concentration. Until the
present time no source has operated at sufficiently high density to
demonstrate an optimum second chamber length due to the collisional
attenuation processes.

An experiment whose purpose is to rationalize the tandem
concept has been performed[14] by injecting electrons of varying
energy into the second chamber and identifying the different H^-
formation processes by their respective energy dependences. The
relative enhancement of H^- yield as a function of the difference of
plasma potential and filament bias is shown in Fig. 3. The negative
ion yield is seen to increase as the effective electron energy is
decreased. This would seem to preclude the polar dissociation
process, the reaction at the bottom of the figure, since its maximum
rate is above 30 eV, and exhibits a threshold near 17 eV. The rise
from 10 eV to 4 eV could possibly be due to the recombinational

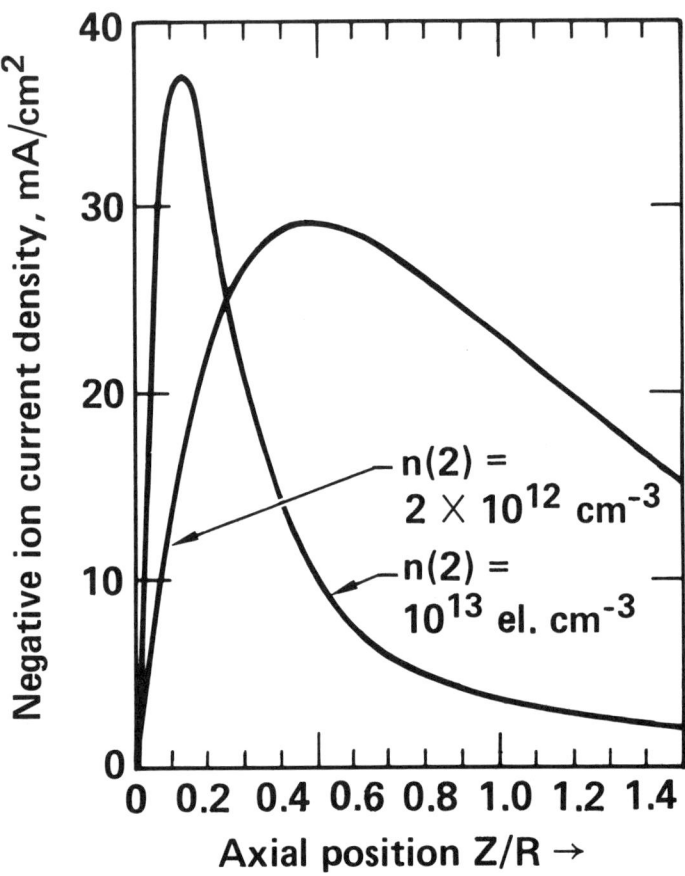

Figure 2. Current density along second chamber.

attachment of electrons to H_3^+ ions,[15] since this process has a maximum cross section in this energy range. The maximum enhancement at the lowest energies, 0.5-1.0 eV, is consistent with either recombinational attachment of H_2 or dissociative attachment of $H_2(v")$.[16,17] If the apparent drop between 1.0 eV and 0.5 eV shown in the figure is not considered to be experimentally significant, the rise at low energies is consistent with H_2^+ recombinational attachment.[18] If the fall from 1.0 to 0.5 eV is taken at face value, the result is consistent only with $H_2(v")$ dissociative attachment.

Since the 1983 BNL Symposium, there has been some clarification of the first chamber optimization. At low densities the principal relaxation process of the vibrational excitation is wall collisions. But as the fast electron density increases, the electron collisional destruction of the vibrationally excited levels becomes the dominant loss of $H_2(v")$, and the active vibrational

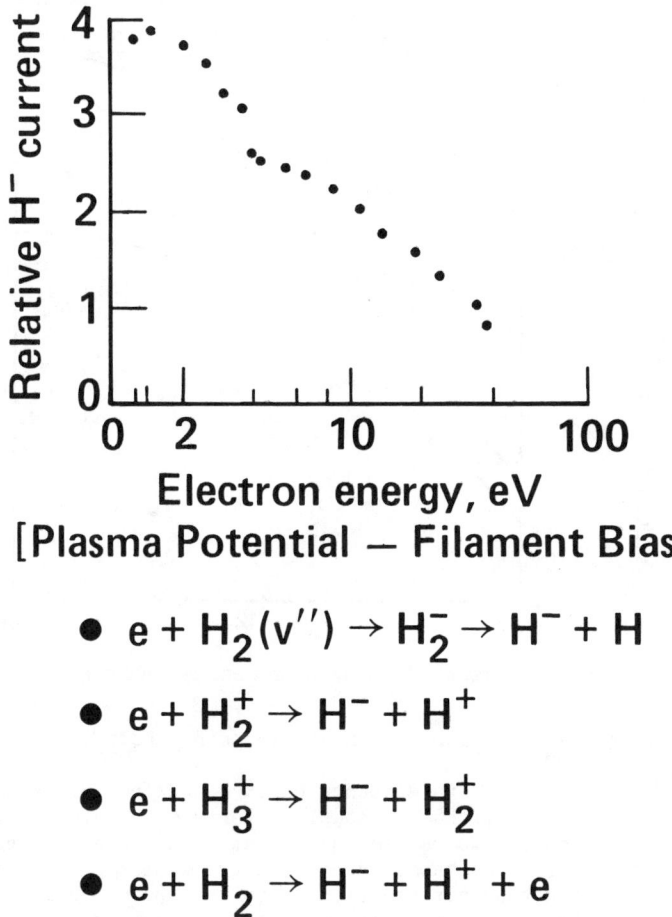

- $e + H_2(v'') \rightarrow H_2^- \rightarrow H^- + H$
- $e + H_2^+ \rightarrow H^- + H^+$
- $e + H_3^+ \rightarrow H^- + H_2^+$
- $e + H_2 \rightarrow H^- + H^+ + e$

Figure 3. Current enhancement vs electron energy.

population approaches an asymptotic value.[19] This is illustrated in Fig. 4 for a system with scale length R = 10 cm and with a ratio of fast (E > 25 eV) to thermal electrons equal to one-tenth. Two solutions are shown: the solid curve for no atomic concentration and the dashed curve for an atomic concentration equal to the molecular concentration.

The performance of the second chamber depends both on the dissociative attachment process and the various loss processes. An extensive array of dissociative attachment cross sections as functions of vibrational and rotational quantum number have been prepared by Wadehra.[17] These calculations are based on a semi-empirical local-potential resonance model developed earlier by Wadehra and Bardsley[20] and fit to the experimental data of Allan and Wong.[16] An alternative model for dissociative attachment has been developed by Gauyacq.[21] In Gauyacq's model the transition is not imagined to proceed through a resonance state, but instead the

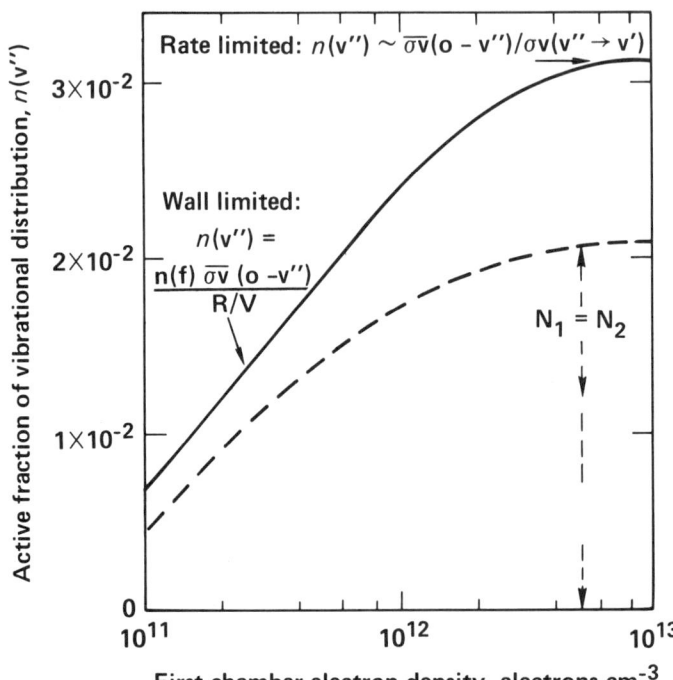

Figure 4. Vibrational population vs electron density.

vibrational excitation and electron capture are induced by the non-adiabatic electronic-vibration coupling terms. In this model a parameter is inserted to provide agreement with the Allen and Wong data up to $v'' = 4$. One has in hand then two semi-empirical models for extrapolating the cross sections into the upper portion of the vibrational spectrum. A comparison of the cross section values for these two models is shown in Fig. 5. The Gauyacq values differ from those of Wadehra and Bardsley for levels $v'' = 5,6,7$ by about 30 to 80%.

An attempt at an essentially exact calculation of dissociative attachment has been initiated by Mundel et al.[22] based upon a full nonlocal-potential resonance model. These calculations have been carried out for the $v'' = 0$, 1, and 2 levels and fall below the experimental values of Allan and Wong. It would be most desirable to have these calculations extended into the upper portion of the spectrum.

In Fig. 6 is plotted the vibrational population distribution generated in chamber one for a high density discharge and the dissociative attachment rates that are appropriate to chamber two. The product of these rates, designated the negative-ion-source-function, is proportional to the negative ion production in the limit of a very short second chamber. This function is also shown in the figure. Inspection of the NISF shows that in excess of 90% of the total negative ion yield is derived from the middle portion of the vibrational spectrum, $5 \leq v'' \leq 11$.

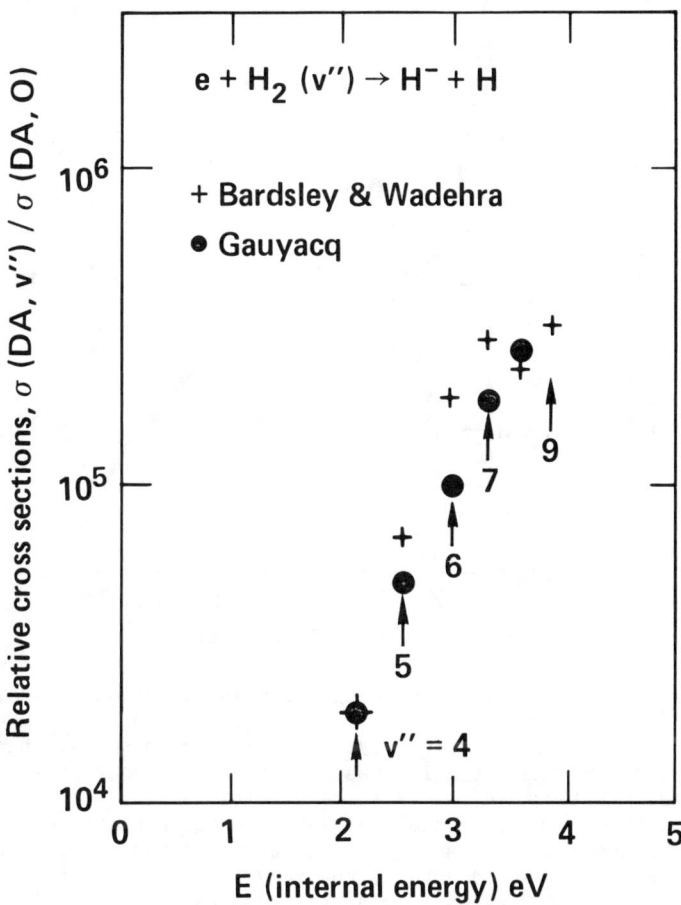

Figure 5. Comparison of dissociative attachment calculations.

The principal negative ion loss processes of concern in the second chamber are ion-ion neutralization,

$$H^- + H^+ \rightarrow H + H \:, \tag{1}$$

and associative detachment[11,12]

$$H^- + H \rightarrow H_2(v'') + e \:. \tag{2}$$

The theoretical and experimental activity with reference to reaction (1) is summarized in Fig. 7.[23-29] Only a brief summary statement of the situation is given here, the reader is referred to the references for an account of the new developments. At collision energies above about 10 eV, the current theoretical and experimental results are in fair agreement. At energies below 4 eV the two theoretical results, Refs. 23 and 29, are in agreement, but lie below the experimental data of Moseley et al.[28] At the collision

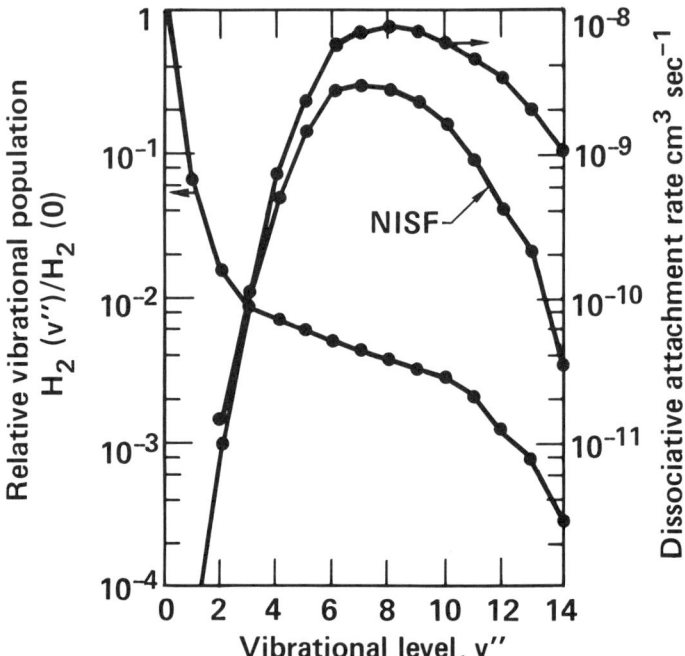

Figure 6. Negative ion source function vs vibrational level.

energy of interest to volume-type negative ion sources, energies near 1 eV, the theoretical cross sections are approximately a factor of three smaller than the experimental values.

With reference to the associative detachment process, reaction (2), the uncertainties are qualitatively different. Here the cross sections are not so uncertain, but very little is known of the concentration of H atoms. Among the greatest uncertainties in evaluating the potential performance of high-density negative ion sources is the almost complete lack of information concerning the atomic concentration.

The variation of negative ion density or current density along the length of the second chamber necessitates a spatial analysis of the ion density in this region. In the low-density, single-chamber discharge where mean free paths are long compared to system dimensions and negative ions are generated uniformly throughout the volume, a spatially independent density analysis will suffice. But in the second chamber of the tandem system a zero dimensional analysis will considerably overestimate the negative ion concentration. This is illustrated in Fig. 8. Here the solid curves are a reproduction of those of Fig. 2, and the upper dashed curve, calculated with a spatially-independent model, is to be compared with the lower solid curve. The spatially-independent model gives a negative ion current density that is almost three times larger than the peak value of the spatially-dependent model. The situation is emphasized further if there are a significant number of free atoms in the discharge. If $N_1 = N_2$ the spatially-dependent solution is reduced to that shown in the lower left-hand

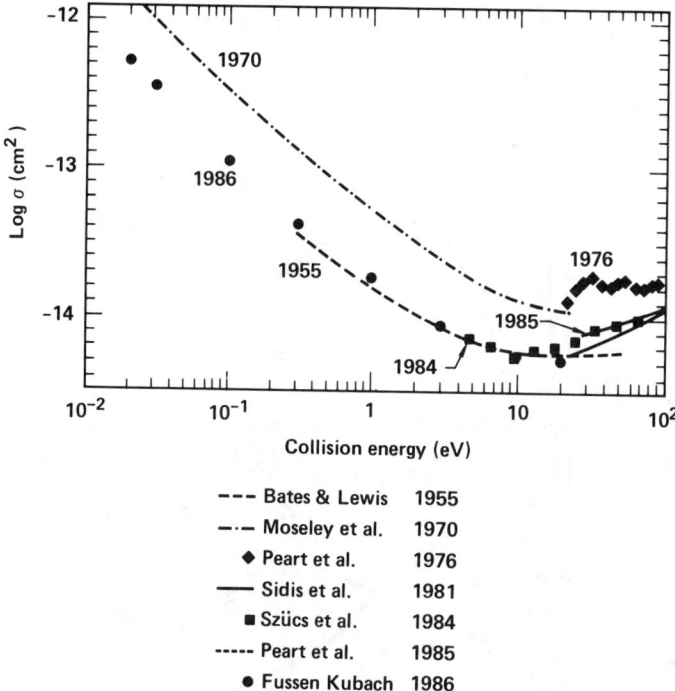

Figure 7. Neutralization cross sections.

corner. The spatially-independent solution is now some six times larger than the peak value.

The scaling law[19] for the tandem system allows one to judge the relative performance of different sources and provides a basis for extrapolating the performance of a particular source to higher current densities. The scaling law can be stated in either of two ways: If one has a particular source geometry with scale length R_0 providing a current density j_0, upon scaling all dimensions downward to a new scale factor R_b and increasing all densities by the factor R_0/R_b while holding the electron energy distribution fixed, the new current density will increase to $j_b = (R_0/R_b) j_0$. Alternatively, if one has optimized the current density j_0 of a system with scale length R_0, then the new optimized system with scale length R_b will emit a current density equal to $j_b = (R_0/R_b) j_0$.

These arguments can be applied to different systems, in a manner shown in Fig. 9. This figure is a plot of extracted current density vs scale length upon which is plotted the data points for three different ion source experiments.[1,30,31] Indicated on the figure is the second-chamber z/R value estimated from the geometry of these different sources. In the source of Holmes et al.[30] (HDG) 6 mA cm^{-2} are extracted from a 10 cm system. If this system were scaled downward to one centimeter the current density would scale up to 60 mA cm^{-2}. The ion source of Jimbo et al.[31] (JELP) is a small 1-cm Penning-type source from which has been extracted a current density of

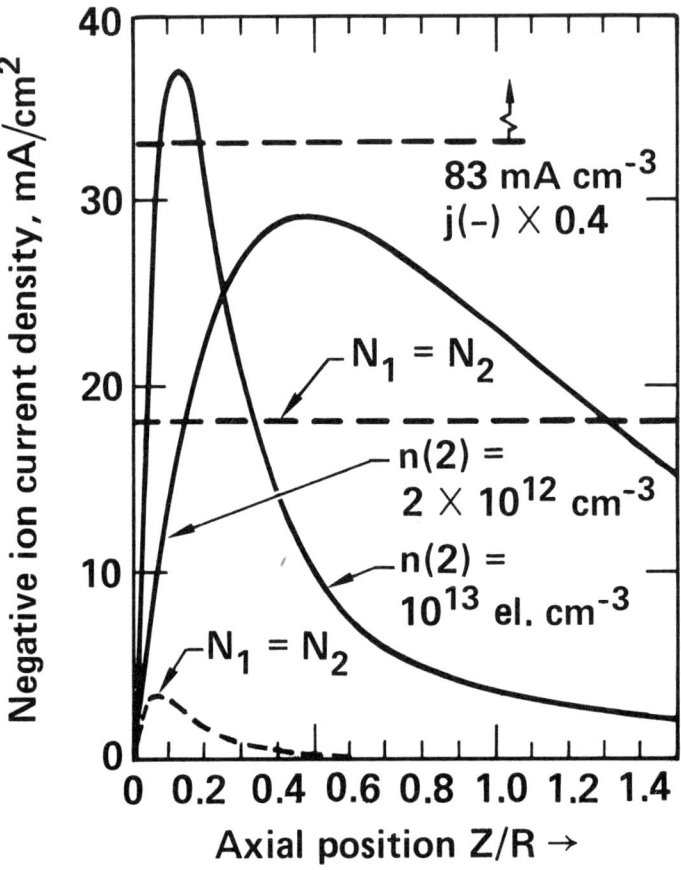

Figure 8. Current density along second chamber.

100 mA cm^{-2}. The ten centimeter source of York et al.[1] (YSLE) would scale to 370 mA cm^{-2} at one centimeter. This scaling tells us that the 100 mA cm^{-2} operation of the Penning source was not an optimized performance, but has the potential for almost a four-fold increase. The ten centimeter systems, Refs. 1 and 30, are known to be non-optimum since the extracted current density was still increasing for increasing discharge current; rather, the operation of these sources was power supply limited. Hiskes et al.[16] have projected an extracted current density of 60 to 80 mA cm^{-2} from a 10 cm system for an atomic concentration equal to one-tenth the molecular concentration; the specific value depending upon the ion drift velocity in the second chamber. This projected value would scale upwards to 600-800 mA cm^{-2} at the one centimeter scale. The high-density tandem system continues to have potential for large increases in the current density.

With the scaling of the current density established, the scaling of the other beam qualities, total current, emittance, and brightness follows immediately.[32] As an example of these scalings consider a cylindrical beam system of radius R_0 scaled downward to a

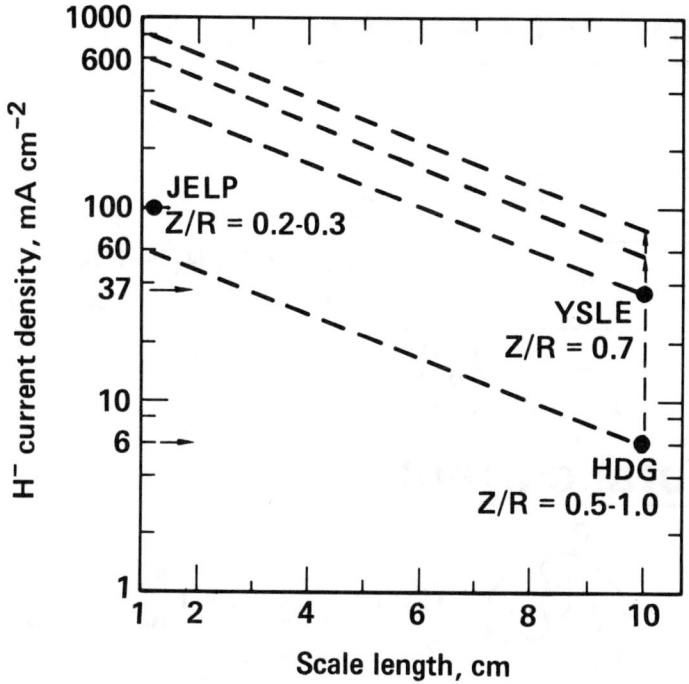

Figure 9. Current density vs scale length.

beamlet of radius R_b to give a beamlet current density $j_b = (R_0/R_b) j_0$. If several of these beamlets are now packed together to fill the original cylinder of radius R_0, one has effectively enhanced the new total current by the scale factor, $J_b = (R_0/R_b) J_0$. The emittance of the new system remains unchanged since the aperture and negative ion energies are unchanged. The source brightness, defined as the ratio of the total current to the square of the transverse emittance, also scales in proportion to R_0/R_b. These scalings are summarized in Fig. 10. The geometric scaling process provides for a favorable enhancement of the different beam qualities, current density, total current, and brightness while the emittance remains unchanged.

III. CONCLUSIONS

The concept of the tandem negative ion generator is now reasonably well defined. The cross section and rate processes for many of the principal underlying atomic and molecular processes are in hand, but certain excited molecular and negative ion destruction processes remain to be clarified. Experimental confirmation of the principal operative mechanisms in the first and second chambers are still largely lacking, notably the demonstration of suitable populations of vibrationally excited molecules in the first chamber, and the concentration of neutral atoms in the overall system and their affect upon the second chamber attenuation.

Current density:

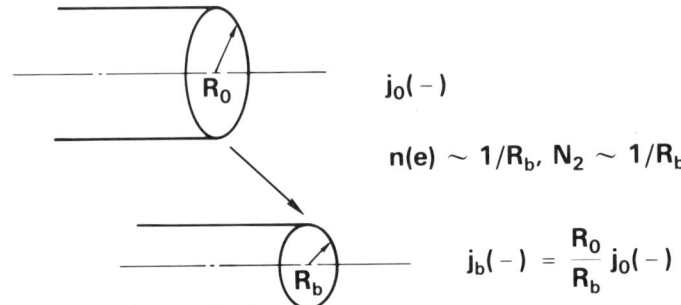

$n(e) \sim 1/R_b$, $N_2 \sim 1/R_b$

$$j_b(-) = \frac{R_0}{R_b} j_0(-)$$

Total current:

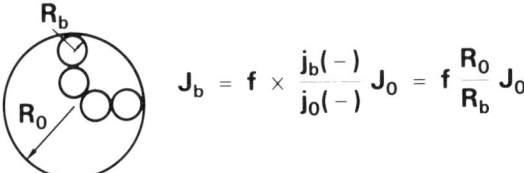

$$J_b = f \times \frac{j_b(-)}{j_0(-)} J_0 = f \frac{R_0}{R_b} J_0$$

Emittance:

$\epsilon \sim R_0 \frac{dr}{dt}(-) \sim$ invariant

Brightness:

$J_b/\epsilon^2 \sim j_b(-)/\left[\frac{dr}{dt}(-)\right]^2 \sim 1/R_b$

Figure 10. Summary of beam quality scalings.

The negative ion current densities that appear to be achievable in a tandem system with a scale length of 10 cm are only marginally interesting for application to neutral beams for magnetic fusion. The scaling of these sources to smaller dimensions will enhance the current density, but at the cost of higher source pressure and power density. The useful scaling that can be achieved represents a compromise between gas density and current density, but the full dimensions of this compromise remain to be clarified.

REFERENCES

1. R. L. York, R. R. Stevens, K. N. Leung, and K. W. Ehlers, Rev. Sci. Instrum. $\underline{55}$, 681 (1984).
2. J. R. Hiskes, Comments on Atomic and Molecular Physics, $\underline{19}$(2) (1987).
3. J. R. Hiskes and A. M. Karo, "Electron Energy Distributions, Vibrational Population Distributions, and Negative Ion Concentrations in Hydrogen Discharges," presented at the NATO Advanced Study Institute on Atomic and Molecular Processes in Contr. Thermo. Research, Palermo, Italy, July 19-30, 1982, Rept. No. UCRL-87779, June 1982.
4. J. R. Hiskes and A. M. Karo, AIP Conf. Proc. $\underline{111}$, 125 (1984).
5. A. M. Karo, J. R. Hiskes, and R. J. Hardy, J. Vac. Sci. Technol. $\underline{A3}$(3), 1222 (1985).
6. J. R. Hiskes, J. Appl. Phys. $\underline{51}$(9), 4592 (1980).
7. J. M. Ajello, S. K. Srivastava, Y. L. Yung, Phys. Rev. $\underline{A25}$, 2485 (1982).
8. W. G. Graham, J. Phys. $\underline{D17}$, 2225 (1984).
9. J. Marx, A. Lebehot, and R. Campargue, J. Physique $\underline{46}$, 1667 (1985).
10. J. R. Hiskes, A. M. Karo, and P. A. Willmann, J. Vac. Sci. Technol. $\underline{A3}$(3) 1229 (1985).
11. R. J. Bieniek and A. Dalgarno, Astroph. J. $\underline{228}$, 635 (1979).
12. R. J. Bieniek, J. Phys. $\underline{B13}$, 4405 (1980).
13. K. N. Leung, K. W. Ehlers, and R. V. Pyle, Rev. Sci. Instr. $\underline{56}$, 364 (1985).
14. K. N. Leung, K. W. Ehlers, and R. V. Pyle, Rev. Sci. Instr. $\underline{57}$, 321 (1986).
15. B. Peart, R. A. Forrest, and K. Dolder, J. Phys. $\underline{B12}$, 3441 (1979).
16. M. Allan and S. F. Wong, Phys. Rev. Lett. $\underline{41}$, 1795 (1978).
17. J. M. Wadehra, Phys. Rev. $\underline{A29}$, 106 (1984).
18. B. Peart and K. T. Dolder, J. Phys. $\underline{B8}$, 1570 (1975).
19. J. R. Hiskes, A. M. Karo, and P. A. Willmann, J. Appl. Phys. $\underline{58}$(5), 1759 (1985).
20. J. M. Wadehra and J. N. Bardsley, Phys. Rev. $\underline{A20}$, 1398 (1979).
21. J. P. Gauyacq, J. Phys. $\underline{B18}$, 1859 (1985).
22. C. Mundel, M. Berman, and W. Domcke, Phys. Rev. $\underline{A32}$, 181 (1985).
23. D. R. Bates and J. T. Lewis, Proc. Phys. Soc. $\underline{A68}$, 173 (1955).
24. J. T. Moseley, R. E. Olson, J. R. Peterson, Case Stud. At. Phys. $\underline{5}$, 1 (1975).
25. B. Peart, R. Grey, and K. Dolder, J. Phys. $\underline{B9}$, L369 (1976).
26. V. Sidis, C. Kubach, and D. Fussen, Phys. Rev. Lett. $\underline{47}$, 1280 (1981).
27. S. Szucs, M. Karemera, M. Terao, and F. Brouillard, J. Phys. $\underline{B17}$, 1613 (1984).
28. B. Peart, M. A. Bennet, and K. Dolder, J. Phys. $\underline{B18}$, L439 (1985).
29. D. Fussen and C. Kubach, J. Phys. $\underline{B19}$, L31 (1986).
30. A. J. T. Holmes, G. Dammertz, and T. S. Green, Rev. Sci. Instrum. $\underline{56}$(9), 1697 (1985).

31. K. Jimbo, K. W. Ehlers, K. N. Leung, and R. V. Pyle, Nucl. Instrum. and Methods A248, 282 (1986).
32. J. R. Hiskes, "Scaling of Current Density, Total Current, Emittance, and Brightness for Hydrogen Negative Ion Sources," Proc. of the NATO Advanced Study Institute on High Brightness Accelerators, Pitlochry, Scotland, July 13-25, 1986.

DISCUSSION

York: If I understand the scaling that you're doing as a function of radius, what's really going to be important is the ratio of surface area to volume in the source.

Hiskes: It depends on the parameter range you're working in. If we're interested in these high density sources, the final parameter that one is concerned with is the actual length from the filter plane to the extraction plane, because that's the dimension over which the ion-ion neutralization and associative detachment is destroying the negative ions. In other words, you want to shorten that length to one mean free path. However, it is true that even at the highest densities, the surface to volume ratio, or the radius of the system, does come in because one is counting on the walls to generate recombination of these fast atoms that are bouncing off the walls and doing the destruction.

York: Based on the recent data from Berkeley which says that yes, this may actually work exactly as you described, is it possible to optimize the size from the calculation? Is it possible to calculate what the optimum surface area to volume or drift length is?

Hiskes: Well, yes, that is possible, but we haven't done that. If you have a configuration of a certain appearance, a cube or a cylinder, the scaling law tells you that if you scale that configuration down, you gain as you go smaller. Now, how one gains as one goes from, say a cylinder to a cube to a cone, is something that requires more detailed calculations. If you go from a cylinder to cylinder, the current density continues to rise as this thing gets vanishingly small. But, of course, you run into trouble.

Wadehra: Since ultimately the interest will be for heavier negative ions like deuterium, for example, is there some scaling law available for the isotopes for various atomic processes?

Hiskes: Strictly speaking, no. And in a strict sense, we expect different results from deuterium and tritium than we would expect for hydrogen. However, a good deal of the reactions that we're concerned with are electronic-type excitations. And, in those type of excitations we would expect the results to be the same for hydrogen or deuterium or tritium, as long as you're exciting the same regions of the energy band. Where you are concerned with the temperatures of the gas, so that the drift velocities of the gas molecules and the excited molecules will depend on the isotopic weight, then you will have some losses, in the case of deuterium with respect to hydrogen. So my guess is that whatever you've done with hydrogen, you're not going to do quite as well with deuterium, and you might do much worse.

Ehlers: John, it seems to me that you're favoring the loss cross section, the loss being largely $H^- + H$, rather than $H^- + H^+$.

Hiskes: I haven't said that but, in fact, that turns out to be true at the present time in some systems.

Ehlers: What is the relative cross-section for $H^- + H^+$?

Hiskes: The $H^- + H^+$ rate is somewhat less than $10^{-7} cm^2/sec$, maybe from a third of that, to 10^{-7}. The associative detachment rate is 10^{-9}, anywhere from a factor of 30 to a factor of 100 smaller. However, the concentrations of atoms are typically 100 to 1000 times larger than ions, so depending on how one can suppress the concentration of atoms, these different rates will shift back and forth. Tomorrow, I think, we'll show that the associative detachment may be the dominant loss process, at least in some of these systems.

Holmes: You used current density, John, toward the end of your talk where, strictly speaking, the rate processes only give you a number density. How do you get current density?

Hiskes: That's a very good question, and what we calculate, in fact, as Andrew points out, is the negative ion density. However, it's easier to talk about the current density, so what we've done is looked at Joe Wadehra's mean negative ion energies that he's calculated, and for these systems we expect the negative ions to be produced somewhere between .2 and .4, maybe .5 eV. So in all the numbers I've shown you, I've taken a mean value of .3 eV for the negative ion energy. However, we've also done calculations on the other end of the range, but rather than clutter up the diagrams, I didn't show them. If you look at the range of energies from, say, .2 to .5 which, I think, is a reasonable range, the numbers shift around 25% up and down.

Holmes: You should actually bring in the concept of drift velocity for negative ions. The random energy essentially gives you zero flux.

Hiskes: Implicit in the optimization of the tandem system is that you have prepared electric and magnetic fields in the second chamber that gather up and sweep the negative ions, as they're formed, towards the emittance plane. So the drift velocity is taken comparable to the negative ion velocity. If you take it much larger you just fool yourself because the negative ions really aren't coming out that fast. But if we're talking about an optimized system, which is what everyone is working for, then I think that this is a reasonable assumption. All the way through here, we're talking about optimized systems, and trying to identify what could be the optimum current density.

ELECTRON ENERGY DISTRIBUTIONS IN MAGNETIC MULTICUSP HYDROGEN DISCHARGES

Wm. F. Bailey and R. G. Jones
Department of Engineering Physics
Air Force Institute of Technology
Wright-Patterson AFB, Oh. 45433

ABSTRACT

Electron energy distributions have been calculated for conditions representative of magnetic multicusp H_2 discharges. A moment approach has been adopted due to the dominance of electron-electron collisions in the highly ionized plasmas considered. Results are presented for a 1-10. A discharge at 40 mTorr. Comparisons are made with numerical solutions of the Boltzmann equation and experiment. Satisfactory agreement is obtained at low energies E < 5.0 eV. Scaling laws with discharge current are presented. Superelastic vibrational collisions are considered and found to have a minor influence on the distribution at all currents. The role of wall losses is examined and related to magnetic filter geometry.

INTRODUCTION

The considerable interest in the production of intense beams of negative ions is driven by applications in low energy kinetic studies, high energy accelerator technology and beam heating in fusion studies. This interest has fostered significant experimental and theoretical studies of the volume production of negative ions in a multicusp geometry.[1-6]

The plasma generators considered are cylindrical chambers with an approximate volume of four liters. The discharge is maintained by thermionic electrons emitted by filaments and accelerated to energies in the range of 100 eV. The cusp geometry is achieved by mounting ceramic magnets on the cylindrical chamber wall and, sometimes, the end plate. A magnetic filter is frequently added to form a tandem discharge.

The understanding and consequent optimization of these sources must be based on a synergistic exchange of experimental diagnostic data and results of theoretical analyses. Based on the wealth of accumulated experimental data[1,2,5] and complex preceding theoretical analyses[3,4,6], we have adopted a much simplified approach to study the electron kinetics. The merits and accuracy of this approach are evaluated and scaling laws are established.

BASIC FORMULATION

The generator is analyzed as an electron-beam excited discharge using the collisional Boltzmann equation. The relevant kinetic processes in molecular Hydrogen are presented in Table I.

Table I Collisional interactions included in the analysis with data source

H_2 CROSS SECTIONS

PROCESS	THRESHOLD	SOURCE
Vibration V(0-1)	.52	For .52<E<2.6 eV, Crompton, Gibson, & McIntosh, Aust J. Phys, (1969), 22, 715. For E>2.6 eV, Kieffer, JILA Information Center Report 13, Sep (1973)
V(0-4) V(0-5)	1.89 2.29	Private communication with J. N. Bardsley & J. M. Wadehra, U. of Pittsburgh, 1979.
Dissociative Attachment V0J0-V4J0	3.70, 3.20, 2.70 2.25, 1.80	Wadehra, J. M. & Bardsley, J. N., Phys Rev Lett, (1978), 41, 1975.
V5J0-V9J0	1.40, 1.03, 0.69 0.36, 0.10	Extrapolated from points received via private communication with J. N. Bardsley, & J. M. Wadehra, U. of Pittsburgh, 1979.
Momentum Transfer		Kieffer, JILA Information Center Report 13, Sep (1973)
Rotation R(0-2) R(1-3)	0.0445 0.075	

These data are supplemented by the corresponding electron-electron specific frequencies characteristic of discharge operation at 1 and 10 A. The vibrational excitation cross sections are shown in Fig. 2.

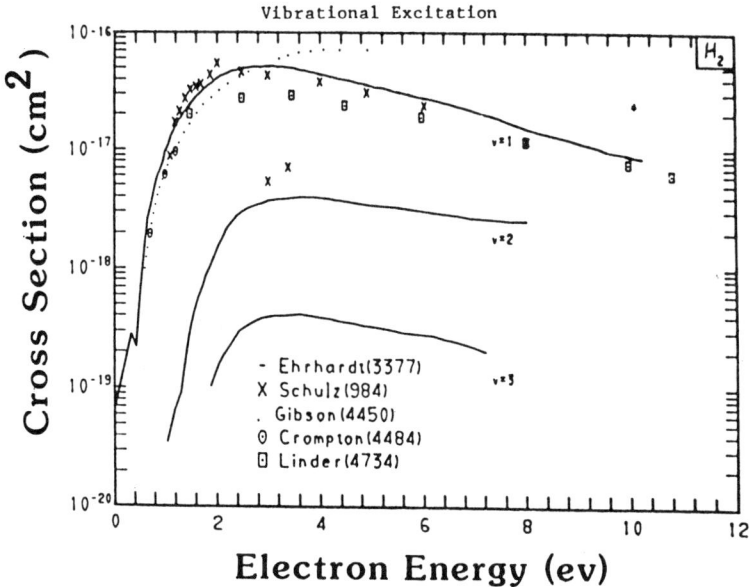

Figure 2. Vibration excitation cross sections in Hydrogen.

We have undertaken calculations with a Maxwellian model based on the high degree of ionization achieved in these devices. This model is based on taking the density and energy moments of the Boltzmann equation, which yield

$$\frac{\partial n_e}{\partial t} = S_o + \nu_{ion} n_e n_o + d n_- n_H$$
$$- \nu_{attach} n_e n_o - \alpha n_e n_+ \quad (1)$$

$$\frac{\partial (n_e <E>)}{\partial t} = <E_s> S_o - \Sigma_j E_j \nu_j n_o n_e$$
$$- <E> \alpha n_e n_+ - 2m/M <E \nu_{MT}> \quad (2)$$

where n_e is the electron density, $<E>$ the average electron energy, S_o the integrated source strength, E_j the threshold energy for process j, α the recombination coefficient, d the detachment coefficient, n_H the density of atomic Hydrogen, M the molecular mass, and n_o is the

Table I (continuation)		
Vibration		
V(0-2)	1.00	
V(0-3)	1.46	
Electronic		Kieffer, JILA Information
E1	8.85	Center Report 13, Sep (1973)
E2	12.00	
Dissociation	8.90	
Ionization	15.43	

A summary presentation of the representative specific inelastic collision frequencies, the product of cross section and velocity, (Q X v), for selected inelastic processes and wall losses is presented in Fig. 1 after Bretagne.[4]

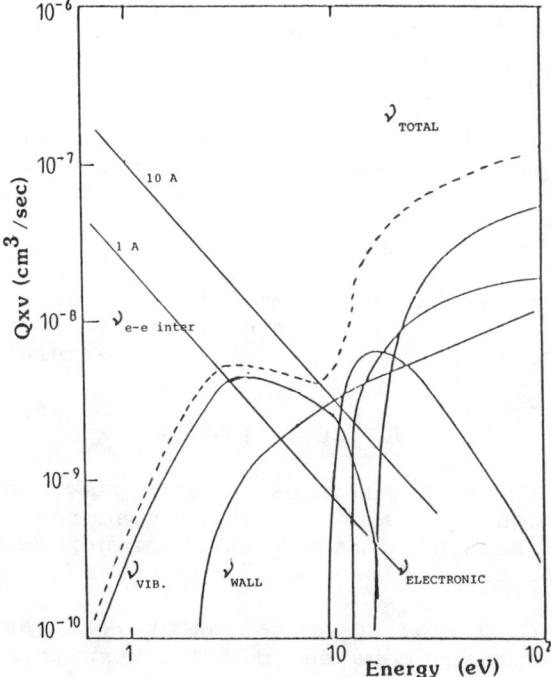

Figure 1. Representative inelastic collision frequencies in H_2 and wall losses (after Bretagne[4]) including electron-electron collisions.

neutral number density. Using the cross section data, Maxwellian averaged collision frequencies, ν_j were generated. The source distribution, $S(E)$, at energy E, is obtained from the classical Thomson theory of ionization, where

$$S(E) = Z(1+Z/E_s)/(E+Z)^2 \qquad (3)$$

Z is the effective energy loss, and E_s is the source energy. The average energy of the source, $\langle E_s \rangle$ is given by

$$\langle E_s \rangle = Z\{(1+x)/x \ln(1+x) - 1\} \qquad (4)$$

$$\text{with } x = E_s/Z.$$

These equations are then complemented with the continuity equations for the positive and negative ions.

$$\frac{\partial n_-}{\partial t} = \nu_{attach} n_e n_0 - d\, n_- n_H - \gamma n_- n_+ \qquad (5)$$

$$\frac{\partial n_+}{\partial t} = S_0 + \nu_{ion} n_e n_0 - \alpha n_e n_+ - \gamma n_- n_+ \qquad (6)$$

The coupled time-dependent moment equations were then solved yielding the electron temperature, T_e, the electron density, n_e, and ion densities, n_- and n_+, at a steady state.

The model was successfully validated in previous studies of excimer laser systems[7] by comparing the temporal evolution of the mean energy and electron density to the results of the detailed calculations of Elliot and Greene.[8]

MULTICUSP RESULTS

In this study of multicusp discharges, the experimental conditions chosen were those reported by Pealat[9]. Comparison is made with the rigorous theoretical results of Bretagne[4] and Gorse[3].

A 40 mTorr discharge operating at 1, 3, and 10 A in pure H_2 is initially considered. The results are summarized in Table II where the calculated values of the electron temperature, and electron and negative ion densities are compared with the data of Pealat[9] and the calculations of Bretagne[4] and Gorse[3].

Table II
Comparison of Discharge Parameters

	(a) Present	(b) Experiment[9]	(c) Theory[3,4]
I(A)	T_e (eV)	N_e (10^{11}cm^3)	N^-/N_e (10^{-3})
1	0.45[a] 0.43[b] 0.32[c]	2.6[a] 1.7[b] 1.9[c]	6.9[a] 7.0[b] 7.0[c]
3	0.62[a] 0.65[b] 0.42[c]	5.0 5.7 3.8	7.0 4.0 4.1
10	0.91[a] 0.85[b] 0.59[c]	10.0 9.8 6.7	6.9 --- 2.7

Excellent agreement is obtained over the entire current range for the values of the electron temperature and density. Given the high degree of ionization, the success of a Maxwellian approach is not surprising. The electron temperature scales at a rate less than $\sqrt{I}$, whereas the electron density varies approximately as $\sqrt{I}$.

The simplicity of this analysis admits interpretation of the discharge scaling laws. The electron losses are recombination dominated. Using the recombination data of Peart and Dolder[10] for H_3^+ yields $n_e \propto \sqrt{I} T_e^{0.37}$. We find the numerical data consistent with this relation. The scaling law for the electron temperature arises from energy balance considerations, which at low T_e can be expressed as

$$\langle E_s \rangle S_o = E_{vib} \, n_e \, n_o \, \nu^o_{vib} \exp(-E_{vib}/T_e) \qquad (7)$$

where ν_{vib} has been approximated as $\nu^o_{vib} \exp(-E_{vib}/T_e)$. Eqn. (7) yields consistent values of T_e. Thus within the approximation of a Maxwellian distribution, both T_e and n_e are analytically specified, and all rates are established.

The Maxwellian model failed to accurately predict the negative ion concentration and its variation with current. This discrepancy arises from the fact that a self-consistent solution of the plasma chemistry and vibrational kinetics was not employed. Instead, those values of H atom concentration and vibrational distribution calculated by Gorse[3] at 1 A were used at all currents. The experimentally observed decrease in the negative ion concentration and vibrational population arises from the increase in H atom concentration at higher currents. To elucidate this sensitivity, detachment due to H atoms was eliminated and the calculation repeated. It was found that T_e and n_e remained essentially unchanged; whereas, the negative ion density increased by a factor of 16, 18, and 18 at 1, 3, and 10 A respectively. The influence of superelastic vibrational collisions was investigated. Calculations performed at vibrational temperatures of 0.0, 0.11, and 0.22 eV, at a fixed rate of dissociative attachment, revealed only an extremely minor influence on the electron temperature and electron number density.

The influence of wall collisions was investigated. When wall losses were significant, the distribution of secondary electrons was modified by a survival probability, P(E), defined by

$$P(E) = 1 - \nu(E)_{wall}/(\nu(E)_{wall} + \sum_j \nu(E)_j) \qquad (8)$$

The wall frequency and total frequency are shown in Fig.1 for a 40 mTorr discharge. This probability distribution was convoluted with the Thomson distribution resulting in a modified source and source mean energy. It was assumed that variations in the wall losses did not significantly alter the plasma potential, which was maintained at 2 eV. For the case of 10 A at 4 mTorr, the electron temperature was increased by a factor of 1.5 from the reported value at 40 mTorr and the electron density reduced by a factor of 9.3. Calculations at 40 mTorr and 10 A, with wall losses characteristic of the filter region of the discharge, yielded a four-fold reduction in both the electron temperature and the electron density. This significant reduction in temperature, although accompanied by a proportional decrease in the electron density, has significant implications with regard to source design and the analysis of magnetic filters used in tandem sources. With regard to sources, the implication is that wall losses can be used to tailor the electron temperature. Conversely, there exists a point of diminishing returns with regard to number and strength of wall and end plate magnets employed. With regard to filter analysis, the increase in wall losses experienced in the vicinity of the filter structure can result in a significant reduction in the electron temperature as the electrons pass through the filter. This temperature reduction would, under appropriate conditions, significantly enhance negative ion production in the vicinity of the filter and also increase the importance of superelastic collisions. Operation of sources at temperatures high enough to optimize vibrational excitation, while yielding low concentrations of negative ions in the source region, could lead to superior production of negative ions in the filter region.

The normalized distribution function, f(E), in pure H_2 at 10 and 1 A is presented in Fig. 3. Here f(E) is given by

$$f(E) = 2\pi/(\pi T_e)^{3/2} \exp(-E/T_e) \qquad (9)$$

As noted previously, the electron temperatures of 0.45 and 0.91 eV are in satisfactory agreement with the calculations of Bretagne[4] (dashed curve); however, the restrictions imposed by the moment approach become apparent at those energies where the electron-electron collisions are no longer dominant and significant departures of the distribution from the Maxwellian characteristic of the low energy regime are observed.

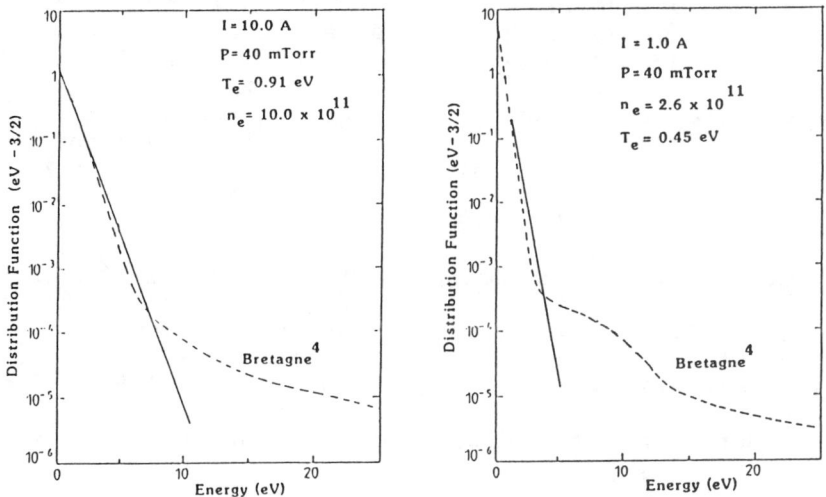

Figure 3. Comparison of normalized electron energy distributions in H_2 at a pressure of 40 mTorr, at currents of 10 A (left) and 1 A (right). Maxwellian model (solid curve), Bretage[4] (dashed curve).

Referring back to Fig. 1, the variation of the electron-electron collison frequency divided by the neutral particle density is shown for the conditions of the 1 and 10 A discharges. When the specific collision frequency for the inelastic processes exceeds this electron-electron term, the Maxwellian character of the distribution will no longer be maintained. We note that this occurs at an energy, E_*, of approximately 3 eV for the 1 A discharge and 7 eV in the 10 A case. This marks the transition point at which we establish our non-Maxwellian distribution. By requiring a local steady state, for $E > E_*$, between source function and excitation loss at each energy, the distribution tail can be approximated. Under these conditions, the excitation rates in the distribution tail will scale as the ratio of the current to the pressure.

CONCLUSIONS

A Maxwellian model of multicusp discharges has proven to be a successful method of characterizing the electron energy distribution. The calculated electron density and temperature are consistent with experiment and compare well with rigorous theoretical solutions[3,4] of the Boltzmann equation. While vibrational temperature has only a minor influence on the distribution in the driver region, superelastic collisions in the filter region take on increased importance. Wall losses are extremely important at pressures less than 10 mTorr. At this and higher pressures, wall losses in the vicinity of the magnetic filter may still be significant and lead to a substantial reduction in the electron temperature. This effect could be used to advantage in tandem source design.

REFERENCES

1. M. Bacal, Physica Scripta, T2/2,467,(1982).

2. A. M. Bruneteau Mordin, "Study on the Hydrogen Negative Ion in Low Pressure Discharges", Dissertation, University of Paris-Sud, (1983).

3. C Gorse, M Capitelli, J Bretagne and M Bacal, Chem Phys, 93, 1,(1985).

4. J Bretagne, G Delouya, C Gorse, M Capitelli, and M Bacal, J Phys. D: Appl. Phys., 18, 811,(1985).

5. K Leung, K Ehlers, J. Appl. Phys., 52, 3905, (1981).

6. J Hiskes, A Karo, NATO Adv. Study Inst. on Atomic and Molecular Processes in Controlled Thermonuclear Research, Palermo, Italy, Report UCRL-87779,(1982).

7. W Bailey, SCEE Rpt, AFAPL, Contract No. F 33615-81-C-2011, (1985).

8. C Elliot, A Greene, J. Appl. Phys., 47,2946, (1976).

9. M Pealat, J Taran, J Taillet and M Bacal, Report ONERA No28/7131 PY, (1981).

10. B Peart, K Dolder, J. Phys.B, 7,1948, (1974).

DISCUSSION

Bacal: Have you thought about how to predict the plasma potential in the calculation so it wouldn't have to rely on the measured values of the plasma potential?

Bailey: In this analysis we assume that the plasma potential remain fixed at the experimentally measured value of 2V. Now, when we're going through temperature changes as dramatic as those that I'm calculating, I would anticipate it changing and we'd have to do that self consistently. So yes, we're initiating some sort of companion studies, looking at diffusion processes and, hence, sheath processes in three-component plasmas. As you're probably aware, the state of knowledge in that area is quite lacking. So what we're looking at, in these very low-pressure discharges, is to analyze the sheath and, consequently, the plasma potential. At this time, I do not have any results I can pass along to you.

Electron-Ion Collision Processes Relevant to H⁻ Ion Sources

J.B.A. Mitchell

Department of Physics and Centre for Chemical Physics
The University of Western Ontario, London, Ontario, Canada. N6A 3K7.

ABSTRACT

Electron ion recombination involving H_2^+ and H_3^+ has attracted considerable attention both from theoreticicans and experimentalists in recent years. The status of these studies will be reviewed with particular emphasis on their role in elucidating negative hydrogen ion source chemistry.

The recombination of H_2^+ and H_3^+ ions are important processes to the chemistry of H⁻ ion sources. Not only are they competing sinks for the population of low energy electrons believed responsible for the formation of H⁻ via dissociative attachment with $H_2(v)$, they are also sources of vibrationally and electronically excited species which also play a role in H⁻ formation. Although these two species represent the simplest diatomic and triatomic molecules respectively their recombination is poorly characterized. In the following, these topics will be addressed with reference to current theoretical and experimental studies. Some of the background to this work has also been discussed in a recent review article (Mitchell, 1986).

H_2^+ DISSOCIATIVE RECOMBINATION

The potential energy curves for H_2 and H_2^+ together with some of the excited states relevant to electron -H_2^+ recombination are shown in Figure 1. It can be seen that the minimum of the ground state of H_2^+ is offset from that for H_2. This means that when H_2 is ionized by electron or photon impact, virticals transition will give rise to a Franck-Condon distribution of vibrational states of H_2^+. Because H_2^+ does not have a dipole moment, these states have very long lifetimes against radiative decay (~10^6secs.). VonBusch and Dunn (1972) have experimentally determined the population of vibrational states in a beam of H_2^+ ions and they found that their measured population was quite close to a Franck-Condon distribution. See table I. For low energy electrons the recombination takes place through initial capture into the $^1\Sigma_g^+$ autoionizing state. This subsequently dissociates so that the excess energy of the electron ⊢ ion system is converted into product kinetic energy thus stabilizing the system against autoionization.

It should be noted that the $^1\Sigma_g^+$ state intersects the H_2^+ ground X $^2\Sigma_g^+$ state in the vicinity of the v=2 level Guberman (1983), Hazi (1983) and (Haroand Sato, 1984). This means that the cross section for the recombination of low energy electrons with H_2^+ (v=0 or 1) via direct capture into this state is expected to be small. It is possible, however, for recombination to occur indirectly through initial capture into vibrationally or rotationally excited Rydberg states of H_2 which lie above the v=0 level of H_2^+. These states

TABLE I
H_2^+
VIBRATIONAL POPULATIONS

v	Von-Busch & Dunn	Franck-Condon	Energy below (ev) Dissociation Limit *
0	0.119	0.092	2.645
1	0.190	0.162	2.374
2	0.188	0.176	2.118
3	0.152	0.155	1.877
4	0.125	0.121	1.651
5	0.075	0.089	1.44
6	0.052	0.063	1.243
7	0.037	0.044	1.059
8	0.024	0.030	0.890
9	0.016	0.021	0.734
10	0.0117	0.0147	0.593
11	0.0082	0.0103	0.465
12	0.0057	0.0072	0.351
13	0.00374	0.0051	0.252
14	0.00258	0.0036	0.168
15	0.00175	0.0024	0.100
16	0.00109	0.0016	0.0491
17	0.00056	0.0008	0.027
18	0.00012	0.0002	0.002

*Taken from Cohen et al, 1960.

can then decay via a transition to the $^1\Sigma_g^+$ state which then dissociates as before.

The inclusion of the indirect mechanism into calculations of the dissociative recombination of H_2^+ has been studied by Giusti et al (1983) using a Multi Channel Quantum Defect approach and more recently by Hickman (1986) who used a Configuration Interaction Method.

Their results are compared in Figure 2. Also shown are experimental results of Auerbach et al (1977) for H_2^+ ions which at the time were believed to be in the v=0, 1 and 2 levels. These ions were produced in a conventional radiofrequency ion source using a hydrogen/helium mixture. The endothermic reaction: -

$$H_2^+ (v>2) + He \rightarrow HeH^+ + H \qquad (1)$$

removes excited H_2^+ ions so that those emerging from the source should have only v=0, 1 and 2 populated. This presupposes however that the H_2^+ ions remain in the source for a sufficient period of time for all the H_2^+ (v>2) to be removed. The rf source used in the experiment of Auerbach et al is calculated to have an average ion residence time of ~35μs. Sen (1985) has modelled the reactions in the rf source and predicted that in fact up to v=6 were populated in Auerbach et al's experiment.

Recently a new ion source has been used in conjunction with the MEIBE experiment at U.W.O. This source, modelled on a design by Teloy and Gerlich (1974) uses a radiofrequency field to trap ions for about 1 millisecond before they escape and are extracted and focussed to form a beam. Sen et al (1986a) have demonstrated that this source is capable of producing H_2^+ ions with only v=0, 1 and 2 populated when a helium/hydrogen mixture is used as the source gas. Using neon instead of helium allows the reaction

$$H_2^+ (v>1) + Ne \rightarrow NeH^+ + H \quad (2)$$

to remove excited H_2^+ ions with v>1. Sen et al were able to demonstrate that in this case the emerging beam was populated predominantly by v=0 and 1 states only.

This source has been installed in the terminal of the Van de Graaff accelerator which is used as an injector into the MEIBE apparatus.

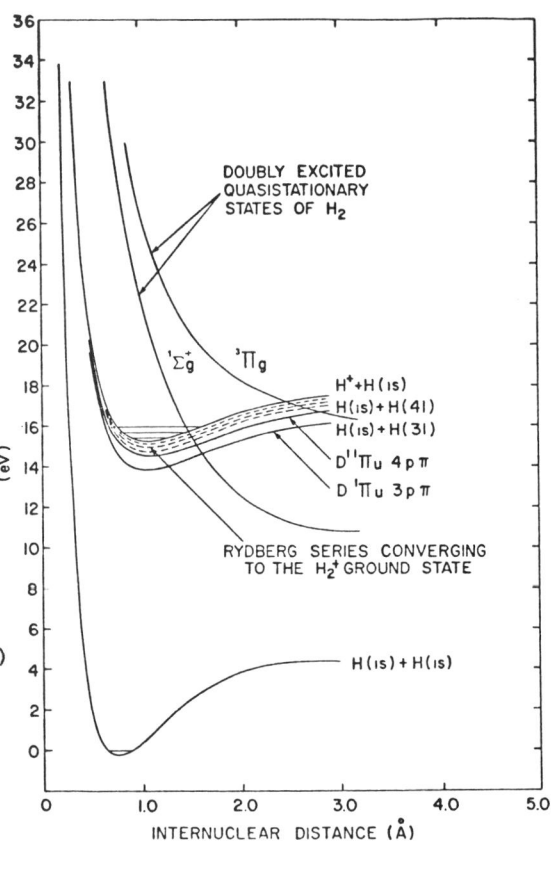

Figure 1. Potential energy curves for H_2^+ and H_2.

Measurements of the recombination of H_2^+ ions formed in the rf trap source are currently in progress. Some preliminary results taken using at 4:1 H_2/He mixture at a source pressure of 30mTorr are shown in Fig. 2. It can be seen that these results are lower than those obtained by Auerbach et al (1977) and indeed they are in better agreement with Giusti-Suzor et al's (1983) calculation than with Hickman's calculation. It must be stressed however that much more experimental work must be performed before a definitive statement can be made regarding H_2^+ recombination. It should also be noted that the calculations show cross sections for a population of states v=0,1,2 in the ratio of 1:2:2. This population assumes that higher vibrational states are removed exclusively by reaction (1), see Table I. Theard and Huntress (1974) however have argued that direct de-excitation to the H_2^+ (v=0) will occur in collisions with inert gases so this would weight the distribution towards lower states. This in turn would lower the calculated cross sections.

Measurements of the dissociative excitation process: -
$$e + H_2^+ \rightarrow H + H^+ + e$$

can be used to determine the internal energy and hence the highest populated vibrational state of the H_2^+ ions used in the recombination measurements. These experiments are also being pursued at this time.

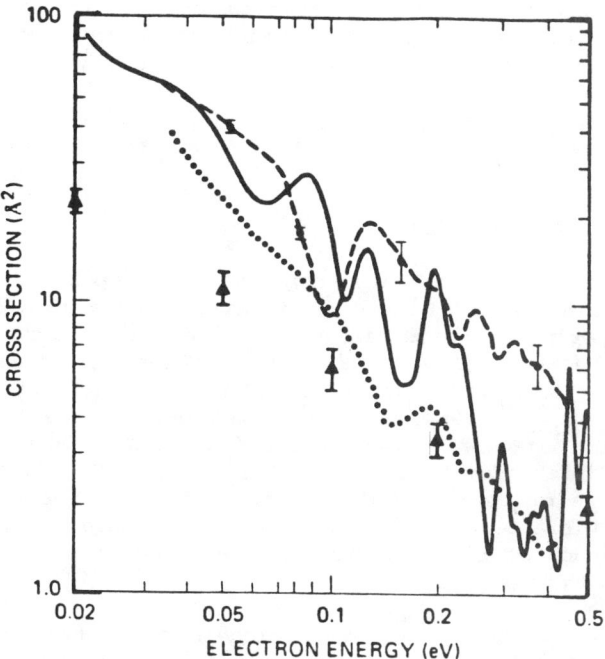

Figure 2. Calculated and measured cross sections for H_2^+ dissociative recombination. --- Hickman, 1986. ... Giusti-Suzor, 1983. ● Auerbach et al, 1977. ▲ Recent experimental.

PRODUCT CHANNELS

The identification of the products of dissociative recombination is a subject of considerable interest. The $^1\Sigma_g^+$ state through which the recombination process proceeds actually goes to the dissociation limit of $H^+ + H^-$ at very large internuclear separation R, O'Malley, (1969). Peart and Dolder (1975) however have measured the cross section for: -

$$e + H_2^+ \rightarrow H^+ + H^-$$

and found that it accounted for a very small fraction of the total recombination cross section. This means that curve crossings must take place between the $^1\Sigma_g^+$ state and excited states of H_2 which subsequently dissociate.

For H_2^+ ions in low vibrational states and low energy electrons the only open channel that can be accessed directly is H(n=2) + H. If more H_2^+ vibrational states are populated and/or if the electron energy is increased then dissociation to H(n = 3, 4, ...) +H becomes possible. Phaneuf et al (1975) and Vogler and Dunn (1975) have measured the partial cross sections for the formation of H(2s) + H and H(n=4) + H respectively using H_2^+ ions with a Von Busch and Dunn vibrational state distribution. In each case they found that the measured partial cross sections accounted for about 10% of the total recombination cross section. Clearly to improve our knowledge of the mechanisms of dissociative recombination it would be desirable to repeat these measurements using state selected H_2^+ ions and lower electron energies.

Long term plans for such a measurement are currently underway for the MEIBE experiment.

H_3^+ DISSOCIATIVE RECOMBINATION

Prior to the last Brookhaven negative ion symposium the dissociative recombination of H_3^+ was in a fairly comfortable state. A number of experimental measurements of rate coefficients or cross sections had been made, Leu et al (1973), Peart and Dolder (1974), Auerbach et al (1977), Mathur et al (1978), MacDonald et al (1984) and Mitchell et al (1984) and these experiments agreed with each other to within a factor of two. Any differences could be explained in terms of differing vibrational distributions. Biondi's measurements and those of Peart & Dolder were believed to have been made using ground vibrational state H_3^+ ions. Those used in the experiments of Auerbach et al and Mitchell et al used ions only some 30% of which were in the v=0 state.

At the last symposium however, Michels and Hobbs (1983) presented a theoretical calculation of the potential energy surface for the autoionizing state through which e - H_3^+ recombination proceeds. This is shown in Figure (3). It can be seen that this state is inaccessible to low energy electrons encountering H_3^+ ions in low vibrational states with $v \leq 3$. Thus direct recombination of electrons with H_3^+ appears to be ruled out in this case. In a subsequent paper, Michels and Hobbs (1984) addressed the problem of indirect dissociative recombination of H_3^+ and concluded that this was also energetically unfavourable. Because of this they predicted that the rate coefficient for the dissociative recombination of ground vibrational state H_3^+ should be two orders of magnitude smaller than that for H_3^+ (v > 3).

Around the same time Adams et al (1984) published experimental results for e - H_3^+ measured using a Flowing Afterglow Langmuir Probe (FALP) apparatus. Because of the high pressures used in the apparatus the authors proposed that the H_3^+ ions would be collisionally quenched to their ground vibrational state. They found that the recombination rate was too small to measure and they placed an upper limit of 2×10^{-8} cm^3 s^{-1} on its value. This has subsequently been revised to 1×10^{-11}cm^{3}s^{-1} (Smith, Private communication 1986). At this point the question is why the other experimental results should be so high or whether the FALP results are incorrect?

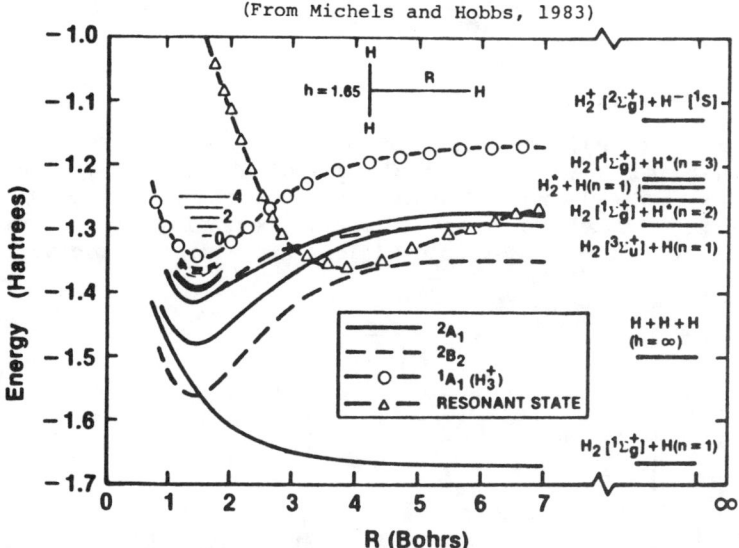

Figure 3. Potential energy curves for H_3^+ and H_3 in C_{2v} symmetry.

Sen et al (1986b) have demonstrated that the radiofrequency trap source mentioned earlier is capable of producing H_3^+ ions with less than 0.5eV of internal energy.

Dissociative recombination and excitation measurements have been performed for H_3^+ ions formed using this source at a variety of pressures and gas mixtures. This series of measurements is not yet complete and so it is too early to make a definitive statement regarding the measured dissociative recombination cross section for completely relaxed ions.

Fig. 4 shows dissociative recombination cross sections for H_3^+ ions prepared in the trap source using pure hydrogen at presures of 25mTorr and 45mTorr respectively. Also included are results taken previously using a conventional radiofrequency ion source. (Mitchell et al, 1983).

It can be seen that the trap ion source results are very much lower than previous measurements The shape of the cross section curves is similar although it appears that at higher energies the curves converge. Dissociative excitation measurements made concurrently with the recombination measurements indicate that the H_3^+ ions which yielded the lower curve had about 1.0 eV of internal energy. Work is in progress to lower this internal energy by using higher source pressures.

When H_3^+ recombines with an electron it can dissociate via two different channels: -

$$e + H_3^+ \rightarrow H_2 + H \quad - - - \text{I.}$$

$$\rightarrow H + H + H \quad - - \text{II.}$$

Mitchell et al (1983) have measured the relative cross sections for each channel and found that for H_3^+ ions of which ~60% were vibrationally excited, channel II dominated over channel I by about a factor of 2. Experiments are currently beginning, to re-examine this ratio for vibrationally cold H_3^+ ions and also isotopic vari iants H_2D^+, HD_2^+ and D_3^+.

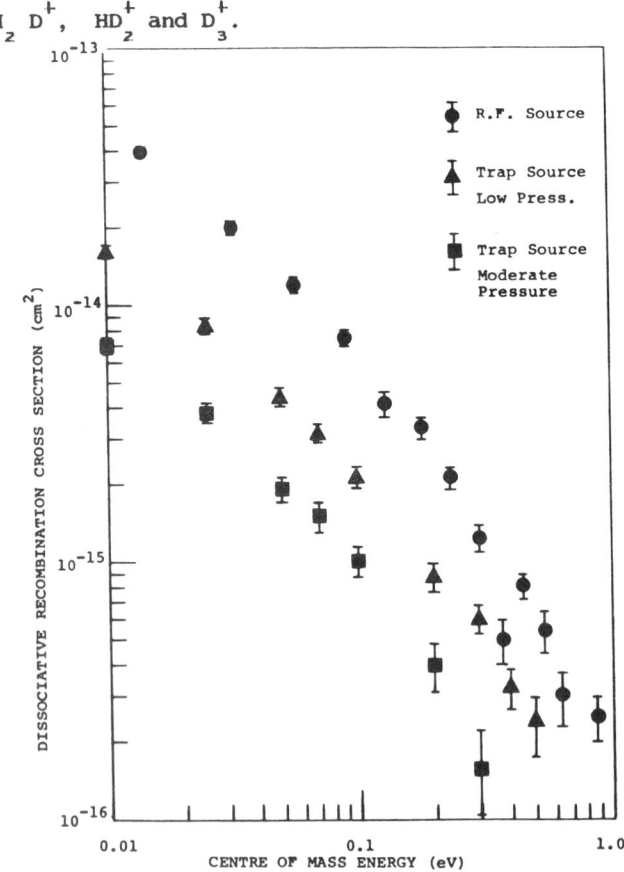

Figure 4. Recent experimental cross sections for H_3^+ dissocciative recombination.

REFERENCES

1. Adams, N.G., Smith, D. and Alge, E., J. Chem. Phys. 81, 1778, 1984.
2. Auerbach, D., Cacak, R., Caudano, R., Gaily, T.D., Keyser, C.J., McGowan, J. Wm., Mitchell, J.B.A., Wilk, S.F.J., J. Phys. B. 10, 3797, 1977.
2a. Cohen, S., Hiskes, J.R. and Riddell, R.J., UCRL-8871- 1960.
3. Giusti-Suzor, A., Bardsley, J.N. and Derkits, C., Phys. Rev. A. 28, 682, 1983.
4. Guberman, S.L., J. Chem. Phys. 78, 1404, 1983.
5. Hara, S. and Sato, H., J. Phys. B. A 27, 101, 1983.
6. Hazi, A.U., Derkits, C. and Bardsley, J.N., Phys. Rev. A 27, 1751, 1983.
7. Hickman, A.P. (Preprint 1986).
8. Leu, M.T., Biondi, M.A. and Johnsen, R., Phys. Rev. A8, 413, 1973.
9. MacDonald, J., Biondi, M.A. and Johnsen, R., Planet Space Sci. 32, 651, 1984.
10. Mathur, D., Khan Su and Hasted, J.B., J. Phys. B. 11, 3615, 1978.
11. Michels, H.H. and Hobbs, R.H., Production and Neutralization of Negative Ions and Beams, AIP Conf. Proc. No. 111, (ed. K. Prelec).
12. Michels, H.H. and Hobbs, R.H., Ap. J. 286, L27, 1984.
13. Mitchell, J.B.A., J.L. Forand, C.T. Ng, D.P. Levac, R.E. Mitchell, P.M. Mul, W. Claeys, A. Sen and J.Wm. McGowan, Phys. Rev. Lett. 51, 885, 1983.
14. Mitchell, J.B.A., C.T. Ng, J.L. Forand, R. Janssen and J.Wm. McGowan, J. Phys. B., 17, L909, 1984.
15. Mitchell, J.B.A. in Phys. of Electron-Ion and Ion-Ion Collisions, NATO ASI, Han-sur-Lesse, Belgium, eds. F. Brouillard and P. Defrance, Plenum, N.Y., 1986.
16. O'Malley, T., J. Chem. Phys. 51, 322, 1969.
17. Peart, B. and K.T. Dolder,, J. Phys. B7, 1948, 1974.
18. Peart, B. and K.T. Dolder, J. Phys. B. 8, 1570, 1975.
19. Phaneuf, R.A., D.H. Crandall, G.H. Dunn, Phys. Rev. A11, 1983, 1975.
20. Sen, A., Ph.D. Thesis, The University of Western Ontario 1985.
21. Sen, A., J.Wm. McGowan and J.B.A. Mitchell, J. Phys. B. (in Press), 1986.
21a. Sen, A. and J.B.A. Mitchell, J. Phys. B. 19, L545, 1986.
22. Teloy, E. and D. Gerlich, Chem. Phys. 4, 417, 1974.
23a. Theard, L.P. and W.T. Huntress, J. Chem. Phys. 60, 2830, 1974.
24. VonBusch, F. and G.H. Dunn, Phys. Rev. A5, 1726, 1972.

ELECTRON EXCITATION OF $H_2(v")$ LEVELS TO YIELD VIBRATIONALLY
EXCITED H_2 MOLECULES

J. R. Hiskes
Lawrence Livermore National Laboratory, Livermore, CA 94550

ABSTRACT

Electron excitation cross sections to produce vibrationally excited H_2 molecules are calculated for molecules initially in the first and second vibrational levels. The pattern of excitation cross sections as a function of final vibrational level v" differs from that found in previous calculations initiated from the ground vibrational level.

I. INTRODUCTION

The electron excitation of hydrogen molecules to form electronically excited states that in turn undergo radiative decay to the different vibrational levels of the ground electronic state provides for an excitation mechanism that is expected to dominate the operation of tandem high-density hydrogen-negative-ion discharges. These excitations provide for a vibrational population distribution that is more enhanced toward the excited vibrational levels, v">0, as the discharge electron density increases. The distributions presented to date however, have been calculated assuming that the initial excitation proceeds only from the lowest vibrational level, v"=0. As the portion of the population distribution in the excited vibrational levels increases to become a larger fraction of the total population, one can expect that the excitations that are initiated from these more highly populated excited levels will bear on the final distributions. Of particular interest is the possibility that these excitations will enhance the final distribution in the region of the spectrum that is principally responsible for negative in formation, $5 \leq v" \leq 11$. To explore this possibility calculations have been performed to evaluate the cross section for excitation to a final vibrational level v" starting from an initial level $\underline{v}"=1$ or $\underline{v}"=2$. These cross sections are indeed found to lead to an enhancement of the final-level cross sections in the range $5 \leq v" \leq 11$ compared to excitations initiated from the v"=0 level.

II. DISCUSSION OF CROSS SECTIONS

This paper is concerned with the excitation of a hydrogen molecule initially in a vibrational level $\underline{v}"$ of the $X^1\Sigma_g$ ground electronic state and excited by energetic electron collisions (E>20 eV) to yield vibrationally excited molecules, $H_2(v")$. These excitations proceed through all members of the singlet electronic spectrum. The subsequent radiative decay to the ground electronic state leaves a residual but vibrationally excited molecule, $H_2(v'')$. The principal singlet excitations proceed via the $B^1\Sigma_u$ and $C^1\Pi_u$ states; for these cases the overall reactions are

$$e + H_2(X^1\Sigma_g, \underline{v}'') \rightarrow e' + H_2(B^1\Sigma_u, C^1\Pi_u)$$
$$H_2(X^1\Sigma_g, v'')) + h\nu \ . \quad (1)$$

The formalism for the effective cross section for the transition, $\underline{v}'' \rightarrow v''$, (1) was derived previously and was evaluated for the lowest vibrational level, $\underline{v}''=0$. These cross sections $\sigma(\underline{v}''=0;v'')$ are reproduced here in Fig. 1.

The cross sections for initial levels $\underline{v}''=1,2$ are shown in Figs. 2,3, respectively. Upon examining the $\sigma(\underline{v}''=1;v'')$ of Fig. 2 and comparing with the $\sigma(\underline{v}''=0;v'')$ of Fig 1 several differences are seen. In Fig. 2 the dominant cross section leads to $v''=1$ rather than $v''=0$. Also, the cluster of cross sections $\sigma(\underline{v}''=1;8,9,10,11)$ are now some considerable factor larger than for $\underline{v}''=0$, while the $\sigma(\underline{v}''1;6,7)$ are reduced. The ordering of the σ's with v'' is now mixed and no longer falls off monotonically with increasing v'' as do those in Fig. 1.

Inspection of the $\sigma(\underline{v}''=2;v'')$ of Fig.3 shows that the largest cross section now occurs for $v''=2$. Here again, the $\sigma(v''=2;5,6,7,8,9,10)$ are enhanced relative to the corresponding σ's of Fig. 1. The pattern of maximum σ for $v''=\underline{v}''$ for all three cases is an interesting one and may have important consequences for negative ion production if this pattern persists into the higher portions of the spectrum.

This work was performed under the auspices of the U.S. Department of Energy by the Lawrence Livermore National Laboratory under contract number W-7405-ENG-48.

REFERENCES

1. J. R. Hiskes, A. M. Karo, and P. A. Willmann, J. Appl. Phys. 58, 1759 (1985).

2. J. R. Hiskes, J. Appl. Phys. 51, 4592 (1980).

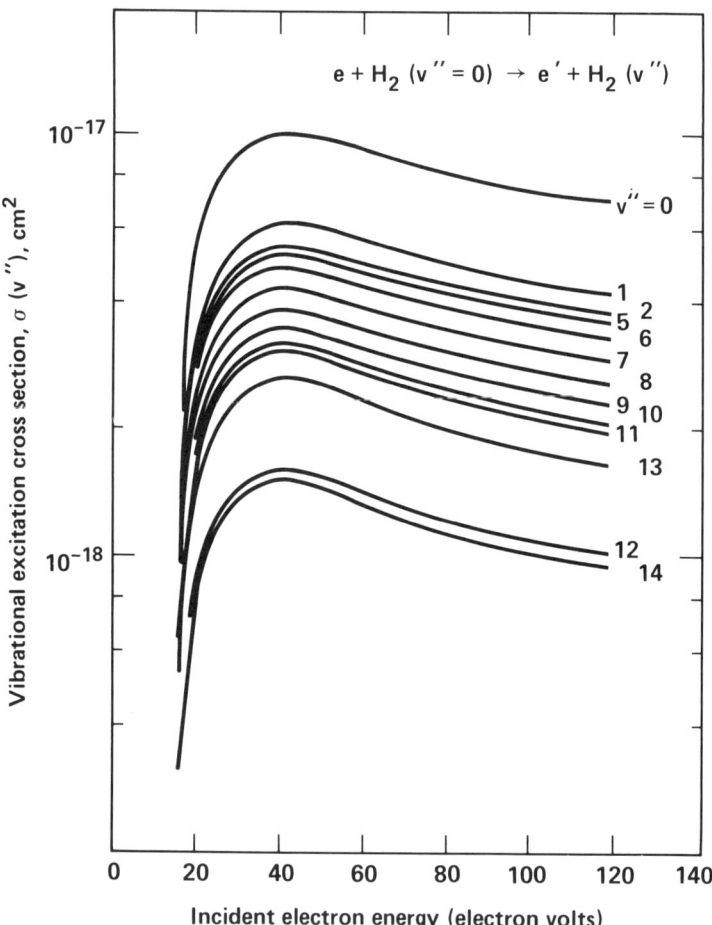

Fig. 1. Excitation cross section to level v" from initial level v"=0.

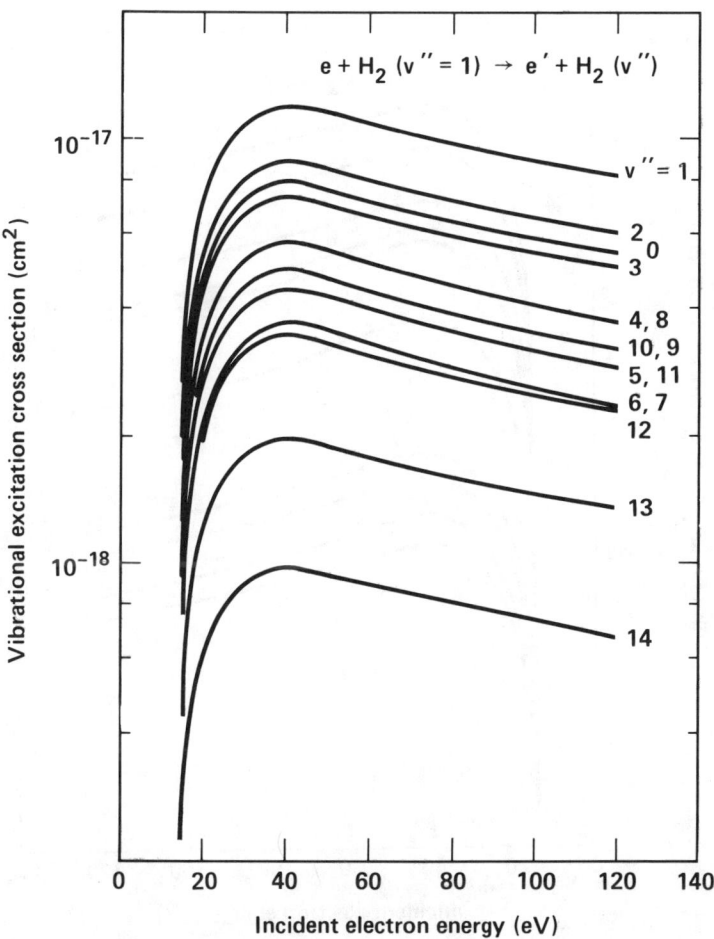

Fig. 2. Excitation cross section to level v" from initial level $\underline{v}"=1$.

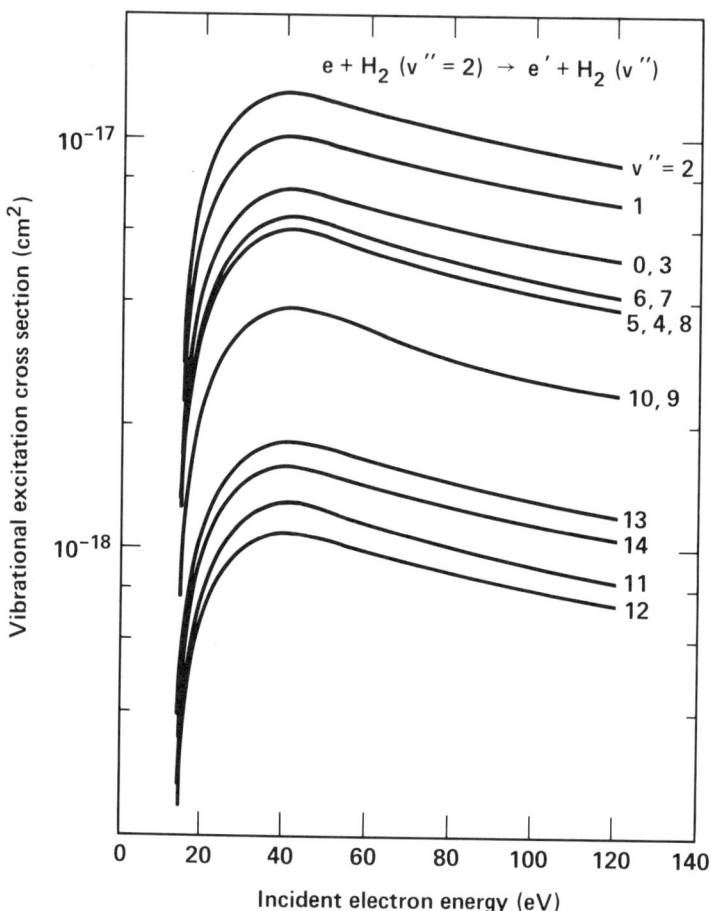

Fig. 3. Excitation cross section to level v" from initial level v"=2.

The Role of H_3^+ Ions in $H_2(v \geq 6)$
Production in Negative Hydrogen Ion Sources

J.B.A. Mitchell
Department of Physics, The University of Western Ontario
London, Ontario, Canada. N6A 3K7

and

W.G. Graham
University of Ulster, Coleraine, Northern Ireland

Abstract

The role of H_3^+ (D_3^+) ions in H_2 ($v \geq 6$) ($D_2(v \geq 9)$) production in $H^-(D^-)$ ion sources has been assessed. It is concluded that at most 4.6% of H_3^+ ions and 7.2% of D_3^+ ions contribute of $H_2(v \geq 6)$ or $D_2(v \geq 9)$ production in these plasmas. Based on previously established theoretical models this implies an additional contribution of about 15% to the overall production of highly vibrational excited molecules.

INTRODUCTION

It is now believed that the principal process leading to the high negative hydrogen ion densities measured in hydrogen plasmas[1], is through electron dissociative attachment to vibrationally excited hydrogen molecules, $H_2(v^*)$ ($D_2(V^*)$)[2,3,4]. Bardsley and Wadehra[5] have calculated that the threshold cross section for dissociative attachment rises by more than a factor of 10^4 as the vibrational quantum number is increased from 0 to 4 in H_2 with even a larger increase in D_2. They also find a distinct isotope effect for $H_2(v<6)$ and $D_2(v<9)$ but none at higher levels. Since no isotope effect was observed in the negative ion densities in the plasmas[1] it is thought that molecules with $v \geq 6$ for H_2 and $v \geq 9$ for D_2 are primarily responsible for the H^- production [2,3,4,6].

Several mechanisms for the production of hydrogen molecules in these vibrational levels have been proposed. Hiskes et al[3] have considered the role of fast primary electron collisions and subsequent thermal electron and molecule collisions on the $H_2(v^*)$ distribution. They conclude that if the $H_2(v^*)$ survives a sufficient number of wall collisions, these processes would lead to a population of highly vibrationally excited hydrogen molecules in the plasma which is almost sufficient to account for the observed H^- density.

A second possible source of $H_2(v^*)$ is through the H_3^+ ions in the plasma. These ions are known to be formed with internal energy and form a major ionic component in most of the hydrogen plasmas studied[7,8]. Since, under the plasma conditions being considered, the rates of ion-ion and ion-electron collisions are small compared with ion-molecule and ion-wall collisions, processes (1) and (2) below are the main loss processes for $H_3^+(v^{1*})$. The internal energy of an H_3^+ ion can be transferred to a hydrogen molecule by resonant proton transfer

$$H_3^+(v^{1*}) + H_2 \rightarrow H_2(v^*) + H_3^+ \text{ - - -} \tag{1}$$

or wall collisions

$$H_3^+(v^{1*}) + \text{wall} \rightarrow H_2(v^*) + H \text{ - - -} \tag{2}$$

In previous modelling [2,3] of H^- ion sources the role of H_3^+ in $H_2(v^*)$ production has been ignored since there was no quantitative information available. In this paper we assess the likely role of processes (1) and (2) in typical H^- ion sources.

PROTON TRANSFER

The resonant transfer of a proton is not the only possible interaction of H_3^+ with H_2. Studies[9] indicate that there may be several, as yet unresolved, interactions. However, it is found that with higher vibrational levels resonant proton transfer dominates.

The $H_2(v=6)$ ($D_2(v=9)$) molecule has an internal energy of 2.63 eV (2.78 eV). Therefore to evaluate the role of H_3^+ in $H_2(v \geq 6)$ it is necessary to estimate what fraction of the H_3^+ (D_3^+) population has internal energy in excess of 2.63 eV (2.78 eV) and how effectively this is transferred to the $H_2(D_2)$ molecules. Smith and Futrell[10] developed a statistical model to predict the internal energy distribution of H_3^+ and D_3^+ ions in a hydrogen plasma. Recently, this model has been elaborated by Anicich and Futrell[11] and these later results will be compared with experimental measurements and their applicability to the present plasma assessed. In this way, an upper limit on the $H_3^+(v^{1*})$ ($D_3^+(v^{1*})$) contribution to $H_2(v \geq 6)$ ($D_2(v \geq 9)$ production in the plasma through process (1) can be established.

(i) $H_3^+(v^{1*})$ formation

The processes leading to H_3^+ production in a hydrogen plasma are: the ionization of the hydrogen gas molecules by the fast primary electrons

$$e + H_2 \rightarrow H_2^+ + 2e \text{ - - -} \tag{3}$$

followed by H_2^+ collisions with H_2 to form H_3^+

$$H_2^+ + H_2 \rightarrow H_3^+ + H \text{ - - -} \tag{4}$$

There is excess energy available in this latter reaction. This is the difference between the dissociative energy of H_2^+, to $H^+ + H$ (2.66 eV) and the proton affinity of H_2 (4.34 eV)[12].

Therefore the excess energy available, if both reactants are in their ground states, is 1.68 eV. This excess energy is in the form of vibrational excitation of H_3^+, with only a small fraction converted to kinetic energy of the product particles[13]. While this energy is insufficient for $H_2(v \geq 6)$ production through processes (1), the H_2^+ ions produced in process (3) will have a distribution of vibrationally excited states and this internal energy will be available for additional vibrational excitation of H_3^+ in process (4).

Von Busch and Dunn[14] have studied the vibrational population of H_2^+ produced in electron-impact ionization of H_2. Their basic assumptions were that the hydrogen molecules are in the ground state and that the ionizing electrons have energies which are high compared to the threshold for the excitation into vibrational states. These conditions are met in the present plasmas since about 98% of the hydrogen molecules are in the ground state, and the primary electron energy is generally greater than 50 eV. Their results indicate that 42% of the H_2^+ ions produced by process (1) will have an internal energy of greater than 1 eV.

Anicich and Futrell[11] have used the results of Von Busch and Dunn[14] to calculate the internal energy distribution of $H_3^+(D_3^+)$ in a hydrogen (deuterium) plasma. They assume a random partitioning of the excess energy amongst all the vibrational and translational states of the product H_3^+ - H system.

Their calculations indicate that in the hydrogen plasma 1.6% of the H_3^+ ions have a vibrational energy of greater than 2.63 eV, i.e. they are in a vibrational state v≥8. (2.98 eV). Those H_3^+ ions in the next lowest vibrational state v=7, have an internal energy of 2.60 eV and account for a further 3.0% of the H_3^+ population.

In a deuterium plasma 3.2% of the D_3^+ are predicted to have a vibrational energy of greater than 2.78 eV, i.e. they are in a vibrational states, v≥ 11 state with an internal energy of 2.87 eV. D_3^+ in the v=10 state accounts for a further 4% of the D_3^+ population.

Anicich and Futrell[11] have examined experimental measurements of the H_3^+ (v^{1*}) population distribution in plasmas and conclude that their model is consistent with these results.

(ii) <u>Internal energy transfer from $H_3^+(v^{1*})$ to $H_2(v^*)$</u>

It is now necessary to consider the likelihood that the internal energy of the H_3^+ (v^{1*}) ($D_3^+(v^{1*})$) ions will be transferred to $H_2(D_2)$ in processes (1) and (2).

Smith and Futrell[9] have studied the energy relaxation of D_3^+ in reactions with H_2. It was found that D_3^+ ions become vibrationally de-excited after, on average, 5-10 collisions with H_2. However, the reaction mechanism is found to depend on the vibrational energy of the D_3^+. While the D_3^+ ions in lower vibrational states gradually lose their internal energy through the formation of intermediate $H_2D_3^+$ complexes, the D_3^+ in the higher vibrational levels undergo resonant deuteron transfer i.e., the D_2 molecule retains the internal energy of the original D_3^+. This rapid decrease in internal energy is supported by the observations of Kim et al[15] and Leventhal and Freidman[16]. Neither report any variation in the reaction rate with different vibrational levels although such effects are inferred in the interpretation of the results of Blakley et al[17].

Therefore it appears that for highly vibrationally excited $H_3^+(D_3^+)$, internal energy is effectively transferred to $H_2(D_2)$ through proton transfer. <u>At most</u> about 4.6% (7.2%) of the H_3^+ (D_3^+) ions in the plasma have sufficient energy to contribute to $H_2(v \geq 6)$ ($D_2(v \geq 9)$), production through this process.

WALL COLLISIONS

When a molecular ion collides with a metal surface it can be neutralized by electron capture processes involving primarily the conduction band electrons. Figure 1 illustrates this process. The capture can occur either resonantly into an excited state of the molecule or via an auger transition in which one electron is captured into the ground state while a second conduction band electron is released into the vacuum[18].

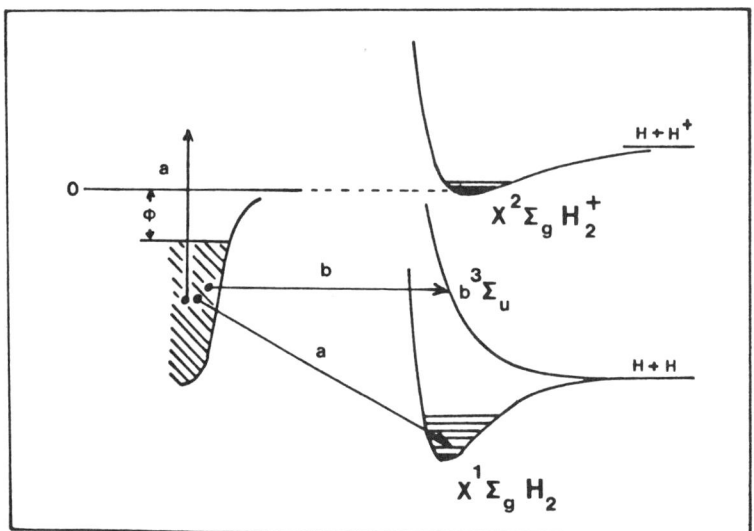

Figure 1. Neutralization processes involving the collisions of a molecule such as H_2 with a metal surface. Auger capture is represented by aa' where one electron is captured into the ground state of the molecule while a second electron is ejected from the metal into the vacuum. Resonant neutralization is illustrated by transition b. Here an electron is captured into an available excited state whose energy overlaps the energy of the electrons in the conduction band. In the case illustrated the excited state is repulsive and subsequently dissociates. As the molecular ion approaches the surface, the induced image potential will shift these states upwards.

It should be emphasized that these neutralization processes are quite distinct from binary free electron recombination processes such as dissociative recombination[19]. With surface neutralization, the electrons are in negative energy states with respect to the initial ion. Free electron recombination involves electrons having positive energies. In order to dispose of this energy the electron-ion system must radiate or dissociate. Radiative recombination occurs when photon emission occurs promptly during the electron-ion collision leading to the stabilization of the recombination. Alternatively an electron having the correct energy can be captured into an autoionizing state lying in the ionization continuum. This state can then decay via photon emission (dielectronic recombination) or in the case of molecules, via dissociation (dissociative recombination). In the former case the recombination energy is removed by the photon, in the latter by conversion to the kinetic energy of the dissociation products. The main point is that the states involved in free electron recombination are not the same as those involved in surface neutralization so that the resulting product states are likely to be different.

The collision of $H_2^+(v)$ with metal surfaces has been discussed by Hiskes and Karo[4], Grannenman et al[20] and more recently by Imke et al[21]. They have shown that H_2^+-wall collisions can indeed produce vibrationally excited H_2 molecules.

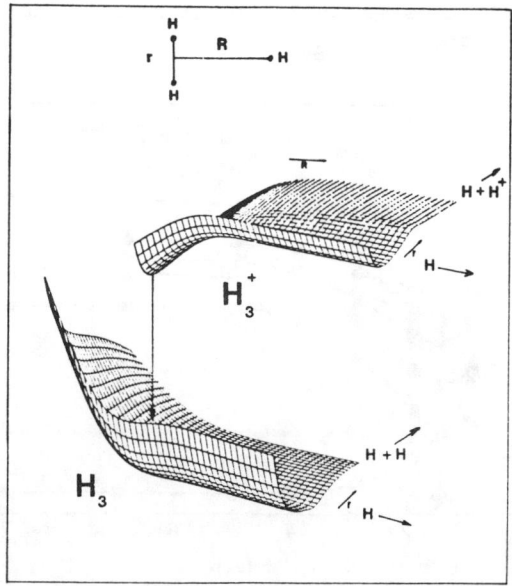

Figure 2 Illustration of the neutralization of H_3^+ to form H_3 via electron capture from a metal surface. If the transition occurs to a point high up on the repulsive side wall of H_3 the resulting H_2 molecule will be vibrationally excited.

The collision of H_3^+ ions with metal surfaces has been studied by Willerding et al[22]. They have shown that the neutralization is dominated by capture into the repulsive $2p^2E^1$ ground state of H_3 which dissociates to give H_2+H. Figure 2 shows a sketch of the curves involved in this capture. The resulting H_2 molecule could be vibrationally excited if the transition was to occur to a point high up on the side wall of the lower state. Unfortunately there is no quantitative experimental data on the vibrational population of the H_2 molecules resulting from H_3^+-wall collisions. Such information can however be obtained from recent theoretical studies on the photodissociation of excited Rydberg states of H_3[23]. This process ocurs via photoemission resulting in a transition from the initial excited state down to the repulsive H_3 ground state.

i.e. $H_3^R \rightarrow H_2 + H + h\nu$.

These Rydberg states are shown in Figure 3. It can be seen that they very closely resemble the H_3^+ ground state. This means that this photodissociation process, which involves a direct Franck-Condon transition from an upper to a lower state, should be very similar to the dissociative neutralization of H_3^+ in wall collisions provided the H_3^+ kinetic energy is low enough that vibrational excitation does not occur via momentum transfer. In an ion source, the H_3^+ ions will have less than one eV of kinetic energy and so this condition is satisfied.

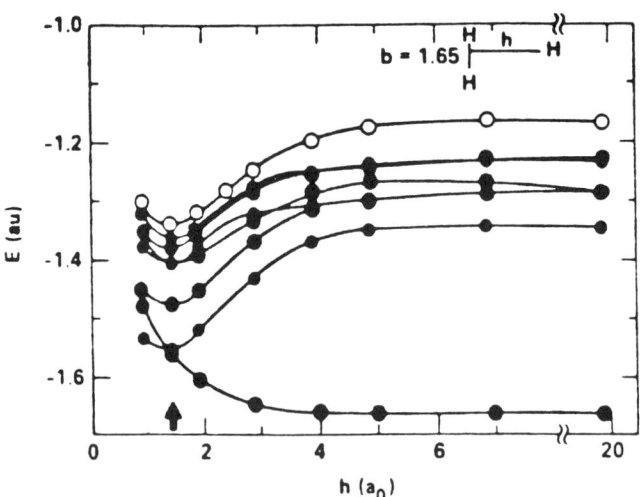

Figure 3. H_3 electronic potential energy curves for C_{2v} nuclear geometries. The large open circles indicate the ground state of H_3^+. Lowest curve is the H_3 ground state. Higher curves are Rydberg-like in the Franck-Condon region which is denoted by the arrow. (From Kulander and Light, 1986).

Kulander and Light[23] have calculated Franck-Condon factors for a number of transitions involving intital vibrational states of H_3^R, namely the (0,0,0), (1,0,0), (0,1,0) and (0,0,1) states to form $H_2(v)$+H. In all cases they found that transitions yielding $H_2(v=0)$ were dominant. It would appear therefore that the probability of producing vibrationally excited H_2 molecules from H_3^+-wall collisions is very small and certainly less than that through proton transfer.

Production of $H_2(v \geqslant 6)$ through H_3^+.

Theoretical models[2,3,4] which have been developed to predict the negative ion density have been concerned almost exclusively with H_2 plasmas. It is assumed that dissociative attachment to $H_2(v \geqslant 6)$ is the dominant production mechanism, and mutual neutralization and/or wall collisions, the main loss mechanism for H^- ions. It has been shown[2] that the density of vibrationally excited molecules n(v) is

$n(v) = An_+$

wehre n_+ is the positive ion density and A is given by

$A = [b \; v(3+)/v(^*)][p(v^*) + \sigma v \; (E,v^*)/\sigma v(2+)]$

where b is the number of wall collisions and $H_2(v^*)$ molecule can survive, $v(3+)$ and $v(^*)$ are the average velocities of the H_3^+ and $H_2(v^*)$ molecules respectively, $\sigma v(E,v^*)$ is the rate at which $H_2(v^*)$ are populated by primary electron collisions and $\sigma v(2+)$ is the rate of wall neutralization of H_2^+ to form $H_2(v^*)$.

The parameter $p(v^*)$ is the fraction of H_3^+ ions converted to $H_2(v \geqslant 6)$ in processes (1) and (2). As mentioned previously these processes have been ignored in previous models. Here we have established that for wall collisions of H_3^+ $p(v^*)$ is essentially zero. For proton transfer $p(v^*)$ has an upper value of about 0.05. Since $\sigma v \; (E,v^*)/\sigma v(2+)$ is approximately 0.33^2, it can be seen that H_3^+ ions will make only an additional 15% contribution to the $H_2(v \geqslant 6)$ production. A similar contribution would be expected in a D_2 plasma. However these processes should still be taken into consideration in detailed calculations of the complete vibrational population distributions in H_2 or D_2 plasmas.

REFERENCES

1. M. Bacal and G.W. Hamilton, Phys. Rev. Lett. 42, 1538 (1979).
2. M. Bacal, A.M. Burneteau, W.G. Graham, G.W. Hamilton and N. Nachman, J. Appl. Phys. 52 1247 (1981).
3. J.R. Hiskes, A.M. Karo, M.Bacal, A.M. Burneteau and W.G. Graham, J. Appl. Phys. 53, 3469 (1982).
4. J.R. Hiskes and A.M. Karon, J. Appl. phys. 56, 1927 (1984).
5. J.N. Bardsley and J.M. Wadehra, Phys. Rev. A20, 1398 (1979).
6. M. Bacal, Physica Scripta T2/2, 467 (1982).

7. E. Nicolopoulou, M.Bacal and H.J. Doucet, J. Phys. (Paris) 38, 1399 (1977).
8. K.W. Ehlers and K.N. Leung, Rev. Sci. Instrum. 54, 1296 (1983).
9. D.L. Smith and J.H. Futrell, Chem. Phys. Lett. 40, 229 (1976).
10. D.L. Smith and J.H. Futrell, J. Phys. B8, 803 (1975).
11. V.G. Anicich and J.H. Futrell, Int. J. Mass Spect. and Ion Proc. 55, 189 (1983).
12. H.H. Harris, M.G. Crowley, T.R. Grossheim, P.J. Woessner and J.J. Leventhal, J. Chem. Phys. 59, 6181 (1973).
13. V.L. Tal'roze, Iz. Akad. Nauk. SSSR Ser Fiz. 24, 1975 (1978).
14. F. von Busch and G.H. Dunn, Phys. Rev. A, 1726 (1972).
15. J.K. Kim, L.P. Thread and W.T. Huntress, Int. J. Mass. Spectrom. Ion Physics 15, 223 (1974).
16. J.J. Leventhal and L. Friedman, J. Chem. Phys. 50, 2928 (1969).
17. C.R. Blakley, M.L. Vestal and J.H. Futrell, J. Chem. Phys. 66, 2392 (1977).
18. H.D. Hagstrum, Phys. Rev. 89, 244 (1953).
19. J.B.A. Mitchell, in Electron-Ion and Ion-Ion Collisions (edited by F. Brouillard and P. Defranc) NATO ASIA, Han-sur-Lesse. Plenum, N.Y. 1986.
20. E.H.A. Granneman, J.J.C. Geerlings, J.N.M. Vanwunnik, P.J. Van Bommel, H.J. Hopman, and J. Los, "Production and Neutralization of Negative Ions and Beams", AIP Conf. Proc. No. 111, American Inst. of Phys., N.Y. p. 206, (1983).
21. U.Imke, K.J. Snowdon and W.Z. Heiland, Phys. Rev. B 34, 41, (1986) and 34, 48, (1986).
22. B. Willerding, K.J. Snowdon and W.Z. Heiland, Phys. B 59, 435, (1985).
23. K.C. Kulander and J.C. Light, J. Chem. Phys. 85, 1938, (1986).

DISCUSSION

<u>Michels</u>: I guess I don't quite understand the analogy of Rydberg transitions to the ground state in H_3^+. After all, you have to have quite a bit of energy for the electron to attach. Don't you think there would be a wall distortion that would totally change this from the pre-gas Franck-Condon factors?

<u>Mitchell</u>: Well, that's true. I didn't mention the image potential, but all I'm really saying is that if you think of it as being a process involving a transition from an H_3^+ down to the neutral state, the Franck-Condon factors for that haven't been calculated, but the Franck-Condon factors for the Rydberg state have been calculated. So I just assumed that they would not be very different because the nuclear geometry is very similar. The shape of the curves is similar.

<u>Michels</u>: But at the point that you have electron attachment, there may be significant distortion of the H_3^+ geometry.

Mitchell: That's true, indeed. The question is, then, how far out from the wall does the actual capture occur? And how much distortion is in there? Now, we're getting into an area where we need a theoretical calculation for it, really, and that's possibly true, that it could make some difference. I think the question is, when does the neutralization occur? And I think that's fairly far out from the wall. The question is, if the image potential shifts the curves up, does it change things very much, or does it just change the energy? Will it change the shape of the curves, or will it just shift them to higher energies?

Peterson: We did some charge-transfer measurements of H_3^+ on cesium and analyzed the fragment energies. From that, I concluded that about 2.6 eV was tied up internally in the H_2. You can have two products from there, and it also goes primarily through a Rydberg level, which is then predissociated, and it can go either to the H + H + H channel, or H_2 + H. From that analysis, I concluded that about 2.6 eV of internal energy was tied up in the H_2. It would be surprising if all that were tied up in rotation. So I don't quite trust your answer.

Mitchell: That's fair enough. But, as I said, I don't know the answer to that, and this was just one thing we came up with. So that's a good reason to do some experimental measurements on this. There aren't any experimental measurements, so at this point, I can speculate.

Bacal: What about the other process which has been proposed and studied for producing directly H^- from H_3^+ in dissociative recombination of H_3^+? It's not your subject, but it is a related subject. Maybe you can tell us what you think about this.

Mitchell: What you're talking about is electrons plus H_3^+ giving H^- + H_2^+?

Bacal: H_2^+, yes, or H^+ + H...whatever...H^-. This was measured experimentally by Peart & Dolder some time ago.

Mitchell: Yes, and they got a very small cross-section for it.

Bacal: Well, but from non-excited H_3^+, or...unclear vibrational excitation of H_3^+, but what does the theory say?

Mitchell: Well the thing about it is, whether Peart & Dolder's H_3^+ was de-excited or not is very much an open question. In fact, it now looks as if it wasn't. It wasn't. In fact, you can never really de-excited H_3^+. You'll always get some vibrational excitation. I think, probably, his results were a lot more excited than what he originally thought they were. Even so, I don't think it's going to make a big difference. The thing is that these curves going to the ion pair - that is actually the main channel for dissociative recombination. But you go through so many curve crossings that go out to the neutral channels that, by the time you got a way out to this state, the process is over.

MODELING OF ELECTRON ENERGY DISTRIBUTION FUNCTIONS AND VIBRATIONAL DISTRIBUTIONS IN VOLUME H$^-$ ION SOURCES*

Jean Bretagne
Laboratoire de Physique des Gaz et des Plasmas
(Associated with the C.N.R.S.)
Université Paris-Sud, 91405 Orsay Cedex (France)

Mario Capitelli and Claudine Gorse
Centro di Studio per la Chimica dei Plasmi (C.N.R.)
Dipartimento di Chimica, Universita di Bari, 70126 BARI (Italy)

ABSTRACT

The modeling of magnetic multicusp H_2 discharges is presented. The model is based on the coupling of the time-dependent Boltzmann equation for the electron energy distribution function with the system of master equations for the H_2 vibrational excitation and kinetic equations for H and H$^-$ densities.

The theoretical results given here are found to be in satisfactory agreement with available experimental ones and indicate, together with previous results, an extended validity range of the model.

INTRODUCTION

Modeling of magnetic multicusp discharges is a problem of current actuality due to the use of these plasmas for the production of intense H$^-$ beams [1,2]. Dissociative attachment from vibrationally excited H_2 molecules ($H_2(v)$) is the accepted mechanism for the production of H$^-$. Optimization of these sources requires enormous theoretical and experimental efforts towards the comprehension of the complicated kinetics acting on the system.

A cooperation between the groups in Bari, Orsay and Palaiseau has been recently undertaken with the aim of a better understanding of some important points such as the electron energy distribution functions (e.e.d.f.) and the vibrational distribution of $H_2(N_v)$ in such plasmas. To this end, two independent computer codes have been developed, one which deals with the self-consistent solution of the e.e.d.f., N_v, N_H (atom number density) and N_{H^-} (negative ion density)[2], the other one which essentially solves the Boltzmann equation for the e.e.d.f.[3]. Detailed descriptions of these two codes have been given elsewhere[2,3]. In this paper we want to illustrate some results linked to the development of the multipole source operating at the Ecole

---

*Work done in cooperation with
Marthe Bacal Group
Laboratoire de Physique des Milieux Ionisés
(Associated with the C.N.R.S.)
Ecole Polytechnique, 91128 Palaiseau Cedex (France)

Polytechnique. Special emphasis will be given to the temporal evolution of N_v, e.e.d.f. and other quantities towards the achievement of quasistationary values.

METHOD OF CALCULATION

The numerical code developed by Gorse et al.[2] simultaneously solves : 1) the Boltzmann equation for e.e.d.f. in electron beam sustained discharges, 2) the vibrational kinetics for the 14 levels of H_2, 3) the dissociation kinetics of H_2, 4) the kinetics of H^- production. Coupling between the different kinetics occurs as a result of the interdependence of some terms in the relevant equations, which are written as follows

$$\frac{\partial n(\varepsilon,t)}{\partial t} = -\left[\frac{\partial J_{el}}{\partial \varepsilon}\right]_{e-M} - \left[\frac{\partial J_{el}}{\partial \varepsilon}\right]_{e-e} + In + Ion + Sup + Rot - S_1 - L_1 \quad (1)$$

$$\frac{dN_v}{dt} = \left[\frac{dN_v}{dt}\right]_{e-V} + \left[\frac{dN_v}{dt}\right]_{E-V} + \left[\frac{dN_v}{dt}\right]_{V-V} + \left[\frac{dN_v}{dt}\right]_{V-T} +$$

$$+ \left[\frac{dN_v}{dt}\right]_{e-D} + \left[\frac{dN_v}{dt}\right]_{w} \quad (2)$$

$$\frac{dN_H}{dt} = S_2 - L_2 \quad (3)$$

$$\frac{dN_{H^-}}{dt} = S_3 - L_3 \quad (4)$$

In Eq. 1, $n(\varepsilon,t)$ is the electron density in the energy range ε to $\varepsilon + d\varepsilon$, while the different terms represent in the order the flux of electrons along the energy axis due to electron-molecule, electron-electron, inelastic, ionization, superelastic and rotational collisions. The last two terms in eq. 1 are the source of electrons emitted by a filament and the loss due to recombination and diffusion. All these terms have been make explicit in ref. 3 (see also ref. 2).

The different terms in eq. 2 refer to the relaxation due to e-V [$e + H_2(v) \rightarrow H_2^- \rightarrow e + H_2(v)$] and E-V [$e + H_2^* \rightarrow e + H_2(w) + h\nu$] processes, to V-V (vibration-vibration) and V-T (vibration-translation), energy exchanges, to electron dissociation and wall processes.

Eq. 3 includes the source term (S_2) of atomic H essentially through dissociation towards the triplet states as well as through ion-molecule reactions, and a loss term (S_2) due to wall recombination. Eq. 4 finally includes the H^- production (S_3) through dissociative attachment from vibrationally excited molecules ($e + H_2(v) \rightarrow H + H^-$) followed by some destruction reactions (L_3). All these terms have been made explicit in our previous works[2-4], the only difference being the inclusion made in this work of dissociative attachment from all levels ($v = 0-14$)[5].

Moreover the V-T terms due to the process

$$H + H_2(v) \rightleftarrows H + H_2(w)$$

makes use of the recent set of rate coefficients calculated by Laganà[6], while the relaxation term due to the wall collisions has been treated as discussed by Hiskes and Karo[1] (see ref. 4). The numerical code developed by Bretagne et al.[3] essentially solves the Boltzmann equation for the e.e.d.f., by taking into account numerous processes. This code can be considered as a guide for understanding many problems connected with the calculation of e.e.d.f. in multicusp H_2 plasmas[3].

SELECTED RESULTS

We start the discussion of our results by considering the coupled kinetics occuring in a multicusp H_2 plasma, the dimension of which (electron loss area A, and volume V) as well as the electrical characteristics (current intensity I, applied voltage V_d and energy spread ΔV_d of the source) are reported in the relevant figures. The reported calculations consider T_H = 4000 K at T_{H2} = 500 K, T_H and T_{H2} being respectively the translational temperature of H atoms and H_2 molecules. For the recombination coefficient γ of atomic H on metallic surfaces we have utilised a γ value equal to 0.1, while the plasma potential V_p which enters in the electron loss term $(L_1)^3$ has been selected according to the experimental values obtained by M. Bacal. In particular values of V_p = 1.15, 1.65, 1.91 V have been respectively used for the pressures of 2, 4 and 8 mTorr.

Fig. 1, which reports the temporal evolution of e.e.d.f., shows the typical contributions in the e.e.d.f. in this kind of plasma i.e. the source electrons (45 $\leq \varepsilon \leq$ 50 eV), a plateau (10 $\leq \varepsilon \leq$ 40 eV) and the thermal component (ε < 10 eV). Moreover we note that the tail of the e.e.d.f. (ε > 15 eV) reaches a quasistationary condition in a time (t $\sim$ 0.5 μs) shorter than the corresponding one (t $\sim$ 0.5 ms) necessary to the low-energy part (ε < 15 eV) to achieve a stationary regime.

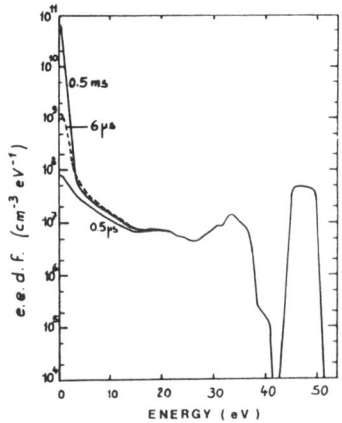

Fig. 1 : Temporal Evolution of the e.e.d.f. for the following conditions : P = 4 mTorr ; I = 5 A ; V = 8.8 l ; A = 830 cm^2 and V_d = 50 V

Figure 2 reports the temporal evolution of N_H and N_{H^-} under the same discharge conditions, these quantities reaching stationary values in a time scale of one millisecond.

Figure 3 reports the vibrational distribution of H_2 at different times as a function of the vibrational quantum number v. The shape of the distribution is the result of the different processes acting in the plasma. In particular the long plateau extending from v = 3 to v = 10-12 is the result of the E-V processes, while the behaviour for v < 3 is dominated by e-V energy exchanges. The small concentration of H atoms as well as their high temperature is such that the deactivation on the plateau is mainly made by wall collisions as well as by $H_2(v)+H_2 \rightarrow H_2(v-1)+H_2$ ones. This last process does not effect the vibrational distribution for v < 10, so that the shape of N_v for v < 3 is the result of the pumping by e-V processes and destruction by wall collisions.

The effect of the pressure on the vibrational distribution is reported in Figs. 4-5 for the same discharge conditions previously reported. We note that the increase of pressure (see fig. 4), increasing the effectiveness of V-T deactivation by molecules, decreases the extension of the plateau, the reverse being true at low pressure (see fig. 5).

The results reported in figs. 3-5 can be considered as representative of discharge conditions characterized by a small production of H atoms. The opposite case has been studied in refs. 2-4. In this last case the high concentration of H atoms is such to prevent the formation of a long plateau in the vibrational distribution of H_2.

Before comparing the numerical results with some available experimental data, we want to discuss some points which can affect the calculations. In particular we want to study the effect of superelastic electronic collisions as well as the choice of the plasma potential on the e.e.d.f. and related properties. These effects will be discussed by using the code developped by Bretagne et al.[3,7].

Figure 6 reports the effect of superelastic electronic collisions i.e. of the processes

$$e + H_2 (a^3\Sigma_g^+, c^3\pi_u) \rightarrow e + H_2 (X^1\Sigma_g^+)$$

on the e.e.d.f. The concentration of excited electronic states has been arbitrarely varied for 10^{-2} to 10^{-6}. We see that the presence of such collisions does generate a peak at about 12 eV, yielding also a decrease of the low-energy portion of electron density. These two points are in qualitative agreement with some experimental findings as discussed in ref. 7. The choice of plasma potential is shown in fig. 7, where e.e.d.f. calculated at V_p = 2 eV and V_p = 5 eV are compared. The increase of the plasma potential is such to strongly increase the electron density, still in agreement with the experimental results[7].

Let us now compare the theoretical results with the experimental ones (see table 1). Unfortunately in the case of the large multipole discussed in this work the experimental values are limited to T_e, n_e or N_{H^-}, lacking informations on N_v and N_H. Keeping in mind, however that n_e and T_e are indicative of the low-energy part of e.e.d.f., while N_{H^-} depends both on e.e.d.f. and on the plateau of N_v, we can conclude that the comparison of the n_e, T_e and N_{H^-} value indirectly samples different zones of the e.e.d.f. and N_v.

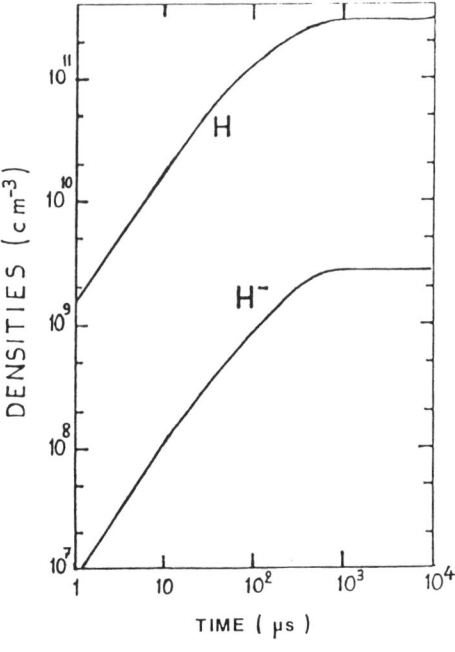

Fig. 2. Temporal evolution of H and H⁻ densities. Same experimental conditions as in Fig. 1.

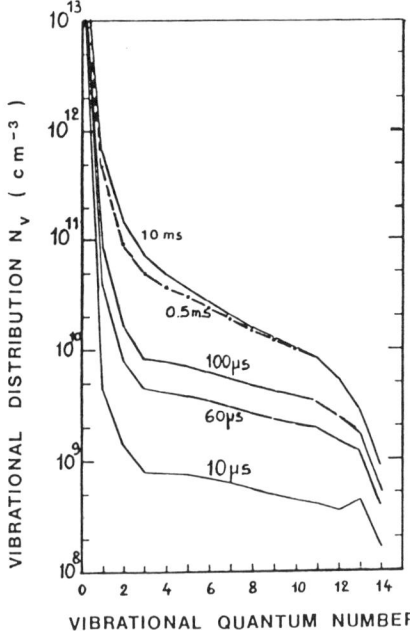

Fig. 3. Temporal evolution of the vibrational distribution for P = 4 mTorr. Other experimental conditions as in Fig. 1.

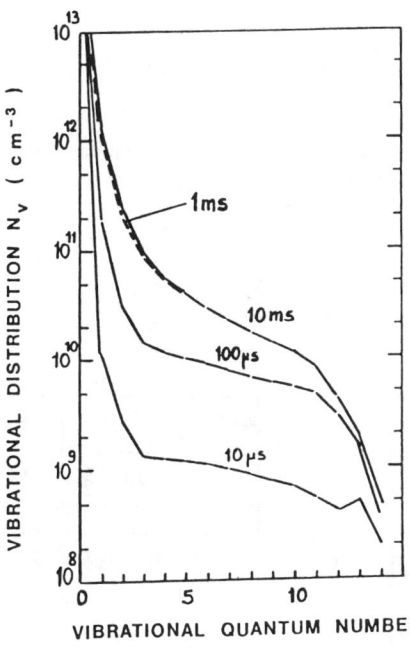

Fig. 4. Temporal evolution of vibrational distribution for P = 8 mTorr. Other experimental conditions as in Fig. 1.

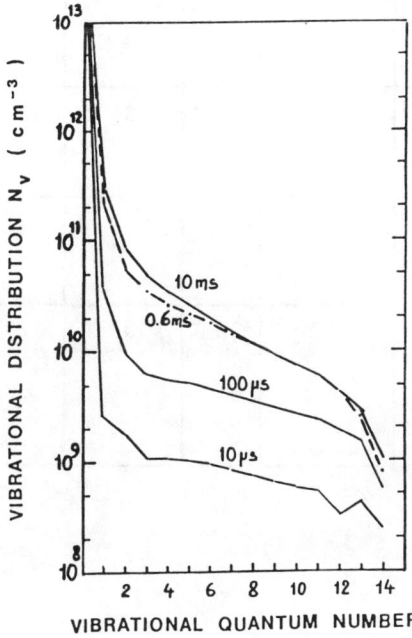

Fig. 5. Temporal evolution of the vibrational distribution for P = 2 mTorr. Other experimental conditions as in Fig. 1.

Table 1

COMPARISON BETWEEN THEORY AND EXPERIMENT
FOR DIFFERENT PRESSURE VALUES

For the experimental results, the two indicated values for n_e and (H^-) are the extreme values obtained in the spatially resolved measurements. The bounds must not be confused with experimental incertainties.

For theoretical results:
    a) results obtained with Gorse et al., code[2];
    b) results obtained by Bretagne et al., code[3].

Pressure mTorr		n_e cm^{-3}	T_e eV	(H^-) cm^{-3}	(H) cm^{-3}	$\dfrac{(H)}{n_e}$	$\dfrac{(H)}{(H)+(H_2)}$	θ_1 K
2	Th. a) b)	3.43×10^{10} 2.5×10^{10}	0.40 0.39	2.3×10^9	1.6×10^{11}	6.7 %	0.40 %	1215
	Exp.	1.1×10^{10} 3.4×10^{10}	0.75	3.2×10^9				
3	Exp.	2.0×10^{10} 6.5×10^{10}	0.37	2.3×10^9 6.3×10^9				
4	Th. a) b)	6.0×10^{10} 4.2×10^{10}	0.35 0.39	2.7×10^9	2.8×10^{11}	4.5 %	0.36 %	1274
5	Exp.	2.0×10^{10} 1.0×10^{11}	0.28	3.0×10^9 4.5×10^9				
8	Th. a) b)	6.4×10^{10} 4.0×10^{10}	0.325 0.40	2.7×10^9	4.9×10^{11}	4.2 %	0.30 %	1258
	Exp.	2.0×10^{10} 1.2×10^{11}	0.37	1.5×10^9 3.8×10^9				

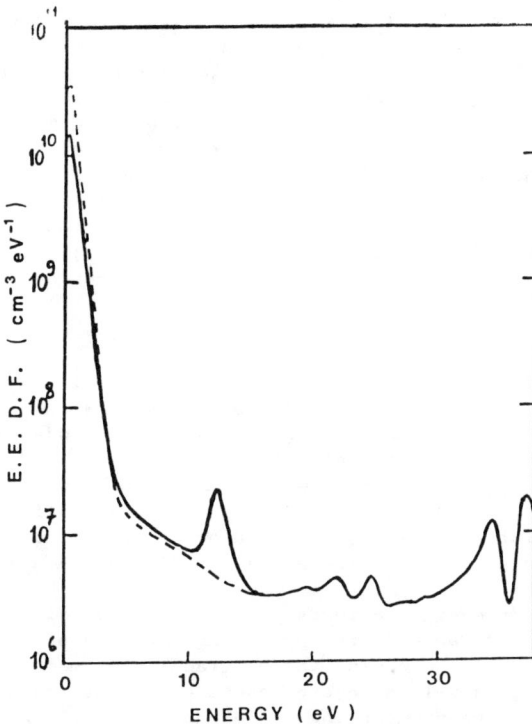

Fig. 6 : Influence of superelastic electronic collisions on the e.e.d.f. for the following conditions : P = 2 mTorr ; Vd = 50 V ; A/V = 0,1 and I = 5 A. The relative concentration m on the triplet states (see text) is : full line, m = 10^{-2} ; dotted line, m = 10^{-6}.

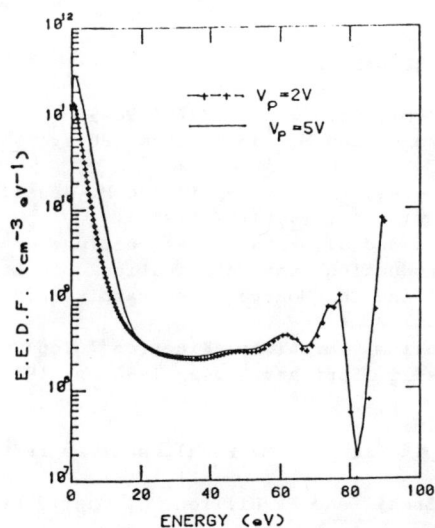

Fig. 7. Influence of the plasma potential V_p.
n_a = 8.5x10^{13} cm^{-3}; I=100 A; E_p = 90 V; A/V = 0.07 cm^{-1}.

First, from table 1, we note a satisfactory agreement between the n_e, T_e theoretical values obtained by the two codes, as already discussed in previous works[3]. A similar satisfactory agreement is found in the comparison of experimental and theoretical results for n_e and T_e. Concerning the concentration of H^-, $[H^-]$, we note that the computer code gives a yield $[H^-]/n_e$ which decreases with increasing the pressure. A similar trend is observed in the experimental results, which anyway are within a factor 2 close to the experimental ones. This point is of particular importance, since, as already mentioned, the production of negative ions depends on both e.e.d.f. and N_v, which in turn reflects the numerons elementary collision processes occuring in the plasma.

CONCLUSIONS

The results reported in this work as well as those presented in previous papers[2-4,7] show that our attempts to model multicusp H_2 discharges for the production of negative ions seem succesfull in reproducing some experimental results.

Of course the final answer to the problem will be given only when new diagnostic tools will produce : 1) the vibrational distribution of N_v in the plateau regime ; 2) the atom and ion concentration as well as their translational temperature ; 3) the e.e.d.f. and related quantities.

The experimental efforts made in this direction by different laboratories[8-12] seem very promising. On the other hand, progress in theory especially concern the interaction of molecules and atoms with surfaces[13,14], as well as energy exchange in gas phase[14], is being made. Cooperation between modeling and diagnostic would give as a result the final answer to the optimization of magnetic multicusp plasma for volume production of H^-.

Acknowledgments. This work has been partially supported by the "Progetto strategico Fusione Nucleare del C.N.R." (year 1986) and Ecole Polytechnique. The numerical results have been obtained with computer means granted by the Conseil Scientifique du Centre de Calcul Vectoriel pour la Recherche (France).

REFERENCES

1. J.R. Hiskes and A.M. Karo, J. Appl. Phys., **56**, 1927 (1984).
2. C. Gorse, M. Capitelli, J. Bretagne and M. Bacal, Chem. Phys., **93**, 1 (1985).
3. J. Bretagne, G. Delouya, C. Gorse, M. Capitelli and M. Bacal, J. Phys. D : Appl. Phys., **18**, 811 (1985) ; **19**, 1197 (1986).
4. C. Gorse, M. Capitelli, M. Bacal and J. Bretagne, Proceedings of the 2nd European Workshop on Production and Application of Light Negative Ions. Ed. M. Bacal and Ch. Mouttet, Laboratoire P.M.I., Ecole Polytechnique, (1986), p. 113.
5. J.M. Wadhera, in "Non-Equilibrium Vibrational Kinetics". Top Curr. Phys., Vol. 39. Ed. M. Capitelli, Springer Verlag 1986, p. 191 and private communication.
6. A. Laganà, same as ref. 4, p. 101.
7. J. Bretagne, C. Gorse, M. Capitelli and M. Bacal, same as ref. 4, p. 55.
8. M. Péalat, J.P.E. Taran, M. Bacal and F. Hillion, J. Chem. Phys., **82**, 4843 (1985).

9. M.P.S. Nightingale, M.J. Forrest and R.Mc Adams, same as ref. 4, p. 123.
10. J.H.M. Bonnie, P.J. Eendhuistra, W. Van Schelt and H.J. Hopman, same as ref. 4, p. 155.
11. W.G. Graham and M.B. Hopkins, same as ref. 4, p. 147, p. 155. ; M.B. Hopkins and W.G. Graham, Rev. Sci. Instrum., 57, 2210 (1986).
12. M. Lefebvre, M. Péalat, J.P.E. Taran, F. Hillion and M. Bacal, same ref. 4, p. 107.
13. A.M. Karo, J.R. Hiskes and R.J. Hardy, J. Vac. Sci. Tech., A3, 1222 (1985).
14. G.D. Billing, M. Cacciatore and M. Capitelli, work in progress.

DISCUSSION

Wadehra: If I understand correctly, one of the basic assumptions that you made in the beginning was that two-term expansion of the distribution function, correct? There was a two-term expansion of the electron energy distribution function in solving the Boltzman equation. Is that right?
Bretagne: We consider the electron distribution function as isotropic and the assumption of the two-term expansion was, in fact, important when we treat the problem of electron energy distribution function calculation in the presence of an electric field. It's not the case here, since we consider that primary electrons injected in the gas at an energy which corresponds to the discharge voltage.
Mitchell: You mentioned recombinations, but passed over it. Am I right in thinking that the recombination is going to compete for the low-energy electrons that are needed for the dissociative attachment to form H^-?
Bretagne: Yes. In fact, what we have assumed in the calculation is that recombination processes are a dissociative recombination of H_3^+ or H_2^+. We are now doing additional calculations involving equilibrium of positive ions. For typical currents that we have here, what we can say is that probably the most important positive ions are H_3^+. In fact, the effect of recombination is to create some saturation of the density of low energy electrons.
Mitchell: But the recombination will remove those low energy electrons. Therefore, if you have too many ions, you would remove all the low energy electrons? Because the cross-section increases as the energy goes to zero.
Bretagne: We have used, for the cross-section for recombination, the results of Peart & Dolder.
Mitchell: That cross-section tends to infinity as the energy tends to zero, therefore it's going to remove the low-energy electrons.
Bretagne: We have assumed that all low-energy electrons can recombine.
Mitchell: But those low-energy electrons are needed for H^- production as well, through the attachment. So the more recombinations, the less attachment?

Bretagne: Yes. In fact, I think it's also a problem, with this calculation, to know exactly the cross-section for dissociative recombination.

Bacal: Well, I would like to make a comment on the question. I mean, we can take into account the cross-section for dissociative recombination of H_3^+, for example. But it appears that this cross-section depends very much on the vibrational excitation of H_3^+. And ground state H_3^+ doesn't recombine at all. That's recent by David Smith. So actually, it's a very difficult question to introduce the right values for the dissociative recombination if there's H_3^+, because we do not know the vibrational distribution of H_3^+.

Mitchell: With regard to the ground state of H_3^+, though, the H_3^+ in the plasma are going to be excited, that's the thing. It's very hard to de-excite H_3^+. So even though the ground state is zero...

Bacal: I think this is a gap in our knowledge which has to be filled - to know what is the vibrational distribution of H_3^+, so we can model it. We can't invent it.

Peterson: I would suggest that you use, at least for the time being, the experimental values that Brian Mitchell has measured in beams because, there, the vibrational distribution of the H_3^+ will be similar to what you'd find in the plasma.

Bacal: Well, okay, we just want to have that value.

Peterson: The cross-sections exist.

Bacal: Okay, good. Very good suggestion.

MUTUAL NEUTRALIZATION – THREE BODY EFFECTS

J. M. Wadehra
Department of Physics and Astronomy
Wayne State University, Detroit, Michigan 48202

ABSTRACT

Computer simulated experiments, using the classical Monte-Carlo trajectories, on the loss of negative ions by the process of mutual neutralization reveal that the neutral ambient gas particles enhance the rate of neutralization. A simple physical picture can explain this enhancement. A review of the traditional mutual neutralization and termolecular recombination processes is also provided.

INTRODUCTION

In dealing with the problems of ion neutralization two processes are commonly discussed. These two processes for the recombination of ions A^+ and B^- are:

$A^+ + B^- \rightarrow A^* + B^*$, two-body mutual neutralization

$A^+ + B^- + M \rightarrow [AB^*] + M$, termolecular recombination.

The asterisk (*) indicates the possibility of excited states of the neutralized products.

In the process of termolecular (also called three body) recombination, the ambient neutral gas particles M serve to assure a simultaneous conservation of total energy and momentum. At moderate to high gas pressures this process is the dominating recombination process since the background gas particles M play a key role at such pressures. It is traditionally suggested that at low gas pressures, when the collisions with the background gas particles are relatively rare, the predominant recombination process is the mutual neutralization. In such discussions it is implicitly implied that the ambient gas particles play almost no role in controlling the process of mutual neutralization. Recent investigations[1], using computer simulated experiments, indicate that in a typical ionized system the processes of mutual neutralization and of three body recombination are in competition in such a fashion that as the rate of mutual neutralization increases, the rate of three body recombination decreases, so that the sum of the two rates is approximately constant. This observation suggests that the background gas particles do indeed affect the rate of recombination via the process of mutual neutralization.

MUTUAL NEUTRALIZATION : TRADITIONAL APPROACH

In the process of mutual neutralization, the total energy released by the transfer of electron from the negative ion to the positive ion is converted either into the excitation energy or the kinetic energy of relative motion of the neutral products. One possible way of achieving it is by means of a pseudocrossing of the initial and the final potential energy curves. Referring to Figure 1, the potential energy curves belonging to the initial pure ionic state (dissociating into $A^+ + B^-$ for asymptotically large internuclear separations R) and the final pure covalent state (dissociating into A^* + B^*) have the same energy at some internuclear separation R_x. If these two molecular states have the same electronic symmetry, their adiabatic potential energy curves cannot cross. The pseudocrossing of two such states at R_x is shown in Figure 1. As the ions approach one another under the influence of mutual Coulomb attraction and reach the pseudocrossing point R_x they either can make a transition with probability p from the ionic adiabatic curve to the covalent adiabatic curve or make no transition with probability 1-p. In the case of latter event as the ions separate out the reverse can happen so that the total probability for mutual neutralization is P = 2p(1-p) which has a maximum value of 0.5.

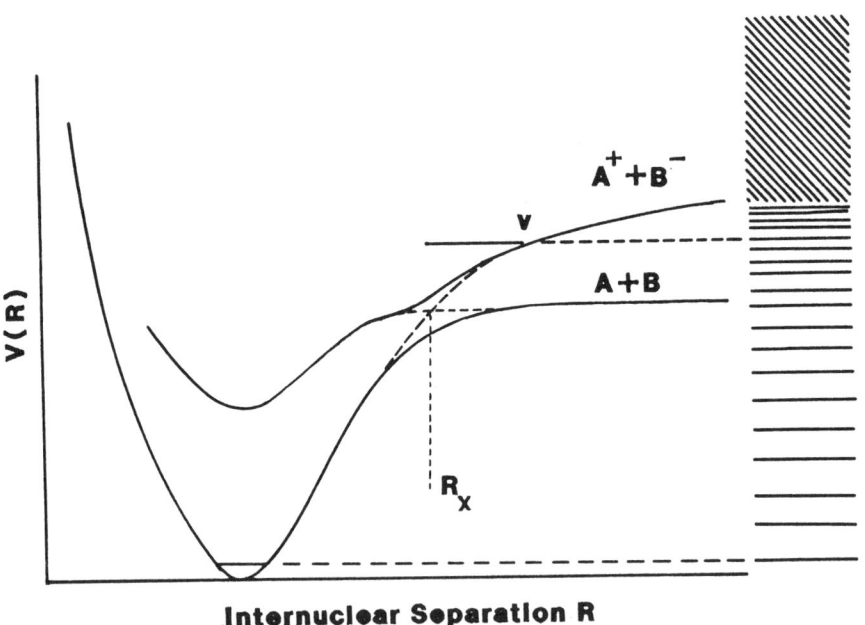

Figure 1. A schematic description of the Landau-Zener curve crossing model for mutual neutralization.

Following Landau[2] and Zener[3] the probability p of transition from one curve to another can be obtained by using the first-order perturbation theory in a two state (ionic and covalent) system where there is interaction only near the crossing point R_x. This leads to the Landau-Zener formula

$$p = \exp(-\eta/v_r) \quad (1)$$

where v_r is the relative radial velocity of the ions at the pseudocrossing point. The Landau-Zener parameter η depends on the slopes of the two potential curves as well as on the energy separation ΔE between the two curves at the pseudocrossing point. In the perturbation treatment, the energy separation ΔE is related to the interaction matrix elements H_{if} coupling the two electronic states. Several schemes exist[4] for approximate evaluation of the coupling matrix elements H_{if} and, therefore, of the Landau-Zener parameter η. Even though the Landau-Zener model has received its fair share of criticism in the literature[5], in practice, it does remain a valuable qualitative tool for estimating the rates of mutual neutralization in an ionic system. The process of mutual neutralization with discussions of theoretical and experimental investigations of various ionic systems has been reviewed in detail[6].

TERMOLECULAR (THREE BODY) RECOMBINATION: TRADITIONAL APPROACH

The classical theory of ion-ion recombination, using the three body mechanism, as developed by Natanson[7] works for any arbitrary gas pressure. This theory successfully bridges the gap between the low pressure model of Thomson[8] and the high pressure model of Langevin[9] for ionic recombination.

The low pressure model of Thomson begins by assuming that the ions are in thermal equilibrium with the ambient gas particles at temperature T. The average energy of relative motion for an ion pair then is $3kT/2$ for large separation distances. As the oppositely charged ions approach under the influence of mutual Coulomb attraction, they gain kinetic energy at the expense of potential energy. The total relative energy remains constant in the absence of any collisions with the neutral ambient gas particles. Now a trapping radius r_T is defined such that if one of the ions makes a collision with the neutral gas particle while separated by a distance less than r_T from the oppositely charged ion, the average loss of kinetic energy will be so large that the total relative energy will become negative. The ions will then form a bound pair and the recombination will be likely. The trapping radius is determined by the condition $3kT/2 = e^2/r_T$, so that

$$r_T = 2e^2/3kT. \quad (2)$$

At room temperature (300 K) the value of r_T is about 370 Å. Note that at high ambient gas pressures, a high probability exists for a subsequent collision with the neutral gas particle that will revert a bound ion pair back into a free ion pair which ultimately will reduce the possibility of recombination. Thus at high pressures the Thomson model overestimates the recombination rate. The second step in the model is to calculate the rate, using the kinetic theory, at which the two ions approach one another within r_T. This rate is enhanced at low pressures by the attractive Coulomb interaction. As the final step, one calculates the probability that one of the ions makes a collision with the neutral gas particle when the ionic separation is smaller than r_T. At low pressures this probability is inversely proportional to the mean free path of the ions and thus directly proportional to the pressure. Thus at low pressures the Thomson model predicts a recombination rate that increases linearly with pressure.

At high gas pressures it is the Langevin model that provides an insight into the process of three body recombination. The drift of the ions is controlled by a delicate balance between the energy gained by the ion pair due to Coulomb attraction and the energy lost by numerous collisions with the ambient gas particles. The rate at which the two ions approach one another is then determined, in terms of diffusion theory, by the mobilities of the ions in the ambient gas. Now since the mobilities vary inversely with the pressure, the Langevin model predicts the recombination rate to decrease, at high pressures, as the pressure increases.

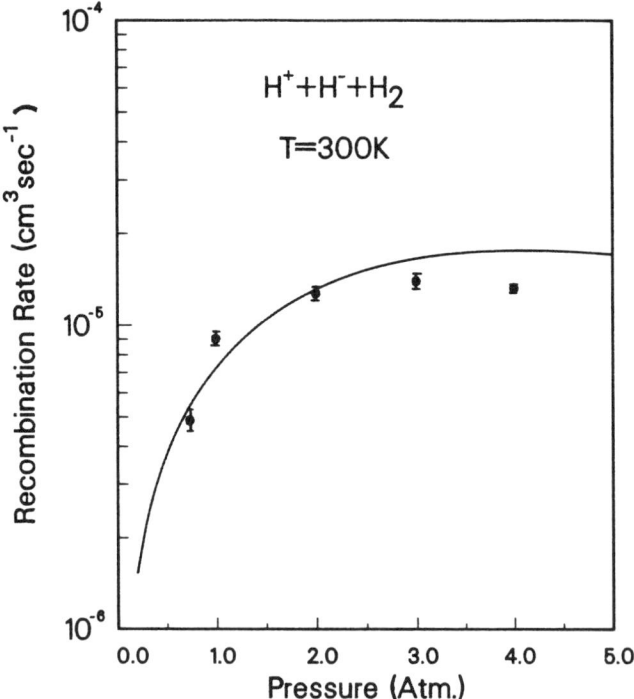

Figure 2. Rates of three body recombination for the system $H^+ + H^- + H_2$ as a function of the gas pressure using the modified Natanson theory (solid line) and the Monte Carlo simulation (circles).

The Natanson theory[7] provides a way of connecting the low pressure Thomson model with the high pressure Langevin model. The original theory was given only for the case of equal mass constituents but since then it has been generalized[10,11] to include the cases when the masses of the ions and the ambient gas particles are all different. Essentially this theory retains the trapping radius concept of the Thomson model, but in this theory the rate at which the ions approach one another is determined by incorporating ideas of both the kinetic theory and the diffusion theory. The Natanson theory predicts a three-body recombination rate that initially increases linearly with pressure, reaches a maximum and then decreases as the pressure increases. Application of a modified Natanson theory[10] to the rate of three body recombination in the $H^+ + H^- + H_2$ system is shown as a function of pressure in Figure 2. For the sake of comparison this Figure also shows the three body recombination rate for this ionic system using the Monte Carlo computer simulation program which will be described below.

COMPUTER SIMULATION "EXPERIMENTS"

A few computer programs have been developed[1,12,13] independently to investigate the processes of ion-ion recombination using the classical Monte Carlo trajectories. A typical "experiment" on the computer begins by observing the motion of an ion pair with relative separation less than a certain value R_o. A good choice for R_o is found to be $2e^2/kT$. The initial velocity components, for both relative and center-of-mass motions, of the ion pair are obtained by generating quasirandom numbers on the computer and by assuming the Maxwellian distribution for velocities. The ions move on the classical Coulomb trajectories and are assumed to have a constant collision frequency. The motion of the ion pair is followed until one of the following conditions is met. First, if the total relative energy after a collision becomes less than some critical predetermined value (usually less than $-5kT$), the three body recombination is assumed to have occured. Second, if the ionic separation becomes larger than the initial separation R_o, the current ion pair is abandoned and a new ion pair is followed. Third, if the ionic separation becomes smaller than the pseudocrossing radius R_x, the probability for curve crossing using the Landau-Zener formula is computed with allowance for multipasses of R_x.

It would be computationally prohibitive to carry out detailed simulations for a wide variety of mixtures of the positive and negative ions and several ambient gases. Furthermore, since the aim is to search for general expected theoretical trends it can be fruitfully realized by restricting the investigations to one representative set of ions and neutral gas particles as far as the masses and interaction potentials are concerned and then cover the wider variety of cases by assigning a set of arbitrary values to the parameters (such as the Landau-Zener parameter η) that determine the mutual neutralization rate. With this computer strategy in mind, the ionic system taken for detailed investigations[1] is $Kr^+ + F^-$ in neutral argon. In the simulations the gas temperature is fixed to be

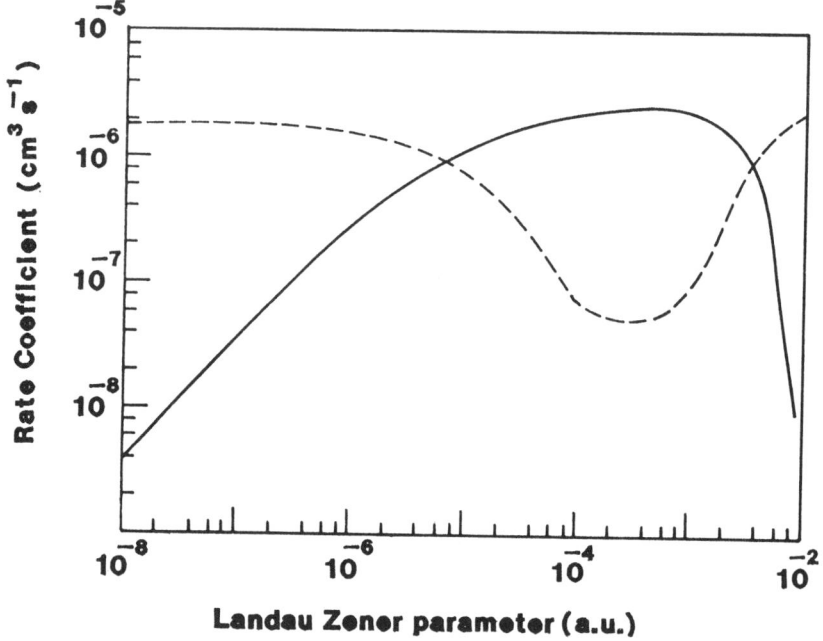

Figure 3. Ion-ion recombination rate as a function of the Landau-Zener parameter η for the $Kr^+ + F^- + Ar$ system. Pressure is 1 atmosphere and the temperature is 300 K. Solid line, mutual neutralization; dashed line, termolecular recombination.

300 K and the gas pressure is varied in the vicinity of 1 atmosphere. At very low gas pressures the computer simulated experiments tend to be quite inefficient due to low collision frequency so that most of the ion pairs separate without having suffered any collisions.

Variation of the mutual neutralization rate with the Landau-Zener parameter η for the $Kr^+ + F^- + Ar$ system is shown in Figure 3. Even though the parameter for this ionic system is estimated, using the semiempirical formula[4], to be about 10^{-11} a.u., in the computer investigations η was allowed to vary arbitrarily. Two important observations can be made from Figure 3. First, when the rate of mutual neutralization increases, the rate of termolecular recombination decreases in such a fashion that the total recombination rate stays approximately the same. Second, around atmospheric pressures the process of mutual neutralization is relatively important only for those systems for which the parameter η lies in the range 10^{-5} a.u. $\lesssim \eta \lesssim 10^{-3}$ a.u.

The pressure dependence of the rate of mutual neutralization in the $Kr^+ + F^-$ system is shown, for a few selected values of the Landau-Zener parameter η, in Figure 4. This Figure also shows the rate of termolecular recombination for the $Kr^+ + F^- + Ar$ system as a

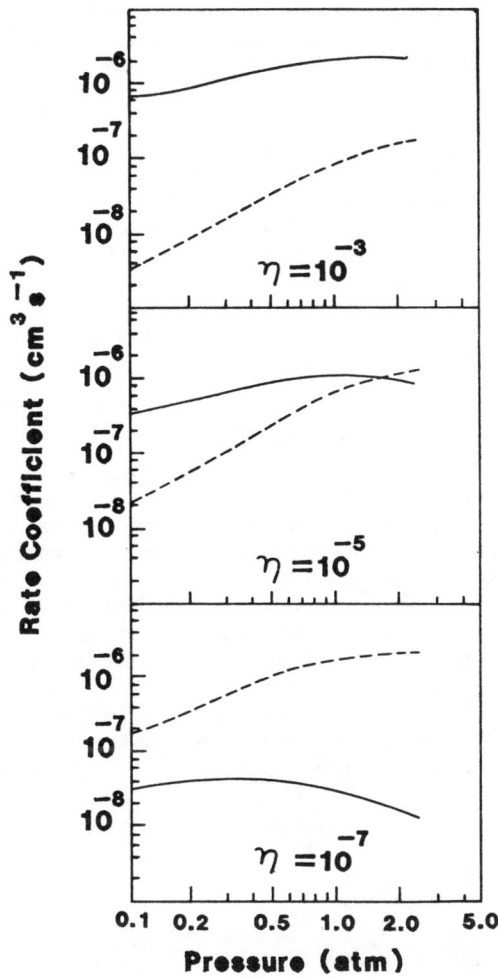

Figure 4. Ion-ion recombination rate as a function of the gas pressure for $Kr^+ + F^- + Ar$ system for various values of the Landau-Zener parameter η. Temperature is 300 K. Solid line, mutual neutralization; dashed line, termolecular recombination.

function of the ambient gas pressure. It is to be noted that the two recombination processes are in competition with one another at all pressures. The fact that the rate of mutual neutralization increases with gas pressure for certain values of η is a clear indication that this rate can be enhanced by the presence of neutral ambient gas. This enhancement, first discussed by Bates[14], can be physically understood rather easily by referring back to Figure 1. The

traditional scenario for the process of mutual neutralization
described earlier assumes that the ionic system $A^+ + B^-$ is in one of
the continuum energy levels (the levels are schematically shown on
the right side of Figure 1) before the mutual approach of the ions.
The ions traverse the pseudocrossing point R_x once on the way in and
then again on the way out as they separate. However, in the presence
of ambient neutral gas particles, it is possible that the ions could
be trapped by collisions in one of the high lying vibrational levels
(labelled v in the Figure) of the ionic potential curve. In such a
situation the ions will have many more opportunities of traversing
the pseudocrossing point R_x which, in effect, would enhance the rate
of mutual neutralization. In fact, once trapped in one of the high
vibrational levels, the ultimate fate of the ion pair in the absence
of any further collisions with the neutral particles will be mutual
neutralization.

CONCLUSIONS

At least a few independently developed[1,15] computer programs,
for simulating the ion-ion recombination processes, have shown that
the rates of both the mutual neutralization and the termolecular
recombination processes are affected by the presence of an ambient
gas and, furthermore, these two recombination processes are in
competition at all pressures. It is also demonstrated that the
enhancement of the mutual neutralization rate due to the presence of
ambient gas is significant only for the systems for which the Landau-
Zener parameter η lies roughly in the range $10^{-5} - 10^{-3}$ a.u. It is not
yet possible to ascertain, for a given ionic system, the degree of
enhancement in quantitative terms though work in this direction is
being pursued[16].

As the final comment, mutual neutralization for the hydrogen
ionic system

$$H^+ + H^- \rightarrow H(1s) + H(n\ell)$$

is of special interest[17] due to its possible application for neutral
beam injection. For this system the values of the Landau-Zener
parameter η, obtained by using semiempirical approach[4], corresponding
to the formation of atomic hydrogen in n = 2, 3 and 4 states are
4.9×10^{-2}, 3.2×10^{-5} and 9.3×10^{-48} a.u., respectively. These values of
η suggest that the rate of mutual neutralization of $H^+ + H^-$ leading
to atomic hydrogen in n = 2 and 3 states could be affected by the
ambient gas. It would be worthwhile to carry out detailed
calculations on this ionic system.

It is a pleasure to thank Professors J. N. Bardsley and M. R.
Flannery for helpful discussions. The support of the Air Force Office
of Scientific Research through Grant Number AFOSR-84-0143 is
gratefully acknowledged.

REFERENCES

1. B. L. Whitten, W. L. Morgan and J. N. Bardsley, J. Phys. **B15,** 319

(1982).
2. L. D. Landau, Phys. Z. Sowjet. $\underline{2}$, 46 (1932).
3. C. Zener, Proc. R. Soc. Lond. $\underline{A137}$, 696 (1932).
4. R. E. Olson, F. T. Smith and E. Bauer, Appl. Opt. $\underline{10}$, 1848 (1971)
5. M. R. Flannery, Atomic Processes and Applications, P. G. Burke and B. L. Moiseiwitsch, eds., p.407 (North-Holland: Amsterdam).
6. J. T. Moseley, R. E. Olson and J. R. Peterson, Case Studies in Atomic Physics, vol. 5, E. W. McDaniel and M. R. C. McDowell, eds., p.1 (North-Holland: Amsterdam).
7. G. L. Natanson, Soviet Phys.-Tech. Phys. $\underline{4}$, 1263 (1959).
8. J. J. Thomson, Phil. Mag. $\underline{47}$, 337 (1924).
9. P. Langevin, Ann. de Chim. et de Phys. $\underline{28}$, 433 (1903).
10. J. M. Wadehra and J. N. Bardsley, Appl. Phys. Lett. $\underline{32}$, 76 (1978)
11. M. R. Flannery, Chem. Phys. Lett. $\underline{56}$, 143 (1978).
12. J. N. Bardsley and J. M. Wadehra, Chem. Phys. Lett. $\underline{72}$, 477 (1980).
13. D. R. Bates and I. Mendas, Proc. R. Soc. Lond. $\underline{A359}$, 287 (1978).
14. D. R. Bates, J. Phys. $\underline{B13}$, 205 (1980).
15. D. R. Bates, J. Phys. $\underline{B14}$, 4207 (1981).
16. M. R. Flannery, private communication.
17. M. Bacal, private communication.

DISCUSSION

Roberts: You said the parameter η depended on ΔE, which depended on the overlap integral. But you never did show how it depends on it.
Wadehra: The factor η, I didn't show because I thought it may not be of interest, but it is given in, I guess standard sources like textbooks. η depends on $1/\Delta E$ and is proportional to the difference between the slopes of the two curves at the crossing point. So, the slope divided by the energy difference.
Mitchell: David Smith, in England, has done a tremendous number of measurements on mutual neutralization in flow tubes and, presumably, over a range of pressures. Has he seen any of these termolecular effects?
Wadehra: That's a question that I was going to almost throw back at the audience. I've seen a lot of papers where people have done mutual neutralization as a function of the energy of ion pairs, relative energy. I have yet to see some graph or some figures being shown as a function of pressure. Maybe I haven't located that, maybe I haven't seen the literature, but I'd like to see that.
Mitchell: Well, the thing with Smith's work is that it's in a flow tube, not a beam experiment. So he can change the pressure, presumably. I don't know whether he's seen it or not.
Michels: When you get into the higher pressure region, don't you have a complexity arising that monomers are driven to dimers, and that dimers are driven to trimers, and you have reactive charge transfer being the dominant process? Certainly in krypton flouride, that's the case.
Wadehra: In the calculation, only two channels were included. And

the channels were of mutual neutralization and three-body recombination. Other processes, in principle, could be included, but were not.

Graham: Two questions. The lowest pressure you showed, I think, was around 76 Torr. Do you know what's happening below that? The other question was, what happens to η if you have H_3^+ instead of H^+? Does that have an effect?

Wadehra: Your first question, lower pressure. As I pointed out, low pressure is very inefficient for carrying out the computer experiments, because of very low frequency of collision. So we didn't do that experiment at low pressure because it just takes too much computer time.

Graham: Have you any feelings of what would happen?

Wadehra: One can, in principle, extrapolate what we already have and that would be my guess because you are, essentially, extrapolating to lower pressure. Still, I wouldn't go to 2 or 3 mTorr, that would be a wild guess. But maybe one can go lower than 76 Torr and see something by extrapolation because the curves seem to be rather smooth here in this area. Coming back to your second question on H_3^+. We have done some calculation on H^+ on H^- which, because I was running out of time, I didn't show you. In this system $H^+ + H^-$ as a function of pressure, this is three-body recombination only, as a function of pressure. The results of Monte Carlo calculations are these dots, and the model of Natanson that I pointed out, which is applicable at all pressures, is the solid curve. Again, at very low pressure it's very difficult to really know what that is, because we cannot go to very low pressure. The only thing we know is that, as far as the three-body recombination rate is concerned, it is supposed to approach zero at zero-pressure, because at zero-pressure there won't be any third body.

Srivastava: In your calculations you did not consider the inverse process of polar dissociation. In an actual plasma, that would be an important process for the mutual neutralization pressure range.

Wadehra: As I pointed out, there were only two processes considered, and that was mutual neutralization versus three-body recombination. The aim was to look at how does one behave in relation to the other.

PRESENT STATUS OF THE MEASURED DISSOCIATIVE ATTACHMENT CROSS SECTIONS

Santosh K. Srivastava
Jet Propulsion Laboratory, California Institute of Technology,
Pasadena, CA 91109

ABSTRACT

A survey of our current knowledge of measured dissociative attachment cross sections for the following processes is presented: i) dissociative attachment of electrons to the ground state of a molecule, ii) dissociative attachment of electrons to the rotationally, vibrationally, and electronically excited states of a molecule, iii) dissociative attachment of electrons to ultra cold molecules and molecular clusters, and iv) dissociative attachment to molecular radical species. Experimental techniques used in the measurement of the cross sections are briefly described.

I. INTRODUCTION

Among a wide variety of processes that can take place when an energetic electron collides with a molecule the dissociative attachment of electrons is of special interest in the development of such devices as particle beam generators and high voltage and high current switches. This process may be represented by the following type of reaction:

$$e^- (E_R) + X_n Y_m \xrightarrow{\sigma(E_R)} X_n Y_{m-1} + Y^-, \qquad (1)$$

where $X_n Y_m$ is a molecule with n number of x atoms and m number of Y atoms, Y^- is a negative ion and $e^- (E_R)$ is an electron with kinetic energy E_R. It has been found that for each molecule there are certain specific number of energies, E_R (some times called resonance energies), at which the dissociative attachment takes place. Although in eq. (1) we have presented the formation of an atomic negative ion, the various molecular negative ions are also produced and have been reported previously for a number of molecules.

The quantity which is of great interest to plasma modellers is $\sigma(E_R)$, the cross section for the process indicated by eq. (1). In addition, a knowledge of the threshold energy at which the dissociative attachment begins to take place and of the electron energy at which the value of $\sigma(E_R)$ is maximum is very important for gaining the information on the potential energy surfaces of the molecular negative ion states.

A simple picture of the process of dissociative attachment is provided by the resonance theory[1]. According to this theory, first of all, an electron with a kinetic energy E_R attaches itself to the parent molecule and forms a temporary molecular negative ion. Then this state either decays to the parent state (auto-ionize) or dissociates into a stable negative ion (dissociative attachment).

A quantity known as "survival probability" determines the value of $\sigma(E_R)$. Details on the factors related to the negative ion formation can be found in two recent books: one by Massey[2] and the other by Smirnov[3]. Several review articles have also been written on this subject during the past two decades. Most recent ones are by Christophorou et al.[4] and by Compton and Bardsley[5]. Reviews on the various theoretical methods have been presented by Bardsley and Mandl[1], Fiquet-Fayard[6], Berry and Leach[7], Hazi[8], Golden[9], Wadehra[10], among many others.

Measurements on the dissociative attachment cross sections date back to 1930 when Bailey and Duncanson[11] reported their results on NH_3, H_2O and HCl. Since then the following aspects of the dissociative attachment process have been studied: i) dissociative attachment of electrons to the ground state of a molecule, ii) dissociative attachment of electrons to the rotationally, vibrationally, and electronically excited states of a molecule, iii) dissociative attachment to ultracold molecules and molecular clusters, and iv) dissociative attachment to molecular radicals. In the present paper our current knowledge about the cross sections for these aspects will be described. Before summarizing the results the various methods that have been used for the cross section measurements will be briefly presented.

II. EXPERIMENTAL TECHNIQUES

Experimental techniques vary according to the type of measurement to be made. Space here does not permit to give a detailed account of each one of them. Therefore, in the following these techniques will be described briefly and references will be given where the reader can find more information.

1. DISSOCIATIVE ATTACHMENT TO THE GROUND STATE OF A MOLECULE. Three different experimental methods have been used in the past for this purpose and will be briefly described in the following:

A) <u>Swarm method</u>: This method involves the use of a swarm of electrons which drifts through the attaching gas under the influence of an externally applied electric field, E. An electron swarm may be defined as a cloud of electrons of density n in a gas of much higher number density, N. This cloud can be generated, for example, by a pulsed UV source at one end of a drift tube and which then drifts and diffuses to the other end in a direction opposite to the electric field.

The attachment coefficient $\alpha(E/P)$ and the drift velocity $w(E/P)$ are measured as a function of the ratio E/P, where P is the pressure of the gas in the drift tube and the drift velocity is defined as the velocity aquired by the center of mass of the electron cloud in an electric field E. Due to collisions suffered by electrons with the gas molecules they acquire a broad velocity distribution which is expressed in terms of a velocity distribution function $f(v, E/P)$. The product $\alpha(E/P) \cdot w(E/P)$ is defined as the absolute rate of electron attachment expressed in units of sec^{-1}

Torr^{-1}. The above mentioned quantities are related to the electron attachment cross section σ(v) by the following relations[12]:

$$\alpha(E/P) \times w(E/P) = N. \int_0^\infty v\sigma(v). f(v, E/P) dv, \quad (2)$$

or

$$\alpha(E/P) \times w(E/P) = N.(2/m)^{1/2} \int_0^\infty E_i \sigma(E_i) f(E_i, E/P) dE_i, \quad (3)$$

where m is the mass and E_i is the energy of the colliding electron. Eq. (3) can also be written in terms of density reduced electric field, E/N (see ref. 4):

$$K(E/N) = (2/m)^{1/2} \int_0^\infty \sigma(E_i) (E_i)^{1/2} f(E_i, E/N) dE_i, \quad (4)$$

where K(E/N) is the attachment rate constant expressed in units of cm^{3}s^{-1}.

Equation (4) is employed for the determination of σ(E_i) for the attaching gases. The various swarm methods are discussed by Christophorou.[12] The procedure is as follows. The molecule or the gas whose cross section is to be determined is mixed in trace amounts with a non electron attaching buffer gas such as Ar or N_2. The function f(E_i, E/N) for the buffer gas is determined accurately in a previous experiment and it is assumed that this function does not change by the addition of a small amount of the attaching gas. The value of K(E/N) is then experimentally determined. With the knowledge of f(E_i, E/N) and K(E/N) one can unfold the values of σ(E_i) by employing the iterative techniques developed by the Oak Ridge National Laboratory Group.[13-16]

B) <u>Electron Beam Method</u>: Whereas in the swarm method the dissociative attachment cross sections, σ(E_i) are determined indirectly by the use of an unfolding technique, in the beam method their values are obtained directly as a function of colliding electron's energy, E_i. In the past, two different types of collision geometries have been employed for this purpose and will be briefly described in the following:

i) <u>Static Gas Collision Geometry</u>: An energy selected beam of electrons is passed through a cell filled with the gas under study. The static pressure of the gas is accurately measured. The energy of the electron beam is then varied. As a result of the process of dissociative attachment of electrons negative ions are produced. From the knowledge of the total negative ion current as a function of electron beam energy, E_i, pressure of the gas, and the value of the electron beam current the absolute value of σ(E_i) is calculated. This method was successfully employed by Rapp and his co-workers. The various experimental details can be found in their paper.[17] Their apparatus did not have the capability of separating the various masses. The experimental arrangement was later modified by extracting the individual negative ion species from the collision region and mass analyzing them by a mass spectrometer.[18,20] However, the drawback of this modification was that it was not suitable for the determination of absolute vlaues of dissociative attachemnt cross sections.

ii) **Crossed Beams Collision Geometry**: In this geometry a beam of molecules whose dissociative attachment properties are to be studied is produced by several methods.[21-23] A detailed account on the various types of beams can be found in ref. 24. Each method depends on the species and on the type of measurement to be made. For example, a hypodermic needle[21] is the simplest tool for making a beam. However, if one desires to specify the spatial density distribution of molecules in a well defined beam then a cepillary array[21] is employed. For species which are liquid or solid at room temperature a high temperature furnace[22] is used. For geometrically well defined beams of molecules in their lowest possible rotational states and for molecular clusters supersonic nozzles[23] have been fabricated.

In the crossed beam geometry[24] the beam of molecules is crossed by an energy selected beam of electrons at 90°. The energy of the electron beam is then varied. At the rosonance energy negative ions are produced. These ions are extracted out of the collision region at right angles to both the electron beam and the molecular beam. Then they are focussed at the entrance aperture of a mass spectrometer. The ion signal is normalized[25] to some known cross section to obtain the cross section values.

C) **Swarm-Beam Method**: Here the two methods, the swarm method and the crossed beam method, are combined to obtain the absolute values of dissociative attachment cross section, $\sigma(E_i)$. Briefly, from the swarm method the absolute rate K(E/N) (eq. 4) is measured for the molecular species under study. The function $f(E_i, E/N)$ for the carrier gas is accurately determined in a previous experiment. From the beam experiment the value of the negative ion yield $I(E_i)$ as a function of electron impact energy E_i is obtained and the negative ion products are identified. With the conditions that the ions involved in the two experiments are the same, one can construct a trial function[12] given below:

$$\sigma_j(E_i) = K_J T_J I(E_i), \qquad (5)$$

where T_J corrects for any uncertainty in the energy scale by shifting the $I(E_i)$ curve along the E_i axis and K_J is a constant for each T_J which transforms the relative values $I(E_i)$ into absolute values of $\sigma_j(E_i)$. The values of T_J and K_J can be obtained by substituting eq. (5) into eq. (3) and performing a double least-squares fitting procedure. Then one obtains the best agreement between the calculated and experimental rates for each value of E/P through the following equation:

$$\frac{d}{dT_J}\left[\sum_i \left\{N(2/m)^{1/2} K_J \int_0^\infty E_i^{1/2} T_J I(E_i) f(E_i) dE_i - \alpha_i W_i\right\}^2\right] = 0, \qquad (6)$$

where the various quantities are defined in section II.1.A. More details of this technique are given by Christophorou and his coworkers.[4,12]

2. DISSOCIATIVE ATTACHMENT TO AN EXCITED STATE OF A MOLECULE. As will be discussed later it has been found that the values of dissociative attachment cross sections increase by many orders of magnitudes when the molecule is in a vibrationally excited state. As discovered by Hiskies[26], this process can play a very important role in a plasma generated for the production of H⁻ beams. Experimental methods for the study of dissociative attachment process for an excited state of a molecule can be divided into two general categories: one for the vibrationally excited molcules and the other for the electronically excited molcules. In the following they will be described very briefly:

A) <u>Attachment to the vibrationally excited states</u>: The experimental arrangements used for this purpose, in general, are similar to ones described above in section II.1 for the swarm and beam methods. In the beam method the source of the target gas is replaced by a high temperature oven (e.g. iridium oven used by Allan and Wong[20]) through which an energy selected beam is passed and the yield of negative ions formed due to the process of dissociative attachment are studied as a function of electron beam energy. Such an oven can be heated to about 2000K and the gas within it then has a vibrational population distribution which can be calculated by the use of the Maxwell-Boltzman equations. Electron swarm methods have also been used for the study of the process of dissociative electron attachment by raising the temperature of the drift tube and then measuring the attachment rates.[27] Due to the limits imposed by the temperatures to which the target gas can be heated the attachment cross sections have been measured only for low vibrational levels of the ground electronic state.

A novel method of studying the attachment process by excited states was demonstrated successfully for CO_2 by Srivastava and Orient.[28] Here two colinear beams of electrons are employed simultaneously. One beam is used to excite the vibrational states and the other beam is employed to generate negative ions. However, this method has two drawbacks. First, for most molecules the absolute values of cross sections for vibrational excitation by electron impact are not known accurately. Second, absolute values of dissociative attachment cross sections can not be obtained at present.

B) <u>Attachment to Electronically Excited States</u>: Experimental measurements on the attachment cross sections for the electronically excited states of molecules are few and have started appearing in the literature only very recently. This is due to the fact that it is difficult to generate enough number of molecules in their excied electronic states for the purpose of performing experiments on the dissociative attachment process. Recently, due to availability of high power excimer lasers, it has been possible to produce appreciable population density of electronically excited states (mostly metastable) in a swarm type of experimental arrangement.[29,30]

3. ATTACHMENT TO RADICAL SPECIES: These species have recently been generated by irradiating a stable molecule by an excimer laser. The photodissociation of the molecule results in a vibrationally or electronically excited radical species. The electron attachment is then studied by a swarm type of experimental arrangement similar to one described in the previous sections. Rossi and his co-workers[31] used a method developed by Gruenberg which consists of a pair of parallel plate electrodes. The gap between them serves the purpose of a swarm drift tube. In another experimental arrangement Ono, Lion and Moseley[32] used an excimer pumped dye laser in the 430-454 nm range to study the attachment properties of NO, OH and SO.

A beam type experimental arrangement was used by McGeoch and Schlier[33] for the study of dissociative attachment of electrons by $(Li)_2$.

III. CROSS SECTION DATA

Although a large number of papers on the process of dissociative electron attachment with molecules have been published since 1930 only very few of them have reported the values of cross sections which can be used for plasma modelling. A survey of cross section data, available up to 1979 can be found in the National Bureau of Standards (NBS) Circular 426 and its supplements (U.S. Department of Commerce, 1976 and 1979). Information regarding the cross sections and a detailed account of the physics involved in the process of dissociative attachment is also given in the most recent review paper by Christophorou and his co-workers[4].

In some molecules there are several resonance energies at which the dissociative attachment takes place. The data presented in this review (Tables I, II, and III) corresponds to the maximum value of the cross section. For example, H^- from H_2 appears at around 3.75 eV, 10 eV, and 14.2 eV. We have given the cross section value for the peak at 14.2 eV which is the largest among the three. Moreover, data are tabulated only for the production of a stable negative atomic ion. More details on the cross sections for the appearance of other species can be found in the references given in the three tables. In most cases data are available from one experimental group only. Wherever more than one measurement has been reported we have tabulated our best judged value. In the following we shall discuss the cross section data for the various types of molcular species:

1. DATA FOR THE GROUND STATE OF A MOLECULE. These data are divided into three parts pertaining to diatomic, triatomic, and polyatomic molecules:

A) <u>Diatomic Molecules</u>: Data on the cross sections for diatomic molecules are pesented in Table I along with the corresponding values of electron attachment energies. We have also indicated the technique used for obtaining the cross sections. These techniques have been described in section II of this paper. It is clear from Table I that the cross sections for most diatomic molecules of interest have been measured in the past.

B) <u>Triatomic Molecules</u>: Table II shows the triatomic molecules for which the values of dissociative attachment cross sections are readily available.

C) <u>Polyatomic Molecules</u>: Dissociative attachment cross sections for polyatomic molecules are usually large and many types of stable negative ions are produced. Christophorou[4] divides them into four categories: i) directly cleaved, ii) complementary, iii) multiple, and IV) rearrangement. The parent polyatomic molecules themself, can be arranged into three general categories: i) hydrogen and deuterium containing molecules such as hydrocarbons, ii) halo carbons, and iii) perfluorocarbons. In most cases swarm methods have been used to derive the attachment rate constants and the cross section values are available only for a few. Christophrou[4] in his recent review article has listed the rate constants for a large number of polyatomic molecules. However, we would like to mention a word of caution. The swarm data for rate coefficients are described by "hydrodynamic" transport parameters which are independent of position and time. They are, therefore, not recommended to be used for non-equilibrium situations. A good survey of swarm data for attaching gases can be found in a JILA information center report number 22 (Aug. 1, 1982) compiled by Gallagher, Beaty, Dutton and Pitchford. In Table III we have presented cross section data only for the production of H^- and D^- from the polyatomic molecules which may be of interest in designing beam sources.

2. DATA FOR THE EXCITED STATES: These can be divided into the following two categories:

A) <u>Vibrationally excited states</u>: Ever since it was demonstrated that the dissociative attachment to H_2 and D_2 can be enhanced by several orders of magnitude if the molecules were vibrationally excited[34-36] and Hiskes[26] discovery of an efficient mechanism of vibrational excitation in a H_2 plasma the number of molecules studied have increased. Due to difficulty in estimating the number density of vibrationally excited states the absolute cross section are known only for a few molecules. The following are the molecules for which the enhancement of dissociative attachment cross section has been observed either by the crossed beam method or by the swarm method: H_2, D_2, O_2, HCl, HF, DCl, N_2O, NO_2, CO_2, SF_6, CCl_4, $CFCl_3$, CH_2Br_2, CH_3I, $CHCl_3$, CF_3Br, CH_3Br, and $1,1-C_2Cl_2H_4$. More recently, Rossi and his co-workers[31] have studied attachment properties of vibrationally excited HCl and HF by using an excimer laser to photo-dissociate C_2H_3Cl and C_3F_3H.

B) <u>Electronically excited states</u>: The first molecule, on which the dissociative attachment to the electronically excited state was reported, was O_2. Burrow[37] repoted cross sections for its metastable state $a^1\Delta g$ which lies 0.98 eV above the ground state. The cross section maximum lies at 6.5 eV and has a value $(4.6 \pm 1.3) \times 10^{-18}$ cm^2 which is about 3.5 times larger than the peak cross section value of 1.3×10^{-18} cm^2 for attachment to the ground state of O_2. However, their cross section value is in a slight disagreement with the most recent measurements by Belic and Hall[38] who found them to be 3.8×10^{-18} cm^2 and 1.8×10^{-18} cm^2 for the $^1\Delta g$ and $X^3\Sigma_g^-$ states, respectively.

With the availability of excimer lasers attachment processes have been studied (experimental method is described in Section II) for the following molecules: $C_6Cl_4O_2$ (Chloronil)[29], C_6F_6[29] (hexafluorobenzene), C_6H_5SH[30] (thiophenol), $C_6H_5SCH_3$ (thioanisole)[30], acetone[29], CH_3COCH_3[29], and 1, 4 benzoquinone.[29]

3. DATA FOR RADICALS: Due to difficulties associated with the production of free radicals for collision studies there are hardly any data available on the process of dissociative attachment. Recently, measurements have been reported on $NO(A^2\Sigma^+)$, $OH(A^2\Sigma^+)$ and $SO(B^3\Sigma^-)$ by Ono et al.[32]

Experimental arrangements are being put together in a few laboratories at the present time to study the radicals and it is hoped that the cross section data will be available in the near future.

4. DATA FOR ULTRACOLD MOLECULES AND MOLECULAR CLUSTERS. As we have already discussed, dissociative attachment cross sections for hot molecules are sometimes much larger than those for cold ones. In addition, the energy threshold for the dissociation shifts towards the lower end. When a molecule is considerably cooled down by a method such as a super-sonic expansion the threshold value increases and the cross sections decrease. Klots and Compton[39] have studied the dissociative attachment process for ultra-cold molecules of N_2O and CO_2. They found that O^- peak at 0.5 eV resulting from the dissociative attachment of electrons with the room temperature N_2O disappeared when it was expanded through a super-sonic nozzle.

By cooling a molecular jet one can form Van der Waals polymers. Klots and Compton have studied dissociative attachment process with clusters of CO_2[39], N_2O[39], and H_2O[40]. For CO_2 the most abundant cluster, which gives rise to O^-, has a general formula $(CO_2)_n$ with $0 < n < 6$. Armbruster et al.[41] have reported the observation of $(H_2O)_n^-$ with $n > 8$.

IV. CONCLUSION

From the above it is clear that the experimental techniques

now exist which can measure the accurate values of cross sections for attachment of electrons to the ground state of a molecule. However, data are available only for few molcules and there is need for more work. The dissociative attachment cross section values for the vibrationally and electronically excited states have started appearing in the literature where the use of excimer lasers has been made. It is hoped that in the near future more research on radicals and clusters will be reported.

Acknowledgement: The work described here was carried out by the Jet Propulsion Laboratory, California Institute of Technology, and was sponsored by AFOSR.

References

1. J. N. Bardsley and F. Mandl, Rep. Pro. Phys. 31, 471 (1968).
2. H. S. W. Massey, "Negative Ions", Cambridge University Press, 1976.
3. B. M. Smirnov, "Negative Ions" (McGraw-Hill, New York 1982).
4. L. G. Christophorou, D. L. McCorkle and A. A. Christodoulides, in "Electron-Molecule Interactions and their Application - Vol. 1" ed. L. G. Christophorou, Academic Press, Inc. (1984), pp. 477.
5. R. N. Compton and J. N. Bardsley, in "Electron-Molecule Collisionπ, ed. I. Shimamura and K. Takayanagi, Plenum Press, N. Y. (1984), pp. 275.
6. F. Fiquet-Fayard, Vacuum 24, 533 (1974).
7. R. S. Berry and S. Leach, Adv. Electronics and Electron Phys. 57, 1 (1981).
8. A. U. Hazi, in "Electron-Molecule Collisions and Photoionization Processes" eds. V. McKoy, H. Suzuki, K. Takayanagi and S. Trajmar, Verlag Chemie Int. (1983), pp. 87.
9. D. E. Golden, Adv. At. Mol. Phys. 14, 11 (1978).
10. J. Wadehra, in "Non-Equlibrium Kinetics" ed. M. Capitelli, Vol. 39, Springer Verlas (1986), pp. 191.
11. V. A. Bailey and W. E. Duncanson, Phil. Mag. 10, 145 (1930).
12. L. G. Christophorou, "Atomic and Molecular Radiation Physics" Viley-Interscience (1971), pp. 426.
13. L. G. Christophorou, D. L. McCorkle and V. E. Anderson, J. Phys. B: Atom. Molec. Phys. 4, 1163 (1971).
14. R. E. Goans and L. G. Christophorou, J. Chem. Phys. 60, 10366 (1974).
15. A. A. Christodonlides, L. G. Christophorou, R. Y. Pai, and C. M. Tung, J. Chem. Phys. 70, 1156 (1979).
16. D. L. McCorkle, A. A. Christodonlides, L. G. Christophorou, and I. Szamrej, J. Chem. Phys. 72, 4049 (1980).
17. D. Rapp and D. D. Briglia, J. Chem. Phys. 43, 1480 (1965).
18. P. J. Chantry and G. J. Schulz, Phys. Rev. 156, 134 (1967).
19. P. J. Chantry, J. Chem. Phys. 51, 3369 (1969).
20. M. Allan and S. F. Wong, J. Chem. Phys. 74, 1687 (1981).
21. R. T. Brinkmiann and S. Trajmar, J. Phys. E 14, 245 (1981).
22. S. Trajmar, W. Williams and S. K. Srivastava, J. Phys. B10,

3323 (1977).
23. R. E. Smalley, D. H. Levy and L. Wharton, J. Chem. Phys. <u>64</u>, 3266 (1976).
24. G. Csanak, D. C. Cartwright, S. K. Srivastava, and S. Trajmar, in "Electron-Molecule Interactions and Their Applications - Vol. 1" ed. L. G. Christophorou, Academic Press, Inc. (1984)μ pp. 1.
25. O. J. Orient and S. K. Srivastava, J. Chem. Phys. <u>78</u>, 2949 (1983).
26. J. R. Hiskes, J. Appl. Phys. <u>9</u>, 4592 (1980).
27. D. L. McCorkle, I. Szamrej and L. G. Christophorou, J. Chem. Phys. <u>77</u>, 5542 (1982). See also: P. G. Datsos and L. G. Christophorou, in "39th Gaseous Electronics Conference, Oct. 7-10, 1986, Madison, Wisconsin", pp. 46.
28. S. K. Srivastava and O. J. Orient, Phys. Rev. A<u>27</u>, 1209 (1983).
29. M. J. Rossi, H. H. Helm, and D. C. Lorents, contributed paper to "39th Gaseous Electronics Conference, October 7-10, 1986, Madison Winsconsin", pp. 44. See also: "38th Gaseous Electronics Conference, 15-18 Oct. 1985, Naval Post Graduate School, Montrey, CA. pp. 46.
30. S. R. Hunter, L. G. Christophoron and L. A. Pinnaduwage, contributed paper to "39th Gaseous Electronics Conference, October 7-10, 1986, Madison, Wisconsin", pp. 45.
31. M. J. Rossi, H. Helm and D. C. Lorents, Abstracts of contributed papers presented in "the 14th International Conference on the Physics of Electronic and Atomic Collisions (ICPEAC), Palo Alto, 11985", ed. M. J. Coggiola, D. L. Huestis, and R. P. Saxon, pp. 284.
32. Y. Ono, H. T. Liou and J. T. Moseley, Abstracts of contributed papers presented in "the 14th International Conference on the Physics of Electronic and Atomic Collisions (ICPEAC)", Palo Aalto, 1985, ed. M. T. Coggiola, D. L. Huestis, and R. P. Saxon, pp. 275.
33. M. W. McGeoch and R. E. Schlier, in "Proceedings of the Third International Symposium on the Production and Neutralization of Negative Ions and Beams", edited by K. Prelec. (American Institute of Physics, New York, 1984), pp. 291.
34. M. Allan and S. F. Wong, Phys. Rev. Lett. <u>41</u>, 1791 (1978).
35. J. M. Wadehra and J. N. Bardsley, Phys. Rev. Lett. <u>411</u>, 11795 (1978).
36. J. N. Bradsley and J. M. Wadehra, Phys. Rev. A<u>20</u>, 1398 (1979).
37. P. D. Burrow, J. Chem. Phys. <u>59</u>, 4922 (1973).
38. D. S. Belic and R. I. Hall, J. Phys. B<u>14</u>, 365 (1981).
39. C. E. Klots and R. N. Compton, J. Chem. Phys. <u>69</u>, 1636 (1978).
40. C. E. Klots and R. N. Compton, J. Chem. Phys. <u>69</u>, 1644 (1978).
41. M. Ambruster, H. Haberland, and H. G. Schindler, Phys. Rev. Lett. <u>47</u>, 323 (1981).
42. M. V. Kurepa and D. S. Belic, J. Phys. B<u>11</u>, 3719 (1978).
43. M. V. urepa, D. S. Babic and D. S. Belic, J. Phys. B<u>14</u>, 375 (1981).
44. O. J. Orient and S. K. Srivastava, Phys. Rev. A<u>32</u>, 2678

(1985).
45. L. G. Christophorou, R. N. Commpton and H. W. Dickson, J. Chem. Phys. 48, 1949 (1968).
46. D. Teillet-Billy, L. Bomby and J. P. Ziesel, "Abstract of contributed Papers" the 13th International Conference on the Physics of Electronic and Atomic Collisions (ICPEAC), Berlin 1983, edts: J. Eichler, W. Fritsch, I. V. Hertel, N. Strolterfoht and U. Wille, pp. 294.
47. O. J. Orient and S. K. Srivastava, Chem. Phys. Lett. 96, 681 (1983).
48. R. N. Compton and L. G. Christophorou, Phys. Rev. 154, 110 (1967).
49. M. Tronc, S. Goursaud, R. Azria, and F. Fiquet-Fayard, Jour. de. Physique 34, 381 (1973).
50. S. K. Srivastava and E. Krishnakumar, to be published (1986).
51. R. Azria, M. Tronc and S. Goursand, J. Chem. Phys. 56, 4234 (1972).
52. J. P. Ziesel, G. J. Schulz, and J. Milhaud, J. Chem. Phys. 62, 1936 (1975).
53. A. C. de AE Souza and S. K. Srivastava, abstract published in "The Tenth International Conference on Atomic Physics (ICAP)" Tokyo, Japan (1986), editors H. Narumi and I. Shimamura. pp. 418.
54. T. E. Sharp and J. T. Dowell, J. Chem. Phys. 46, 1530 (1967).
55. R. Azria, J. Phys. (Paris) 33, 663 (1972).
56. R. N. Compton, Phys. Rev. 180, 111 (1969).
57. A. Chutijian and S.H. Alajajian, submitted to Phys. Rev. Lett. (Private communication).

Table I. Recommended maxium values of cross sections for dissociative attachment of electrons to diatomic molecules. The corresponding electron attachment energies in eV are given in brackets.

Molecule	Technique	σ_{max} ($\times 10^{-19} cm^2$)	Author (year)
$H_2(H^-)$	e-beam	0.195(13.9)	Rapp and Briglia (1965)[17]
$D_2(D^-)$	e-beam	0.0941(14.0)	''
HD(Total)	e-beam	0.137(13.95)	''
$O_2(O^-)$	e-beam	14.06(6.5)	''
CO(Total)	e-beam	2.02(9.19)	''
$NO(O^-)$	e-beam	11.16(8.15)	''
Cl_2 (Cl^-)	e-beam	2.01×10^5(0.00)	Kurepa and Belic (1978)[42]
Br_2 (Br^-)	e-beam	1.77×10^2(0.00)	Kurepa et al. (1981)[43]
F_2 (F^-)	krypton photoionization	infinite at threshold (s-wave), and 2.72×10^4 (0.01)	Chutjian and Alajajian (1987)[57]
$HCl(H^-)$ (Cl^-)	e-beam	9.3 (9.05 eV) 265.9(0.85eV)	Orient and Srivastava(1985)[44]
$HBr(Br^-)$	swarm-beam	2.7×10^3(0.28)	Christophorou et al.(1968)[45]
$HI(I^-)$	swarm-beam	2.3×10^5(~0/00)	''
$DCl(Cl^-)$	swarm-beam	1.4×10^2(0.81)	''
$DBr(Br^-)$	swarm-beam	1.87×10^3(0.28)	''
$DI(I^-)$	swarm-beam	1.4×10^5(~0.00)	''
Li_2	beam	$\sim 3 \times 10^3$(~0.4eV)	McGeoch and Schlier (1984)[33]
Na_2	beam	$40 + 20 cm^2 \cdot ev$ (0.05+.05eV)	Teillet-Billy et al.(1983)[46]

Table II. Recommended maximum values of cross sections σ_{max} for dissociative attachment of electrons to tri-atomic molecules. The corresponding attachment energies in eV are given in brackets.

Molecule	Technique	σ_{max} ($\times 10^{-19} cm^2$)	Author (year)
CO_2	e-beam	4.48 (8.2)	Orient and Srivastava (1983)[47]
$D_2O(D^-)$	swarm-beam	52 (6.5)	Compton and Christophorou (1967)[48]
$D_2S(D^-)$	e-beam	9 (5.35)	Tronc et al. (1973)[49]
$H_2O(H^-)$	e-beam	69.9 (6.6)	Srivastava and Kumar (1986)[50]
(O^-)		3.2 (11.8)	
$H_2S(H^-)$	e-beam	12 (5.35)	Azria et al. (1972)[51]
$CS_2(S^-)$	e-beam	3.7 (3.35)	Ziesel et al. (1975)[52]
N_2O	e-beam	85.9 (2.2)	Rapp and Briglia (1965)[17]
$SO_2(O^-)$	e-beam	80.8 (4.3)	Orient and Srivastava (1983)[25]
(S^-)		3.1 (4.0)	
$OCS(S^-)$	e-beam	2.9×10^2 (1.35)	Ziesel et al. (1975)[52]

Table III. Dissociative attachment of electrons to some hydrogen and deuterium containing polyatomic molecules. Maximum values of cross sections, σ_{max}, are given and corresponding electron impact energies in eV are presented in the brackets.

Molecule	Technique	σ_{max} ($\times 10^{-19} cm^2$)	Author (year)
$CH_4(H^-)$	e-beam	0.96 (10)	DeSouza and Srivastava (1986)[53]
$CD_4(D^-)$	e-beam	0.95 (9.5)	Sharp (1967)[54]
$C_2H_2(H^-)$	e-beam	0.31 (7.45)	Azria (1972)[55]
$NH_3(H^-)$	e-beam	57.4 (5.65)	Compton (1969)[56]
$ND_3(D^-)$	e-beam	29.5 (5.86)	''
$SiH_4(H^-)$	e-beam	2.2×10^3 (9.47)	DeSouza and Srivastava (1986)[53]

DISCUSSION

Azria: You showed the results from cross-section from Christophorou, et. al. on DBr and DCℓ and so on, and I always see these results. Everybody knows that these results are completely wrong. By a factor of 5, 6, or more. And new values have been published now for many years by our group.

Srivastava: Well, I checked the results from your group, but I didn't find the absolute numbers.

Azria: There are absolute numbers for the isotope effects in DCℓ and DBr.

Srivastava: The isotope effects, they are the ratios. But what are absolute numbers?

Azria: Well, you measure the absolute value for Cℓ^- from HCℓ. And without a computer you can get the cross-section for DCℓ..

Srivastava: We found these, actually, but unfortunately they were all ratios and I did not know what the absolute numbers were.

Azria: I made this comment because even in recent reviews they are taking all these old values, but theoretically and experimentally we know that these values are wrong.

Bacal: Well, I think it's a good thing you are here so you can provide the absolute values, and please complete the table with those.

ELECTRON-STIMULATED DESORPTION NEGATIVE IONS FROM CONDENSED MOLECULES

R. Azria* and L. Sanche
MRC Group in the Radiation Sciences
Faculty of Medicine, University of Sherbrooke
Sherbrooke, Quebec, Canada J1H 5N4

ABSTRACT

The formation of negative ions in electrons scattering from a number of molecules (O_2, CO, NO, N_2O) condensed on a refrigerated polycrystalline platinum surface is reported in the energy range 0-20 eV. Two mechanism namely, dissociative electron attachment and polar dissociation account for the observations. The behavior of the negative ion signal with thickness average as well as comparison with gas-phase observations is discussed.

INTRODUCTION

The formation of negative ions by electron impact is a process of fundamental and practical importance in many areas of physics and chemistry.[1,2] There exist two mechanisms by which isolated molecules dissociate and form negative ions by colliding with electrons: polar dissociation and dissociative attachment. For a diatomic molecule AB these processes may be represented, respectively, by the equations

$$e + AB \rightarrow AB^* + e \rightarrow A^+ + B^- + e \qquad (1)$$

$$e + AB \leftrightarrows AB^{-*} \rightarrow A^* + B^- . \qquad (2)$$

The first reaction proceeds via an excited electronic state AB^* and the second via a transient molecular negative ion state AB^{-*}. Reaction (2) has been intensively investigated in a large number of isolated (gas phase) electron-molecule systems and is now quite well understood. Energy dependence of negative ion intensity, spatial and kinetic energy distributions of the dissociating products are currently available and provide insight into the dissociation mechanism (i.e., coupling), the symmetry of the resonant state and the electron-molecule interaction potential, the final state of the fragments and the dissociation dynamics of the resonant state.

Very recently, experiments with physisorbed O_2[3] on polycrystalline platinum have demonstrated the presence of dissociative attachment mechanism responsible for producing negative ions below the energy threshold of polar dissociation. These results will be reminded here and similar results on series of molecules presented. Particular features of electron scattering in the condensed phase such as the influence of neighboring polarizable targets and the possibility of multiple collisions will be considered as well as the modification of ions receeding from the surface.

© American Institute of Physics 1987

EXPERIMENT

The apparatus consists of an electron gun, a refrigerated target, and a mass spectrometer housed in an UHV system reaching pressures below 5 X 10^{-11} Torr. Electrons are accelerated to an electrically isolated polycrystalline platinum ribbon press fitted on the cold end of a closed-cycle refrigerated cryostat. The angle of incidence is 20° from the surface. Condensed layers are grown on the ribbon and the film thickness estimated, within 50% accuracy.[4] Negative ions produced by electron impact on the condensed film are measured as a function of electron energy by a quadrupole mass spectrometer equipped with ion lenses and positioned at 70° from the film surface. The energy spread of the electron beam was 0.3 eV. Its energy was calibrated within ± 0.4 eV with respect to the vacuum level by measuring the onset of electron transmission through the films and by comparing the energy of the structure in the transmission spectra with those obtained at high resolution.[5] The target ribbon was cleaned by repeated resistive heating up to 1500 K. After cooling, the target temperature could be monitored by a gold-iron thermocouple located behind the platinum ribbon. The data were recorded with incident currents of about 10^{-8} A, target temperature, ranging between 15 and 40 K according to the molecule under study and film thickness varying from submonolayer coverage to about 15 "monolayer".

RESULTS AND DISCUSSION

Figures 1, 2, 3, and 5 show energy dependence of negative ion yields produced by electron impact on films of condensed O_2, CO, NO and N_2O, respectively. Many pronounced structures are observed. In order to observe electron stimulated desorption of negative ions from a neutral multilayer film condensed on a metal surface, electron must be provided either by the metal surface or by the impinging electron beam. Since back donation of an electron from the metal substrate to a dissociating molecule located at or near the surface of a multilayer insulating film is a very unlikely process,[6] the structure in the energy dependence of negative ions yields below the threshold for dipolar dissociation (~ 16 eV) is understood as the decay of transient anions into the dissociative electron attachment channel.

These transient anions occur, in the condensed phase, at lower energies than in the gas, mainly because of electronic polarization of neighboring molecules.[7] However, in the dissociative attachment process only negative ions which overcome this polarization force can escape the solid so that only the higher-energy portion of the negative ion distribution (produced by higher-energy electrons) is observed in the vacuum. The two effects have a tendency to cancel each other and it is expected that if a resonance decays in the dissociating channel in both gas phase and condensed phase, the peak position of negative ion formation should be nearly the same, however with an energy range more confined in the condensed phase. In condensed O_2 (Fig. 1) the 6.6 eV peak is associated with the negative ion state $^2\Pi_u$, which is known to lead to the 6.5 eV O^- peak in

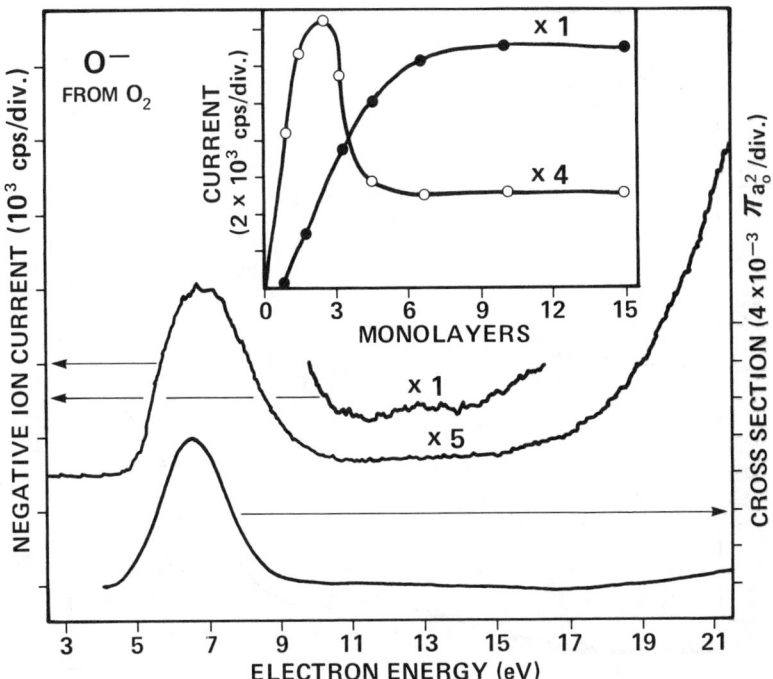

Fig 1. Energy dependence of the O⁻ yields produced by electron impact on a six-"monolayer"-thick film of condensed O_2 and on gaseous O_2 (bottom curve; from Ref. 9). The inset shows the film-thickness dependence of the O⁻ signal produced by 7-eV (full circles) and 21-eV (open circles) electrons.

the gas (bottom curve of figure 1).[8] Furthermore, in condensed O_2 this resonance has recently been observed to cause a strong enhancement of vibrational excitation in the 6-8 eV region.[7] The O⁻ peak in the condensed state looks broader than that in the gas phase because it contains contributions from electron which suffer single and multiple vibrational or electronic losses.

The behavior of the O⁻ intensity as a function of film thickness is shown in the inset of Fig. 1. The solid circles represent O⁻ ions arising from electron bombardment with 7 eV impact energy. Values for the open circles were recorded at incident energy of 21 eV where dipolar dissociation is usually the dominating process. These curves indicate that the dissociative attachment process does not seem to be influenced by the platinum surface, whereas the dipolar dissociation mechanism is enhanced at low coverages where escaping ions are more likely to arise from molecules physisorbed on the metal. The small "hump" at 13 eV may arise from the decay of another repulsive O_2^- state.

In condensed CO (Fig. 2), C⁻ and O⁻ ions are observed. The energy dependence of O⁻ and C⁻ yields is different from that found in the gas phase[8] where dissociatve attachment reactions had a sharp

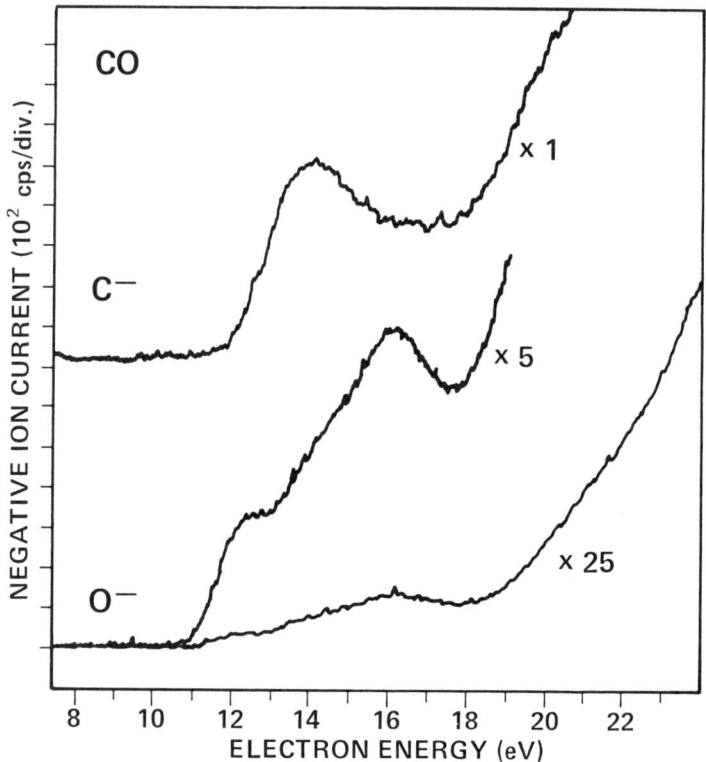

Fig. 2. Energy dependence of C⁻ and O⁻ yields produced by electron impact on a six-"monolayer"-thick film of condensed CO.

onset in O⁻ production at 9.65 eV followed by a maximumm near 9.8 eV and a C⁻ signal with two maxima around 10.5 and 11 eV. Here, O⁻ production sets in at 10.6 eV and maximizes near 12 and 16 eV. The C⁻ signal, completely absent in the range 10-11.5 eV, exhibits a maximum near 14 eV. Its intensity is about an order of magnitude smaller than that of the O⁻ maximum at 16 eV. The O⁻ intensity in the 12-14 eV range is found to be two orders of magnitude smaller than that found in the 10-12 eV range for isolated CO, when normalized to the O⁻ intensity at 35 eV impact energy. These features maybe understood by considering a reduced lifetime of the transient anions in the condensed phase and the probability that escaping ions can lose the additional electron by collisions with neutrals forming the film.[3]

NO molecules dimerize in a trapezoid conformation (N_2O_2) by forming a weak chemical bond between nitrogen atoms upon condensation. O⁻ ion currents measured as a function of electron energy are shown in Fig. 3 for a four "monolayer" thick film of condensed NO and for NO gas.[8] Structure is much more abundant in the condensed phase where broad peaks in O⁻ production are located at 6.8, 9.1, 13.5 and 16.9 eV. We noticed that the amplitude of the signal doubles when the temperature of deposition is lowered from 40 to

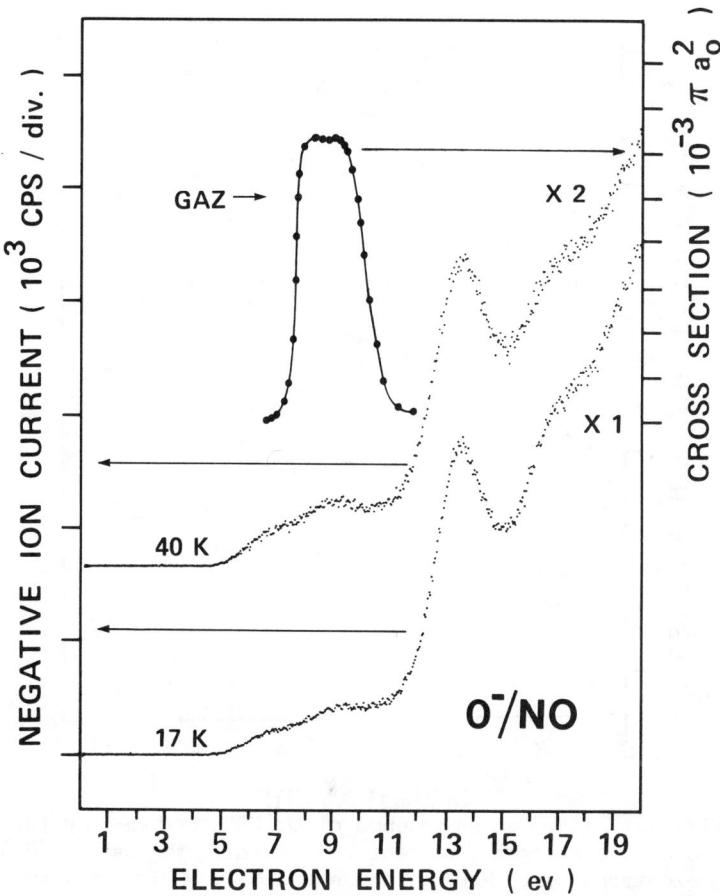

Fig. 3. The middle and lower curves represent the energy dependence of O- ESD from a multilayer film of condensed NO deposited and held at 40 and 17 K, respectively. The cross section for producing O- by electron impact on gaseous NO is shown on top.

17 K.

In gaseous NO the dissociative attachment reaction is generally accepted to give $O^-(^2P)$ ions and $N^*(^2D)$ excited neutral fragments. According to comparison in figure 3 the peak around 9 eV in the condensed phase correlates to the process observed in the gas phase thus leading to O- and N_2O^* excited neutral fragments. The 13.5 eV peak is tentatively associated with $N_2O_2^-$ transient anion in the ground state.[9] Other $N_2O_2^-$ quasibound states also contribute to O- signal as evidenced by the two other peaks at 6.8 and 16.9 eV in figure 3.

The thickness dependence of the O- signal emerging from NO condensed at 17 K and bombarded with 13.5 and 8.0 eV electrons is represented by the two upper curves of figure 4. Beyond a single

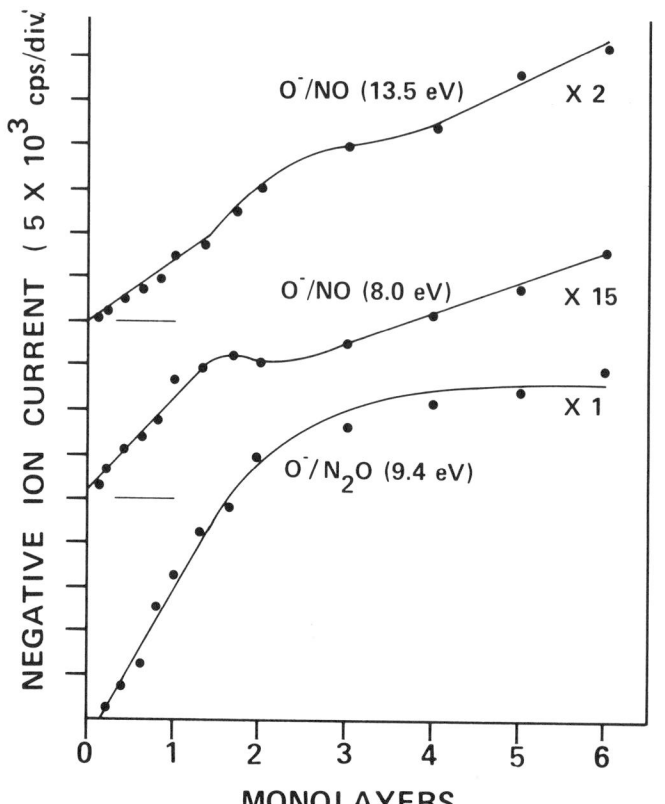

Fig. 4. Thickness dependence of O- ESD produced by 13.5 and 8.0 eV electrons incident on condensed NO and by 9.4 eV electrons incident on condensed N_2O. The platinum substrate temperature was 17 K.

layer coverage, the signal from a first order reaction should saturate as a function of thickness due to the increasing attenuation of the anions formed deeper in the bulk; the steady increase in the O-/NO signal is puzzling and a new mechanism may have to be invoked to explain this finding.

Figure 5 shows the O- ion current measured as a function of electron energy from a multilayer film of N_2O. This curve exhibits two maxima located at 9.4 and 16.4 eV and bears little resemblence with that obtained in gas-phase N_2O,[8] where a strong O- peak is observed near 2 eV and a much smaller one near 11 eV. The thickness dependence of O- signal at 9.4 eV is shown on the lower curve of figure 4. The 9.4 eV peak in Fig. 5 can be correlated with a maxima near 10.6 eV in the gas phase dissociative attachment cross section, but this comparison does not appear justified since the intensity of that latter peak has been shown to depend quadratically on pressure indicating an indirect process. The linearity of the thickness dependence of the O-/N_2O signal at submonolayer coverage (Fig. 3) indicates the presence of a first order reaction.

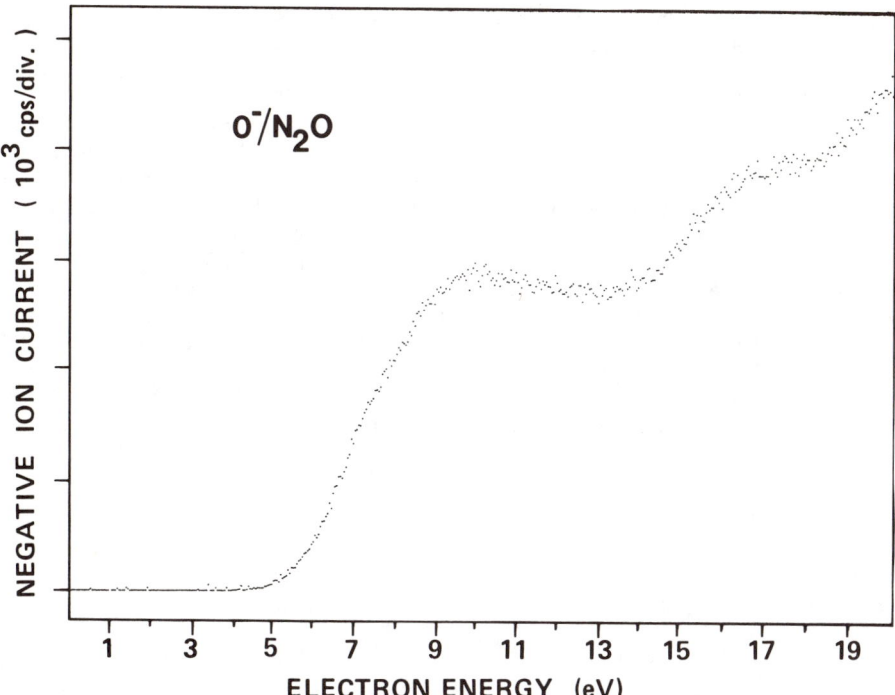

Fig. 5. ESD of O⁻ from a multilayer film of N₂O condensed at 17 K on a polycrystalline platinum surface.

CONCLUSION

Electron stimulated desorption of negative ions in the range 5-20 eV from O_2, CO, NO and N_2O - covered platinum substrate is reported. the structure in the negative ion yields could be attributed to the decay of transient molecular anion states into the dissociative attachment channel. When compared to the gas phase dissociative attachment, these results exhibit more structure and consequently allowed for the detection of a larger number of quasibound anions. Electron stimulated desorption provides a new method by which electron resonances can be detected on metal surfaces.

* L.C.A.M. Bât. 351, Université Paris Sud, 91405, Orsay, France.

REFERENCES

1. For a review of the processes leading to negative ion formation by electron impact in gases and their applications, see H.S.W. Massey, "Negative Ions" (Cambridge Univ. Press, London, 1976).
2. For a review of the processes leading to negative ion formation by electron impact on molecules or atoms chemisorbed on surfaces see N.H. Tolk, M.M. Traum, J.C. Tully, and T.E. Madey, "Desorption Induced by Electron by Electronic Transitions" (Springer, New York, 1983).

3. L. Sanche, Phys. Rev. Lett, 53, 1638 (1984).
4. L. Sanche, J. Chem. Phys. 71, 4860 (1979).
5. G. Bader, G. Perluzzo, L.G. Caron, and L. Sanche, Phys. Rev. B30, 78 (1984).
6. For a molecule chemisorbed on a metal surface the additional electron may be provided by the metal. Dynamical charge transfer from the metal to an adsorbed molecule is also possible as shown in surface-enhanced Raman scattering (see A. Otto, in "Light Scattering of Solids", edited by M. Cardona and G. Guentherodt (Springer, Berlin, 1984), Vol. 4, p. 289).
7. L. Sanche and M. Michaud, Phys. Rev. Lett, 47, 1008 (1981).
8. D. Rapp and D.D. Briglia, J. Chem. Phys. 43, 1480 (1963).
9. L. Sanche and L. Parenteau, J. Vac. Sci. Technol. A3, 1240 (1986).
10. P.J. Chantry, in Proceedings of the 22nd Gaseous Electronic Conference, Gatlinburg, TN, 1969, p. 38.

ENERGY AND ANGULAR DISTRIBUTION OF ELECTRONS DETACHED IN H^--He COLLISIONS (50eV-2keV)

F.Penent, J.P.Grouard, R.I.Hall and J.L.Montmagnon.
Groupe de Spectroscopie par Impact Electronique et Ionique[†], Université Pierre et Marie Curie, 4 place Jussieu, T12-E5, 75252 Paris cedex 05, France

ABSTRACT

The energy spectra of electrons detached in H^--He collisions for H^- energies up to 2keV have been measured directly at several angles using an electrostatic energy analyser. There is a marked difference between these results and those obtained by both time of flight techniques and theory. At small angles spectacular structure is observed resulting from the decay of excited states of H^- (resonances).

The prototype system H^--He is a test bed for experiment and theory on electron detachment and we now have a fair understanding of the mechanisms. This system is particularly interesting because it is a two state problem and deals with the interaction between one bound state and one continuum of the molecular collision complex. Accurate potential curves can be obtained allowing a detailed study of the collision dynamics. In particular, theory has brought to light the important role played by the coupling of the electron to the continuum by the nuclear motion (dynamic coupling). On the experimental side, accurate total detachment cross sections have been measured[1] and more detailed information in the form of detached electron energy distributions are becoming available and provide a stringent test for theory[2]. These are obtained either from time of flight (TOF) measurements[3] of the H atom or from direct observation of the electrons[4]. The TOF technique performs an impact parameter (classically speaking) study of the detachment process but summing over all electron ejection angles and suffers from low resolution. On the other hand direct observation of the electrons sums over impact parameters but benefits from good resolution allowing a more direct comparison with theory. The latter method was used in the present work.

The experimental lay-out is shown in fig 1. The H^- ions, produced in a simple discharge source, are mass selected in a Wien filter before impinging on the target gas beam effusing from a hollow needle. The detached electrons are analysed in energy and angle by a cylindrical electrostatic filter. The detached electrons have energies in the region of $\sim$ 1eV and present a distribution in the form of a broad peak with a rapid rise at low energies and a slow decay to high ones. The width of the distribution increases with H^- energy (fig 2). An accurate measurement of the distribution presents serious experimental problems due to the low energy of the electrons. These are imposed by the transmission characteristics of

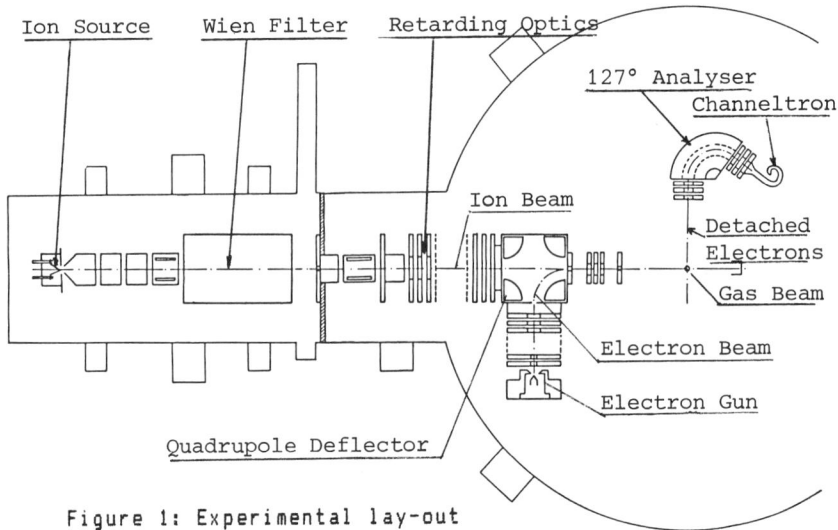

Figure 1: Experimental lay-out

the optics of the electron energy analyser which can vary rapidly with electron energy. This difficulty can be circumvented by calibrating the optic transmission against known electron energy distributions produced when He is ionised by electron impact near threshold.

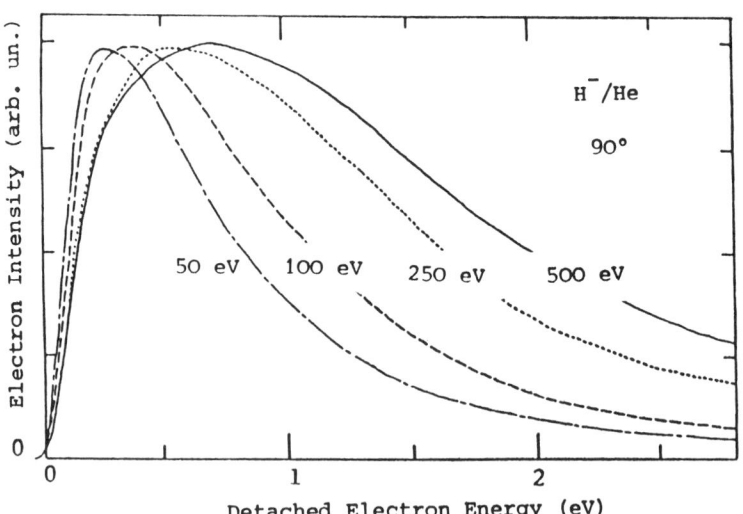

Figure 2: Detached electron energy distributions.

In the previous study of H⁻-He collisions the experimental set-up was such that the negative ion and electron beams entered the collision region at angles separated by 45°. This implied that it was difficult to make the collision volume 'seen' by the electron analyser to be identical in both cases i.e. for the calibration procedure on one hand and for the detachment observations with the negative ion beam on the other. We have done a reappraisal of these observations after rebuilding part of the instrument so that the ion and the electron beams are coaxial thus producing identical collision volumes in each case.

Figure 3 shows the energy location of the half maximum on the

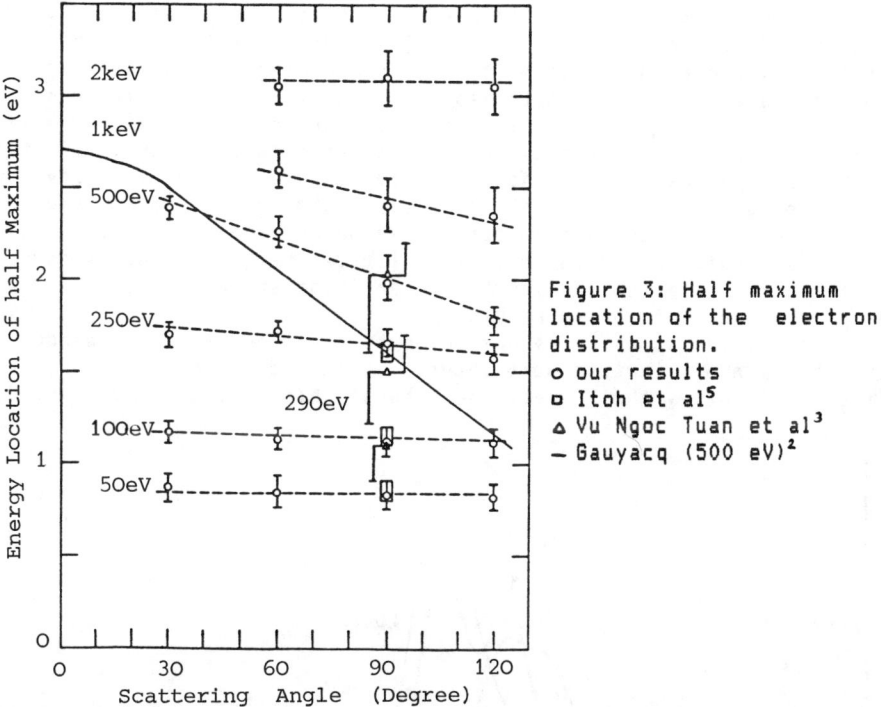

Figure 3: Half maximum location of the electron distribution.
o our results
□ Itoh et al[5]
△ Vu Ngoc Tuan et al[3]
— Gauyacq (500 eV)[2]

high side of the electron distribution plotted against angle for H⁻ energies up to 2keV. These results are now in good agreement with the observations of Itoh et al[3] at 90° who used the same calibration technique. The theoretical result of Gauyacq[2] at 500eV and 90° is much narrower than our observation but agreed well with our previous results . The triangles are representative of the observed half maximum energies at neutral scattering angles making the main contribution to the total detachment cross section. The bars then represent the range of energies observed for all scattering angles. These TOF measurements are broadened by instrumental effects but once these taken into account are in good agreement with theory and would then be in disagreement with our present observations. (The electron energy distributions deduced from TOF results are presented at 90° where kinematic effects are negligible for the electron spectra). This large difference between the electron energy spectra observed directly and deduced from TOF measurements is very puzzling. With regard to the prototype role played by the $H^- - He$ system it is important that this discrepancy be elucidated. This will involve perhaps more sophisticated experiments and help from more elaborate theories.

The evolution of the energy of the half maximum with observation angle depends on the velocity of the emitter which one can imagine to range from that of the center of mass of the $H^- - He$ system to that of H^- alone. At 50eV the kinematic distortion of the electron spectra is negligible whether one considers the H^- frame or the center of mass frame as the emitter. At 500eV the distortion is still weak for an electron emitted by the center of mass but becomes considerable when H^- emits as is shown by the calculations of Gauyacq. The present results at 500eV would indicate a regime between these two extremes. At 2keV the molecular picture of the collision process should no longer be valid as the H^- velocity is much greater than the orbital velocity of the outer electron and one would expect H^- to be the emitter, however these results would indicate that there is no or very little change in the electron distribution as a function of the observation angle. This is a most surprising and intriguing result.

When observations were performed at small angles new phenomena in the form of stucture were observed. This structure is shown in figure 4 with the direct detachment background substracted.

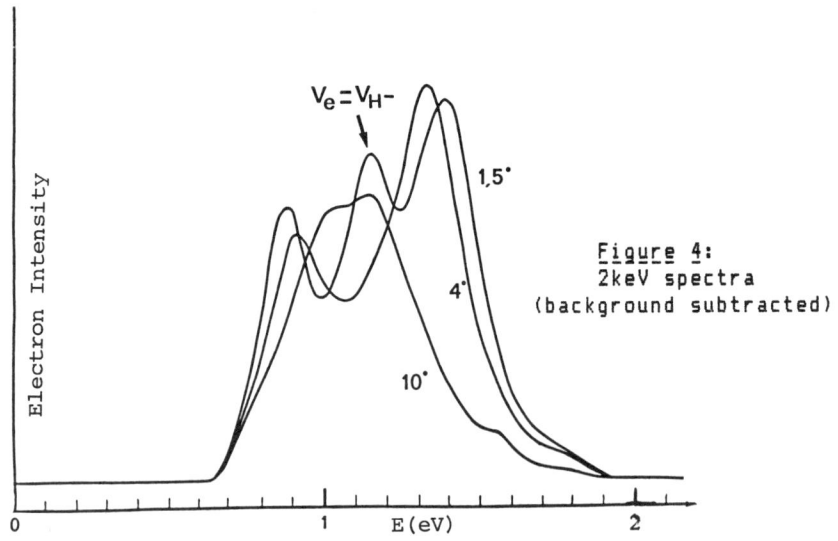

Figure 4:
2keV spectra
(background subtracted)

The structure is located at electron energies corresponding to the translational energy of H⁻ i.e. around Ve=V_{H⁻} thus the electron energy in the H⁻ frame is very small <.1eV. We interpret the structure as being partly due to the decay of the 2s2p $^1P^0$ shape resonance of H⁻ to the n=2 state of H and can be understood by refering to the velocity diagram shown in figure 5. The resonance emits an electron with the velocity Ve in all directions. When the ion velocity $V_{H⁻}$ is greater than Ve, then at the lab. angle θ_L, two peaks will be observed in the spectrum at lab. velocities V_L^+ and V_L^-. This would be the situation for the 4° spectrum of figure 4.

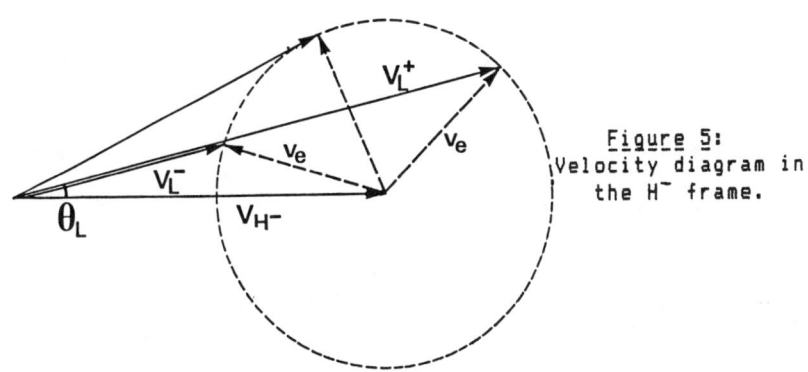

Figure 5: Velocity diagram in the H⁻ frame.

As θ_L increases the two peaks merge giving the 10° curve. The electron energy determined from these observations is 17±2meV which is in excellent agreement with the energy of the $^1P^0$ resonance determined by Bryant et al[6]. As θ_L is decreased to 1.5° a third peak appears at exacly Ve=$V_{H⁻}$ i.e. at zero energy in the H⁻ frame. A similar phenomenon has been observed by Duncan et al[7] and theory would relate it to the formation of H n=2 and the long range interaction of the detached electron with the dipole formed by this state. This interaction leads to a finite cross section for electron impact excitation of the n=2 state due to resonances and consequently to the breakdown of the Wigner threshold law. Thus the central peak is the manifestation of zero energy electrons in the H⁻ frame and the galilean transformation of the double differential cross section from the H⁻ frame to the lab. frame i.e.

$$(d\sigma_L/d\Omega_L dE_L) = (V_L/V_{CM})(d\sigma_{CM}/d\Omega_{CM} dE_{CM}).$$

‡ Associé au CNRS (UA 774)

References.
1. Huq M.S., Doverspike L.D., Champion R.L. and Esaulov V.A., J.Phys.B., 15, 951, (1982)
2. Gauyacq J.P., J.Phys.B., 13, 4417, (1980)
3. Vu Ngoc Tuan, Gauyacq J.P. and Esaulov V.A., J.Phys.B., 16, L95,

(1983)
4. Esaulov V.A., Grouard J.P., Hall R.I., Landau M., Montmagnon J.L., Pichou F. and Schermann C., J.Phys.B., 17, 1855, (1984)
5. Itoh Y., Hege Y. and Linder F., J.Phys.B., (1986)
6. Bryant H.C. et al , Phys.Rev.A, 27, 2889, (1983)
7. Duncan M.M., Menendez M.G. and Hopkins J.L., Phys.Rev.A, 30, 655, (1984)

COMPUTER SIMULATIONS OF PARTICLE-SURFACE DYNAMICS*

Arnold M. Karo and John R. Hiskes
Lawrence Livermore National Laboratory
Livermore, California 94550

and

Thomas M. DeBoni**
University of Texas at Austin
Austin, Texas 78751

ABSTRACT

Our simulations of particle-surface dynamics use the molecular dynamics codes that we have developed over the past several years. The initial state of a molecule and the parameters defining the incoming trajectory can be specifically described or randomly selected. Statistical analyses of the states of the particles and their trajectories following wall collisions are carried out by the code. We have carried out calculations at high center-of-mass energies and low incidence angles and have examined the survival fraction of molecules and the dependence upon the incoming trajectory. We report also on preliminary efforts that are being made to simulate sputtering and recombinant desorption processes, since the recombinant desorption of hydrogen from typical wall materials may be an important source for vibrationally-excited hydrogen in volume sources; for surface sources the presence of occluded hydrogen may affect the concentration of atomic species.

INTRODUCTION

Computer molecular dynamics (CMD) is a powerful technique for examining the microscopic details of a wide variety of phenomena. During the past few years we have reported on the successful application of CMD to the study of a number of important technological problems relevant to an understanding of the plasma-wall interaction zone of fusion plasmas. The problems of interest involve particle-surface dynamics, where experimental data is extremely difficult, as well as costly, to obtain. As we have described in more detail elsewhere,[1] CMD involves the numerical solution of Newton's Second Law equations of motion for all the atoms, thus providing a detailed time history of the coordinates and velocities of the particles. The initial positions and velocities of the particles represent the initial conditions on these equations, and it then remains only to define the forces acting on each of the particles. This has been done

---

*Work performed under the auspices of the U. S. Department of Energy by the Lawrence Livermore National Laboratory under contract number W-7405-ENG-48.

**Consultant to Lawrence Livermore National Laboratory.

in our present calculations of particle-surface grazing-angle collisions by means of the usual sum-of-pair-potentials approximation (Figure 1). It should be noted that realistic simulations of recombinant desorption and sputtering will require a more general treatment of the forces acting on the particles, and our work in these areas will be briefly described.

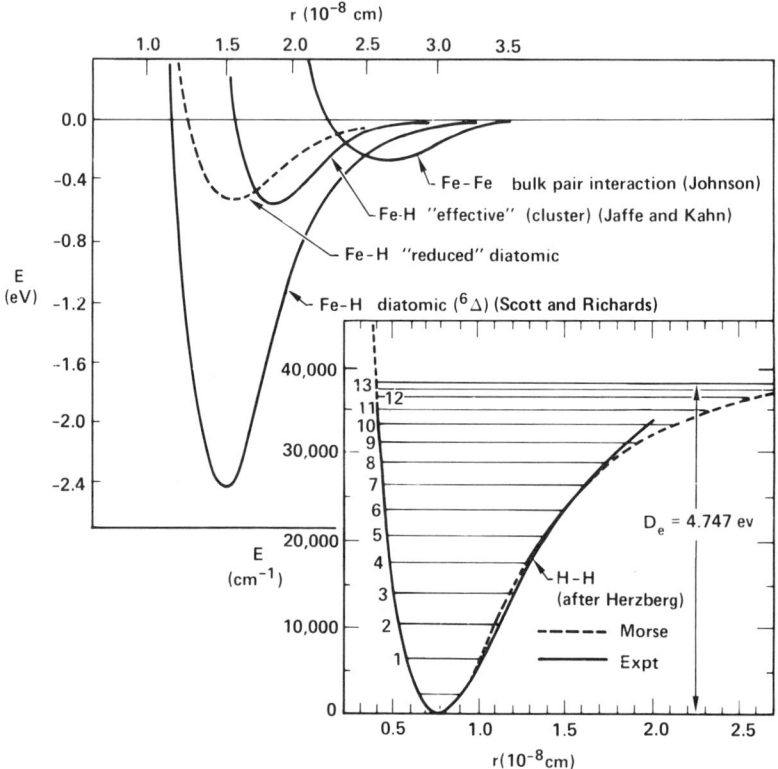

Figure 1. Pair potentials for the interactions between atoms (from ref. 1).

For the development of either volume or surface sources of intense negative hydrogen ion beams, it is crucial to understand the possible mechanisms that can affect both the atomic concentrations and the concentrations of vibrationally-excited molecules in the plasmas. In volume sources, several processes are known to form highly-vibrationally-excited molecules which then by means of dissociative attachment with low-energy electrons form H^- ions from the decay of an intermediate H_2^- state.[2] Our work to date has emphasized the excitation of H_2 molecules to high vibrational levels by fast-electron collisions or from H_2^+ ions that have been accelerated across the discharge-wall potential, undergoing Auger neutralization immediately before impact with the discharge chamber wall.[1,3] As we have discussed previously, at high densities and depending on the discharge parameters, either electronic excitation or H_2^+ Auger neutralization will be the dominant source of vibrational excitation.[3,4]

In order to examine both the creation and depletion of vibrationally-excited molecular hydrogen, we began our early calculations with the de-excitation and re-excitation of highly-vibrationally-excited H_2 molecules, initially at thermal (500K) velocities, undergoing repeated wall collisions.[5] By statistically analyzing a large number of trajectories, we obtained qualitative estimates of the average loss of vibrational energy as a function of the number of wall collisions and also obtained qualitative estimates of the rates of energy exchange and the extent of equipartition among the degrees of freedom of the molecular system, as well as estimates of the average transfer of energy between the molecules and the wall. Later, in addition to studying low-energy, thermal wall collisions, we carried out studies of collisions where the translational energies ranged up to 50 to 100 eV.[1] The purpose of these calculations was to determine the survival probability of vibrationally-excited H_2 molecules after undergoing high-energy collisions with a wall composed of metal atoms and to determine as well the fraction of survivors that remain in, or are excited to, high-lying vibrotational states. The results showed that the process of the wall recombination of H_2^+ to form H_2 ($v" \geq 6$) could make an important contribution to the total vibrationally excited population and to the negative ion yield, comparable to that from the fast-electron process.[3]

The application of these results to the interpretation of ion-source operation has proven successful and has provided a basis for extrapolation to high-power systems. At the same time because of the unusual nature of the initial conditions defining the state of the molecule, experimental results with which to compare the results of the simulations are for all practical purposes nonexistent. However, as we noted in our earlier study of the generation of vibrationally-excited H_2 molecules by H_2^+ wall collisions, we found a substantial fraction of surviving neutrals for translational kinetic energies ranging from 1 to 20 eV. Subsequently, experimental verification of the prediction of the survival of a large neutral fraction of H_2 for beams of H_2^+ incident on a metal surface has been reported.[6] These experimental studies and the current interest in molecular-ion scattering have encouraged us to undertake a series of grazing-angle trajectory calculations in which the great advantage of using computer experiments to modify the initial conditions can be utilized. These calculations will be discussed in the following section.

In addition, in order to understand the plasma-wall interaction zone in more detail, we have been designing simulations that can (1) describe the recombinant desorption of hydrogen molecules from metal surfaces containing a partial monolayer of hydrogen atoms upon which a flux of energetic protons is directed and (2) describe the sputtering and possible recombination of hydrogen atoms lying just below a heavy metal surface when a flux of light ions (e.g., H^+) or heavy ions (e.g., Cs^+) is directed toward such a prepared surface.

Our preliminary studies of these processes will be described in the third section.

LOW-ANGLE SCATTERING OF MOLECULAR HYDROGEN FROM METAL SURFACES

In Figure 2, we show schematically a typical starting configuration for an H_2 molecule set in motion with some velocity "v" toward a heavy-atom wall and with an initial phase angle α and an angle θ with respect to the direction perpendicular to the surface. Vibrotational states of H_2 have been selected that represent vibrationally-excited species with $v" = 2, 8$, and 12, corresponding

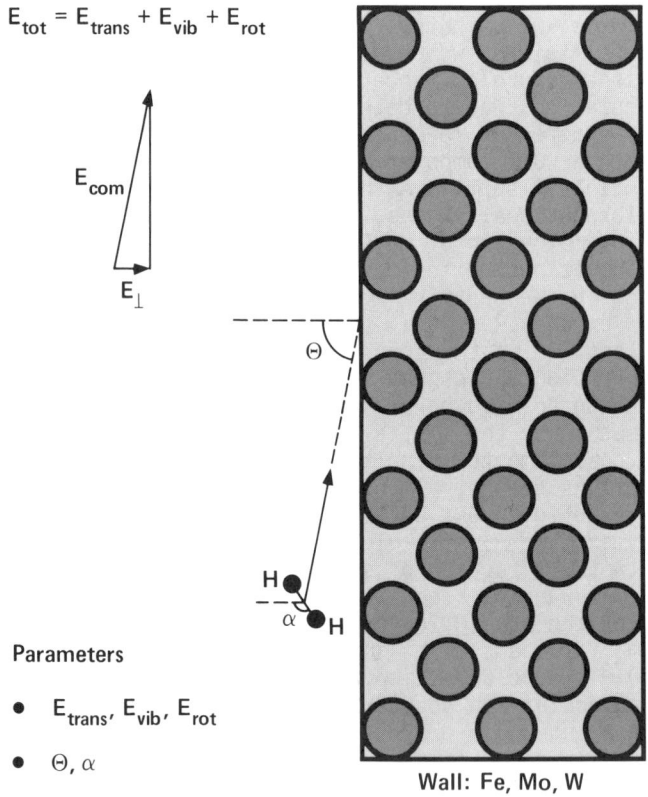

Figure 2. Schematic representation of a typical starting configuration for an H_2 molecule moving at a low incidence angle toward a wall. Initial conditions include the velocity v, the phase angle α, and the angle θ from the normal, as well as the wall temperature T.

to energies ranging from 10,200 cm^{-1} to 35,250 cm^{-1} (1.266 eV to 4.371 eV) above the potential energy minimum (cf. Figure 1). The rotational quantum number J was set equal to 1 in each case, corresponding to a plasma temperature of 500 K where this rotational state predominates. The center-of-mass velocities have been given values so that at each grazing angle the normal component of center-of-mass translational kinetic energy will be 1, 4 or 10 eV. This provides a correspondence with our earlier work[1] where the trajectories were selected to be normal or near-normal with the perpendicular components of the translational energies ranging from 1 to 50 eV. Results from these earlier calculations indicated that survival fractions and excitations were essentially statistically independent of the angle of approach of the incoming trajectory when the normal component of energy was held constant. For the grazing-angle collisions, the mechanism we have previously described for the near-surface neutralization of an energetic beam of H_2^+ ions[3] will result in a range of vibrotational states corresponding to those indicated above. For the interatomic potentials, which are shown in Figure 1, we have used a modification of the two-body potential of Scott and Richards[7] for FeH, where we have reduced the strength of the interaction by a factor of 5, thereby simulating the weaker pair interaction between a hydrogen molecule and a cluster of Fe atoms. (An FeH effective "cluster" two-body potential obtained from <u>ab initio</u> calculations on Fe_4H clusters has also been shown to give statistically very similar results.) The Fe wall potential does not play a significant role in the dynamics, and we have continued to use the two-body potential of Johnson[8] obtained from elastic constant data. Finally, as shown in Figure 1, a Morse fit to the experimental H_2 potential curve is very accurate up to v" = 8 or 9. The effect of a somewhat more slowly-rising potential for v" = 12 will very likely be subtle and less important than other approximations involved in the calculations.

The results to date of more than a thousand trajectories are summarized in Table I. It can be seen that the percent of surviving neutrals ranges from slighly less than 10 percent to more that 50 percent. As the normal component of beam energy changes from 1 to 10 eV, the fraction of surviving neutrals, taking into account statistical error, remains about the same for initial vibrational excitations v" = 2 and 8; for v" = 12 the neutral fraction appears to increase, although this may be a spurious result because of the relatively small number of trajectories examined so far. Additional computer studies are in progress to decrease the statistical uncertainly. More importantly, the surviving fraction of neutrals with large vibrational excitation (i.e., $f_1 f_2$) remains significant and of the order of 10 to 20 percent throughout the range of v" and $E_\perp$ that we have examined. A comparision with results from our earlier calculations of the survival fraction after normal and near-normal surface collisions,[1] shown in Table II, indicates for both the earlier and the present calculations that the survival rate is consistent with a strong dependence on the normal velocity component of the incoming molecule; however, the dynamic processes leading to vibrational excitation are seen to be enhanced by grazing collisions with the wall. Thus, for v" = 2 and $E_\perp$ = 2 eV, there is

Table I. Fraction f_1 of molecules surviving a wall collision at high impact energies. The fraction of the surviving molecules with $v'' \geq 6$ is f_2; the grazing angle is 85°.

E_{com} (eV)	$E_\perp$ (eV)	\multicolumn{9}{c}{v''}
		\multicolumn{3}{c}{2}	\multicolumn{3}{c}{8}	\multicolumn{3}{c}{12}
		f_1	f_2	$f_1 f_2$	f_1	f_2	$f_1 f_2$	f_1	f_2	$f_1 f_2$
131.6	1.0	0.40	0.20	0.08	0.55	0.42	0.23	0.08	0.38	0.03
526.6	4.0	0.29	0.45	0.13	0.29	0.72	0.21	0.19	0.87	0.16
1316.6	10.0	0.33	0.17	0.06	0.22	1.00	0.22	0.28	0.77	0.21

Table II. Fraction f_1 of molecules surviving a wall collision at higher impact energies. The fraction of the surviving molecules with $v'' \geq 6$ is f_2.

$E_\perp$		\multicolumn{9}{c}{v''}
		\multicolumn{3}{c}{2}	\multicolumn{3}{c}{8}	\multicolumn{3}{c}{12}
		f_1	f_2	$f_1 f_2$	f_1	f_2	$f_1 f_2$	f_1	f_2	$f_1 f_2$
1 eV	0°	1.00	0.00	0.00	1.00	0.47	0.47	0.60	0.50	0.30
	15°	1.00	0.00	0.00	1.00	0.38	0.38	0.56	0.54	0.30
4 eV	0°	0.92	0.13	0.12	0.38	0.32	0.12	0.64	0.28	0.18
	15°	0.94	0.16	0.15	0.48	0.54	0.26	0.64	0.22	0.14
10 eV	0°	0.42	0.57	0.24	0.20	0.83	0.17	0.18	0.89	0.16
	10°	0.27	0.73	0.20	0.15	0.67	0.10	0.14	0.86	0.12

now more than sufficient translational energy to excite low-lying vibrational states of the molecule during the grazing collision to high vibrotational levels. In addition to adding many more trajectories to reduce the statistical uncertainty, we are attempting to understand the dynamical events leading either to molecular survival or destruction at these large total energies.

ATOMIC AND RECOMBINANT DESORPTION OF HYDROGEN FROM METAL SURFACES

Other recent simulations have been designed to study the recombinant desorption of hydrogen molecules from metal surfaces containing a partial monolayer of hydrogen atoms upon which a flux of energetic protons is directed. Computer experiments have also been designed to look at the sputtering and possible recombination of hydrogen atoms lying just below a heavy metal surface when fluxes of light or heavy ions are directed toward such a prepared surface. These are processes affecting both the atomic concentrations and the concentrations of vibrationally-excited molecules in the plasmas. The recombinant desorption of hydrogen could be an important source of vibrationally-excited hydrogen in volume sources; for surface sources, the presence of occluded hydrogen in the wall could affect the concentration of atomic species and provide an additional mechanism for the creation of excited molecular hydrogen.

Figures 3 and 4 show schematic representations of possible initial configurations for the simulations described above. For the calculation of recombinant desorption of hydrogen from surfaces, an Fe surface was formed upon which we placed a partial monolayer of hydrogen atoms. Surface temperatures were set equal to 500 K. Fluxes of atomic hydrogen with a random velocity distribution equivalent to a plasma temperature of 1000 K and with translational velocities ranging from 2-20 eV were directed toward the surface. Within a few picoseconds in these calculations, dynamical events are found to occur leading to the desorption of hydrogen atoms and molecular hydrogen. Although results from these initial calculations seem to indicate that the recombinant desorption of hydrogen from

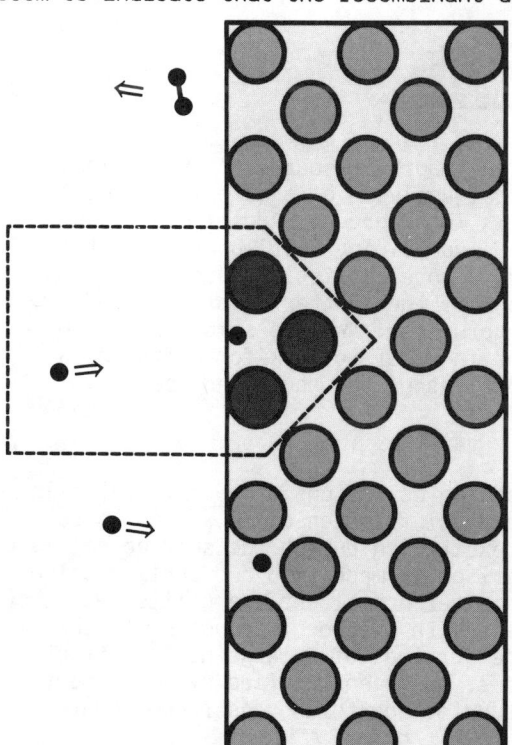

Figure 3. Schematic representation of an assembly of particles used to study recombinant desorption of hydrogen from a metal surface. Multi-body force terms are included in describing the dynamics of the cluster within the dashed region.

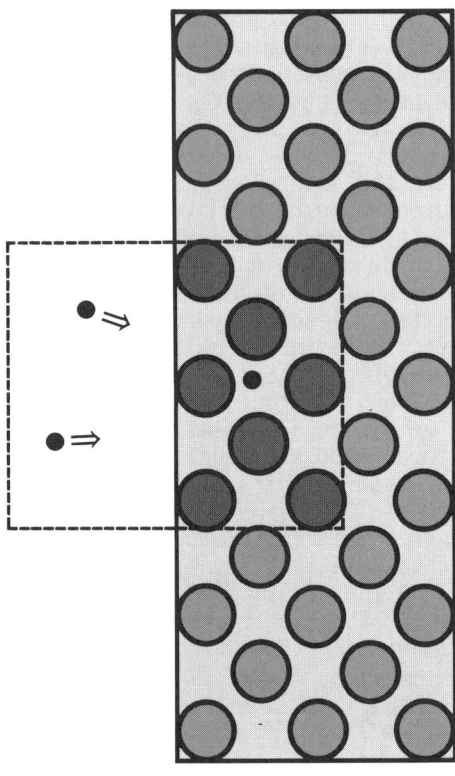

Figure 4. Schematic representation of an assembly of particles used to study sputtering and possible recombination of hydrogen atoms placed at equilibrium positions just below a heavy metal surface. The incoming flux may be composed of either light (as shown) or heavy ions. Multi-body force terms are included in describing the dynamics of the cluster within the dashed region.

typical wall materials can be an important source for vibrationally-excited hydrogen, the statistical analysis of a much larger number of runs is needed, and this work is in progress. Additionally, an improved molecular dynamics code containing multi-body force terms is being written that will provide a more realistic approach to these dynamic processes. Calculations of the sputtering of occluded hydrogen are also currently in progress. We are simulating both heavy and light ion bombardment in order to look for differences in the energy and angular distributions of the sputtered species.

SUMMARY

In the simulations of glancing collisions of H_2 with metallic surfaces, we have found the survival fraction of molecules to be significant and vibrotational excitation during the surface collision to be very likely. When comparison is made with our earlier calculations of the survival fraction after normal surface collisions, the survival rate appears consistent with a strong dependence on the normal velocity component. The survival of a large neutral fraction, as predicted by our calculations, has been verified by experiment. Simulations directed toward studying sputtering and recombinant desorption of H and D atoms from metal surfaces have shown the ready formation of atomic and excited molecular species, thus identifying mechanisms affecting their concentrations in plasmas.

REFERENCES

1. A. M. Karo, J. R. Hiskes, and R. J. Hardy, J. Vac. Sci. Technol. A3(3), 1222 (1985).
2. J. R. Hiskes, A.M. Karo, M. Bacal, A. M. Bruneteau, and W. G. Graham, J. Appl. Phys. 53 (5), 3469 (1982).
3. J. R. Hiskes and A. M. Karo, AIP Conference Proceedings No. 111, (American Institute of Physics, New York, 1984), p. 125.
4. J. R. Hiskes and A. M. Karo, AIP Conference Proceedings No. 111, (American Institute of Physics, New York, 1984), p. 3.
5. A. M. Karo, J. R. Hiskes, K. D. Olwell, T. M. DeBoni, and R. J. Hardy, AIP Conference Proceedings No. 111, (American Institute of Physics, New York, 1984), p. 197.
6. B. Willerding, W. Heiland, and K. J. Snowdon, Phys. Rev. Lett. 53, 2031 (1984).
7. P. R. Scott and W. G. Richards, J. Chem. Phys. 63, 1690 (1975).
8. R. A. Johnson, Phys. Rev. 134A, 1329 (1964).

DISCUSSION

Mitchell: With regard to this hydrogen recombination on the wall, are you aware of the interstellar chemistry modeling that has been done on that, where atomic hydrogen collides with a cold grain, which is about 4°K, sticks there, and then another atomic hydrogen comes in, recombines with the first atomic hydrogen, and the recombination energy gives it enough energy to kick it off the grain and back into the environment? This is the only mechanism for forming molecular hydrogen in colloids that's been known. And yet, the observations have shown that colloids are predominantly molecular hydrogen, so it's a very, very efficient process, indeed.

Karo: Very good. Your question is: Am I aware of that? I don't think I'm aware specifically, so I would appreciate it if you have a reference of that.

Azria: I don't know if I missed something, but how do you characterize your surface? You imagine that you have a clean surface, or something absorbed on this?

Karo: The characterization of the surface? I had started with a surface on which I had placed a monolayer of hydrogen atoms. It was started with a prepared surface, and then I brought the flux in.

DIFFUSION AND FREE FLOW THROUGH A MAGNETIC FILTER

R. G. Jones and Wm. F. Bailey
Department of Engineering Physics
Air Force Institute of Technology
Wright-Patterson AFB, Oh. 45433

ABSTRACT

A parametric study of charged particle transport through magnetic filters in multi-cusp discharges was conducted. Among the phenomena discussed are the free and diffusive flow of species through the filter, a cooling mechanism for electrons, and the dependence of ion flow with mass and with the strength and profile of the magnetic field.

INTRODUCTION

The magnetic filter in a multicusp ion source permits the transport of ions and electrons[1] from the source to the extraction chamber (see Figs. 1 and 2). The ion flow generally decreases with increasing magnetic field strength.[2] If the maximum magnetic field strength is lowered while the line integral of the magnetic field from the source to the extraction chamber is kept constant, the magnetic filter becomes more transparent.[3] We will derive a general expression for the free flow of charged particles across a triangular magnetic field profile, compare the magnitude of the free flow with the diffusive flow, and explain the observed variations in ion flux into the extraction region for all three ion species.

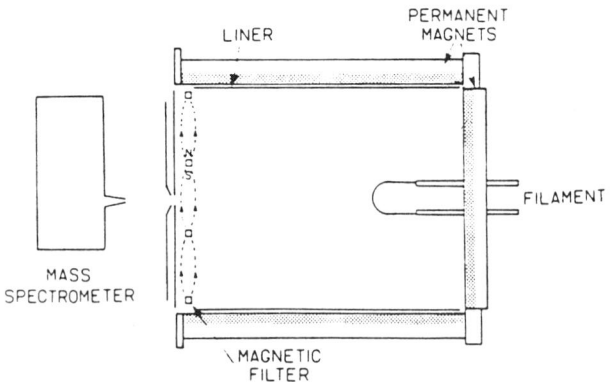

Fig 1. Ion Source (from Leung, Ehlers and Pyle[4])

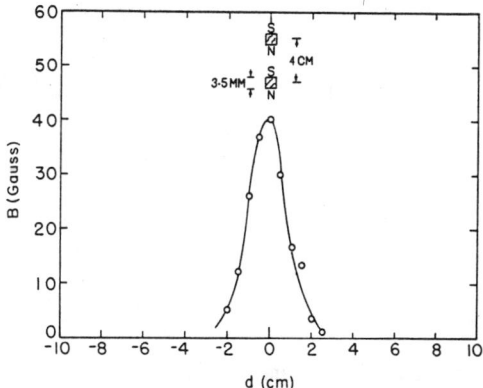

Fig 2. Typical field profile (after Leung and Ehlers[2])

FREE FLOW

We calculate the flux of charged particles across a magnetic filter. The initial conditions on particle velocity that permit transit are determined from the Lorentz force equations. Figure 3 shows the coordinate system employed. The "y" equation parametrized in x, yields

$$V_y(x) = V_y^o - e/(mc) \int_0^x dx' \, B(x') \qquad (1)$$

The "x" equation then can be written as

$$V_x \frac{dV_x}{dx} = \frac{e}{mc} \left[V_y^o - \frac{e}{mc} \int_0^x dx' \, B(x') \right] B(x) \qquad (2)$$

The solution is

$$V_x^2(x) = V_x^{o\,2} + 2\,\frac{e}{mc} \int_0^x dx' B(x')$$
$$- 2e^2/(mc)^2 \int_0^x dx' B(x') \int_0^{x'} dx'' B(x'') \qquad (3)$$

If we model the magnetic field profile as a triangle

$$B(x) = \begin{cases} \frac{2B}{\Delta} x & 0 \leq x \leq \Delta/2 \\ 2B(1 - x/\Delta) & \Delta/2 \leq x \leq \Delta \end{cases} \qquad (4)$$

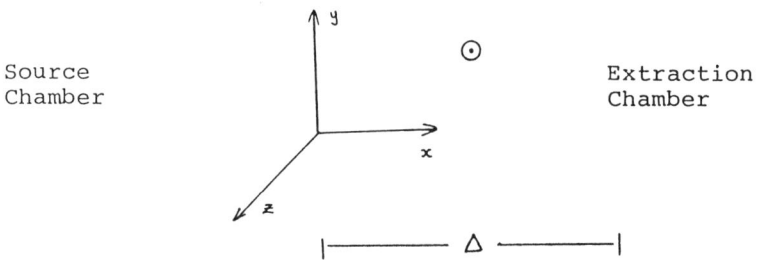

Fig 3.   The coordinate system

we find that the x-component of velocity at $x = \Delta$ is

$$V_x^2(\Delta) = V_x^{o2} + \omega_b \Delta V_y^o - \omega_b^2 \Delta^2/4 \tag{5}$$

where $\omega_b = \frac{eB}{mc}$. The condition for transit is $V_x(\Delta) \geq 0$, so flux through the filter is

$$\Gamma_x = N \overline{V_x^F} = \iiint d^3 v^o \, f(\vec{V_o}) V_x^o \tag{6}$$

subject to the condition

$$V_x^{o2} + \omega_b \Delta V_y^o - \omega_b^2 \Delta^2/4 \geq 0$$

Assuming a Maxwellian distribution function, we find that

$$\overline{V_x^F} = \left(\frac{kT}{2\pi m}\right)^{1/2} \left[1 - \mathrm{erf}\left(\left[\frac{m}{2kT}\right]^{1/2} \frac{\omega_b \Delta}{4}\right)\right] \tag{7}$$

DIFFUSION ANALYSIS

Particles with velocities too low to traverse the filter may diffuse across the filter magnetic field. Assume this flow is ambipolar and proceeds with a directed velocity[5]

$$\overline{V_x^D} = -kT \left[\frac{\nu}{M_e W_e^2}\right] \left(\frac{1}{n}\right) \frac{\partial n}{\partial x} \tag{8}$$

where ν is the total collision frequency, M_e the electron mass, and W_e the average electron cyclotron frequency. In the discussion that follows, it is assumed that the electron temperature is 1 eV, and that each ion species' temperature is low in comparison. The effective value of T in eqn. (8) is then the electron temperature.

We model the ion flux to the extraction chamber as the sum of the free flux, eqn. (7), and the diffusive flux, eqn. (8). The free flux is a function of $\int B dr \propto \omega_b \Delta/2$. Since $(1/n) \partial n/\partial x$ is approximately equal to $1/\Delta$, the diffusive flux varies as
$$(B\Delta/2)^{-1} B^{-1} = [\int B dr]^{-1} [B^{-1}].$$
Given accurate values of the ion number densities and temperatures, it is possible to predict the change in each ion species' concentration in the extraction chamber as B and $\int B dr$ are varied. But without that information, the model can explain observed variations in ion concentrations which cannot be explained assuming diffusive flux alone.

K. N. Leung and K. W. Ehlers[3] measured the effective transparencies of four magnetic filters (their data are reproduced below as Table 1). Filters 1 and 4 had the same value of $\int B dr$. The H^+ and H_3^+ percentages are higher using filter 1, while the H_2^+ percentage is higher using filter 4. On the basis of diffusion alone, each species' velocity should decrease by the same factor when B is increased. (Variations in plasma grid potential will alter H_2^+ percentages[3], but this effect should be of minor importance for these filters.)

Filter	Magnet cross section (mm^2)	Magnet spacing (cm)	Maximum B field (G)	$\int B dr$ (G cm)	Effective transparency (%)	Plasma grid floating potential (V)	$H^+:H_2^+:H_3^+$
filter	—	—	—	—	100	−56	24:35:41
1	3.5 × 3.5	4	40	89	68	−15	30:17:53
2	4.5 × 4.5	4	76	166	43	−2	41:15:44
3	4.5 × 4.5	6	35	104	70	−10	35:19:46
4	4.5 × 4.5	8	20	89	80	−25	26:25:49

Table 1. Filter data from Leung and Ehlers.

However, if the flux through the filter is the sum of a diffusive term and a free flow term for each ion species, it is possible to model the changes in species concentrations as $\int B dr$ varies. The high H^+ percentage of filter 2 can be understood if the H^+ collision frequency is large compared with the collision frequencies of H_2^+ and H_3^+. The general decrease in H^+ percentage with decreasing $\int B dr$ results from an increase in H_2^+ and H_3^+ free flux. Figure 4 illustrates the rise in H_3^+ free flux for decreasing $\int B dr$. For an ion temperature of 1000 °K, the H^+ free flux is unimportant above $\int B dr = 100$ G-cm.

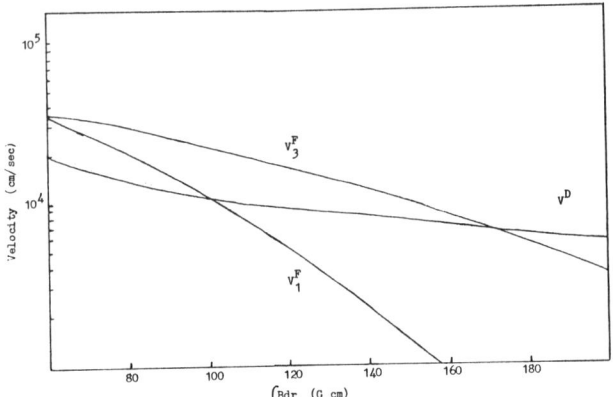

Fig 4. Diffusive, V^D, and Free Flow, V^F, Velocities for Ion Temperatures of 1000 °K.

The experimental variations between Filters 1 and 4, discussed above, will occur under the following conditions. The condition for a decrease in H^+ percentage as B is decreased is

$$\left(\frac{\nu_2}{\nu_1} - \frac{V_2^F}{V_1^F}\right) N_2 + \left(\frac{\nu_3}{\nu_1} - \frac{V_3^F}{V_1^F}\right) N_3 < 0 \quad (9)$$

where ν_1 is the total collision frequency for H^+ and V_1^F is the free flux velocity of H^+ at $\int Bdr = 89$ G cm. Similarly, the H_2^+ flux increase if

$$\frac{V_2^F}{V_1^F}\left(N_1 + \frac{\nu_3}{\nu_1} N_3\right) > \frac{\nu_2}{\nu_1}\left(N_1 + \frac{V_3^F}{V_1^F} N_3\right), \quad (10)$$

and the H_3^+ flux decreases if

$$\frac{V_3^F}{V_1^F}\left(N_1 + \frac{\nu_2}{\nu_1} N_2\right) < \frac{\nu_3}{\nu_1}\left(N_1 + \frac{V_2^F}{V_1^F} N_2\right). \quad (11)$$

If the number densities, collision frequencies and velocities satisfy these conditions, the diffusive plus

free flow model of flux correctly accounts for the
observed variations of the ion flow through the filter.

FILTER EFFECTS ON THE ELECTRON DISTRIBUTION FUNCTION

For electrons, the free flow flux is essentially
zero. Diffusion, however, remains significant. In
additions, the electron energy distribution is
significantly effected in the vicinity of the filter.
The specific collision frequencies of relevance in
molecular Hydrogen are sketched in Fig 1 of reference 7.
The specific wall loss curve given, which scales directly
with loss area to volume ratio and inversely with
pressure, is representative of a discharge operating at
40 mTorr with a loss area to volume ratio of 0.07. In
the filter region, this ratio can increase dramatically.
Using a Maxwellian model of the electron energy
distribution[7], the electron number density and
temperature have been calculated for a 10 A, 40 mTorr
discharge (Fig 4). The calculated electron temperature
and density in the source region are 0.9 eV and 9.89×10^{11}
cm^3. When the wall losses dominate the specific
collision frequencies, the distribution tail is
truncated. Calculations at an area to volume ratio of
1.0 yielded an electron temperature of 0.23 eV and number
density of 2.6×10^{11}. This significant reduction in
electron temperature suggests that negative ion
production in the filter region is sustained by diffusion
and may be enhanced due to wall losses.

CONCLUSION

A simple model of charged particle transport through
a magnetic filter may explain observed variations in ion
percentages in the extraction chamber for different
magnetic field configurations. Investigation of the
electron energy distribution in the filter region reveals
that the increased area to volume ratio in the vicinity
of the magnetic filter results in a lower electron
temperature. This may contribute to increased negative
ion production in the filter region.

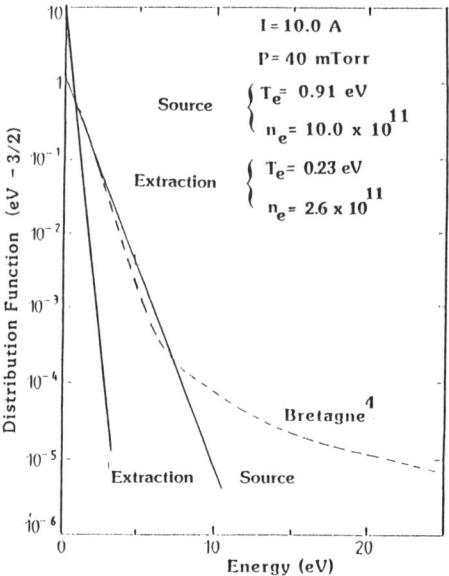

Fig. 4. Electron energy distribution function in source and filter regions.

REFERENCES

1. K. N. Leung, "Proceedings of the Symposium on the Production and Neutralization of Negative Hydrogen Ions and Beams", 1977.

2. K. W. Ehlers and K. N. Leung, Rev. Sci. Instrum., 52, 1452 (1981).

3. K. W. Ehlers and K. N. Leung, Rev. Sci. Instrum., 53, 1423 (1982).

4. K. N. Leung, K.W. Ehlers, and R.V. Pyle, Appl. Phys. Letters, 47(3),227 (1985).

5. V. E. Golant, A. P. Zhilinsky, I. E. Sakharov, and S. C. Brown, Fundamentals of Plasma Physics (John Wiley & Sons, New York, 1980).

6. J. Bretagne, G. Delouya, C. Gorse, M. Capitelli, and M. Bacal, J Phys. D: Appl. Phys., 18, 811, (1985).

7. W. Bailey, R. Jones, Fourth International Symposium on the Production and Neutralization of Negative Ions and Beams, (1986).

COMMENTS ON H⁻ VOLUME PRODUCTION IN CS-SEEDED ION SOURCES

J. R. Peterson
Chemical Physics Laboratory
SRI International
Menlo Park, CA 94025

Considerable interest was generated at the IAEA Negative Ion Beam Workshop in Grenoble, France, in March, 1985, by a report from the Kurchatov Institute on the development of a 2-ampere steady-state H⁻ ion source, in which the ions were volume-produced in a discharge in H_2, seeded with Cs vapor.[1] The mechanism primarily responsible for this remarkably high current from a volume production source was not yet understood, but it was tentatively presumed to involve the collisional energy transfer from electronically excited Cs 6p atoms into H_2 vibrations. In any case, it was apparently different from the surface-plasma interactions that have been assumed to control the H⁻ production in the Dudnikov-Dimov type sources.

BACKGROUND

Following the experiments of Allan and Wong, and theoretical developments by Wadehra and Bardsley,[2,3] it has become commonly accepted that vibrationally excited H_2 plays a critical role in the efficient generation of H⁻ in a hydrogen plasma as first observed by Bacal et al. The "volume production" sources now being developed at LBL and elsewhere based on pure H_2 plasmas have shown promise for high-brightness in tests at LANL, nevertheless the reported high-current results from the Kurchatov source stimulated a consideration of the mechanisms involved in it.

The theoretical calculations[2,3] of dissociative attachment in H_2,

$$e^- + H_2(v) \rightarrow H^- + H,$$

show that the maximum cross section increases by 10^5 between $v = 0$ and $v = 6$, with 10^4 of this occuring by $v = 4$. Because the maximum cross section for attachment via excitation of the H_2^- $^2\Sigma_u^+$ resonance occurs near the threshold energy,[2,3] which decreases as v increases, the attachment rates will increase even more dramatically for high v's if the bulk of the electrons have kinetic temperatures under a few eV.

EXCITATION OF H_2 VIBRATIONS BY EXCITED CESIUM ENERGY TRANSFER

Because dissociative attachment with H_2 is the primary H^- production mechanism in H_2 plasmas, we first consider the possibility of increasing the vibrational temperature of the gas to improve the efficiency of the ion source, and in particular for a method of reaching v = 4 or higher. Collisional energy transfer from the 6p ^2P resonant state of Cs, which could be "resonantly trapped" in the volume is the most likely mechanism. There have been no measurements of energy transfer from this reaction, but the similar reaction between Na $3p^2P$ and H_2 has undergone fairly extensive study. Reiland, Tittes and Hertel[4] found that Na 3p transfers a about 1 eV (max. probability) of the total 2.2 eV available, to either H_2 or D_2, yielding vibrational distributions of: 60%, 13%, 61%, and 26% and 0%, for v = 0 → 4, respectively. Cs 6p has about 1.6 eV available, and using a similar relative distribution for E-V transfer in H_2 to that of Na, one can conclude that the distribution from a single collision would peak at v = 1.

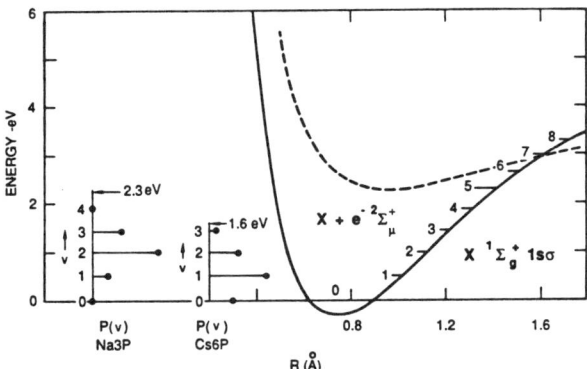

Figure 1 Potential energy diagram of the $H_2(X\ \Sigma_g^+)$ ground state and the location (Ref. 2) of the broad $H_2^-\ \Sigma_\mu^+$ resonance that is responsible for dissociative attachment of low-energy electrons. Bar graphs to the left indicate relative excitation of $H_2(v)$ by thermal collisions with Na 3^2P (Ref. 4) and Cs 6^2P (estimated).

Figure 1 shows the distribution in $H_2(v)$ found by Reiland et al.[4] for Na(3p)-H_2 collisions, and, by analogy, one that could be predicted to be excited by Cs(6p) collisions. These distributions shown at the lower left, indicate the relative excitation of the vibrational levels of the $H_2\ X^1\Sigma$ ground state, whose energies are shown on the potential energy diagram. Arrows above the distributions at 2.1 eV and 1.6 eV indicate the energies of the Na and Cs excited states, the maximum that can be transferred in a single collision.

Although the reaction rates (for Na) are reasonably high ($>10^{-10}$ cm^{3}s^{-1}), it is obvious that single collisions with Cs(6p) cannot excite much above v = 1. Excitation of higher v by multiple collisions with excited Cs is quite unlikely, thus it appears that excitation of H_2(v ≥ 5) by cascade radiation induced by the high energy electrons, examined by Hiskes,[5] would be faster than by excited Cs.

Other reactions that could involve excited Cs were considered, but none look promising. However, there are two beneficial effects that ground state Cs could have.

POSSIBLE ENHANCEMENT OF H$^-$ VOLUME PRODUCTION DUE TO GROUND-STATE Cs

1. Electron Cooling.

The introduction of Cs into the volume would have a cooling effect on the electron temperature, due to excitation and ionizing collisions. This effect would

(a) increase the dissociative attachment rates, as well as
(b) reduce the H$^-$ loss by electron collisional detachment.

The magnitude of the effect would depend on the amount of the Cs present and cannot be estimated, but the ionization and excitation cross sections of Cs by electron impact are all large, and because there are no comparable inelastic cross sections in H_2 below about 10 eV, the presence of Cs could be important in cooling the electrons. Other aspects have been considered by Ehlers and Leung.[6]

2. Production of H_2(v) by H_3^+ + Cs Charge Transfer.
We have found[7] that about two-thirds of these reactions,

$$H_3^+ + Cs \rightarrow H_3^* \rightarrow H_2^\ddagger + H + Cs^+,$$

between Cs and the main positive ion, H_3^+, lead directly to 3.6 ± 1.5 eV internal (rovibrational) excitation of H_2, which is equivalent to v = 9 for pure vibration. The remaining 1/3 of the reactions yield H + H + H products.[7] The total cross sections[8] are very large, $\geq 150\text{Å}^2$ and independent of energy from 1000 eV down to 50 eV, the lowest energy examined (See Figure 2), thus they are

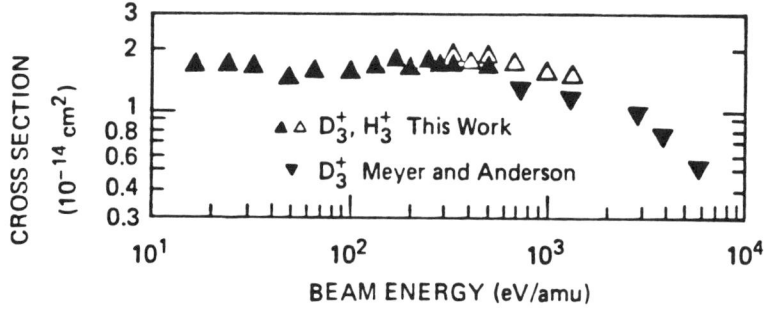

Figure 2 Cross sections for electron capture of H_3^+ in Cs (from Ref. 8).

still large near thermal energies. And because the reaction goes to near-resonant Rydberg states of H_3, (which subsequently predissociate), the final states should be nearly independent of collision energy.

Of course, substantial rotational excitation can also be expected, however Wadehra[9] has shown that combined rovibrational energy is more effective than pure vibration in enhancing dissociative attachment in H_2. In fact, his rates for electron temperatures of 1 eV reach a maximum of about 10^{-8} cm^3/sec for a total internal energy of about 3.8 eV. This reaction with ground-state Cs is fast and effective, and does not require Cs excitation. It thus is, with little doubt, more important than E-V excitation of $H_2(v)$ by Cs 6p. Whether it is as important as vibrational excitation by electron impact is not known.

It is impossible to determine its actual effect without knowing the actual conditions and modeling the plasma. However it can at least be noted that if $n_e \sim 10^{12}$ cm^{-3}, which is reasonable for a negative ion source, and if the dissociative attachment rate coefficient is 10^{-8} cm^3/s (the calculated value for 3.8 eV of rovibrational energy), then an H_2^* will undergo dissociative attachment in 10^{-4} s. Our results[6] also show that about 5 eV is released as kinetic energy in the H_2^* + H products, of which about 1.5 ev would go to the H_2^*. These would have a speed of about 10^6 cm/s, and would leave the ~ 10 cm of the plasma volume in 10^{-5} s, and would thus have about 10% chance of forming H^-. This is a substantial probability, and shows that the mechanism is worth considering.

Finally it should be noted that H_3^+ and D_2^+ ions are vibrationally excited when formed from $H_2^+ + H_2$ and $D_2^+ + D_2$ reactions at thermal energies. From the analysis of the limited energies of the D + D + D products, we concluded that the D_3^+ ions in our beam had internal energies of about 2.2 eV on the average, with half maxima at about 0.5 eV and 3.2 eV. This finding within uncertainties, is consistent with an analysis by Smith and Futrell,[10] who concluded that the D_3^+ is expected to have the excitation energies that at least 69% of of the D_3^+ ions are formed with 0.6 eV or more. The conditions in these experiments are typical of those that exist in any ion source operating at pressures where two-body collisions dominate reactions. The internal excitation in H_3^+ and D_3^+ will also affect the rates and final product states of e + H_3^+ dissociative recombination, another source of $H_2(v)$.

SUMMARY

It does not appear likely that energy transfer from excited Cs atoms can excite H_2 vibrationally enough to be responsible for the high efficiency of the ion source. On the other hand, charge transfer with H_3^+ the neutral Cs is efficient, and leaves the H_2^* products highly excited. Cursory considerations indicate that these can have a substantial probability of forming H⁻ before leaving the plasma volume, and thus the reaction may aid volume production. The presence of Cs in the volume could also have an important cooling effect on the electrons, which would increase the dissociative attachment rate on $H_2(v)$ and also reduce H⁻ loss by electron collisional detachment.

ACKNOWLEDGEMENTS

This work was supported by the Air Force Office of Scientific Research under Contract No. F49620-85-K-0017, and by the Department of Energy under Contract No. DE-AT03-80ER53091.

REFERENCES

1. See S. P. Antipov, L. I. Elizarov, M. I. Martynov, and V. M. Chesnokov, Pribori I Texnika Eksperimenta No. 4, pp. 42-44, July-August, 1984, transl. by Consultants Bureau, New York.
2. J. N. Wadehra and J. N. Bardsley, Phys. Rev. Lett. $\underline{41}$, 1795 (1978).
3. J. N. Bardsley and J. N. Wadehra, Phys. Rev. A $\underline{20}$, 1398 (1979).
4. W. Reiland, H. U. Tittes, and I. V. Hertel, Phys. Rev. Lett. $\underline{48}$, 1369 (1982).
5. J. R. Hiskes, J. Appl. Phys. $\underline{51}$, 4592 (1980).
6. K. W. Ehlers and K. N. Leung, Proc. 3rd Intl. Symp. Production and Neutralization of Negative Ions and Beams, Brookhaven, 1973, p.227.
7. J. R. Peterson and Y. K. Bae, Phys. Rev. A $\underline{30}$, 2807 (1984). Although this experiment measured D_3^+ + Cs product energy distributions, similar results for the H_2 + H channel were obtained in H_3^+ + Cs experiments at the FOM Institute AMOLF, Amsterdam, by D. de Bruijn and coworkers.
8. Y. K. Bae, M. J. Coggiola, and J. R. Peterson, Phys. Rev. A $\underline{31}$, 3627 (1985); F. W. Meyer, C. J. Anderson, and L. W. Anderson, Phys. Rev. A $\underline{15}$, 455 (1977).
9. J. M. Wadehra, Phys. Rev. A $\underline{29}$, 106 (1984).
10. D. L. Smith and J. H. Futrell, J. Phys. B $\underline{8}$, 803 (1975).

SOURCE DIAGNOSTICS

DIAGNOSTIC TECHNIQUES FOR NEGATIVE ION SOURCES

M. Bacal
Laboratoire de Physique des Milieux Ionisés
Laboratoire du C.N.R.S., Ecole Polytechnique
91128 Palaiseau Cedex, France.

ABSTRACT

The following diagnostic techniques, useful in characterizing negative ion sources, are reviewed: the photodetachment technique for measuring the negative ion density and velocity distribution and the spectroscopic techniques for characterizing the neutral species.

INTRODUCTION

Design of efficient H^- negative ion beam sources requires a full understanding of the mechanisms of H^- production, decay and extraction in source plasma. The most interesting parameter of all is, therefore, the spatially resolved H^- density. Normal spectroscopic techniques are impossible since H^- has only a single bound state. The most frequently used techniques to date for measuring the H^- density, the photodetachment technique and the electrostatic probe technique, will be discussed. Proposed, but not yet demonstrated, direct optical methods for the observation of H^- ions by Thomson scattering and by a variant of laser induced fluorescence in the VUV, and methods for measuring H^- ion velocity distribution will also be described.

The knowledge of other negative ion source parameters, such as the electron density and temperature, the electron energy distribution function, the density and temperature of atomic and molecular hydrogen, the positive and negative ion mass spectra, is also very important. Several conventional techniques are therefore useful in negative ion source work, such as electrostatic probes and Stark broadening of atomic lines for the measurement of electron density and temperature, as well as mass spectrometry for positive and negative ion analysis and separation of the extracted electron and negative ion components. The techniques described in other papers at this Symposium are not discussed in this review.

PHOTODETACHMENT TECHNIQUE.

The photodetachment technique, proposed by Taillet[1] for the study of an oxygen plasma, has been developed at Ecole Polytechnique for use in volume H^- ion sources.[2-4] In this technique the extra electron is detached from the H^- ion by means of a pulsed laser beam:

$$H^- + h\nu \rightarrow H(n) + e \qquad (1)$$

and the resulting increase in electron density, equal to the negative ion density, is measured.

A well-known H^- photoabsorption continuum occurs for $h\nu > 0.754$ eV with the maximum cross section ($4.1 \times 10^{-17} cm^2$) at about 1.5 eV. When the photon energy is less than ~ 10.94 eV ($\lambda > 1130$ Å) the hydrogen atom resulting from photodetachment is in the ground state, n=1. Above 10.94 eV additional channels involving H(n) open up. The calculations of Hyman et al[5] indicate that photons with energy higher than 10.94 eV can produce a noticeable fraction of hydrogen atoms in excited states. However, existing diagnostic techniques use low energy photons and therefore, the detection method is based solely on the measurement of the electrons resulting from photodetachment.

The photon energy was chosen larger than the electron affinity of hydrogen (0.75 eV), but smaller than the electron affinity and ionization potential of impurity atoms. At Ecole Polytechnique we use a Nd:Yag laser (1.2 eV); at this photon energy the cross section for photodetachment, σ, of H^-/D^- is near its maximum value.

The density of electrons created per unit volume by laser photodetachment depends upon σ (cm^2), the laser fluence (number of photons per pulse per unit area) I (cm^{-2}) and the initial density of negative ions per unit volume, n_- (cm^{-3}):

$$\Delta n_e = n_- [1 - \exp(-\sigma I)] \qquad (2)$$

or, in terms of laser pulse energy per unit area, E:

$$\Delta n_e = n_- [1 - \exp(-\sigma E/h\nu)] \qquad (3)$$

$h\nu$ is the photon energy.

Eq. 3 exhibits saturation. At low laser pulse energy Δn_e is proportional to E, but becomes independent of it at high E, because all the negative ions are photodetached. At Ecole Polytechnique we chose to measure the saturation value of Δn_e, at high E. This is accessible with a pulsed Nd:Yag laser: Δn_e saturates for $E \simeq 30$ mJ/cm^2. In this condition:

$$\Delta n_e = n_- \qquad (4)$$

and the relative density of H^-, n_-/n_e, can be measured from the relative increase of the electron density, $\Delta n_e/n_e$. The change in electron density can be measured by probes or microwaves, depending upon the geometry and plasma density. With this aim we have used a cylindrical probe, coaxial with the laser beam. The relative negative ion density n_-/n_e can be obtained by comparing the amplitude of the transient current pulse (Δi^-), following the laser pulse, with the current drawn from the plasma by the same positively biased probe under stationary conditions (i^-_{dc}). The ratio $\Delta i^-/i^-_{dc}$ yields directly n_-/n_e when the entire region around the probe from where electrons are collected, is illuminated. The diameter of this region has been determined experimentally[6] and was found to be of the order of ten Debye lengths.

The technique can be authenticated by changing the laser pulse energy and ascertaining that the measurement of Δn_e is consistent with Eq. 3. A separate measurement of n_e is made using a disk probe.

The first experiments[4] using the photodetachment technique were made with a fixed probe, situated at the center of the plasma generator. Recently this technique was used for obtaining spatially resolved H^- density, namely the variation of H^- ion density along the axis of the cylindrical hybrid source[7,8]. A cylindrical probe (0.5 mm in diam, 1.5 cm long) coaxial with the laser beam, can be moved along the source axis. The probe is biased 20 V positive with respect to the wall. Figure 1 shows the axial variation of n_- when a stray magnetic field (of 20 Gauss at most) is present in front of the plasma electrode. When the plasma electrode is biased positively, the negative ion density in front of it increases dramatically, while the electron density is very much reduced. Further away from the plasma electrode the effect of the plasma electrode bias is very weak. This spatially resolved measurement of H^- ion density helped to explain the previously observed effect of the plasma electrode bias upon the variation of the extracted negative ion current[7,8]. It was shown that this effect was essentially due to the presence of a small stray transverse magnetic field in front of the plasma electrode.

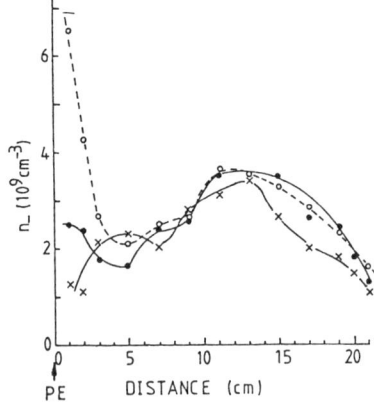

Fig. 1. Axial variation of H^- density measured by photodetachment in the hybrid multipole (50V - 5A - 2mTorr), for different values of the plasma electrode bias, V_b:
● - $V_b=0$; o - $V_b=+1V$; x $V_b=+3V$
A stray magnetic field of at most 20 Gauss is present in front of the plasma electrode.

It is also possible to use photodetachment for measuring the negative ion density at a laser pulse energy lower than that required for saturation. In this case the knowledge of the photodetachment cross section and laser pulse energy are necessary, in order to determine what negative ion fraction has been destroyed (Eq. 3). This can be useful if there are no powerful lasers available in the wavelength range required for the photodetachment of the studied negative ion species. Gottcho and Gaebe[9] measured the density of Cl^- in a rf discharge through Cl_2. A N_2 laser (337.1 nm) with maximum laser pulse energy of 0.4 mJ was used. This work is also interesting because it demonstrates the possibility of (a) time-resolved measurements and (b) opto-galvanic detection of transient signals in the external circuit (requiring the knowledge of the interaction volume in order to evaluate n_b/n_e).

<u>Laser induced fluorescence technique for negative ion density measurement.</u> The development of laser sources in the VUV^{10} opens the possibility to use photodetachment at photon energies where a fraction of the resulting atoms is in an excited state[6], and can radiate Lyman-α or Lyman-β radiation. A particularly favorable situation could be at the wavelength corresponding to the upper of the two recently observed resonances[11] (the shape resonance) in the photodetachment cross section near 11 eV, where this cross section is very large. Although most of the excited atoms are in the 2p state[6] and decay radiatively right away, collisions and electric fields in the plasma will quickly lead to emission of Ly-α by the 2s atoms also. However, to the best of our knowledge, the Ly-α radiation following photodetachment via this resonance has not yet been observed.

This variant of laser induced fluorescence applied to H^- detection could be useful in cases when the detection of the change in charged particle currents is difficult. This is the case in the study of H^- formation in pulsed magnetically insulated diodes. These diodes operate in vacuum, therefore self-absorption of VUV radiation by neutral atoms and molecules is not a problem. The small width of the shape resonance $(23\pm6 meV)^{11}$ could be used to authenticate the Ly-α fluorescence due to photodetachment of H^- from other sources of Ly-α radiation, and allow the determination of the H^- velocity distribution.

<u>Photodetachment diagnostics of negative hydrogen ion velocity distribution.</u> Two new applications of photodetachment are proposed for determining the negative ion velocity distribution. The first concept, proposed by R.A. Stern, is a variant of the 'optical tagging' technique which has been

developped recently for studying positive ion velocity distributions in plasmas[12]. A short laser pulse is beamed across the source and illuminates the volume to be diagnosed. The Nd:Yag laser radiation is sufficiently intense to destroy by photodetachment most of the H^- ions within the beam volume. As soon as the laser pulse is off, H^- ions flowing into the beam volume from the surrounding plasma begin to replenish the H^- density. A second laser pulse, delayed from the first, is now switched on into the diagnosed volume. It will photodetach the fastest H^- ions which have migrated into the volume within the time delay between the two pulses. The photodetachment destroys the H^- ions and liberates instead an equivalent number of electrons, whose arrival at a nearby positively-biased probe is detected rapidly. Thus the rate of H^- inflow can be determined by varying the delay between the laser pulses, and comparing the ensuing electron signals. This yields the relative density of H^- ions with various speeds and maps out the H^- velocity distribution, a function which underlies most quantities of importance such as H^- temperature, density and flow velocity. These quantities are critical for the H^- ion source emittance.

The second possibility to measure the velocity distribution of the H^- ions takes advantage of the existence of the lower of the two recently observed resonances near 11eV[11] (Feshbach resonance). The very small width of this resonance would allow to determine the velocity distribution of the H^- ions by scanning the VUV laser wavelength, while detecting the electrons released by photodetachment. It should be verified, however, that the resonance is not destroyed by the electric fields present in the plasma.

PROBE MEASUREMENT OF THE NEGATIVE ION DENSITY

In spite of the existence of the photodetachment diagnostics for the negative ion density, electrostatic probes continue to be used in negative ion source work, because they are inexpensive, easy to handle and necessary by all means for measuring other plasma parameters.

In the experiments performed before 1977 at Ecole Polytechnique[13] the negative ion density was measured by the technique proposed by Doucet[14], based on the analysis of the characteristic of a thin cylindrical probe. The comparison[3] of the data obtained by probes with those obtained by photodetachment has indicated that thin cylindrical probes give only a qualitative indication of the presence of negative ions.

Recently Hopkins and Graham[15] discussed whether thick cylindrical probes could be used to determine n_- from the difference between n_+ and n_e and reached a negative answer.

They observed in hydrogen plasma that the measured n_+ is a factor of 1.4 to 1.6 larger than the measured n_e. When the discharge was run with argon, the value of n_+ is again consistently a factor 1.6 higher than the value of n_e. Since no long-lived negative ions of argon are known, they concluded that this discrepancy in n_+ and n_e is not due to the presence of large densities of negative ions.

Holmes et al[16] derived n_-/n_e from the ratio of the negative particle (electrons and H^- ions) and positive ion saturation fluxes to a planar probe, which has a guard ring. A modified version of this technique has been used recently by Holmes et al[17]. The use of these probe techniques is subject to the usual limitations of probe measurements, and can only give a qualitative indication, especially in the presence of magnetic fields.

THOMSON SCATTERING

Burgess[18] discussed the possibility of developping a purely optical means of detecting H^- ions via the modifications to ordinary ion-feature Thomson scattering, occurring due to the presence of negative ions in the plasma. One of the difficulties is the photoionization of H^- by the incident laser pulse itself, thereby limiting the choice of incident laser pulse wavelength and energy. According to Burgess[18] scattering experiments should, however, be possible either using a CO_2 laser wavelength of 10.6 μm, i.e. well below the H^- photoionization limit, or by illuminating the plasma with a CW laser of sufficiently low power. To the best of our knowledge, experiments verifying this possibility have not yet been performed.

TRANSLATIONAL EMISSION SPECTROSCOPY

The very large associative detachment and vibrational deactivation cross sections due to hydrogen atoms justify the need for measuring the atomic hydrogen density and temperature. When the atomic hydrogen density is sufficiently large the charge exchange of H^- and H can lead to a significant heating of the H^- ion population. Keller and Smith[19] assumed that under such conditions the H^- ions have the temperature of the H atoms and the measured atomic temperature can be useful for calculating emittance. Ref. 19 also contains a very interesting discussion of the use of various Balmer-series lines for measuring the hydrogen atom temperature from Doppler broadening and the electron density from Stark boradening.

The analysis of the Hα and Hβ line shapes can provide the relative density of atomic and molecular hydrogen[20], $n(H)/n(H_2)$.

This method is based on recent crossed-beam studies[21,22] of dissociative excitation of H_2

$$e + H_2 \longrightarrow e + H^*(n) + H \tag{5}$$

in which the emitted Balmer line profiles have been detected (Fig. 2). These studies have shown that a large fraction of the excited atoms produced by dissociative excitation have energies comprised between 2 and 8 eV; they form two 'fast groups' which are at the origin of the broad wings of the observed Balmer line. A slow group (translational energy from 0 to 2 eV) forms the narrow central peak (Fig. 2). The excitation function of the slow group has a threshold at 17.1 eV, while the threshold for the excitation of the fast groups are, respectively, 24 and 27 eV.

The $H\beta$ (λ = 4861.3 Å) line profile measured from the plasma of a 90V-4mTorr multicusp discharge is shown in Fig. 3 for two values of the discharge current (1 and 10 A)[23]. The multicusp plasma generator, described in Ref. 20, is a conventional multipole, with one of the end plates unmagnetized. The line profile observed from this plasma is similar to the one observed in the crossed-beam experiment, since it presents broad wings; however, the height of the narrow central peak is much larger in the plasma (Fig. 3) than in the crossed-beam experiment (Fig. 2). This is due to the large contribution, in the case of plasma, of light emission following the excitation of ground state hydrogen atoms by electrons:

$$e + H \longrightarrow e + H^*(n) \tag{6}$$

The proposed method[20] for determining $n(H)/n(H_2)$ is based upon the possibility of separating by high resolution spectroscopy the fast (E≥2eV) excited hydrogen atoms in n=3, 4 or 5, from the slow ones. The slow excited atom component contains the following two contributions:
 (a) from excitation of ground state atomic hydrogen (Eq. 6);
 (b) from the slow group resulting from dissociative excitation of molecular hydrogen (Eq. 5).

The fast component is identified as the fast group produced by dissociative excitation of H_2 (Eq.5), provided reaction 5 dominates over other dissociative processes.

The emission intensities due to atomic hydrogen I(H) and to molecular hydrogen (fast group) If(H_2) are determined from the line spectrum. Their ratio is related to the relative atomic hydrogen density as follows:

$$I(H)/If(H_2) = [n(H)/n(H_2)] \times [N(H)/Nf(H_2)] \tag{7}$$

Since the pumping rate N(H) for direct excitation of atomic hydrogen is much larger than the pumping rate N(H_2) for

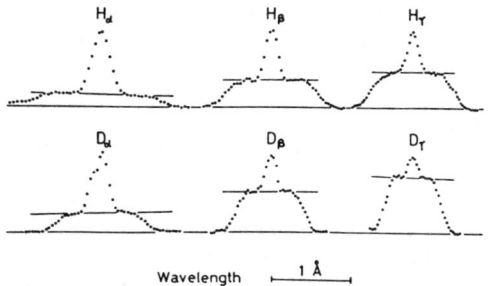

Fig. 2. High resolution (0.03-0.041 Å) spectra of the Balmer-α, β, γ lines produced by electron impact (100eV) on H_2 and D_2. The horizontal line separates the slow (above) and fast (below) H*(D*) groups. From Higo et al[22].

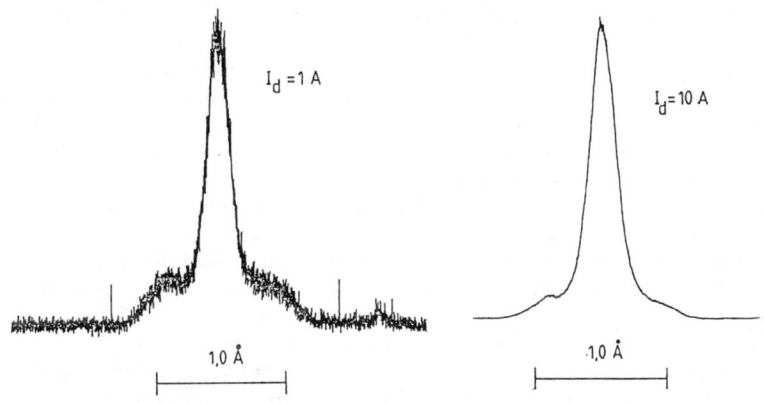

Fig. 3. Balmer-β line shape from a 90V-40mTorr multicusp discharge, for two values of the discharge currents (1A and 10A). From Bruneteau et al[23].

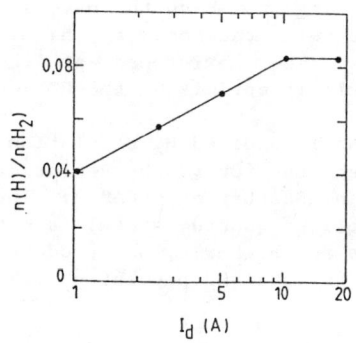

Fig.4. Dependence of the relative atomic and molecular hydrogen fraction upon the discharge current in a 90 V - 40 mTorr multicusp discharge.
From Bruneteau et al[23].

excitation of the fast component in dissociative excitation of molecular hydrogen (50 times in the studied case[20]) this method can be useful when the molecular hydrogen fraction is not too low.

Recent measurements[23] in a 90V-40 mTorr discharge indicate that $n(H)/n(H_2)$ varies from 0.040 at 1 A to 0.085 at 18A (Fig. 4).

This method can be applied to any particular discharge provided (a) mechanisms (5) and (6) dominate the emission of the Balmer-series line; (b) the electron energy distribution is known and contains electrons with energies above 27 eV (c) the relative density of atomic hydrogen is not too high.

Nightingale et al[24] reported Hα and Hβ line shapes observed in the driver section of a tandem multipole; they did not observe clear evidence of wings. A possible reason for this could be the very high degree of dissociation of the gas.

Observation of molecular lines can be an indication of the presence of H_2 molecules. In the experiments of Ref. 23 molecular lines were observed. The ratio between the intensity of Hβ and that of the molecular line at 4856.55 Å was 140 in a 90V-10A-40mTorr discharge.

After subtracting from the measured line spectrum the total molecular contribution, one obtains the line spectrum due to the excitation of atomic hydrogen by electrons. The shape of this line is Gaussian. The Doppler contribution is determined from the difference between the observed F.W.H.M. and the other contributions (instrument profile F.W.H.M., determined to be 50 mÅ using a He Ne laser, fine structure - 77 mÅ for Hβ), assuming they add quadratically. The derived excited H-atom temperature varies with discharge current from 1400 K at 1 A, to 3080 K at 18 K. These values are lower than those reported in Ref. 20, where the mentioned corrections were not made.

LASER DIAGNOSTICS OF NEUTRAL ATOMIC AND MOLECULAR HYDROGEN

The measurement of rovibrationally excited populations of H_2 is important in order to test the basic hypothesis of the volume production mechanism, according to which the negative ions are produced by dissociative attachment to highly vibrationally excited molecules[4]. The existence of the corresponding diagnostics could help in optimizing the H$^-$ ion production.

The measurement of rovibrationally excited H_2 populations in a multicusp H$^-$ ion source has been done for v≤3 by Péalat et al[20] by Coherent Anti-Stokes Raman Scattering (CARS). The sensitivity of this method (10^{12} mol/cm^3/quantum state) is not sufficient for measuring populations for higher v, but produced important information concerning the total H_2 population (the

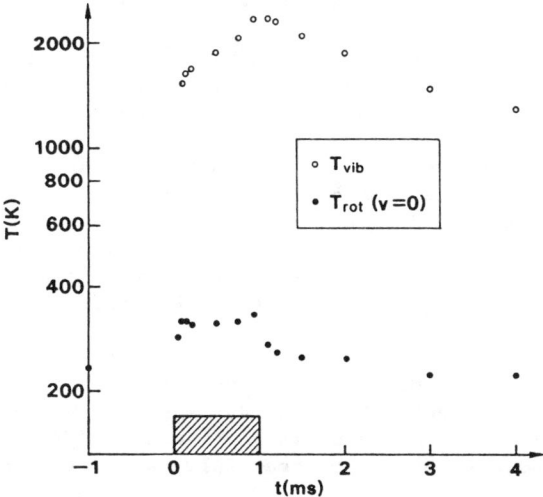

Fig. 5. Rise and decay of rotational temperature of v=0 and vibrational temperature derived from the populations of v=1 and v=2. Measurements made using CARS at the center of the plasma. Boxcars beam arrangement. Plasma conditions: 90V-10A-4mTorr. From Lefebvre et al[25].

decrease in H_2 under plasma conditions, due to dissociation, was measured), rotational and vibrational temperatures, with both space and time resolution[25]. Recent time-resolved measurements of rovibrational populations by CARS have led to the following conclusions[25]:
1. For a copper wall, covered with deposits resulting from the operation of thoriated tungsten filaments, the relaxation time for vibration at low pressure, is 1.4 msec (Fig. 5). If one considers that the time it takes a molecule to reach the wall is 200 μsec (as deduced from the time of rotational deexcitation, see Fig. 5), then one concludes that about seven wall collisions are required in order to deexcite a molecule $H_2(v=1)$. This was found in the case of low pressure operation (4 mTorr), typical for volume H$^-$ ion sources.
2. The contribution of the filaments in maintaining both the vibrational and rotational temperatures is significant, as was observed in the absence of the main discharge.

Recently other laser diagnostics have been considerd[26-28] for measuring atomic and rovibrational molecular populations, such as resonant multiphoton ionization[26], absorption[27] and laser induced fluorescence (LIF) resulting from two-photon and single-photon absorption[28]. The results obtained with resonant multiphoton ionization are discussed in another paper at this Symposium[26]. Ground-state hydrogen-atom density in H$^-$ sources will be measured[27] by the attenuation of Ly-β radiation produced by frequency tripling of light from a dye laser. The plasma to be studied is expected to be optically thin for Ly-β. Two-photon and single-photon absorption LIF techniques are investigated at SRI International[28]. The two-photon absorption technique requires laser wavelengths from 193-260 nm. Single-photon LIF technique, although simpler in concept, requires tunable laser radiation in the region 120-160 nm. The production of this VUV radiation is currently an important state-of-the art research area in nonlinear optics[11]. Two difficult problems need to be solved: production of the VUV radiation and the development of an efficient VUV photon collection system.

Acknowledgments. This work was supported by Ecole Polytechnique and Centre National de la Recherche Scientifique (France). Enlightening discussions with R.I. Hall, Ady Hershcowitch and R.A. Stern are gratefully acknowledged.

REFERENCES

1. J. Taillet, C.R. Acad. Sci. Paris, 269, série B, 52 (1969)
2. M. Bacal, G.W. Hamilton, A.M. Bruneteau, H.J. Doucet and J. Taillet, Rev. Sci. Instrum., 50, 719 (1979)
3. G.W. Hamilton, M. Bacal, A.M. Bruneteau, H.J. Doucet, and J. Taillet, Second Int. Symp. on the Production and Neutralization of Negative Hydrogen Ions and Beams, October 6-10, 1980. Report B.N.L. 51304, p. 90, Th. Sluyters Editor (1981)
4. M. Bacal, Physica Scripta, 2/2, 467 (1982)
5. H.A. Hyman, V.L. Jacobs and P.G. Burke, J. Phys. B, 5, 2282 (1972)
6. M. Bacal, A.M. Bruneteau and M. Nachman, J. Phys. Lettres, 42, L-5, (1981)
7. M. Bacal, P. Devynck and F. Hillion, Proc. 2nd European Workshop on the Production and Application of Light Negative Ions, March 5-7, 1986, Ecole Polytechnique, Palaiseau, p. 75.
8. M. Bacal, J. Bruneteau, P. Devynck and F. Hillion, this Symposium.
9. R.A. Gottcho and C.E. Gaebe, IEEE Trans. on Plasma Science, PS-14, 92 (1986)
10. F. Rostas, Proc. XIIIth Summer School on Quantum Optics, organized by the University of Gdansk, Frombork (Poland), Sept. 2-8, 1985, edited by J. Fiutak and J. Lizerski, World Scientific (Singapour), p. 328 (1986).

11. H.C. Bryant, B.D. Dieterle, J. Donahue, H. Sharifian, H. Tootoonchi, M. Wolfe, P.A.M. Gram and M.A. Yates-Williams, Phys. Rev. Lett., 38, 228 (1977)
12. R.A. Stern, Europhysics News, 15, N° 5, 2 (1984)
13. M. Bacal, E. Nicolopoulou and H.J. Doucet, Proc. Symp. on the Production and Neutralization of Negative Hydrogen Ions and Beams, Sept. 26-30, 1977, Report BNL 50727, K. Prelec Editor, p.26.
14. H.J. Doucet, Phys. Lett., 33A, 283 (1970)
15. M.B. Hopkins and W.G. Graham, Rev. Sci. Instrum., 57, 2210 (1986)
16. A. Holmes, G. Dammertz and T. Green, Rev. Sci. Instrum., 56, 1697 (1985)
17. A.J.T. Holmes, L.M. Lea, A.F. Newman, and M.P.S. Nightingale, Proc. 2nd European Workshop on the Production and Application of Light Negative Ions, March 5-7, 1986, Ecole Polytechnique, Palaiseau, p. 1.
18. D.D. Burgess, Plasma Physics and Controlled Fusion, 27, 349 (1985).
19. R. Keller and H.V. Smith, Jr., IEEE Trans. Nucl. Sci., NS-32, 1736 (1985) and this Symposium.
20. M. Péalat, J.P.E. Taran, M. Bacal and F. Hillion, J. Chem. Phys., 82, 4943 (1985)
21. R.S. Freund, J.A. Schiavone, D.F. Brader, J. Chem. Phys., 64, 1122 (1976)
22. M. Higo, S. Kamata, T. Ogawa, Chem. Phys., 73, 99 (1982)
23. A.M. Bruneteau, G. Hollos, F. Hillion and M. Bacal, Private communication.
24. M.P.S. Nightingale, M.J. Forrest and R. McAdams, Proc. 2nd European Workshop on the Production and Application of Light Negative Ions, March 5-7, 1986, Ecole Polytechnique, Palaiseau, p. 123.
25. M. Lefebvre, M. Péalat, J.P.E. Taran, F. Hillion and M. Bacal, Proc. 2nd European Workshop on the Production and Application of Light Negative Ions, March 5-7, 1986, Ecole Polytechnique, Palaiseau, p. 107.
26. J.H.M. Bonnie, P.J. Eenshuistra, W. van Schelt and H.J. Hopman, Proc. 2nd European Workshop on the Production and Application of Light Negative Ions, March 5-7, 1986, Ecole Polytechnique, Palaiseau, p.147 and p. 155, and this Symposium.
27. G. Stutzin, K.N. Leung, A.S. Schlachter, J.W. Stearns, W.B. Kunkel, R. Stevens, G. Worth, P. Gohil, W.G. Graham, Bull. Am. Phys. Soc., 31, Nb. 9, p.1506 (1986), Paper 5P 1.
28. W.K. Bischel, D.R. Crosley and J.R. Peterson, Private Communication.

DISCUSSION

Hiskes: When you have the discharge current at 10A, what is the plasma density?

Bacal: Well, if it was at 40 mTorr it was 10^{12}. And the electron temperature, less than 1 eV. If it was at 4 mTorr, it was like 4×10^{11}.

Michels: In this laser excitation of H^-, why isn't it simply photodetached in the singlet channel? Why would you expect Lyman-α?

Bacal: Just because the photon is very energetic, so there is energy for remaining in an excited state. I don't anticipate that, that's theory which anticipates that. And there have been calculations indicating that above the threshold of 10.9 eV, most of the negative ions destroyed will be in n=2.

Peterson: There is a continuum line down below that, that goes to the ground state. But related to your point, the shape resonance ends up with a 2s H atom, not a 2p.

Bacal: There are a mixture of both.

Hershcovitch: They're a mixture of three states, and the ratio is about 50% H(2p), 13% H(2s), and the rest H(1s). This is the theoretical work of Hyman, Jacobs, and Bruk; although this result seems to be somewhat controversial.

Bacal: There is, first, theory, which predicted this situation. And recently, in '77, experiments at Los Alamos have indicated that, indeed, there existed two - a shape resonance and a Feshbach resonance. Well, they confirmed theory.

Peterson: That's true, but the shape resonance goes to the 2s - it's a p-wave.

Hershcovitch: Only a small part of it.

Bacal: Okay. I'm not ready to explain. I'm just believing that we have to try to find this radiation.

Holmes: You say you actually measured the drift velocity using the laser dissociation?

Bacal: No, I haven't. I will. I wanted to improve my talk by telling some things which haven't been done, because you already heard the photodetachment, so I just tried to tell you also things which haven't been done, but could be done. We plan to do it in a short time. Indeed, we need a second laser, so for the moment, we are seeking this laser beam. But, it is not a difficult experiment, in principle. And since you say "drift velocity", I just want to say, we have seen some things which we couldn't understand from the correlation of negative ion density with the extracted current, which can have two explanations: either the temperature of the negative ions is very high, or we don't have a temperature which is high, but a high drift velocity. So, all these phenomena should be clarified by measuring the negative ion velocity. That's what we propose to do, and we'll try, but didn't do yet.

Holmes: I have one comment. There's a possibility that negative ions move supersonically in the plasma. So the drift velocity does not, necessarily, correlate with the temperature.

Bacal: Sure, but we will measure the velocity, whatever it is.

EXPLORATION OF A HYDROGEN DISCHARGE USING RESONANT MULTIPHOTON IONIZATION

J. H. M. Bonnie, P. J. Eenshuistra and H. J. Hopman
Association EURATOM-FOM
FOM Institute for Atomic and Molecular Physics, Kruislaan 407,
NL-1098 SJ Amsterdam, The Netherlands

ABSTRACT

We report the first successful application of resonant multiphoton ionization to diagnose a plasma. The principles of this technique and the way it is implemented in the apparatus are discussed. In principle the technique is capable of measuring concentrations of atoms and molecules, as well as determining their internal energy. We demonstrate its possibilities with the measurement of the rotational temperatures of H_2 in the ground state and the metastable $c^3\Pi^-_u$ state, which are in the range of 400 to 900 K. The density of the metastable molecules is determined using a single-photon ionization process. This yields a value of $3*10^9$ cm^{-3} at a discharge current of 30 A.

INTRODUCTION

The existence of high densities of hydrogen negative ions in discharges[1] has been attributed to dissociative attachment of low energy electrons to vibrationally excited hydrogen molecules[2,3]. One way to test this hypothesis is to measure the vibrational distribution of hydrogen molecules in a discharge and compare it with theory. Pealat et al.[4] have tried to measure this distribution with CARS. They measured the concentration of the $\upsilon = 0-3$ levels. We have set up an experiment to measure this distribution with <u>R</u>esonant <u>M</u>ultiphoton <u>I</u>onisation (RMI) which promises[5] to give a much lower detection limit of $\approx 10^8$ molecules cm^{-3}.

THE PRINCIPLE OF RESONANT MULTIPHOTON IONIZATION

In multiphoton ionization, the energy of the photons used is such that a molecule or atom has to absorb more than one photon to overcome the ionization potential. Such a process only occurs if the light intensity is sufficiently high: after absorption of one photon the molecule is excited to a socalled virtual state which is not an eigenfunction of the molecular Hamiltonian. In accordance with the uncertainty principle, the molecule can be in this virtual state only for a short time ($\approx 10^{-15}$ sec). If the light intensity is high enough, the excited molecule will absorb a second photon and a second virtual state can be reached. In this way the molecule can pick up a sufficient number of photons to become ionized. In resonantly enhanced multiphoton ionization or RMI, the energy of the photons is chosen such that one of the intermediate states corresponds to a real state which has a much longer lifetime ($\approx 10^{-8}$ sec) than a virtual state. The cross section for this process is much larger than the cross section for non-resonant multiphoton ionization. The energy jump associated with the absorption of a certain number of photons matches the energy difference between two states only at one particular wavelength. This apect of RMI makes state-selective detection possible.

In Fig. 1 a potential energy diagram is given for the hydrogen molecule, in its ground state and in some of its electronically excited states. The RMI principle is indicated by the arrows which represent photons with a wavelength of about 290 nm and the dotted lines representing the virtual states. Three photons are used to excite the molecules from the $X^1\Sigma^+_g(v"=0)$ ground state to the $C^1\Pi_u$ ($v'=2$) electronically excited state, and the absorption of one additional photon causes ionization, a process conveniently abbreviated as (3+1) RMI. The laser power density required to drive the three photon excitation step is $\approx 10^{14}$ Wm^{-2}. Of course all vibrational levels sketched are subdivided into rotational levels, which means that at one particular wavelength only molecules in one specific quantum state may be ionized. By measuring the amplitude of the ionization signal as a function of wavelength, information is obtained about the population of the different quantum states in the ground state.

THE DIAGNOSTIC SYSTEM

A schematic set-up of our experiment is given in Fig. 2. A laser beam is focused approximately 5 mm in front of a hole in the discharge chamber out of which a hot gas jet effuses. A magnetic field of ≈ 1 T at the laser focus, generated by a magnetic circuit which is only partially shown, prevents charged plasma particles from reaching the laser focus. The right hand side of the magnetic circuit can be biased to a positive potential of 4 kV providing an extraction field of ≈ 1 MV/m. This accelerates the laser-produced ions into a periodic focusing/defocusing lens system which takes care of the transport of the ions to a parallel plate energy analyzer. As the laser is focused in the middle of the gap in the magnetic circuit, the laser-produced ions are accelerated to an energy of 2 keV, whereas photo-desorption ions, generated by plasma light on nearby surfaces, get an energy of 0 or 4 keV. With the energy analyzer it becomes possible to select only the laser-produced ions. The fact that the discharge is a copious source of UV radiation makes this precaution necessary. The complete particle transport system is in fact a time of flight mass spectrometer which enables us to distinguish between H^+ and H_2^+ ions and impurities. Because the RMI-principle is also applicable to hydrogen atoms, this aspect of the system makes it possible to directly monitor changes in atomic density with changing discharge parameters. The detector is a Johnston electron multiplier employed in current mode. After amplification, the signal is fed into a gated, current-integrating analog to digital converter. Both H^+ and H_2^+ signals and laser pulse energy, as measured with a photo diode, are recorded for every laser shot. The whole experiment is controlled by a micro computer.

Tunable UV radiation between 285 and 305 nm is generated by frequency-doubling the output of an excimer pumped dye laser. Typically 3 mJ UV laser pulse

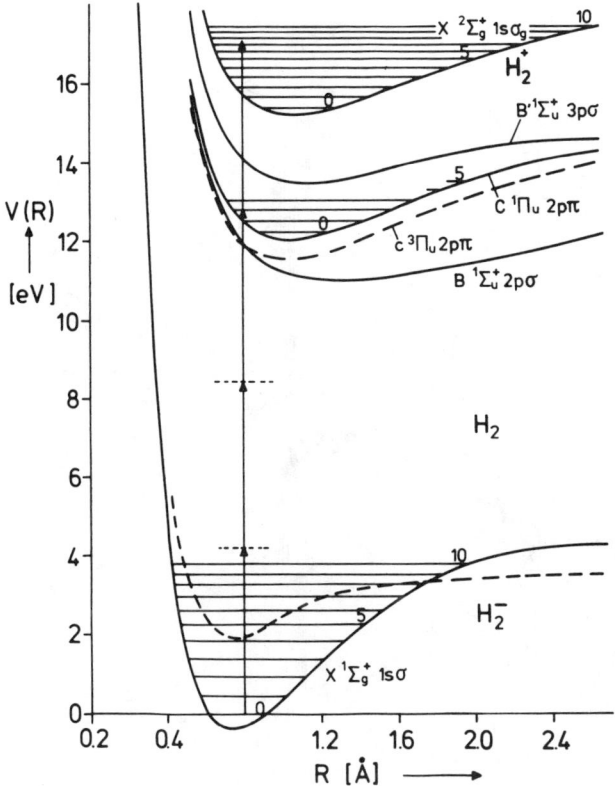

Figure 1. The potential energy diagram of the hydrogen molecule in its ground state and in some of its electronically excited states. The arrows indicate an example of the three photon excitation, one photon ionization scheme we use in the RMI process.

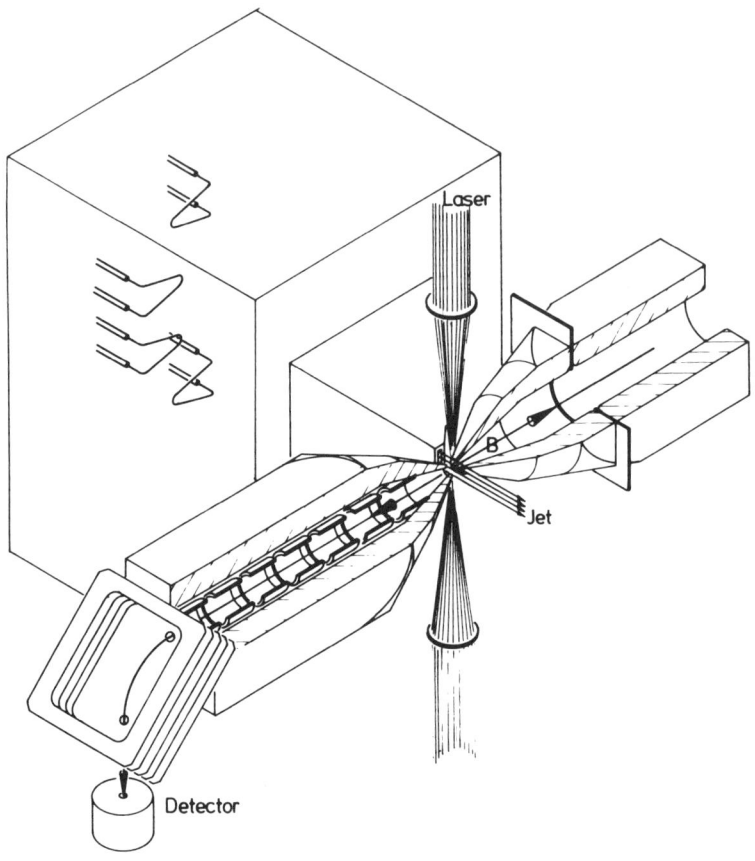

Figure 2. Schematic set-up of the detection system for ions resulting from the RMI-process. Indicated are the discharge chamber; the hot gas jet effusing from the plasma; the laser beam which is focused in this jet; part of the magnetic field generating circuit; the periodic focusing/defocusing lens system; the parallel plate energy analyzer and the electron multiplier.

energy is focused to a spot of ≈ 40 μm diameter by means of a lens with a focal length of 8 cm. Laser pulse duration is ≈ 20 nsec.

The ion source is a magnetic multipole bucket source. The dimensions of the rectangular source chamber are $14*14*19$ cm^3. The walls are made of oxygen-free copper and are water-cooled. Except for the front plate, all walls are equipped with cobalt-samarium permanent magnets, which generate a magnetic cusp field of ≈ 0.07 T on the inside of the wall. The frontplate is electrically insulated from the rest of the source and can be biased to a separate potential. In the back plate 6 tungsten filaments with a diameter of 1.5 mm are mounted on insulated, water-cooled feedthroughs. The source can be operated up to 30 A, 200 V dc, or in a pulsed mode with a top current of 200 A, pulse length of ≈ 2 msec and a maximum repetition rate of 40 Hz. Working pressure in the source is between 0.1 and 5 Pa.

RESULTS

Rotational temperatures

On starting the measurements, initially without discharge so that all molecules are in $\upsilon = 0$, we discovered a dissociation process leading to H$^+$, superimposed on the RMI-process. Allthough the appearance of ions always coincided with a resonance to one of the electronically excited states of H$_2$, we found strong variations of the number of H$^+$ to H$_2^+$ ions with the vibrational quantum number υ' of the intermediate state.[6] As this effect is not immediately relevant to the measurements presented here, it will not be discussed further.

In Fig. 3 we give the (3+1) RMI signal around 293 nm, obtained for a dc discharge with discharge current $I_d = 20$ A, voltage $V_d = 100$ V and pressure $P_d \approx 11$ Pa. The lines are due to the $X^1\Sigma^+_g(\upsilon''=2) \to B'^1\Sigma^+_u(\upsilon'=0)$ transition, for which we observe an R-branch (change in rotational quantum number $\Delta j = +1$) and a P-branch ($\Delta j = -1$). The numbers between parentheses in Fig. 3 give the rotational quantum number in the $X^1\Sigma^+_g(\upsilon''=2)$ state. A plot of the normalized line height, corrected for the degeneracy of the rotational levels and nuclear spin states, versus energy separation of the levels, a so-called Boltzmann plot, yields the rotational temperature T_r. Figure 4 shows such a plot. The slope of the line through the data points corresponds to $T_r = 500$ K. It is seen that the point measured for j = 5 falls below this line, indicating a higher occupation of this level than according to a thermal distribution. Similar deviations from thermal equilibrium have been observed by Pealat et al.[4] as well.

The more or less wavelength-independent part of the H$_2^+$ signal in Fig. 3 is caused by the one-photon ionization of hydrogen molecules in the metastable $c^3\Pi^-_u$ state. These molecules have a radiative lifetime of ≈ 1 msec.[7] At a thermal speed of $\approx 3*10^3$ m/s they can thus travel ≈ 3 m which is much longer than the distance from the edge of the discharge to the laser focus (≈ 1 cm). Specific quantum states of $c^3\Pi^-_u$ were detected by (1+1) RMI measurements in the wavelength region from 570 to 605 nm. We found one-photon excitations from $c^3\Pi^-_u$ to all four n=3 triplet gerade states in hydrogen. In addition, many of the peaks in the spectra we recorded

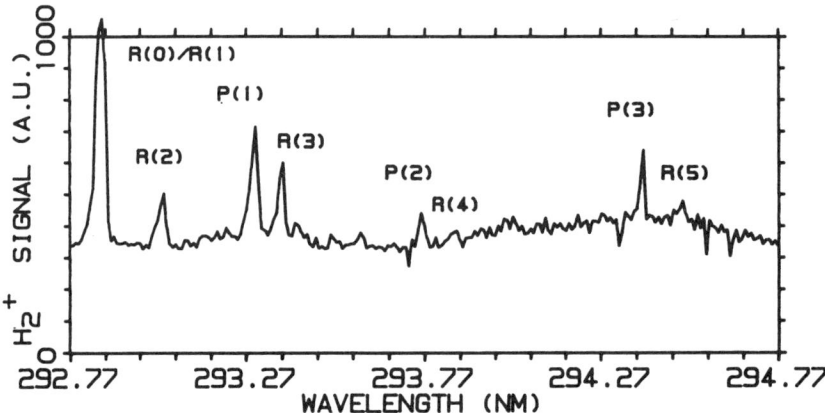

Figure 3. H_2^+ signals obtained for (3+1) RMI measurements on the $X\,^1\Sigma^+_g(v''=2) \rightarrow B'\,^1\Sigma^+_u(v'=0)$ transition. The numbers in the parentheses give the rotational quantum numbers in the $X\,^1\Sigma^+_g(v''=2)$ state. Discharge parameters are $I_d = 20$ A, $V_d = 100$ V and $P_d = 1$ Pa.

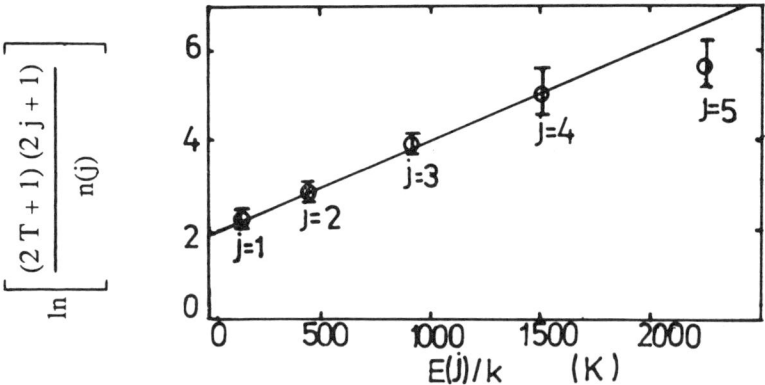

Figure 4. Boltzmann plot for the data from Fig. 3. The line fitted through the data point corresponds to a Boltzmann distribution with $T_r = 500$ K.

correspond to the electric dipole forbidden transition $c^3\Pi^-_u \to d^3\Pi_u$. In Fig. 5 we give the Q-branches ($\Delta j = 0$) for the (0-0), (1-1) and (2-2) vibrational bands of this transition. Most of the P and R lines for these bands were also found in the spectra. We investigated the nature of this forbidden transition and found that it is of the enforced-dipole type,[8] induced by the electric field of our detection system. This electric field may couple the $c^3\Pi^-_u$ or the $d^3\Pi_u$ state to another state as a consequence of which the selection rule no longer holds. For instance, calculations[9] show that electric fields of the order of 1 MV/m are already strong enough to couple the $c^3\Pi^-_u$ and the $a^3\Sigma^+_g$ states. This coupling reduces the life-time of $c^3\Pi^-_u(v"=0)$ by a factor of ≈ 3. However, quenching of $c^3\Pi^-_u$ by this field can be neglected, as the molecules drift from the middle of the discharge to the detection volume in ≈ 30 μsec.

We decided to deduce T_r from the forbidden transitions as these peaks are well resolved in contrast to the peaks of the allowed transitions. In addition, the rotational branches for the forbidden transition appear in conveniently small wavelength regions.

The results showing the variation of T_r for $c^3\Pi^-_u(v")$ and $X^1\Sigma^+_g(v")$ with arc current are given in Table I.

<u>Table I.</u>   Rotational temperatures [Kelvin] for $X^1\Sigma^+_g(v = 0,1,2)$ and $c^3\Pi^-_u(v = 0,1,2)$ for different arc currents [Ampere]. Other discharge parameters are $V_d = 100$ V and $P_d = 0.4$ Pa in case of $c^3\Pi^-_u$ and 1.2 Pa in case of $X^1\Sigma^+_g$.

Arc Current	10		20		30	
State	$X^1\Sigma^+_g$	$c^3\Pi^-_u$	$X^1\Sigma^+_g$	$c^3\Pi^-_u$	$X^1\Sigma^+_g$	$c^3\Pi^-_u$
$v = 0$	460	460	530	500	600	570
$v = 1$	560	440	680	500	750	570
$v = 2$	640	360	750	380	900	450

<u>Densities</u>

To obtain the total $c^3\Pi^-_u$ density in the discharge we used photons at $\lambda = 337$ nm. These photons have enough energy to ionize $c^3\Pi^-_u$ molecules in a one-photon process, regardless of their quantum state. At this wavelength the dye laser delivers a high enough pulse energy to saturate this ionization process. Furthermore, we need to know the size of the acceptance volume of the detector, the current corresponding to one single ion and the overall efficiency of the detection

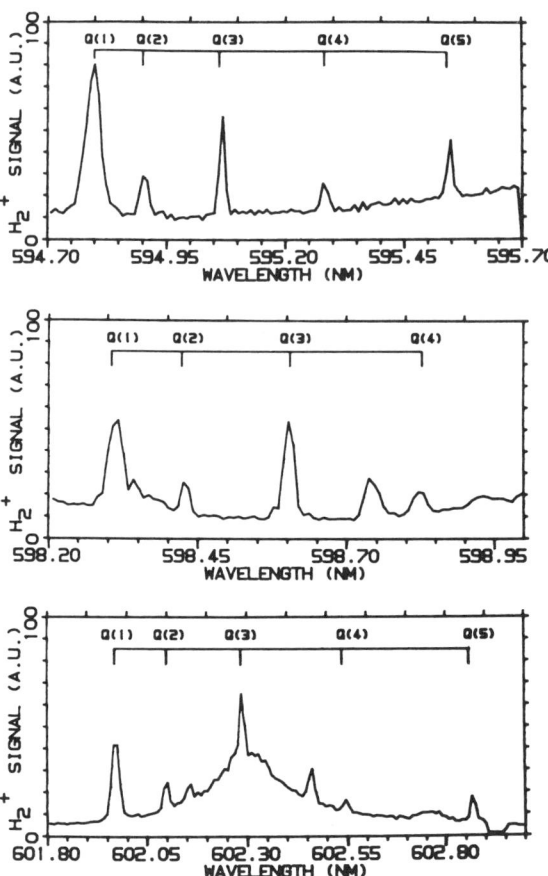

Figure 5. H_2^+ signals obtained when employing (1+1) RMI. Indicated are the line positions of the Q-members of the (0-0), (1-1) and (2-2) vibrational bands (from above to below respectively) for the $c^3\Pi_u^- \to d^3\Pi_u$ transition. Discharge parameters are: $I_d = 30$ A, $V_d = 100$ V and $P_d = 0.4$ Pa.

system. We determined the size of the acceptance volume, i.e. the volume from which ions can reach the detector, by scanning the laser focus through this volume and detecting the (3+1) RMI signal, from cold hydrogen gas, as a function of focal position. The (3+1) RMI signal emanates from a small volume ($\approx 10^{-3}$ mm^3) determined by the laser focus. The acceptance volume was found to be ≈ 1 mm^3. For the measurements at 337 nm the laser beam was only slightly focused in order to be sure that this whole volume was illuminated. By decreasing laser and discharge power to reach the regime where ion counting becomes possible, the detector signal from a single ion could be measured. Computer calculations on the transmission of the detection system show[11] that it can be taken as 100%. According to the manufacturer's specifications, the quantum efficiency of the detector exceeds 90% if the ion energy is larger than 3 keV, as is the case in our experiment. We therefore take the overall efficiency of the detection system, which is the product of transmission and quantum efficiency, to be equal to 1. Note, however, that the error that might be introduced in this way causes an underestimation of the density of $c^3\Pi^-_u$. Finally, as the hydrogen molecules flow from the discharge through the aperture towards the laser beam, the density decreases by a certain factor. This factor was determined experimentally to be ≈ 30, for ground state molecules when the discharge was turned off. At the pressures we use, this factor does not depend on the gas temperature and therefore we assume that it is the same for $c^3\Pi^-_u$ when the discharge is on.

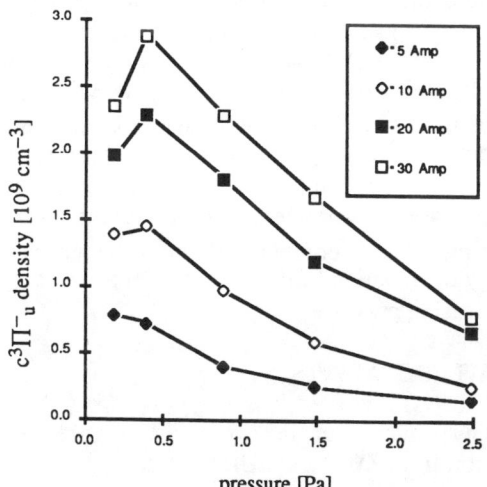

Figure 6. Variation of $c^3\Pi^-_u$ density with neutral gas pressure for different arc currents. Discharge voltage is 100 V.

Part of the measurements on the scaling laws are presented in Fig. 6. In view of the discussion in the previous paragraph the values on the vertical axis should be regarded as approximate, and in fact are lower limits for the $c^3\Pi^-_u$ density. The relative values, however, are accurate within $\approx 10\%$.

DISCUSSION

Rotational temperatures

The temperatures for $X^1\Sigma^+_g(\upsilon)$ as obtained by Pealat et al.[4] are in the same range as ours. However, we note that Table I shows a peculiar difference in the variation of T_r with υ: at constant arc current and increasing υ, T_r increases for $X^1\Sigma^+_g(\upsilon)$, whereas it decreases for $c^3\Pi^-_u(\upsilon)$. Apparently, metastable hydrogen is rotationally colder than the ground state for $\upsilon > 0$. Rotational cooling has been observed[10] in the excitation of H_2^+ from $X^1\Sigma^+_g$, and is associated with the reduction in the rotational constant. The behaviour in our experiment is not yet completely understood. It may include an indication that $X^1\Sigma^+_g$ and $c^3\Pi^-_u$ are involved in different reaction mechanisms. Further measurements are needed to clarify this point.

Densities

The variation of the $c^3\Pi^-_u$ density with neutral gas pressure and arc current can be understood qualitatively using a simple model in which $c^3\Pi^-_u$ is created by electron excitation of groundstate molecules and destroyed as a consequence of collisions with walls, plasma electrons and gas molecules. Plasma electrons may excite $c^3\Pi^-_u$ to the close-lying, short-lived (≈ 10 nsec) $a^3\Sigma^+_g$ state, a process for which the cross section is expected to be quite large ($\approx 10^{-13}$ cm^2), due to the small energy difference between these two states.[12] Collisions with gas molecules may convert $c^3\Pi^-_u$ into $c^3\Pi^+_u$, which decays rapidly by predissociation through the $b^3\Sigma^+_u$ state.[13] At low pressures, the loss rate is dominated by wall collisions and therefore the $c^3\Pi^-_u$ density should increase with pressure and arc current. This is seen in Fig. 6 below 0.4 Pa. At higher pressures, a combined effect of the other two loss mechanisms for $c^3\Pi^-_u$ and energy degradation of the primary electrons causes the $c^3\Pi^-_u$ density to decrease with increasing pressure.

ACKNOWLEDGEMENTS

This work is part of the research program of the association agreement EURATOM–FOM, with financial support from ZWO and EURATOM.

REFERENCES

[1] M. Bacal, Physica Scripta T2/2, 467 (1982).

[2] J. R. Hiskes and A. M. Karo, J. Appl. Phys. 56, 1927 (1984).

[3] C. Gorse, M. Capitelli, J. Bretagne and M. Bacal, Chem. Phys. 93, 1 (1985).

[4] M. Pealat, J-P. E. Taran, M. Bacal and F. Hillion, J. Chem. Phys. 82, (1985) 4943.

[5] E. E. Marinero, C. T. Rettner and R. N. Zare, Phys. Rev. Letters 48, 1323 (1982).

[6] J. H. M. Bonnie, P. J. Eenshuistra, J. Los and H. J. Hopman, Chem Phys Letters 125, 27 (1986); J. H. M. Bonnie, J. W. J. Verschuur, H. J. Hopman and H. B. van Linden van de Heuvell, Chem Phys Letters 130, 43 (1986).

[7] C. E. Johnson, Phys. Rev. A 5, 1026 (1972).

[8] R. H. Garstang, in *Atomic and Molecular Processes*, edited by D. R. Bates (Academic Press, New York-London, 1962).

[9] Robert P. Freis and John R. Hiskes, Phys. Rev. A 2, 573 (1970).

[10] D. P. de Bruijn and W. Koot, private communication.

[11] J. H. M. Bonnie, E. H. A. Granneman and H. J. Hopman, Proceedings of "Workshop on Basic and Advanced Fusion Plasmas Diagnostic Techniques, Varenna, Italy (1986) to be published.

[12] J. R. Hiskes and A. M. Karo, "Electron Energy Distributions, Vibrational Population Distributions and Negative Ion Concentrations in Hydrogen Discharges," presented at the NATO Advanced Study Institute on Atomic and Molecular Processes in Contr. Thermo. Research, Palermo, Italy, July 19-30, 1982, Rept. No. UCRL-87779, June 1982 (unpublished).

[13] D. P. de Bruijn, J. Neuteboom and J. Los, Chem. Physics 85, 233 (1984).

DISCUSSION

Hiskes: I wonder if you have an estimate of the cross section to go from the triplet state to the singlet Π state.

Bonnie: No, I don't have an estimate. It's just a process that might be of importance.

Azria: About the sensitivity of the method. Up to which vibrational level do you think you would be able to go?

Bonnie: Well, up till now we have measured up to v=3, and the reason we were not able to detect to higher v's is because of these metastable molecules. Because they spoil our signal.

Peterson: In the wavelength dependence of the resonant (3+1), there seem to be some dips as well as these peaks. Do you understand what causes this?

Bonnie: We sometimes have breakdown to this 4 kV part of the magnetic field generating system, and then the signal drops for a short time.
Peterson: Those are not consistent, though?
Bonnie: No. If you do a second scan, then they would appear at another position, or not at all.
Bretagne: Two questions. The first question: What kind of laser do you use? Second question: Have you made a study of RMI signals as a function of power of the laser?
Bonnie: Answer of your first question: We have a Lambda Physik excimer pumped frequency doubled dye laser. As far as your second question is concerned, this dependence is, indeed, of great importance, because the signal is extremely dependent on the laser power, because you have the three photon excitation step. The dependence is not as much the laser pulse energy as the true laser intensity. Right now, we are working on a system with which we can measure the true laser intensity in the focus, and then we will perform this measurement you are speaking about.
Bretagne: Are you sure that there is no saturation effect in the RMI signal due to the fact that all levels from which you start are depopulated by the laser?
Bonnie: Saturation? Because you depopulate the lower levels? Well, to be sure of that we have to do this measurement you asked about. We did some variation by measuring the laser pulse energy, so not the true intensity, the laser pulse energy. And from that we got the impression that we are not depleting the lower states.

VISIBLE AND V.U.V. EMISSION MEASUREMENTS IN A TANDEM MULTICUSP ION SOURCE

W.G. Graham and M.B. Hopkins, Physics Department
University of Ulster, Coleraine
BT52 1SA. Northern Ireland.

INTRODUCTION

Light emission measurements provide a non-perturbative method of obtaining information about plasmas and since the spectra observed arise from atomic processes, the measurements will emphasize that particular aspect of the plasma behaviour.

There are several factors which complicate the interpretation of light emission measurements, the plasma may be optically thick to some lines, the measured emission comes from the line of sight of the spectrometer and therefore possibly from different regions of the plasma. Also the spectral response of the instrument must be known, this can be a particularly difficult problem with a vacuum ultraviolet spectrometer.

In H^- ion sources the vacuum ultraviolet (V.U.V.) emission is of particular interest since the present hypothesis is that H^- production is through dissociative attachment of thermal electrons to vibrationally excited molecules, $H_2(v*)$[1,2]. The principal source of $H_2(v*)$ in the plasma is thought to be from singlet excitation of H_2 by fast electrons[1,2].

$$e + H_2(X^1\Sigma g, v''=0) \rightarrow e + H_2(B^1\Sigma u, C^1\pi u)$$

followed by radiative decay to vibrationally excited ground state H_2,

$$H_2(B^1\Sigma u, C^1\pi u) \rightarrow H_2(X^1\Sigma g, v'') + h\nu \quad (1)$$

The transition from the $B^1\Sigma u$ and $C^1\pi u$ levels to the ground state, $X^1\Sigma g$, of H_2 are termed the Lyman and Werner Bands respectively[2] and are found in the V.U.V. region.

Atomic hydrogen is also thought to play an important role in H^- ion sources both in atomic de-excitation of $H_2(v*)$[3] and in the destruction of H^- in associative detachment collisions[3]. Atomic hydrogen emission lines are found in the V.U.V. (Lyman series) and visible (Balmer Series) regions. The Lyman Series is of particular interest since the decay terminates in the ground state of the atom.

In this paper data based on relative measurements

of V.U.V. and visible emission from a tandem multipole H^- ion source is presented. In particular the change in emission intensity with plasma parameters and viewing region has been studied. The relative measurements side step some of the experimental problems mentioned above. Additionally the production rate of $H_2(v \geq 5)$ has been estimated.

APPARATUS

The plasma source used in the present experiments has been described in detail elsewhere and is similar to the H^- "small source" in use at the Culham Laboratory[4]. The hydrogen discharge is magnetically confined using permanent magnets in a line cusp geometry. A "virtual" filter is created by breaking the multipole geometry on two adjacent sides. Magnetic field lines cross the source dividing it into two regions: the driver region where primary electrons from two heated filaments produce the plasma and vibrationally excited molecules and the extractor region which has no primary electrons and a low electron temperature, since the filter prevents energetic electrons from crossing from the driver region.

Two Hilger and Watts 1 m normal incidence grating spectrometers were used to study the light emission from the plasma. One, a vacuum spectrometer equipped with a 600 line mm^{-1} grating blazed at 150 nm, was used to study emission at wavelengths from 90 to 180 nm and the other, a visible spectrometer with a 1200 line mm^{-1} grating blazed at 500 nm, was used to study wavelengths from 300 to 800 nm. Both spectrometers could be connected to ports which viewed either the centre of the driver or the extractor region. Particular care was taken to ensure that the optical geometry was such that when studying the extractor region the spectrometers did not have direct line of sight to the filter or driver regions.

Under normal operating conditions the discharge current, I_d, could be varied from 1 to 20 A, the discharge voltage, V_d, from 50 to 120 V and the gas pressure, P, from 0.5 to 10 m Torr. The plasma parameters were measured using cylindrical Langmuir probes and an interactive real-time analysis of the probe characteristics[5] using Laframboise theory[6]. A maximum electron density of a few 10^{12} cm^{-3} could be obtained in the driver region.

EMISSION SPECTRA

A typical V.U.V. spectrum from the driver region of the present H^- ion source is shown in Figure 1. The spectral features as a whole appear unchanged with variation of the plasma operating conditions. The spectral features are similar to those observed in a diffusion dominated H_2 plasma and which have discussed in detail previously[7]. The main feature is the Lyman (Lyα) line of atomic hydrogen at 121 nm. Other lines of the Lyman series (Lyα, Lyβ and Lyγ) can also be identified. The other features are due mainly to emission from molecular hydrogen[7]. The extent of the Werner and Lyman bands and the regions in which strong transitions leading to $H_2(v≥5)$, through process 1, are expected are indicated in Figure 1.

Although the resolution is presently restricted to about 0.2 nm it is possible to identify emission for particular molecular transitions, for example the structure at 118 nm can be identified with a C(v=4) to X(v=9) transition and that a 161 nm with a B(v=5) to X(v=12) transitions. The intensity of these two particular features is taken as representative of the intensity of their respective series.

A typical visible spectrum is shown in Figure 2. In the extractor region the visible region is dominated by radiation from the hot filaments. Aside from this the main feature is the Balmer (Hα) line of atomic hydrogen at 656 nm. The Hα and Hβ lines of atomic hydrogen can also be identified. There are other line features in the spectra, some can be assigned to possible impurity atoms, however there are others which as yet have not been identified.

RATE OF FORMATION OF $H_2(v≥5)$

In a previous paper the rate of production of vibrationally excited hydrogen molecules with $v ≥ 5$ was estimated from V.U.V. emission measurements in a diffusion-type plasma[7]. In such plasmas the fast electrons are lost rapidly to the walls and the high energy electron energy distribution function can be assumed to be almost monoenergetic with the energy determined by the discharge voltage.

In the present multicusp ion source the fast electrons are confined and the electron energy distribution function is more complex but has now been determined experimentally[8]. The rate of excitation of molecules, leading to light emission is given by $R = \int_0^\infty n(\Sigma)\sigma(\Sigma)vdv$ where $\int n(\Sigma)d\Sigma$ is the electron energy distribution function, (which has been measured) and, $\sigma(\Sigma)$ is the cross section for excitation.

The cross section for the production of vibrationally excited molecules with $v ≥ 5$, $\sigma(v≥5)$ can be obtained from the calculations of Hiskes[9]. The cross section for the production of Lyman α (Lyα) has just been remeasured at 100 eV by Woolsey et al[10], previous measurements[11] at other energies have been scaled by the amount suggested by Woolsey et al[10].

These cross sections can be combined with the measured electron energy distribution function to calculate rate constants for excitation and it is found that $R(v \geq 5)/R(Ly\alpha) = 3.7$.

This can be compared with the measured intensity of light emission at Lyman α, Int(Lyα), and the integrated intensity of light from the Werner and Lyman Bands where strong transitions leading to $v \geq 5$ are expected to be found (corrected for the Lyman contribution), Int($v \geq 5$). It is found that Int.($v \geq 5$)/Int(Lyα) = 3.3±0.4. In excellent agreement with the theoretically calculated ratio. (The measurements were obtained with a discharge current of 5A, gas pressure of 5×10^{-3} T and discharge voltage of 60V).

By normalizing Int(Lyα) to the calculated value R(Lyα) it is possible to estimate that the rate of production of vibrationally excited hydrogen molecules with $v \geq 5$ is approximately 3×10^{15} cm^3 s^{-1}, under the operating conditions presented above.

VARIATION OF EMISSION INTENSITY WITH Id

In a previous paper[12] we have shown that for the present plasma operating conditions the fast electron density (n_{fe}) in the driver region varies as $Id^{1.2}$ where Id is the discharge current, while the thermal electron density (n_e) depends also on the thermal electron temperature (kT_e) and that the product $n_e kT_e$ varies linearly with Id. For the conditions under which the present emission measurements were obtained n_e is found to vary at $Id^{0.66}$.

In Figures 3 and 4 the emission intensity of Hα and Lyα lines of atomic hydrogen are shown as a function of Id in both the driver and extractor regions. The dependence of other emission lines under the same plasma conditions are given in Table I.

In any region the atomic hydrogen lines have within the experimental uncertainties approximately the same Id dependence, with the dependences somewhat more pronounced in the extractor region than the driver region. The results for Hα are in good agreement with the $Id^{1.4}$ reported by Nightingale et al[13] in both regions of a similar ion source.

As discussed below one of the major findings has been the considerable amount of visible and V.U.V. emission detected in the extractor region. A more pronounced dependence on the discharge current is found not only in the H atomic lines but also in V.U.V. H_2 lines. This appears to indicate that a similar process is responsible for all light emission in the extractor region.

TABLE 1

Variation of light intensity with I_d

Light intensity I_d^α, value of exponent given below.

	V.U.V. Lyα	Atomic Lyβ	Lyγ	Visible Hα	Atomic Hβ
Driver	1.3	1.2	1.2	1.3	1.3
Extractor	1.4	1.4		1.6	1.5

	V.U.V. C-X(118nm)	Molecular B-X(161nm)
Driver	0.8	0.8
Extractor	1.1	1.1

	n_e	n_{fe}
Driver	0.7	1.2
Extractor	0.7	

COMPARISON OF INTENSITY IN THE DRIVER AND EXTRACTOR REGION

In our measurements of the electron energy distribution function in these sources[8], the fast electron density is found to drop by at least a factor of 100 across the virtual filter. It is therefore surprising to find a large amount of visible and V.U.V. emission in the extractor region since this is clearly not driven by electron excitation.

Table 2 shows the ratio of light emission in the driver region

TABLE 2

Comparison of light intensity in driver and extractor region.

	Driver/Extractor		Driver/Extractor
Ly	3.5-3.0	C-X(118nm)	2.6-1.4
Ly	3.0-2.5	B-X(161nm)	2.6-1.4
H	17-10	n_e	3.0-3.5
H	23-17	n_{fe}	>100

to that in the extractor region. As shown in Figures 3 and 4, this ratio changes somewhat with discharge current so the ratio at I_d = 1A and 20 A are given in Table 2. A strong visible emission from the extractor region has also been reported by Nightingale et al[13] from a similar plasma source. They have been unable to identify a particular mechanism for n = 3 production.

It is interesting to note that the fractional production in the extractor is greater in the V.U.V. than in the visible and that atomic and molecular lines are reduced by approximately the same amount. This may indicate that light emission may be from the same basic process in both the extractor and driver region, that is, the dissociation of molecular hydrogen.

Our electron energy distribution function measurements[8] exclude the possibility of fast electrons being present in the extractor region and therefore it appears that the energy in this light emission must be carried across the filter as internal energy of either metastable molecules or excited ions. Here we speculate on two possible sources of such energy, $H_2 C^3 \pi_u$ and H_3^+.

$H_2 c^3 \pi u$ is a metastable, electronically excited[3] state of H_2. It is produced in fast electron collisions with H_2, with a threshold at an electron energy of about 11 eV and a peak production cross sectin at about 14 eV of 1.4×10^{-16} cm^2.[14] The main destruction mechanism is through slow electron collisions creating $H_2 (a^3 \Sigma g)$[15] which quickly decays to the H_2 repulsive state $b^3 \Sigma_u^+$,[16] the light emission is seen as the H_2 continuum, a featureless emission in the wavelength region 140 to 170 nm. The $H_2 b^3 \Sigma_u^+$ state yields two ground state hydrogen atoms.

The cross sections for this destruction cross section has been calculated[15], for example, for 0.25 eV electrons the calculated destruction cross section is 4×10^{-13} cm^2. This would give the $H_2 c^3 \pi u$, under typical plasma conditions, a lifetime of about 10^{-7} s, with the resulting states making no significant contribution to the atomic and molecular features observed in the V.U.V. spectra.

H_3^+ ions are created in the plasma with internal energy.[17] Dissociative recombination of an electron and ground vibrational H_3^+ is exothermic by 9.3 eV to form H_2 + H. There is sufficient internal energy available for the resultant H_2 to be created in the $B^1 \Sigma u$ or $C \pi u$ states, which would then decay to the H_2 ground state generating the observed molecular spectra. Likewise the atomic hydrogen could be created in an excited state.

It has been shown that there are many reaction product channels[18] in dissociative recombination of electrons with H_3^+, to our knowledge the possibility of electronically excited H_2 being formed from electron or wall collisions has not been explored. We intend to explore this possibility further by a more detailed study of the spectra in the extractor region and installing mass spectrometry to correlate the H_3^+ density with V.U.V. emission.

CONCLUSION

Spectroscopic investigations can provide important information on the atomic processes occurring in present H^- ion sources, for example, the V.U.V. emission has been used to estimate the production rate for vibrationally excited molecules.

Importantly a model of the atomic processes occurring in these sources should be able to explain the spectroscopic observations. At present there is no satisfactory explanation for the relatively intense light emission from the extractor region of tandem multicusp ion sources. Understanding these processes may aid our understanding of H^- production and destruction mechanisms in the plasmas.

REFERENCES

1. M. Bacal, A.M. Bruneteau, W.G. Graham, G.W. Hamilton and M. Nachman, J. Appl. Phys. 52, 1247 (1981).
2. J.R. Hiskes, A.M. Karo, M. Bacal, A.M. Bruneteau and W.G. Graham, J. Appl. Phys. 53, 3469 (1982).
3. J.R. Hiskes and A.M. Karo, J. Appl. Phys. 56, 1927 (1984).
4. A.J.T. Holmes, G. Dammertz, T.S. Green and A.R. Walker, Proc. Int. Ion Engineering Congress. Kyoto, Japan, Sept. 12-16, p.71 (1983).
5. M.B. Hopkins and W.G. Graham, Rev. Sci. Instrum. 57, 2210 (1986).
6. J.G. Laframboise, Univ. of Toronto, Instit. for Aerospace Studies Report No. 100 (1966).
7. W.G. Graham, J. Phys. D. Appl. Phys. 17, 2225 (1984).
8. M.B. Hopkins and W.G. Graham. (These proceedings.)
9. J.R. Hiskes. J. Appl. Phys. 51, 4592 (1980).
10. J.M. Woolsey, J.L. Forand and J.W. McConkey. J. Phys. B. 19, 1493 (1986)
11. E.C. Zipf and M.J. Mumm, J. Chem. Phys. 55, 1661 (1971).
12. W.G. Graham and M.B. Hopkins. Proc. 2nd European Workshop on the Prodtn. and Applic. of Light Negative Ions. (Ecole Polytechnique) 1986 p.61.
13. M.P.S. Nightingale, A.J.T. Holmes, M.J. Forrest and D.D. Burgess, J. Phys. D. Appl. Phys. 19, 1707 (1986).
14. N.J. Mason and W.R. Newell, J. Phys. B.19, L587 (1986).
15. T.N. Rescigno and A. Orel, Private Communication.
16. J.L. Terry, J. Vac. Sci. Technol. A1, 831 (1983).

17. J.B.A. Mitchell and W.G. Graham (These proceedings.)
18. H.H. Michels and R.H. Hobbs Astrophysical J. **286**, L27 (1984).

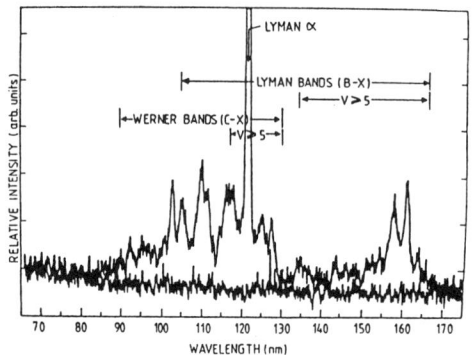

Fig. 1 A typical V.U.V. spectrum obtained from the driver region with plasma operating conditions Id=5A, Vd=60V and gas pressure = 5×10^{-3} Torr.

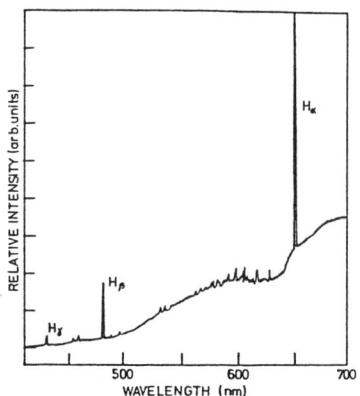

Fig. 2 A typical visible spectrum obtained from the extractor region with plasma operating conditions Id=5A, Vd=60V and gas pressure = 1×10^{-3} Torr.

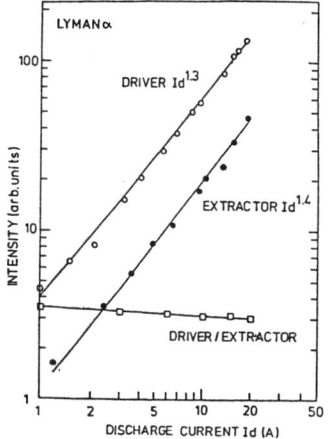

Fig. 3 Variation of intensity of Lyman α emission with discharge current Id, for Vd=60V and gas pressure = 5×10^{-3} Torr.

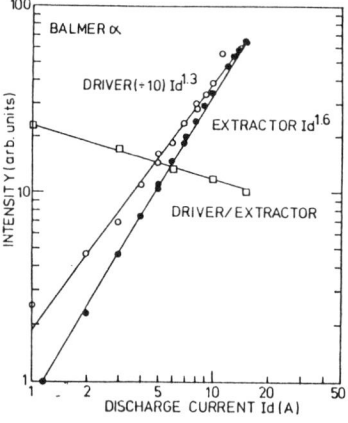

Fig. 4 Variation of intensity of Balmer α emission with discharge current, Id, for Vd=60V and gas pressure = 5×10^{-3} Torr.

DISCUSSION

Hiskes: Bill, do you have any H_3^+ in that extractor region that might recombine and give you vibrational excitation?

Graham: We absolutely do, yes. And that's obviously a contender. Whether or not there's going to be enough energy around....one of the problems is that you essentially reproduce, again, the entire vacuum ultraviolet spectra, so you do get the atomic and the molecular lines. It's not as if you're only seeing the atomic lines, but you also see the molecular lines. One of the things we're still concerned about is that there might be reflection, or something like that, and we really haven't nailed that down yet.

Hiskes: In the case of atomic ions, what about other atoms? What about resonant scattering?

Graham: Well, I don't see how that would reproduce the entire spectra, that's the problem. It would give the atomic part, maybe, but it's explaining why the molecular part is there as well. Vacuum UV shouldn't reflect that well, we shouldn't be getting 30% reflection efficiency, so...

Bacal: I wanted to ask about the atomic lines in the VUV. Why not self absorption?

Graham: Well, it's a possibility, but it doesn't explain the molecular lines.

Bacal: No, no, but it explains the atomic lines. I wouldn't say there is anything which can't be explained by self absorption in the atomic lines.

Graham: I agree. That's a possibility. If you look at what you expect the absorption cross-section to be, though, I think there's too much atomic light there. It's only down a factor of three, 30% of what we see in the driver region.

Mitchell: Bill, when the electrons recombine with H_3^+, that's exothermic by 9 eV to the H_2 channel; so would that not give you the molecular radiation?

Graham: It depends what state is formed when you get that recombination. I don't think it's clear at all that you're going to get an electronically excited molecule created. I don't know.

Holmes: I don't think there's a lot of H_3^+ around, actually.

Graham: Well, of course, once you go to the Culham sources, the high intensity sources, that's certainly the case.

SPECTROSCOPIC MEASUREMENTS OF ATOMIC SPECIES IN VOLUME SOURCES

M. P. S. Nightingale and M. J. Forrest
Culham Laboratory, Abingdon, Oxon OX14 3DB, England.

(1) INTRODUCTION

The use of spectroscopic techniques in the diagnosis of conditions within plasmas, and in particular in the investigation of atomic and molecular processes, has long been established as one of the principle means of obtaining data without perturbing the plasma under investigation. The use of such measurements in the field of negative ion source development has arisen via the need:
(a) to identify the mechanism responsible for H^- production
(b) to measure the H^- density directly
(c) to measure the densities and temperatures of all atomic, ionic and molecular species of relevance to the optimisation of H^- production.

Atomic hydrogen is one constituent species that has been of interest to the H^- source community for some time, since it
(i) may be responsible for destruction of both H^- and vibrationally excited molecules via the reactions:
$$H^- + H \rightarrow 2H + e$$
$$H_2 (\nu^*) + H \rightarrow H_2 (\nu<\nu^*) + H$$
(ii) atomic hydrogen is one species for which accurate destruction, loss and excitation cross-sections are known, and its density is a useful guide to the reliability of theoretical models describing the atomic and plasma physics of such sources.

The atomic hydrogen density has been obtained experimentally in one source by Nightingale, Holmes, Forrest and Burgess[1] using laser fluorescence to measure the $n = 2$ excited state population and a numerical model to derive the total atomic population, and by Péalat, Taran, Bacal and Hillion[2] using a model for the electron energy distribution for their discharge to analyse the lineshape of the $H\alpha$ emission from their source.

In addition the origin of the atomic spectrum emitted from the extraction zone beyond the filter field has recently attracted some interest. Nightingale et al [1] published spectra from both source regions and commented that no atomic emission should be visible from the extraction region since this does not contain the fast electrons responsible for excitation, and that no obvious explanation for the spectrum's existence was forthcoming. Similar behaviour has since been reported by Graham and Hopkins[3] using visible and VUV spectroscopy to show that the fractional variation in hydrogen $n = 2$ population across the filter is likely to be even lower than that of the $n = 3$ and 4 populations derived from the $H\alpha$ and $H\beta$ emission.

(2) DC EMISSION MEASUREMENTS

Given these questions, high resolution $H\alpha$ and $H\beta$ lineshapes have been measured for the emission spectrum from the Culham small negative ion source (described by Nightingale et al[1]) for both

source regions in order to
(i) use the method used by Péalet et al to derive the atomic hydrogen density for comparison with that obtain using laser fluorescence
(ii) obtain atomic temperatures from the measured Doppler widths.

The measurements were made using a one-metre Ebert monochromator equipped with a 316-line/mm Echelle grating and an optical multichannel analyser providing a resolution of 165 mÅ (measured using a He/Ne laser). Hα profiles were measured for varying arc currents ranging from 10A to 40A and source filling pressures between 1.75 mT and 10.8 mT, and the resultant spectra numerically analysed to deconvolve the instrument profile to provide the atomic lineshape. An example of a resultant Hα lineshape from the discharge region is shown in figure 1. The effect of the Hα fine structure must be taken into account in order to derive an atomic temperature. This is performed using a numerical calculation of the profile FWHM in terms of the atomic temperature (figure 2) given the profiles (figure 3) calculated using the relative fine structure component intensities provided by Tallents (private communication). These measurements showed
  (i) no clear evidence for the presence of the profile shoulders observed by Péalat et al[2]
 (ii) less than 8% variation in linewidth with varying discharge conditions (factor of four in arc current and six in pressure)
(iii) a profile FWHM of 312 mÅ corresponding to a temperature of 2750 K (0.24 eV) after correcting for the fine structure (note that similar results were obtained using Hβ).

The low Doppler temperature, coupled with the lack of variation with arc conditions, suggests that the width may not be a true Doppler width at all. Indeed such profiles were observed previously by McNeill and Kim[4], who suggested that the observed profile arises purely from the Frank Condon energy associated with molecular dissociation by electrons with energies in the range 15 - 28 eV, and bears no relation to the temperature of the thermalised atomic population. This would suggest that the lack of shoulders seen in the present work might be due to the source geometry used. Should the 120 V electrons emitted by the filaments suffer sufficient energy degradation that their energy is less than 28 eV when they reach the volume observed in the emission experiment, then the observed lineshape will not contain the shoulders induced by high energy electron collisions.

To investigate the possibility that fast electron dissociation is responsible for the observed profiles, further measurements were made in the extraction region, which contains a negligible fast electron population (T_e = 0.3 - 0.5 eV). A typical result is shown in figure 4. The average measured profile width was 286 mÅ, corresponding to a temperature of 1811 K (0.16 eV) after removal of the fine structure contribution. Again little dependence of the profile width on arc conditions was observed (when the arc current and pressure were both varied by factors on three). Although the measured temperature is lower than the corresponding temperature for the discharge volume, the similarity in profiles observed in the two regions must cast doubt concerning the proposed excitation mechanism involving fast electrons.

In addition to the above question, one extra anomaly arises from the $H\alpha$ lineshapes observed in the two regions. All the spectra appear to show two fine structure peaks (as in figures 1 and 4). The numerical profiles shown in figure 3, however, would suggest that no fine structure should be resolvable at these temperatures. This behaviour was also seen by McNeill and Kim, who observed some contribution to their lineshape from cold atoms. The observation of the fine structure might indeed be a demonstration of a profile composed of emission from atoms of varying temperature. Since the emission measurements are line integrated, cold atoms formed by the recombination of positive ions at the walls might contribute strongly to the line profiles. Warmer hydrogen in the centre of the discharge would then be responsible for the broad base of the profiles.

These measurements therefore not only failed to explain the origin of the atomic spectrum emitted from the extraction region, but produced extra anomalies for both regions via the constancy of linewidths observed. This suggests that fast electron collisions may not be responsible at all for atomic excitation in either source region. Alternative excitation mechanisms might be

(i) Mutual neutralisation $\quad H^+ + H^- \to H + H^*$

(ii) Thermal electron excitation of metastable atoms $\quad H(2s) + e \to H(n>2) + e$

(iii) Thermal electron collisions with metastable molecules $\quad H_2(C^3\Pi) + e \to H^* + H + e$

(iv) Recombination $\quad H^+ + e \to H^* + h\nu$

(v) Dissociative recombination $\quad H_2^+ + e \to H^* + H$

Clearly a method of unambiguously determining whether fast electrons do indeed play a role in the production of excited atoms is required. The first results from such an experiment are now described.

(3) PULSED EMISSION MEASUREMENTS

The density of fast electrons (T > 10 eV) in the discharge region of the Culham source is typically 5% of the thermal electron density and approximately given by a Maxwellian electron energy distribution of central temperature around 20 eV. If the source arc current is cut off rapidly (less than 5 µsecs), the subsequent containment time for these fast electrons will be of order

$$\tau_p = \frac{4V}{A_p} \left(\frac{m_e}{8kT_p}\right)^{1/2}$$

where V is the plasma volume, and A_p the primary loss area given by the cusp losses. Assuming that the cusp width is given by the hybrid Larmor diameter, it can be shown that the above containment time should lie in the range 0.5 - 3 µsecs, a value which has been verified elsewhere experimentally (see for instance Hopkins and Graham - this workshop). Should the atomic excitation indeed proceed via fast electron collisions after all, the $H\alpha$ emission will decay within a similar timescale.

The Culham small negative ion source has therefore been run in the configuration shown in figure 5, with the arc voltage pulsed for 1.6 msec at 0.8% duty cycle producing an arc current decay time of

1.0 μsec (1/e) as shown in figure 6.

Prior to recording the Hα traces, planar Langmuir probe measurements were performed in both source regions for a 25 A. 120 V. 3 mTorr discharge. Probe and Hα traces were taken at various values of the bias voltage applied between the beam forming electrode and the anode, and a sample of the results is shown in figure 7. When the probe decays are plotted on log/lin scales it is found that all the probe traces decay exponentially, with one change in decay rate at a time of 40 - 90 μs after the arc switching, with the decay times given in table 1 (τ_+ for the probe ion saturation current decay time and τ_- for the electron saturation decay time, and (a) and (b) used to denote the early and late decays). These results show that the initial exponential decay time drops with increasing bias, but that the later decay times decrease very rapidly with bias. The mechanism for this behaviour is not yet clear, but may well help clarify the role that the bias potential plays in H^- production.

The Hα decays are similarly exponential, and an example is given in figure 8. Only a single exponential decay is seen until very late times, and these are also listed in table I.

Two immediate conclusions can be drawn:
(a) The decay times are far longer than those corresponding to atomic excitation via fast electron collisions.
(b) The emission decays faster than either of the corresponding probe traces.

Although this experiment cannot as yet demonstrate the precise mechanism responsible for atomic excitation, it should be noted that the Hα decay times given in table 1 are in fairly good agreement with those also listed in table 1 derived from the expression

$$\frac{1}{\tau_\pm} = (\frac{1}{\tau_+} + \frac{1}{\tau_-})$$

Such a decay rate would apply if the atomic excitation proceeds via recombination of a positive and a negative particle, as would be the case for recombination, dissociative recombination and mutual recombination. Note that this agreement would also apply if the high bandwidth of the Hα filters used (18 Å) allows sufficient continuum radiation to reach the photomultiplier that this emission dominates that from the atomic line. Measurements have therefore been with the 18Å bandwidth filter replaced by a 1 metre Monospek with the slit widths set to produce a resolution of 1.45Å (FWHM). These measurements showed little change in Hα decay time. In addition, the emission of a detuning of 2.1Å from the Hα line showed no pulse shape at all, thus confirming that there is no contribution to the measured decay times from the underlying continuum.

During the course of these measurements, the Hα emission from the discharge region was also recorded at (zero bias only). This shows an Hα intensity that drops more rapidly (with a 1/e time of between three and eight μs) to a value of 33% of the DC level, and then decays at a similar rate to the corresponding case in the extraction region (40.5 μs). Finally the arc current was raised to 100 A, but no change was observed in the Hα decay rate for the extraction zone at zero bias to within the accuracy of the measurements.

These measurements therefore appear to demonstrate that atomic excitation does not take place via fast electron collisions. Perhaps more importantly they demonstrate that afterglow studies may well offer a valuable insight into the physics of source operation.

ACKNOWLEDGEMENT

This work is partially funded by the United States Air Force Office of Scientific Research under contract co. F49620/86/C/0064.

TABLE I Measured decay times

Bias (V)	τ_+ (µs) (a)	τ_+ (µs) (b)	τ_- (µs) (a)	τ_- (µs) (b)	$\tau_{H\alpha}$ (µs)	$\tau_\pm$ (µs) (a)
0	73 ± 5	109 ± 10	56 ± 2	94 ± 3	32 ± 3	31.7 ± 3.3
0.5	61 ± 7	101 ± 10	37 ± 10	76.5 ± 2	35.5 ± 7	23.0 ± 9.0
1.0	59 ± 5	38 ± 3	47 ± 1	34 ± 1	30.5 ± 5	26.1 ± 2.8
1.5	40 ± 4	21 ± 2	38 ± 6	9 ± 1	25 ± 3	19.5 ± 5.0

REFERENCES

1. M. P. S. Nightingale, A. J. T. Holmes, M. J. Forrest and D. D. Burgess, J. Phys. D: Appl. Phys. 19, 1707, 1986.
2. M. Péalat, J. P. E. Taran, M. Bacal and F. Hillion, J. Chem. Phys. 82, 4943, 1985.
3. W. G. Graham and M. B. Hopkins, Proc. 2nd Eur. Workshop on Production and Application of Light Negative Ions, Palaiseau, 1986 p.61.
4. D. H. McNeill and J. Kim, Phys. Rev. A 25, 2152, 1982.

Figure 1

H$_\alpha$ PROFILE (AFTER DECONVOLUTION)
DISCHARGE REGION
25A/130V/5·6mT

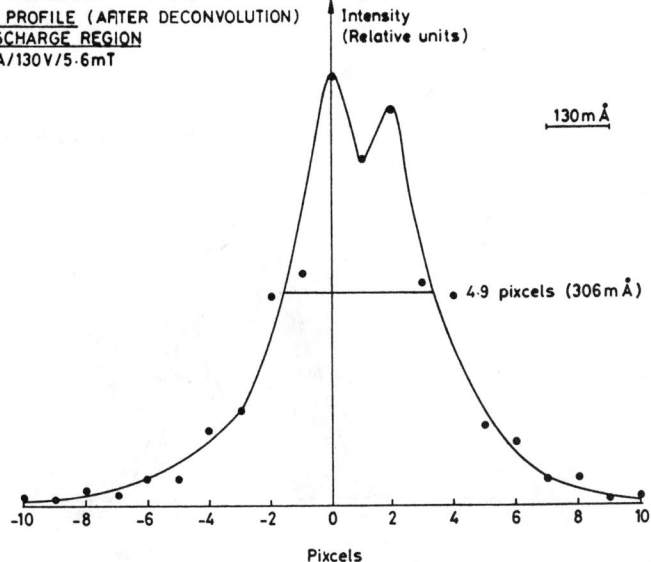

Figure 2

Total Profile Width as a function of Doppler Width α

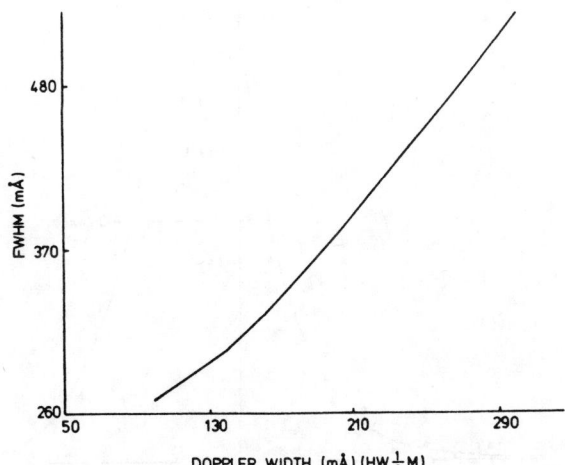

Figure 3

Hα Lineshape as a function of Doppler HW$\frac{1}{e}$ M α

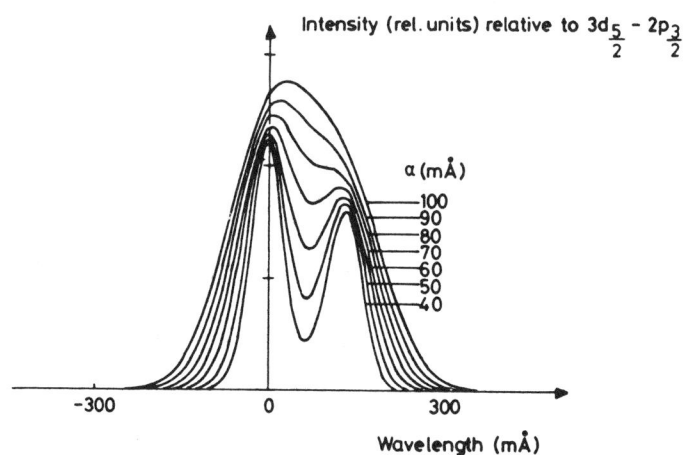

Figure 4

Hα PROFILE (AFTER DECONVOLUTION)
EXT. REGION
25A/130V/5.6mT

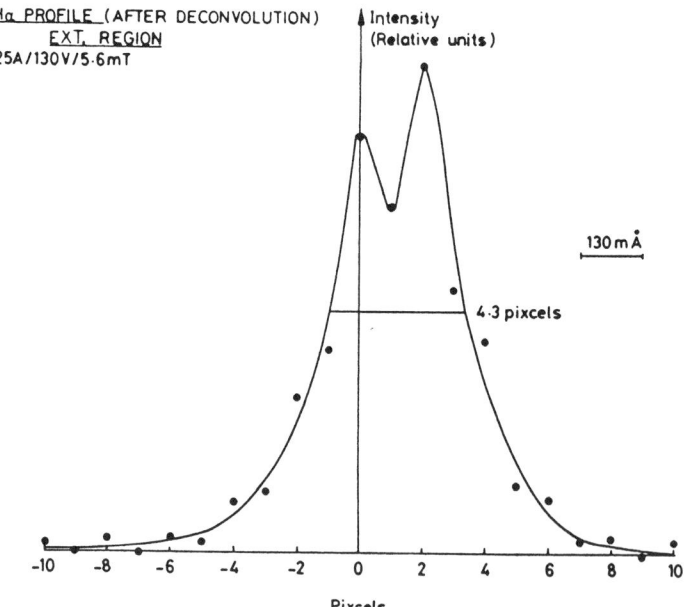

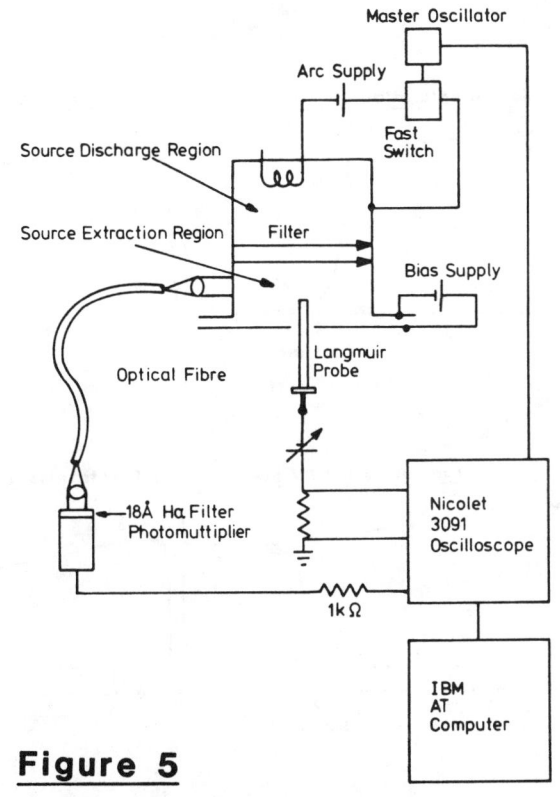

Figure 5

Schematic Layout for the pulsed Hα Spectroscopy Experiment.

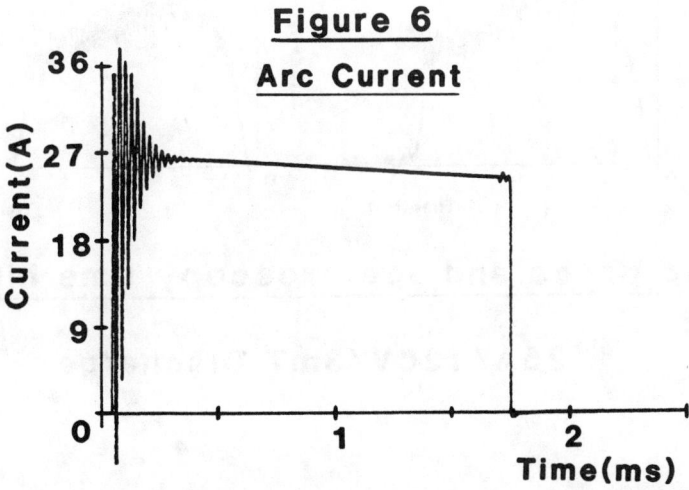

Figure 6
Arc Current

Figure 7

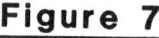

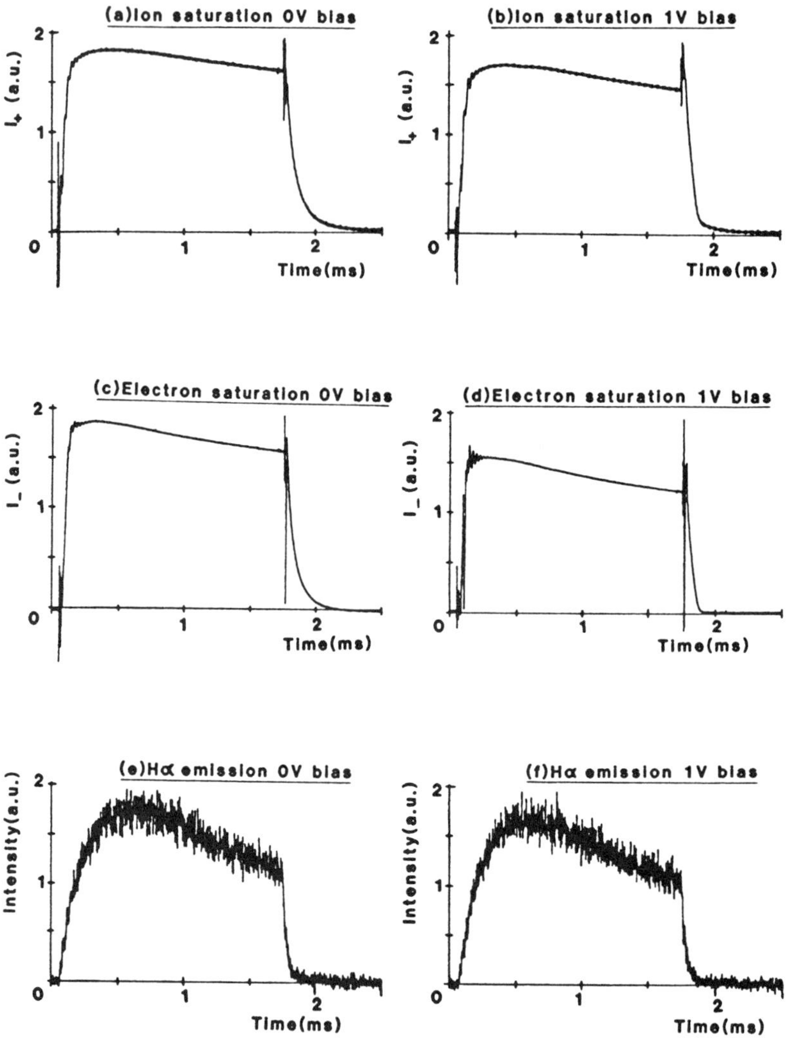

Pulsed probe and spectroscopy time histories

25A/120V/3mT Discharge

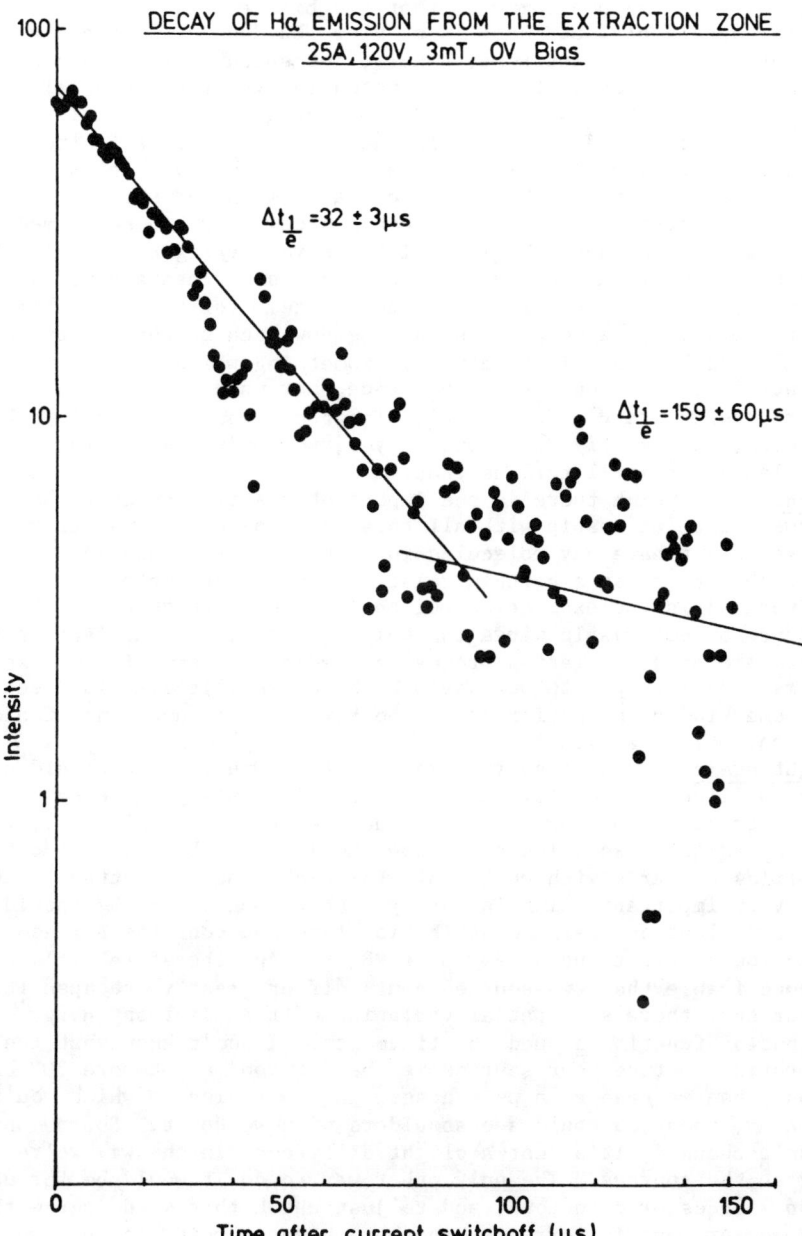

DISCUSSION

Ehlers: I have one comment, that the bias effect that you see could conceivably, at least in my view, be due to the speed at which you dump the cold electrons; namely, the speed at which you dump them is a function of the plasma potential vs. the potential on that grid. And if it remains at anode and the plasma potential remains high, you trap cold electrons in there for possibly a finite time which, in turn, could dump more rapidly if you have a bias on it.

Nightingale: I think there is a certain amount of source physics that this particular experiment could show. I don't see immediately where we go from here, but yes, I think you're right. I think this could be very valuable insight into the loss mechanisms for such things as cold electrons within the source. For the decay time, one of the problems we do have is knowing how much of that decay we see is H^-, and how much is electrons. Something which I think we can attack is on the ion saturation trace, how much of the ions that we're seeing are H^+, H_2^+, and H_3^+. The guess would be that, probably late on, it's mostly H_3^+. But, as yet, we don't have a diagnostic available that will tell us that.

Bacal: I think there is one important question which we have to answer in relationship with all this spectroscopy - whether we have or we do not have any molecules in this type of discharge. So the fact that there is a certain relationship between intensities in the driver and in the extraction region is a separate question, but the absence of observable wings can only be related to the fact that there are no dissociation processes leading to formation of excited atoms. Another way to see whether there are molecules is observing at some kind of molecular line. So here is the question: Could you see any molecular line?

Nightingale: I can give you two answers. One is that we did not see any line in the visible that we could identify as being a molecular structure; we saw no band structure in our emission spectra. As far as the discussion about how the lack of shoulders on our profiles compares with what Ecole Polytechnique has gotten, I think one very important point is that you should only see the shoulders if your electron energy distribution function contains a reasonable fraction of electrons above about 28 eV. And therefore, it's not inconceivable that two sources could differ greatly, because it is clear that there's a spatial variation of the electron energy distribution function around the filaments. I don't know what the comparison between our sources is, but if you've got more 30V electrons than we have as a percentage, in the region in which you're looking, then you could see shoulders when we don't. Source physics hasn't changed, it's just a slight difference in the way we're looking at the sources. The only other way to do it would be for us to swap sources or detectors, and to just check that we do agree that the measurement is correct. Another relevant point is that we do expect, in the very near future, to try and do the emission work again, but this time with a laser pumped system, so we use the laser flourescence setup but measure the line profile of the flourescence.

The important point about that is that it would not be line integrated. We would be able to do spatially resolved measurements in the center of the plasma.

<u>Mitchell</u>: That change in the slope of the decay time of H_α - could that be because of some new process switching on, that would be particularly effective in producing H_α?

<u>Nightingale</u>: I certainly can't rule that out. Perhaps it's slightly unfortunate to have picked this trace. Most of the H_α traces don't show a second slope, and you could even argue that it isn't here on this one. The signal to noise gets quite low there. One point I haven't made, that perhaps ought to come in, is that we have repeated this set of measurements at higher resolution. There was some fear that what we were seeing was the underlying continuum changing, because this dependence of the slope observed for the ion and electron saturation currents, would also be explained if we were looking at continuum radiation from free-free and free-bound. So we used a monospec instead of the 18Å filter and tuned off the line and proved that everything we see in this decay is quite definitely H_α. It might be argued, though, that you're starting to see filament emission. I did a very quick calculation that suggested the filaments are going to change across fractions of a second. It may well be that we're just coming up against the filament emission, that's a slight scatter off the walls.

SPATIALLY AND TEMPORALLY RESOLVED EEDF MEASUREMENTS IN A HYDROGEN DISCHARGE

M. B. Hopkins and W. G. Graham
Physics Department
University of Ulster
Coleraine BT52 1SA, Northern Ireland

ABSTRACT

Measurements of the spatial dependence of the electron energy distribution function, eedf, in the driver and extraction region of a tandem multicusp negative hydrogen ion source are presented. These results and the temporal evolution of the plasma parameters, including the eedf, in the discharge and post discharge regime of a pulsed plasma provide an insight into the basic processes occurring in the discharge.

INTRODUCTION

The production of intense beams of negative ions requires as a source a uniform dense plasma which is normally formed in a magnetic multicusp discharge by electrons emitted from filaments and accelerated by an applied voltage, V_d[1,2,3]. The energetic ionizing electrons are confined in the plasma by the magnetic fields at the walls resulting in a high ionization efficiency[4].

In explaining the properties of these multicusp plasmas it cannot be assumed that the electron energy distribution function, eedf, is Maxwellian. This has important consequences in calculating the rate coefficients of electron induced processes such as ionization and excitation. Although theoretical[5,6] calculation have given invaluable insight into the development of the eedf in the central magnetic field free region of the multicusp device, the spatial variation of the eedf throughout the source is of crucial importance in establishing the actual volume averaged rates of electron processes. The spatial distribution of the eedf can be affected by weak external magnetic fields leading to a greatly altered ion source behaviour and efficiency[7]. A similar problem arises from internal magnetic fields, such as those produced by the filament current[8]. Furthermore, spatially resolved measurements of the eedf are important in investigating the nature of operation of the filter field[1,2,3] and clearly demonstrate the effectiveness of this field in preventing high energy electrons from entering the extraction region.

By operating the source in a pulsed mode, temporal information is obtained on the development and decay of the eedf during and after the discharge pulse. The eedf in the post discharge regime is shown to be Maxwellian and decays to near the gas temperature with times of the order 10^{-3} S. It is believed that a coupling between the vibrational energy of the H_2 molecules and the kinetic energy of the electrons is

© American Institute of Physics 1987

responsible for the long decay times[9]. Therefore, measurements of the eedf in the post discharge or afterglow may provide information on the vibrational energy of the H_2 molecules.

The eedf is obtained from the second derivative of a Langmuir probe current-voltage, I(V) characteristic according to the Druyvesteyn equation[10]. Langmuir probes are particularly useful in multicusp plasma where the confining magnetic fields ensure an isotropic velocity distribution even for high energy electrons. Previous attempts at measuring the eedf using gridded energy analysers[11] located at the anode of the discharge, only analysed the velocity component, normal to the wall, of electrons escaping the plasma and are thus unlikely to reflect the energy distribution of the confined electrons. A further advantage of Langmuir probes is that a movable probe may be used to spatially resolve the eedf without greatly perturbing the plasma.

APPARATUS

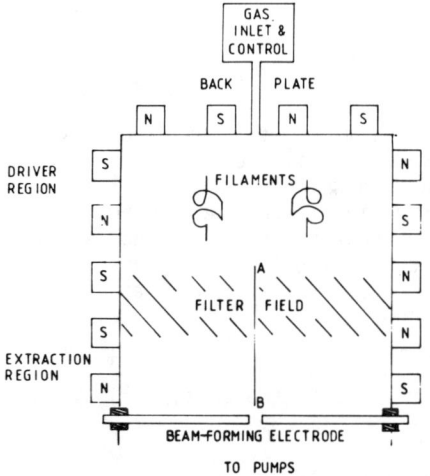

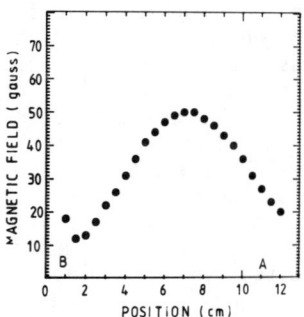

FIG.1 : Schematic diagram of the multicusp source with filter.

FIG.2 : Variation of the magnetic field strength as a function of position through the filter field.

Figure 1 shows a schematic of the source which consists of a stainless steel box with dimensions 19 × 19 × 24 cm. Permanent magnets with pole strengths of 3.6 KG are spaced 4.5 cm apart on the outside walls in the normal line cusp symmetry. This symmetry is broken on opposing walls near the extraction end of the source allowing a one-dimensional filter field to cross the source dividing it into two regions. The magnetic field strength is shown in figure 2 as a function of position from the beam forming electrode (BFE) towards the driver region along the line marked A/B in figure 1. The discharge is produced by electrons emitted from two tantalum filaments heated by a d.c. current. The filaments are biased negatively with respect to the vessel wall which forms the anode.

The Langmuir probes consist of a tungsten wire 0.33 mm in radius, shielded by a 1.9 mm diameter ceramic sleeve. The wire extends typically 10 mm beyond the ceramic to form the probe. A more detailed description of the apparatus has been published in an earlier paper[12].

The probe I(V) characteristic is obtained using a high speed large dynamic range data acquisition system[13]. The second derivative is obtained by direct differentiation of the stored I(V) characteristic. Noise is reduced by ensemble averaging data points, typically 10^6 points per characteristic and smoothing is performed with an energy half-width varying from .5 eV at low energies to 4 eV at high energies.

MEASUREMENT OF THE EEDF

At a typical operating pressure of 2 mTorr it is assumed that the confining magnetic fields will produce an isotropic but non-homogeneous electron distribution function which is stable, it is possible to define the electron distribution in terms of energy and position only $f(\varepsilon,\underline{r})$. In the pulsed discharge the eedf is not stable but may be represented by the periodic function $f(\varepsilon,\underline{r},\tau)$, where τ is the delay time. In the present results τ varies from 0 to 2×10^{-3} S and thus $f(\varepsilon,\underline{r},\tau)$ has a frequency of 500 Hz. The discharge pulse begins at $\tau = 0$ and ends at $\tau = 2 \times 10^{-4}$ S. The post discharge or afterglow begins at $\tau = 2 \times 10^{-4}$ S and ends at $\tau = 2 \times 10^{-3}$ S.

The second derivative I''(V) is related to the eedf, $f(\varepsilon)$ by the Druyvesteyn[10] formula.

$$n(\varepsilon) = n_e f(\varepsilon) = 2I_e''(2m\varepsilon/e)^{\frac{1}{2}}/e\, S$$

where the energy $\varepsilon = V - V_p$, I_e is the electron current to the probe, the primes indicate differentiation with respect to ε, S is the probe surface area, m and e are the mass and charge of the electron. The notation $n(\varepsilon)$ is used to denote $n_e f(\varepsilon)$.

In order to reduce noise in the eedf it is possible to ensemble average a large number of eedf's for periods up to several hours. The general plasma parameters are monitored over the integration period to ensure a sufficiently stable plasma.

The detailed analysis[12] of the characteristic using Laframboise theory[14] gives a reasonably accurate determination of the ion current dependence on probe voltage, $I_+(V)$. Therefore, it is possible to define more accurately the electron probe current,

$I_e(V) = I(V) - I_+(V)$. The contribution of the ion current to the second derivative of the probe current is normally insignificant. However, it is important to have a reasonable estimate of the potential error due to ion current in the eedf measurement. The measured value of V_p is used to transform the probe voltage to particle energy, $\varepsilon = V_p - V$.

RESULTS AND DISCUSSION

It has been proposed[4] that the high ionization efficiencies achieved in multicusp plasmas is the result of the magnetic confinement of the electrons emitted from the filaments. These electrons are often assumed to be monoenergetic with an energy of eV_d and referred to as primaries. It has been shown that this concept of primary electrons is valid in low pressure, low current discharges[15,16]. However, at normal source operating pressures of ~ 2 mTorr the primaries are degraded in energy by collisions and will be referred to as fast electrons.

Spatial Variation of the eedf

In figure 3 the concept of fast electron confinement is shown. The electron density per unit energy $n(\varepsilon)$ at ε = 30 eV is plotted as a function of position in a plane which cuts through the centre of the source (see figure 1). The spatial distribution of 30 eV electrons is seen in figure 3 to follow the structure of the confining and filter magnetic fields. In the centre of the driver region the density is basically homogeneous with a value of 2×10^8 cm^{-3} eV^{-1}. However, there are 4 anomalous regions of high fast electron concentration. These peaks are positioned at points where the plane cuts close to the filaments. The peaks are probably due to confinement of the fast electrons by local magnetic fields produced by the filament current[8]. When the source is operated on one filament only, the two peaks associated with that filament disappear. Outside the central driver region it is apparent from figure 3 that the fast electron density falls by an order of magnitude per centimeter. Probe measurements show that there is not a similar drop in plasma density. The plasma production region is confined to a driver volume estimated to be 1.5 L or 17% of the total source volume.

The idea of confining the fast electrons is again used in the concept of a magnetic filter field[3]. In the present device the fast electron free region is desired to provide a region of enhanced negative ion production and reduced destruction. To illustrate the effect of the filter field on the full range of electrons energies figure 4 shows the variation of the eedf, $n(\varepsilon)$ as a function of position on a line through the filter field. The spatial scan goes from just inside the driver region, through the filter, to within 1 cm of the beam forming electrode and is illustrated by the line marked A/B in figure 1. The discharge conditions are identical to those of figure 3. Figure 4 shows that in the driver region (10-11 cm) there is no primary peak at eV_d but that the primaries are highly

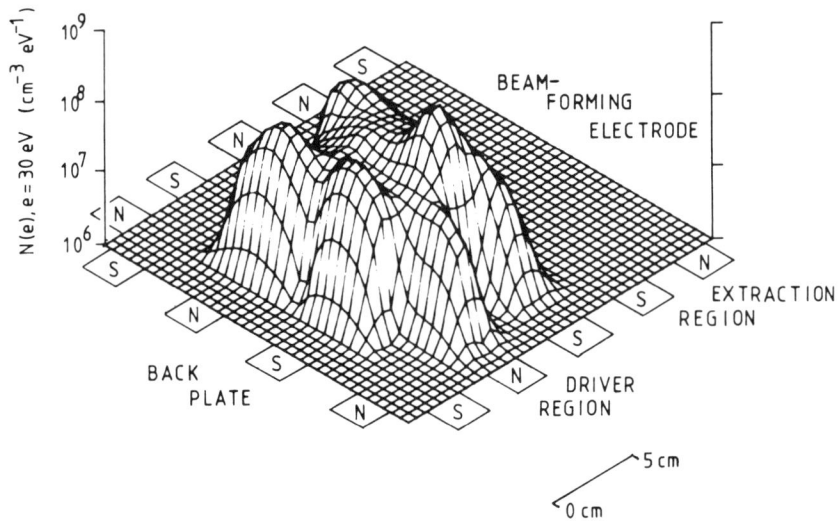

FIG.3 : Spatial distribution, in two dimensions, of 30 eV electrons within the source. Discharge conditions I_d = 10A, V_d = 60V, P = 2 mTorr.

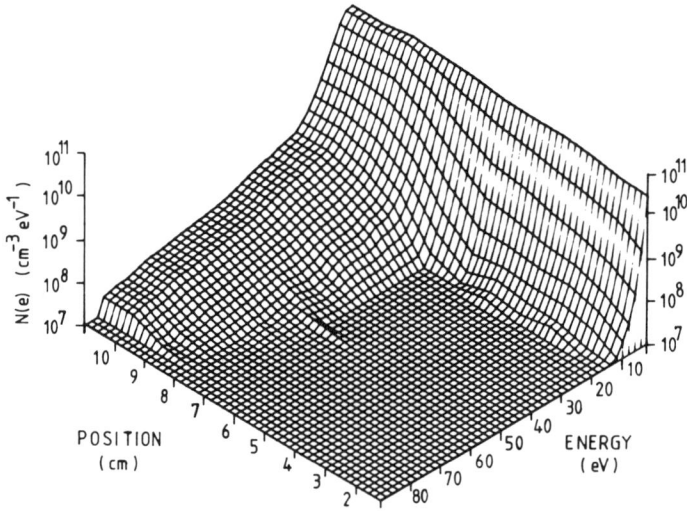

FIG.4 : Spatial variation, in one dimension, of the eedf across the filter field along the line marked AB in figure 1. Discharge conditions as figure 3.

degraded and spread along the energy axis. This produces a non-Maxwellian eedf. As the filter field (6-9 cm) is approached the fast electron portion of the eedf is most affected by the magnetic field. In the extraction region (1-6 cm) the fast electrons are effectively removed and the eedf is Maxwellian. There is a fall in total electron density of a factor of ~ 3 in crossing the filter field. The other major change in crossing the filter into the extraction region is the cooling of the bulk electron temperature by a factor of ~ 3.

Temporal evolution of the eedf

It is convenient to define two time regimes. Firstly the discharge regime, $0 < \tau < 2 \times 10^{-4}$ S, where the discharge voltage, $V_d = 60$ V and secondly the post discharge regime, $2 \times 10^{-4} < \tau < 2 \times 10^{-3}$ S, where $V_d = 0$.

(a) Discharge Regime

The importance of the present studies of the discharge regime lies in establishing the time necessary for the plasma to reach steady state conditions. Furthermore, to verify that the probe techniques already developed successfully in the steady state discharge can be applied to the pulsed source. It will then be possible to extend the eedf measurements to current higher than those possible with our present apparatus in the steady state.

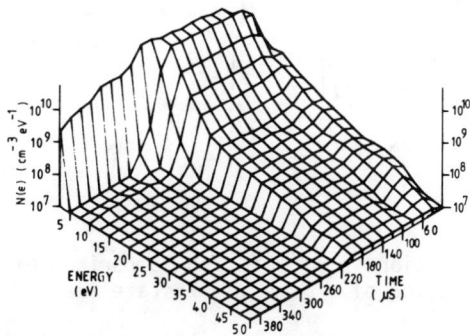

FIG.5 : Temporal variation in the driver region of the eedf in the discharge and early post discharge regimes. Discharge conditions I_d = 2A, V_d = 60V, P = 2 mTorr.

In the pulsed mode V_d has a total rise time of less than 2×10^{-6} S and a total decay time of less than 1×10^{-5} S. In figure 5 the temporal evolution of the eedf is shown in both the discharge and the post discharge regimes, $0 < \tau < 4 \times 10^{-4}$ S. The eedf's are measured in figure 5 at intervals in τ of 2×10^{-5} S. In the discharge regime the fast electron component established quickly at $\tau = 2 \times 10^{-5}$ S, but the bulk electron component requires a much longer period. Until a high plasma density has established the discharge current is unstable. This is reflected in the oscillations of the fast electron component of the eedf in the period $0 < \tau < 8 \times 10^{-5}$ S. In fact, as figure 6 clearly shows n_e reaches a

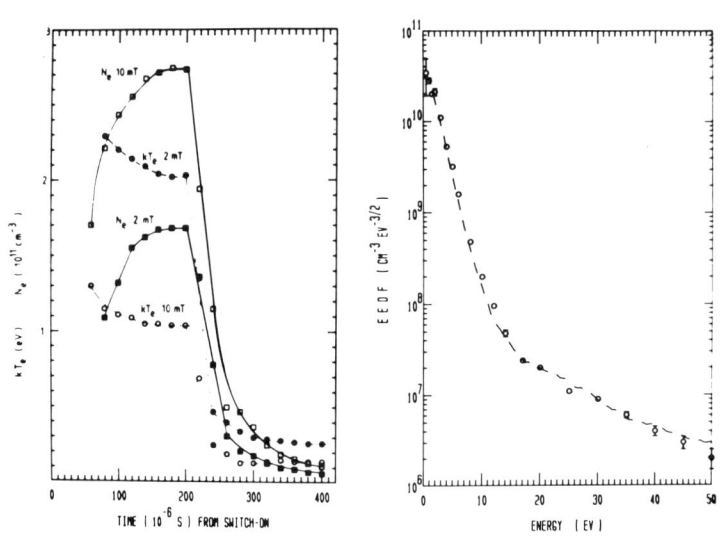

FIG.6 : Temporal variation in the driver region of the bulk electron temperature and density in the discharge and early post discharge regimes. I_d = 4A, V_d = 60V, P = 2 mTorr.

FIG.7 : Comparison of the steady state (dashed line) with the pulsed discharge regime (open circles) at τ = 1.8×10^{-4} S. I_d = 2A, V_d = 60V, P = 2 mTorr.

steady state value at $\tau = 1.6 \times 10^{-4}$ S. The values of kT_e and n_e at

$\tau = 1.8 \times 10^{-4}$ S agree within a few percent with the measured steady state values. The fast electron density n_{fe} and temperature kT_{fe} (not shown) reach steady state values at $\tau = 4 \times 10^{-5}$ S. To further illustrate the agreement between the pulsed and steady state plasmas figure 7 gives a direct comparison between the eedf at 1.8×10^{-4} S (as shown in figure 5) with the steady state eedf. Furthermore, the eedf's in figure 7 are expressed as $n(\varepsilon)\,\varepsilon^{-\frac{1}{2}}$ in which form a Maxwellian eedf should give a straight line on a semi-logarithmic plot indicating that the eedf's are non-Maxwellian but can be approximated by a bi-Maxwellian distribution. This bi-Maxwellian approximation is used to analyse the probe characteristics[12].

It has been noted that the electron emission from the filaments is initially space charge limited. The plasma potential V_p is highly negative with respect to the anode at $\tau = 2 \times 10^{-5}$ S and only approaches its steady state value after $\tau = 4 \times 10^{-5}$ S. This leads to an unstable plasma at early values of τ as is reflected in the eedf measurements in figure 5.

(b) Post discharge regime

It is believed that the main mechanism for H^- production in a multicusp plasma is through the dissociative attachment of slow electrons to highly vibrationally excited H_2 molecules. This process is encouraged in the H^- ion source by having two distinct regions. A driver region where the filaments are located which produces the plasma and populates the vibrational levels by fast electron impact on H_2 molecules. An extraction region into which the plasma and vibrationally excited molecules can diffuse but where the fast electrons are prevented from entering (see figure 4). Although the low energy electrons in the extraction region may vibrationally excite the H_2 molecules directly through a resonant mechanism involving the H_2^- state (e-V process), it is thought that the dominant mechanism, at least for vibrational levels $v > 5$, involves the electronic excitation by fast electron impact on the H_2 molecule followed by radiative decay (E-V process). The E-V process is confined to the driver region and the vibrationally excited molecules produced must then diffuse into the extraction region.

In the post discharge regime the electrons quickly lose energy in collisions with the background gas with characteristic times of the order of 10^{-5} S at 2 mTorr. At times of the order of 10^{-4} S, the energy return from super elastic collisions of electrons with vibrationally excited H_2 molecules will partially balance this electron energy loss[9]. At pressures of 4 mTorr the vibrational "temperature" has been recently measured by CARS as 2.2×10^3 K, with a relatively long relaxation time of 1.5×10^{-3} S[18]. Therefore, it is believed that the temporal evolution of the eedf for times $\tau > 3 \times 10^{-4}$ S will be strongly correlated with the vibrational "temperature" in the discharge regime and by inference in the steady state discharge. The two decay regimes of kT_e are seen in figure 6, at 2 mTorr kT_e decays rapidly reaching a quasi-stationery value, at $\tau = 3.6 \times 10^{-4}$ S, having a much slower decay rate. At 10 mTorr the quasistationary value is reached much quicker at $\tau = 2.8 \times 10^{-4}$ S. The values of kT_e

in figure 6 are derived from a direct analysis of the probe characteristic. The more accurate second derivative technique shows that the eedf relaxes at 2 mTorr with a characteristic time of 3×10^{-5} S upto $\tau = 2.8 \times 10^{-4}$ S and then much more slowly. The initial relaxation rate of kT_e at 10 mTorr was faster than the techniques temporal resolution of 2×10^{-5} S could measure.

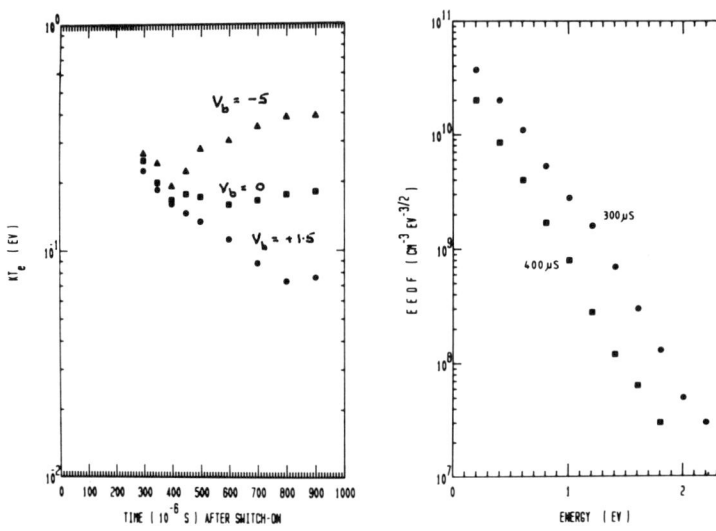

FIG.8 : Evolution of the electron temperature with different filament bias voltages (see text) $I_d = 2A$, $V_d = 60V$, P = 2 mTorr.

FIG.9 : The eedf in the post discharge regime (driver region) $I_d = 4A$, $V_d = 60V$, P = 2 mTorr.

However, there is an experimental problem in that the discharge is produced by electrons emitted from hot tantalum filaments. These filaments are still hot in the post discharge regime with a voltage drop of 8 V and may emit electrons with energies of several eV. In order to investigate the effect of the emitted electrons on the eedf in the post discharge, the filaments are biased with a variable bias V_B relative to the anode. The bias is applied to the negative leg of the filaments. When V_B is negative the filaments emit electrons, as V_B is made more positive the energy of the emitted electrons is reduced. When V_B is greater than V_p (~ 1.5V) the filaments are no longer emitting electrons but tend to deplete the plasma. To illustrate the effect of V_B on the eedf figure 8 shows a plot of kT_e as a function of τ, in the driver region, at three different values of V_B. As can be seen the values of kT_e diverge drastically at $\tau > 4 \times 10^{-4}$ S. The negatively biased filament clearly heats the eedf, while the effect is small when the filaments are biased at anode potential ($V_B = 0$). The optimum bias appears to be when the filaments are biased at the plasma potential (~ 1.5V). This prevents electrons being emitted into the plasma but the greatly enhanced loss of electrons to the filaments may have a cooling effect on the eedf. To reduce the depletion caused by biasing the filaments at V_p the filament area was reduced by operating the source on one filament only. In figure 8 when τ is in the period $3 \times 10^{-4} < \tau < 8 \times 10^{-4}$ S, kT_e is seen to decay with a characteristic rate of 3×10^{-4} S ($V_B = 1.5V$). At $\tau = 8 \times 10^{-4}$ S, kT_e reaches another quasistationary value linked to the equilibrium gas temperature. In the period $8 \times 10^{-4} < \tau < 2 \times 10^{-3}$ S (not shown), kT_e stabilizes at a temperature of 0.07 eV or 800 K. In using kT_e to describe the eedf in the post discharge regime, it is assumed that the eedf is Maxwellian. In order to show that this is a valid assumption figure 9 shows the measured eedf, $n(\varepsilon)\varepsilon^{-\frac{1}{2}}$, at $\tau = 3 \times 10^{-4}$ S and $\tau = 4 \times 10^{-4}$ S. The linear plots indicate that the eedf's are indeed Maxwellian, in the driver region, in the post discharge regime. At $\tau = 4 \times 10^{-4}$ S it can be reasonably assumed, in the pressure range 1 to 10 mTorr, that the major source of heating the eedf is through super elastic collisions with the vibrationally excited H_2 molecules. Considering the long relaxation times measured for the vibrational temperature and the Maxwellian nature of the eedf, it is expected that the electron temperature kT_e at $\tau = 4 \times 10^{-4}$ S is strongly correlated with the vibrational temperature of the H_2 molecules in the discharge regime. Furthermore, two major mechanisms for heating the vibrational temperature are expected to operate in this type of source. The E-V process is a function of fast electron density n_{fe} and operates in the driver region only. The effectiveness of the e-V process is a function of bulk electron density and operates throughout the source. It has already been established[16] that the fast electron density is an inverse function of pressure while the bulk electron density increases with pressure in the range 1 to 10 mTorr. Figure 10 shows the dependence of kT_e at $\tau = 4 \times 10^{-4}$ S on pressure at different positions in the plasma. The positions are measured from the BFE along the line marked A/B in figure 2. These results confirm the expected trends. the vibrational temperature decreases with pressure in the driver (12 cm), whereas in the extraction region (1 cm) the vibrational temperature increases with pressure. At 4 cm and 7 cm in figure 10 mixing occurs at low pressures indicating that vibrationally excited molecules diffuse from the driver region into the extraction region. This mixing

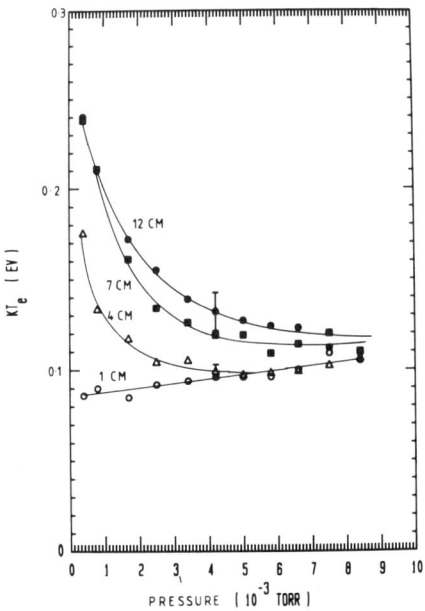

FIG.10 : The electron temperature at $\tau = 4 \times 10^{-4}$ S (which reflects the vibrational "temperature" of the H_2 molecules) as a function of pressure at different positions in the source.

vanishes in the case of the 4 cm position at 4 mTorr. This requires a diffusion time of 4×10^{-4} S over 6 cm at 4 mTorr, which is close to the expected value[18].

In conclusions the measurement of the eedf using Langmuir probes is a flexible and useful technique leading to a better understanding of the basic operation of these plasmas. The extention of the eedf measurement into pulsed discharges offers new and intriguing possibilities, one of which is the measurement of a characteristic energy or temperature of the vibrationally excited molecules in this type of source.

REFERENCES

1. K. N. Leung, K. W. Ehlers and M. Bacal, Rev. Sci. Instrum., 54, 56 (1983).
2. M. Bacal, F. Hillion and M. Nachman, Rev. Sci. Instrum., 56, 649 (1985).
3. A. J. T. Holmes, G. Dammertz and T. S. Green, Rev. Sci. Instrum., 56, 1607 (1985).
4. K. N. Leung, N. Hershkowitz and K. R. Mackenzie, Phys. Fluids, 19, 1045 (1976).
5. J. Bretagne, G. Delouya, C. Gorse, M. Capitelli and M. Bacal, J. Phys. D: Appl. Phys., 18, 811 (1985).
6. J. Bretagne, G. Delouya, M. Capitelli, C. Gorse and M. Bacal, J. Phys. D: Appl. Phys., 19, 1197 (1986).

7. O. Kaneko, Y. Oka, K. Sakurai and T. Kuroda, Rev. Sci. Instrum., $\underline{57}$, 67 (1986).
8. S. Tanaka, M. Akiba, H. Horrike, M. Matsuoka, Y. Ohara and Y. Okumeura, Rev. Sci. Instrum., $\underline{57}$, 145 (1986).
9. C. Gorse, M. Capitelli and A. Ricard, J. Chem. Phys. $\underline{82}$, 1900 (1985).
10. M. J. Druyvesteyn, Z. Phys., $\underline{64}$, 781 (1930).
11. A. P. H. Goode, T. S. Green and B. Singh, J. Appl. Phys., $\underline{51}$, 1896 (1980).
12. M. B. Hopkins and W. G. Graham, Rev. Sci. Instrum., $\underline{57}$, 2210 (1986).
13. M. B. Hopkins, W. G. Graham and T. J. Griffin, Rev. Sci. Instrum., (accepted for publication).
14. J. G. Laframboise, University of Toronto, Institute for Aerospace Studies Report No.100 (1966).
15. H. Amemiya, Jap. J. Appl. Phys., $\underline{25}$, 595 (1986).
16. M. B. Hopkins and W. G. Graham, Proc. Euro. Workshop on the production and application of light negative ions, Ecole Polytechnique, Palaisean, France, March (1986), p.69.
17. J. R. Hiskes and A. M. Karo, J. Appl. Phys., $\underline{56}$, 1927 (1984).
18. M. Lefebvre, M. Pealat, J. P. E. Taran, F. Hillion and M. Bacal, Proc. Euro. Workshop on the production and application of light negative ions, Ecole Polytechnique, Palaisean, France, March (1986), p.107.

This work is supported by the Science and Engineering Research Council, Great Britain.

DISCUSSION

Bretagne: Mike, could you show again the electron distribution function after you switched off the discharge? There is some appearance of super-elastic electronic collisions in the electron distribution function. About 10 eV.
Hopkins: There could be; the resolution of this technique is about 10^7. This is just a small glitch here that could be real or could not. I wouldn't like to say that's real. As a matter of fact, I prefer to say at the moment, it's not, until such time as we can investigate that further. We had hoped to see contributions from excited states, etc. in our eedf work, but we haven't been able to obtain the resolution necessary to do that. So yes, we are thinking of looking into that.
Holmes: Have you looked at the variation of this asymptotic electron temperature as a function of discharge?
Hopkins: Yes. It's a slow increasing function of discharge current.
Holmes: Doesn't that suggest something else, that you could have gas heating, for example, and electrons are coupling to the temperature of the gas?
Hopkins: That's exactly what we're seeing.
Holmes: Okay. That's not the same as vibrational temperature.
Hopkins: We're talking about this discharge. The vibrational temperature is quite high, it's been measured to be quite high. Gorse has shown, in her calculations, that if the vibrational temperature is very high, in those circumstances the electron temperature will go to the vibrational temperature rather than to the translational temperature. But I'm not saying that this electron temperature is the vibrational temperature; it's not. But it's strongly coupled to the vibrational temperature.
York: You showed the plasma distribution of the electron temperature as you cross from the production region into the extraction region. Did you ever look at that as a function of the bias placed on what we commonly call the plasma electrodes to see if it changed the distribution of electrons in the extraction region?
Hopkins: We have done some work on that, but we haven't done a systematic study. For instance, if you bias the beam-forming electrode very negative, then you effectively push the plasma potential up, because the electrons can't get out. And you see a build-up of all the electrons in the tail. We bias, for instance, at -20V, and you could see a large build-up in eedf right up to 20V and then it dropped off. So you could look at wall losses that way quite easily.
York: What happens if you bias it positive?
Hopkins: If you bias it positive, you lose plasma, but I can't give you, offhand, any systematic variation. We could look at it.
Ehlers: I remember, very well, some of the work we did with Los Alamos, and Ralph Stevens can perhaps verify this. During the time the arc was switched off, in that particular case it was a period of perhaps 10 or 20 µs, at which point the fast electrons disappeared. The extracted H$^-$ rose, like 30%, for a very short instant, and then,

of course, tailed on down. So you get 30% more during that time that the H⁻....

Hopkins: If you look at the comparison here between these two plots, I think you'll see that they're very similar. So basically, what you have is a temporal filter. You're removing all the fast electrons temporally rather than by doing it spatially.

Lietzke: The spatial resolution of the electron distribution function showed peaks which you attributed to a trapping by the filaments. Did I see it correctly that there seemed to be two regions of peaks between the back of the source and the filter? And does that mean that you have two sets of filaments in there, some close to the filter?

Hopkins: In that plot, there are actually two filaments. The filaments actually cover quite a large area. This peak would be associated with close to one leg of filament, and this peak would be close to the other leg. Now, these three peaks are not, in fact, peaks; they're extensions into the cusp field. This is a true peak, and it has more to do with how close the plane cuts to the filaments. When you're scanning, if the filament is slightly bent, you'll go closer. These peaks only occur within about 1 cm or less of a filament, so it depends on the geometry of the filament and the geometry of your scan.

Lietzke: I noticed that's very close to the filter region. Is that normal for this source, to have those hot primaries so close?

Hopkins: The filter removes the hot primaries.

Lietzke: I understand, but you're placing very hot electrons close to the filter, which is something that we do not do at Berkeley yet.

Hopkins: Well, I think we've shown that the filter effectively removes all these fast electrons. In theory, a lot has been said about the different types of sources, the tandem source, and the other sources where the filaments are in the cusp field, but really, they're all quite similar in their action. Basically, the magnetic field will prevent fast electrons from crossing it.

Lietzke: I think the significance here, in terms of trying to explain the higher H⁻ that Culham is getting compared to what we've been getting at Berkeley, is that we do not have our filaments close to the filter. And what I see here is that you do. I think that may be very significant.

Hopkins: This source is not a Culham source. It's a copy of a Culham source. How close it is, we don't know.

Lietzke: It's closer to a Culham source than ours is.

Hopkins: We can move the filaments.

Hershcovitch: Has anybody tried to correlate the various excitation lines to the H⁻ output?

Hopkins: We haven't actually measured any extracted current, so...

Hershcovitch: My question was to all the previous four or five speakers. I mean, that's supposed to be the bottom line in this work, is how does this correlate?

Nightingale: Perhaps I can make a comment on that one. We've been looking at the pulsing of the source to produce H⁻ in pulsed applications. It's difficult to know exactly what decay time you get with H⁻ because, to a large extent, with the facility we're using, it's given by the particular power supplies that we've got. However, we can show that if we, in principle, could take the limiting resistor we have in the power supply line to zero, we would get a non-zero decay time. It does look to be a physical limit. It does look to be about 20 µs.

Hopkins: The life time of fast electrons is, say, 1 to 2 µs, or something like that.

Nightingale: It does look like the H$^-$ would decay on its own with roughly the time scale we're seeing for probe traces and H$_\alpha$ which agrees very much with what Mike's seen for thermal distribution.

Holmes: I had one last comment, actually, on this particular drawing. This suggests that the primary electrons actually disappear as you move away from the filaments, your intermediate valley. You are localized around the filament. This is something we see in our larger negative ion sources as well, except....

Hopkins: The primary electrons are localized around the filament. That would depend on the position. The filaments are situated there 6 cm from this wall, 6 cm from that wall, and 6 cm from the filter, but they cover by 4 cm. If you move them towards this wall, I presume those peaks will become very localized. And all the excitation will go on in there. By the way, when this is suspended on the source, this volume is only about 15% of the total source volume, so all the electronic excitation and vibrational excitation through electron excitation occurs in about 15% of the total source volume.

Nightingale: If I understood it correctly, you're saying that as you change the filament bias, you may either be changing the heating of the plasma due to the filaments, not the arc voltage, or you may be increasing the loss area for electrons. Have you tried repulsing that bias potential late on, say 50-100 µs into your decay, pulsing it back up from +2V up to -2V to see whether you get an increase in temperature again, which would fit with the filament heating, or nothing at all, which would agree with the fact that it's just extra loss area?

Hopkins: We haven't tried that, but the loss area doesn't explain the heating. I mean, we actually saw electron temperature go up. I don't think you can explain that in terms of loss area. That's obviously due to the filament electrons heating the background eedf. As for whether the loss area cools the electron temperature, I think most people find that if you bias the beam-forming electrode, you'll cool the electrons.

SPECTROSCOPIC INVESTIGATION OF H⁻ AND D⁻ ION SOURCE PLASMAS*

H. Vernon Smith, Jr., Paul Allison, and
Roderich Keller†, AT-2, MS H818
Los Alamos National Laboratory, Los Alamos, NM 87545

ABSTRACT

Several H I (Balmer), Cs I, Cs II, and Mo I lines emitted by the small-angle source and 4X source plasmas are studied. After correcting for Stark broadening, the H_α line width gives the H-atom temperature kT_{HO}. After correcting for Doppler broadening, the H_β and H_δ line widths give the electron density n_e. For pulsed operation of both sources, kT_{HO} is 1.5 to 2 eV and n_e is 1 to 2 x $10^{14}/cm^3$, with kT_{HO} and n_e scaling approximately with the square root of the discharge current. For the 4X source operated on D_2, kT_{DO} and n_e are near the values of kT_{HO} and n_e obtained for H_2 operation. Assuming that the H⁻/D⁻ ion temperature equals the H/D-atom temperature, we deduce a lower limit to the H⁻/D⁻ beam emittance.

INTRODUCTION

The Penning surface-plasma source (SPS) provides a bright H⁻ ion beam for accelerator applications.[1] Knowledge of the plasma parameters of this source may prove quite valuable in developing a theoretical model of it and in understanding its performance limits. In our initial study[2] of the plasma parameters of the Penning SPS discharge using quantitative optical spectroscopy, we looked only at the 4X source[3,4] discharge. The approximate plasma density, H-atom temperature, and electron temperature for only one set of discharge parameters for each of the three, pulsed-source plasma modes were determined. In this work we look in more detail at two of the 4X source discharge modes by varying the discharge parameters over a range of conditions, plus we study the small-angle source[5] (SAS) over a wide range of its operating parameters. We also take a brief look at the operation of the 4X source on deuterium. A preliminary report of the present work is given elsewhere.[6]

EXPERIMENTAL APPARATUS AND METHOD

Because the experimental method is discussed in Ref. 2, only a brief summary, including changes in the procedure, will be given here. A schematic of the experimental apparatus is shown in Fig. 1. The discharge light emitted from the source passes through a quartz vacuum window and is imaged with a lens onto the monochromator entrance slit. The monochromator slits are set for 0.01-nm full

---

*Work performed under the auspices of the U.S. Dept. of Energy and supported by the U.S. Army Strategic Defense Command.
†Permanent address: GSI, Darmstadt, West Germany.

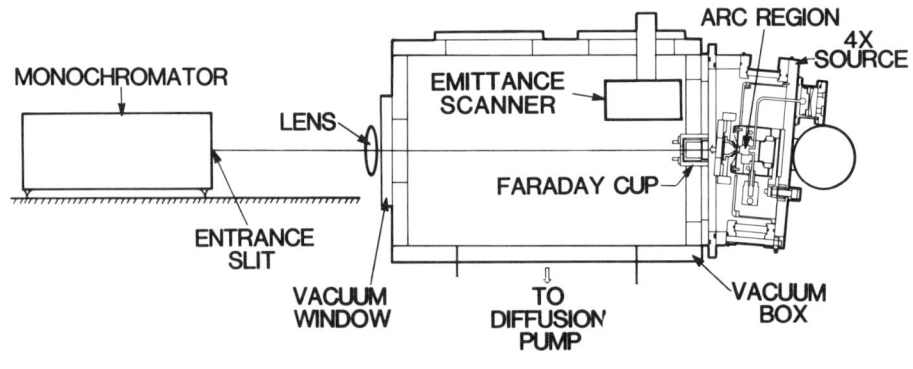
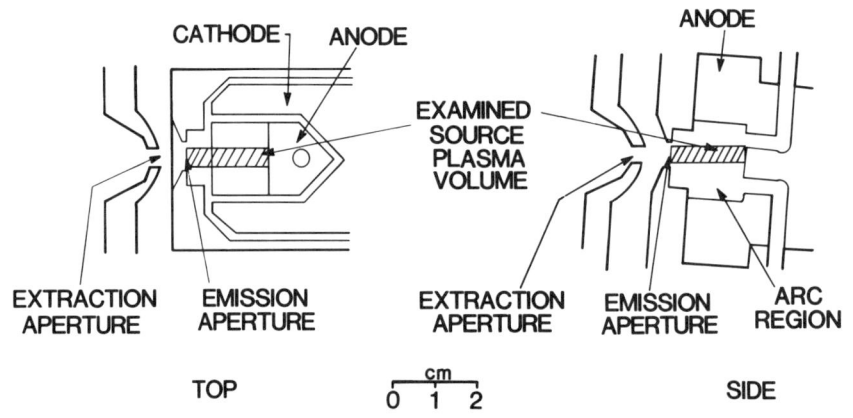

Fig. 1. Upper: experimental arrangement of the 4X source and the 1-m monochromator on the ion source test stand (not to scale). The distance from the emission aperture to the lens is 67 cm; to the monochromator, 107 cm.

Lower: horizontal (top) and vertical (side) sections of the 4X source plasma volume. Only a small portion (hatched area) of the arc region is examined with the monochromator.

half-width instrument broadening. The photomultiplier tube (PMT) current I_{PMT} is measured at a preselected time during the 1-ms-long discharge pulse by using a gated sample-and-hold circuit. This signal is then fed to the input of a strip-chart recorder. Because the strip-chart recorder and monochromator grating drive are driven at known constant speeds, the horizontal (time) axis is easily converted to wavelength. Sample recorded H_α, H_β, H_γ, and H_δ line shapes are shown in Fig. 2. The time required to record each curve in Fig. 2 was 1 min. The source pulsed-discharge repetition rate is 5 Hz; therefore, each curve is constructed from 300 discharge pulses.

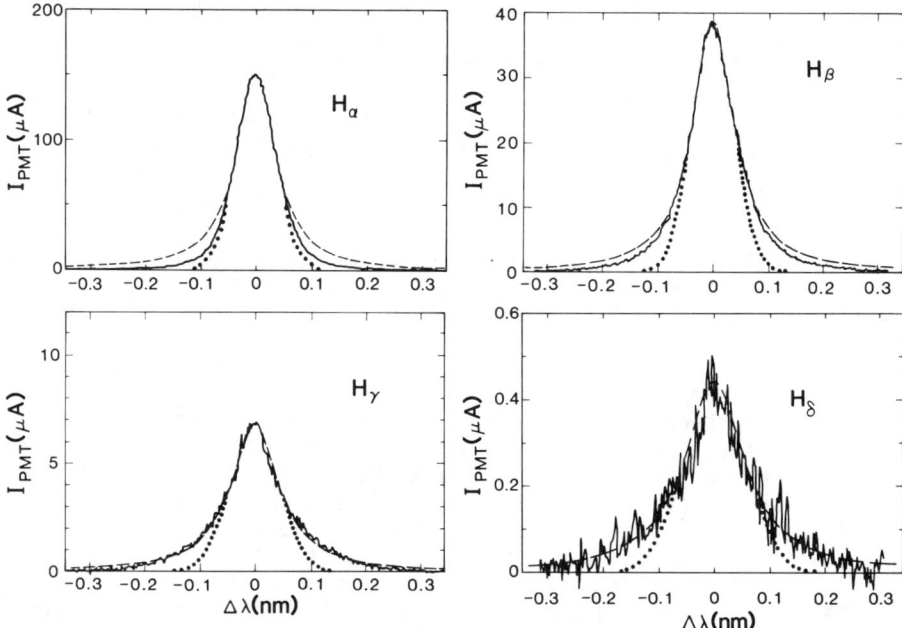

Fig. 2. H_α, H_β, H_γ, and H_δ Balmer line profiles recorded for a 152 A, Mode II discharge in the 4X source. The Lorentzian (dashed) and Gaussian (dotted) curves have the same half-widths as the measured profiles and are included for comparison purposes only.

The 4X source will operate in three distinct pulsed H_2 discharge modes (Table I of Ref. 4). Mode I has discharge voltage $V_D \simeq 300$ V, with a 1500-G magnetic field and 2.2 Tl/s H_2 gas flow. At 29-keV extraction voltage, Mode I produces a maximum H^- beam current of 110 mA, having ±25% beam noise; Mode II, $V_D \simeq 120$ V, 1500 G, 2.9 Tl/s, 120 mA of H^- at 29 keV with ±20% beam noise; and Mode III, $V_D \simeq 500$ V, 700 G, 3.3 Tl/s, 70 mA of H^- at 29 keV with $\leq\pm1\%$ beam noise. We study two of these 4X source modes, Modes II and III, in this work. When the 4X source pulsed discharge is run on D_2 gas, it operates in Mode II. The SAS, with its fixed arc magnetic field, also operates in Mode II.

For each set of discharge conditions studied, we recorded the line profiles of the H_α, H_β, H_γ, and H_δ lines with 0.01-nm instrument resolution and the integral line intensities of several Cs I, Cs II, and Mo I lines (Table I of Ref. 2) with 0.05-nm instrument resolution. The electron temperature kT_e (determined from one ionization state of an element) is calculated from

$$kT_e = (E_j - E_i)/\ln[(A_j g_j \lambda_i I_i K_i)/(A_i g_i \lambda_j I_j K_j)] \qquad (1)$$

where E_i and E_j are the transition energies for two different lines; A the transition probabilities; g the statistical weights; λ the wavelengths; I the recorded integral intensities; and K the calibration factors. More details of this calculation are given in Ref. 2.

The measured hydrogen line profiles result from a convolution of Stark broadening, related to the electron density (ion-dynamical effects[7] further increase the Stark broadening of H_α); Doppler broadening, caused by the motion of radiating atoms; instrument profile; and the fine-structure splitting of the sublevels. For H_α profiles, a fine-structure correction curve is calculated following the procedure of Ref. 8, considering all seven sublevel positions. The fine-structure splitting is negligible for the other H-Balmer series profiles.

To separate Stark and Doppler broadening, a general tendency can be exploited: Doppler broadening scales proportionally to the wavelength of the observed line[9]; thus, H_α shows the largest Doppler broadening effect of the lines we examine. Stark broadening is larger for H_β and H_δ than for H_α and H_γ (Ref. 9). Therefore, H_α and H_γ should be better suited for atomic temperature determinations; H_β and H_δ, for electron density measurements. However, we find that H_γ has anomalous Lorentzian broadening (see below), so we do not use H_γ in any of our plasma-parameter determinations.

The actual unfolding procedure uses the tabulated Voigt function parameters.[10] Knowing the total width and the width of either a Gaussian or a Lorentzian profile, one can immediately look up the width of the other profile. Hydrogen Stark profiles have nearly Lorentzian shapes, whereas Doppler- and instrument-broadening effects produce Gaussian shapes. The instrument profile half-width is subtracted from the Doppler-broadening half-width by assuming they add quadratically. For the two 4X source deuterium measurements, we use the deuterium Doppler broadening, and we use the H_α fine-structure correction for D_α. For the D_α, D_β, and D_δ Stark-broadening corrections, we use the H_α (Ref. 11), H_β (Ref. 9), and H_δ (Ref. 9) values, respectively.

An iterative procedure is used to unfold the H-atom temperature and electron density from each set of H_α and H_δ measurements. Typically, four passes are required for kT_HO and n_e to converge. The kT_HO deduced in this manner is used to correct the H_β line for Doppler broadening, and the resulting H_β Stark broadening is used to obtain another value for n_e. Only the H_δ n_e values are used for the SAS (Fig. 6) because only in a few cases did we record the SAS H_β line shape. The rms deviation of the two values for n_e found from H_β and H_δ is 5%. Thus, we have good confidence in our values for n_e deduced from the hydrogen Balmer line shapes.

We have somewhat lower confidence in our kT_HO values because of the uncertainty in the theoretical knowledge of the Stark broadening of H_α. It is well known[7] that treating the ion motion in the static approximation leads to underestimating the Stark broadening of H_α for $n_e \sim 10^{14}/cm^3$ by about a factor of 3. To account for the ion motion (the "ion-dynamical effect"), Stehlé and Feautrier[11] use the impact approximation: we use their results in correcting our recorded H_α

line shapes for Stark broadening. However, Kelleher[12] has performed a calculation of ion dynamical broadening for H_α using a computer code that calculates the ion-generated perturbing fields using Monte Carlo techniques. For $kT_{H^+} = kT_e = 1$ eV and $n_e = 1 \times 10^{14}/cm^3$, Kelleher's prediction for the FWHM of H_α is 0.0089 nm, about 54% of Stehlé and Feautrier's 0.016 nm. If Kelleher's prediction is used, our values of kT_{H^0} deduced from the H_α line increase by 30%, the values of n_e deduced from H_β decrease by 15%, and the values of n_e deduced from H_δ decrease by 5%. Thus, the theoretical uncertainty in the Stark broadening of H_α causes ~30% uncertainty in our values for kT_{H^0} and ~10% uncertainty in our values for n_e. We note that the H-atom temperatures we report here (deduced from H_α) are for the n = 3 excited state. We assume that the H-atom temperatures for the ground state and n = 3, 4, and 6 excited states are identical. We do not use the recorded H_γ line shapes for any determination of the source plasma parameters because every H_γ line shape we record is almost entirely Lorentzian in shape (Fig. 2), despite its expected Doppler (Gaussian) shape. From Griem's tabulation of Stark widths,[9] H_γ is expected to have a very small Lorentzian half-width (about 0.01 nm). However, we observe H_γ has approximately 0.1-nm Lorentzian width. Evidently ion-dynamical effects are very important for H_γ. An investigation of this possibility is under way.[12] In our analysis, we assume local thermal equilibrium (LTE) and Maxwellian particle velocity distributions. These assumptions may not be valid.

RESULTS AND DISCUSSION

The deduced H-atom temperature vs discharge current for the 4X source is shown in Fig. 3; for the SAS, in Fig. 4. The electron density vs discharge current for the 4X source and SAS is shown in Figs. 5 and 6, respectively; the electron temperature vs discharge current, in Figs. 7 and 8. One of the most striking features of the results is the wide variation of the electron temperatures calculated from the Mo I, Cs I, Cs II, and H I integral line intensities. For the 4X source (Fig. 7), there is little variation of kT_e with I_d, and the average values for kT_e descend from 0.7 eV, for the Mo I results; to 0.6 eV, for Cs II; to 0.5 eV, for Cs I; to 0.2 eV, for H I. This general trend seems to hold true for the SAS as well (Fig. 8). This variation of kT_e with emitting species (hydrogen and cesium are purposely added to the source discharge, whereas the molybdenum comes from the sputtering of the molybdenum cathode material) probably indicates that the electron temperature distribution is not a simple Maxwellian.

The SAS has one primary mode of operation characterized by the discharge voltage being about 100 V--this corresponds most closely to Mode II for the 4X Source. The electron density vs I_d measurements for the SAS (Fig. 6) are all for this 100-V discharge mode. The SAS electron density varies approximately as the square-root of I_d. The same data for the 4X source, plotted in Fig. 5, are primarily for what we have termed Modes II and III. For completeness, on Fig. 5 we have included the four points we reported in Ref. 2. There appears to be a real difference between the plasma state of

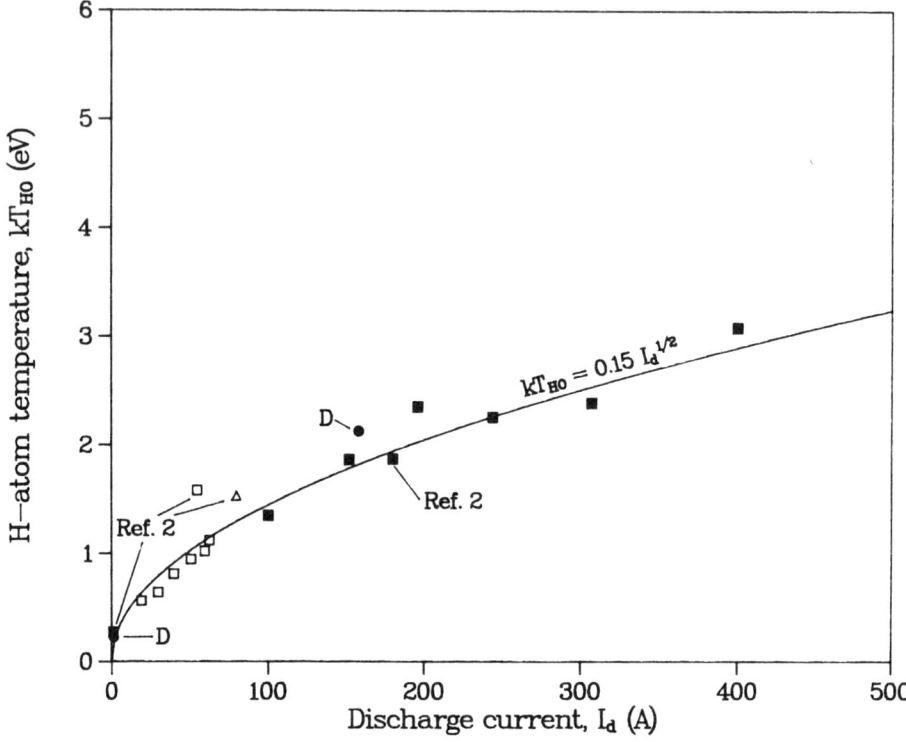

Fig. 3. H-atom temperature vs discharge current for the 4X source. The triangle is for Mode I, filled squares are for Mode II, and open squares are for Mode III. The filled circles are for a Mode II deuterium discharge. Four points from Ref. 2, corrected as discussed in the text, are included in this figure. The curve labeled $kT_{H0} = 0.15\, I_d^{1/2}$ is included to guide the eye.

Modes II and III. Mode III appears to have a nearly linear dependence of n_e on I_d, as one would naively expect. For Mode II, n_e appears to have about a $I_d^{1/2}$ dependence, which is not understood at present. We observe ±40% fluctuations on the discharge voltage and ±20% fluctuations on the H⁻ current from Mode II, but <±1% fluctuations of these parameters for Mode III. This leads us to speculate that plasma turbulence effects may exist in Mode II but not in Mode III (Ref. 2). These hypothetical plasma-turbulence effects may also cause anomalous plasma losses, hence the nonlinear dependence of n_e on I_d for Mode II. The $I_d^{1/2}$ dependence of n_e for the SAS suggests that plasma-turbulence effects may be important in that source also. Comparing the kT_e and kT_{H0} vs I_d curves for the SAS (Figs. 8 and 4, respectively), at the lowest I_d, $kT_e \gtrsim kT_{H0}$, as it is for most plasmas. However, for discharge currents ≥ 40 A, $kT_{H0} > kT_e$, difficult to understand unless our determination of kT_e

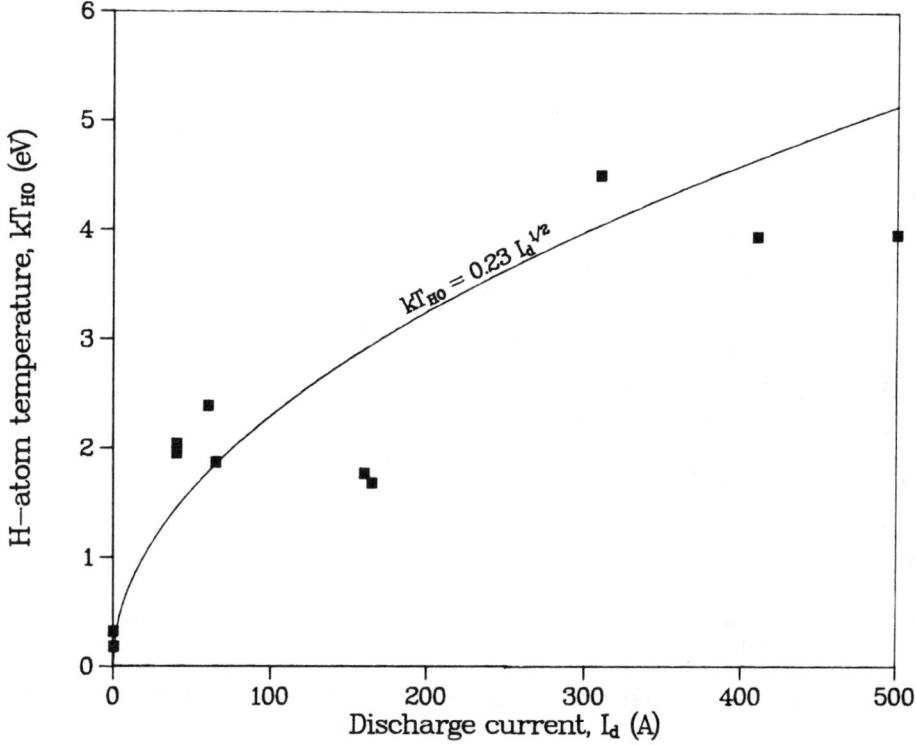

Fig. 4. H-atom temperature vs discharge current for the SAS. The curve labeled $kT_{HO} = 0.23\ I_d^{1/2}$ is included to guide the eye.

is inaccurate or there is anomalous heating of the atoms such as turbulence heating.

Despite the apparent difference in plasma state between Modes II and III, the 4X source H-atom temperature appears to depend as $I_d^{1/2}$ on the discharge current (Fig. 3), with the points from Modes II and III lying near the same curve. Use of the Stehlé correction[11] to the Stark broadening of H_α leads to the pulsed results in Ref. 2 being lowered an average of 28% for kT_{HO} and raised an average of 17% for n_e. The corrected kT_{HO} and n_e values are plotted on Figs. 3 and 5. The SAS kT_{HO} vs I_d curve (Fig. 4) also appears to follow approximately an $I_d^{1/2}$ dependence, but the scatter in the data is larger than that for the 4X source.

On Figs. 3 and 5 we have also included the results of two measurements on deuterium discharges in the 4X source. One point is for a low arc current, cw discharge (I_d = 1.3 A); the other for a high arc current, pulsed discharge (I_d = 158 A). The D-atom temperatures deduced from these two measurements are close to the H-atom temperature curve shown in Fig. 3. The maximum D⁻ current we have

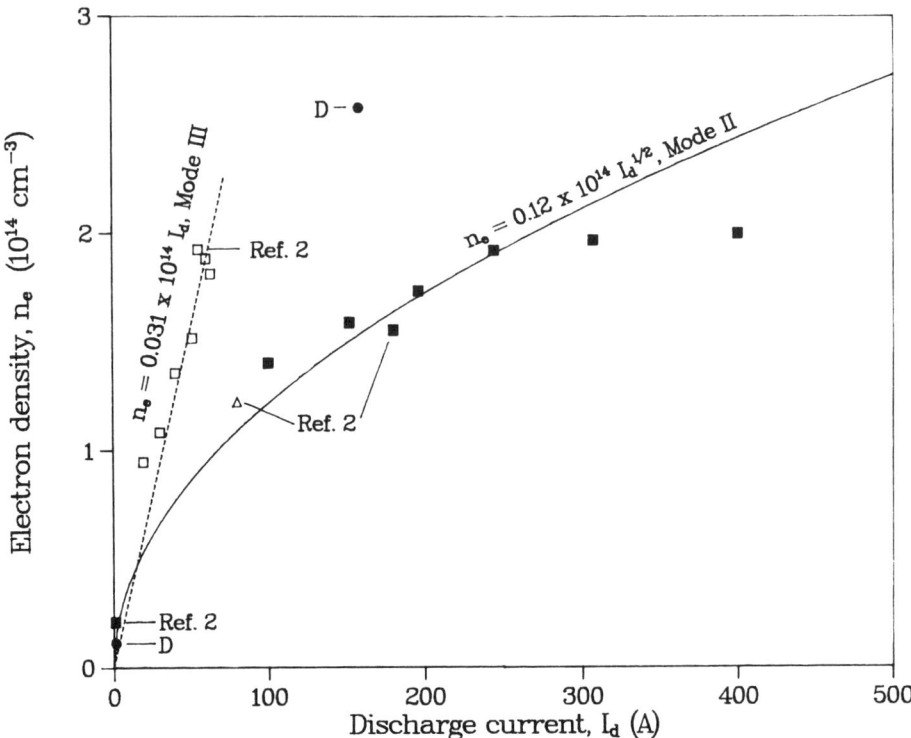

Fig. 5. Electron density vs discharge current for the 4X source. The legend for the points is given in the caption for Fig. 3. Four points from Ref. 2, corrected as discussed in the text, are included in this figure. The dashed straight line through the Mode III points, as well as the curve labeled $n_e = 0.12 \times 10^{14} I_d^{1/2}$, is included to guide the eye.

obtained from the 4X source at 29-keV extraction voltage is 55 mA for arc conditions that are not too different from those for which the 158-A deuterium spectroscopy measurements were made.

For pulsed operation of both the SAS and the 4X sources at their normal discharge currents of 150 A, the H-atom temperature is about 2 eV, considerably larger that the ~0.3 eV reported for the conventional duopigatron[13] and the bucket source.[13,14] Perhaps the relatively high H-atom temperature for the SAS and 4X sources is due to turbulence heating.

If the H$^-$ ion temperature equilibrates with the H-atom temperature,[2] then we can compare the H$^-$ ion temperature deduced from the spectroscopy measurements with that deduced from the SAS and 4X source[4] emittance measurements. To get the effective H$^-$ temperature from the two-dimensional, rms emittance ε_{rms} we use[15]

$$\varepsilon_{rms} = R(kT_{H^-}/Mc^2)^{1/2}/2 \qquad (2)$$

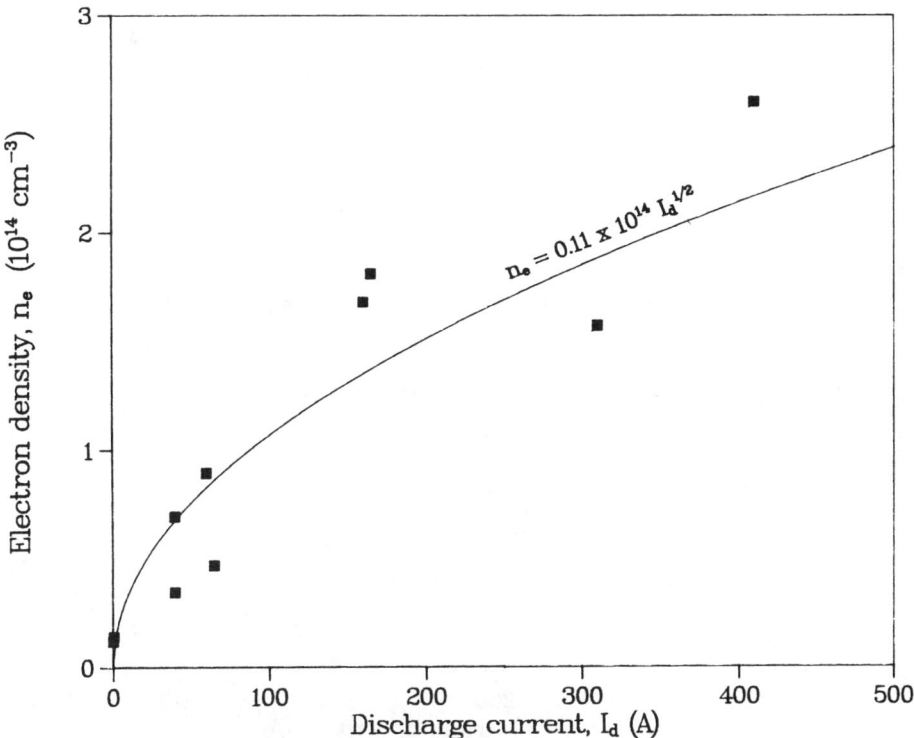

Fig. 6. Electron density vs discharge current for the SAS. The curve $n_e = 0.11 \times 10^{14} I_d^{1/2}$ is included to guide the eye.

where R is the emission aperture radius and M is the negative ion mass. The H^-/D^- current, current density, and ion temperature values are given in Table I. We note that the kT_{H^-} values deduced from the spectroscopy measurements are considerably below the values deduced from the emittance (phase-space) measurements. This difference can arise from either the H^- ion temperature being much larger than the H-atom temperature (nonequilibrium, despite our estimate) or from effects other than temperature contributing to ε_{rms}. These effects can include aberrations in the extraction system, perveance fluctuations in the extraction gap, dispersion of the H^- ions in the source magnetic field, and nonlinear space-charge compensation effects in the H^- beam transport. We think that the first two effects are likely to be much more important than the latter two.[2,4] Using the spectroscopy value for kT in Eq. (2), we determine a lower limit to ε_{rms} for the 4X source and the SAS. These limits are 0.0044 and 0.0093 π•cm•mrad respectively, considerably below our lowest measured ε_{rms} for these sources, 0.011 and 0.018 π•cm•mrad.

Finally, although we have not made emittance measurements of the D^- beam from the 4X source, from the spectroscopy measurements we deduce that kT_H0 and kT_D0 are approximately equal. If $kT_H0 = kT_{H^-}$ and $kT_D0 = kT_{D^-}$, then we would expect ε_{rms} for D^- to be ~0.7 ε_{rms} for H^-.

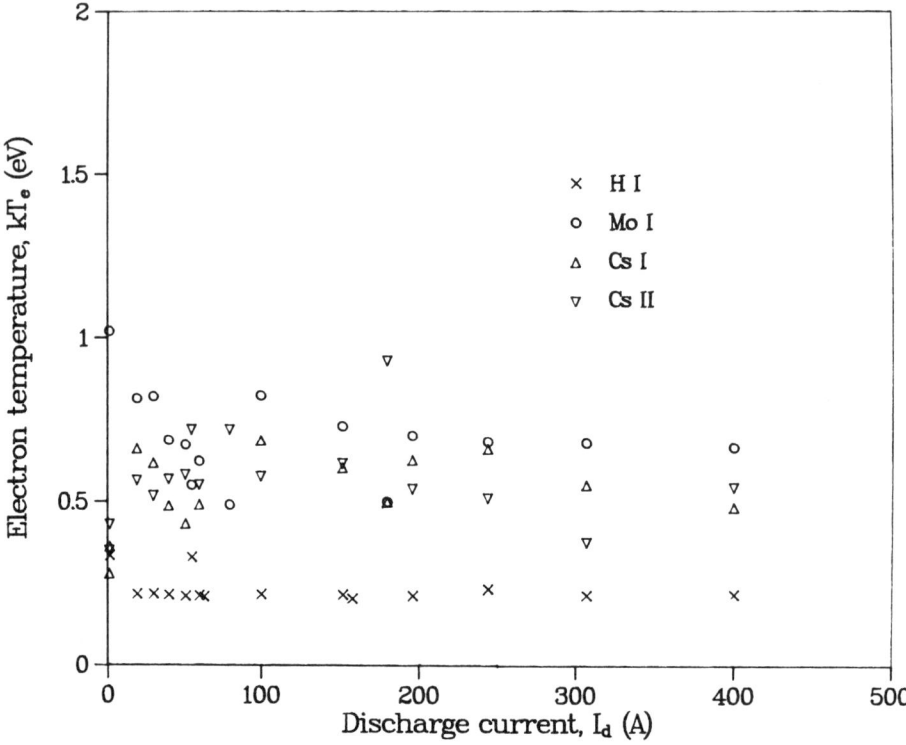

Fig. 7. Electron temperature vs discharge current for the 4X source. The legend for the points is given on the figure.

TABLE I Current, current density, and ion temperature values

Ion	Source (Discharge Mode)	I_- (mA)	j_- (mA/cm^2)	kT_H- Spectroscopy Value[a] (eV)	kT_H- Phase-Space Value[b] (eV)
H^-	4X (I)	110	480	1.5	15
H^-	4X (II)	100	437	1.9	25
D^-	4X (II)	50	218	2.1	--
H^-	4X (III)	67	293	1.0	6.2
H^-	SAS (II)	140	2800	2.0	6.6

[a] Assumes kT_H^0 equals kT_H^-.

[b] Calculated from $kT = 4\, \varepsilon_{rms}^2 Mc^2/R^2$ for 4X source and $kT = 3\, \varepsilon_{rms}^2 Mc^2/a^2$ for the SAS (Ref. 15).

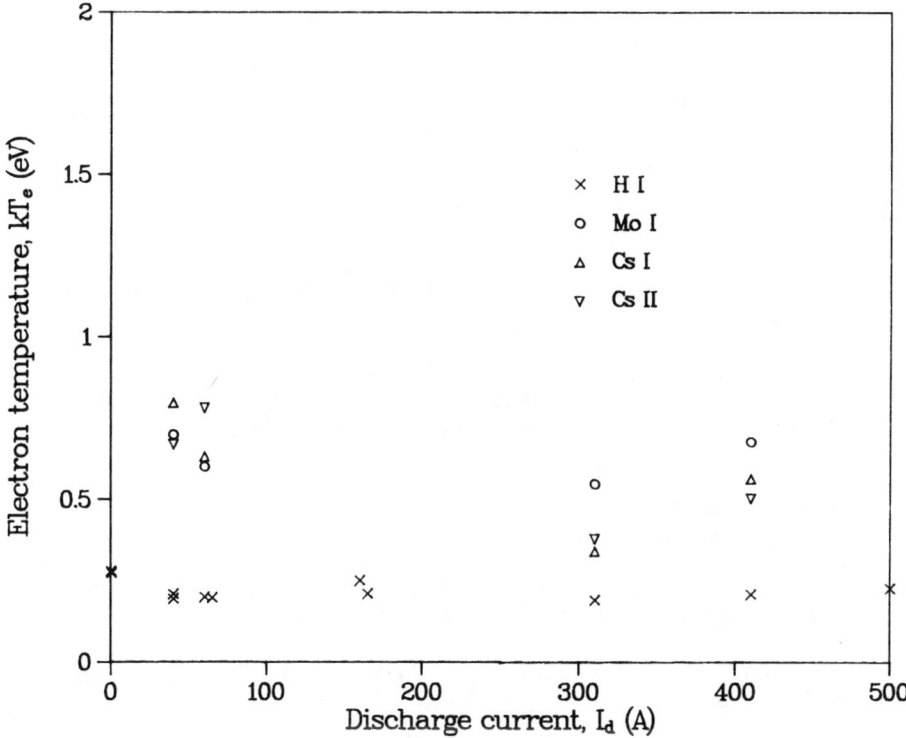

Fig. 8. Electron temperature vs discharge current for the SAS. The legend for the points is given on the figure.

CONCLUSIONS

For pulsed operation of the 4X and small-angle sources, kT_{H0} is 1.5 to 2 eV and n_e is 1 to 2 x 10^{14} cm^{-3}. If kT_{H0} gives a true indication of the H$^-$ ion temperature in the source plasma, then a factor of 1.8 to 3.6 reduction in the H$^-$ beam emittance may be possible.

REFERENCES

1. G.E. Derevyankin and V.G. Dudnikov, "Production of High Brightness H$^-$ Beams in Surface-Plasma Sources," AIP Conf. Proc. No. 111, AIP, New York, 376 (1984).

2. R. Keller and H.V. Smith, Jr., IEEE Trans. Nucl. Sci. 32, (5) 1736 (1985).

3. H.V. Smith, Jr., P. Allison, and J.D. Sherman, "A Scaled, Circular-Emitter Penning SPS for Intense H$^-$ Beams," AIP Conf. Proc. No. 111, AIP, New York, 458 (1984).

4. H.V. Smith, Jr., P. Allison, and J.D. Sherman, IEEE Trans. Nucl. Sci. 32, (5) 1797 (1985).

5. P. Allison and J.D. Sherman, "Operating Experience With a 100-keV, 100-mA H⁻ Injector," AIP Conf. Proc. No. 111, AIP, New York, 511 (1984).

6. R. Keller, H.V. Smith, Jr., and P. Allison, Bull. Am. Phys. Soc. 30, 1608 (1985).

7. J. Seidel, in Spectral Line Shapes, edited by B. Wende (de Gruyter, New York, 1981), p. 3.

8. M.W. Grossman, "H⁻ Ion Source Diagnostics," Proc. of Symp. on the Production and Neutralization of Negative Hydrogen Ions and Beams, Brookhaven National Laboratory report BNL-50727, 105 (1977).

9. H.R. Griem, Spectral Line Broadening by Plasmas, (Academic Press, New York, 1974)

10. G. Traving, in Plasma Diagnostics, ed. W. Lochte-Holtgreven, (North Holland, Amsterdam, 1968), Chap. 2.

11. C. Stehlé and N. Feautrier, J. Phys. B: At. Mol. Phys. 17, 1477 (1984).

12. D.E. Kelleher, National Bureau of Standards, Washington, D.C., 20234. Private communication.

13. D.H. McNeill and J. Kim, Phys. Rev. A 25, 2152 (1982).

14. M. Pealat, J-P.E. Taran, M. Bacal, and F. Hillion, J. Chem. Phys. 82, 4943 (1985).

15. P. Allison, J.D. Sherman, and H.V. Smith, Jr., "Comparison of Measured Emittance of an H⁻ Ion Beam with a Simple Theory," Los Alamos National Laboratory report LA-8808-MS (June 1981).

DISCUSSION

Peterson: Could you go over again what you think the causes are for the much larger phase space apparent in the diagnosed temperature?
Smith: Okay. We think that it could be ion optic aberrations; it could be perveance variations in our extraction gap; it could be fluctuations in the plasma density, either spatially or else with time, and this would cause a rotation of our phase space diagram which would have the effect of having an increased emittance.
Nightingale: That analysis in which you looked at the atomic temperature and your negative ion temperature assumes that, in both cases, they're in equilibrium.

Smith: Yes, that's right.

Nightingale: Why is there any reason to believe they will be? Because surely, in the atomic case, H_α and H_β decay so quickly that the temperature you get is connected with the formation process for the excitation. What you see is actually given by the formation process; how fast the electrons are that produce the excitation or, if it's mutual neutralization, what energy that gives to the H atom. Not the atomic temperature itself. It's not in equilibrium with the background gas or the electrons or, even more so, the H^-. Is not the H^- temperature in your source given by the mechanism for H^- production, rather than equilibrium value with background gas or atomic species?

Smith: Well, one of the things that has been speculated on is that in our source there is a large resonant charge exchange that goes on between surface-produced H^- ions from the cathodes, and the H atoms that are in the plasma. So perhaps there would be an opportunity for equilibrium to exist between the atoms and the negative ions.

Nightingale: What sort of atomic density do you then think applies to your discharge?

Smith: We have not measured that. We can only speculate that it would be on the order of $10^{15}/cm^3$.

Allison: Well, 10^{14} would be enough.

Nightingale: Atomic?

Smith: That's only a speculation. We have not made a measurement. We would like very much to be able to do that.

Nightingale: That's 2 or 3 orders of magnitude higher than volume sources we talked about earlier. If you've got that much enhancement, you may be running into a problem that certainly H_α, and possibly H_β and H_γ, may be optically thick.

Smith: We have made an estimate of that. We have never seen any indication that we are optically thick for H_α or H_β. We think we're about an order of magnitude below that point.

Peterson: A related question in that instance: Is it at all possible that the phase space increased temperature is due to charge transfer that occurred in the extraction region?

Allison: If there were charge exchange in the extraction region, then you'd have a tremendous energy variation in the beam. And that would show up in dispersion of the beam and the bending magnet. We see no such thing.

Lietzke: If the enhanced emittance is due to rapid rotations of the emittance diagram, due to fluctuations, you should be able to see a difference in emittance between the noisy mode and the quiet mode. Do you see it?

Smith: Yes, indeed we do. The emittance for the noisy mode is roughly double that of the quiescent mode.

DISCHARGE CHARACTERISTICS OF A PLASMA GENERATOR FOR SITEX AND VITEX ION SOURCES*

C. C. Tsai, W. K. Dagenhart, W. L. Stirling, G. C. Barber, H. H. Haselton,
P. M. Ryan, D. E. Schechter, J. H. Whealton
Oak Ridge National Laboratory, Oak Ridge, TN 37831

J. J. Donaghy
Washington and Lee University,
Lexington, VA 24450

ABSTRACT

Surface Ionization with Transverse Extraction (SITEX) and Volume Ionization with Transverse Extraction (VITEX) ion sources are being developed to produce intense beams of light negative ions for neutral particle beam applications. The salient feature of these ion sources is their ability to form intense negative-ion beams. With the objective of improving the performance of these sources, an experimental study of their plasma properties has been conducted. The effects of various electrodes in the plasma generator were investigated. Low electron and ion temperatures (below 1 eV) and positive plasma potential up to +6 V have been measured. The measured distributions of plasma density and potential reveal the existence of multichamber characteristics in the source plasma. The significant discharge characteristics and the plasma properties associated with the performance of SITEX and VITEX ion sources are discussed.

---

*Research sponsored by the Department of Defense under Interagency Agreement DOE No. 40-1442-84 and Army No. W31RPD-53-A180 with the Office of Fusion Energy, U.S. Department of Energy, under Contract No. DE-AC05-84OR21400 with Martin Marietta Energy Systems, Inc.

INTRODUCTION

The two principal methods for producing H^-/D^- ions in ion sources[1-11] are surface conversion and volume generation. These negative-ion production mechanisms have been studied extensively, and recent reviews are available.[12-18] In a surface conversion source (such as the SITEX source), positive ions of the source plasma impinge on cesiated metal surfaces, where they backscatter and are converted into negative ion, or release negative ions from the converter surface. In a volume generation source (such as the VITEX source), negative ions are generated essentially by a dissociative attachment process in which vibrationally excited molecules collide with and attach to low-energy plasma electrons and are converted into negative ions via spontaneous dissociation. Normally the vibrationally excited molecules in an intense arc discharge are produced from molecules excited by energetic electron collisions and/or from molecular ions neutralized and excited by colliding with the arc chamber walls. In a plasma generator having molecules such as CH_4 and H_2, the production of H^- ions could also be effected by the process of polar dissociation.

EXPERIMENTAL ARRANGEMENT

Figure 1 shows the plasma generator, electrical connections, and setup for the plasma property study. The plasma generator for the SITEX and VITEX sources consists of a molybdenum-coated graphite arc chamber, one or two tantalum filaments, one converter, and one plasma grid. The applied uniform magnetic field (B_y) is parallel with the arc column between the filaments and is variable up to 6000 G. The arc collimation slots in this experimental setup are rectangular, with a 2.5- by 20-mm cross section and 20-mm length. The working gas, hydrogen or deuterium, is admitted into both collimation slots adjacent to the two filaments and also into the region behind the converter. Normally the plasma is studied during a discharge with constant applied magnetic field, continuous gas feed, and continuous filament heating. The

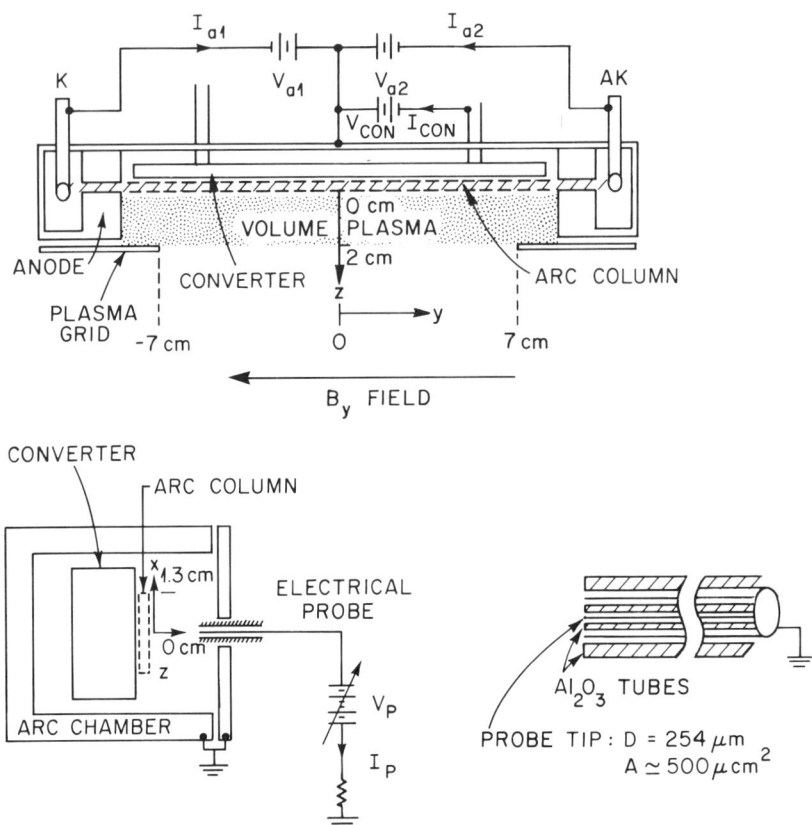

Fig. 1. Arrangement used to study plasma properties of SITEX and VITEX sources.

discharge is initiated by applying a pulsed negative potential to the filaments with the arc chamber at ground potential. An intense electron beam is formed by the primary electrons emitted from the hot filaments and accelerated through the collimation slots (which act as anodes). This beam ionizes the hydrogen gas to create an intense plasma in the arc column. The efficiency of electron ionization is improved by the reflection of energetic electrons at both ends of the arc column. Such a discharge is similar to the so-called Penning, PIG, and reflex discharges. However, a negatively biased converter works as a large ion collector and can be used to control source plasma density. In addition, the secondary electrons emitted from the converter surface can enhance the discharge. Thus the converter voltage can be used to control the plasma density and plasma potential.

The conditions for each arc discharge are usually set by adjusting the arc voltage V_a, which is variable up to 200 V; the arc current I_a, which is variable up to tens of amperes; and the gas feed Q, which is variable up to 200 sccm. With a gas conductance of 100 L/s through the aperture of the plasma grid, the gas pressure in the source plasma is variable up to 20 mTorr. In addition, the filament heating current of about 300 A can be adjusted to control the arc current and hence the intensity of the arc discharge. Based on this experimental study, we observed that the arcs interact with each other and enhance the plasma density.

The plasma properties were measured using an electrical probe that can be scanned both parallel to the magnetic field (left-right) and perpendicular to the converter surface (in-out). The probe tip is 0.25 mm long and 0.25 mm in diameter. In some experimental studies, the probe tip is recessed into a ceramic tube housing with the objectives of minimizing the probe's disturbance of arc discharges and measuring local plasma properties. To facilitate studies of the plasma properties near the arc column, the probe was located by a scanner with a position resolution of 0.05 mm.

EXPERIMENTAL RESULTS

Under normal operations, each arc can be raised above 10 A. However, during this study the arcs were limited to 3 A to prevent the probe tip from melting. Figure 2 reveals that both arcs produce plasmas with similar density and uniformity. When both arcs are turned on simultaneously, however, the plasma density triples and the plasma nonuniformity increases. Depending upon the arc conditions, the plasma density and uniformity can be quite different from those shown in the figure. In fact, a floating hot filament can degrade the arc discharge of the other filament. A speculation to explain this observation is that the electron cloud of the floating hot filament partially shorts it to the

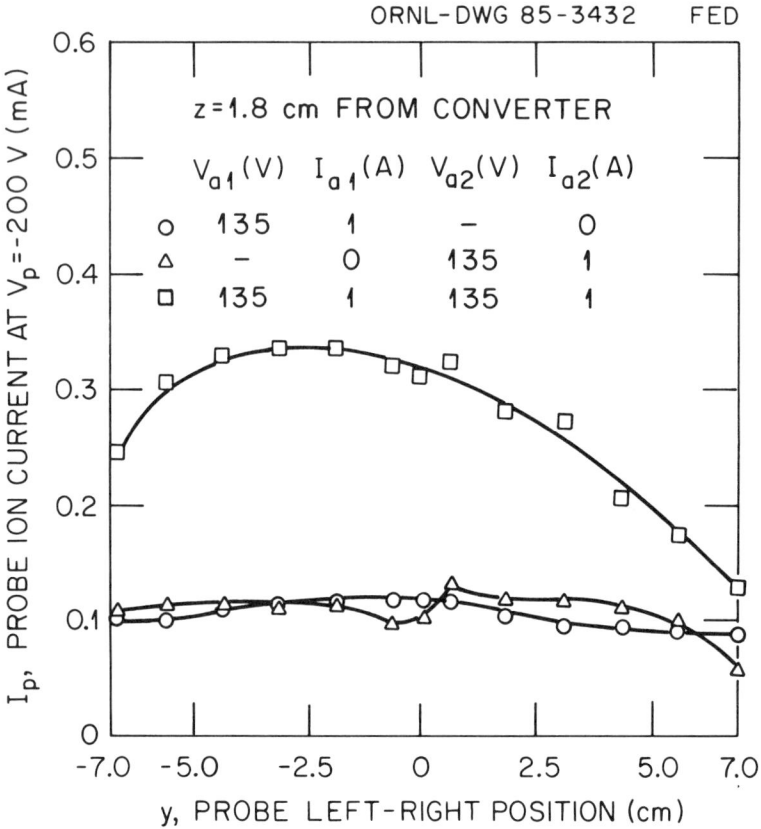

Fig. 2. Effect of arc discharge on plasma density and uniformity.

anode, making it unable to reflect the energetic electrons and weakening the discharge. We also note that the arc intensity can be maximized by optimizing the applied magnetic field and adjusting the gas feed.

Figure 3 shows a sharp decrease of plasma density away from the arc column toward the plasma grid. Here, $z = 0$ is the edge of the arc column. With a negatively biased converter, the measured plasma density decreases as the converter voltage increases. We also note that the plasma potential near the arc column is proportionally decreased. The measured electron temperature is below 1 eV, and the plasma potential is several volts positive in the region near the plasma grid. A

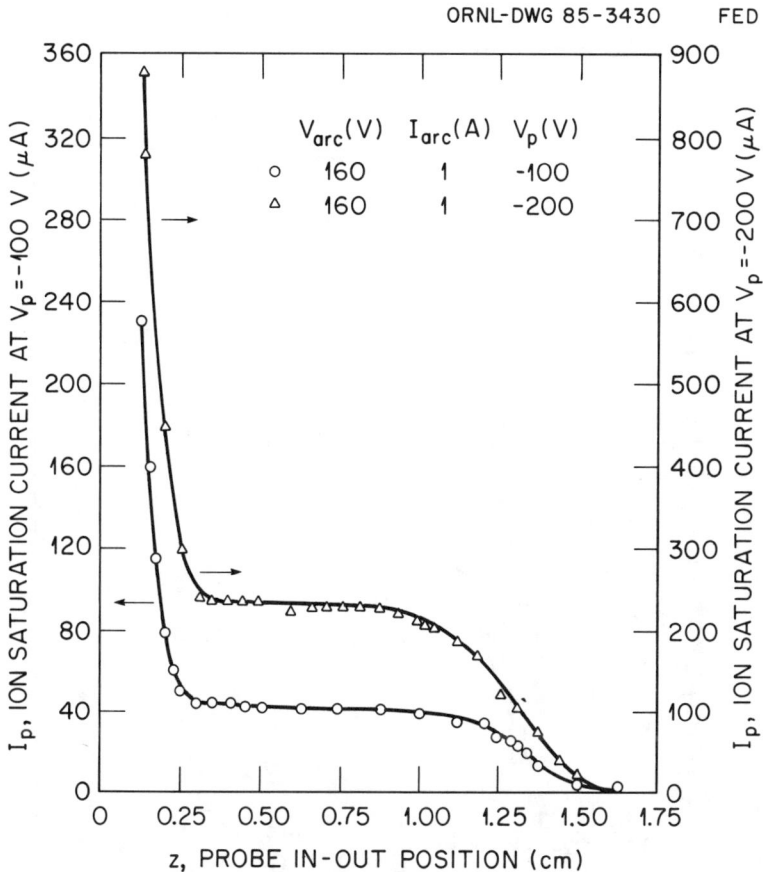

Fig. 3. Sharp variation of plasma density in the region near the arc column.

typical distribution of probe floating potential is shown in Fig. 4. This figure reveals that the floating potential near the arc column is always lower than the converter potential. For this particular experiment, the probe tip was not extended outside the ceramic tube. The ceramic tube could be charged by energetic electrons which impinged on the ceramic tube surfaces near the probe tip. These results suggest the existence of multichamber characteristics in the source plasma, which are useful for enhancing volume production of negative ions, as emphasized by Hiskes.[15]

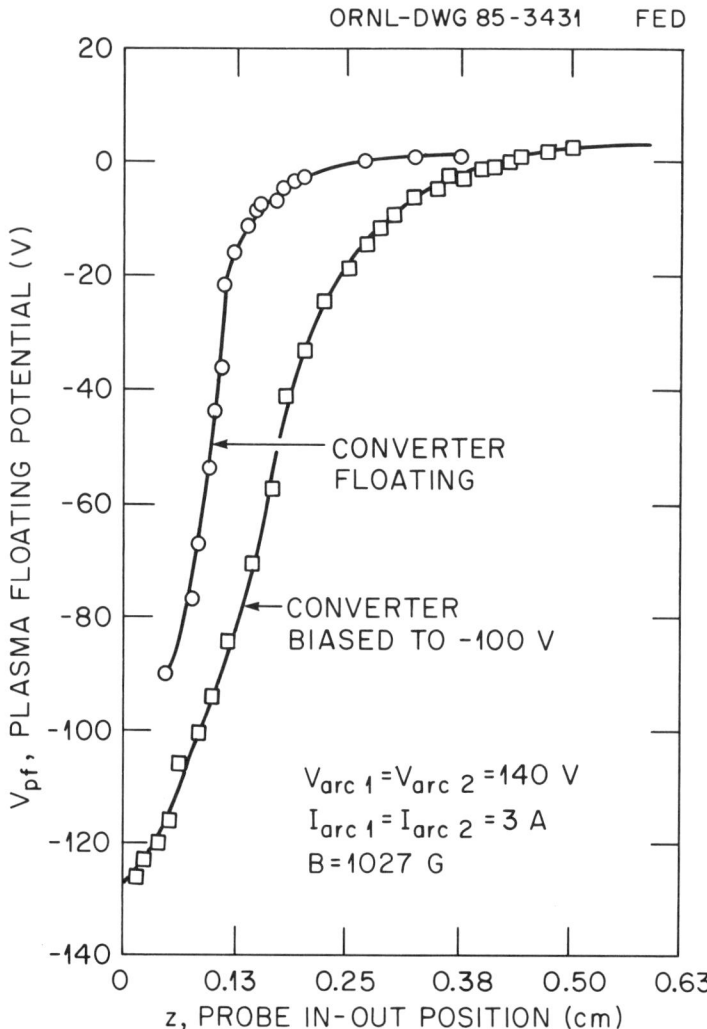

Fig. 4. The plasma floating potential in the region near the arc column.

DISCUSSION

Based on the potential distribution in the source plasma and on the energies of the primary and the plasma electrons, it should be possible to reach a high negative-ion production rate. However, the positive plasma potential up to +6 V with respect to the plasma grid has been measured by both a hot and a cold electrical probe. Such a potential hill will make it difficult for the volume negative ions to be extracted. In order to increase the output of negative ions, we tested ideas for increasing either the transmission efficiency of volume negative ions toward the extraction region or the volume production rate in that region. One idea is to change the biasing potential of the plasma grid. We observed that the plasma floating potential changing by less than 2 V when the plasma grid potential is changed by 10 V. This reveals the shielding effect of the plasma sheath adjacent to the plasma grid. We observed that the plasma density was decreased slightly. However, we also observed that the negative-ion output increases about 10% as the plasma grid is biased negatively. Under such conditions, the volume negative ions should be repelled from the plasma grid and positive ions attracted toward the grid. Thus, the increase of negative-ion output may be due to the increase of the volume production rate in the extraction region. This observation suggests that the ions of the VITEX beams in these experiments are dominated by those created in the extraction region.

Based on Hiskes' hypothesis, the negative-ion production rate can be increased if the concentration of vibrationally excited molecules and low-energy plasma electrons can be increased. For the case with the plasma grid biased negatively, the molecular ions are attracted toward the plasma grid and converted into excited neutrals. Moreover, the electric field so established in the extraction region will increase the density of low-energy electrons. Thus, the observed increase of negative-ion output in that experiment reveals the important role of both the surface excitation and neutralization and the dissociative attachment processes in the VITEX source.

In an ion source, the current density of negative-ion beams will increase with the increase of the production rate and the decrease of the destruction rate. Usually the production rate is a positive function of the density of the excited molecules and the density of low-energy electrons. On the other hand, negative ions are destroyed by the processes of ion-ion neutralization collision and ion-atom associative detachment collision. Hence, in order to increase negative-ion output, the source has to be operated under conditions that will increase the molecular species but decrease the atomic species. In the VITEX ion source, effective ways to achieve such desired operating conditions are to increase anode gas feed and decrease cathode gas feed, to increase the applied magnetic field, to bias the plasma grid negatively with respect to the anode, and to shorten the distance between the arc column and the plasma grid. Negative hydrogen ion beams of output current greater than 20 mA and with a current density above 120 mA/cm^2 have been successfully formed by applying such methods..

Another effort to increase H$^-$ output of the VITEX source involved feeding heavy gases such as argon, helium, and CH$_4$ into the anode region. We observed that the H$^-$ output only increases about 10% for argon and helium. However, with CH$_4$, the H$^-$ output is comparable to that for a hydrogen gas feed. The highest H$^-$ output was achieved when the cathode gas feed was turned off. At this earlier experiment, the source delivered 12 mA/cm^2 H$^-$ beams from either CH$_4$ plasma or H$_2$ plasma. The fact that CH$_4$ could be used as a working gas for H$^-$ production suggests that the process of polar dissociation[18] may contribute to H$^-$ production in the VITEX source.

In a SITEX ion source, the cesiated converter surface is subjected to bombardment of positive ions of hydrogen and cesium. The negative ions are created essentially by surface conversion processes. However, as in a VITEX source, negative ions can also be created by volume generation processes. Comparison of the production rate of negative ions at the cesiated converter surface with that in the extraction region indicates that the negative ions of SITEX beams should be dominated by those originating from the converter surface. This statement is supported by the following experimental results. The negative-ion output increases with the converter bias potential. The maximum output can be quadruple that of a floating converter. Moreover, the temperature of SITEX ions is estimated from parallel emittance measurements to be about 5 eV, instead of about 0.5 eV as for the VITEX beams. Our experimental studies confirmed that SITEX sources are capable of producing long-pulse intense deuterium negative-ion beams with beam current density above 48 mA/cm^2 (Ref. 5).

ACKNOWLEDGMENTS

The authors wish to thank their colleagues for their assistance in keeping the test facility running: S. C. Forrester and F. Sluss for their work on the mechanical hardware of the facility, including ion sources, and B. E. Argo and D. O. Sparks for their work on power supplies and electronics.

REFERENCES

1. W. K. Dagenhart et al., "Short-pulse operation with the SITEX negative ion source," Proc. Third Int. Symp. on Production and Neutralization of Negative Ions and Beams: AIP Conf. Proc. **111**, 353 (1984).
2. W. L. Stirling et al., "Normalized emittance of SITEX negative ion source," Proc. Third Int. Symp. on Production and Neutralization of Negative Ions and Beams: AIP Conf. Proc. **111**, 450 (1984).
3. J. H. Whealton et al., "2-D accelerator deisgn for SITEX negative ion source," Proc. Third Int. Symp. on Production and Neutralization of Negative Ions and Beams: AIP Conf. Proc. **111**, 524 (1984).
4. W. L. Stirling et al., Volume H^- Generation with the SITEX Source in the VITEX Reflex Mode Without Cesium, Report No. ORNL/TM-9753, Oak Ridge, National Laboratory, 1986.
5. W. K. Dagenhart, C. C. Tsai, W. L. Stirling, P. M. Ryan, D. E. Schechter, J. H. Whealton, and J. J. Donaghy, these proceedings.
6. K. Prelec, "A Penning source for negative hydrogen ions," Nucl. Instrum. Methods **144**, 413-421 (1977).
7. K. T. Ehlers, Nucl. Instrum. Methods **32**, 309-316 (1965); K. Jimbo et al., Nucl. Instrum. Methods in Physics Research **A248**, 282-286 (1986).
8. M. Bacal et al., "Production of negative hydrogen ions in low-pressure hydrogen plasmas," pp. 26-34 in Proc. Symp. on Production and Neutralization of Negative Ions and Beams: Report No. BNL-50727, Brookhaven National Laboratory, 1977.
9. K. Leung and K. W. Ehlers, in Proc. Symp. on Production and

Neutralization of Negative Ions and Beams: Report No. BNL-51304, Brookhaven National Laboratory, 1980; K. Leung and K. W. Ehlers, Rev. Sci. Instrum. **53**, 1423 (1982); **52**, 1452 (1981).

10. P. W. Allison, "Experiments with a Dudnikov-type H^- ion source," p. 119 in Proc. Symp. on Production and Neutralization of Negative Ions and Beams: Report No. BNL-50727, Brookhaven National Laboratory, 1977.

11. M. Bacal et al., "Progress in developing a volume D^- ion source," in Proc. 10th Symp. on Fusion Engineering, 1983 (IEEE, New York, 1984).

12. A. Pargellis and M. Seidl, "Formation of H^- ions by backscattering thermal hydrogen atoms for a cesium surface," Phys. Rev. B **25**, 4356-4361 (1982).

13. M. Seidl and A. Pargellis, "Production of negative hydrogen ions by sputtering absorbed hydrogen from a cesiated molybdenum surface," Phys. Rev. B **26**, 1-6 (1982).

14. J. R. Hiskes et al., "Generation of H^-/D^- ions on composite surfaces with application to surface/plasma ion source systems," J. Vac. Sci. Technol. **A2**(2), 670-674 (1984).

15. J. R. Hiskes et al, "Optimum extracted negative-ion current densities from tandem high-density systems," J. Appl. Phys. **58**, 1759-1764 (1985).

16. J. R. Hiskes et al., "Hydrogen vibrational population distributions and negative ion generation in tandem high hydrogen discharges," J. Vac. Sci. Technol. **A3**, 1229-1233 (1985).

17. A. M. Karo and J. R. Hiskes, "Vibrational relaxation in H_2 molecules by wall collisions: applications to negative ion source processes," J. Vac. Sci. Technol. **A3**, 1222-1228 (1985).

18. S. K. Srivastava and O. J. Drient, "Polar dissociation as a source of negative ions," Proc. Third Int. Symp. on Production and Neutralization of Negative Ions and Beams, AIP Conf. Proc. **111**, 56 (1984).

DISCUSSION

Hershcovitch: You said you got 122 mA/cm^2 on VITEX?
Tsai: Yes.
Hershcovitch: What was the total current and the extraction aperture?
Tsai: The extraction aperture is .2 cm^2, and the current is ~ 24 mA.
Bacal: You mentioned the measured energy of the ions was 0.5V. How did you measure that?
Tsai: This is from the emittance measurement. Tomorrow, Kelly Dagenhart will give more details about that.
Lietzke: In VITEX you still show the converter. What role does the converter play in VITEX?
Tsai: We are still in the process of study. But if the channel of surface neutralization excitation is important, then the position of the converter is still very important. At the moment we are not using it, but we intend to adjust the position to study how it changes.
Lietzke: Does it depend on the converter voltage?
Tsai: At the moment we leave it floating, but are you thinking about the plasma grid? From most reports up to now, you usually hear of 1V or 2V bias, where you can get improvements by a factor of 3 or 5, but I never observed that. I only observe about a 20% increase. But it's not 1 or 2V. In fact, it's 6-10V.

H^- SOURCES

A MODEL FOR H⁻ VOLUME PRODUCTION ION SOURCES

T. S. Green, A. J. T. Holmes and M. P. S. Nightingale

(1) INTRODUCTION

Sources of H⁻ ions are being developed at the Culham Laboratory with the objective of establishing the feasibility of constructing megawatt neutral injectors for fusion experiments operating at high particle energies[1,2,3], and as injectors into high energy accelerators. Plasma discharges operating in hydrogen have been shown to provide more copious flux of H⁻ ions than originally expected following the pioneer work of Bacal[4]. This work and subsequent theoretical discussions have led to the recognition that the most probable route to H⁻ production is a two-stage process - excitation of H_2 molecules into vibrational and rotational states followed by dissociative attachment of electrons to these excited molecules. Further it has been shown that the reaction rate for each of these processes is optimised by having different electron distribution functions in each case and that a two-stage, or tandem, discharge can produce separate plasma regions allowing spatially separated electron distributions to be created[5,6].

Fig. 1 shows schematically such a tandem or filter source. In one region electrons emitted from the hot cathode with high energies ionise the gas and produce significant molecular excitation (this region we call the ionisation region). The second region, near the plane through which ions are extracted, is magnetically isolated from the first. Only cold electrons can diffuse into this region. As a result no ionisation or excitation takes place in this region. But the electron temperature may be such that dissociative attachment is probable. This region is called the extraction region.

In this paper we consider a simple model of this type of discharge, which allows us to make predictions of the parametric dependence of H⁻ density on arc conditions in such sources. The model, described in section 2, is an extension of a model developed to describe parametric dependences of positive ion production rates in sources by Green et al[7]. Essentially we consider that the electron distribution is the sum of two components: a primary component sufficiently energetic to ionise and excite, and a thermal component which is not. The effective temperatures of these components are considered to be essentially constant, independent of current and gas density, but their densities are not.

This approach differs from those used by Hiskes & Karo[6] and Gorse et al[8], who calculate in detail the energy distribution functions. It has the advantage of giving some simple insight into the variation of performance with source design. In addition, processes involving atomic hydrogen are neglected here as they do not appear to influence the negative ion density in the extraction region[9].

(2) DISCHARGE MODEL

2.1 Ionisation Region

In the ionisation region we have a situation analogous to that discussed by Green et al[7]. Primary electrons are injected from the cathodes; they are either lost to the anode or degraded by inelastic collisions into thermal, non-ionising, electrons. Their density, n_p, is given by the relation

$$\frac{I_{ARC}}{eV} = n_p N_o S_{IN} + \frac{n_p}{\tau_p} \qquad (1)$$

I_{ARC} = arc current,
V = discharge volume,
e = electron charge,
N_o = gas density,
S_{IN} = reaction rate coefficient for inelastic collisions,
τ_p = lifetime for loss of primary electrons to the anode.

Positive ions are created by ionisation with a rate coefficient S_{ION}, and their density, n_{+1}, in this region is given by

$$\frac{n_{+1}}{\tau_+} = n_p N_o S_{ION} \qquad (2)$$

Thermal electrons are created by ionisation and as a result of energy degradation of the primaries. They are sufficiently hot that they do not produce H⁻ by dissociative attachment significantly, and further give a large destruction rate of any H⁻ which may be formed, by collisional detachment. Consequently we may neglect negative ions in this region. Hence the thermal electron density is given by the charge neutrality equation:

$$n_{+1} = n_p + n_{e1}$$

and, since generally $n_p \ll n_{e1}$,

$$n_{+1} = n_{e1} \qquad (3)$$

We can now take the original analysis one step further. We suppose that we can characterise the population of vibrationally excited molecules by a density N* which will vary rapidly with parameters, whilst the distribution between the different states varies slowly. This is in reasonable agreement with detailed computations of the distribution[6,8]. The value of N* can be derived from a balance equation

$$\frac{N^*}{\tau^*} + N^* n_p S_D = N_o n_p S^* \qquad (4)$$

where τ^* is the time constant for de-excitation of the molecules by wall collisions.
S_D is the reaction rate coefficient for dissociation of an excited molecule by collisions with primaries.

S* is the reaction rate coefficient for vibrational excitation of molecules (this includes all processes between an electron and a molecule which may lead to excitation of the molecule[6].

It follows that

$$N^* = N_0 \frac{S^*}{S_D + \frac{1}{n_p \tau^*}}$$

Substituting for n_p from equation (2) we have

$$N^* = N_0 \frac{S^* n_{+1}}{S_D n_{+1} + \frac{N_0 S_{ION} \tau_+}{\tau^*}} \qquad (5)$$

2.2 Extraction Region

In this region there are no primary electrons[5], only cool thermal electrons of density n_{e2}. H^- ions are produced by dissociative attachment. Following Bacal[4], we take the dominant processes for H^- loss to be diffusion to the wall and mutual recombination with positive ions. The balance equation which determines the negative ion density, n_-, is thus:

$$\frac{n_-}{\tau_-} + n_- n_{+2} S_{ii} = n_{e2} N^* S_{DA} \qquad (6)$$

where τ_- is the life time for H^- for loss to the walls
n_{+2} is the positive ion density in region 2
S_{ii} is the reaction rate coefficient for mutual recombination
S_{DA} is the reaction rate coefficient for dissociative attachment.

From plasma neutrality we have:

$$n_- + n_{e2} = n_{+2} \qquad (7)$$

Finally we introduce a parameter α which relates the positive ion density on the two sides of the filter, which depends on the physics of motion across the filter.

$$\alpha = \frac{n_{+2}}{n_{+1}} \qquad (8)$$

We now have the equation to relate the observables n_- and n_{+2}:

$$\frac{n_{+2}}{n_-} - 1 = \left[\frac{S_{ii}}{S_{DA}} + \frac{1}{\tau_- S_{DA} n_{+2}}\right]\left[\frac{S_D}{S^*}\frac{n_{+2}}{N_0} + \frac{S_{ION}}{S^*}\frac{\tau_+}{\tau^*}\alpha\right] \qquad (9)$$

If we neglect the term arising from diffusion of H^-, which is only significant at very low densities, then we may write equation (9) in the form:

or its equivalent

$$\frac{1}{f} = \frac{n_{+2}}{n_-} = 1 + A + Bn_{+2} \tag{10a}$$

$$\frac{1}{n_-} = B + \frac{1+A}{n_{+2}} \tag{10b}$$

where
$$A = \frac{S_{ii}}{S_{DA}} \frac{S_{ION}}{S^*} \frac{\tau_+}{\tau^*} \alpha \tag{10c}$$

$$B = \frac{S_{ii}}{S_{DA}} \frac{S_D}{S^*} \frac{1}{N_o} \tag{10d}$$

A further form of this equation is to relate n_- to the arc current using equation (1) as well:

$$\frac{1}{n_-} = \frac{(1+A)\left(N_o S_{IN} \tau_p + 1\right) eV}{\alpha \, I_{ARC} \, \tau_+ \, \tau_p \, N_o \, S_{ION}} + B \tag{10e}$$

(3) IMPLICATIONS OF THE MODEL

The two forms of equations (9) and (10) are useful separately. Equation (9) shows that there is a maximum value of the negative fraction 'f' equal to $(1 + A)^{-1}$. Equation (10) shows that n_- saturates with increasing arc current at a value of B^{-1}, whilst currents n_- increases linearly with n_{+2}.

3.1 Limiting Negative Ion Fraction

As shown, the limiting value of f is

$$f = \frac{1}{1+A} = \frac{1}{1 + \frac{S_{ii}}{S_{DA}} \frac{S_{ION}}{S^*} \frac{\tau_+}{\tau^*} \alpha} \tag{11}$$

The term A depends on reaction rate coefficients which in turn depend on the electron energy distribution. It also depends on source design via the factor $\frac{\tau_+}{\tau^*} \alpha$. One expects τ_+/τ^* to be of the order of, or less than, unity, although its exact value depends on how well the ions are contained. The parameter α is less than unity and is typically 0.25 or less. Decreasing α, i.e. using stronger filters, increases f but may result in a lower negative ion density. Optimisation of α and f is part of the source design.

3.2 Saturation Density

The value of B^{-1} is the limiting value of n_-. As above, it equals $\frac{S_{DA}}{S_{ii}} \frac{S^*}{S_D} N_o$. This value depends only on reaction rate coefficients and gas filling pressure but not on source design.

3.3 Efficiency of H⁻ Production below Saturation

Equation (10e) shows that n_-/I_{ARC} equals:

$$\frac{\alpha}{(1+A)} \frac{\tau_+}{(N_0 \frac{\tau_p}{S_{IN}} \frac{N_0 S_{ION}}{\tau_p} + 1) eV}.$$

We see clearly that we have two aspects of this relation: the role of region 1 as an efficient ionisation system, and the role of the filter. As in Green et al[7], we see that high values of τ_p, i.e. good primary electron containment, lead to n_p being inversely proportional to N_0. The consequence in this limiting case is that

$$\left.\frac{n_-}{I_{ARC}}\right|_{max} = \frac{\alpha}{1+A} \frac{\tau_+}{eV} \frac{S_{ION}}{S_{IN}}$$

The ratio $\frac{S_{ION}}{S_{IN}}$ increases with arc voltage[7], so that below saturation the source is a more efficient producer of H⁻ as the arc voltage increases.

The term $\frac{\alpha \tau_+}{1+A}$ can be altered to $\frac{\alpha \tau_+}{1+C\alpha\tau_+}$, since A is proportional to $\alpha\tau_+$. This shows that when we alter source design to obtain maximum n_-/I_{ARC}, by changing the filter which varies α directly and τ_+ as a consequence, we get the maximum when $C\alpha\tau_+$ equals unity. This corresponds to f equal to 0.5/C.

(4) COMPARISON WITH EXPERIMENT

Fig. 2 shows data reported in reference 9 for variation of n_- with the extracted positive ion current at several pressures. The data clearly show saturation. In Fig. 3 we plot $1/n_-$ versus $1/j_+$ as a test of equation (10b).

The data are in good agreement with the prediction in the range of values plotted. (We note that at low currents significant departure is observed, due we believe to the influence of the diffusion loss term for H⁻ at low densities.)

For lower pressure discharges there are discrepancies due to large variations in T_e.

We have measured separately n_-, n_{+2} and n_{e2} at saturation for several pressures. According to equation (6) these are related by the equation:

$$n_- n_{+2} S_{ii} = n_{e2} N^* S_{DA} = n_{e2} N_0 S_{DA} \cdot \frac{S^*}{S_D} \qquad (12)$$

when we neglect diffusive loss of H⁻.

To check this directly we plot the term $\frac{n_- n_{+2}}{n_{e2}}$ versus N_0 for this data, as in Fig. 4. The data agree well with the theoretical relationship.

Further evidence for the validity of this model is provided by measurements reported by Holmes et al$^{(9)}$. These authors have shown that the particle densities in the extraction region can be varied, without changing conditions in the ionisation region, by biasing the extraction electrode (fig. 1) relative to the anode.

Data have been obtained for variation of n_{+2}, n_{e2} and n_- at the saturation limit for constant gas density and variable bias. As in equation 12

$$n_- n_{+2} = n_{e2} \cdot \frac{S_{DA}}{S_{ii}} \times \frac{S^*}{S_D} \times N_o$$

The data are plotted in the form of the product of n_- and n_{+2} versus n_{e2} in Fig. 5. They follow the linear variation predicted above.

(5) CONCLUSIONS

We have presented a model for a two-stage discharge in which H$^-$ ions are produced, as an extension of previous models for positive ion sources. The model explains qualitatively a range of data which show saturation of the density of H$^-$ ions, as arising from saturation in the density of ro-vibrationally excited molecules. This does not preclude the possibility that saturation and even a decrease in H$^-$ density may occur if the electron temperature rises to a level at which the H$^-$ production rate falls or the destruction rate due to electron collisions increases.

An important implication of this model is that the attainable level is independent of source design as it affects particle confinement times. On the other hand, the approach to saturation does depend on source design.

This model is essentially zero-dimensional and does not address the spatial variation in density observed experimentally or the relation between particle density or current density, issues which are under investigation at present.

ACKNOWLEDGEMENT

This work was supported by the United States Air Force Office of Scientific Research under contract no. F49620/86/C/0064.

REFERENCES

(1) Neutral Injection Heating of Toroidal Reactors, Culham Report CLM-R112 (1971).
(2) D. R. Mikkelsen and C. E. Singer. Proc. Non-Inductive Current Drive in Tokamaks Workshop, 1, II, CLM-CD (1983), Culham Laboratory.
(3) W. S. Cooper, O. A. Anderson, D. A. Goldberg and J. Finke. Bull. Am. Phys. Soc. 27, 1142 (1982).
(4) M. Bacal. Physica Scripta, T2/2, 467 (1982).

(5) A. J. T. Holmes. Rev. Sci. Instr., 53, 1523 (1982).
(6) J. R. Hiskes and A. M. Karo. J. Appl. Phys., 56 1927 (1984).
(7) T. S. Green, A. R. Martin, C. Goble and M. Inman. 7th European Conf. on Controlled Fusion & Plasma Physics, 93 (1975).
(8) C. Gorse, M. Capitelli, J. Bretagne and M. Bacal. Chem. Phys., 93, 1 (1985).
(9) A.J. T. Holmes, L.M. Lea, A. Newman and M. P. S. Nightingale. Submitted to Rev. Sci. Instr.

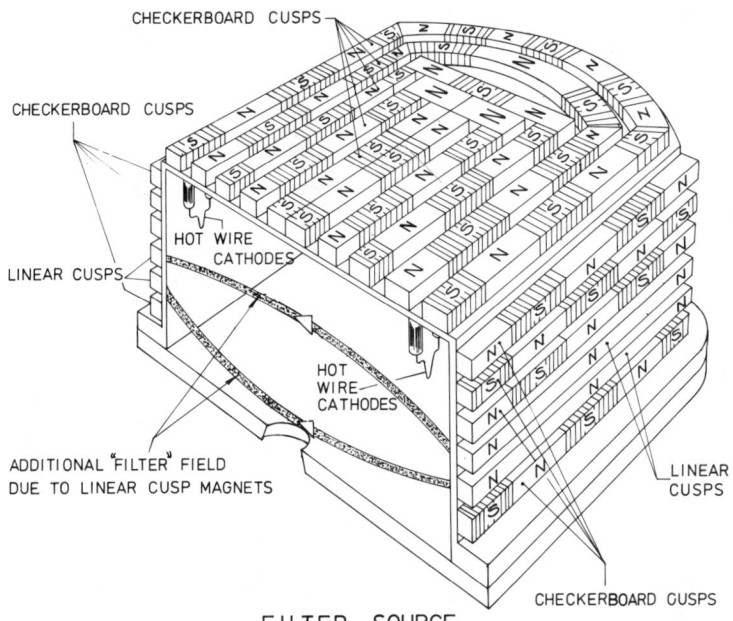

Figure 1

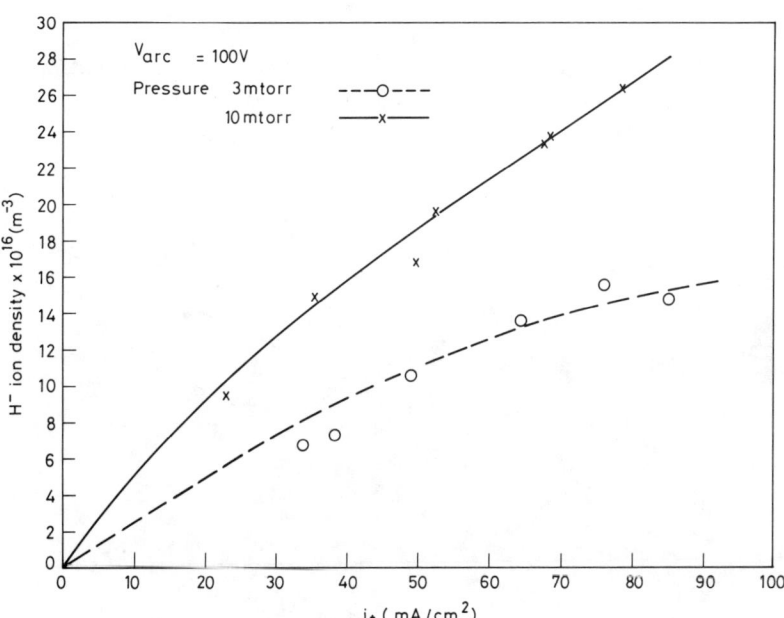

Figure 2

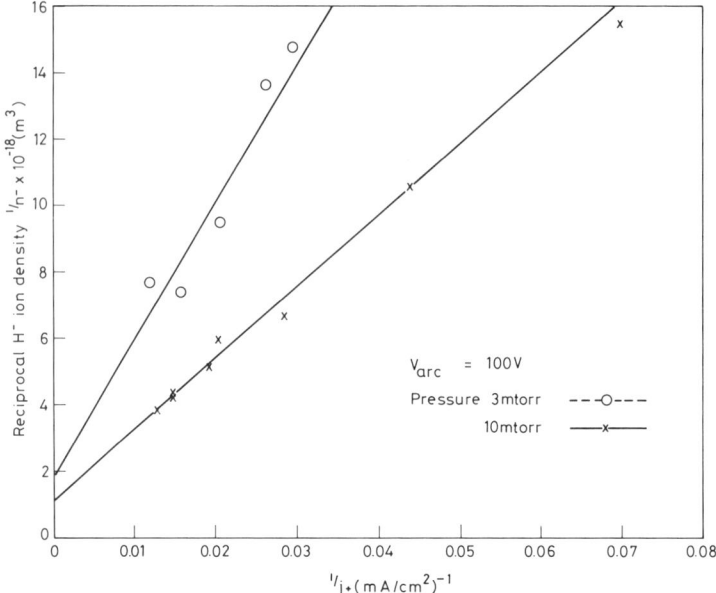

Figure 3

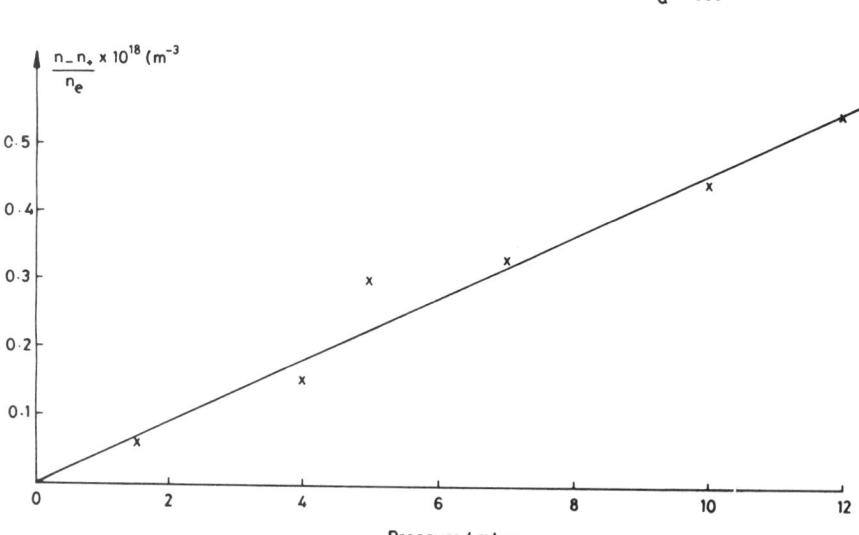

Figure 4

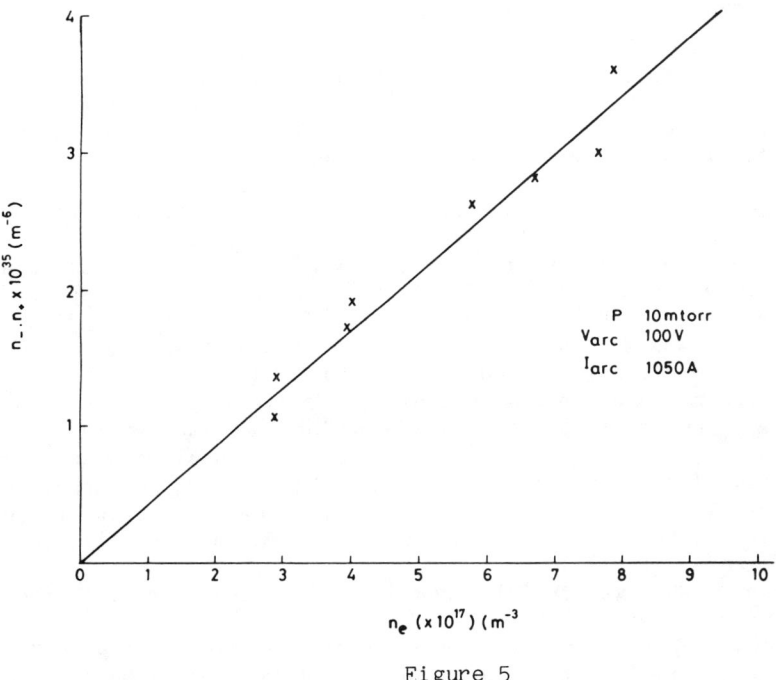

Figure 5

DISCUSSION

<u>Graham</u>: The modeling you did for the hydrogen atoms was for intense discharges where H^+ was the dominant ionic species. Do you think that would be the same in less intense discharges, where other molecular ionic species would be present?

<u>Holmes</u>: I don't know. It's a difficult question, because you have chemistry coming in. You have the presence of the molecular ions, which add extra channels for vibrational molecule production, over and above the modeling I've been talking about here. Essentially, what we've done in this model is to simplify the atomic physics as far as we can, by making the assumption of very intense discharges, where we have one ion species. There is a problem, that when we go to the one-dimensional modeling, we will have to keep the number of ion species to a minimum, because every extra ion species means an extra 3 moments in the Boltzman equation, and it rapidly gets totally out of control. The minimum plasma I can deal with is electrons, positive ions, i.e. H^+, and H^-. If I have H_2^+, H_3^+ and other ions or other molecular ions that you could dream up, it gets very, very difficult, indeed, and I could not see a way of actually, then, creating a 1-D model.

<u>Hiskes</u>: What was the electron density in the primary chamber when you were operating?

<u>Holmes</u>: In this source, the electron density in the driver or production region, for the intense discharge, would be about $4-5 \times 10^{12}$.

<u>Hiskes</u>: And you said, as far as you could tell, there was no atomic concentration in the discharge.

<u>Holmes</u>: Within these error bars on the measurements, yes.

<u>Hiskes</u>: What is your upper limit for the atomic concentration compared to the molecular concentration?

<u>Holmes</u>: I can't give you an exact answer to that one, John. All I can say is that if you look at the graphs, you can see there is a small scatter in a number of points. So I suppose I could say that the ratio of this quantity to the error bars on the points is about 10%. So, I would say from that, the atomic concentration effects must be less than 10%, just from the error bars on the measurements.

<u>Hiskes</u>: Can you put that in terms of the ratio of the atomic concentration to the molecular concentration?

<u>Holmes</u>: Not at the moment. The only thing we've done is the measurements which Mark Nightingale mentioned yesterday, which suggest that the atomic concentration is very small, in a different source, admittedly. He measured 2% for a discharge current of about 25A. This is a much bigger source and a much bigger current, but it would probably indicate that a few percent, maybe 10%, is the sort of number we would be talking about. I can't be more precise than that at the moment.

A NUMERICAL MODEL OF THE MIRROR ELECTRON
CYCLOTRON RESONANCE MECR SOURCE

Göran Hellblom, The Royal Institute of Technology,
Stockholm, Sweden

ABSTRACT

Results from numerical modelling of a new type of ion source are presented. The plasma in this source is produced by electron cyclotron resonance in a strong converging magnetic field. Experiments have shown that a well-defined plasma column, extended along the magnetic field (z-axis) can be produced. The electron temperature and the densities of the various plasma particles have been found to have a strong z-position dependence. With the numerical model, a simulation of the evolution of the plasma as a function of z is made. A qualitative agreement with experimental data can be obtained for certain parameter regimes.

1. INTRODUCTION

A new type of H^- ion volume source based on plasma generation by electron cyclotron resonance with an imposed high frequency field of 10 GHz has recently been reported by the author[1]. The plasma is confined by a strong axial magnetic field of 0.3-0.6T. The magnetic field is adjusted so that the electron cyclotron resonance occurs in the axially directed wave-guide through which the high frequency power is injected as well as the gas. The wave-guide ends in a larger diameter cylindrical cavity shortly after the resonance region. The magnetic field strength increases from the resonance region over the length of the cavity, forming a magnetic mirror field (figure 1).

A beam of plasma thus moves from the resonance region into the magnetic mirror and reaches the extractor. A number of interesting experimental observations have been made on this configuration. A narrow plasma column with a cross-section given by the wave-guide and the converging magnetic field lines was observed to be formed. A positive ion current density of 160 mA/cm^2 was measured. The plasma density can be estimated from this figure giving around 1.0×10^{12} ions per cm^3. A ratio of extracted negative and positive drain currents as low as 5 was found, implying that a large fraction of the negatively charged particles are H^- ions rather than electrons.

Electrical probe measurements[2] indicate a decreasing electron temperature towards the extraction. Electron

© American Institute of Physics 1987

temperatures of 1-4 eV in the cavity close to the wave-guide and 0.2-1.0 eV at the extraction region have been measured. The gas pressure in the source was estimated to be 0.5-3.0 mtorr which corresponds to $0.2-1.0 \times 10^{14}$ molecules per cm^3.

2. THE ADOPTED THEORETICAL MODEL

The purpose of this paper is to model the plasma observed in the experiments described above. A number of approximations and idealizations have been made. According to experimental results the plasma is stable, permitting us to neglect collective particle effects, and to concentrate on microscopic binary particle collisions. One of the most important assumptions in the model is that the plasma drifts away from the source region along the magnetic field with a constant velocity v_z. This velocity can be estimated from the strength of the plasma source and by assuming that an equal amount of plasma goes either way from the source along the magnetic field. The strength of the source is taken as 2 amperes[3] of ions and electrons for 1 kW of HF-power absorbed in the plasma. Further assumptions on which the model is based are the following:

a. The plasma has the form of a square-shaped bar, with DxD cm^2 end area, and being extended in the z-direction.
b. The plasma is quasi-neutral.
c. The hydrogen gas density is constant, and represents a parameter in the computation.
d. The density of vibrationally excited hydrogen molecules (H2") along z is constant, thus being an additional parameter in the computation. There is a consumption and a de-excitation of these molecules in the plasma. On the other hand, the plasma that recombines on the plasma electrode might act as a source of excited milecules, resulting in the assumed approximately constant density of H".
e. The electrons have a Maxwellian energy distribution, and they are distributed in 20 energy groups, from 0 to 20 eV.
f. The ions are mono-energetic.
g. The electron temperature decreases in the positive z-direction according to the formula $T_e = T_{eo} + T_{max} \exp(-0.2z)$, where T_{eo} and T_{max} are parameters in the computation.
h. The plasma bar is divided into NZ segments in the z-direction, each with the length D_z=total length/NZ. The ion densities, the electron temperature and the electron energy group densities are constant within a given segment.

i. The particle densities and the electron energy group populations are adjusted when passing from one segment to the next. The adjustment terms for the various particle species are given by Production-Loss within a given segment.
j. The atomic hydrogen has an extra particle loss term given by $n_H*2Dz/(D+2Dz)$ when the drift velocity is approximately equal to the thermal velocity $\approx v_z$.

The following parameters are fixed

1. The composition of the plasma at Z=0 is 60% H2+, 20% H+, 20% H3 and no H-.
2. The total length in Z (Z_{total}) is 16 cm.
3. The atomic hydrogen density is 20% of the gas density at Z=0.

The reactions included in the model are presented in Table 1, where also the ion-ion and ion-neutral cross-sections are given. The electron collision cross-sections are given in Table II. The details of the computation of the adjustment terms for the various species in the plasma that have been considered are described in the Appendix. The relative populations of the different electron energy groups as a function of z and the electron temperature for some combinations of T_{eo} and T_{max} are shown in Figure 3.

3. A DISCUSSION ON THE CROSS-SECTION

Before moving to the results of the computation, some comments on the most important cross-sections will be made. The plasma is produced by high energy electrons (>20 eV) that ionize the gas. The model does not include the hot electron region which is the source of plasma in the model. The cross-sections for ionization of the gas (reactions 7 and 15) are included in the model. The cross-sections for ionization of atomic hydrogen (reaction 18) are enhanced compared to those of ground state atoms to include the effect of a large fraction being in the 2S level. The H2+ has a large probability of giving H3+ in a gas collision (reaction 10 in the model). The H3+ ion can in a successive gas collision give a proton according to reaction 11. The absorption cross-sections of slow electrons on hydrogen molecule ions are high, as seen by the cross-sections for the reactions 8 and 14. The proton fraction is therefore enhanced as the plasma moves through a low electron temperature region.

The two most important cross-sections from the negative ion point of view are C1 and C2, which represent the production and destruction of H- ions by electron collisions. With highly vibrationally excited hydrogen molecules ($H_2"$) there exists an electron energy domain of 0.5-3 eV where the production probability of H^- is high and the destruction probability due to electron collision is low [4,5,6]. When vibrationally excited molecules are present in a low temperature hydrogen plasma one can therefore expect an important H^- population to build up.

In the model the density of vibrationally excited hydrogen molecules $H_2"$ is a parameter in the calculation. An interesting outcome of the simulation is the $H_2"$ density needed to explain the experimental observations made. The mutual neutralisation cross-section between negative ions and protons, given by C_3, is very high, but the limited collision frequency gives a mixture of H^- and H^+, a certain survival time.

4. RESULTS

With the assumptions specified in Section 2 and with the reactions and cross-sections given in Tables 1 and 2, a computation with the method described in the Appendix is performed. The evolution of the H^-, H^+, H_2^+, H_3^+ and atomic hydrogen H densities as a function of z is shown in Figure 3 for a specified set of parameters. Three different step lengths or numbers of segments have been used. One sees that even 10 segments give good accuracu. The curve obtained with 20 segments will be our reference case. Here the initial plasma density was 1×10^{12} cm^{-3}, the drift velocity 2×10^6 cm/sec, the gas density 1×10^{14} molec./cm^3, the $H_2"$ density 1×10^{12} molec./cm^3, and the electron energy distribution according to case 2 as shown in Figure 3 where T_{eo}=1.0 eV and T_{max}=5.0 eV. The effect of varying the electron energy distribution with respect to the reference case is shown in Figure 4. The Figures 5 and 6 show the effect of varying the drift velocity V_z and the density of $H_2"$ with respect to the reference case.

5. DISCUSSION AND CONCLUSION

The outcome of the calculations based on the adopted numerical model can be made to agree with experimental observations for certain parameter values. The gas pressure, the plasma density, the plasma composition at Z=0 and the electron temperature used in the model have some experimental ground. The density of $H_2"$ is not known. In the calculations a result coinciding with experimental observations is obtained when $n_{H_2"}$ is roughly equal to 1×10^{12} molecules per cm^3, which is the plasma density and 1% of the gas density. In a calculation done by J.R. Hiskes et.al[12] a fraction of $H_2"$ in the H_2-gas of about 0.5% is found. The important conclusion from the modelling, is that with realistic assumptions and parameter values a qualitative agreement with experiment can be obtained.

ACKNOWLEDGEMENT

The author wants to thank Dr. Rolf Pauli for helpful discussions on details in the model, Dr. Bo Lehnert for reading and commenting the manuscript and John Tonks for his linguistic corrections of the manuscript.

REFERENCES

1. G. Hellblom and C. Jacquot, Extraction of Volume Produced H⁻ Ions From a Mirror Electron Cyclotron Resonance Source, Nuclear Instuments and Methods, A 243 (1986)255-259.
2. G. Hellblom, Discussion on Hot Probe Measurements in a Negative Ion Source, The Institute of Technology, Sweden, TRITA-PFU-85-03.
3. P. Sermet, Source intense d'atomes de deuterium à très basse énergie en vue du Chauffage de Plasmas d'intérêt thermonucléaire.Thèse de doctorat ès sciences physiques, soutenu à l'Institute National Polytechnique de Grenoble, March 1978.
4. M. Bacal, A.M. Bruneteau, and M. Nachman, J. Appl. Phys. Vol. $\underline{55}$ p 15-24 (1984).
5. A.M. Bruneteau Mordin, Thèse de doctorat ès science physiques, soutenue à l'université de Paris-Sud le 5 Juillet 1983, P.M.I Report 1310, 1983.
6. J.M. Wadehra, Appl.Phys.Lett. Vol.$\underline{35}$ p 917-919 (1979).
7. C.F. Barnett, et al., Atomic Data For Controlled Fusion Research, Oak Ridge Nat.Lab., Feb. (1977), ORNL-5206 (Vol.$\underline{1}$).
8. C.F. Barnett, et al., Atomic Data For Controlled Fusion Research, Oak Ridge Nat.Lab., ORNL-5207, (Vol.$\underline{2}$ of ORNL-5206).
9. H. Tawara, T. Kato and M. Ohnishi, Ionization Cross Sections of Atoms and Ions by Electron impact, Institute of Plasma Physics, Nagoya University, IPPJ-AM-37.
10. Culham Laboratory CLM-R175.
11. W.R. Gentry, D.J. McClure and C.H. Douglass, Rev.Sci. Instrum. $\underline{46}$ p 367-375 (1975)
12. J.R. Hiskes, A.M. Karo, and P.A. Willman, J. Appl. Phys., Vol.$\underline{58}$ p 1759.1764, (1985).

Table I. Reactions included in the model, with the ion collision cross-sections given in cm^2.

Reaction	Cross-section	Reference
e + H2" --> H- + H	C1(i)	Ref.5 p59
e + H- --> H + 2e	C2(i)	Ref.5 p74
e + H2+ --> H- + H+	C6(i)	Ref.5 p43
e + H2 --> H+ +...	C7(i)	Ref.9 p15
e + H2+ --> 2H	C8(i)	Ref.10
e + H2+ --> H+ + H + e	C9(i)	Ref.10
e + H3+ --> H2 + H	C13(i)	Ref.10
e + H3+ --> 3H	C14(i)	Ref.8 p7.6
e + H2 --> H2+ + 2e	C15(i)	Ref.9 p14
e + H3+ --> H- + H2+	C17(i)	Ref.5
e + H --> H+ + 2e	C18(i) 50% H(2S)	Ref.9 p16,17
H+ + H- --> 2H	C3=5.0x10^{-14}	Ref.7 A8.3
H2+ + H- --> H2 + H	C4=5.0x10^{-14}	Ref.5 p87
H2 + H2+ --> H3+ + H	C10=1.0x10^{-15}	Ref.11
H2 + H3+ --> H+ +2H2	C11=5.0x10^{-17}	
H3+ + H- --> 2H2	C16=5.0x10^{-14}	Ref.5 p89

Table II. Cross-sections for electron collision processes ($\times 10^{-16}$ cm^2) with the electron energy ranging from 1 to 20 eV.

ENERGY GROUP	C1	C2	C6	C7	C8	C9	C13	C14	C15	C17	C18
0-1	1	0	0.08	0	15	0	0	20	0	0	0
1-2	5	0	0.05	0	5	0	0	9	0	0	0
2-3	4	0	0.02	0	2	0	0	5	0	0	0
3-4	2	3	0	0	0.5	0	0	4	0	0.5	0
4-5	1	10	0	0	0	0	0	3	0	1	0
5-6	0.5	15	0	0	0	0	0	2	0	1.5	0
6-7	0	20	0	0	0	0	0	2	0	2	0
7-8	0	25	0	0	0	0	0	2	0	2	1
8-9	0	30	0	0	0	0	0	2	0	2	2
9-10	0	35	0	0	0	0	0	1	0	1.5	2
10-11	0	35	0	0	0	0.5	0	0	0	1	2
11-12	0	35	0	0	0	2	0	0	0	0.5	2
12-13	0	35	0	0	0	4	0	0	0	0	2
13-14	0	35	0	0	0	3	0	0	0	0	2
14-15	0	35	0	0	0	3	0	0	0	0	2
15-16	0	35	0	0	0	2	3	0	0	0	2
16-17	0	35	0	0	0	2	3	0	0	0	2
17-18	0	35	0	0	0	2	3	0	0.05	0	2
18-19	0	35	0	0.01	0	2	3	0	0.1	0	2
19-20	0	35	0	0.01	0	2	3	0	0.2	0	2

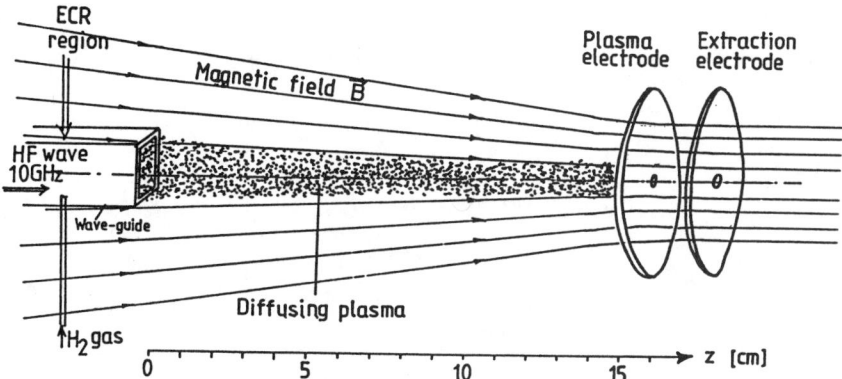

Figure 1. A schematic drawing of the apparatus.

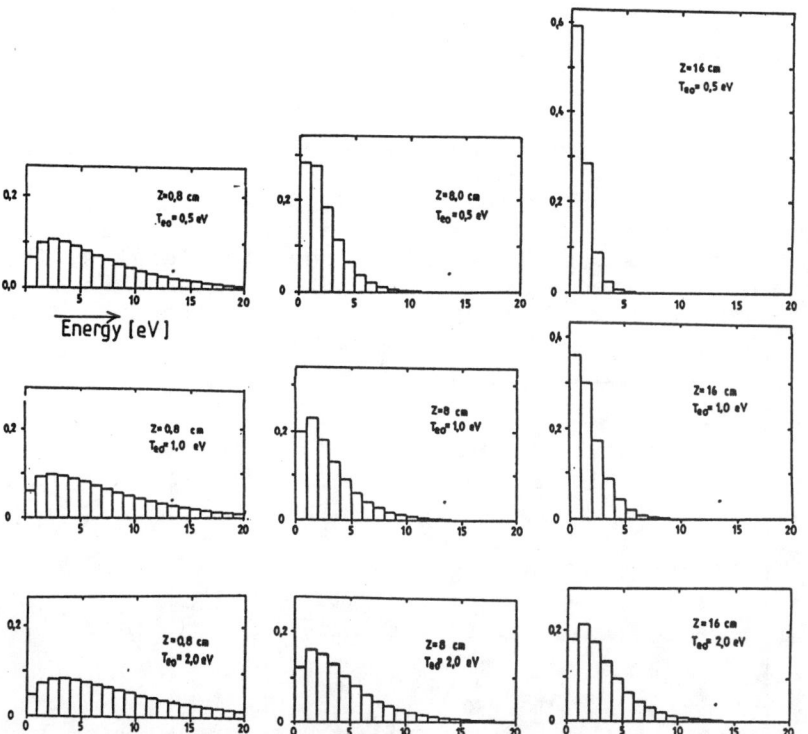

Figure 2. The relative electron population in each energy group 1-20 eV with respect to the total number of electrons. The electrons have a Maxwellian velocity distribution. The electron temperature decreases in the positive z-direction according to the formula $T_e = T_{eo} + T_{max} \exp(-z*0.2)$. The value of T_{max} is fixed and equal to 5.0 eV, and T_{eo} is varied from 0.5 to 2.0 eV.

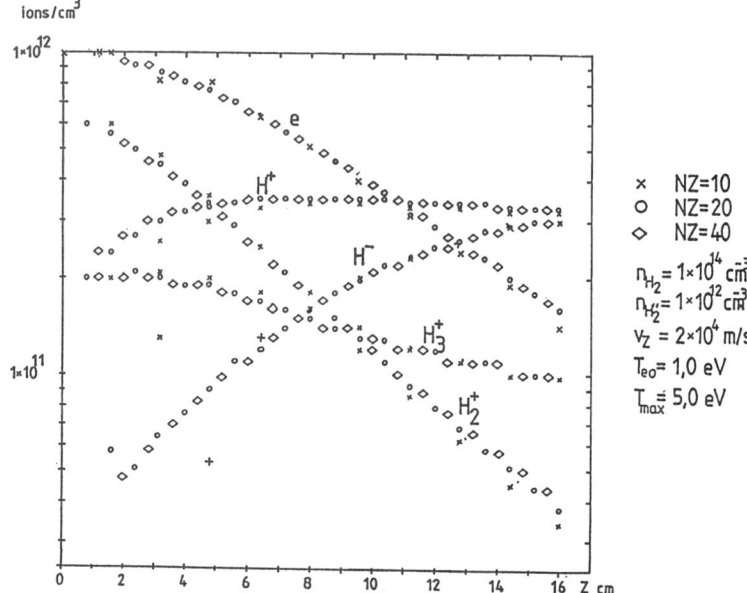

Figure 3. The evolution of the different particle species in the plasma as a function of z. The computation has been made with three step-lengths. The result obtained with the parameters given in the figure and with 20 steps will be our reference case.

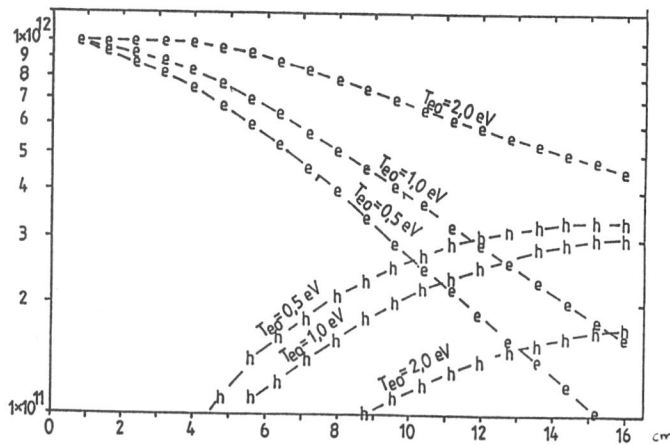

Figure 4. The evolution of the H^- ion and the electron densities as a function of z. The electron energy distribution is varied according to cases 1-3 in Figure 1. The other parameters are given by the reference case in Figure 3.

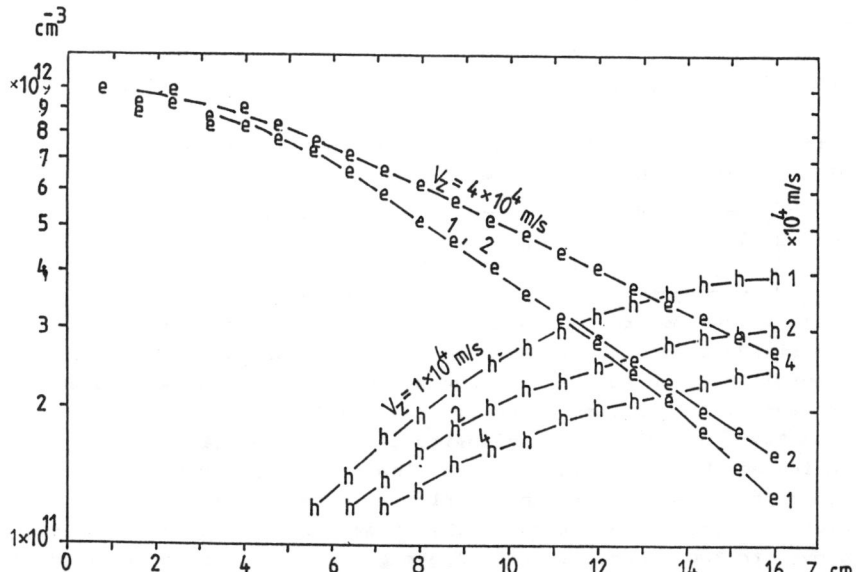

Figure 5. The evolution of the H⁻ ion and the electron densities as a function of z. The ion drift velocity v_z is varied from $1*10^4$ to $4*10^4$ m/s. The other parameters are given by the reference case in Figure 3.

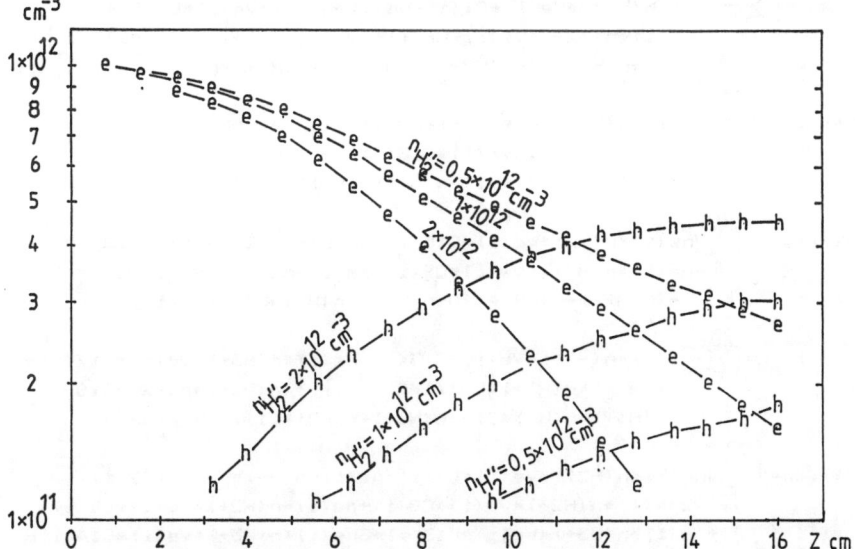

Figure 6. The evolution of the H⁻ ion and the electron densities as a function of z. The density of vibrationally excited hydrogen molecules $n_{H''}$ is varied from $0.5*10^{12}$ to $2*10^{12}$ cm⁻³. The other parameters are given by the reference case in Figure 3.

APPENDIX

Details of the numerical model.

The rectangular plasma column discussed in the text, is divided into NZ segments as shown by the following figure:

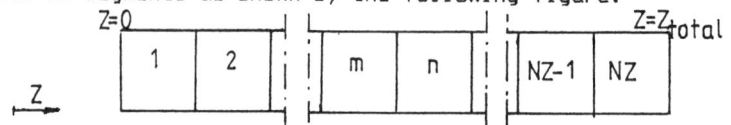

The plasma is assumed to move with a constant drift velocity V_z in the positive Z-direction. The fractional composition of the plasma in segment 1 is given as an input parameter. While the plasma moves through segment 1 in a time given by $(Z_{total}/NZ)/V_z$, reactions takes place with an intensity given by the collision frequencies and by the cross-sections. The reactions and their cross-sections considered in the model are listed in Tables 1 and 2. The original particle densities in the first segment are adjusted when moving over to the second segment. The magnitude of these adjustments depends on the intensity of reactions in the proceeding segment.
In the general case, when moving from segment m to segment n and all particle densities being those of segment m, we have for the various species

$$(\Delta n_{H-})_{m-n} = \left[\sum_{1}^{20} [ne(i)*n(H2")*Ve(i)*C1(i) - ne(i)*n(H-)*Ve(i)*C2(i) + ne(i)*n(H2+)*Ve(i)*C6(i) + ne(i)*n(H3+)*Ve(i)*C17(i)] \right.$$
$$\left. - (n(H+)*n(H-)*C3 + n(H2+)*n(H-)*C4 + n(H3+)*n(H-)*C16)VI \right] \Delta t$$

$$(\Delta n(H+))_{m-n} = \left[\sum_{1}^{20} [ne(i)*n(H2+)*Ve*C9(i) + ne(i)*n(H2)*Ve*C7(i) + ne(i)*n(H)*Ve(i)*C18(i)] \right.$$
$$\left. + (n(H3+)*n(H2)*C11 - n(H+)*n(H-)*C3)VI \right] \Delta t$$

$$(\Delta n(H2+))_{m-n} = \left[\sum_{1}^{20} [ne(i)*n(H2)*Ve(i)*C15(i) - ne(i)*n(H2+)*Ve(i)*C8(i) - ne(i)*n(H2+)*Ve(i)*C9(i) - ne(i)*n(H2+)*Ve(i)*C6(i)] \right.$$
$$\left. - (n(H2+)*n(H2)*C10 + n(H2+)*n(H-)*C4)VI \right] \Delta t$$

$$(\Delta n(H3+))_{m-n} = \left[\sum_{1}^{20} [-ne(i)*n(H3+)*Ve(i)*C13(i) - ne(i)*n(H3+)*Ve(i)*C17(i) - ne(i)*n(H3+)*Ve(i)*C14(i)] + (n(H2+)*n(H2)*C10 \right.$$
$$\left. - n(H3+)*n(H-)*C16 - n(H3+)*n(H2)*C11)VI \right] \Delta t$$

$$(\Delta n(H))_{m-n} = \left[\sum_{1}^{20} [ne(i)*n(H2")*Ve(i)*C1(i) + ne(i)*n(H-)*Ve(i)*C2(i) + 2ne(i)*n(H2+)*Ve(i)*C8(i) + ne(i)*n(H2+)*Ve(i)*C9(i) + ne(i)*n(H3+)*Ve(i)*C13(i) + 3ne(i)*n(H3+)*Ve(i)*C14(i) - ne(i)*n(H)*Ve(i)*C18(i)] + (n(H+)*n(H-)*C3 \right.$$
$$\left. + n(H2+)*n(H-)*C4 + n(H2)*n(H2+)*C10)VI \right] \Delta t$$

where $n_e(i)$ = the electron density from energy group i.
       $V_e(i)$ = the electron velocity corresponding to the kinetic energy of energy group i.
       VI = the ion velocity V_I, approximately equal to V_z.
       Δt = the time that the plasma particles stay in a segment, the time is given by $(Z_{total}/NZ)/V_I$.
       $n(H2")$ = the density of vibrationally excited hydrogen molecules.
       $n(H2)$ = the hydrogen gas density.
       $n(H+)$ = the proton density.
       $n(H2+)$ and $n(H3+)$ = the molecular ion densities.
       $C1(i), C2(i), C6(i), C7(i), C8(i), C9(i), C13(i), C14(i), C15(i), C17(i)$ and $C18(i)$ are the electron collision cross-sections for electrons from energy group i.
       $C3, C4, C10, C11$ and $C16$ ion-gas or ion-ion cross-sections.

DISCUSSION

Ehlers: I heard your estimate of density and your positive ion current density, but I failed to hear your expected negative ion current density.
Hellblom: We have a problem there in pinpointing the negative ion density that we have because we have been extracting at low energy, typically a few kV. And with the current density we have, we have a strong blowup of the beam. So the spectrometer that we have put in has gotten much too low a current density.
Ehlers: What kind of drain current do you read? Did you quote that number?
Hellblom: The positive drain current was 160, and the negative was 5 times larger. The spectrometric readings that we have, though, do indicate 20 mA/cm^2 of H$^-$. It would be safe to mention that figure at least.
Forrester: I'm still puzzled about the same question Ken was asking. You say the negative drain is 5 times as great as the positive drain. You also had curves showing that the electron density, at the plasma electrode, would be very low compared to the negative ion density at that electrode. Putting these things together, are you saying that you suspect that there's nearly 5 times as many negative ions as positive ions? I'm puzzled by the whole thing.
Hellblom: This drain current ratio implies that there is a large majority of H$^-$ ions over electrons.
Forrester: That means 1/2 A/cm^2 of negative ions...?
Hellblom: No, no, because the drain current is mostly electrons, even at small electron density....
Forrester: Because the electron velocity is still so much larger?
Hellblom: Yes.
Holmes: You describe, in your model, a gradient in electron temperature. What do you think caused this gradient in temperature? Because your field runs parallel to the plasma axis, not transverse, as in most negative ion sources.
Ehlers: It's mirroring, Andy. It's sort of like a mirror.
Hellblom: I could not give a satisfying answer to that, other than that the primary electrons, or the fast electrons from the electron cyclotron resonance are mirrored. My colleague, Jacob Brynolf, is working on a model on the electron energy distribution, and this is a rather difficult question.
Holmes: I was wondering, because we have the same problem, but ours is transverse fields, so we can argue that it's because of the low mobility of electrons across the field lines; but in your case...
Hiskes: You said you had assumed an atomic density ratio of about 20%. Could you say something about how you arrived at that number?
Hellblom: That was taken from a reference by Pierre Sermet, I think, on ECR sources. This was just to have an initial value in the calculation. I should say that I failed to include this atomic hydrogen reaction with H$^-$ with the cross-sections that you showed yesterday. This is a discrepancy in the model.

ANALYSIS AND INTERPRETATION OF A HIGH DENSITY
TANDEM NEGATIVE ION SOURCE

J.R. Hiskes
Lawrence Livermore National Laboratory, Livermore, CA. 94550

A. F. Lietzke and C. Hauck
Lawrence Berkeley Laboratory, Berkeley, CA.

INTRODUCTION

In the last few years the development of tandem-discharge hydrogen-negative-ion-source systems has proceeded along both experimental and theoretical lines. To some extent these developments have proceeded independently, either the available theoretical model was inadequate to account for a specific geometrical configuration, or the experimental data was not sufficient to provide adequate input parameters for calculation. In the tandem system described here the electron temperature, electron density, and other relevant parameters have been obtained for a high-density system whose electron densities range up to $3 - 5 \times 10^{12}$ electrons cm^{-3}. The model calculation for the atomic processes has been extended to include both electron density and electron temperature spatial variations through the second chamber. These spatial variations are essential for an adequate interpretation of tandem systems where steep density gradients may occur beyond the magnetic filter region. In this paper we shall combine the experimental density data with the new spatially dependent atomic model for the purpose of attempting a correlation of the observed and calculated current densities.

EXPERIMENTAL

The experiment was performed on the Test Stand 1 (TS-1) facility (Fig. 1) at the Lawrence Berkeley Laboratory. For this experiment, we limited TS-1 to accelerate one beam, from a 1x2.5 cm^2 rectangular slot, with an energy of 6 kV. The beam was detected electrically by saturated +ion collection and calorimetrically by water-flow calorimetry, 60 cm downstream, on a beam dump. The electrical signal was calibrated against the current inferred from the steady-state calorimetry whenever the pressure or accelerator voltage was changed.

The $\bar{H}$ plasma chamber consisted of a standard "cusp SmCo magnetic bucket" (22x22x22 cm). Four rods (separated by 6.4 cm) of water-cooled magnets (SmCo, 4.7x4.7 mm), were inserted in the plasma chamber at three axial locations: 2.5 cm., 6.9 cm., and 11.3 cm. upstream from the beam-forming electrode. (Fig.2) The magnet rods separated the plasma chamber in two regions: 1) a hot electron region where the filaments were located and 2) a cold electron region which was terminated by the beam-forming electrode. Visualization of these two chambers is aided by visualizing the B-field lines. The relative electron density for the exit chamber was controlled by the "strength" of th magnetic filter and by the voltage of the beam-forming electrode. For this experiment, we chose source parameters which maximized the H⁻ delivery to the beam dump, i.e. optically overdense and on the verge of beam-clipping by the exit electrode.

The parameters for the data used in the fitting are listed in Table 1. Density and temperature information is reported elsewhere at this conference.(1)

FILTER PLANE SERIES

SOURCE PARAMETERS:		FILTER LOCATION		
		2.5cm	6.9cm	11.3cm
Arc	0-200 A			
	100 V			
Gas Pressure	10 mTorr			
Rect. Aperture	1 x 2.54cm			
J⁺ bucket (mA/cm²)		313	370	(460)
$V_{\text{filter frame}}$	@anode			
4 SmCo filter rods	@anode			
Ext. Mags. (SmCo)	none			
B_{G1}	≈0			
V_{G1}(Bias adj.for max. H⁻@200A)		+2.5	+4.0	+4.0
V_{accel}	6.0kV			
B_{G2}	T(N)			
V_{G2} Bias (volts)	0			
I_{G2} (mA)		260	500	730
$I_{\text{dump[sec.]}}$ (mA)		31	17.5	12.5±1%
I_{H^-} [∂=1.8] (mA)		17.2	9.7	6.9
J_{H^-} [Area=2.54cm²] (mA/cm²)		**6.8**	**3.8**	**2.7**

THEORETICAL DISCUSSION

The array of atomic and molecular processes that determine the vibrational population distribution and the negative ion concentration in a tandem discharge has been discussed in previous works.[2,3,4] A principal parameter that remains uncertain is the atomic concentration. Some estimates for the atomic concentrations in high-power hydrogen discharges designed for the optimum extraction of protons have been made by Chan et al.[5], and by Vella[6]. In a positive ion source discharge the species mix favors high proton ratios provided the discharge contains a high concentration of atoms[5]. In hydrogen discharges intended for negative ion extraction however, a high atomic concentration in deleterious, causing negative ion destruction via the associative detachment process

$$H^- + H \rightarrow H_2(v'') + e \qquad (1)$$

To gain a perspective of the relative atomic concentrations we calculate the ratio of atomic to molecular density for two limiting cases that are expected to represent lower and upper limits, respectively. We obtain a lower limit assuming the source of atoms is limited to molecular collisional dissociation, process (1) in Table II, and dissociation of H_2 molecules impinging on the hot filament, process (7). The energetic atoms ($E \simeq 2.5eV$) that result from collisional dissociation and the lower energy thermally dissociated products are assumed to be completely absorbed upon striking the chamber walls ($\gamma = 1$). For an upper limit we have taken into account the sum of processes listed in the Table. The relative atomic contributions from these different processes are shown on the right. The recombination of positive ions on the chamber walls, process (2), is seen to be an important source of atoms. The rates for these processes are summarized by Chan et al., who performed a similar calculation. The mean atomic energy for this upper limit case is taken to the 0.35eV, derived from spectroscopic data[7]. Inspection of Fig. 4 shows a possible range of values for these two limiting values that runs form one percent to thirty percent at electron densities near $3 \times 10^{12} cm^{-3}$.

TABLE II

Relative contributions to the atomic concentration at $n=3.3 \times 10^{12}$ electrons cm^{-3}.

	Process	Contribution
1.	$e + H_2 \rightarrow H + H + e$	36%
2.	$H^+, H_2^+, H_3^+ + WALL \rightarrow H + \ldots$	28%
3.	$H_2^+ + H_2 \rightarrow H + H_3^+$	8.8%
4.	$e + H_2 \rightarrow e + H_2^* \rightarrow H + H + e$	6.8%
5.	$e + H_3^+ \rightarrow H + H_2 \,/\, H + H + H$	6.3%
6.	$e + H_2^+ \rightarrow H + H$	5.2%
7.	$H_2 + Filament \rightarrow H + H$	5.1%
8.	$e + H_2^+ \rightarrow H + H^+$	2.9%
9.	$e + H_3^+ \rightarrow 2H + H^+ + e$	<1%

The model calculation for the negative ion concentration and current density to described here is based upon the schematic shown in Fig. 5. In the first chamber the electron and gas densities do not vary with position along the chamber. The first chamber is defined as that region to the left of the magnetic filter where fast electrons (E > 25eV) are present. The density of fast electrons is calculated using the model for the electron energy distribution described in Ref. 2. The second chamber is defined as that region where no fast electrons are present.

In the second chamber the electron densities, electron temperatures, vibrationally-excited-molecule densities, and negative ion densities are known to vary with axial position from the region of the magnetic filter up to the extraction plane. To account for these spatial variations in the model we have divided the second chamber into four zones, with the representative electron density and electron temperature variations shown in the lower portion of the figure.

With the experimental data described above the necessary input parameters for the calculation are available but with the exception of the atomic concentration. Here we treat the atomic concentration as a free parameter that is adjusted until the calculated current density agrees with the extracted current density. This fit is attempted at a filter-extraction plane separation of $z = 2.5$cm. For a first-chamber electron density of 3.3×10^{12} electrons cm^{-3}, the extracted current density is observed to be 6.8mA cm^{-2}. In the earlier analysis of the high-density positive-ion-source discharge Chan et. al., arrived at an H_2 gas temperature of approximately 1500°K. In the multipole system under investigation here, the required discharge current per unit gas pressure to achieve the same electron density is lower, and we might expect the gas temperature also to be lower.

On Fig. 4 is shown the atomic ratio necessary to reproduce the observed 6.8 mA cm^{-2}, and calculated for two gas temperatures, 900°K and 1500°K. The atomic ratios are close to, but less than, the projected upper limit values.

It is found experimentally that the extracted current density of 6.8mA cm^{-2} is a maximum as a function of gas pressure about a pressure of 10 mtorr for fixed discharge current. We have varied the gas pressure by a factor of two in the calculation so as to examine the pressure dependence. Increasing the gas density will decrease proportionately the number of high energy electrons, and shift the ratio of atomic to molecular densities along a line parallel to the upper-limit line shown in Fig. 4. The results of these gas density variations at a temperature of 1500°K are shown in Fig. 6. The calculations confirm the current density maximum near 10mtorr.

Having found the atomic ratio parameter at the $z = 2.5$ cm filter position we use this value as a basis for predicting the extracted current densities at the $z = 7$ and $z = 11.2$ cm filter positions, respectively. In moving the filter to $z = 7$ or 11.2 cm the volume of the first chamber in reduced and the electron density is increased. Moving to the $z = 7$ cm position causes the volume to be reduced by a factor of 1.3 but the electron density is observed to increase by about a factor 1.2. At the $z = 11.2$ position the volume is approximately halved compared to the original $z = 2.5$ position, and the electron density increase here is extrapolated upward by a factor of 1.5. These density increases case the atomic/molecular ratio to increase toward larger values, along the two temperature lines shown in Fig. 4. These calculated values projected from the $z = 2.5$ position data are shown in Fig. 7 and compared with the observed extracted current densities. The model calculation provides for a reasonably good comparison with the experimental values.

These model results are clarified by examining the negative ion density variations along the tandem system. The negative ion density variations for the three filter positions, z = 2.5, 7, and 11.2 cm, respectively, are shown in Fig. 8. For these calculations we have taken the negative ion energy to be equal to 0.3eV, a value suggested by the dissociative attachment calculations of Wadehra.[8] The calculated negative ion densities are multiplied by the negative ion velocity appropriate to 0.3 eV to provide the ordinate current density scale shown in the figure.

Inspection of the figure shows a constant negative ion density in the first chamber, then rising and falling in the first zone of the second chamber. The rise in this first zone is due to the removal of the effect of H collisional destruction caused by the energetic electrons; the subsequent fall is due to attenuation of the vibrationally excited molecules as they move out of chamber one into chamber two. In zone two the negative ion concentration rises sharply due to the more optimum electron temperature for dissociative attachment and the reduction of collisional detachment. The densities rise to a maximum and begins to fall because of the increased attenuation of the excited molecules moving through the second chamber. The attenuation of the negative ions at the larger distances, toward z = 7 and 11.2 cm, is due principally to the associative detachment, process (1).

The thermal electron density, n, the fast electron density, n(f), the atomic density, N_1, and the molecular density, N_2, used in these calculations are summarized in Table III. The values for the various rate processes are those given in Refs. 2,3,4. For the ion-ion collision process

$$H^- + H^+ \rightarrow H + H, \qquad (2)$$

we have used the recent values of Fussen and Kubach[9].

TABLE III

Density values used for the solutions of Figs. 7 and 8

	z = 2.5 cm	z = 7. cm	z = 11.2 cm
$n(cm^{-3})$	3.3×10^{12}	4.0×10^{12}	5.0×10^{12}
$n(f)$ (900°K)	1.4×10^{11}	1.7×10^{11}	2.1×10^{11}
(1500°K)	2.2×10^{11}	2.7×10^{11}	3.4×10^{11}
N_1 (900°K) →	2.7×10^{13}	3.6×10^{13}	4.7×10^{13}
(1500°K)	2.7×10^{13}	3.6×10^{13}	5.0×10^{13}
N_2 (900°K) →	1.3×10^{14}	1.3×10^{14}	1.3×10^{14}
(1500°K)	0.8×10^{14}	0.8×10^{14}	0.8×10^{14}

CONCLUSIONS

A spatially-dependent model calculation for the negative ion density in a high-density tandem-negative-ion generator has been applied to a specific ion source apparatus. The calculated extracted current densities agree qualitatively and quantatively with the observed current densities. The atomic concentrations inferred in the system are within a factor of two of the upper limits for the atomic concentrations expected for a high-density hydrogen discharge.

REFERENCES

1. A. F. Lietzke and C. Hauck, this conference.
2. J. R. Hiskes, A. M. Karo, and P.A. Willmann, J. Appl. Phys. 58, 1759 (1985).
3. J. R. Hiskes, A. M. Karo, and P. A. Willmann, J. Vac. Sci. Technol. A3(3), 1229 (1985).
4. J. R. Hiskes, and A. M. Karo, J. Appl. Phys. 56, 1927 (1984).
5. C. F. Chan, C. F. Burrell, and W. S. Cooper, J. Appl. Phys. 54, 6119 (1983).
6. M. C. Vella, Private Communication.
7. C. F. Burrell, Private Communication.
8. J. M. Wadehra, Phys. Rev. A 29, 106 (1984).
9. D. Fussen and C. Kubach, J. Phys. B 19, L31 (1986).

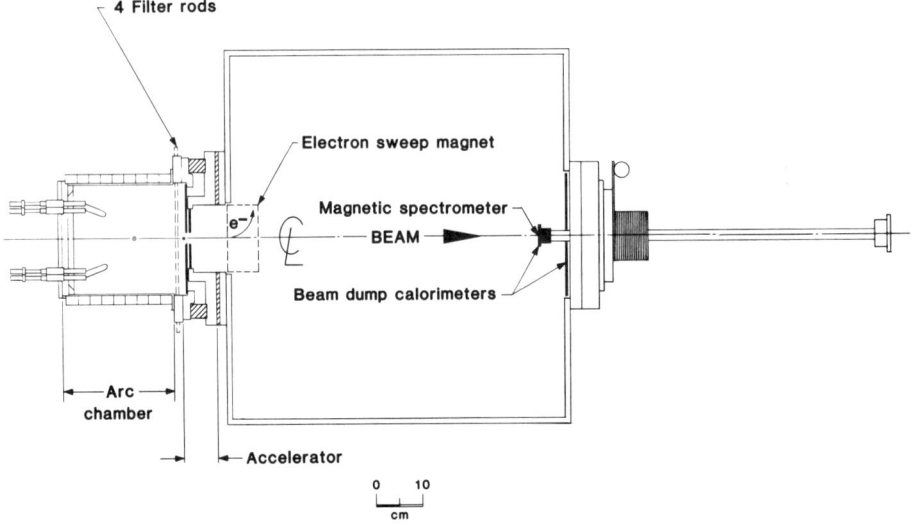

Figure 1.   SCHEMATIC OF TS-1

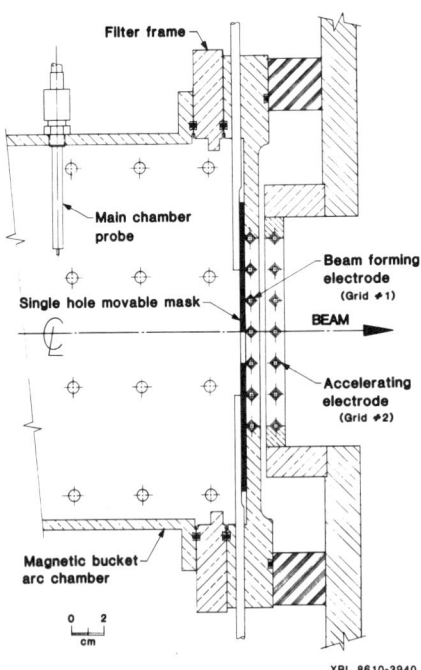

Figure 2.   DETAILS OF FILTER AND EXTRACTION REGION

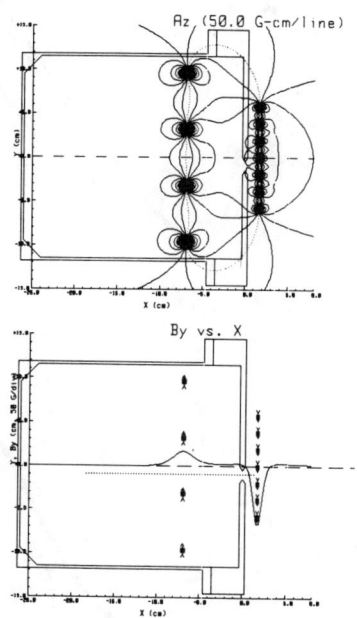

FIGURE 3

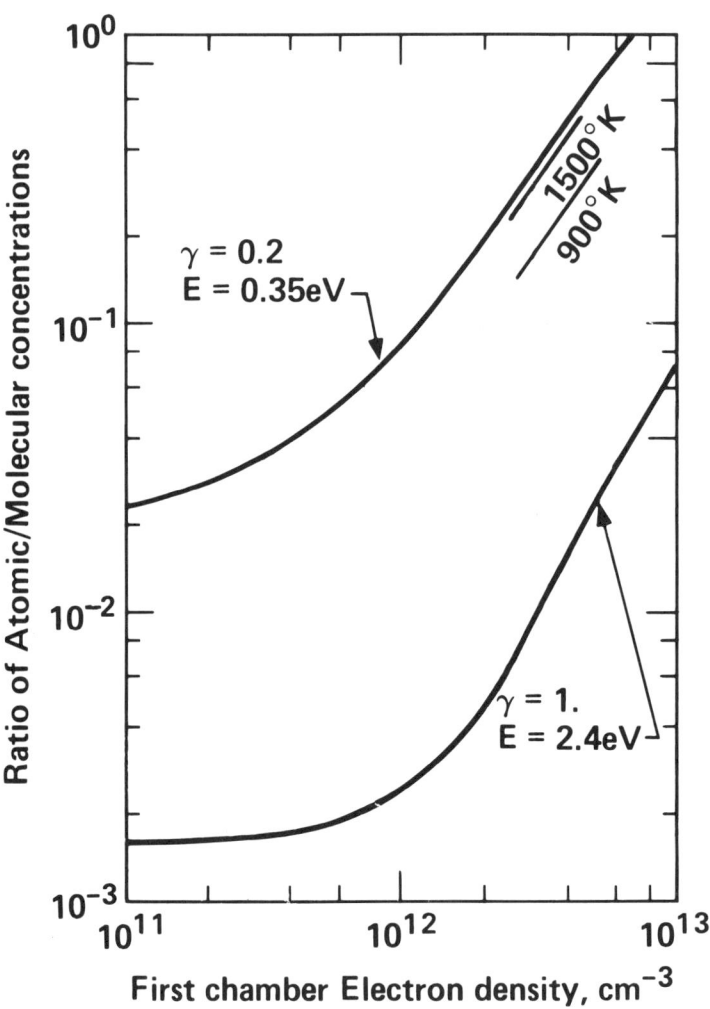

Figure 4.

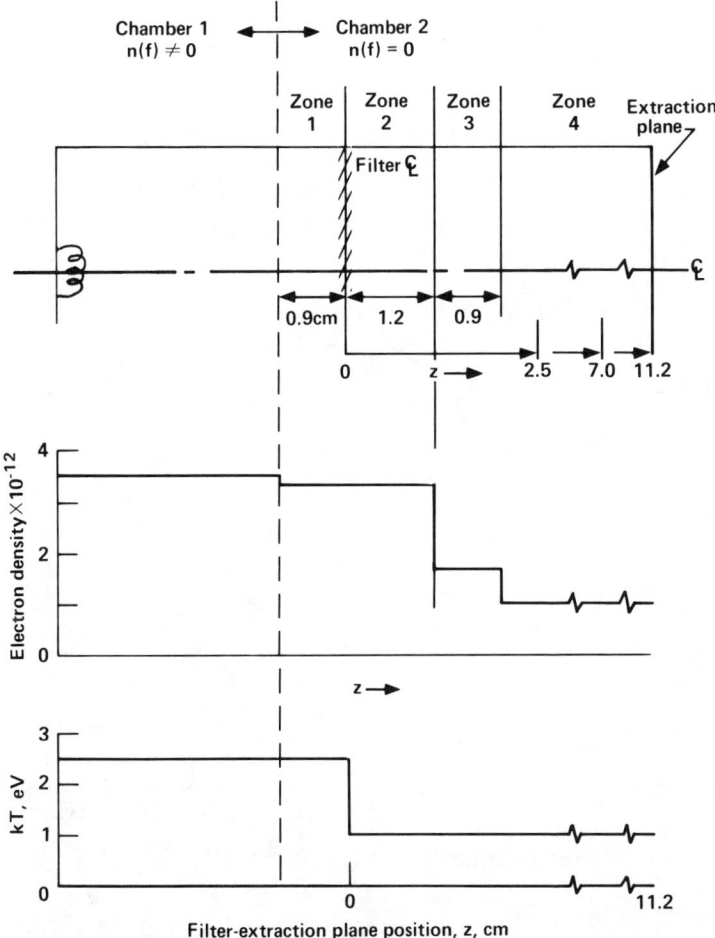

Figure 5.

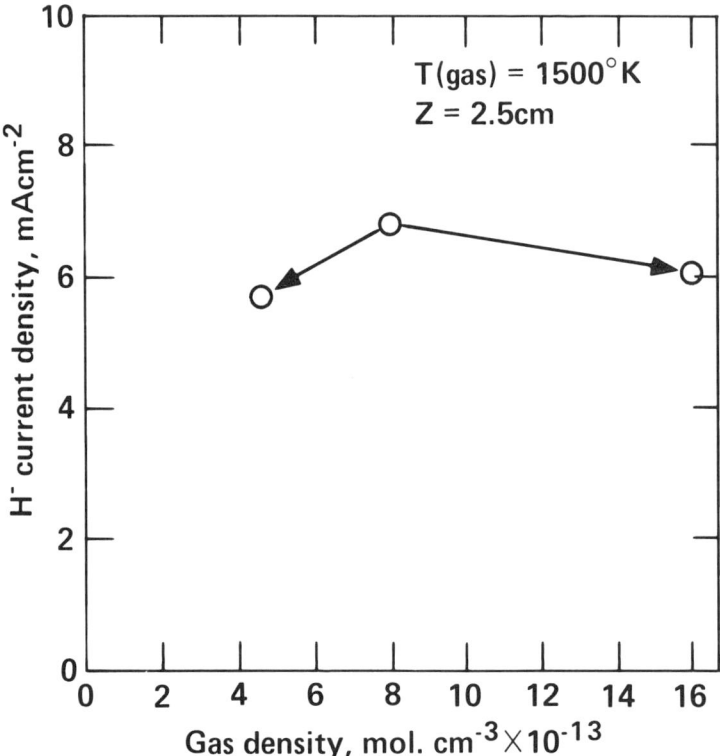

Figure 6

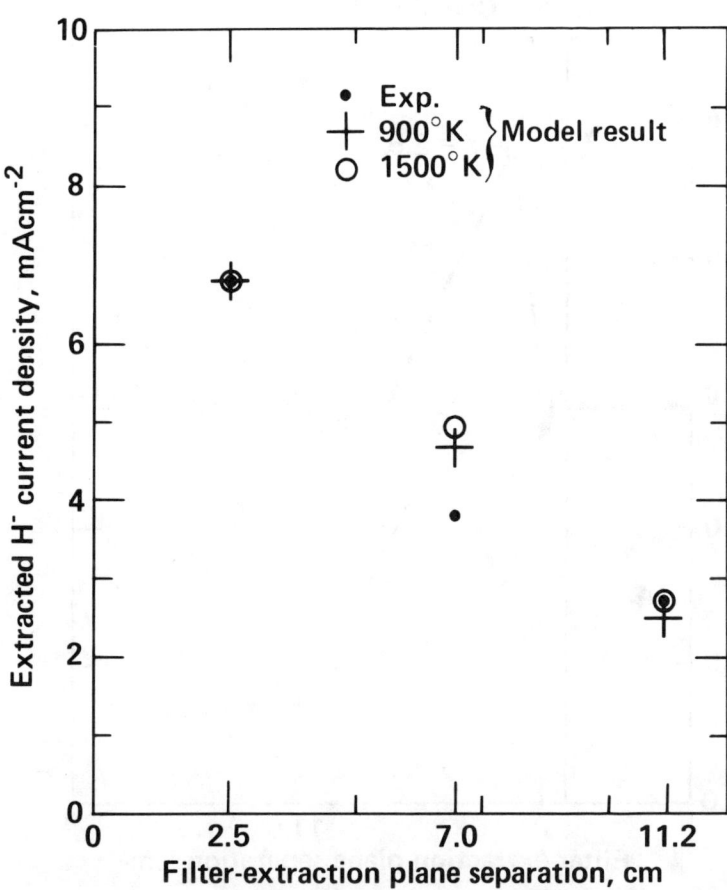

Figure 7

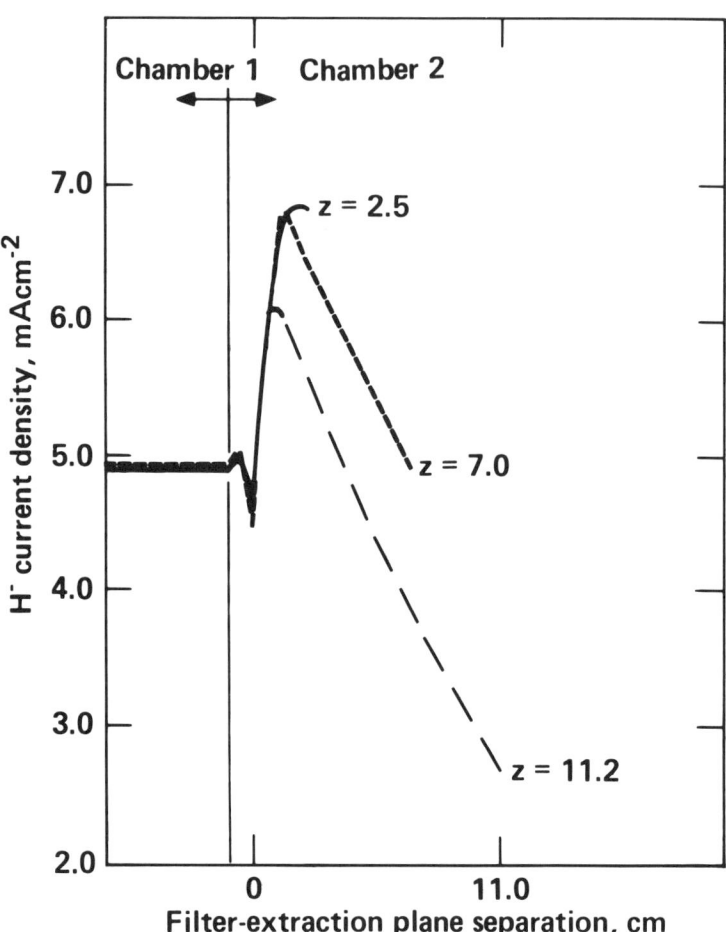

Figure 8

DISCUSSION

Graham: Can you reconcile the two statements we've just heard? Andrew Holmes made a statement that atomic hydrogen doesn't play any part in his source, and you're telling us that it's a primary consideration in your model.

Hiskes: We can't reconcile it at this moment, because I have not analyzed Andrew's discharge, and he has not analyzed our discharge; but I'm sure that by the next meeting, there will have been some reconciliation.

Ehlers: I might add, there's another conundrum. Namely, Paul says that they're making their negative ions by charge exchange, in which they use the atomic hydrogen to a benefit. So we've got to find out whether atomic hydrogen is a friend or an enemy. I'm more inclined to believe it's an enemy. It's hard to have friends in this business.

Holmes: I find it interesting that we seem to be starting off with virtually the same density in the plasma production region, about $3-5 \times 10^{12}$. And the only thing I can think of, which might be an answer to Bill Graham's question a few minutes ago, is that what may happen is, that the ion-ion recombination rate slowly rises as you lower the ion temperature. It may possibly be that if you have a low enough ion temperature, the ion-ion recombination rate always wins.

Hiskes: Well, you would have to get the ion temperature, the atom-ion relative energy, to well below 1/10 eV for that to be the case. Well below. And, from the positive ion source work, the atom energies were about 0.35 eV, and the positive ion energies are typically around 1 eV. So I can't accept that possibility, at least for these discharges at the present time.

VOLUME PRODUCTION OF H⁻ IONS AT ECOLE POLYTECHNIQUE.
A METHOD FOR EXTRACTING VOLUME PRODUCED NEGATIVE IONS.

M. Bacal, J. Bruneteau, P. Devynck and F. Hillion

Laboratoire de Physique des Milieux Ionisés
Laboratoire du C.N.R.S., Ecole Polytechnique
91128 Palaiseau Cedex, France.

ABSTRACT

This paper is dedicated to report the conclusions reached from examining the extracted (H⁻ and electron) currents and the H⁻ and electron densities near the extractor, in the hybrid multicusp source. The extraction of H⁻ is greatly improved by the presence of a small magnetic field parallel to the plasma electrode, which magnetically insulates this electrode from the bulk plasma. When biased positive, the plasma electrode depletes the electron population in the neighbouring region. To maintain plasma neutrality, negative ions from the main plasma replace the electrons in this region. Thus a high fraction of negative ions builds up in front of the plasma electrode. The implication of these phenomena for negative ion beam formation are discussed.

INTRODUCTION.

There has been considerable interest in the development of intense negative ion sources for use in high-energy neutral beam heating and diagnostic systems for nuclear fusion plasmas. The initial work on volume production as a source of H⁻ ions[1,2] has suggested that the dominant production mechanism is through dissociative attachment of low energy electrons (<1 eV) to highly vibrationally excited molecules.

Previous work[3] at Ecole Polytechnique, performed using the photodetachment technique for measuring the negative ion density[4], has shown that the central region of the hybrid magnetic multicusp source contains a high density and a high proportion of H⁻ ions (∼10%). More recently negative ion beams were extracted from this source; the extracted negative ion and electron currents appeared to be sensitive to the bias potential applied to the electrode of the extraction system in contact with the plasma (the so-called plasma electrode, PE)[3,5]. No change in the negative ion and electron densities is observed in the center of the source when the PE bias potential (denoted V_b) is varied[6].

These observations suggested that the plasma region located next to PE is very important for negative ion

© American Institute of Physics 1987

extraction and deserves a detailed study. We have established, by measuring the negative ion density, n_, at 2.2 cm from PE in a smaller volume hybrid multipole, that the optimization of the extracted negative ion current, obtained by biasing positive the plasma electrode, is correlated with an accumulation of negative ions in front of this electrode[6]. In order to understand the processes ruling negative ion extraction, we had to measure the spatial variation of the negative ion density and of other plasma parameters.

It is the purpose of this paper to report the systematic study of the axial variation of the plasma parameters in meantime with studying the extracted charged particle currents. Preliminary results[7] have shown that the stray magnetic field present in front of PE, had a large effect upon the extracted negative particle currents and their densities in the neighbourhood of PE[8-10].

EXPERIMENTAL SET-UP.

The source studied in these experiments has been described in detail elsewhere[3,5,6,8,11]. A cylindrical stainless steel chamber is surrounded by ten Samarium-Cobalt magnets (with a surface magnetic field of 3500 Gauss), with the north and south poles alternatively facing the plasma (Fig. 1). The end plates are not magnetized. The bottom end is bound by the plasma electrode of the extractor, PE, which is 9 cm in diam, and by an annular grid, G, which is grounded.

The primary electrons are produced by ten thoriated tungsten filaments biased 50V negative with respect to the chamber walls. These filaments are located in the upper part of the chamber, in the multicusp magnetic field; each filament is located in the radial plane passing through the saddle point of the multicusp magnetic field (in between two wall magnets). Due to this choice of the filament position, the primary electrons are trapped and confined in the neighbourhood of the cylindrical sidewall. Few energetic electrons escape into the central, magnetic field-free region.

<u>Extraction system.</u> In our investigation on the extraction of volume produced H^- ions from the hybrid source, we were interested in separating the electrons from the negative ions and in measuring separately their currents. We used an extraction system consisting essentially of three electrodes[5], shown on Fig. 2. The first electrode, in contact with the plasma, PE, has a circular extraction aperture, 0.8 cm in diameter. The second electrode, called 'separator', is located at 0.62 cm from PE and has also an opening 0.8 cm in diameter. A pair of Sm-Co magnets located in the separator, just behind the opening, create a transverse magnetic field (300 Gauss)

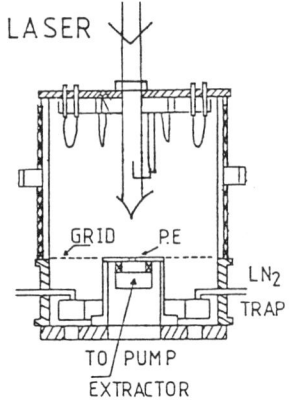

Fig. 1. Hybrid multicusp plasma generator.

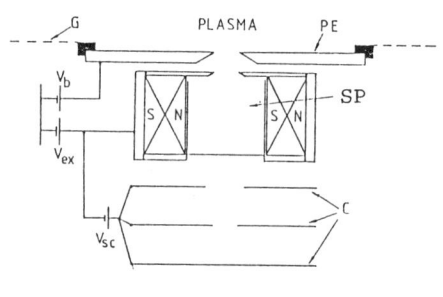

Fig. 2. Extraction system.
PE: plasma electrode;
SP: separator; G-grid;
C: collector.

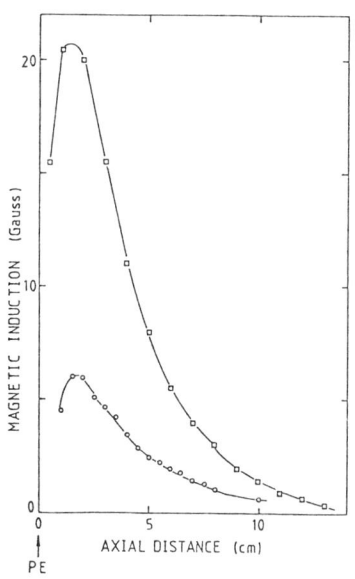

Fig. 3. Axial variation of the magnetic induction in front of the plasma electrode.
o : B-reduced;
□ : B-normal.

strong enough to deflect the accelerated electrons onto the separator, when the extraction voltage does not exceed 5 kV. The H⁻ ions are little affected by the presence of this field, which causes only a small lateral displacement of the H⁻ ion flux reaching the collector. The entire separator electrode is made of soft iron, in order minimize the stray magnetic fields. In some experiments this stray field was further reduced by shielding the back ends of the magnets with soft iron plates. We found that the value of the stray magnetic field was affecting the extracted negative ion and electron currents.

Stray magnetic field near the plasma electrode. The measurements were made with a Hall probe and showed that the largest magnetic field component is parallel to PE. Fig. 3 presents the dependence of the stray magnetic field induction upon the distance from PE for the two investigated situations:

a) without additional shielding of the back ends of the extractor magnets ('B-normal'), and

b) with soft iron plates added to the back ends of the magnets ('B-reduced'). As can be seen, magnetic screening reduces the maximum field induction in front of PE from 20 Gauss to about 6 Gauss.

Measurement of the H⁻ ion density. The H⁻ ion density was measured using the photodetachement technique[4]. In this technique the electron is detached from the H⁻ ion by means of a pulsed laser beam and collected by a cylindrical tungsten probe (0.5 mm in diam, 1.5 cm long) biased at +20 V relative to the anode. This creates a probe current pulse, Δi^-, whose height is proportional to the H⁻ ion density. The dc current to the probe, i^-_{dc}, is proportional to the electron density. Therefore the measured ratio $\Delta i^-/i^-_{dc}$ gives the relative negative ion density n_-/n_e. An independent measurement of n_e is necessary to determine n_-. We used a disk probe, 4 mm in diam. The laser beam was provided by a Nd:Yag laser (380 mJ/pulse, 15 ns pulse duration), with a photon energy of 1.2 eV. The laser beam enters the plasma through a small orifice in the upper flange and crosses the source along its axis. The cylindrical photodetachement probe is coaxial whith the laser beam and can be moved along the source axis. In this way, it is possible to obtain an axial profile of the H⁻ ion density.

RESULTS AND DISCUSSION

Fig. 4 shows that the dependence upon the PE bias, V_b, of the extracted negative ion, I^-, and electron, I_e, currents, is affected by the stray magnetic field in front of PE. The results correspond to a 50V-10A-2.5mTorr discharge. I^- exhibits the most pronounced maximum and attains the largest value, and I_e the most considerable drop, when the stray magnetic field in

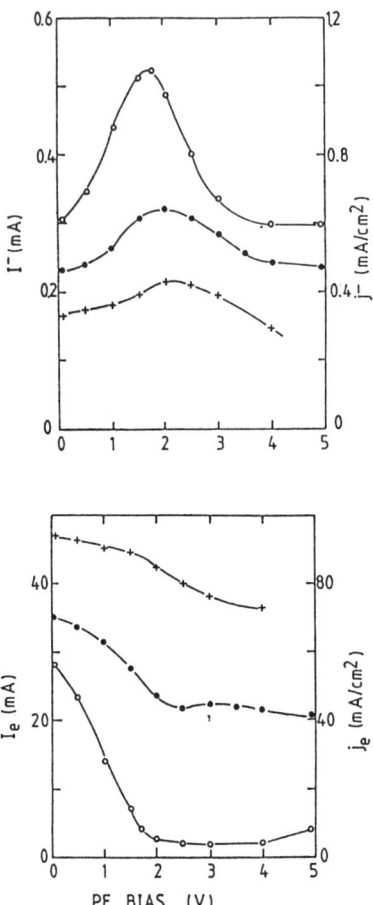

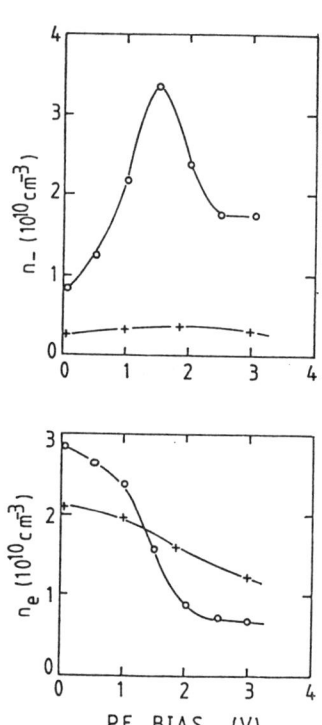

Fig. 4. Dependence of the extracted negative ion and electron currents upon PE bias, with different magnetic fields in front of PE.
o : B-normal (largest B)
● : ceramic magnets in the separator (medium B)
+ : B-reduced (lowest B)
50V-10A-2.5mTorr discharge.
V_{ex}=+1kV; Vgr=0.

Fig. 5. Dependence upon PE bias of the negative ion and electron densities at 1 cm from PE. 50V - 10A - 3mTorr discharge.
o : B-normal;
x : B-reduced.

the plasma is the strongest.

Figure 5 shows that the negative ion and electron density dependence upon V_b, measured on the axis of the source, at 1 cm from PE, with B-normal and B-reduced. Note that at this location n_- varies in a similar way to I^- and attains the maximum values at the same V_b value (+1.5 V). Although there is a similarity in the variation of the electron current and electron density, the relative values of the electron current and density are different: the extracted electron current, with B-reduced, is larger than with B-normal in the studied range of V_b, while the thermal densities are comparable. This different behaviour is discussed in Ref. 12, where it is shown that it is related to the higher amount of fast electrons present in the case of B-reduced.

Fig. 6 shows, for B-normal and B-reduced, the pressure dependence of I^- and I_e, with optimum V_b and with $V_b = 0$. Note that I^- is largest with B-normal in the whole pressure range studied. However, the increase of I^- due to optimizing V_b is largest at low pressure and almost vanishes when the pressure is increased.

Figure 7 shows the effect of V_b upon the plasma potential, electron temperature and density in the center of the plasma. The transition from B-reduced to B-normal makes the plasma parameters at the plasma center less sensitive to PE bias potential. With a weak magnetic field near PE (B-reduced), the plasma potential and electron temperature goes up, while the electron density goes down when V_b is increased. With a stronger magnetic field (B-normal), V_p, n_e and kT_e are not affected by V_b. This indicates that B-normal efficiently insulates the PE from the plasma electrons. Note the large difference in the electron temperature (a factor of two), related to the different values of the magnetic field near the PE. This change in kT_e has no major effect upon n_- and n_-/n_e in the center of the plasma.

Figure 8 compares the axial profiles of the electron density, n_e, and negative ion density, n_-, in the neighbourhood of PE, with B-normal and B-reduced. These profiles are shown for values of V_b from 0 to +3 V. When B-reduced is applied, n_e goes down with increasing V_b in the whole plasma. However when V_b is increased with B-normal applied, the decrease of n_e is localized mainly in the magnetized region, within 3 cm from PE. The electron density gradient depends strongly on V_b (Fig. 8).

Fig. 8 also presents the behaviour of n_- in front of PE; when B-reduced is applied, n_- only slightly increases with V_b. With B-normal, n_- increases by a factor of 2.6 at 1 cm from the PE, when $V_b = +1$ V is applied, and goes down when V_b increases further.

The role of the stray magnetic field, which can be

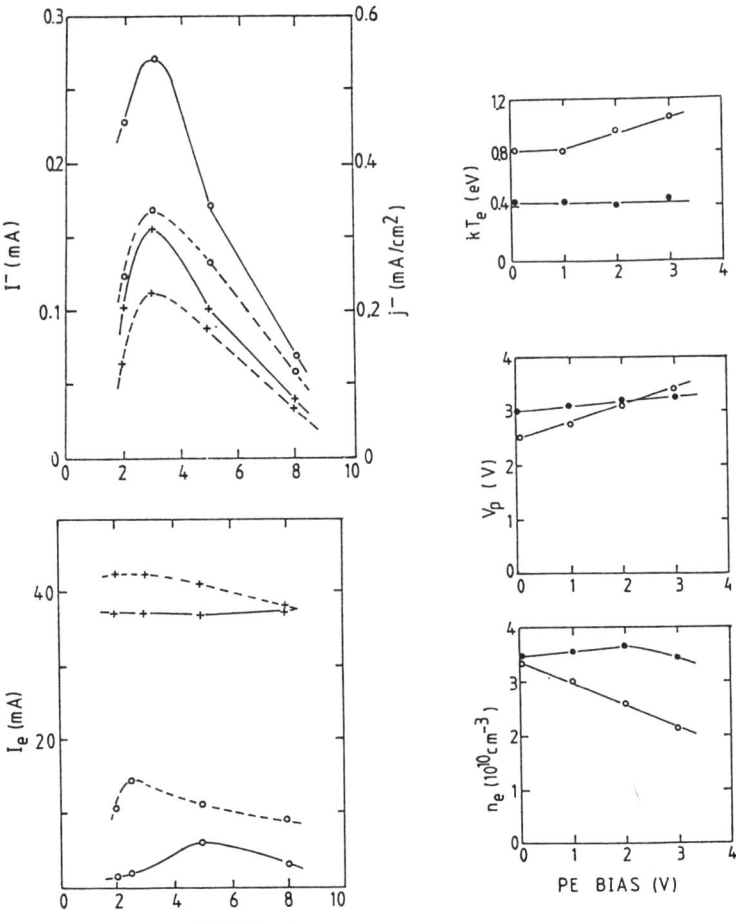

Fig. 6. Dependence of the extracted negative ion and electron currents upon H_2 pressure, with different magnetic fields in front of PE.
o : B-normal;
+ : B-reduced.
50V - 5A discharge.
V_{ex}=+1kV.
Full line: optimum V_b.
Dotted line: V_b=0.

Fig. 7. Dependence upon the PE bias of the plasma potential, electron temperature and density. Center of the source. 50V - 5A - 2mTorr discharge.
● : B-normal;
o : B-reduced.

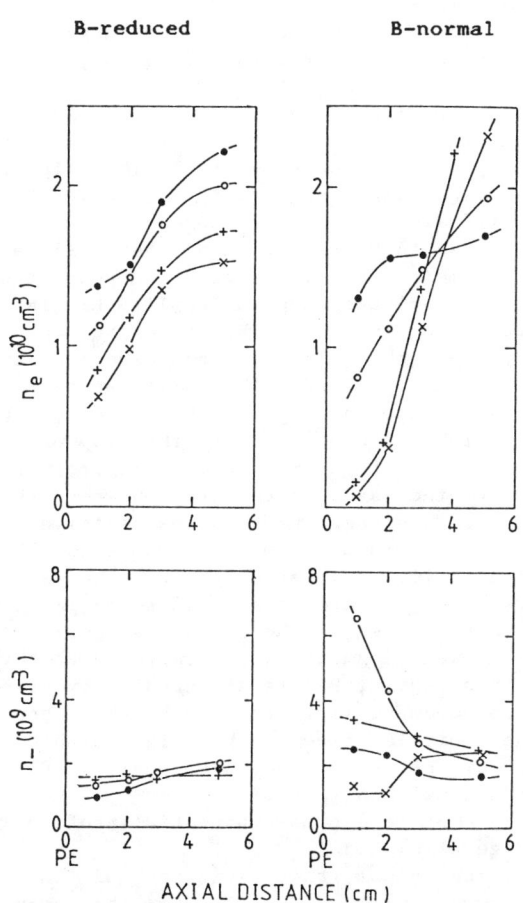

Fig. 8. Dependence of the electron and negative ion densities upon the axial distance from PE. 50V - 5A - 2mTorr discharge.
● : $V_b = 0$;   o : $V_b = +1V$;   + : $V_b = +2V$;   x : $V_b = +3 V$.
Left side: B-reduced.   Right side: B-normal.

idealized as parallel to PE, is to magnetically insulate PE from the plasma. It is shown in Ref. 9 that, in a collisionless situation, with B-normal, the electrons present in the main plasma (E<50eV) are unable to cross this transverse magnetic field and to attain PE; with B-reduced, electrons with E>15eV can reach PE. This explains the observed large difference in the extracted electron currents (Fig. 4), in spite of comparable thermal electron densities (Fig. 5).

When biased positive, PE collects a large electron current, denoted I_b. Figure 9 shows the variation of this current versus V_b for the case B-normal. The electron density present in the magnetized region in front of PE has to be attributed to diffusion processes across the magnetic field. Since the region close to PE region cannot be easily attained by electrons from the main plasma, the electron population in this region is strongly depleted when the applied PE bias potential allows the electrons to flow up to PE, as can be seen on Fig. 8.

Note on Figure 9 that the extracted electron current goes down with increasing V_b. Its variation is opposite to that of I_b. Indeed the increase of I_b is the primary cause of the reduction of n_e, which in turn, leads to the drop of I_e.

The new important fact indicated by Figure 8 is that the negative ion density increases in the region where the electron density is reduced. This suggests that the negative ions arrive from the main plasma into the magnetized region close to PE, where plasma neutrality is perturbed by the action of the magnetic field, which blocks the electrons. As a matter of fact, the heavier (compared to the electrons) positive ions are unaffected by the weak magnetic field employed, but their flow to the wall (in this region, PE) is affected by two factors:
1. the electron movement is restrained by the magnetic field; this modifies the ambipolar flow in the neighbourhood of PE.
2. the change of PE potential; the positive PE would even repel the positive ions when $V_b > V_p$.

The evolution of the plasma potential profile with V_b is illustrated by Figure 10.

Figure 9 also shows the variation of the extracted positive ion current versus V_b. It can be seen that this current goes down when the positive bias potential of PE is increased. Note that the negative ion current goes up to a maximum value, which is close to the value of the positive ion current, and then both currents decrease, conserving very close values. This occurs because the plasma neutrality in this region requires equal positive and negative ion densities, since n_e is very low.

Conclusion. The described results show that in the present experimental configuration the extracted H⁻ current is

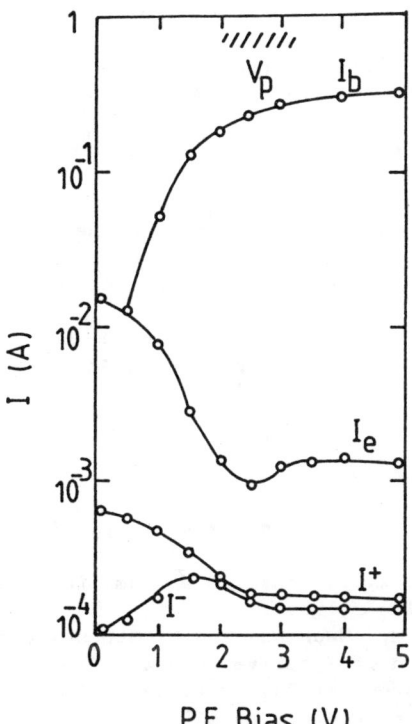

Fig. 9. Dependence upon PE bias of the extracted currents (positive ion I^+, negative ion, I^-, and electron, I_e) and current to PE, I_b. The range of plasma potential variation is indicated on the top of the figure. 50V-5A-2mTorr discharge. V_{ex}= +1kV. B-normal.

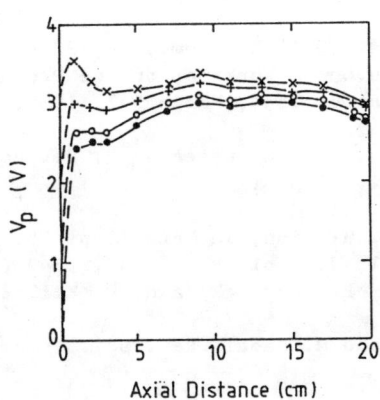

Fig. 10. Dependence of plasma potential upon the axial distance from PE, for four values of V_b:
● : V_b= 0; o : V_b = +1V;
+ : V_b = +2V; x : V_b=+3V.
50V-5A-2mTorr discharge. B-normal.

conditionned by the joint action of the magnetic field in front of PE and of the applied PE bias potential. One finds here a certain resemblance with the tandem or magnetically filtered configuration[13]. This resemblance is only apparent: in the described here experiments the magnetic field in front of PE acts only upon extraction, by concentrating in the extraction region the negative ions formed in the main plasma. In the tandem configuration the role of the magnetic filter is to create the target plasma region, the only where the negative ion formation is possible.

We propose to use the observed properties of the magnetic field in front of PE to optimize extraction of volume produced ions and considerably reduce the extracted electron component.

ACKNOWLEDGEMENTS. This work was supported by Ecole Polytechnique and Centre National de la Recherche Scientifique (France).

REFERENCES

1. M. Bacal and G.W. Hamilton, Phys. Rev. Lett., 42, 1538 (1979)
2. M. Bacal, A.M. Bruneteau, W.G. Graham, G.W. Hamilton and M. Nachman, J. Appl. Phys., 52, 1247 (1981)
3. M. Bacal, F. Hillion, M. Nachman and W. Steckelmacher, Production and Neutralization of Negative Ions and Beams, (3rd Int'l Symposium, Brookhaven 1983), A.I.P. Conference Proceedings N° 111, edited by K. Prelec, p. 419.
4. M. Bacal, Proceedings of this Symposium.
5. M. Bacal, F. Hillion and M. Nachman, Rev. Sci. Instrum., 56, 649 (1985)
6. M. Bacal and F. Hillion, Rev. Sci. Instrum., 56, 2274 (1986)
7. Proceedings of the 2nd European Workshop on the Production and Application of Light Negative Ions, March 5-7, 1986, Ecole Polytechnique,
8. M. Bacal, A.M. Bruneteau, J. Bruneteau, P. Devynck, F. Hillion and J. Bretagne, in Ref. 7, p. 17.
9. J. Bruneteau, in Ref. 7, p. 81.
10. M. Bacal, P. Devynck and F. Hillion, in Ref. 7, p. 75.
11. A.M. Bruneteau and M. Bacal, J. Appl. Phys., 57, 4342 (1985)
12. M. Bacal, J. Bruneteau, P. Devynck and F. Hillion, in Ref. 7, p. 201.
13. K.N. Leung, K.W. Ehlers and M. Bacal, Rev. Sci. Instrum., 54, 56 (1983)

DISCUSSION

Ehlers: I think it's important to, again, state Marthe's words. The magnetic filter does, indeed, allow us to generate a small region, in the case of hydrogen, of atomic plasma composed of positive ions and negative ions. And that's something I never thought would ever be done.

Forrester: In your very first slide, you showed what I think was a cylindrical, multicusp magnetic filter. I didn't quite understand that geometry.

Bacal: I tried to make a schematic representation, instead of a real one. You have magnets around the device as in any bucket source. We have vertical magnets parallel to the axis, alternatively facing north and south. This is the usual multicusp bucket configuration. What I'm saying is that our filaments are placed in between two magnets in this region of high transverse magnetic field. The dotted line I showed is just a kind of a similitude of a magnetic filter. We don't put anything there, but there is a magnetic field acting, in the same way as the filter acts in Andrew Holmes' configuration or in Kao Leung's configuration. I just tried to make a parallel between these two devices.

Holmes: I noticed that, in some of your results, you actually have supersonic negative ion velocities travelling faster than the sound speed of the positive ions. These drift velocities will then correlate with the density gradient or temperature gradient in the plasma. I was wondering whether you're at low pressures when you have these very high negative ion drift velocities where you see the highest plasma density gradients, or not.

Bacal: Well, sure. In the region next to the extractor.

Holmes: Is it higher in the high pressure?

Bacal: I think so, yes. Because at 8 mTorr we can't make these very low electron densities.

Whealton: Is there any hint of noise in the source, particularly when you get near the purely ionic plasma regime? Ion acoustic or 100 kilocycle-type noise?

Bacal: No, I don't think we have seen any sign of noise in this situation. There is no difference between the optimum and the non-optimum situation.

Schmor: I'm wondering if, at the low plasma potentials where you see your H^- current dropping, whether you looked at the extraction voltage as an optimizing parameter.

Bacal: Of course the extraction voltage is important, but we couldn't really notice an effect of the extraction voltage upon the plasma potential. However, we could notice, in some conditions, that the optimum plasma electrode bias is effected by the value of the extraction voltage. You see, this is a dependence of the extracted current on the discharge current, which seems to saturate when the voltage is low, and doesn't saturate so obviously at higher extraction voltage. So, obviously, the extraction voltage goes into the plasma in some way, and collects negative ions from farther and farther. Having a purely ionic plasma is a good move, because then we can apply very high extraction voltages. In a state where we don't have it, we are then limited to not very high extractor values just to be able to get rid of the electron component in some part of the extractor. Obviously, we are interested in working with higher and higher extraction voltages to take out the negative ions from all this region which contains them.

Nightingale: First, I have a comment on that last point. I think that the talk that Roy McAdams will be giving does include data from the Culham program that might explain why you have a dependence on the extraction voltage. Now, could you just clarify, again, what magnetic field you're referring to as normal? What is its value?

Bacal: The maximum value was 20G and the product (distance by magnetic field) is 96G-cm.

Nightingale: You made the comment that that made your hybrid discharge essentially different from the tandem ones. I think that we would claim that since we have magnets in our extraction plate anyway, for electron suppression, in fact we have a very similar....

Bacal: Yes, but you have other magnetic fields everywhere in the source, so I don't know. I can't make a comparison with your source. Maybe in Culham you have more experience on that than anybody else because of the positive ion work. You know the effect of the stray fields from the extractor; I think there was some problem with that. Anyway, you probably have this field. We have to be conscious of its existence, that's all I mean. I can't analyze your device, because you have to try to do experiments with a strong field in an extractor and with a weak field in an extractor. Just try to see what it is without the other fields. That would be interesting. Kao has no magnetic fields from the extractor in some of their experiments, when he moved the magnetic filter to the plasma electrode, and he did see an increase in the negative ion current. He didn't have this little magnetic field. When I did it, I didn't see any effect of moving the filter, just because I already had this magnetic field there and the filter didn't add anything.

H⁻ ION SOURCE SCALING STUDIES AT LBL*

A.F. Lietzke and C.A. Hauck
Lawrence Berkeley Laboratory, Berkeley, Ca. 94720

ABSTRACT

Four experiments are reported: 1) Constant arc voltage operation was compared with constant arc current operation over a similar range of plasma density. The higher arc voltage required in the constant arc current operation produced more H⁻, but it was at the expense of an even larger increase in the accelerated electron current. 2) A comparison of different magnetic filter locations showed the highest accelerated H⁻ and the lowest e⁻ content at the closest filter location. 3) Beam-forming electrode aperture comparison for round apertures from 2mm to 10mm diameters and a 11x27mm oval slot showed a decreasing H⁻ and e⁻ with increasing aperture area. 4) Increasing accelerator voltage produced an increasing plasma potential and electron temperature and decreasing electron density in the exit chamber of the plasma source.

INTRODUCTION

In our H⁻ ion source scaling studies we have observed that many parameters affect the production and/or delivery of volume-produced H⁻ ions to the accelerator: gas pressure, location of filaments, arc current, arc voltage, aperture size, filter location, filter strength, type of electron feed or heater, the relative potential of the filter rods, the relative potential of the beam-forming electrode (Grid 1), and the size of the plasma chamber. One experiment reported here compared the accelerated H⁻ and e⁻ yields over a range of main chamber plasma density (a two chamber system) for two styles of operation... constant current vs. constant voltage. This experiment was a more methodical version of earlier unreported work that showed pessimistic results for high arc voltage operation. Reports from Stevens and York[1] at LANL prompted us to extend the earlier work to higher voltages. We also wanted to display a comparison in a more device-independent manner. We chose "J^+_{bucket}" (the main chamber saturated +ion current density) as an appropriate independent variable that "measured" the main chamber "excitation" level.

In the second experiment we extended the low density work of Leung,

*This work was supported by U. S. Doe under Contract No.DE-AC03-76SF00098.

et al.[2] to higher density. We compared the H⁻ and e⁻ yields (over the range of our power supplies) at three filter locations relative to Grid 1. Data from this experiment has also been used for calibration of Hiskes' model of H⁻ production and is presented elsewhere in this conference.[3]

A new beam-forming electrode was installed which gave a new accelerator gap spacing and permitted delivery of the total beam at higher current densities. The design also allowed changing the aperture size of the beam-forming electrode without venting the system.

The fourth parameter we investigated was the effect of the accelerator extraction voltage on the plasma potential and electron density and temperature in the exit chamber of the ion source.

EXPERIMENTAL SETUP

Fig.1 is a schematic of LBL's TS-1 facility. The plasma chamber (22x22x22 cm, cubic) was a standard longitudinal multi-cusp magnetic bucket (1.3x1.9x22 cm, SmCo, 4 cm- spaced side-wall cusps parallel to the beam direction). It was divided into two chambers by four magnetic rods

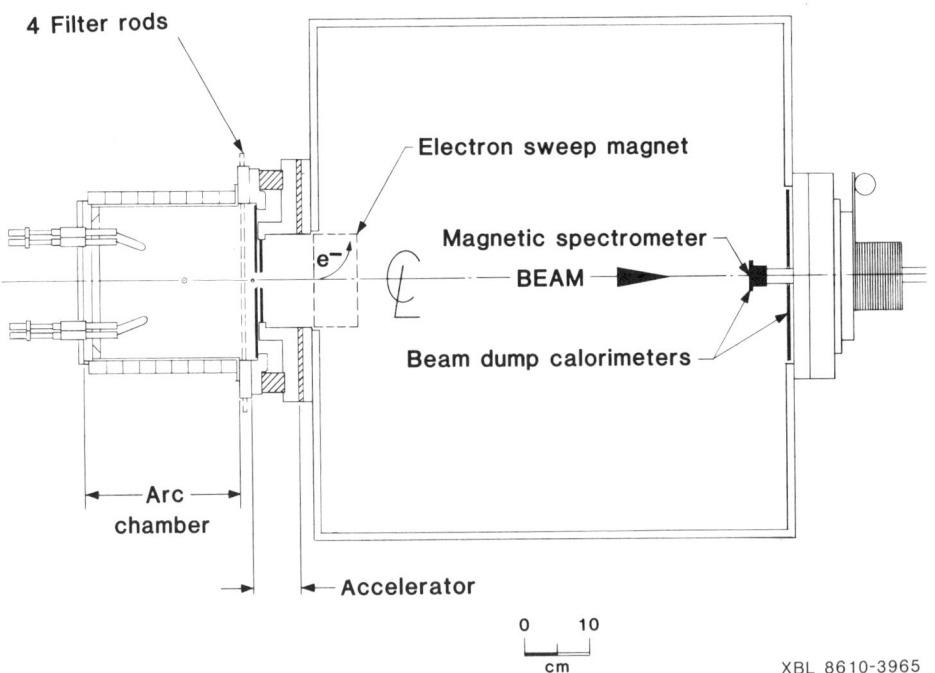

Fig. 1 Schematic view of LBL's Test Stand-1 facility.

(4.7x4.7x300mm SaCo with 6.4cm spacing) (Fig. 2) that were inserted into the bucket at either of three planes(2.5, 6.9, 11.3 cm) upstream from Grid 1 extraction plane. The peak filter field is 90 Gauss; and the peak Grid 2 field is 350 Gauss (Fig. 3). The filter thickness is 300 G-cm (Fig. 4). Eight 1.5mm diam. filaments injected up to 200 amperes of ionizing electrons into the main (hot) chamber at energies up to 200eV.

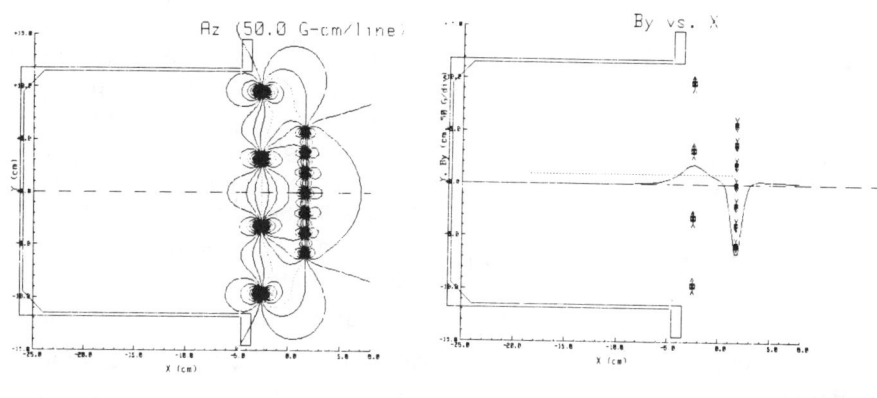

Fig. 2   Magnetic structure: 4 filter rods @ 2.5cm up-stream from the beam-forming electrode; 7 Grid #2 magnets interact.

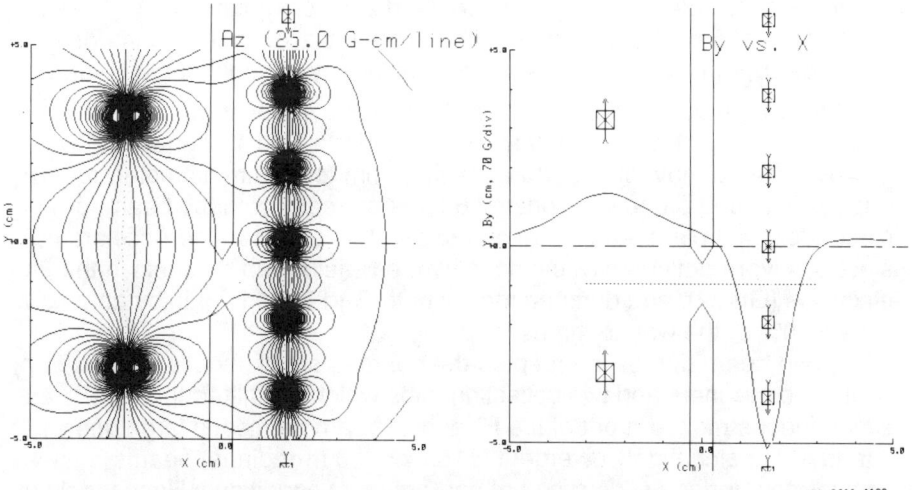

Fig. 3   Magnetic structure near the beam-forming electrode.

The arc/filament power supply system could be operated with either constant current or constant voltage. The hydrogen gas flow for all

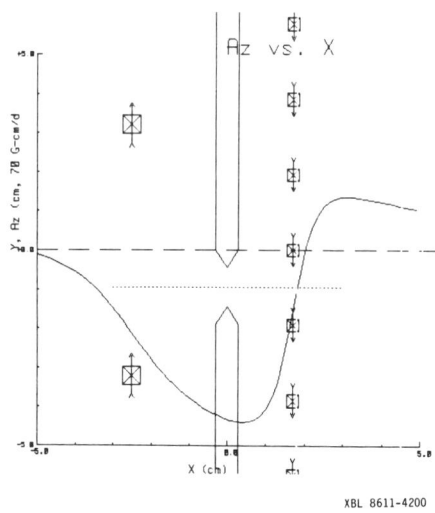

Fig. 4 Vector potential near the beam-forming electrode.

experiments was the previously determined optimum which gave a cold (arc-off) chamber pressure of 10 mtorr. One water-cooled cylinder-probe monitored the center of the main plasma chamber; and a cup probe (for good low-density saturation characteristics) monitored the exit (low temperature) chamber 8mm upstream from the beam-forming electrode's mid-plane, and 5mm off the aperture's center-line. One of the probes was routinely swept (usually the one in the exit chamber) to ascertain the dependence of the plasma potential (V_p), the electron temperature (T_e), and the electron density (N_e) upon the parameter under investigation. These probe characteristics for the main chamber, when the filter location was at 2.5 cm, are shown in Fig.5.

For the first two experiments we had a one gap accelerator (Fig. 6) with one aperture exposed (10x25 mm, rectagular) that had an experimentally determined overdense (Grid 2 clipping) limit of $J/V^{1.5}$=45 nano-pervs/ 2.5 cm^2 hole. No magnets were installed in the beam-forming electrode ($B_{Grid\ 1}$=0)(Fig. 3), so as to reduce any impediment to H$^-$ or e$^-$ flow toward the accelerator hole. For the last two experiments, the filter location was 1.7cm upstream from Grid 1 and the Grid 1 aperture (11x27 mm, oval) had a movable aperture mask (2 mm to 10 mm diameter, round) (Fig.7). The cup probe was located 6 mm upstream from the beam-forming electrode's mid-plane, and 5 mm off the aperture's center-line. Accelerated electrons were deflected twice: once by the B-field from magnets in the 2nd electrode (Grid 2), and dumped mostly onto Grid 2, and again by an auxiliary B-field 10% of the way to the beam dumps.

Two beam dumps intercepted the beam: a small, mobile one (6x9cm) contained magnets and two collecting cups which permitted the simultaneous measurment of the H$^-$ and e$^-$ beam profiles; a large one was intended to catch highly divergent (±160 x ±350 m-radians) beams accelerated under a wide range of non-optimum conditions. Hence, only at the lowest currents, and (with inadequate voltage) at the highest currents would the dumps under-estimate the beam. A window, permitting visual beam observations from the side (the aperture's long direction and the beam

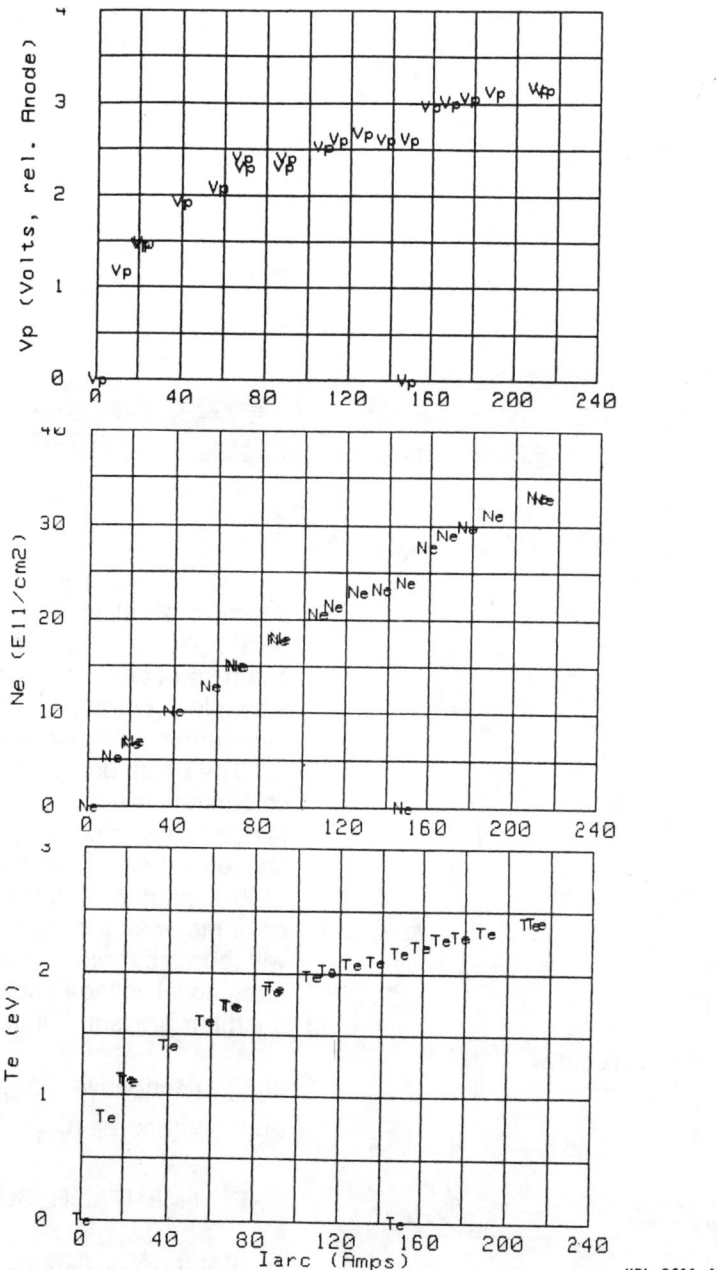

Fig. 5 Main chamber plasma potential, electron density and temperture vs. I_{arc} @ V_{arc}=100V, pressure=10mtorr (cold), filter rods @2.5cm.

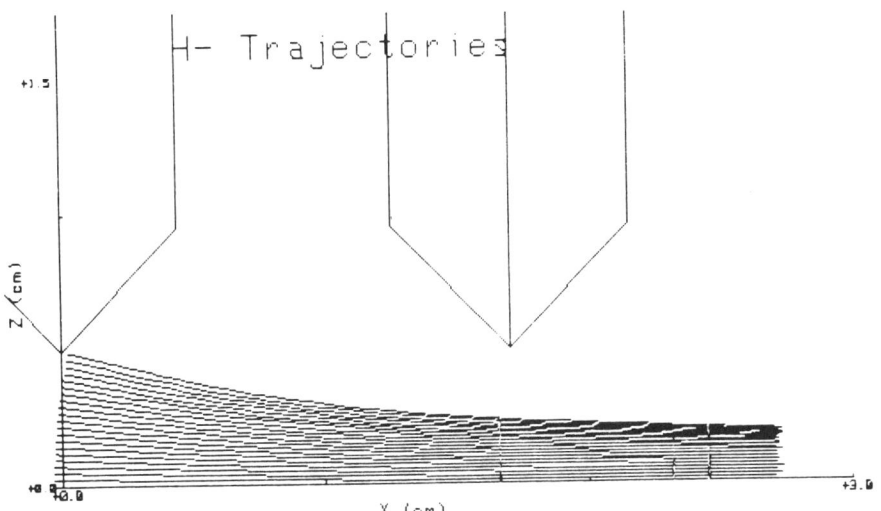

Fig. 6 H⁻ trajectories in the round aperture approx.@ $J(H^-)$=8 mA/cm^2, V_{accel}=6000 V.

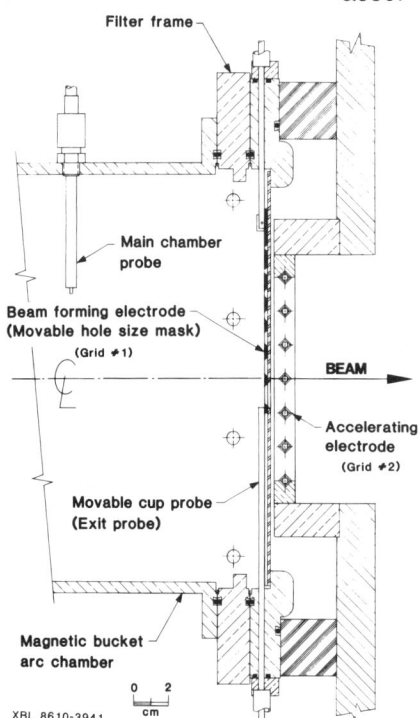

Fig. 7 Schematic view of arc chamber and accelerator, with movable aperture mask.

beam observations from the side (the aperture's long direction and the beam short direction), and the movable spectrometer could be used to determine the size of the beam.

The beam dumps were monitored calorimetrically and electrically (saturated +ion current). The more convenient electrical signal was calibrated against the water-flow calorimetry and found to be linear with beam current, but dependent upon accel voltage and dump chamber pressure. In this paper

$J(H^-)$= (Heat power) /V_{accel}/A_{hole}; and J_e=I(grid 2)/ A_{hole}.

EXPERIMENTAL RESULTS

For the V_{arc}/I_{arc} comparison, the arc voltage was varied from 60-170 volts at two currents (50A,100A); and the current was varied from 0-200A at

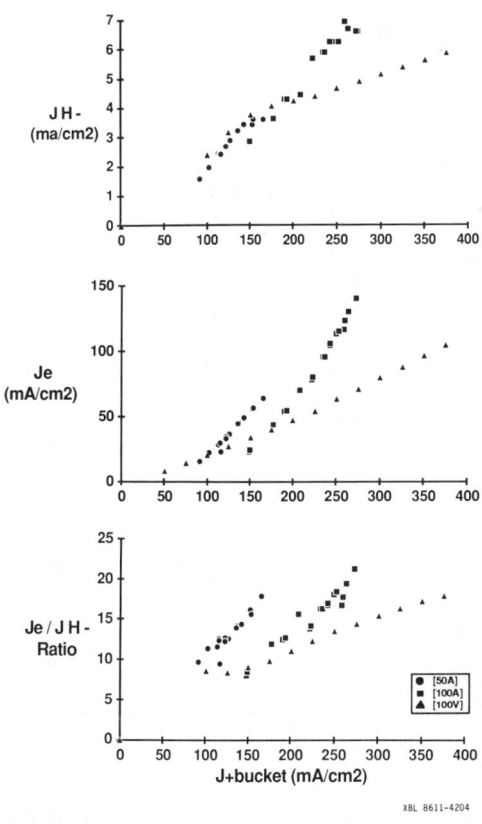

Fig. 8 I_{arc} vs. V_{arc} variation.

one voltage (100V). An 11kV accel voltage was used to avoid Grid 2 beam-clipping at the top end. At the low power end, all the J (H⁻) data overlaps (Fig. 8); But at high power, the high voltage (constant current) operation produced more H⁻, avoiding the first knee characteristically observed with a variation in arc current. Unfortunately this increase came at the expense of more electrons and a higher e⁻/H⁻ ratio than obtained with 100V operation.

In the filter location experiment (V_{arc}=100V, 0<I_{arc}<200A, V_{accel}= 6 kV), some Grid 2 beam-clipping was observed as a rollover in J (H⁻) at the top end (Fig. 9). The closest filter delivered the highest H⁻ and the lowest e⁻ (consistent with the lower density results of Leung[2]). The data with the filter at 11.3cm was even more extreme, but was not plotted because the main chamber probe was part way into the filter and read ≈2x too low (indicating that the density falls going into the filter).

The aperture size scaling study (V_{arc}=100V, I_{arc}=75A, filter @ 1.7cm, V_{G1} @ anode, V_{accel}=12kV) indicated a factor of 1.25 decrease in J(H⁻) (calorimetrically) and 3x smaller accelerated electron current density as the aperture size was increased from 2 mm to 10 mm diam. (Fig. 10). The majority of the data clustered around J(H⁻)=7mA/cm². The 3cm² (oval) aperture data showed a proportionately similar decrease (0.65 of the 2mm intensity). At another operating condition (higher H⁻, lower e⁻), the qualitative trend was the same. The plasma potential and electron temperature increased, and the electron density decreased in the exit chamber, with larger aperture diameters (Fig.11). Most of the changes were observed only for the largest apertures. The arc-only measurements were used for reference purposes.

The results of the V_{accel} study (V_{arc}=100V, I_{arc}=75A, filter @ 1.7cm, V_{G1} @ anode, D=8mm) showed that the density of the electrons in the exit chamber decreased as the accel voltage increased (Fig. 12). This was also observed for larger apertures. Most of the variation was observed between 0-2kV.

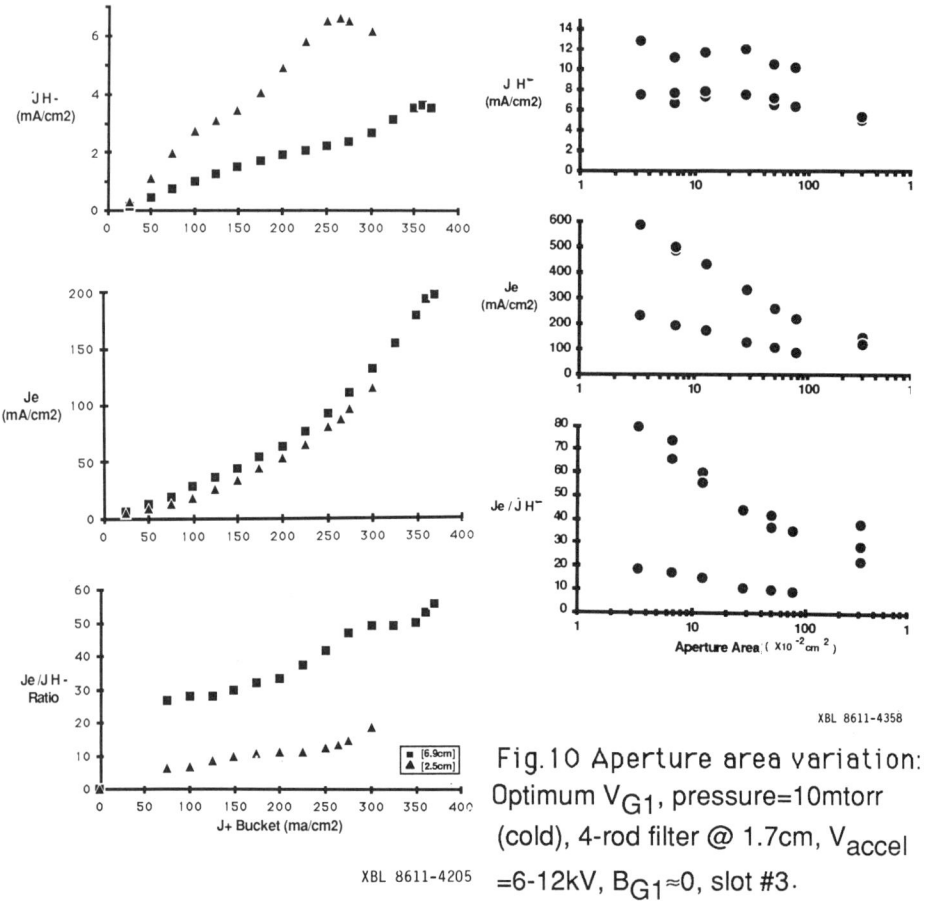

Fig. 9 4-rod filter location: Optimum V_{G1}, pressure=10mtorr (cold), V_{accel}=6kV, $B_{G1} \approx 0$, slot #3 (2.5cm^2).

Fig. 10 Aperture area variation: Optimum V_{G1}, pressure=10mtorr (cold), 4-rod filter @ 1.7cm, V_{accel}=6-12kV, $B_{G1} \approx 0$, slot #3.

DISCUSSION

The higher H⁻ observed in the high V_{arc} operation may support either of two arguments: 1) the more energetic primary electrons would likely increase the density of electrons that are energetically capable of vibrationally exciting molecules close enough to the filter to survive and

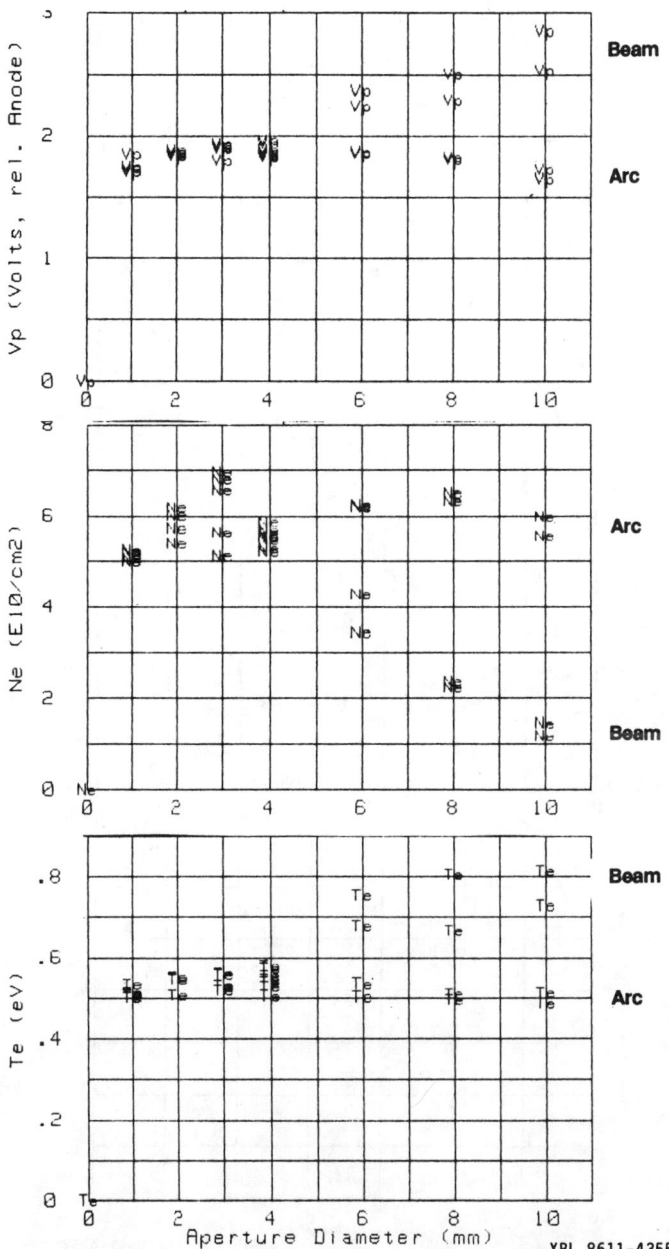

Fig. 11 Exit chamber plasma potential, electron density and temperature vs. aperture diameter @ V_{arc}=100V, I_{arc}=75A, pressure=10mtorr (cold), 4-rod filter @ 1.7cm, V_{G1}=0.

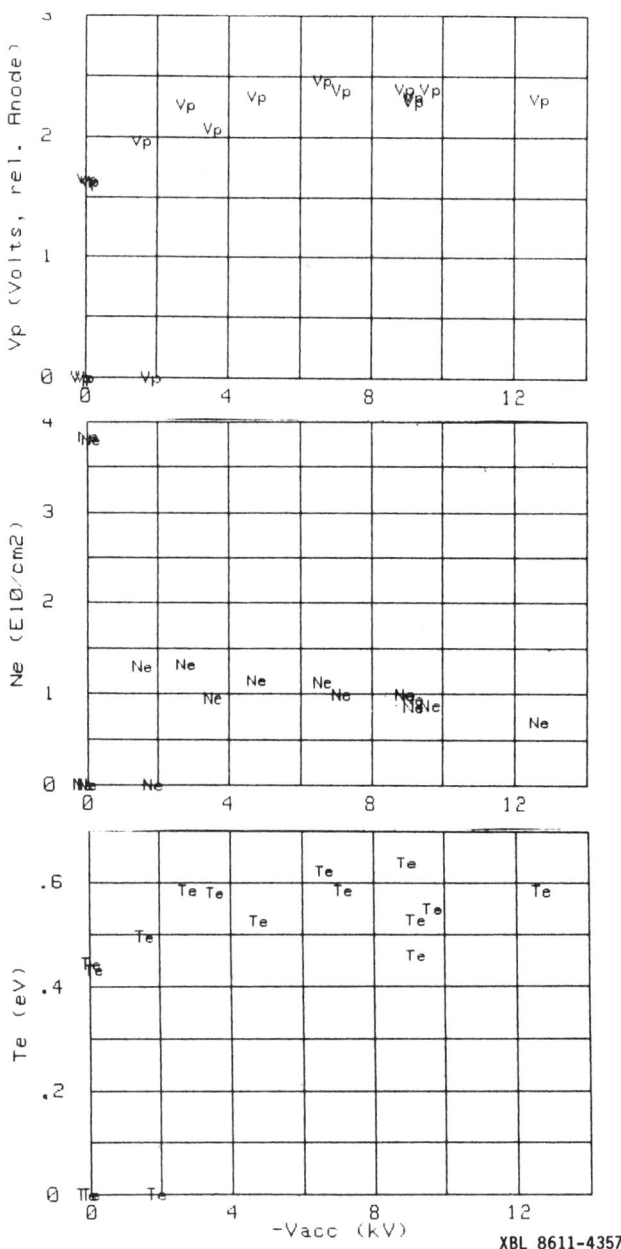

Fig. 12 Exit chamber plasma potential, electron density and temperature vs. V_{accel} @ V_{arc}=100V, I_{arc}=75A, pressure=10 mtorr (cold), 4-rod filter @ 1.7 cm, V_{G1}=0.

produce extractable H⁻. 2) The higher e⁻ density observed in the second chamber more efficiently utilizes the available molecules.

What is not so clear is why the exit electron density increased (relative to the first chamber density) with higher arc voltage. Perhaps the more energetic primary electrons decouple from the body of the electron distribution. This would result in a lower main chamber electron temperature and faster diffusion across the filter. Unfortunately, we did not measure the main chamber temperature. The exit temperature is always lower and increases more slowly with arc power, thus, we can not support this conjecture. It is also not clear as to the applicability of these results to more practical systems having stronger electron-control B-fields at the beam-forming electrode.

The second experiment produced results in realistic agreement with Hiskes' model (see Hiskes and Lietzke, this conference). According to that model, the vibrationally excited molecules from the edge of the main plasma volume do not survive well in the exit chamber (due to destruction by atomic hydrogen generated in the main chamber). The H⁻ ions are similarly fragile. As the filter, which controls the electron temperature transition, is moved upstream, the dominant source of the H⁻ moves away from the beam-forming electrode and less H⁻ is accelerated.

In the aperture size experiment, larger H⁻ and e⁻ currents were obtained with larger apertures, but, when normalized by their respective areas, lower current densities were observed. The dependence was more pronounced for the electrons and was opposite to earlier (unpublished) long slit data. The anomaly is unresolved.

The accelerated H⁻ was approximately correlated to a decrease in Ne at the exit probe (8mm upstream)(Fig.10, Fig.11). Noticeable changes began to appear for D≤5mm which implied that the plasma parameters were affected L≤1.5 diameters upstream. The reduction in H⁻ with increasing diameter is believed to be due to a partial depletion of cold electrons by acceleration through the aperture. The removal of cold electrons is possible (B≈0) and consistent with the observed increase in the exit plasma potential and electron temperature. Electron heating was precluded because the exit potential was smaller than the main chamber potential.

The last experiment indicated that these changes (Ne,Te,Vp) are relatively insensitive to the plasma sheath position which must move semi-qualitatively as $J \sim V^{1.5}/d^2$. Most of the dependence existed below 2kV. This was interpreted to mean that the scale length for upstream plasma perturbation greatly exceeded the distance of plasma boundary movement in this experiment (1-2mm).

CONCLUSIONS

Four experiments have been reported. They show that higher arc voltage operation generally yielded more H⁻ at the same main chamber density, but this increase occurred at the expense of higher e^-/H^- ratios. The closest filter plane (2.5cm from the beam-forming electrode) yielded the highest H⁻ output intensity and the lowest e^-/H^- ratio. The larger apertures delivered larger currents but had smaller H⁻ current densities; and they had smaller e^-/H^- ratios, contrary to previous slit experiments. Larger apertures also produced greater perturbations on the plasma immediately upstream from them, but only when some small accelerating voltage was applied. Most of the voltage dependence occurred when $V_{accel}<2kV$.

REFERENCES

1. R. R. Stevens Jr. and R. L. York, private communication (1968).
2. K. N. Leung, K. W. Ehlers, and R. V. Pyle, Rev. Sci. Instr. 56, 364 (1985).
3. J. R. Hiskes and A. F. Lietzke, This proceedings.

OPERATION OF A MAGNETICALLY FILTERED MULTICUSP VOLUME SOURCE*

Ralph R. Stevens, Jr., R. L. York
Los Alamos National Laboratory, Los Alamos, NM 87544

and

K. N. Leung, and K. W. Ehlers
Lawrence Berkeley Laboratory, University of California
Berkeley, California 94720

ABSTRACT

The results of experimental studies on an optimized version of the Lawrence Berkeley Laboratory (LBL) volume source are presented. Negative ion yields and emittance data were obtained for operation with both H^- and D^- beams. At high arc power, H^- beam currents up to 10 mA with rms normalized emittances of $0.0080\,\pi \cdot$ cm $\cdot$ mrad were obtained from a 6.3-mm-diam emission aperture. The yields of D^- beams were approximately half those of H^- beams, and the normalized emittances were 1.7 times smaller at the same current density. The results of these studies indicate that the present operation is limited by the extraction system rather than the ion source.

INTRODUCTION

For the past year there has been an ongoing program at the Los Alamos National Laboratory to test a magnetically filtered multicusp volume source for production of H^- and D^- beams at high duty factor for accelerator applications. An optimized version of the LBL volume source has been operated in pulsed mode up to 5% duty factor at high arc power to determine the ion beam yields and emittances that can be produced. This source has been operated on both a 100-keV test stand at LAMPF and a 30-keV test stand on the ion source test stand (ISTS) facility. The results of these tests indicate that both H^- and D^- beams can be produced with sufficiently high brightness to be of use for high duty factor accelerator operation.

---

*Work performed under the auspices of the U.S. Dept. of Energy and supported by the U.S. Army Strategic Defense Command.

EXPERIMENTAL APPARATUS

The optimized LBL volume source[1] is shown in Fig. 1. The source chamber is 20 cm in diameter and 22 cm long. It is divided into a driver region and an extraction region by a set of filter rods containing samarium-cobalt magnets; the filter strength is 256 G-cm. The filter rods in this optimized source are located almost at the extractor plane. The source chamber is made of oxygen-free copper and is surrounded by ten columns of samarium-cobalt magnets. These columns are connected by four additional columns of magnets on the end plate to form (together with the filter rods) a full-line cusp configuration for magnetic confinement of the plasma. The primary electrons in the driver region were produced by three 0.15-cm-diam

Fig. 1. The LBL magnetically filtered multicusp volume source.

tungsten filaments. The gas pressure in the source was varied from 2 to 20 mtorr, whereas the pressure in the experimental chamber varied from 2×10^{-6} to 2×10^{-5} torr.

Beam tests were carried out both on the LAMPF test stand and on the ISTS facility. The tests at LAMPF employed a tetrode extraction system similar to that used on earlier tests with the original LBL volume source.[2] The beams were accelerated to 72 keV, focused with a solenoid lens, mass analyzed with a 45° bending magnet, and then transported to an emittance scanner. This scanner is a conventional slit and collector system and has a spatial resolution of 0.2 mm and angular resolution of 1.5 mrad. Beam currents were measured using both toroids and Faraday cups on both the analyzed and the unanalyzed sections of the transport line. A more detailed description of this test stand is presented in Ref. 2.

The tests carried out on the ISTS facility employed a high perveance, accel-decel extraction system designed for accelerating beam currents up to 140 mA at 29 keV. A schematic diagram showing the volume source mounted on ISTS is shown in Fig. 2. Electron loading precluded operation at H^- beam currents over 10 mA; therefore, smaller emission apertures were installed in the plasma electrode for the present tests. A pair of samarium-cobalt magnets was also installed in the plasma electrode to reduce electron drain current. Operation of the system was then possible with a 6.3-mm-diam emission aperture. The tests on ISTS were carried out at a relatively low voltage (10 to 20 keV) but with electron drain currents up to 0.5 A. The

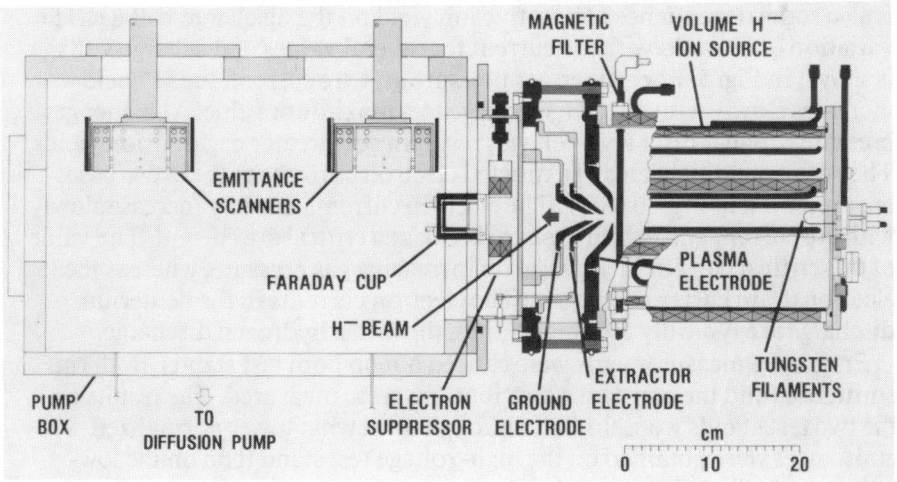

Fig. 2. The LBL volume source on the ISTS facility.

beam was extracted and subsequently transported through a double dipole magnet system having sufficient ∫ Bdl (800 G-cm) to deflect the extracted electrons onto a cooled beam stop while introducing only a 1.0-mm offset in the central trajectory of the negative ion beam at 20-keV beam energy. The negative ion beam currents were then measured using a magnetically suppressed Faraday cup. An electrostatic sweep emittance scanner was employed on the ISTS facility and was typically operated with a spatial resolution of 0.6 mm and an angular resolution of 0.5 mrad.[3]

In both the LAMPF and ISTS tests, the source was operated at high arc power; arc voltages up to 400 V and arc currents as high as 500 A were employed. The beam pulse lengths were typically 800 μs, and the pulse repetition rates varied from 0.5 Hz (ISTS) to 60 Hz (LAMPF). Moreover, in both cases, the electron loading on the high-voltage power supply rather than the ion source limited operation at still higher arc power.

EXPERIMENTAL RESULTS

The negative ion currents obtained from a volume source depended primarily on the source pressure and the arc current. For a given gas pressure, the negative ion yield increases with increasing arc current to a broad maximum value and then decreases slowly. For sufficiently high arc current, the negative ion yield increases linearly with increasing source pressure and again exhibits a broad maximum before decreasing at high gas pressures. The variation of H^- and D^- beam currents obtained for several values of gas flow are shown in Figs. 3 and 4. The H^- yields obtained here are in agreement with results obtained at TRIUMF with a similar source.[4] There is also some dependence of negative ion yield on the discharge voltage. The variation of H^- yield with arc current for several values of discharge voltage is shown in Fig. 5. For a given gas pressure and arc current, the H^- yield increases slowly with discharge voltage to a maximum value. At higher gas pressures, higher discharge voltages result in still greater negative ion yields. The corresponding yields of extracted electron current for these two cases are presented in Figs. 6 and 7. The electron currents initially increase slowly with arc current and exhibit a sharp break at a critical arc current. The value of this critical current increases with increasing gas pressure, whereas the electron drain current decreases. The electron currents in the deuterium discharge are typically 50% higher than those in a hydrogen discharge.

Emittance measurements were carried out on both test stands. Both rms emittances and the emittance distributions were measured. The results on the two test stands were similar, although somewhat lower normalized emittances were obtained on the high-voltage test stand than on the low-voltage test stand. Typical data showing the dependence of H^- and D^- normalized emittances as a function of beam current density are presented

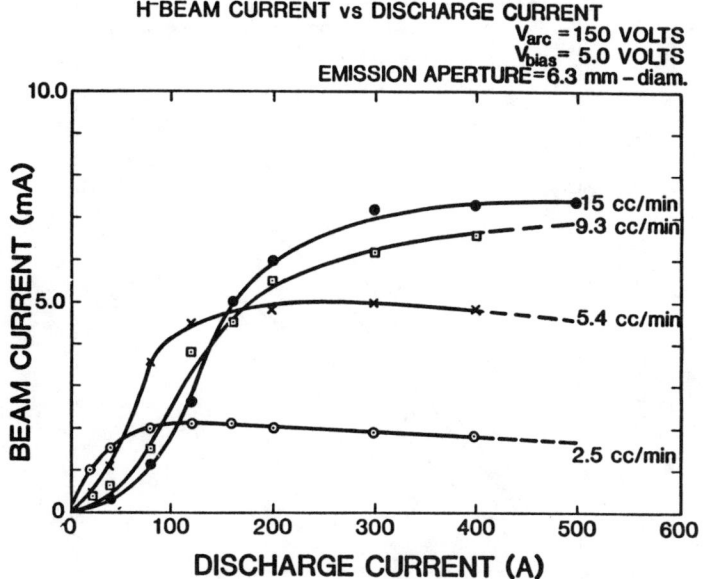

Fig. 3. H⁻ beam current vs discharge current.

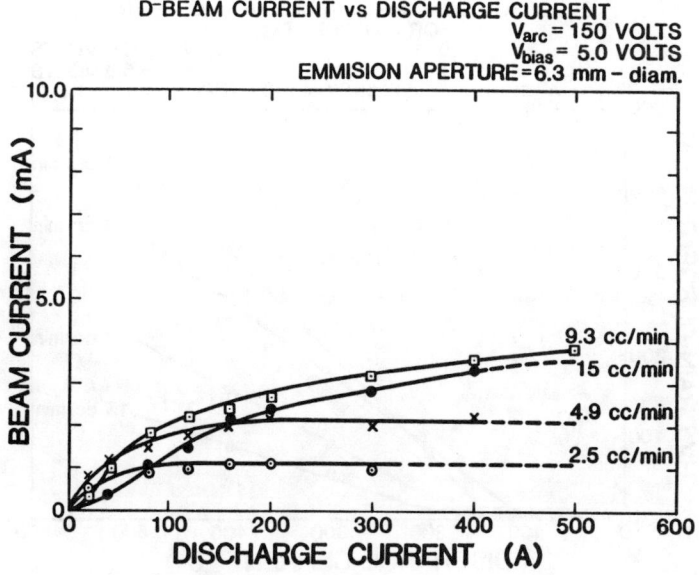

Fig. 4. D⁻ beam current vs discharge current.

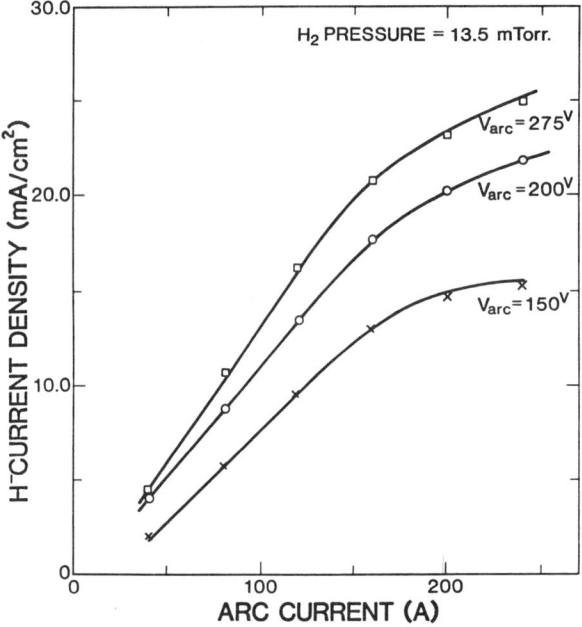

Fig. 5. Dependence of H⁻ current density vs arc current on arc voltage.

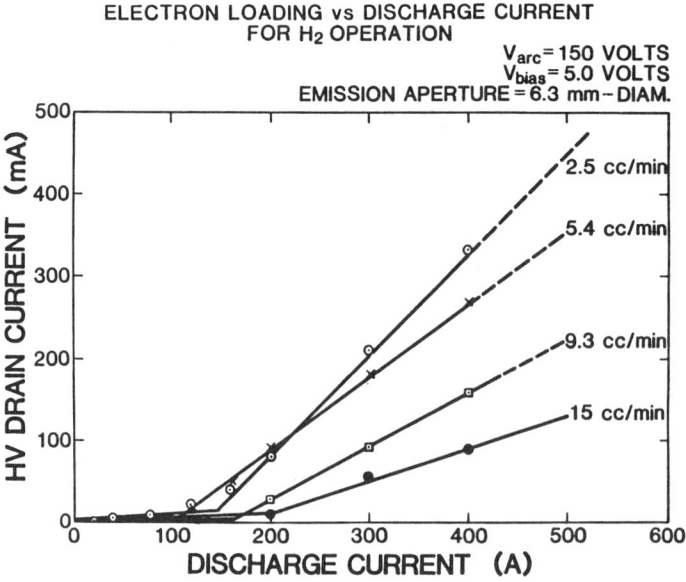

Fig. 6. Electron drain current vs discharge current for a hydrogen operation.

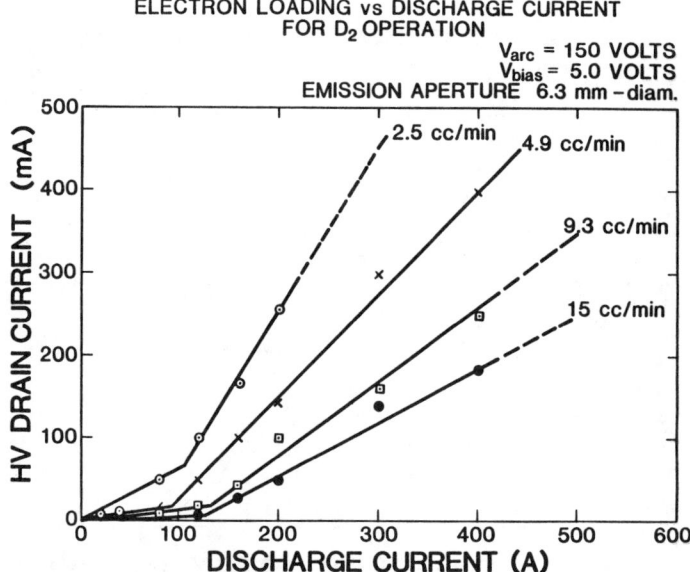

Fig. 7. Electron drain current vs discharge current for a deuterium operation.

in Fig. 8. The data for D⁻ beams were limited to half the current density as that for H⁻ beams because of the lower yields obtained for deuterium operation in this source.

In considering the emittance distribution of a beam, it is useful to plot the marginal emittance E(F) for a given beam fraction F as a function of $\ln[1/(1-F)]$. A beam having a Gaussian distribution function in transverse phase space will have a linear graph in this plot; departures from Gaussian behavior, if any, are then readily apparent. To determine the rms emittance for the total beam, we have used an estimator based on the rms emittance of the marginal distribution $E_{rms}(F)$, i.e., on the rms emittance of that fraction F of the total beam above a specified threshold. For a Gaussian distribution characterized by ε_{rms}, we have[5]

$$E_{rms}(F) = \varepsilon_{rms}\ (1 + (1-F)\ln[(1-F)/F]) \ .$$

For arbitrary beam distribution, this equation defines the rms emittance of an equivalent Gaussium distribution

$$\varepsilon_{rms} = E_{rms}(F)/1 + (1-F)\ln[(1-F)/F] \ .$$

Now, by extrapolating the value of ε_{rms} as we let F→1, we estimate the rms for the total beam from the marginal distribution without having to obtain

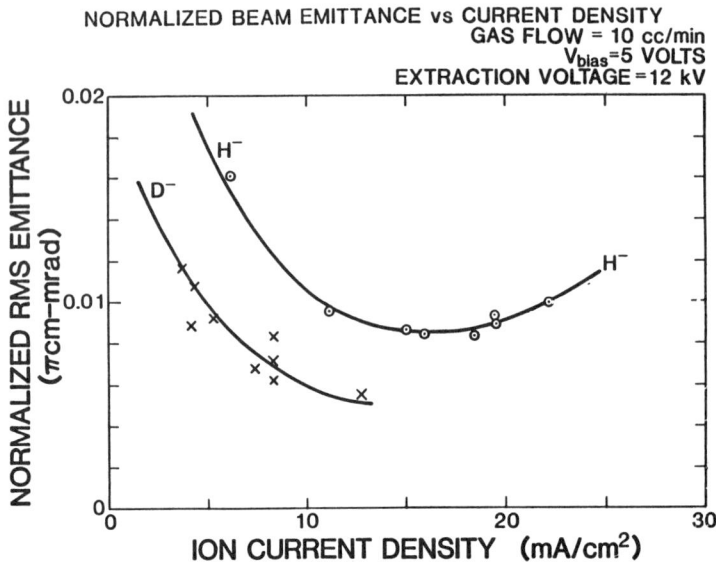

Fig. 8. Normalized rms emittance vs current density.

E(F) at F = 1. If, in fact, the beam distribution is Gaussian, the values obtained from this estimator will be independent of F. The variation of the estimator values with F is a measure of the departure of the beam distribution from a Gaussian distribution. This procedure has been used to determine the rms emittances given in this paper.

The emittance distribution of the beams obtained from this volume source in general could not be characterized by a single Gaussian function. The plots of E(F) vs ln[(1-F)/F] usually have two distinct linear regions, one for small values of F and the other for F values close to 1. In Fig. 9, a plot of E(F) vs ln[(1-F)/F] is presented for several beams obtained when the ion source parameters and the total beam energy were kept constant and the extraction voltage was varied. As the extraction voltage was increased, the emittance corresponding to larger F values increased, whereas the emittance values for the smaller values of F were essentially independent of extraction voltage. The higher emittance values produced with increasing extractor voltage are associated with filamentation in the emittance distribution and, thus, the rms emittance of the total beam increases. The emittance values for lower values of F (the core of the beam) are unchanged and presumably reflect the contribution from the thermal energy spread in the ion source. Similar behavior has been observed in these distribution functions for variations in the plasma bias voltage.

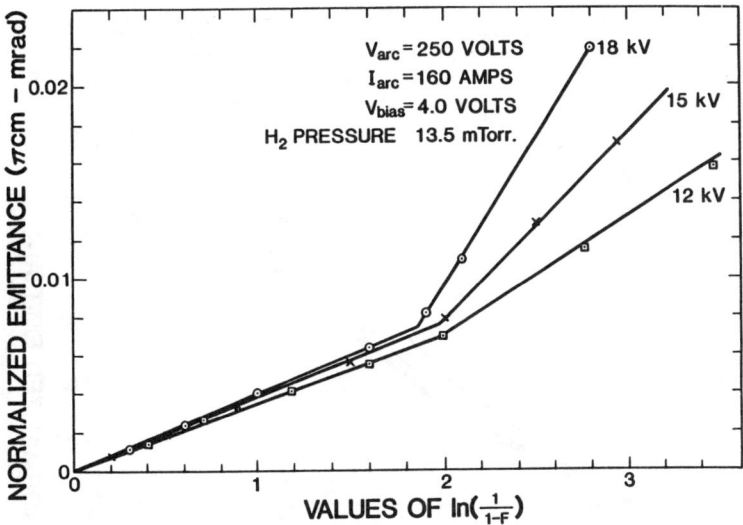

Fig. 9. Dependence of beam emittance vs ln[1/(1-F)] on extraction voltage.

However, studies carried out on the effect of the decel gap voltage in the accel-decel accelerator showed a different effect on the emittance distributions. The results obtained for a 6.0-mA H⁻ beam are shown in Fig. 10. As the decel voltage was increased from zero to several tens of volts, the entire emittance distribution curve was reduced. When the decel voltage exceeded 100 V, there was no further effect on the emittance distribution. This behavior suggests that the electrostatic well produced by the decel gap is reducing the emittance growth in the beam in the transport line, presumably by changing the neutralization of the beam. Similar behavior has been seen with a surface converter source[6] and with the TRIUMF volume source.[7]

DISCUSSION

In general, operation with hydrogen and with deuterium is similar; the D⁻ ion currents and emittances are less, whereas the electron loading is higher. The negative ion yield curves (Figs. 3 and 4) increase approximately quadratically at low arc currents and then saturate. The quadratic behavior is consistent with the two-stage mechanism generally believed to account for the negative ion production in volume sources, i.e., formation of vibrationally excited molecules from excitation by fast electrons in the driver region followed by disassociative attachment with thermal electrons in the extraction region. For sufficiently high arc currents, the negative-ion yield saturates and depends directly on source pressure because the principal

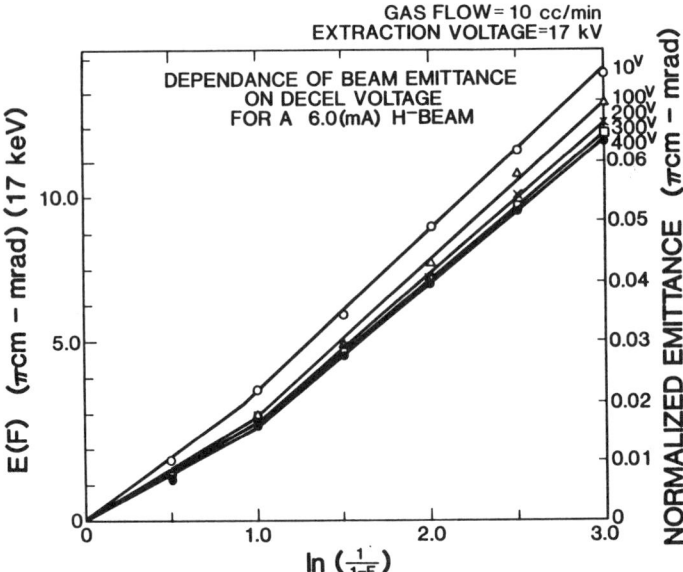

Fig. 10. Dependence of beam emittance vs ln[1/(1-F)] on decel voltage.

production mechanism (electronic excitation) and the principal destruction mechanism (collisional detachment) of the vibrationally excited molecules depend on the same fast electron density.

For very high arc currents, the negative ion yields begin to decrease because of the increasing importance of other destruction mechanisms (mutual destruction with positive ions and associative detachment with hydrogen atoms). For sufficiently high gas pressure, the negative ion yield again saturates. Estimates of stripping loss in the system are not large enough to account for this decrease, and this behavior is most likely due to a decrease in the population of the relevant excited states caused by collisions with gas molecules.

The current densities obtained in the present study (30 mA/cm^2) with a 6.3-mm-diam aperture are somewhat less than what was obtained previously[2] with a smaller emission aperture. The present operation, however, entailed tens of milliamperes of beam impingement on the extractor electrode, which was several times the transmitted current. The high beam loss suggests that the yield reduction is probably due to negative ion loss on the extractor electrode because of ion optical effects rather than a failure of ion source scaling with emission aperture area. Further tests are needed to resolve this question.

The emittance values obtained in this present work also do not scale exactly as expected from the previous work. The rms emittance values obtained for H⁻ beams with a 6.3-mm-diam emission aperture had a minimum value of $0.0080\,\pi \cdot$ cm $\cdot$ mrad, whereas previously emittance values of $0.0032\,\pi \cdot$ cm $\cdot$ mrad were obtained with a 3-mm-diam aperture, which would imply $0.0067\,\pi \cdot$ cm $\cdot$ mrad with a 6.3-mm aperture. The strong dependence of emittance on beam perveance (Fig. 8) and on extraction voltage (Fig. 9) in the present study indicates that the apparent emittance growth is also due to extraction optics. In the previous studies these difficulties were in large part obviated by the use of the tetrode accelerating column, which permitted independent control of the extraction voltage and the beam energy.

CONCLUSIONS

The tests on the optimized LBL volume source have demonstrated that H⁻ current densities of 30 mA/cm² can be obtained from a 6.3-mm-diam emission aperture with normalized rms emittances of $0.0080\,\pi \cdot$ cm $\cdot$ mrad. The yield of D⁻ ions are one-half those of H⁻ ions for the same ion source condition while the D⁻ emittances are 1.7 times smaller for the same ion current density. The scaling of beam current and emittance with emission aperture shows some apparent deterioration at this larger (6.3-mm) emission aperture diameter. Higher quality extraction systems are now needed to verify whether this scaling failure is a property of the ion source or of the extraction system now being used.

ACKNOWLEDGMENTS

We would like to thank P. Schaftstall and T. Dorr for technical assistance. This work was performed under the auspices of the Department of Energy and supported by the U.S. Army Strategic Defense Command.

REFERENCES

1. K. N. Leung, K. W. Ehlers, and R. V. Pyle, Rev. Sci. Instrum. **56**, 364 (1985).
2. R. L. York, R. R. Stevens, Jr., K. N. Leung, and K. W. Ehlers, Rev. Sci. Instrum. **55**, 681 (1984).
3. P. W. Allison, J. D. Sherman, and D. B. Holtkamp, IEEE Trans. on Nucl. Sci., **30** (4), 2204 (1986).
4. K. R. Kendall, M. McDonald, D. R. Mosserop, P. W. Schmor, and D Yuan, Rev. Sci. Instrum. **57**, 1277 (1986).
5. P. W. Allison, "Some Comments on Emittance of H⁻ Ion Beams," these proceedings.
6. J. Kwan, Lawrence Berkeley Laboratory, private communication, 1986.
7. P. W. Schmor, TRIUMF Laboratory, private communication, 1986.

CHARACTERISTICS OF A SMALL MULTICUSP H⁻ SOURCE*

K. W. Ehlers, K. N. Leung, R. V. Pyle and W. B. Kunkel
Lawrence Berkeley Laboratory
University of California
Berkeley, CA 94720

ABSTRACT

It is shown that the H⁻ current extracted from a magnetically filtered multicusp source can be enhanced by optimizing the extraction chamber length, by employing the proper chamber wall material, by mixing hydrogen with xenon gases in the discharge and by injecting very low energy electrons into the filter and extraction regions. A large improvement in H⁻ yield can be achieved by using a multicusp source with a much reduced plasma volume. In this arrangement, H⁻ current densities as high as 240 mA/cm^2 have been obtained from a 1-mm-diam extraction aperture.

INTRODUCTION

H⁻ and D⁻ ions are used to generate efficient neutral beams with energies in excess of 150 keV.[1] It has been demonstrated that volume-produced H⁻ ions, extracted from a filter-equipped multicusp source can provide high quality H⁻ beams with sufficient current density (40 mA/cm^2) to be useful for both neutral beam heating of fusion plasmas and accelerator applications.[2] In order to produce this high H⁻ current density, it was necessary to operate the source with a discharge current as high as 350 A. For long pulse or dc operations, it is desirable to improve this relatively low arc efficiency so as to reduce source cooling requirements and to prolong cathode lifetime.

Several methods to improve the efficiency of the filter-equipped H⁻ source have been investigated. By optimizing the extraction chamber length, a substantial improvement in the H⁻ output has been achieved.[3] Experimental results have demonstrated that the H⁻ yield can be enhanced by choosing aluminum or copper as the chamber wall material[4] or by mixing hydrogen and xenon gases in the source discharge.[5] A substantial increase in H⁻ yield also occurs when very low-energy electrons (E ≈ 1 eV) are added into the filter or extraction regions.[6] Most recently, a small multicusp source has been fabricated and operated successfully to generate volume-produced H⁻ ions.[7] From this new source, H⁻ current densities as high as 240 mA/cm^2 have been extracted from a

---

*This work is supported by the Air Force of Scientific Research and by the U.S. DOE Contract No. DE-AC03-76SF00098.

LBL-22385

1-mm-diam aperture. The increase in H⁻ output is mainly due to an increase in the source plasma density.

EXPERIMENTAL SETUP

A schematic diagram of the ion source used for generating volume H⁻ ions is shown in Fig. 1. The stainless-steel source chamber (20 cm diam by 24 cm long) is surrounded externally by 10 columns of samarium-cobalt magnets which form a longitudinal line-cusp configuration for primary-electron and plasma confinement. A samarium-cobalt magnet filter divides the entire chamber into an arc discharge and an extraction region. Detailed description of this filtered multicusp source arrangement has been reported previously.[2,8] In brief, the filter provides a limited region of transverse magnetic field which is strong enough to prevent the energetic primary electrons from entering the extraction zone. However, both positive and negative ions, together with cold electrons are able to penetrate the filter and they form a plasma in the extraction region.

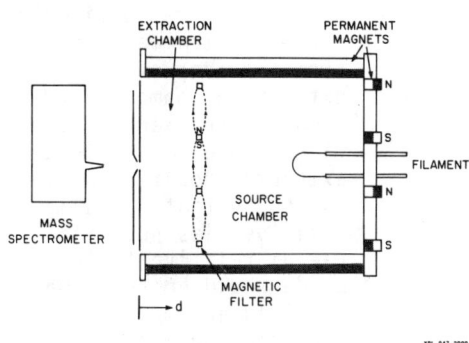

Fig. 1 Schematic of the multicusp ion source equipped with a magnetic filter.

The open end of the source chamber is enclosed by a two-electrode acceleration system. Positive or negative ions were extracted from the source through a small 0.1 x 1.0 cm² slot. A steady-state hydrogen plasma was produced by primary electrons emitted from two 0.05-cm-diam tungsten filaments. The entire chamber wall, together with the filter rods, served as the anode for the discharge. In order to optimize the H⁻ output, the first accelerator (or plasma) electrode was biased at a potential more positive than the chamber wall.

Plasma parameters were obtained by small planar Langmuir probes located at the center of the source and extraction chambers. A compact magnetic deflection mass spectrometer,[9] located just outside the extractor was used for relative measurement of the extracted H⁻ ions. In addition, a permanent-magnet mass separator was used with a Faraday cup to measure the extracted H⁻ and electron current. With this arrangement, it is possible to measure the ratio of H⁻ to electron current as well as the extracted H⁻ ion current density.

EXPERIMEMTAL RESULTS

(a) Optimization of Extraction Chamber Length

In order to determine the optimum H⁻ yield as a function of the length of the extraction chamber, a moveable extraction and spectrometer system was fabricated.[3] Figure 2 summarizes the data that were obtained as the extractor was moved from the edge of the extraction chamber (d=0) towards the plane of the filter (d=6 cm). At each extractor position d, it was found that a small positive bias potential (+4 V) relative to the anode would enhance the H⁻ yield accompanied by a reduction in electron current. The data in Fig. 2(a) show that a substantial increase in H⁻ yield (about a factor of 6) occurs when the extractor is moved from the edge of the source to a position near the filter. This result seems to indicate that a sizable portion of the extracted H⁻ ions are formed in the filter region.

The largest part of the extractor power supply drain current I^- is composed of electrons. Figure 2(b) and 2(c) illustrate that this drain current I^- and the extracted positive ion current I^+ behave similarly when d < 3 cm. When d > 3 cm, I^- decreases but the positive ion current drain does not, indicating that the electrons are drifted away by the E x B motion in the

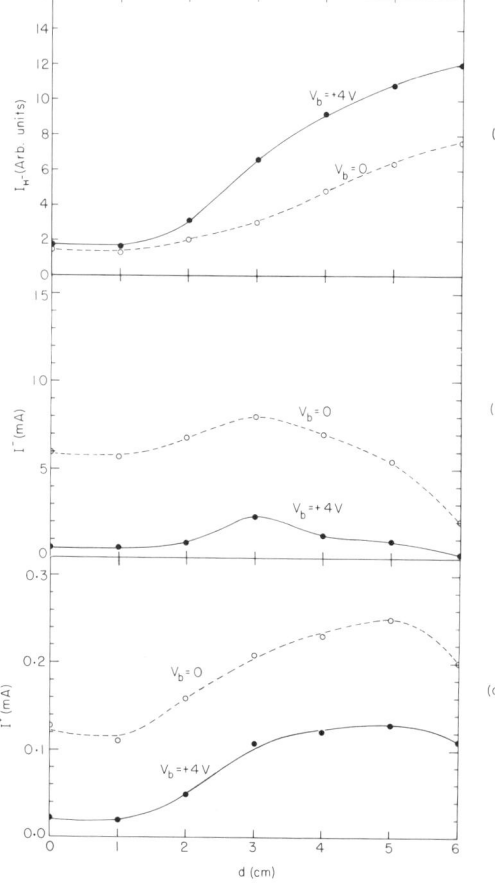

Fig. 2 H⁻ yield and the extracted electron and positive ion current as a function of the extractor position.

acceleration gap, or more readily collected on the biased plasma electrode.

With the extractor located at d=0, the extractable H⁻ current density for 1 A of discharge current in this test source geometry is 0.12 mA/cm² and the ratio of I_e/I_{H^-} is 200. When the extractor is moved very close to the filter (d=6 cm), the H⁻ current density for the same discharge current increases to 0.7 mA/cm² and the ratio of I_e/I_{H^-} is less than 2.

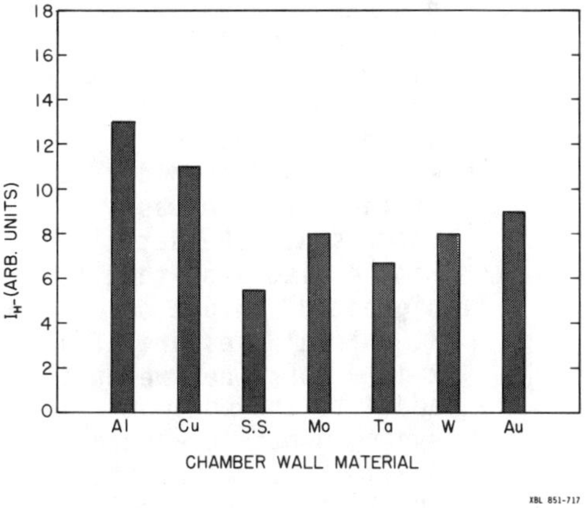

(b) <u>Optimization of Chamber Wall Material for H⁻ Production</u>

To study the effect of wall materials on the H⁻ yield, thin (0.13 mm thick) cylindrical metal liners were installed on the chamber wall.[4] These liners were cleaned in an ultrasonic alcohol bath before installation. To ensure good thermal and electrical contact with the source chamber, two stainless-steel rings were used to force the liner to lie flush against the vessel wall. Figure 3 shows the magnitude of the H⁻ output signal when seven different metal liners are compared under the same plasma conditions. The results show that aluminum and copper generate the highest H⁻ yield while stainless-steel produces the lowest. The H⁻ output of other metal liners such as Mo, Ta, W, and Au falls between those of Al and stainless-steel.

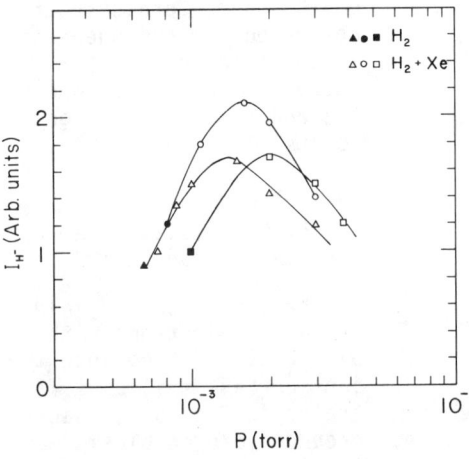

Fig. 3 Relative yield of H⁻ ions when the source is operated with different liners.

Fig. 4 The H⁻ yields as a funstion of total hydrogen and xenon pressure for three different hydrogen base pressures (indicated by ▲●■). The discharge is maintained at 80 V, 3 A.

(c) <u>Effect of Mixing H_2 with Xe on H⁻ Production</u>

The effect on the H⁻ yield by adding xenon gas into the hydrogen discharge[5] is illustrated in Fig. 4. The

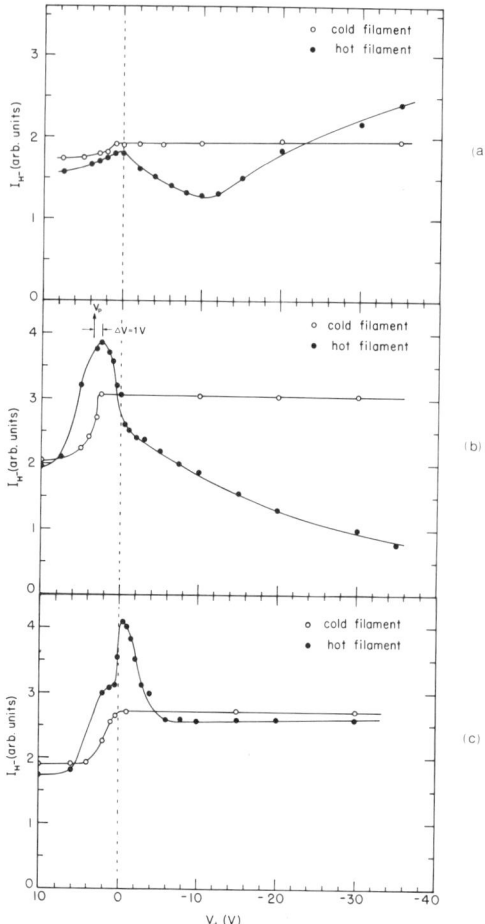

Fig. 5 H⁻ yield as a function of the discharge voltage of the second filament when it is installed inside (a) the source chamber, (b) the extraction chamber, and (c) the filter region. The data shows the H⁻ yield when the filament is operated with the without electron emission.

source was initially operated with only pure hydrogen. As xenon gas was introduced into the discharge, and with the bias potential V_b on the plasma electrode optimized, the H⁻ output first increased, reached a maximum, and then decreased as the pressure was increased. The data in Fig. 4 also show that the highest H⁻ output occurs at a total pressure of 1.5×10^{-3} Torr when xenon is added to the optimum hydrogen base pressure of 8×10^{-4} Torr. At this point, the increase in H⁻ yield is more than 75% for a constant discharge power of 80 V, 3 A. If the ionization gauge readings on the source pressure are corrected for hydrogen and xenon, then the real increase in pressure due to the addition of xenon gas is just 13%.

(d) <u>Enhancement of H⁻ Production by Cold Electron Injection</u>

The effect on H⁻ yield by adding electrons with different energies into the ion source[6] is illustrated in Fig. 5. A background hydrogen plasma was initially produced by a dc discharge of 80 V, 2 A from one set of filaments located in the source chamber. Additional primary electrons are then injected into the source chamber, the extraction chamber or the filter region from a second set of filaments. The emission current from the filament of set 2 was maintained at approximately 1 A. The H⁻ output as a function of the discharge voltage of the second filament set when it was positioned in the three regions is presented in Fig. 5. The data in Fig. 5(a) show that there is a reduction in H⁻ yield when electrons with energies lower than the ionization energy of H_2 are introduced into the source chamber. However, an increase in H⁻ output was obtained when electrons with energies ≈ 1 eV are added into the filter or extraction region (Fig. 5(b) and 5(c)). This result indicates that only very low-energy electrons can enhance the formation of the H⁻ ions in the filter or extraction chamber region. Since both dissociative attachment of vibrationally excited H_2 molecules and dissociative recombination of H_2^+ ions have the highest reaction rate for forming H⁻ at these low electron energies, further investigation is required in order to determine the exact process which is responsible for generating the H⁻ ions.

(e) <u>H⁻ Generation from a Small Filtered Multi-cusp Source</u>

In order to generate intense H⁻ beams, a high density source plasma is required. For this reason, a small multicusp source (7.5 cm diam by 8 cm long) has been constructed to generate volume H⁻ ions.[7] This source (Fig. 6) is equipped with a strong neodymium magnet filter and the magnetic-field-free discharge volume is small (2.5-cm-diam by 4 cm long). To enhance the H⁻ yield, the chamber wall of the source is fabricated from copper. The first electrode of the accelerator is placed very close to the filter plane so as to reduce the extraction chamber length.

CBB 856-3985

Fig. 6 A small multicusp H⁻ source.

At low discharge power, the plasma density in the source region is about 5 times higher than that of the larger multicusp source described in the previous sections. The spectrometer output signal shows that the H⁻ yield also increase by about the same amount. This small source can be operated with either tungsten or directly-heated LaB_6 cathodes. Since the entire source chamber wall together with the surrounding dipole magnets are cooled by a water jacket, the source is capable of long pulse or steady state operation at high discharge power.

This source has been operated at short pulse lengths (~ 7 ms) to generate a small H⁻ beam (extraction aperture diameter = 1 mm) for arc current as high as 120 A. Figure 7 is a plot of the extracted H⁻ current density J_{H^-} versus the

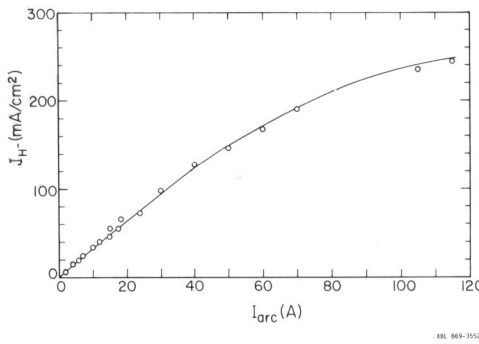

Fig. 7 Extracted H⁻ current density versus arc current.

discharge current for a constant arc voltage of 150 V. The result demonstrates that H^- current densities as high as 240 mA/cm^2 can be obtained from the source with an arc current of about 120 A. The ratio of extracted electron current of H^- current is about 4. Experiments are now planned to enlarge the aperture size so that higher H^- current can be extracted.

REFERENCES

1. K. H. Berkner, R. V. Pyle, and J. W. Stearns, Nucl. Fusion 15, 249 (1975).
2. R. L. York, Ralph R. Stevens, Jr., K. N. Leung, and K. W. Ehlers, Rev. Sci. Instrum. 55, 681 (1984).
3. K. N. Leung, K. W. Ehlers, and R. V. Pyle, Rev. Sci. Instrum. 56, 364 (1985).
4. K. N. Leung, K. W. Ehlers, and R. V. Pyle, Appl. Phys. Lett. 47, 227 (1985).
5. K. N. Leung, K. W. Ehlers, and R. V. Pyle, Rev. Sci. Instrum. 56, 2097 (1985).
6. K. N. Leung, K. W. Ehlers, and R. V. Pyle, Rev. Sci. Instrum. 57, 324 (1986).
7. K. N. Leung, K. W. Ehlers, R. V. Pyle, and W. B. Kunkel, (to be published).
8. K. N. Leung, K. W. Ehlers, and M. Bacal, Rev. Sci. Instrum. 54, 56 (1983).
9. K. W. Ehlers, K. N. Leung, and M. D. Williams, Rev. Sci. Instrum. 50, 1031 (1979).

EXTRACTION OF NEGATIVE ION IN
REFLEX TYPE SHEET PLASMA NEGATIVE ION SOURCE

T. Kuroda, K. Sakurai, Y. Oka and O. Kaneko
Institute of Plasma Physics, Nagoya University
Nagoya 464, Japan

ABSTRACT

A sheet plasma negative ion source is designed to study development of high current negative hydrogen ion source for NBI. The characteristics of plasma source and the preliminary results of beam extraction are described. The plasma parameter, especially electron density and amount of fast electron, is changeable with discharge condition. In the preliminary experiment of extraction of beam, the negative ion current density is estimated to be 6 $\mu A/cm^2$ at 0.75 kW discharge.

1. INTRODUCTION

Since advanced neutral beam injector must meet requirements of high energy in excess of 200 keV, high current density and long pulse operation, negative ion based neutral beam injector is a critical issue for future neutral beam injection heating technology on Tokamak. To develop a source capable of providing such a high current hydrogen or deuterium negative beams, experimental studies of negative ion production and extraction of beams have been done in various type plasma source, i.e., magnetron type source[1] bucket source[2] penning discharge source[3] sheet plasma source[4] etc., concerning to surface production and volume production methods. Recently several experiments for direct-extraction volume type source have demonstrated successful results of beam current density of around 20 mA/cm^2. On the other hand, theoretical and computational studies have been done for generation of hydrogen negative ion in discharge and they, especially tandem high density hydrogen discharge model[5] seems to be consistent with the results of the experiments. From these results, direct-extraction volume-type negative ion source is considered to be promising one and it has some advantage compared to surface production source with cesium-activated convertor.

A sheet plasma experiment shows an attractive results for possibility of application to high current negative ion source[5] However, it has not enough data base to apply itself as a prototype high current negative ion source for neutral beam injector. The mechanism of generation of H$^-$ ions in the sheet plasma seems also to be explain by the tandem discharge model. To search the possibility of developing the sheet plasma for a large negative ion source of neutral beam injector and to get much data base for the design, a new type of negative source (a reflex type sheet plasma negative ion source) is designed on the basis of principle of the tandem discharge model and the results of the sheet plasma. This paper reports source design

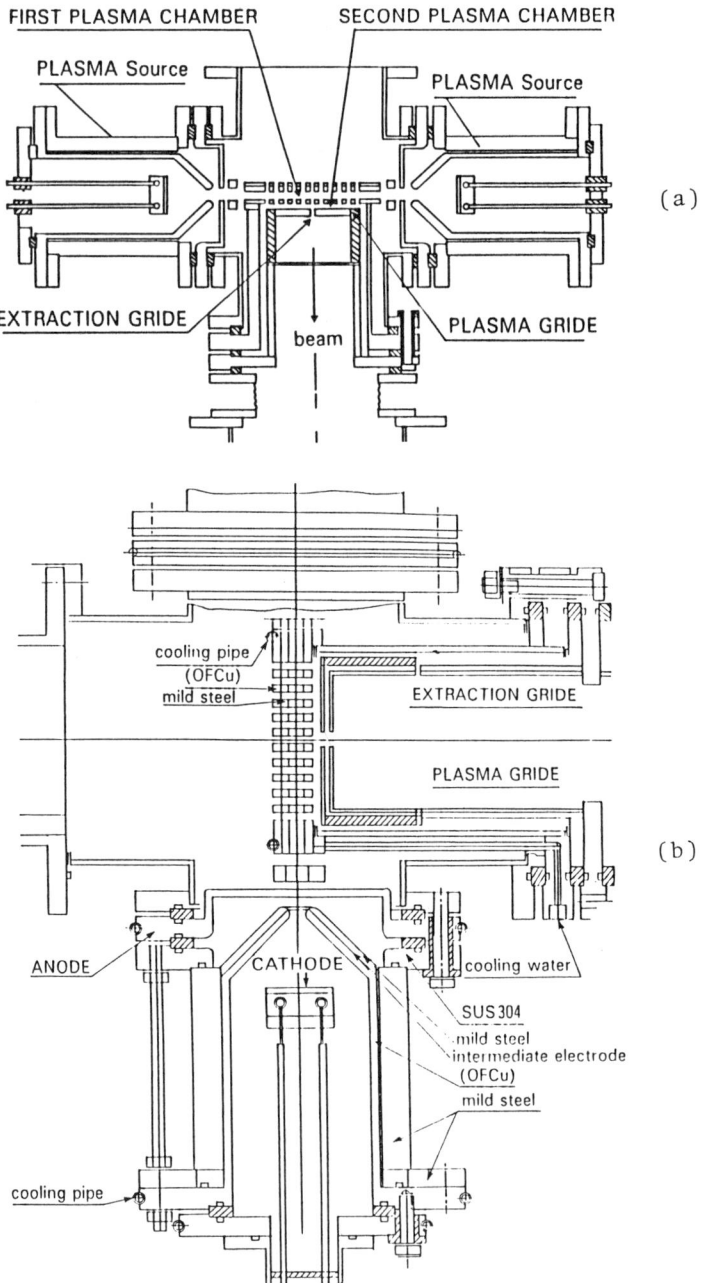

Fig.1 Schematic diagram of experimental apparatus.
(a) Experimental arrangement.
(b) Source construction.

and the preliminary experimental results for the source. The discharge characteristics, plasma parameters, beam extraction are described.

2. EXPERIMENTAL APPARATUS

A schematic diagram of the experimental apparatus is shown in Fig.1. The original design and the principle of the source are described in ref.6. The source consists of the discharge region provides the high temperature plasma (first chamber) and the region provides the diffused thermal electron plasma (second chamber). The first chamber consists of two plasma sources on the both end and a magnetic channel with axial guiding magnetic field. It is bounded by the array of mild steel square shape plates with a slite of 1 cm by 5 cm. The distance between each square plates is 5 mm. The axial magnetic field is produced by an electric magnet. The calculated pattern of magnetic field is shown in Fig.2. The magnetic field is weaker in the plasma source than in the magnetic channel. The lines of force pass through the cathode surface is pressed at the anode and extends to the magnetic channel.

The plasma source is a type of duoplasmatron. The anode and the intermediate electrode have a slit of 1 cm by 5 cm. The anode is made of copper with cooling and the intermediate electrode is made of copper covered with mild steel to guide the magnetic field. The oxide cathodes of 3 cm in diameter are used in this experiment.

The second chamber is a region between the array of the square plates and the plasma grids. The depth of the second chamber is adjusted by changing the position of the plasma grid.

Plasma produced in the source flows into the magnetic channel passing through the narrow anode slit. The primary electrons emitted from the cathode also pass through the anode slit to the magnetic channel. They are confined in the magnetic field, reflected by the cathode on the both side and oscillated between two plasma source. During oscillation, they enhance ionization in the magnetic channel and diffuse across the magnetic field to the second chamber losing their energy. The plasma and the primary electrons generated in the plasma source are squeezed into the magnetic channel to make a thin sheet plasma with high concentration ratio of the primary electrons in the first chamber.

The plasma in the magnetic channel diffuses into the second chamber across the

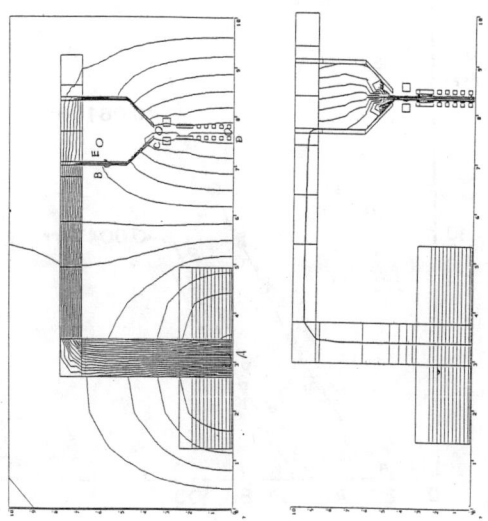

Fig.2 Calculation of magnetic lines of force.

magnetic field. In addition to the diffuse plasma, plasma is also produced by high energy electron in the second chamber. So the depth of the second chamber is important parameter to optimize the plasma density as well as the generation of negative ion. To optimize the generation of negative ions by minimizing the loss due to some collisional process, the thickness must be short. On the other hand, from a point of generation view of the enhancement of the dissociative attachment process, it must be thick. Hiskes et al. give the relation between ratio of negative ion concentration and second-chamber thermal electron density? The electron temperature in the second chamber must be low enough to suppress electron collisional detachment.

To test a preliminary beam extraction, the extraction electrode consists of plasma grid and earth grid, each of which has five holes of 4 mm in diameter on the center line. The extraction electrode with slit (1.5 mm x 20 mm) is also tested. The gap between the two grid is 2 mm. The plasma grid is biased negatively up to 6 kV and the extraction current is measured. The extraction beam current is detected by a collector cylinder.

The source is operated as a following conditions; the discharge voltage, V_d is ranged from 26 V to 100 V, the discharge current I_d up to 15 A, magnetic field B 20 Gauss to 160 Gauss at the center axis of the first chamber and 45 Gauss to 180 Gauss at the second chamber, gas pressure P in the magnetic channel 0.006 Torr to 0.06 Torr.

3. PLASMA PARAMETERS IN THE SOURCE

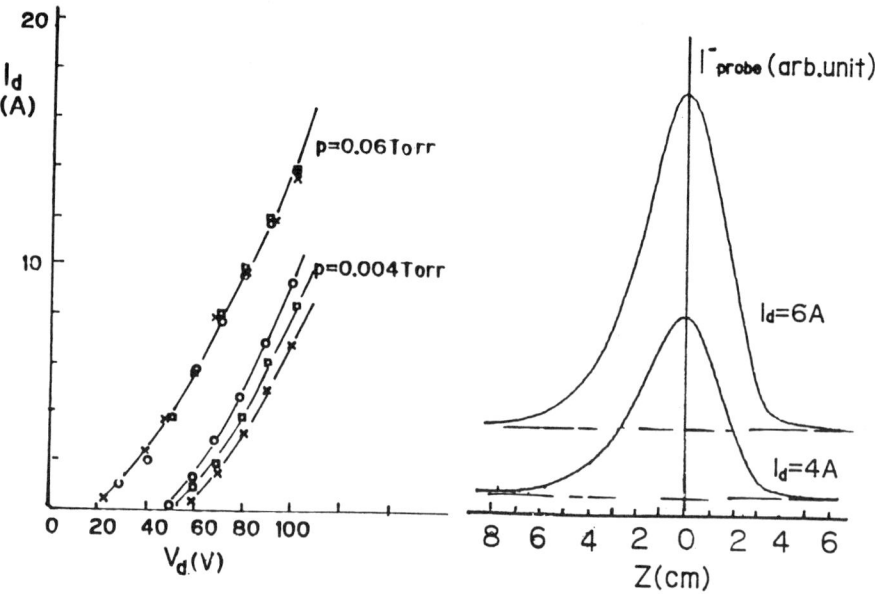

Fig.3 Discharge characteristics of plasma source.

Fig.4 Plasma density profile.

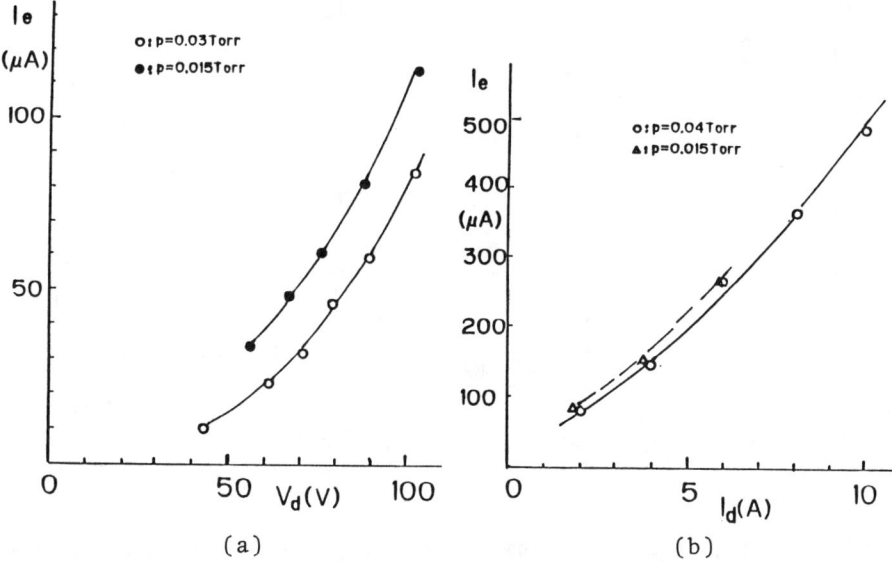

Fig.5 Plasma density as a function of discharge voltage (a) and discharge current (b).

Typical discharge characteristic of a single plasma source are shown in Fig.3. The discharge of both source is operated by a single power supply. When the array of square plates was connected to the anode, V-I characteristics did not change.

Discharge of the plasma source was tested for two cases of different distance between the anode and cathode. The discharge voltage is higher in the longer distance case than in the short distance case to get same discharge current I_d. Discharge voltage was adjusted by changing cathode heater current, that is, cathode temperature. As the cathode is located at a weaker magnetic field in the case of longer distance discharge, backheating to the cathode by plasma ion is decreased and as the results the constant discharge condition can be easily obtained.

The plasma parameters are measured by a movable cylindrical Langmuir probe along the thickness of sheet. Figure 4 shows the plasma density profile along the thin direction of the sheet. The half width of the profile is about 5 mm. It changes slightly by a potential of array of the mild steel plate and with changing the magnetic field and asymmetry of the profile is caused by the existence of extraction grid. The width of the sheet in the vertical direction is nearly equal to the cathode diameter but it is expanded in the case of weak magnetic field.

Figure 5 shows the electron density at the first chamber as a function of V_d and I_d. The density increases both with increasement of the discharge and the discharge current. Electron density as a function of V_d is shown in two different gas pressure. Electron

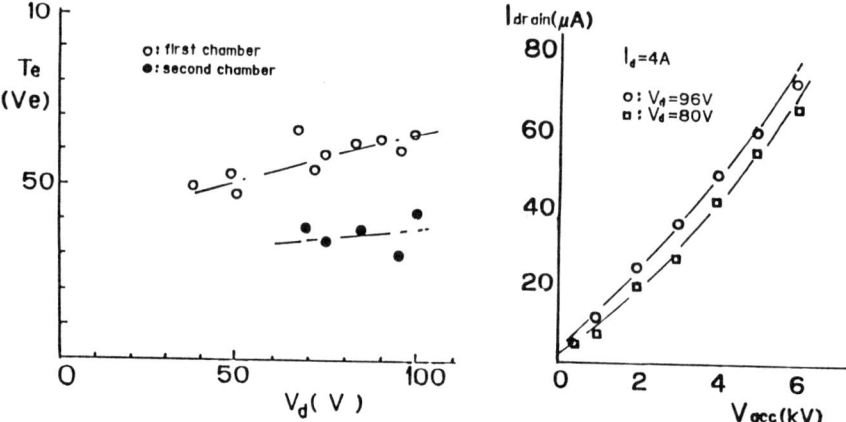

Fig.6 Electron temperature as a function of discharge voltage.

Fig.7 Extraction current vs. extraction voltage.

density is higher in lower than in higher pressure discharge at same discharge power. But in higher discharge voltage the ratio of that, $n_e(p = 0.037)/n_e(p = 0.015)$ is increased. The plasma density also depends on the operating gas pressure and its optimum range of operating pressure exists.

The electron temperature is shown in Fig.6 as a function of the discharge voltage V_d at the first and the second chamber respectively. The temperature is in the range of 5 eV to 10 eV at the first chamber and the probe characteristics vurce shows the existence of fast electrons. The electron temperature at the second chamber is much lower than at the first chamber. It is in the range of 1 eV to 5 eV. These results shows that the plasma is produced in the magnetic channel with fast electrons. The plasma density in the center of sheet varies from 10^8 to 10^{10} in the range of the present experiment parameter.

4. EXTRACTION OF BEAMS

The beam extraction is examined in the voltage range of up to 6 kV for three cases of the depth of the second chamber.

The characteristic of beam extraction is shown in Fig.7. The extraction current also increases with the increasement of the discharge voltage.

The extracted beam current to the vollector is shown in Fig.8 as a function of extraction voltage in the case of shortest depth of the second chamber. Electron is cut off at a distance of more than 1.5 cm in magnetic field of 180 Gauss. The beam current depends on the discharge voltage and in the same extraction current it is larger in the higher discharge voltage. The dependence of beam current to the collector on V_d is shown in Fig.8 for same discharge parameters. Figure 9 shows relation between the drain current of extraction and the collector current. The ratio of the collector current to the

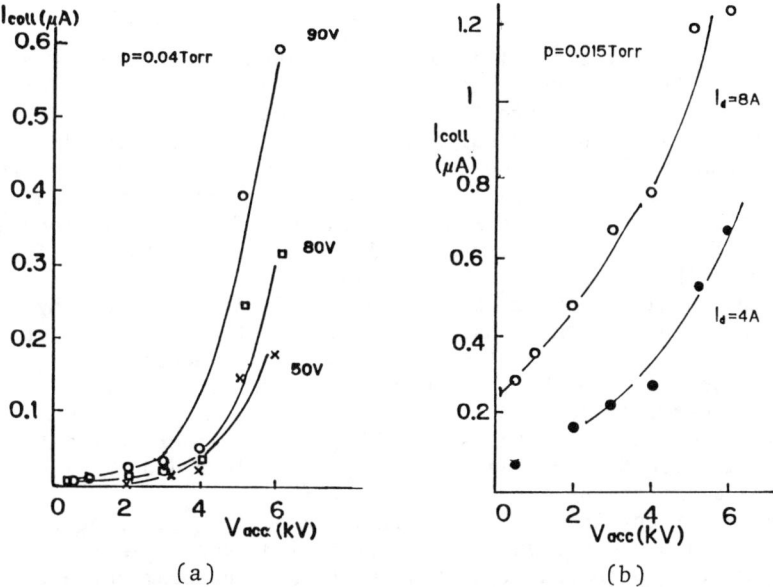

Fig.8 Beam current to collector vs. extraction voltage.
(a) Dependence on discharge current.
(b) Dependence on discharge voltage.

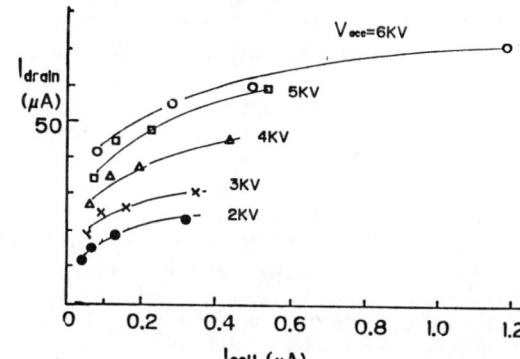

Fig.9 Beam current to collector as a function of drain current.

drain current is increased with increasing the discharge current. It also increases with the increasement of discharge voltage. This results show that beam current can be increases with increasing the plasma density.

Beam extraction in the different depth of the second chamber was examined. Beam current is larger in the short second chamber than in the long at same discharge voltage. The plasma density becomes higher in the short second chamber, so it is not clear whether the increasement of beam currents is due to the increasement of negative ion production rate and the decrease of the destruction of H⁻ ion due to the dissociative detachment and the electron collisional detachment.

5. SUMMARY

The characteristic of a reflex type sheet plasma source has been examined. The distance between two plasma source determines the critical discharge voltage to produce plasma at the center of the extraction grid. These results show that enough density plasma will be obtained by choosing the cathode position, operating pressure and discharge voltage for much larger discharge chamber with longer distance between the anode and the cathode. The density along axial magnetic field is not measured but it seems to be rather difficulty to produce the axial uniform plasma in this source.

The negative beam current of up to 1.5 µA is obtained by the collector. The current density is estimated to be 6 µA/cm^2 at 0.75 kW discharge. But the analysis of the mass is necessary to qualitatively discuss although it is supposed that electrons are eliminated by the magnetic field.

The short depth of the second chamber is rather better to get much beam current. The reason of it is not clear and the more detailed examinations are needed in future.

Beam current increases with simultaneously increasing the discharge voltage and current. More powerful discharge in the plasma source will be prepared at next state to get the beam current of more than beam current of 1 mA.

REFERENCES

1. K. Prelec, Proc. 3rd Symp. on the Production and Neutralization of Negative Ions and Beams, Brookhaven, p.333 (1983).
2. K. A. Leung, K. W. Ehlers and M. Bacal, Rev. Sci. Instrum. 54, 56 (1983).
3. G. E. Derevyankiu and V. G. Dudnikov, Proc. 3rd Symp. on the Production and Neutralization of Negative Ions and Beams, Brookhaven, 376 (1983).
4. J. Uramoto, Institute of Plasma Physics, Nagoya University research report, IPPJ-731 (1985).
5. J. R. Hiskes and A. M. Karo, J. Appl. Phys. 56, 1927 (1984).
6. T. Kuroda, H. Okamura and O. Kaneko, Proc. 2nd Symp. on the Production and Neutralization of Negative Hydrogen Ion and Beams, Brookhave, p.240 (1980).

DISCUSSION

Lietzke: I perceive that you have a double sheath on the electron emission and you're collecting electrons across the field, and that's normally observed to have a high level of fluctuation. What kind of level of fluctuation do you see in your beam current?
Kuroda: In the design of this source, we wanted to have less noise in the plasma. So it was designed with a rather weak magnetic field in the center of the plasma. The Penning discharge has a lot of noise because, in the Penning discharge, the magnetic field strength is rather high and the particles have to diffuse to the anode by crossing the magnetic field. That situation in a plasma is very noisy, so I want to avoid that. So we chose the duoplasmatron source so as not to have much oscillations, or noise.
Ehlers: Less than 1% you said?
Kuroda: Yes.
Tsai: How big is the extraction grid aperture?
Kuroda: 4 mm in diameter.
Holmes: I notice your beam current doesn't saturate when you increase the beam energy for fixed discharge conditions. It should saturate, like a probe. The beam current should saturate at some value, because the number of ions or electrons that leave the discharge is limited by the plasma density. And your curves just keep on going up.

PRODUCTION AND FORMATION OF INTENSE H⁻ BEAMS

R. McAdams, A. J. T. Holmes, M. P. S. Nightingale,
L. M. Lea, M. D. Hinton, A. F. Newman and T. S. Green
Culham Laboratory, Abingdon, Oxon OX14 3DB, England.

ABSTRACT

Experiments on the extraction of intense d.c. H⁻ beams have been carried out on a volume source with a triode accelerator. The maximum beam currents so far obtained have been 145 mA at 83 keV and 126 mA at 89 keV. An estimate of the fraction of this beam which is composed of electrons has been made and found to be 6%. A study has also been carried out on the effect of the first gap voltage of the accelerator on the extraction of a beam and a model of the extraction physics is proposed.

INTRODUCTION

There are two H⁻ sources in operation at Culham. Both are volume discharge sources which use external magnets to confine the plasma and also to form a filter field across the source. One is a large source, and up until now this has been used to study H⁻ and D⁻ densities in volume discharges. Beam extraction experiments have only been performed using a pinhole accelerator[1]. A small source has been used to investigate beam extraction using a triode accelerator. This operated in a d.c. mode and beams of up to 33 mA from a 16 mm diameter aperture have been extracted[2,3].

Now, the large source has been attached to a triode accelerator with the initial aim of achieving 100 mA of H⁻ at an energy of 100 keV in d.c. operation. The main purpose of this paper is to discuss the results achieved from this facility. We will also describe our work, from both sources, on the dependence of the intensity of the extracted current on the first gap voltage, and propose a model for the observed behaviour.

EXPERIMENTAL APPARATUS

In Fig. 1 we show a schematic diagram of the large source. A checkerboard magnet configuration is used to confine the plasma within a volume of dimensions $55 \times 30 \times 20$ cm³. The dipole filter field is formed by two unipolar rows of magnets along the longest dimension on each side of the source. There are 24 tungsten filaments which drive the discharge.

The accelerator is shown in Fig. 2. It is a triode with a thick second electrode and is also a 3/2 times replica of the accelerator used for the small Culham source[2]. Each electrode contains a number of magnets. Those in the beam forming electrode adjacent to the plasma reduce, but do not eliminate, the electrons extracted through the aperture. The first set of magnets in the second electrode are then used to deflect the extracted electrons into a trap region in the second electrode. H⁻ ions are re-steered onto the beam axis by the remaining sets of magnets in the second and third electrodes. The aperture in the beam forming electrode has a diameter of 24 mm.

The current accelerated to full voltage is measured by the drain current on the high voltage power supply and we also measure the current collected on the second grid. In the ideal situation, the former is the extracted H⁻ beam current and the latter is the extracted electron current but this never leaves the accelerator. The system works in a quasi d.c. mode where beam pulses of 30 seconds duration are extracted.

Various diagnostic measurements are made on the beam once it leaves the beam forming electrode. At present these consist of calorimetry and profile measurement. A schematic diagram of the setup is shown in Fig. 3. Within the accelerator, thermal power loading measurements are made on the second and third electrodes. After leaving the accelerator the beam is allowed to drift in a large cryopumped tank. The beam halo is removed by a scraper and the amount of power falling on the scraper is measured. Further downstream is a wire scanner to monitor the beam profile. Finally the beam is stopped by a calorimeter. This has two purposes. Firstly the total power falling on the calorimeter is measured, and secondly the beam profile in two orthogonal directions is measured by rows of nickel pins containing thermocouples which lie behind holes in the front plate of the calorimeter.

An emittance diagnostic has now been added and other diagnostics, such a d.c. current monitors, a beam potential measurement probe and a plasma ion temperature probe, are at present under development.

RESULTS

The major result of our work to date is shown in Fig. 4, where the extracted H⁻ current is plotted as a function of arc current for a constant arc voltage and gas flow. The maximum extracted current is 145 mA at a beam energy of 83 keV. For extracted currents less than 60 mA with a floating first electrode, the ratio of the electron current measured at the second grid to the extracted current is approximately four. For currents higher than this the ratio rises to close to five. However these values are still less than those observed in the small Culham source (ratio ~ 7) and are probably attributable to the higher magnetic flux in the beam forming electrode of the larger source compared to that in the small source.

Application of a bias voltage between the anode and the beam forming electrode can reduce the number of electrons extracted into the second electrode. This is shown in Fig. 5. In fact, at bias voltages of ≥ 2.5V the ratio of extracted electrons to H⁻ ions becomes less than unity. However, at least a 50% reduction in the H⁻ flux is observed.

Using the calorimetric measurements from the beam stop calorimeter, the scraper ring, and the third electrode full power accountablility is achieved to within a few percent. At low currents the beam stop and scraper ring account for 85% of the beam power. At high currents this value is decreased, the cause thought to be overheating of the electrodes and the magnets, resulting in increased power to the third electrode deposited by ions with distorted trajectories. In the previous section it was stated that under ideal conditions the extracted current consisted only of H⁻ ions and that all the electrons were collected in the trap region of the second electrode. We have made an estimation of the fraction of the high voltage power supply drain current which is composed of electrons.

Assume that a fraction, g, of the electron current collected on the second electrode, I_e, leaks through the trap and is accelerated to full beam voltage, V_B. Then we have

$$I_D = I_{H^-} + g\, I_e \qquad (1)$$

where I_D is the high voltage power supply drain current and I_{H^-} is the H⁻ current. Now assume also that these electrons which are accelerated to full voltage are trapped by the various magnetic fields and deposited on the third electrode. Then the power loading on this electrode is

$$P_3 = [f\, I_{H^-} + g\, I_e]\, V_B \qquad (2)$$

where f is the fraction of the H⁻ current striking the electrode.

Combining equations (1) and (2) gives

$$\frac{P_3}{I_D} = V_B \left[g\, \frac{I_e}{I_D}(1-f) + f \right] \qquad (3)$$

Thus a plot of P_3/I_D against I_e/I_D should be a straight line and the slope and intercept enable an estimation of the important fractions f and g to be made. This plot is shown in Fig. 6. I_e/I_D was varied by changing the bias voltage. It can be seen that the data shows a linear relationship between P_3/I_D and I_e/I_D indicating that the model described above is a good one. We find that f = 0.16 and g = 0.0154. Thus for the case of the extraction of 145 mA of beam current, 136 mA of this is H⁻ ions and the remainder electrons. Also the results imply that approximately 22 mA of the H⁻ current strikes the third electrode. Only one such estimation has been made so far and it is clear that others, under different conditions, are necessary.

In Fig. 7 we show a beam profile as measured by the beam stop calorimeter at 4 m from the source. From the lower graph, which is for one side of the profile, it is clear that the beam is composed of two Guassian components: an inner core, and an outer halo. The 1/e radius of the beam is 26 mm.

We have also carried out a number of experiments to investigate the extraction of H⁻ beams from the plasma sources. What we have found can be summarised briefly as follows. The intensity of the extracted beam is highly dependent on the first gap accelertion voltage and the beam is focussed at an approximately constant value of the ratio of beam voltage to first gap voltage. The second gap voltage acts only as a focussing element. Fig. 8 shows how the extracted current depends on the first gap voltage in the case of the large source. As V_1 is increased so is the extracted current until a value of V_1 is reached where the extracted current saturates - the so-called 'S' shaped curve. The electron current remains constant. A family of such curves for the small source, at different arc currents, is shown in Fig. 9.

This behaviour is completely different from positive ion sources where the extracted current is essentially independent of V_1 and where there is a unique value of $I/V^{3/2}$ in order to have a collimated beam. The model proposed by Green and Holmes suggests that the plasma boundary is fixed. Then the situation is somewhat analogous to an electron diode with a fixed quasi-planar cathode. At low values of V_1 the current extracted is given by the Langmuir-Blodgett equation.

$$I_{H^-} = C \frac{a^2 k^2}{d^2} V_1^{3/2} \qquad (4)$$

where a is the cathode radius, d the diode gap and $C = 4\pi \varepsilon_0/9 (2e/m_i)^{1/2}$. The term k includes the curvature correction which is approximately $(1 - 0.4 \, d/R)$ where R is the boundary curvature radius. At high values of V_1 the extracted current is limited by that available from the cathode. However we must modify equation (4) slightly to allow for the effect on the space charge in the extraction gap due to the electrons also extracted. Then we have

$$I_{H^-} \left[1 + \left(\frac{m_e}{m_i}\right)^{1/2} \frac{I_e}{I_{H^-}} \right] = C \frac{a^2 k^2}{d^2} V_1^{3/2} \qquad (5)$$

In order to test the model we have calculated the value of d/ak from equation (5) and plotted it against the extracted current normalised to the arc current. This is shown in Fig. 10 for the small source data. Over a wide range of currents the value of d/ak remains fixed. At large currents where the negative ion current saturates, the plasma boundary can no longer remain fixed and the value of d/ak increases rapidly. The model does not work at low currents where the value of d/ak collapses.

Over the range for which d/ak is constant we obtain d/ak ~ 1.8 for the small source and d/ak ~ 1.3 for the large source. Since the large source is a 3/2 times replica of the small source we would have expected these values to be equal. Furthermore neither of these values agree with the mechanical value of d/ak ~ 1. The explanation may lie in the differing magnetic fields in each source, the flux being higher in the large source, and also in the effect of this field on the plasma boundary leading to different values of R. These values of d/ak may also be responsible for beam aberrations and the large power loading on the third electrode (≥ 4 kW at 83 kV, 140 mA) is testimony to this.

CONCLUSIONS

We have reported measurements on a volume source and an accelerator which can produce d.c. H^- beams of up to 145 mA at 83 keV. The electron fraction of the beam is only of the order of a few percent. In future it should be possible to produce beams of even greater intensity. Further work is needed at present to completely characterise the beam and this is to be carried out in the near future.

We have also shown that a simple model can provide a good description of the physics of the extraction of H^- beams from plasma volumes. A better model would be of great importance in understanding how the magnetic field geometry affects the position and shape of the plasma boundary ultimately leading to a better knowledge of the factors relating to beam quality.

ACKNOWLEDGEMENTS

This work was supported by the United States Air Force Office of Scientific Research under contract number F49620/86/C/0064.

REFERENCES

1. A. J. T. Holmes, L. M. Lea, A. F. Newman & M. P. S. Nightingale, Proceedings of the 2nd European Workshop on the Production and Application of Light Negative Ions, p1, 1985.
2. A. J. T. Holmes, M. P. S. Nightingale & T. S. Green, Proceedings of the 2nd European Workshop on the Production and Application of Light Negative Ions, p215, 1985.
3. Unpublished result.

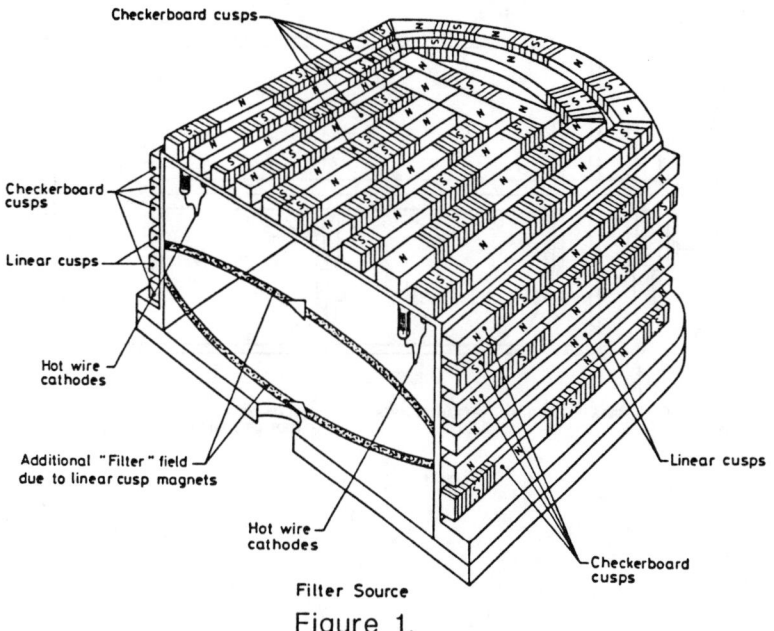

Figure 1.

Filter Source

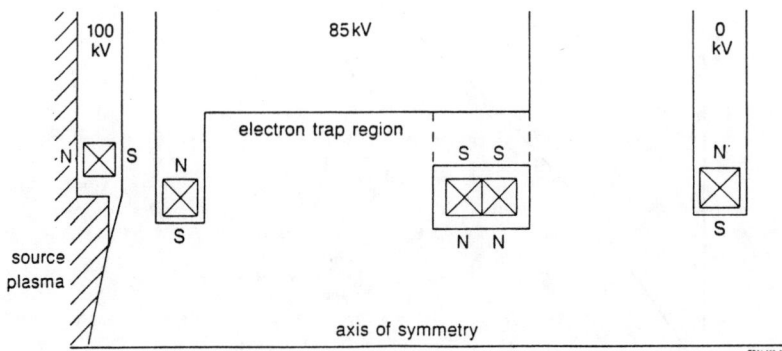

Figure 2.

View of H⁻ Accelerator

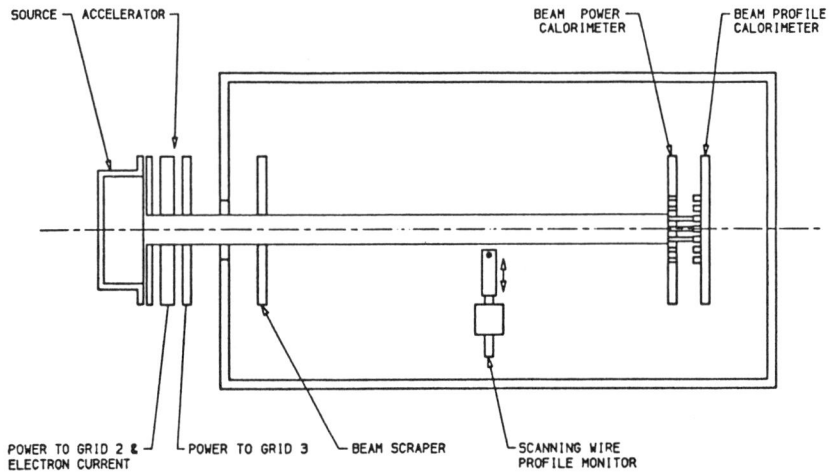

Figure 3.

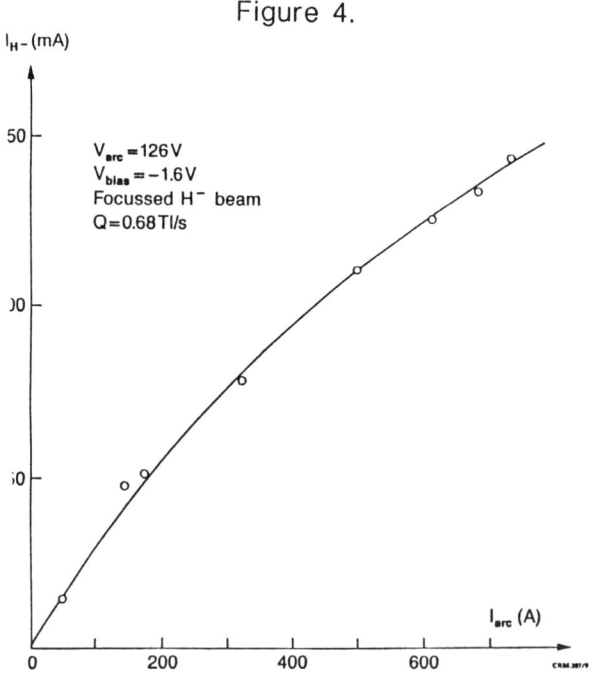

Figure 4.

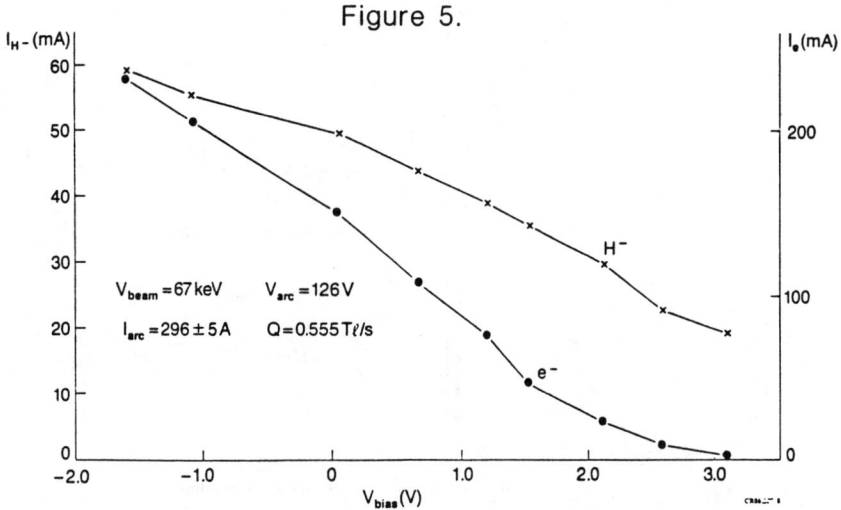

Figure 5.

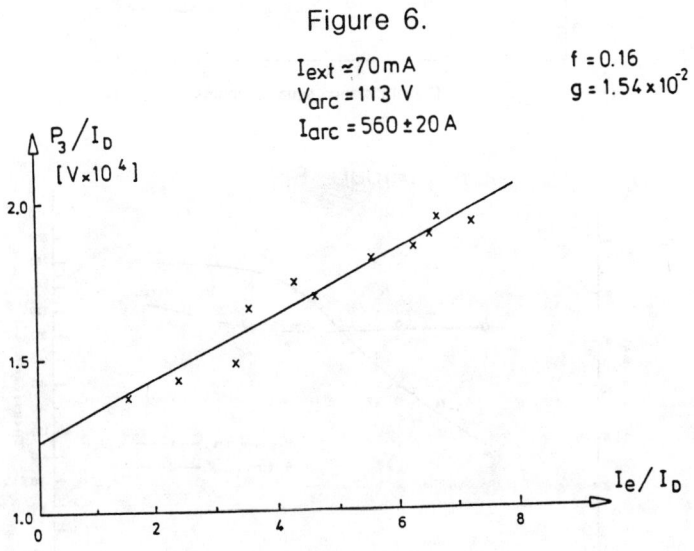

Figure 6.

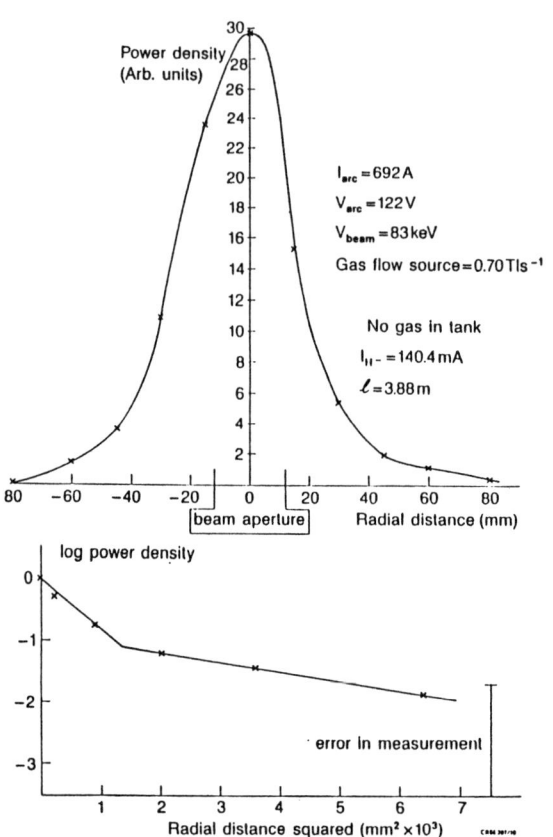

Figure 7.

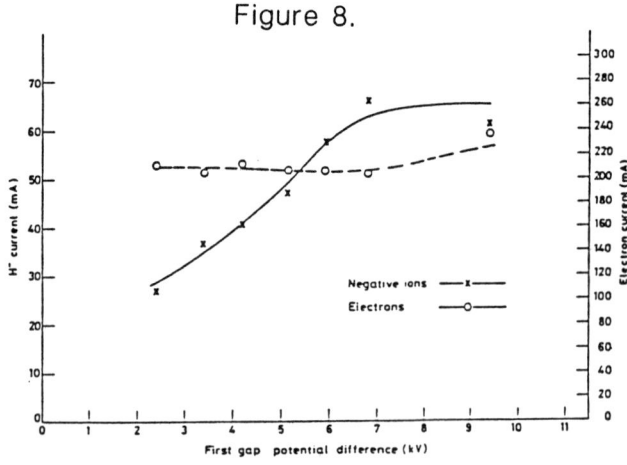

Figure 8.

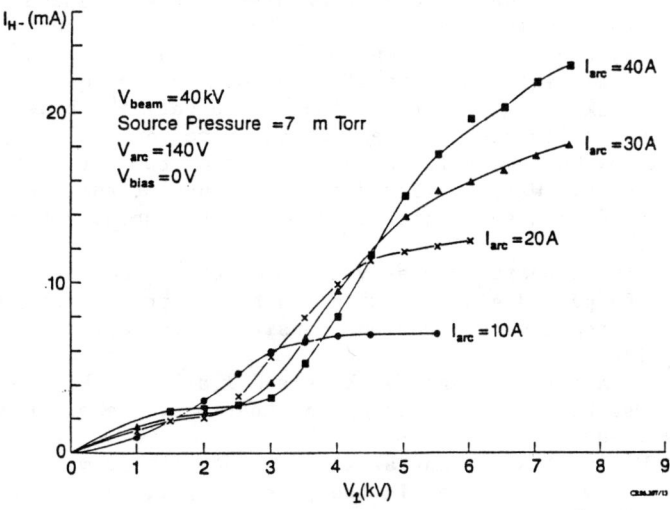

Figure 9.

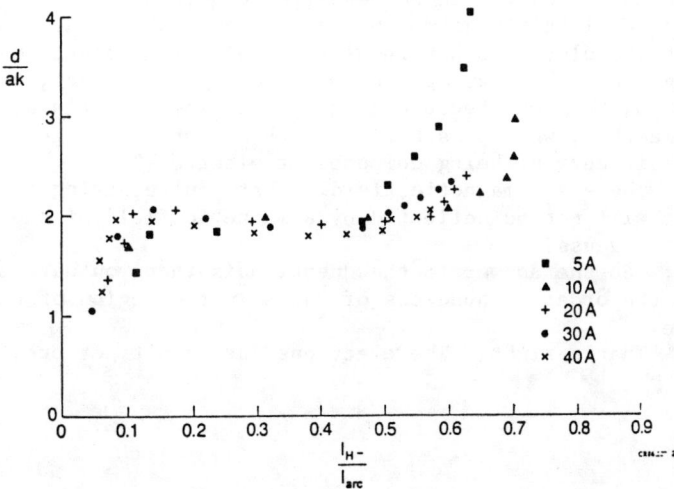

Figure 10.

DISCUSSION

Tsai: (question on the effect of biasing the plasma electrode).

Holmes: Can I comment on that? All the volume sources with true accelerators, where we have tens of kilovolts and try and form a collimated beam, show the sort of behavior that Roy described. There's no peak in the H⁻ production as you vary the bias potential. When you have a low energy accelerator, where you don't try and form a beam - a few hundred volts or a kilovolt or two - then this peak seems to appear. We've observed some evidence of that from our small source work when we didn't have a proper accelerator, in the early days.

Tsai: With a positive bias you would get the small peak there?

Holmes: The peak disappears when you have a proper accelerator. You see, the negative ion current density decrease with increasing positive bias.

Tsai: I have tried negative bias and I see a peak.

Holmes: Negative bias? That's the other way. We never looked in that direction.

McAdams: We let that potential float up negative sometimes, but it's limited by some diodes in the power supply, so it will only go to just over -1V.

Hershcovitch: On our hollow cathode discharge source, we had a problem, initially, of detecting electrons as H⁻. After we modified the extraction, to verify that we had no electrons, we switched our discharge to argon, which forms no negative ions. Have you ever considered doing something similar?

McAdams: We'll consider it. I've never tried that, personally.

Holmes: We have tried it in the pinhole accelerator. We used an argon discharge and we had no negative ions then.

Whealton: Will Stirling brought up the point about the approximation that the electrons hit the last accelerating electrode. I'm wondering about the next obvious question. If, in fact, the electrons follow the beam because of gas focussing potentials, or something like this, what possibility is there for some higher fraction of the drain current being composed of electrons?

Holmes: There's a magnetic field. What you're saying is, the electrons will not be deflected by a magnetic field of the order of hundreds of gauss.

Whealton: So the answer to the question is that you have a magnetic field on the order of hundreds of gauss in the region of that last electrode.

Holmes: That's right. The electrons just can't get out.

A HIGH CURRENT VOLUME H$^-$ ION SOURCE WITH MULTI-APERTURE EXTRACTOR

Y. Okumura, H. Horiike, T. Inoue, T. Kurashima, S. Matsuda,
Y. Ohara, and S. Tanaka
Japan Atomic Energy Research Institute, Naka-machi, Naka-gun,
Ibaraki-ken, 311-02, Japan

ABSTRACT

We describe a high current volume H$^-$ ion source, which produces more than 1 A H$^-$ ion beams. H$^-$ ions formed by volume processes in a magnetically filtered multicusp plasma generator have been extracted through a multi-aperture extractor. The extractor consists of four grid, each has 209 apertures of 9 mm diam. within a rectangular area of 12 x 26 cm^2. The ion beam current was measured calorimetrically and it was confirmed that H$^-$ ion beams of up to 1.26 A were extracted with a current density of 9.5 mA/cm^2 at 21 keV for 0.2 s. Impurity concentration in the beam was less than a few percent according to momentum mass analysis.

INTRODUCTION

High current and high energy negative deuterium (hydrogen) beams will be required for plasma heating and for current drive in future fusion reactors. For instance, 100 A/ 500 keV D$^-$ ion beam is required in the conceptual design of 20 MW negative-ion-based injector[1] for FER (Fusion Experimental Reactor), which is now being designed at JAERI. Out of several approaches to produce D$^-$ or H$^-$ ion beams[2], volume production seems to be the most favorable method to produce such a high current negative beams, because the structure is simple enough to scale up and the produced ion beams have good beam qualities. Several laboratories have been investigating the volume H$^-$ source[3-7] and have succeeded in producing H$^-$ ion beams with high current densities[8,9]. Up to the present, however, the total current of H$^-$ ion beams generated by those sources has been an order of mA, while more than 1 A beams have been already generated by surface production[10] and double charge exchange methods[11].

At JAERI, our objective is to develop a high current negative ion source which is directly applicable to the neutral beam injectors for future fusion reactors. Initial studies on the volume production have been concentrated on the optimization of source geometry and magnetic configurations so as to increase the H$^-$ ion yield[12] and 15 mA/cm^2 H$^-$ beam was extracted through a 4 mm diam. aperture[13,14]. Then we developed multi-aperture four-grid extraction system with 12 apertures of 10.3 mm diam. and a H$^-$ ion beam of 0.1 A was extracted with a like amount of electrons[15]. In order to demonstrate >1 A beam extraction in volume ion source, we have fabricated a large multi-aperture extractor which has total extraction area of 133 cm^2. In the present paper, we describe the design of the 1 A volume source and present some experimental results on initial extraction test of the source.

ION SOURCE

A schematic of the ion source and the electrical connection of power supplies are shown in Fig. 1. The plasma generator is a rectangular stainless steel chamber, 21 x 36 cm^2 in area by 15 cm deep, surrounded by three rows of SmCo magnets on each side wall and five rows on end plate of the chamber. These magnets form continuous line cusps perpendicular to the source axis. An additional couple of line cusps are installed near the surface of plasma grid. The discharge is created by electron emission from eight tungsten filaments near the end plate. The entire wall of the chamber serves as an anode.

About 3 cm above the plasma grid is a magnetic filter. The filter is an array of five water cooled copper tubes of 8 mm diam. in which small SmCo magnets are inserted. The spacing distance between the tubes is 4 cm and the filter strength ($\int B d\ell$) is about 0.2 kG.cm. We can reduce the filter strength by subtracting the magnets from two tubes out of the five. Throughout the present paper 'weak filter' denotes this reduced strength filter, while 'strong filter' denotes the original filter. The filter makes the transverse magnetic field in the plasma which modifies the electron energy distribution[4,16] and enhances the H^- yield[17].

The extractor is isolated from the plasma generator by a thin insulator inserted between the magnetic filter and the plasma grid. The extractor comprises four grids called plasma, extraction, suppression, and acceleration grids, respectively. Each grid has 209 apertures of 9 mm diam. within the rectangular area of 12 x 26 cm^2. The H^- ions are extracted together with a large amount of plasma electrons by applying a negative potential of up to 5 kV to the plasma grid with respect to the extraction grid. In the extraction grid, small SmCo magnets are inserted between two adjacent rows of apertures so that the integrated transverse magnetic field ($\int B d\ell$) becomes to be zero. The electrons are deflected by the magnetic field and strike the extraction grid. The H^- ions receive only a small net displacement by the magnetic field and are accelerated to 21 keV by the potential between the extraction and the acceleration grids. The suppression grid should be held slightly negative with respect to the extraction grid in order to prevent the secondary electrons, which is produced by both secondary electron emission on the extraction grid and beam-gas interactions, from leaking out to the acceleration gap[15]. Throughout the present experiment, however, the suppression grid was connected electrically to the extraction grid, since our interest was in producing high current H^- beams. All the grids except for the suppression grid are made of oxygen free copper and have water cooling channels between two adjacent rows of apertures. The suppression grid is made of molybdenum.

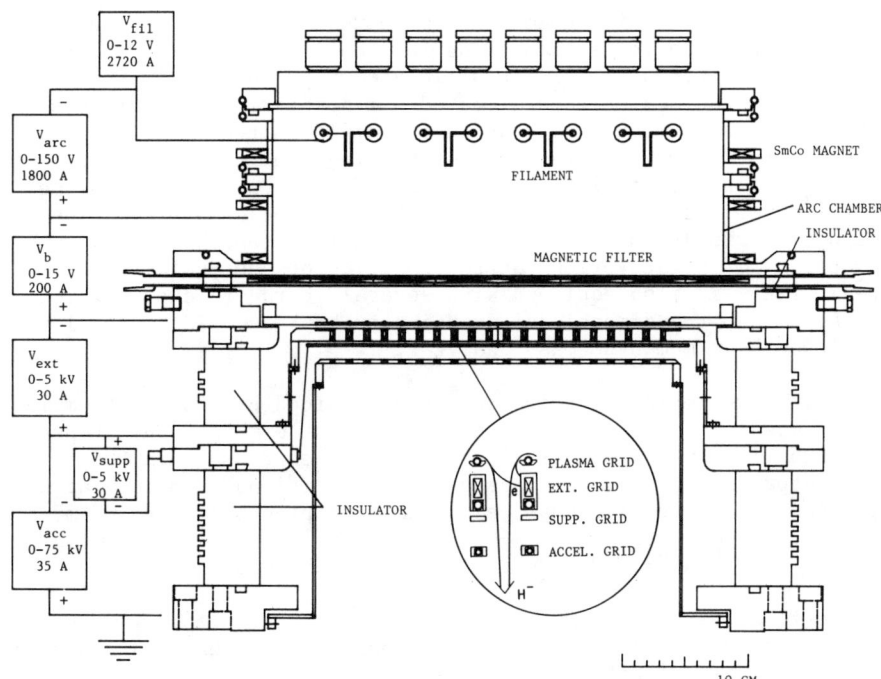

Fig. 1 Cross section view of >1 A volume H⁻ ion source and electrical connection of power supplies.

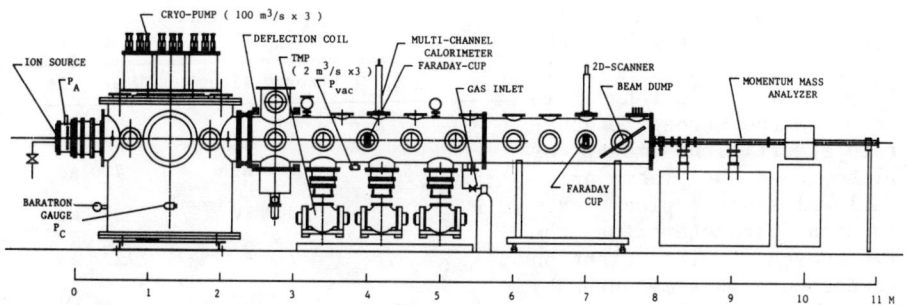

Fig. 2 A schematic of entire beam line.

EXPERIMENTAL SET UP

A view of entire beam line is shown in Fig.2. The beam line length from the ion source to an ion beam dump is about 7.5 m. The H$^-$ ion beam current and its profile are measured calorimetrically by a multi-channel calorimeter located 4 m downstream of the ion source. In order to remove electrons completely from the beam, a magnetic field of 20-30 Gauss is applied by using deflection coils, whose center is placed at 2.5 m downstream of the ion source. Since the total power received by the calorimeter did not change even when the magnetic field was applied, we may conclude that the beam is wholly composed of negative ions at the position of the calorimeter. The beam line is also equipped with several Faraday cups, which are mounted on two-dimensional scanning mechanism. The beam line is followed by a momentum mass analyzer.

The vacuum in the beam line is maintained by four turbo molecular pumps. The total pumping speed is about 6 m^3/s. The pressure in the beam line was typically 0.25 Pa when the pressure in the plasma generator was 1.0 Pa. If necessary, an additional pumping can be produced by three cryopumps, each has a pumping speed of 100 m^3/s.

EXPERIMENTAL RESULTS AND DISCUSSIONS

Magnetic Filter

We have tested several types of magnetic filters and found that the 'weak filter' offers the highest H$^-$ ion current. In case of the original 'strong filter', the current of electrons extracted with the H$^-$ ions ($I_{ext}-I_{H^-}$) was small compared with that for the 'weak filter'. Thus the heat dissipation in the extraction grid is small, but the H$^-$ current was also smaller than that for the 'weak filter' by about 30 % for the same arc discharge power. It seems that the transverse magnetic field formed by the 'strong filter' is too strong to generate a sufficient amount of plasma near the plasma grid. Table I shows typical extraction parameters for the two cases. The electron temperature was measured by a Langumuir probe, which is placed 7 mm apart from the plasma grid, and was found to be typically 0.65 eV for 'strong filter' and 0.75 eV for 'weak filter' at an arc power of 70 V/ 700 A.

Table I Extraction parameters for two types of magnetic filter.

	'Strong Filter'	'Weak Filter'
V_{arc}	70 V	70 V
I_{arc}	700 A	700 A
V_{ext}	3.4 kV	3.4 kV
I_{ext}	7.2 A	12.2 A
V_{acc}	18 kV	18 kV
I_{acc}	1.25 A	1.74 A
H$^-$ Current*	0.82 A	1.09 A

* Measured by calorimeter

Dependence on Arc Voltage

Figure 3 shows the dependence of H^- ion current on the arc voltage for the two types of magnetic filter. The arc current was kept to be 200 A by regulating filament temperature. Although the optimum arc voltage that gives the highest arc power efficiency is almost the same for both types of the filter, the power efficiency decreases rapidly with the arc voltage in the case of 'weak filter'. This may be considered as follows; since the energy of the primary electrons increases with the arc voltage, those energetic electrons are apt to escape from the filter in the case of 'weak filter' and give undesirable effect on H^- production. The dependence of the power efficiency on the arc voltage is weak in the case of 'strong filter'. The optimum arc voltage was about 70 V and the maximum arc power efficiencies were 0.026 A/kW and 0.019 A/kW for the 'weak' and 'strong filter', respectively.

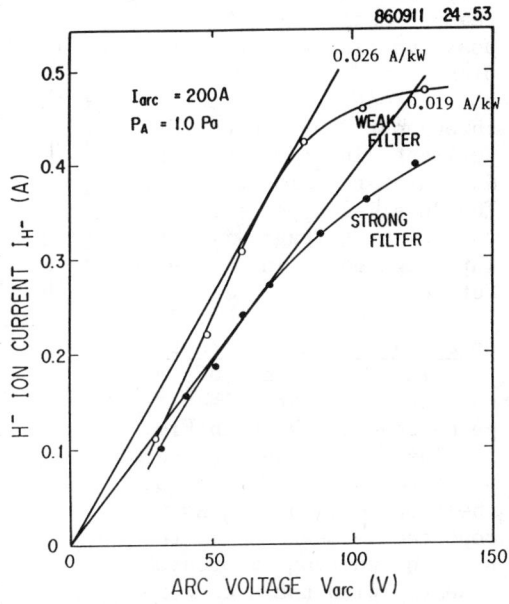

Fig. 3 H^- ion beam current as a function of arc voltage for two types of magnetic filter

Pressure Dependence

Since the confinement of the primary electrons is fairly good in the plasma generator, the arc discharge occurs at an extremely low pressure of 0.05 Pa. However, H^- ions are scarcely formed in such a low pressure region. The H^- ion current increases with the pressure in the arc chamber and has a maximum at 0.5-1.2 Pa. Figure 4 shows the H^- ion current as a function of the pressure for various arc discharge current. It should be noted that the optimum pressure increases with the arc current; i.e. higher arc current (or higher H^- current) requires higher pressure.

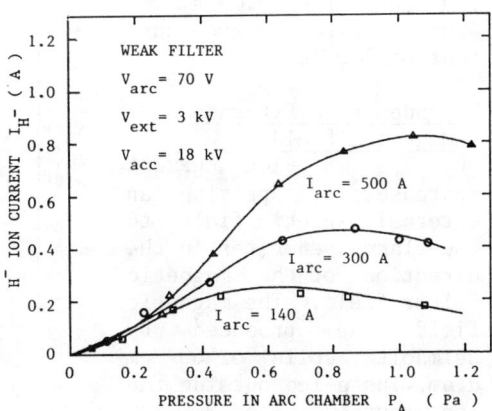

Fig. 4 H^- ion beam current vs pressure in arc chamber for various arc discharge currents.

Bias Voltage

The plasma grid can be

biased positively or negatively with respect to the anode. The H^- ion and the electron current vary with the bias voltage; the electron current decreases significantly when the grid is biased positively, while the H^- ion current is almost constant for negative bias voltage but decreases slightly at high positive bias voltage. The plasma grid is connected to the anode ($V_b = 0$) throughout the present experiment in order to produce higher H^- ion current. The detailed dependence on the bias voltage will be investigated in near future.

Dependence on Arc Current

The dependence of H^- current on the arc discharge current is shown in Fig. 5. The arc voltage and the pressure were kept to be their optimum value. As we reported elsewhere[12], the H^- ion current increases linearly with the arc current or plasma density for low arc discharge but is gradually saturated at higher arc current. The H^- ion current of 1.09 A was achieved with the arc current of 700 A.

Dependence on External Magnetic Field

The H^- ion current increases by applying an external magnetic field to the plasma generator in the direction of the magnetic filter field. The magnetic field was produced by Helmholtz coils of 60 cm diam. installed outside the ion source. The H^- ion current is shown in Fig. 6 as a function of the magnetic field strength at the position of the plasma

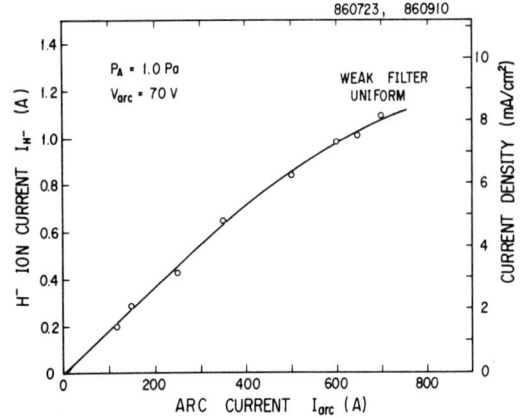

Fig. 5 H^- ion beam current vs arc discharge current.

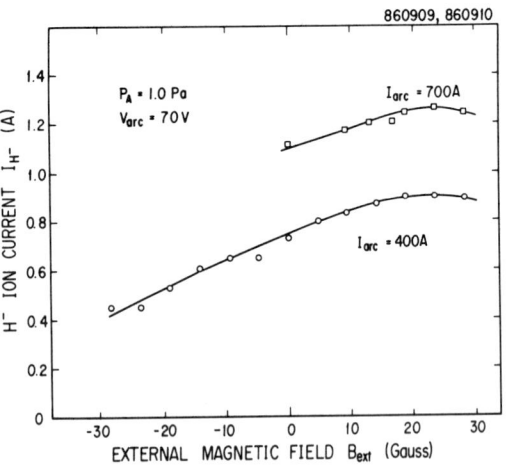

Fig. 6 Variation of H^- ion beam current with external magnetic field.

grid. The H^- ion current increases with the magnetic field and has a maximum value of 1.26 A at B_{ext} = 25 Gauss and I_{arc} = 700 A. The reason why the H^- ion current increases is not clear, but this result suggests there is still some room for improvement in magnetic field configuration.

Beam Profile

An example of power density profile of the H^- ion beam is shown in Fig. 7. The e-folding widths are 25 cm and 30 cm for horizontal and vertical profiles, but it should be noted that the lengths of the sides of the extraction area are 12 cm and 26 cm in horizontal and vertical directions, respectively. Considering the extraction area, we can conclude that the horizontal beam divergence is worse than the vertical one. This is mainly due to the displacement of the beamlets caused by the magnetic field in the extraction grid and partly due to density variation of the source plasma in horizontal direction. By masking all apertures except a single aperture in the plasma grid, we have measured the beamlet divergence by the Faraday cup and found that the divergence are 0.8 deg and 0.9 deg in horizontal and vertical directions, respectively.

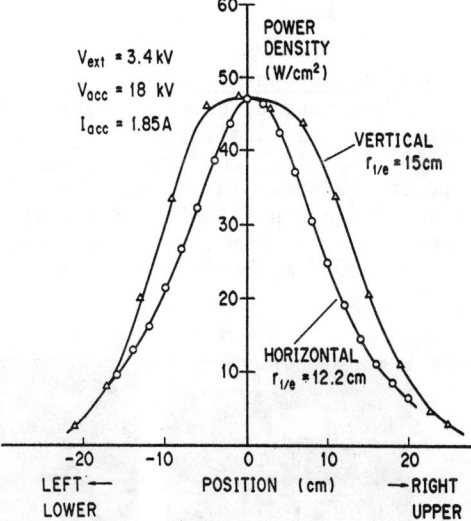

Fig. 7 An example of power density profile of the H^- ion beam measured by multi-channel calorimeter at 4 m apart from the ion source.

Space Charge Expansion in a High Vacuum

By using this ion source, we have investigated the space charge effect of H^- ion beam in a high vacuum. All the apertures except central nine apertures are masked in order both to reduce gas flow rate and to clarify the change of the divergence. The beam line was evacuated to 1×10^{-3} Pa by operating one or two cryopumps. Figure 8 shows the beam divergence as a function of the pressure in beam drift region. The pressure was varied by introducing hydrogen gas directly into the beam drift region. Although the beam divergence of 0.85 deg was obtained at pressures higher than 5×10^{-3} Pa, the divergence becomes worse at lower pressures due to the space charge expansion. Further study of the effect on beam radius and beam current density is now planned.

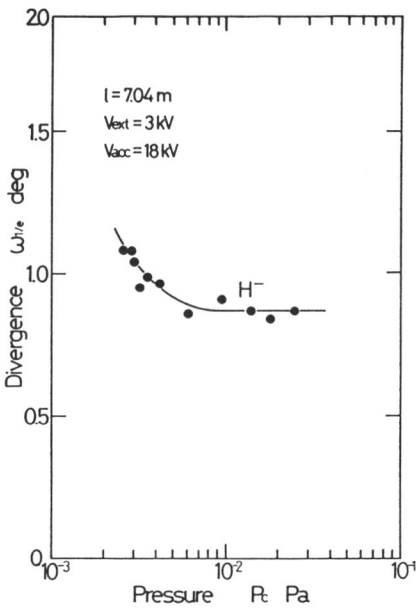

Fig. 8 Dependence of beam divergence on the pressure in beam drift region.

Momentum Mass Analysis

The impurity concentration in the beam has been measured by the momentum mass analyzer. The maximum mass number detectable is $M=240$ for ion energy of 50 keV[18]. An example of the mass spectra is shown in Fig.9, where H^- ion beam current is 0.3 A and beam energy is 21 keV. Since the line density of the residual hydrogen gas from the ion source to the mass analyzer is as high as 5×10^{16} molecules/cm^2, the negative ions are converted to

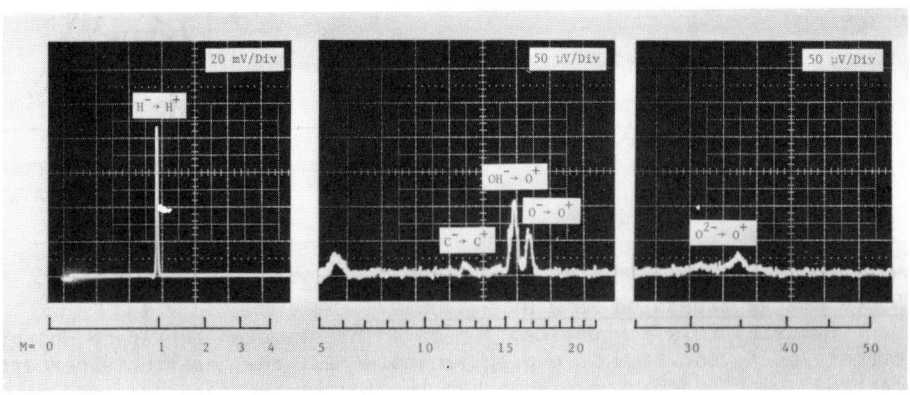

Fig. 9 An example of mass spectra of the H^- ion beam.

positive ions. The impurity concentration in the original beam was estimated by taking into account the conversion efficiency. It was found that the beam contains a few percent light impurities (C^-, O^-, OH^-) at the beginning of the experiment, but the concentration decreases to a level of one percent by discharge cleaning of the source. We have scanned the mass number up to 200, but no metal impurities have been observed.

CONCLUDING REMARKS

In order to clarify the difficulties in scaling up the volume H^- ion source and to demonstrate >1 A H^- ion beam production, we have fabricated the large magnetically filtered multicusp plasma generator and the multi-aperture extractor whose extraction area is 133 cm^2. The following remarks should be noted;

1) There is no big problem in scaling up the volume H^- ion source.
2) 1.26 A/ 21 keV/ 0.2 s H^- ion beam has been obtained with a current density of 9.5 mA/cm^2.
3) Impurity contents in the beam was no more than a few percent. No metal impurities have been observed.
4) The extracted beam has a good beam quality. The beamlet divergence was 0.8-0.9 deg at 21 keV.
5) There is an optimum pressure which makes the H^- ion current maximum. The optimum pressure increases with the arc discharge current.
6) The optimum arc voltage was about 70 V.

ACKNOWLEDGMENTS

The authors are indebted to Dr. K. Shibanuma and the members of his group for assistances in operating the cryopumps. They would like to thank their co-workers, the members of plasma heating laboratory, for their valuable discussions and assistances. Thanks are also due to Drs. H. Shirakata, M. Tanaka, K. Tomabechi for support and encouragement.

REFERENCES

1. H. Horiike, Y. Ohara, Y. Okumura, T. Shibata, and S. Tanaka, Japan Atomic Energy Research Institute Report, JAERI-M 86-064 (1986) (in Japanese).
2. K. Prelec, in Proc. Linear Accelerator Conf., SLAC, Stanford, California (1986).
3. M. Bacal and F. Hillion, Rev. Sci. Instrum. 56 2274 (1985).
4. K. N. Leung, K. W. Ehlers, and M. Bacal, Rev. Sci. Instrum. 54 56 (1983).
5. A. J. T. Holmes, G. Dammertz, and T. S. Green, Rev. Sci.

Instrum. $\underline{56}$ 1697 (1985).
6. K. R. Kendall, M. McDonald, D. R. Mosscrop, P. W. Schmor, and D. Yuan, Rev. Sci. Instrum. $\underline{57}$ 1277 (1986).
7. J. Uramoto, Institute of Plasma Physics, Nagoya, Research Report IPPJ-731 (1985).
8. R. L. York, R. R. Stevens, Jr., K. N. Leung, and K. W. Ehlers, Los Alamos National Lab. Report LA-9931 (1984).
9. T. S. Green, in Proc. of 11th Symp. on Fusion Engineering, Austin, Texas, (1985) p.103.
10. J. W. Kwan, G. D. Ackerman, O. A. Anderson, C. F. Chan, W. S. Cooper, G. J. DeVries, A. F. Lietzke, L. Soroka, and W. F. Steele, Rev. Sci. Instrum. $\underline{57}$ 831 (1986).
11. E. B. Hooper, P. Poulsen, P. A. Pincosy, Appl. Phys. $\underline{62}$ 7027 (1981).
12. Y. Okumura, Y. Ohara, H. Horiike, and T. Shibata, Japan Atomic Energy Research Institute Report, JAERI-M 84-098 (1984).
13. Y. Okumura, H. Horiike, H. Inami, Y. Ohara, and T. Shibata, in Proc. of 9th Symp. on Ion Source and Ion-Assisted Technology, Tokyo,(1985) p.121.
14. T. Shibata, H. Horiike, H. Inami, S. Matsuda, Y. Ohara, Y. Okumura, and S. Tanaka, in Proc. of the IAEA Technical Committee Meeting on Negative Ion Beam Heating, Grenoble (1985).
15. Y. Okumura, H. Horiike, H. Inami, S. Matsuda, Y. Ohara, T.Shibata, and S. Tanaka, in Proc. of 11th Symp. on Fusion Engineering, Austin, Texas, (1985) p.113.
16. A. J. T. Holmes, Rev. Sci. Instrum. $\underline{53}$ 1523 (1982).
17. J. R. Hiskes and A. M. Karo, J. Appl. Phys. $\underline{56}$ 1927 (1984).
18. Y. Okumura, Y. Mizutani, Y. Ohara, and T. Shibata, Rev. Sci. Instrum. $\underline{52}$ 1 (1981).

VOLUME PRODUCED H⁻, D⁻ ION SOURCE FOR PROTON ACCELERATOR AND THERMO-NUCLEAR FUSION RESEARCH BY SHEET PLASMA

Joshin Uramoto
Institute of Plasma Physics
Nagoya University, Nagoya, Japan

ABSTRACT

In order to apply a sheet plasma type of volume produced H⁻, D⁻ ion source to injection of proton accelerators and neutral beam injection of thermo-nuclear fusion researches, a large sheet plasma of width 20-25 cm and length 30 cm is produced under a discharge current of 175A, a H_2 or D_2 gas pressure of about 2.5×10^{-3} Torr and gas flow of (2.0-2.5) Torr ℓ/sec in a magnetic field of (60-110) Gauss.

H⁻ or D⁻ ion current is extracted with a current density up to 19 mA/cm² or 16 mA/cm², which is uniform within 10% over a large area (over 250 cm²) of the sheet plasma surface, and at an acceleration voltage up to 4.5 kV.

The electron current accelerated with the H⁻ or D⁻ ions is negligible beyond a distance of 2 cm from the final acceleration electrode.

INTRODUCTION

As H⁻, D⁻ ion sources, it is obvious that the volume production method[1,2,3,4,5,6] is superior to the methods[7,8] using cesium, with respect to handling and impurities. However, the volume production method has many problems: low H⁻, D⁻ ion current density, extraction area, suppression of electron current, discharge power efficiency, gas efficiency and measurement separating electron current from H⁻, D⁻ ion current. For these problems, we have already reported seven basic papers concerning the development[9] of an efficient sheet plasma, the application[6] of it to volume production ion sources, a method of extraction[10] of H⁻, D⁻ ion current over large area while suppressing electrons, research[11] on an effective vacuum chamber for volume produced H⁻, D⁻ ions in the sheet plasma, a simple method[12] to measure total H⁻, D⁻ ion current while separating electron current by using a magnetic field applied to the sheet plasma itself, and physical merits[13] of the sheet plasma as a volume produced H⁻, D⁻ ion source, in comparison with the multicusp plasma.[5,14] In this paper we will try to determine a sheet plasma type of volume produced H⁻, D⁻ ion source useful for proton accelerators[15] and thermo-nuclear fusion research.[16,18]

The main vacuum chamber extracting H⁻, D⁻ ions is expanded from the previous[11] size $19 \times 18 \times 23$ cm³ to a size $28 \times 25 \times 30$ cm³. Thus, through increments of the sheet plasma width and length, the H⁻, D⁻ extracting area will be increased practicably (above 50 cm² for proton accelerators and 500 cm² for thermo-nuclear fusion researches). Second, by increasing the discharge current, the H⁻,

D^- ion current density will be increased practicably (above 5 mA/cm^2 for proton accelerators and 15 mA/cm^2 for thermo-nuclear fusion researches). Third, the extracted electron current will be reduced to a level negligible compared to the extracted H^-, D^- ion current. Finally, the discharge power efficiency, gas efficiency for the extracted H^-, D^- ion current, and the heat load of extracting electrodes for the H^-, D^- ion will be discussed for the applications.

PRODUCTION OF A LARGE SHEET PLASMA

Schematic diagrams for production of a practicable sheet plasma and extraction of a H^-, D^- ion current in a high current density are shown in Figures 1a, 1b, 1c and 1d. To produce a "noshi-mochi plasma"[9] larger than the previous one,[11] two strong rectangular permanent magnets are used, as seen in Figures 1a and 1b. Thus, an efficient sheet plasma with 20-25 cm width, about 1.0 cm thickness, and 30 cm length over the extraction region (V.C.) of H^-, D^- ions is produced, whose surface area for the extraction is up to 600-750 cm^2.

The magnetic field B_z along the sheet plasma flow is about 110 G in the center (y=0) and about 60 G near the extracting electrode L (y≈15 cm), as seen in Figure 1b.

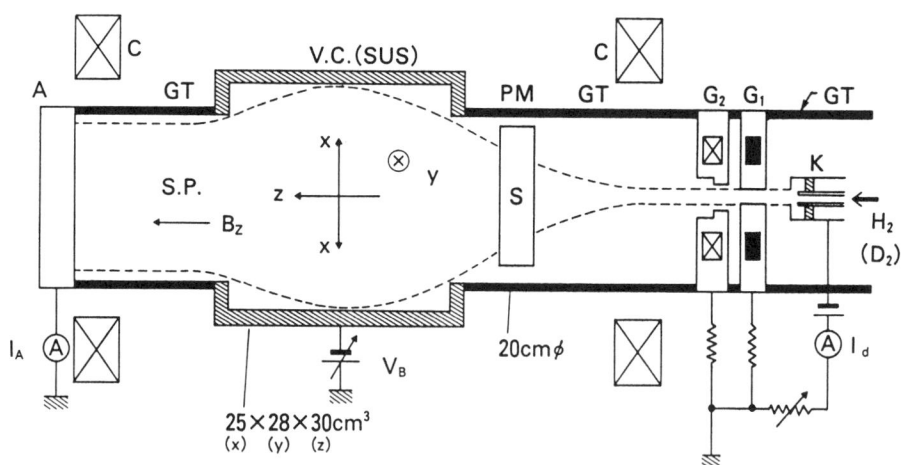

Figure 1a. Schematic diagram of the experimental apparatus (sheet plasma surface).
S.P.: sheet plasma
GT: glass tube
V.C.: main vacuum chamber (SUS)
I_A: discharge anode current
A: discharge anode
K: discharge cathode
V_B: bias voltage of V.C.
I_d: total discharge current
x, y, z: orthogonal coordinates
G_1, G_2: 1st, 2nd intermediate electrode
C: magnetic field coils for axial magnetic field B_z

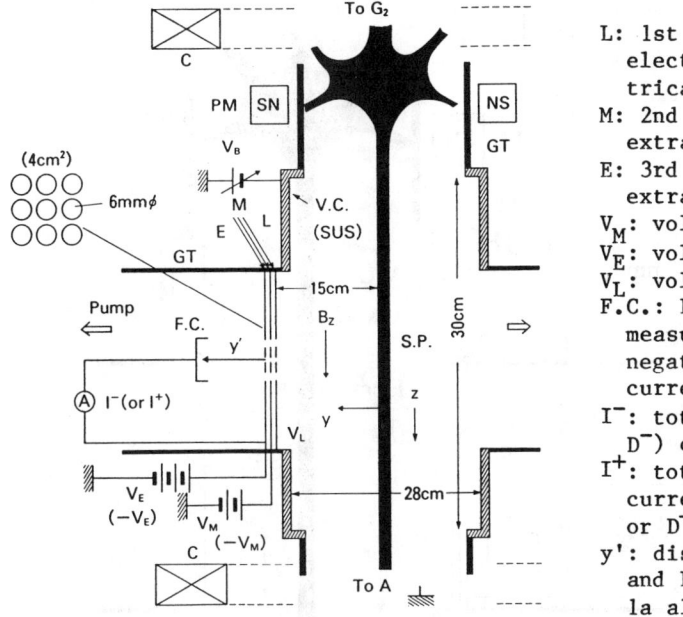

Figure 1b. Schematic diagram of the experimental apparatus (measurements of accelerated negative or positive ions).

L: 1st extraction electrode (electrically floated);
M: 2nd or intermediate extraction electrode;
E: 3rd or final extraction electrode;
V_M: voltage of M;
V_E: voltage of E;
V_L: voltage of L;
F.C.: Faraday Cup measuring total negative or positive current;
I^-: total negative (H^-, D^-) current;
I^+: total positive current (H^+, H_2^+, H_3^+ or D^+, D_2^+, D_3^+);
y': distance between E and F.C. (see Figure 1a also).

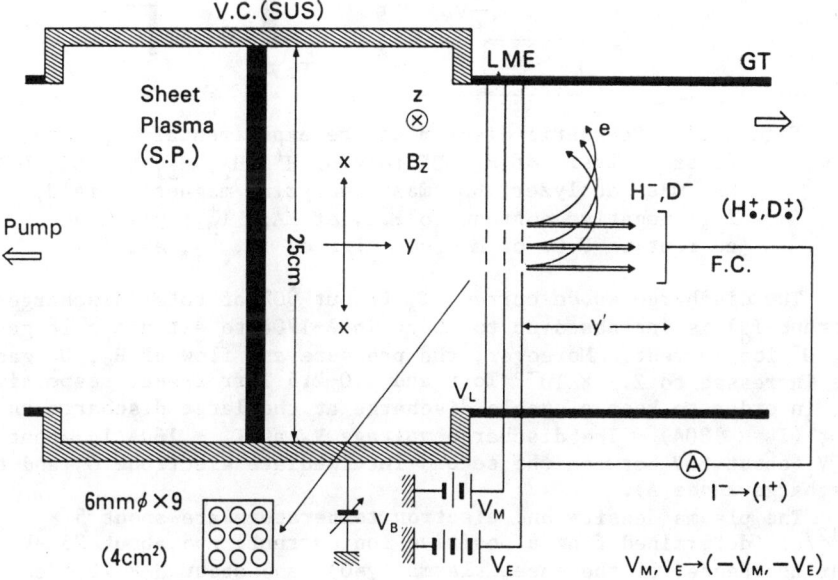

Figure 1c. Schematic diagram of the experimental apparatus (measurement separating electron current from H^-, D^- current). e: electron current (see Figures 1a and 1b also).

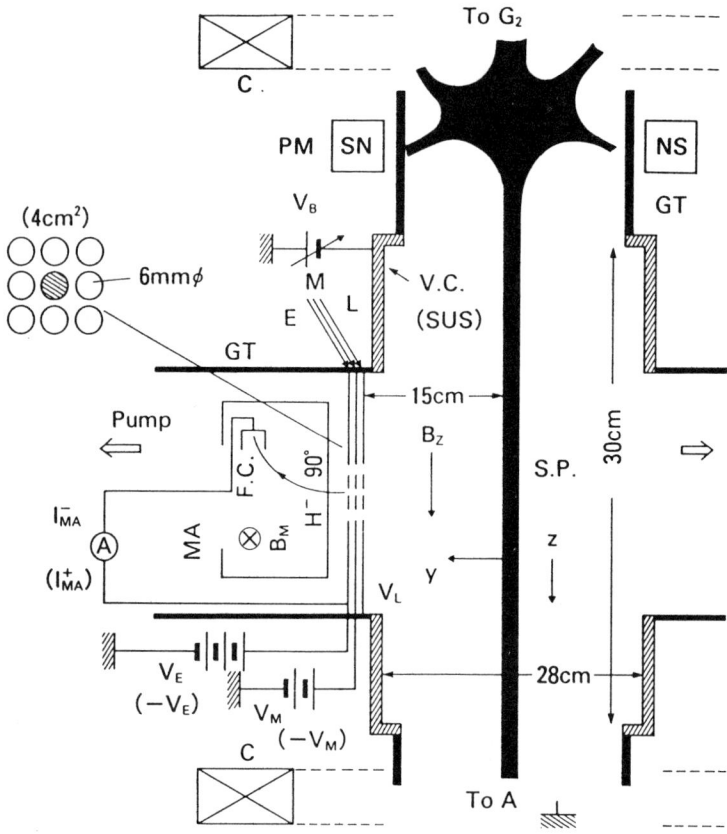

Figure 1d. Schematic diagram of the experimental apparatus (mass analysis of H^-, D^- ions or H^+, H_2^+, H_3^+; D^+, D_2^+, D_3^+). MA: mass analyzer; B_M: mass analyzing magnetic field; I_{MA}^-: negative current to F.C. of MA; I_{MA}^+: positive current to F.C. of MA (see Figures 1a, 1b, and 1c).

The discharge anode current I_A (about 90% of total discharge current I_d) is increased up to about 160A-170A to extract a large H^-, D^- ion current. Moreover, the pressure and flow of H_2, D_2 gas are increased to 2.5×10^{-3} Torr and 2.0-2.5 Torr ℓ/sec, respectively, in order to keep a stable discharge at the large discharge current ($I_d \approx 180A$). The discharge voltage V_d at $I_A \approx 160A$ is about 120V (about 65V between the second intermediate electrode G_2 and the discharge anode A).

The plasma density and electron temperature are about 5×10^{12}/cc (determined from a positive ion current) and about 25 eV near the center of the sheet plasma (y=0), and about 3×10^{11}/cc (determined from a positive ion current), and about 2.0 eV near the first extracting electrode L. The distributions of plasma density $n(y)$ and electron temperature $V_e(y)$ are shown in Figures 2a and 2b, respectively.

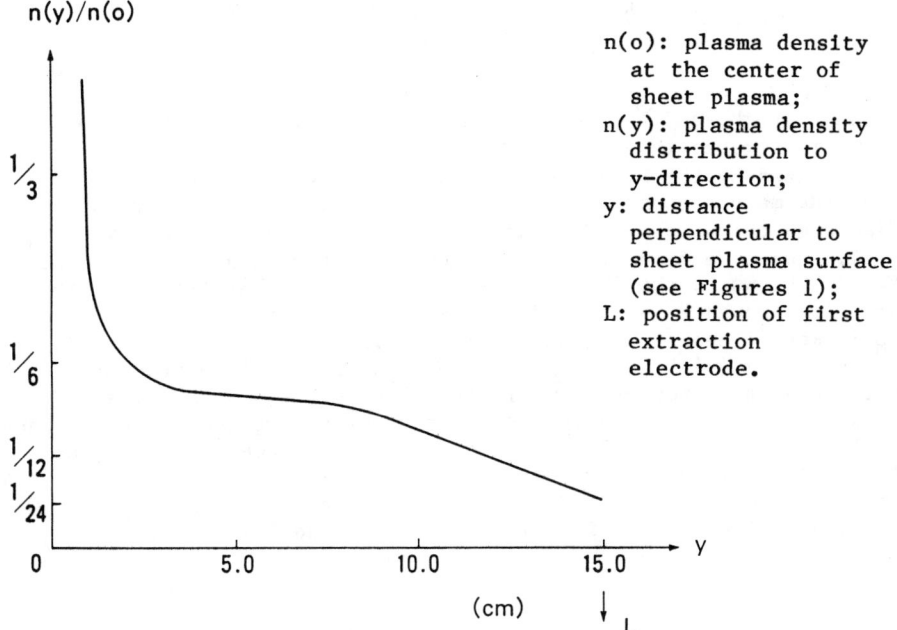

Figure 2a. Normalized plasma density distribution n(y)/n(o) (determined from positive ion current density).

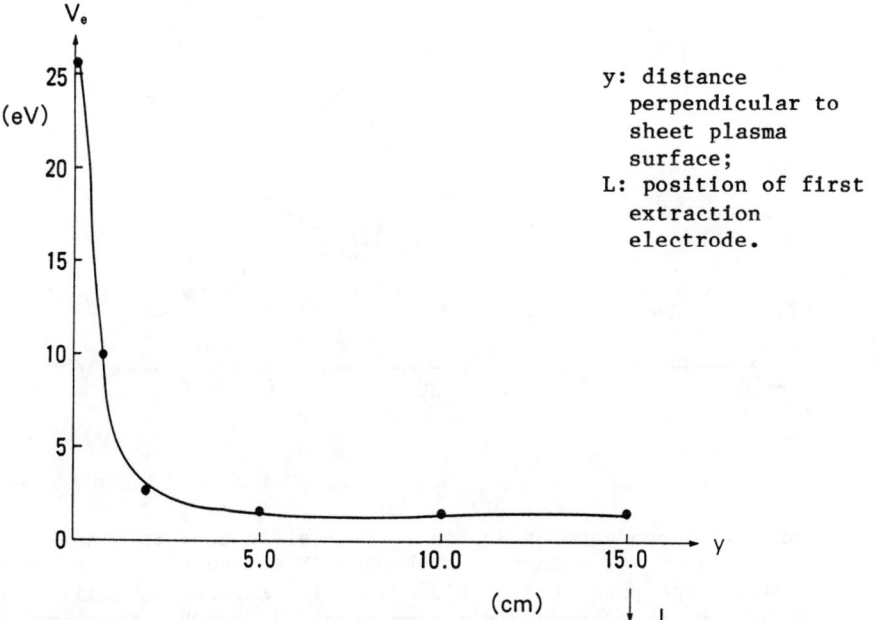

Figure 2b. Electron temperature V_e distribution.

DISCHARGE CURRENT AND EXTRACTOR VOLTAGE
FOR TARGETS OF H⁻, D⁻ ION CURRENT

The H⁻ ion current is extracted by three extraction electrodes[10] L, M, and E from outside (y≈15 cm) of the sheet plasma, as shown in Figures 1b and 1c. First, dependence of the H⁻ ion current (normalized by the H⁻ ion current at V_B = -50V) on a bias voltage V_B of the main vacuum chamber V.C. is shown in Figure 3 for large discharge currents (> 100A) keeping a constant discharge power ($I_d V_d$ ≈ 12.4 kW). We find that the H⁻ ion current reaches a maximum near V_B = -50V, while the first extraction electrode L is floated electrically around V_L = -5V and the second extraction electrode voltage V_M is kept[10] near V_M ≈ 0.2 V_E for the final extraction electrode voltage V_E = 4-4.5 kV. Here, the H⁻ ion current is extracted through 9 apertures of 6 mmϕ in the three electrodes L, M, and E, as shown in Figures 1b and 1c and as previously reported.[10] Dependence of the H⁻ ion current I⁻ on the distance y' between the final extraction electrode E and Faraday Cup F.C. is shown in Figure 4 for a final extraction voltage V_E = 4 kV (and a magnetic field B_z ≈ 60 G), and the H⁻ ion current is compared with the positive hydrogen ion

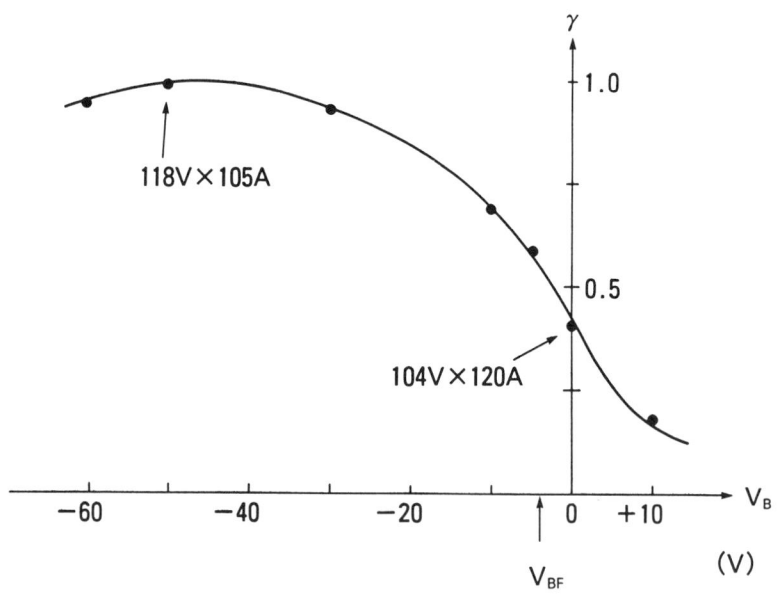

Figure 3. Dependence of H⁻ ion current efficiency on bias voltage V_B of main vacuum chamber (V.C.) under a constant discharge power (12.4 kW). γ: H⁻ ion current normalized by a H⁻ ion current in a case of of V_B = -50V. Discharge conditions: voltage V_d ≈ 104V and current I_d = 120A at V_B= 0; V_d = 118V and I_d = 105A at V_B = -50V.

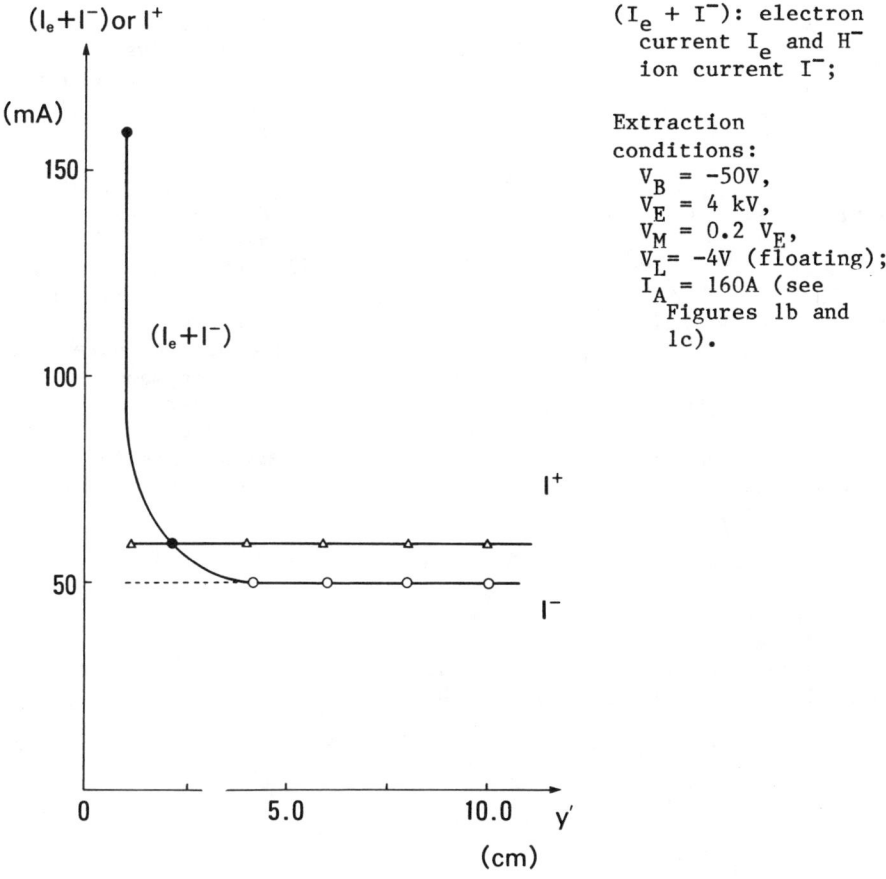

Figure 4. Dependence of H⁻ ion current I^- and (H^+, H_2^+, H_3^+) ion current I^+ on distance y' between F.C. and E.

(H^+, H_2^+, H_3^+) current by changing signs of the M electrode and E electrode voltages from plus to minus. It is seen in Figure 4 that the electron current I_e, extracted with the H⁻ ion current, decreases abruptly for the distance y' > 1.5 cm, while the positive hydrogen ion current I^+ is constant independent of y'. This fact shows that the electron current I_e is separated from the H⁻ (or D⁻) ion current beyond y' = 1.5 cm by the magnetic field outside of the sheet plasma.

Dependences of the total H⁻ ion current I^- and the corresponding H⁻ ion current I_{MA}^- to the mass analyzer (Faraday Cup) on the discharge anode current I_A are shown in Figures 5a and 5b with a final extraction voltage V_E = 4 kV, where I^- is nearly proportional to I_{MA}^-.

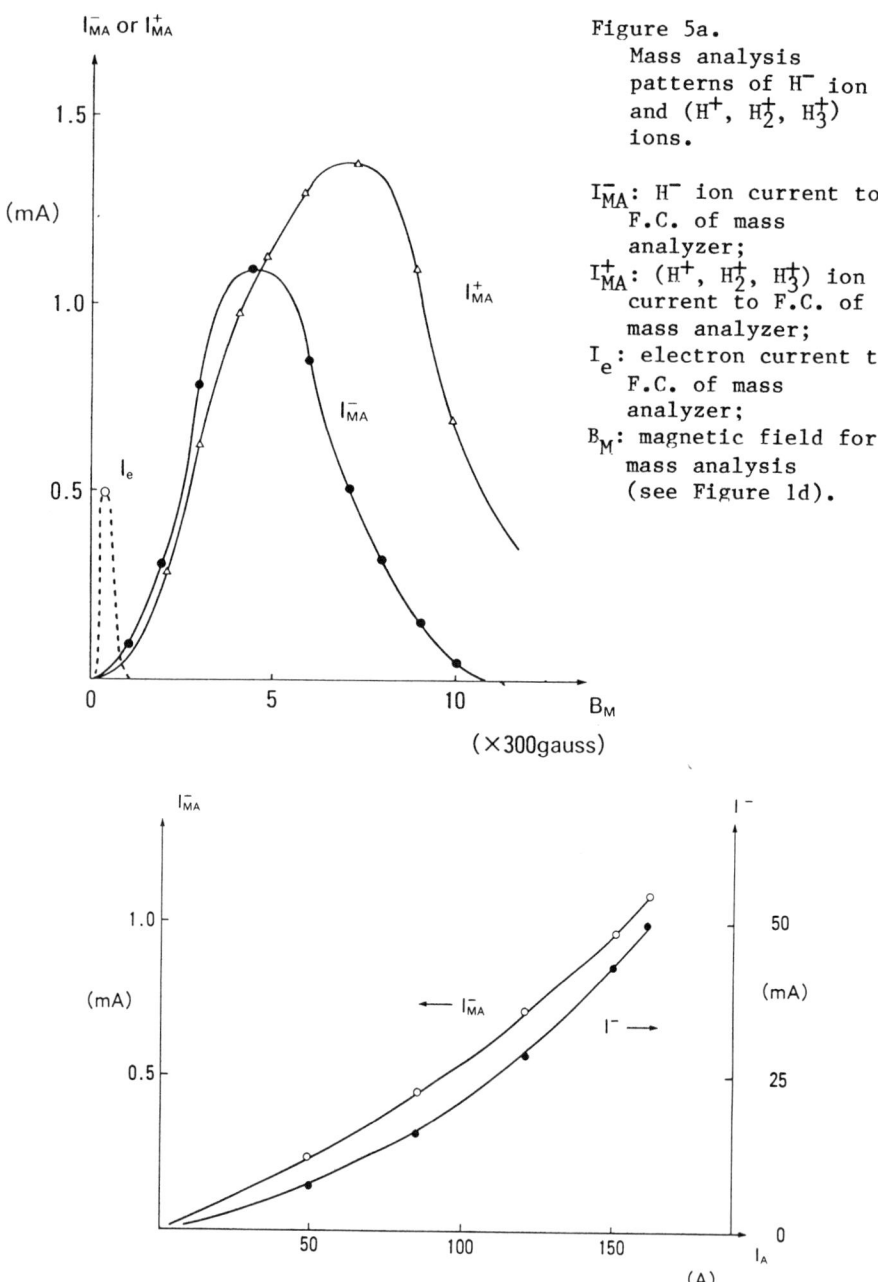

Figure 5a. Mass analysis patterns of H^- ion and (H^+, H_2^+, H_3^+) ions.

I_{MA}^-: H^- ion current to F.C. of mass analyzer;
I_{MA}^+: (H^+, H_2^+, H_3^+) ion current to F.C. of mass analyzer;
I_e: electron current to F.C. of mass analyzer;
B_M: magnetic field for mass analysis (see Figure 1d).

Figure 5b. Dependences of total H^- ion current I^- and maximum H^- ion current I_{MA}^- to F.C. of mass analyzer on discharge anode current I_A. Extraction voltages: V_E = 4 kV; V_M = 0.2 V_E; V_L = -4V (floating); V_B = -50V.

Characteristics of the total H⁻ ion current I⁻ versus the final extraction voltage V_E are shown in Figure 6 under various discharge anode currents I_A, in which the Faraday Cup F.C., for determination of the total H⁻ ion current, is set at y' ≥ 4 cm (where the electron current is negligible). The experimental $I^- - V_E$ characteristics can be compared with a theoretical space-charge limit law for H⁻ ion current (similar to H⁺ ion current),

$$I^-_{TH} = 5.4 \times 10^{-8} V_E^{3/2} S/d^2, \tag{1}$$

where I^-_{TH}, V_E, S and d are the theoretical H⁻ ion current (A), the final extraction (acceleration) voltage (V), total extraction area (cm²), effective acceleration distance (cm) for H⁻ ions, respectively. From Eq. (1), we obtain the theoretical curve TH of Figure 6 under the experimental conditions: S(6 mmφ × 9) ≈ 2.5 cm², d ≈ 0.8 cm. However, it is difficult to compare the theoretical curve TH with the experimental ($I^- - V_E$) characteristics exactly, because of the space-charge effect of the electrons.

Similarly, dependences of the positive hydrogen ion (H^+, H_2^+, H_3^+) current I^+ on V_E are shown in Figure 7 and the experimental ($I^+ - V_E$) characteristics is compared with two theoretical space-charge limit laws TH for H^+ or H_3^+ ions. We find that the experimental

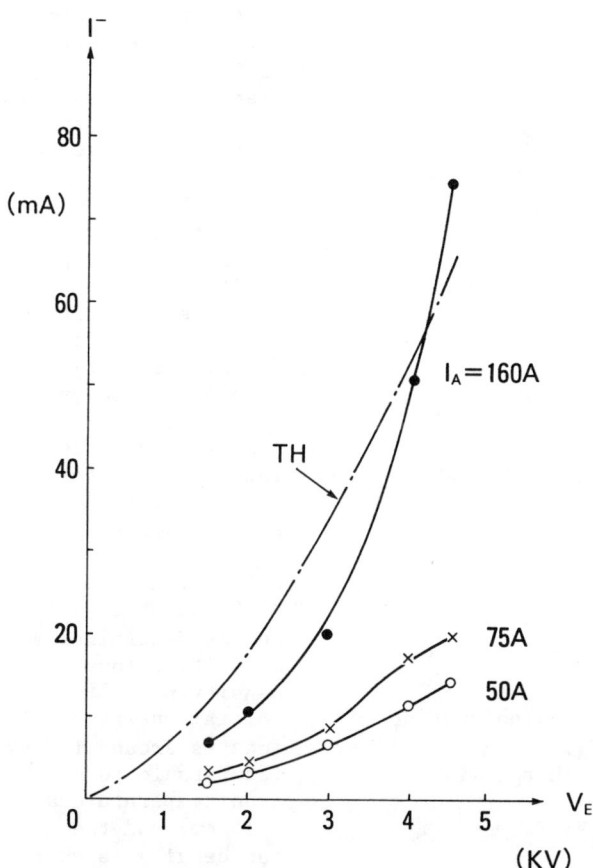

Figure 6. Dependences of H⁻ ion current I⁻ on extraction voltage V_E under various discharge anode currents I_A. TH: theoretical space-charge limit curve for H⁻ ion.
$V_M = 0.2\ V_E$; $V_L \approx -4V$; $V_B \approx -50V$.

values of I^+ at I_A = 160A are about twice larger than the corresponding theoretical value TH (H_3^+), while a large part of the positive hydrogen ions are composed of H_3^+ ions, as understood from Figure 5a. This difference may be explained by space-charge neutralization due to electrons.

From these experimental results, it is concluded that an H^- ion current with a current density of 13-19 mA/cm^2 (50 mA/4 cm^2 - 75 mA/4 cm^2), is extracted (y ≈ 15 cm) from this sheet plasma with a discharge anode current I_A ≈ 160A (total discharge current I_d ≈ 175A) and an extraction voltage V_E ≈ 4-4.5 kV. It is noted that the H^- ion current density reaches 20-30 mA/cm^2 if it is estimated by only the total aperture area S = 2.5 cm^2 (6 mmϕ × 9 apertures). Moreover, this H^- ion current density can be related to a plasma density near the first extraction electrode L (y = 15 cm), which is near 3 × 10^{11}cc at I_A ≈ 160A and is determined as a positive ion density n^+. If the initial energy of H^- ions is around 1.0 eV (comparable to electron temperature at y = 15 cm) and the H^- ion density is nearly equal to the positive ion density n^+, the H^- ion current density j_E^- is estimated by

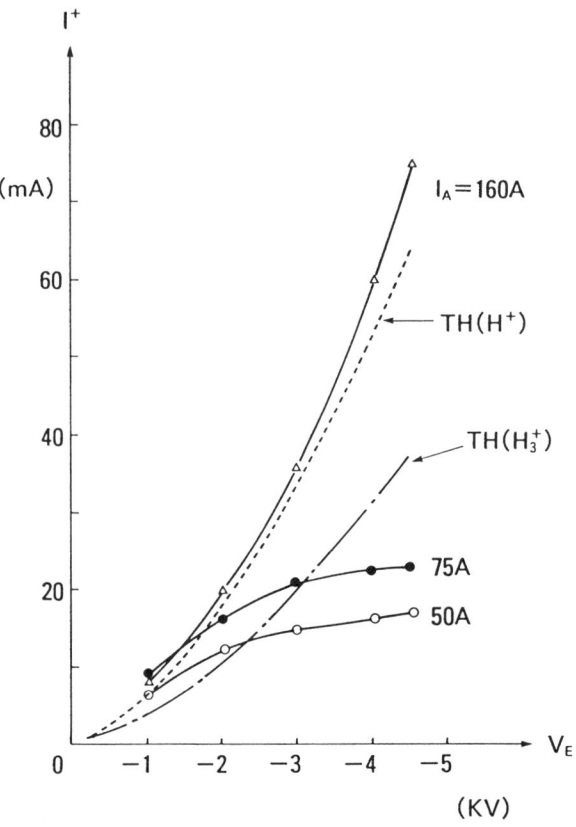

Figure 7. Dependences of H+ ion current I^+ on V_E for various I_A.
TH: theoretical space-charge limit curve for H^+ ion.
$V_M = -0.2|V_E|$; V_L ≈ -4V; V_B ≈ -50V.

$$j_E^- = en^- v^- \approx 67 \text{ mA/cm}^2, \qquad (2)$$

where e, n^- and v^- are electron charge, H^- ion density and initial velocity, respectively. Thus, we find that the corresponding experimental H^- ion current (20-30 mA/cm^2) as described already, is much lower than this value j_E^- estimated in Eq. (2). We consider that this difference may be due to electrons. That is, the exact plasma neutralization is determined by

$$n^- = n^+ - n_e, \qquad (3)$$

where n_e is an electron density near the first extraction electrode L.

For volume produced D^- ions in this sheet plasma, dependence of the D^- ion current I_D^- on the discharge anode current I_A (in D_2 gas) is shown in Figure 8 for an extraction voltage $V_E \approx 4.5$ kV. We find that the D^- ion current density reaches 15 mA/cm^2 (24 mA/cm^2 for only the extraction area of 6 mmϕ × 9 apertures) at I_A = 170A.

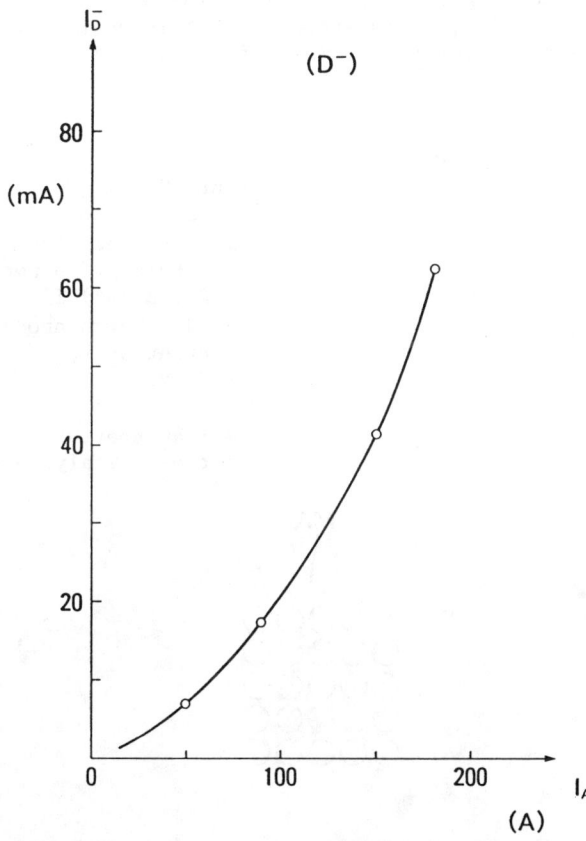

Figure 8. Dependence of D^- ion current I_D^- on discharge anode current I_A at extraction voltage V_E = 4.5 kV.

I_D^-: D^- ion current from apertures of 6 mmϕ × 9/4 cm^2.
$V_M = 0.2\ V_E$;
$V_L \approx 5V$ (floating);
$V_B = -50V$.

Thus, we may conclude that a sufficient H^- or D^- ion current density is obtained for use in injection of proton accelerators and thermo-nuclear fusion research neutral beam injector (NBI) by this large sheet plasma.

EXPANSION OF THE EXTRACTING AREA AND UNIFORMITY OF H^-, D^- ION CURRENT

For an NBI, extraction of H^-, D^- ion current over a large area is required, with uniformity of the H^-, D^- ion current density. An expansion of the extracting area is accompanied by a large flow of H_2, D_2 gas from the sheet plasma region to the H^-, D^- ion beam region through apertures of the three extraction electrodes. Thus, the vacuum pumps with large pumping speeds are required. The extraction area in this experimental apparatus, constructed with small vacuum pumps, is restricted to 6 mmϕ × 25 apertures (total aperture area ≈ 7 cm^2) in about 13 cm^2, as shown in Figure 9. An H^- ion current is extracted from the 25 apertures in the extraction electrode L, M, E, and dependence on the discharge anode current I_A is shown in Figure 9 at the extraction voltage V_E = 4.5 kV. The H^- ion current is about 200 mA at I_A ≈ 170A, which shows that it is nearly proportional to the extracting area (the number of extracting apertures).

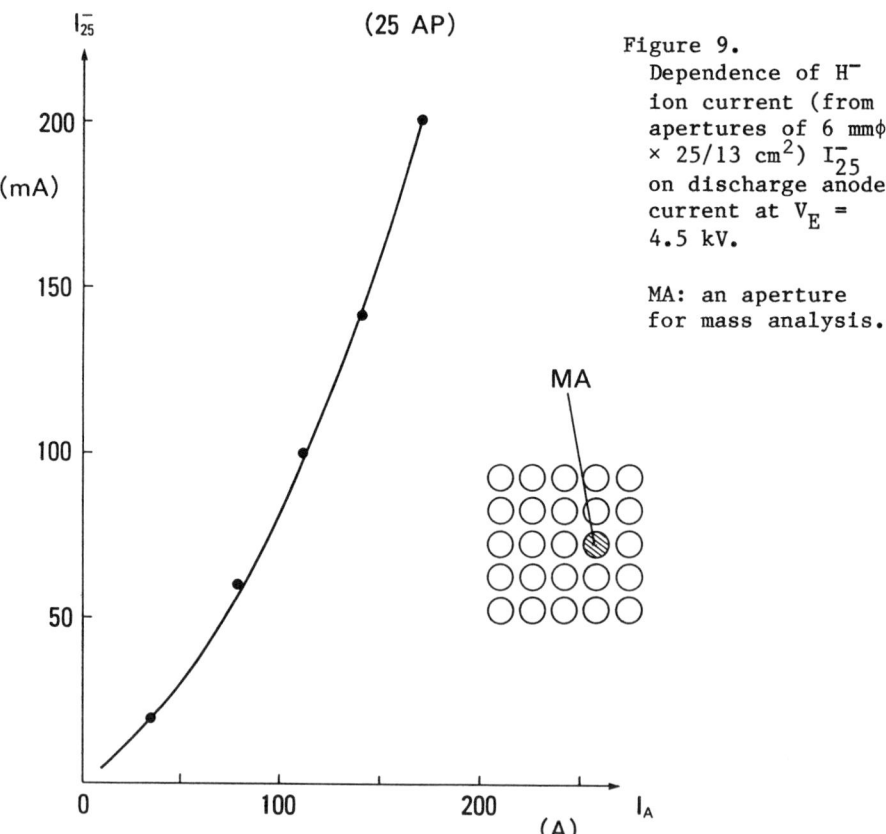

Figure 9. Dependence of H^- ion current (from apertures of 6 mmϕ × 25/13 cm^2) I_{25}^- on discharge anode current at V_E = 4.5 kV.

MA: an aperture for mass analysis.

Finally, the uniformity of the H⁻ ion current over the surface of the sheet plasma is investigated by rotating L, M, E electrodes with eccentric extraction apertures, as shown in Figure 10. We find that variation of the extracted H⁻ ion current versus the rotational angle θ is within 10% over the surface area of about 250 cm² in x-z plane of V.C.. Thus, we estimate that the H⁻ ion current density is nearly uniform over the surface of the sheet plasma.

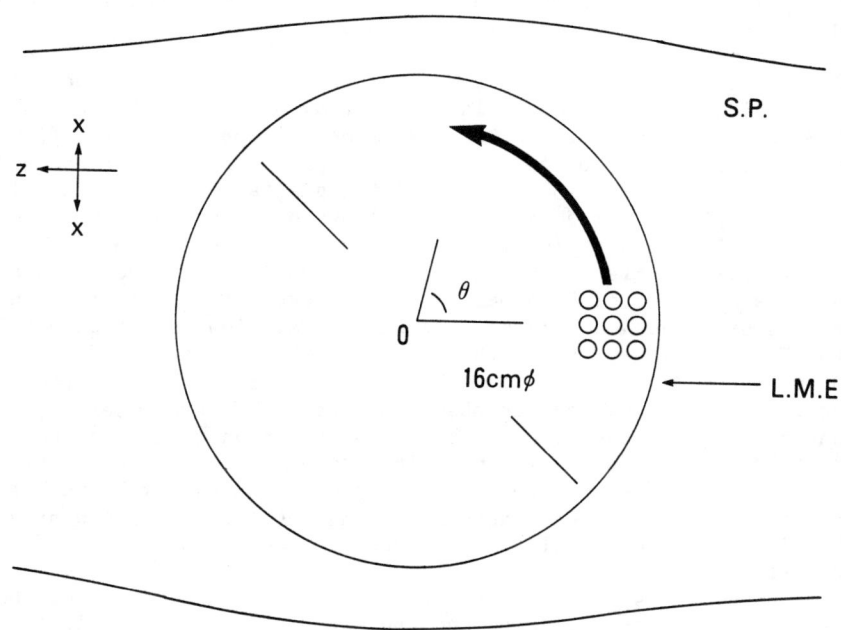

Figure 10. H⁻ ion extraction from various positions of sheet plasma surface.
S.P.: sheet plasma (x - z plane);
L, M, E: three extraction electrodes;
0: center of three electrodes;
θ: rotational angle of three electrodes (H⁻ ion currents are determined at θ = 0, 90°, 180° and 270°).

APPLICATIONS TO ACCELERATOR AND THERMO-NUCLEAR FUSION RESEARCH

From these experiments, we can conclude that the H⁻, D⁻ ion source with this efficient sheet plasma is applicable for injection of proton accelerator[15] and NBI.[17,18]

However, for NBI applications, and to exceed a positive ion based NBI[17] in efficiencies, some problems must be investigated further. First, the heat load of the final extraction electrode E, due to an electron current, is mentioned. For example, the current into E or M electrode (which we estimate to be electron currents mainly) reaches about 0.4A ($V_E \approx 4.5$ kV) or 0.15A ($V_M \approx 1$ kV) when the maximum H^- ion current of 75 mA is extracted from 9 apertures of L, M, E electrodes, as seen in Figure 1b. That is, the electrical power to E or M electrode is each about 1.8 kW (650W/cm^2 for the extraction area 4 cm^2) or 150W (38W/cm^2). These electrical powers (heat loads) to E and M become serious when H^-, D^- ion current is extracted over a larger area in an NBI. Therefore, the electrons entering the final extraction electrode E must be greatly reduced by improvements of experimental conditions.

Second, pumping of the H_2, D_2 gas flow is mentioned. In order to produce a stable sheet plasma with a large width (x-direction) in a high discharge current, a large gas flow and a high basic gas pressure are required. In this experimental apparatus, the flow and the pressure of H_2, D_2 gas in the sheet plasma region are about 2.5 Torr ℓ/sec and 2.5×10^{-3} Torr. If the gas is pumped out in only the sheet plasma region and the gas flow to the extraction regions or H^-, D^- ion beam region is negligible, the required speed of the main vacuum pump is around 10^3 ℓ/sec, which is easily achieved by using usual commercial vacuum pumps. However, if the extraction area is expanded and a large part of the gas flow is pumped out in the H^-, D^- ion beam stage (through the extraction electrode apertures) where a pressure around 2.5×10^{-5} Torr may be required, the required pump speed reaches about 10^5 ℓ/sec. In practical H^-, D^- ion sources, the gas flow through the extraction electrode apertures can be greatly reduced by increasing the number of the extraction electrodes. Therefore, the required pumping speed may be much smaller than 10^5 ℓ/sec. We consider pumping out the gas in two stages, that is, in the sheet plasma stage and in the extracted H^-, D^- ion beam stage.

Third, it is important to expand the width and length of the sheet plasma, that is, to expand the extraction area in relation with the gas efficiency and discharge power efficiency.

Similarly, the technique of focussing H^-, D^- ion current (beams) may be important to reduce the heat loads on extraction electrodes and to improve the gas efficiency.

REFERENCES

1. J.M. Wadehra, Appl. Phys. Lett. 35, 917 (1979).
2. M. Bacal and G.W. Hamilton, Phys. Rev. Lett. 42, 1538 (1979).
3. M. Allan and S.F. Wong, Phys. Rev. Lett. 41, 1971 (1978).
4. K.W. Ehlers, Nucl. Instr. Meth. 32, 309 (1965).
5. K.N. Leung, K.W. Ehlers and M. Bacal, Rev. Sci. Instrum. 54, 56 (1983).
6. Joshin Uramoto, Research Report of Institute of Plasma Physics, Nagoya University, Nagoya, Japan IPPJ-645 (1983).
7. K.W. Ehlers and K.N. Leung, Rev. Sci. Instrum. 51, 721 (1980).
8. K.N. Leung and K.W. Ehlers, Rev. Sci. Instrum. 53, 803 (1982).

9. Joshin Uramoto, Research Report of Institute of Plasma Physics, Nagoya University, Nagoya, Japan IPPJ-574 (1982).
10. Joshin Uramoto, IPPJ-666 (1984).
11. Joshin Uramoto, IPPJ-731 (1985).
12. Joshin Uramoto, IPPJ-730 (1985).
13. Joshin Uramoto, IPPJ-728 (1985).
14. K.N. Leung, K.W. Ehlers and M. Bacal, Rev. Sci. Instrum. $\underline{55}$, 338 (1984).
15. A. Takagi, et al, KEK Preprint 83, January 29, 1984 A (Japan), Proc. of the 3rd Symp. on the Prod. and Neutral. of Neg. Hydr. Ions and Beams (Brookhaven, 1983).
16. Kouichi Jimbo, Lawrence Berkeley Lab. Report No. LBL-13769 (Ph.D. thesis).
17. K.W. Ehlers, et al, Proc. 2nd Symp. on Ion Sources and Formation of Ion Beams (Berkeley, 1974), P.I - 5 - 1.
18. F. Schwirzke, Lawrence Berkeley Lab. Report No. LBL-6827.

PRODUCTION OF NEGATIVE IONS BY DOUBLE CHARGE EXCHANGE AND THEIR ACCELERATION

N.N. Semashko, V.V. Kuznetsov, A.I. Krylov, P.S. Firsov
I.V. Kurchatov Institute of Atomic Energy, Moscow, U.S.S.R.

INTRODUCTION

The need for negative hydrogen ion beams with currents of a few tens of amperes and higher, for applications in controlled fusion, has emerged in the seventies. By that time, sources of positive ions with beam currents of up to a hundred amperes had been successfully developed in many laboratories throughout the world. Such sources have served as a reliable basis for the development of the double charge exchange of protons in alkaline metal vapor targets as a technique for negative ion production. Sources used previously yielded negative ion beams with the currents of a few hundred milliamperes, at best. It was necessary to demonstrate experimentally not only the production of intense H⁻ beams, but also the post-acceleration of such beams. The MIN-facility[1] was constructed for this purpose at the Kurchatov Institute of Atomic Energy and experimental results obtained there will be given below.

FACILITY

A sketch of the MIN-facility is given in Figure 1. The IBM-5 and IBM-6 gas discharge sources[2] with positive ion currents up to 12A and 55A, respectively, in the energy range 3-9 keV, and with the pulse duration 10-25 ms, were used. The volume of vacuum chambers ($8~m^3$ each) in the facility allows one to increase, if necessary, the pulse duration by an order of magnitude without additional pumping. Sodium vapor serves as the charge-exchange target.[3] The vapor crosses the ion beam path in the form of a supersonic jet. The gas from the ion source can pass into the negative ion transport section only through the vapor jet, and this decreases considerably the gas load in this part of the facility.[4,5,6] The neutral hydrogen pressure drops from 10^{-4}-10^{-3} Torr at the ion source side to 10^{-6}-10^{-5} Torr downstream of the jet; the cross section of ion beam apertures is 22 × 48 cm². Then, beyond the charge-exchange target, there is a cylindrical insulator with an internal diameter of 80 cm, where flat electrodes for further acceleration are located. The apertures in the electrodes are 14 × 65 cm. The size of the aperture in front of the first electrode can be reduced with a movable limiter.

Two movable beam dumps measure the total beam current and the current density distributions of charged particles and neutrals in the beam. A low transverse magnetic field is used to separate electrons from negative ions in the post-acceleration part of the system. The energy spectrum and angular spread in the beam are measured by an electrostatic analyzer. Secondary plasma parameters at different positions along the path are studied by using flat, hot, and capacitive movable probes and a mass-analyzer. The parameters of the charge-exchange target are measured with calorimeters, and via the intensity of the resonance sodium emission induced by a special electron beam, with the sodium vapor flux sensors located along the path.

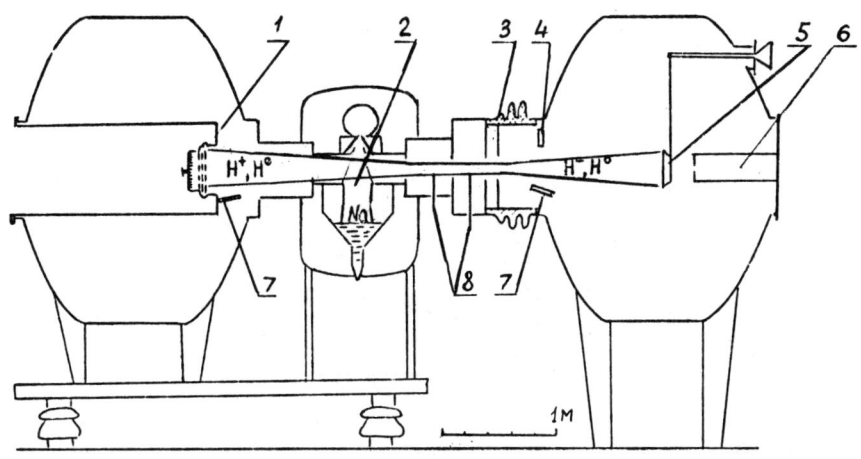

Figure 1. MIN-Facility.
1 - positive ion source; 2 - charge-exchange target;
3 - negative ion acceleration system; 4 - high frequency mass analyzer of secondary ions; 5 - ion collector;
6 - electrostatic energy analyzer; 7 - sodium flux sensors; 8 - probes.

NEGATIVE ION BEAM PRODUCTION

The primary beam of positive ions and neutrals, atomic and molecular, is passed through the charge-exchange sodium target with a line density of $(0.5-2.0) \times 10^{15} cm^{-2}$. After passing through the target the beam consists of about 10% of negative hydrogen ions, the rest being the atoms and neutral molecules of hydrogen; the fraction of positive ions is less than 1%. The current density of negative ions is measured by means of enclosed secondary emission probes. The total current of negative ions is found by integration over the current density. The neutral current is determined by measurements with an open secondary emission probe. The total negative ion current vs. the total primary positive ion current (as measured in the ion source power-supply circuit) is given in Figure 2. This dependence is close to a linear one. To maintain the optimal focusing conditions the energy of particles in the primary beam has to increase according to the 3/2 power law if the beam current extracted from the source is increasing (so that the maximum current corresponds to the energy 9 keV). Therefore, the initial positive ion beam composition, as well as charge-exchange conversion efficiencies of protons in the initial beam and of protons and atoms produced

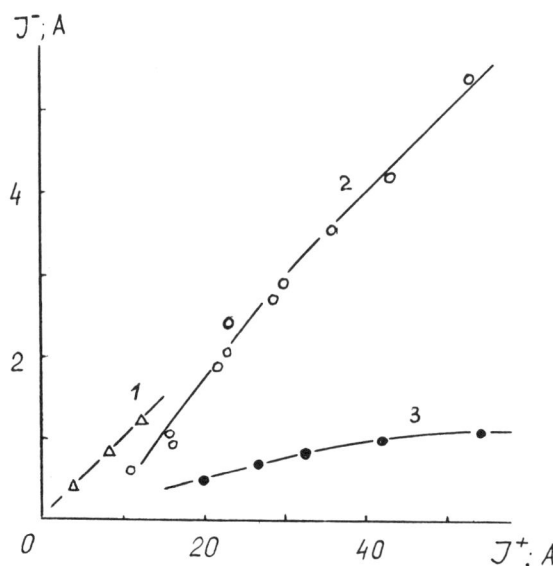

Figure 2. Negative ion current vs. positive ion current.
Curves 1 & 2: H^- ions, obtained with two different positive ion sources;
Curve 3: He^- ions.

after dissociation of two-atomic and three-atomic ions, will change. The composition of negative ion beams for two ion sources is shown in Figure 3. From 50% to 75% of negative ions are produced out of the molecular ions, which have a rather large dissociation cross-section in sodium.[6] The overall effect of these and other factors (stripping in the plasma, in the target, in the residual gas) is an almost-linear rise of the negative ion current with the primary beam current, within the range of measurements. It is possible that it would rise even further with the primary beam current. The energy of particles in the primary beam is limited to 9 keV in our experiment.

The role of the plasma produced in the vapor jet target by the beam should be particularly considered. Initially, we were alarmed by the expected limitations of the charge-exchange technique, namely, by the plasma production in the target and, due to this, by the reduction in the ion yield. The experimentalists in Novosibirsk observed a similar phenomenon:[7] atomic yield from a magnesium jet target decreased with an increase in the current density of protons. From two to four slow ions per each fast particle can be produced in the target under our conditions. The maximum density of the secondary plasma in the target was measured to be $6 \times 10^{12} cm^{-3}$; this corresponds to an ionization degree in the jet equal to 6%. Under such conditions a reduction in the negative ion yield is only a few percent.

However, it turned out that the secondary plasma has showed its effect in another way. The interactions between the sodium plasma

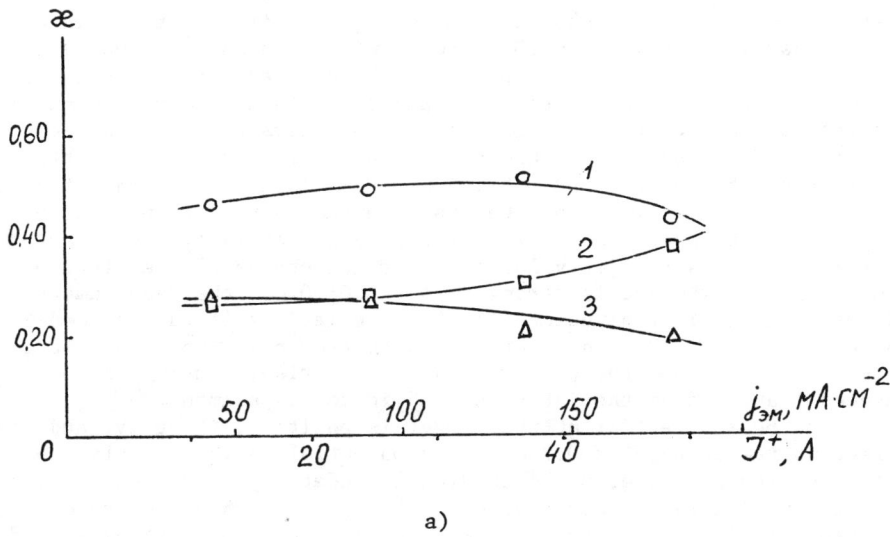

a)

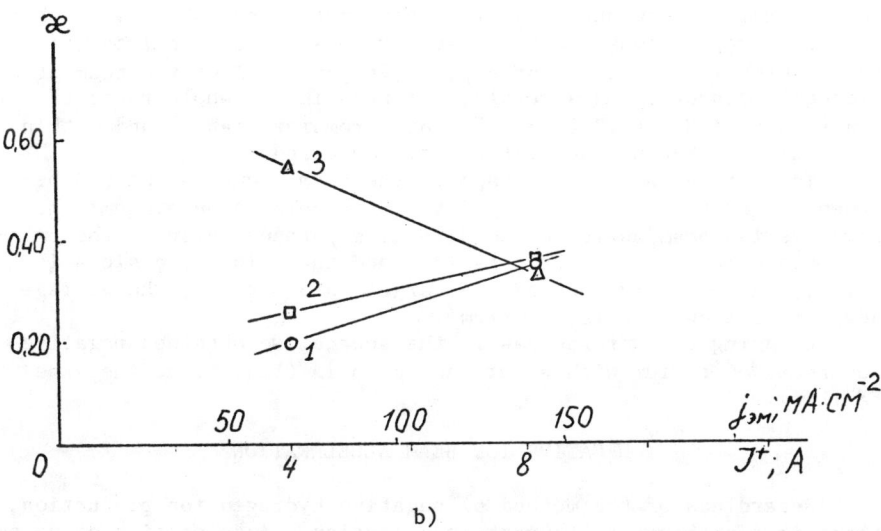

b)

Figure 3. Negative hydrogen ion beam composition.
a) for IBM-6 source; b) for IBM-5 source.
The energy of negative ions is equal to the initial primary energy for curves 1; to 1/2 of it for curves 2; and to 1/3 of it for curves 3.

and the sodium jet result in a strong deceleration of the supersonic
jet due to a resonance charge exchange. It manifests itself in a
non-linear increase (up to 10^{16} atoms/cm^{2}s at maximum currents) of
the leakage rate of sodium vapor (co- and counter-directed with the
beam) from the jet during the beam passage. The hydrodynamic model
of the interaction between the beam and a supersonic jet has been
developed.[8] According to this model, the intensity of interaction
depends on the parameter $\alpha = I\sigma_1/V_0 e$, where I is the current of fast
particles per unit width of the charge-exchange target in A/cm; σ_1
is the ionization cross-section of the target atoms by these fast
particles; V_0 is the jet velocity; e is the charge of the electron.
The effect on the jet is strong at $\alpha \geqslant 0.03$-0.05; the Mach number
drops from its initial value (in the calculations it was assumed M_0
= 00) down to M $\leqslant$ 2. In accordance with a drop in the Mach number,
the temperature and the pressure in the jet rise, causing an enhanced leakage from the jet, observed in the experiment.

The beam potential at the target is positive, about 4V, and the
electron temperature is 0.7 eV. The potential in the negative ion
beam is also positive, not exceeding 1V under any vacuum conditions,
(including initial conditions before the pulse), when the hydrogen
pressure is lower than the critical one,[9] which equals $(2-4) \times 10^{-5}$
Torr under our conditions. Mass spectra measurements of the secondary plasma with the high frequency mass-spectrometer have shown
that it mainly consists of sodium ions (more than 70%). The density
of sodium vapor along the beam path exceeds the critical one, 10^{10}
cm^{-3}, which is the cause for a positive potential of the beam at any
hydrogen pressure. As a result, the beam in the whole range of
study (0.5-5A H$^-$, (0.2-2) $\times 10^{-5}$ Torr) remains stable, and within
the band 0-10 MHz no oscillations are observed.

To improve the beam transport, the beam focus is located between the target and the entry into the acceleration system; the
negative ion beam angle in the direction perpendicular to the emissive slots in the ion source is ±3°, and that along the slots is
±0.5°. The beam size beyond the target is 5 × 40 cm, the average
negative ion current is 30 mA/cm^2.

Changing the working gas in the source, we obtained negative
ion beams of helium with a current up to 1A (Fig. 2) at the same
facility.

NEGATIVE ION BEAM ACCELERATION

Regardless of the method of negative hydrogen ion production,
there is a problem of its post-acceleration. Acceleration of negative ion beams produced by the charge-exchange technique has some
peculiar features. As the fraction of ions in the beam produced
from molecular positive ions is large (see Fig. 3) it is desirable
that all three components be accelerated with similar efficiency,
i.e. the acceleration system should have a long focal length. Calculations[10] have shown that a system of flat electrodes satisfies
this requirement. If negative ions are not separated from neutrals
after the passage of the beam through the target (which is the case
in our system), the acceleration process is complicated by the

presence of the neutral beam, as well as of the vapor and electrons from the target.

First attempts to accelerate the beam resulted in breakdowns in the accelerating gap. They occurred at a certain delay after switching the beam on, depending on the beam current and the accelerating voltage (Fig. 4, curves 1-5). Studies have shown that short-circuiting of an accelerating gap occurs due to the initiation of a self-sustained discharge by the beam of ions and neutrals. The discharge is initiated due to the sodium film accumulated on the electrodes and on other surfaces in the acceleration system in intervals between the pulses. Sodium is knocked out from these surfaces during the beam passage, as well as entering the accelerating system from the jet target due to the perturbation by the beam. Increased pressures of hydrogen and sodium, produced by other means in the accelerating gap, do not result in ignition. The discharge continues to burn even after switching the beam off, and the arc duration and the current are determined by the parameters of the power-supply system.

We have managed to avoid the breakdown by placing a limiter, 10 × 30 cm, in front of the input electrode. This limiter intercepts 10-20% of the current from the beam periphery. Thus, the dependence on the accelerating voltage is eliminated (Fig. 4, curve 6). Heating the limiter and the electrodes up to 100-120°C eliminates the dependence on the current, and the time of acceleration becomes equal to the original beam duration (curves 7 and 8). This allows us to accelerate negative ion beams with currents greater than 4A,

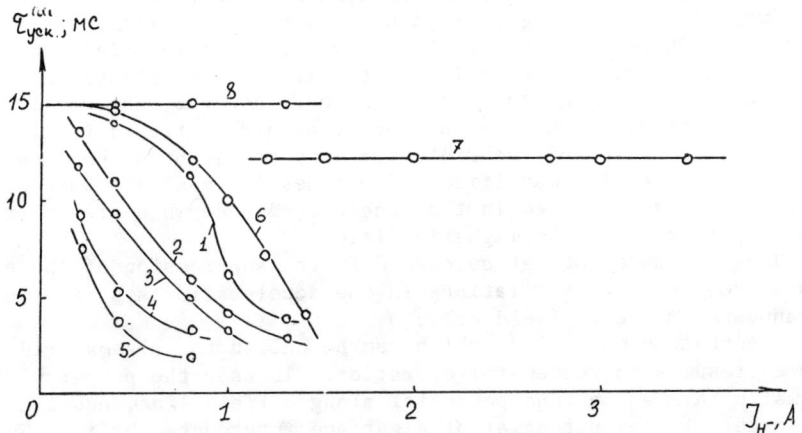

Figure 4. Pulse length of accelerated negative ion beam vs. its current, for different methods of suppressing breakdowns in the system.
<u>Curves 1-5</u>: without limiters and heating; energies of acceleration for curves 1-5 are 10, 20, 30, 40, and 50 keV, respectively;
<u>Curve 6</u>: in the presence of a limiter (all energies);
<u>Curves 7 & 8</u>: limiter and heating (all energies).

up to an energy of 80 keV.

A strong rise in the current occurs during the pulse due to the acceleration of electrons from the secondary plasma, and some oscillations are observed. The main source of electrons is the charge-exchange target, where 20-40 electrons per each negative ion are produced. However, this number is not greater than 0.5 in the drift region. A positive potential (≥ 1V) of the beam helps the diffusion of electrons from the target into the accelerating gap.

The presence of oscillations observed between the target and the accelerating gap is related to the acceleration of electrons. The spectrum of frequencies has two characteristic regions: near 10 kHz and 100 kHz. In the primary beam, as well as in the accelerated one, the oscillations in the band 0-10 MHz are not observed.

Dc and ac components of the electron current are given in Figure 5, a and b, curves 1.

Similar oscillations during the passage of an electron current through a plasma having parameters close to those of our secondary plasma ($n_e \simeq 10^8 - 10^9 cm^{-3}$) were observed by a number of experimentalists.[12,13]

A negative potential barrier at the grid is one of the possible techniques for suppressing the accompanying electrons and oscillations in the acceleration system. Oscillations are suppressed, and the electron current is reduced to a level determined by the secondary electron emission from the grid under the beam bombardment. However, this method has evidently no future for a steady-state, high current injector.

Some experiments on the suppression of electrons by a transverse magnetic field were carried out. The maximum field was at a distance of 20 cm from the input electrode. When the field was 85 Oe at the beam axis, a reduction in the level of oscillations by almost an order of magnitude was observed. However, the complete suppression of electrons does not occur at high currents (Fig. 5, curves 2). Doubling of the field improves the suppression; however, oscillations are also magnified a few times (curves 3). Weak oscillations are also observed in the accelerated beam when the suppression is performed by the magnetic field.

Thus, we have not yet succeeded in the suppression of the electron current nor of oscillations in the acceleration system by using a transverse magnetic field only.

A certain combination, which can be called the plasma grid method, results in successful operation. It uses the property of a plasma to have a constant potential along a field line, equal approximately to the potential of a surface intersected by it. The plasma grid located between the poles of a magnet is an isolated frame, 4 mm wide, which surrounds the beam. The magnetic field at the center of the gap is 85 Oe. When a negative potential is applied to the frame, electrons and oscillations are effectively suppressed (Fig. 5, curves 4). A cut-off potential of 10-20V is sufficient for beam currents below 2A. However, one needs 400V at an H^- current of 3.5A. Note that the oscillations appear at 2 ms and disappear at 6 ms, without reappearing again.

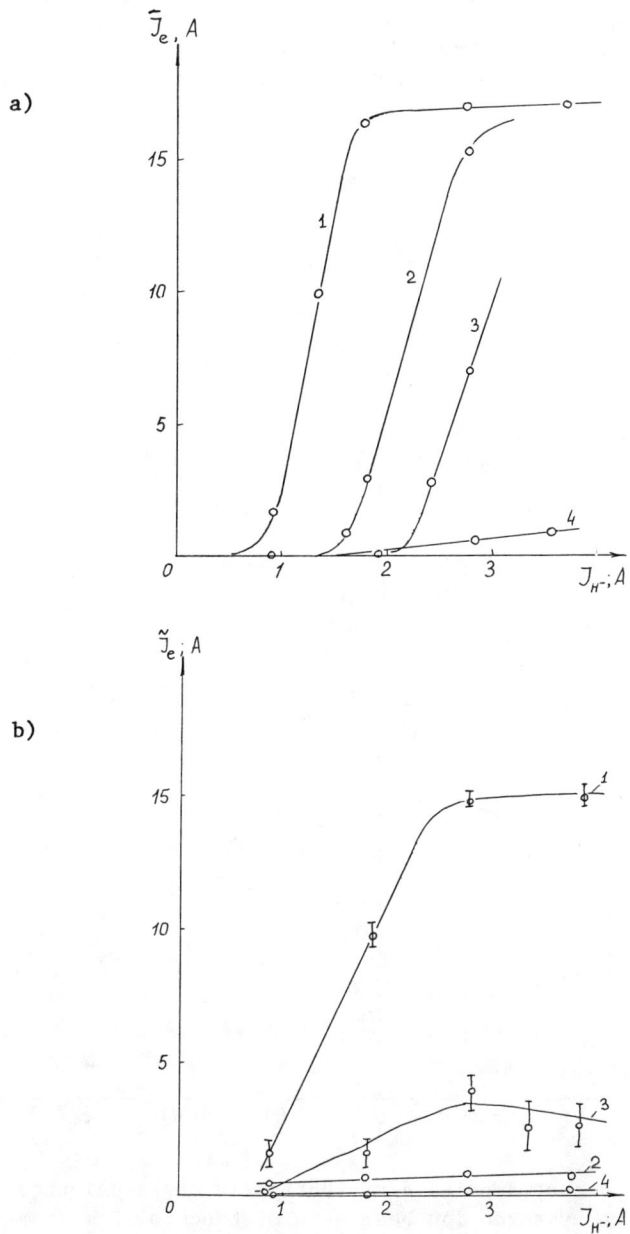

Figure 5. DC (a) and AC (b) components of the electron current vs. accelerated negative ion current for different methods of suppression.
1 – without suppression; 2 – with the magnetic field 85 Oe; 3 – with the magnetic field 165 Oe; 4 – suppression with a plasma grid.

The current density distribution across the accelerated beam of negative hydrogen ions at a distance 2 m from the acceleration system is shown in Figure 6. One can see reduction in the beam width with the accelerating voltage due to an increase in the longitudinal velocity.

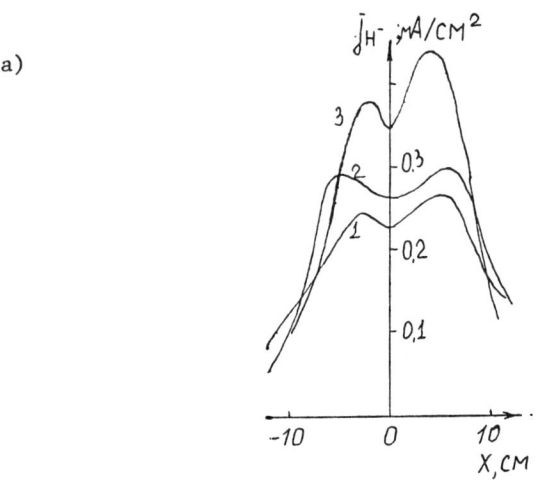

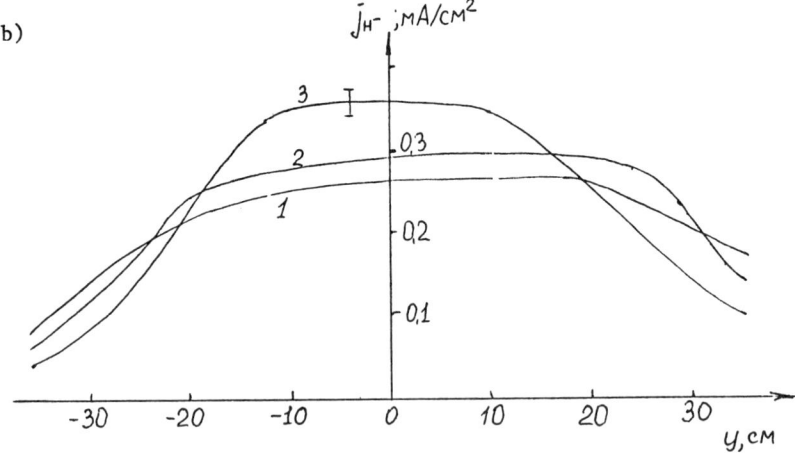

Figure 6. Current density distribution in the accelerated negative hydrogen ion beam at a distance of 2 m from the acceleration system:
a) directed along the emission electrode slots in the source;
b) directed across the slot. The H^- beam current is 1A; initial primary energy is 3.5 keV; ion energy after acceleration: 1 - 30 keV; 2 - 40 keV; 3 - 60 keV.

DISCUSSION

The studies made have shown the possibility of multiampere H⁻ beam production by double charge exchange in sodium vapor. Analogous results (H⁻ current of 2A) have been obtained in Ref. 14. They have shown some peculiarities of this technique. Further increase in the negative ion beam current is possible either by increasing the primary source current density or by increasing the larger dimension of the source emitting surface. The experiments have shown the possibility of accelerating such ribbon beams with a relatively small electron load and at a low level of oscillations. Operational stability of the acceleration system in the studied range does not depend on the accelerating voltage. Therefore, one can hope to accelerate high power beams of negative ions to considerably higher energies, where some other physical limitations may become important. Pulse length and negative ion acceleration limits are related to the characteristics of the positive ion source used. An H⁻ beam of 3A, 0.3 s was produced.

Advantages and shortcomings in the production of negative ions by the double charge exchange depend, to a great extent, on the neutral beam system. A possible system design is shown in Figure 7.

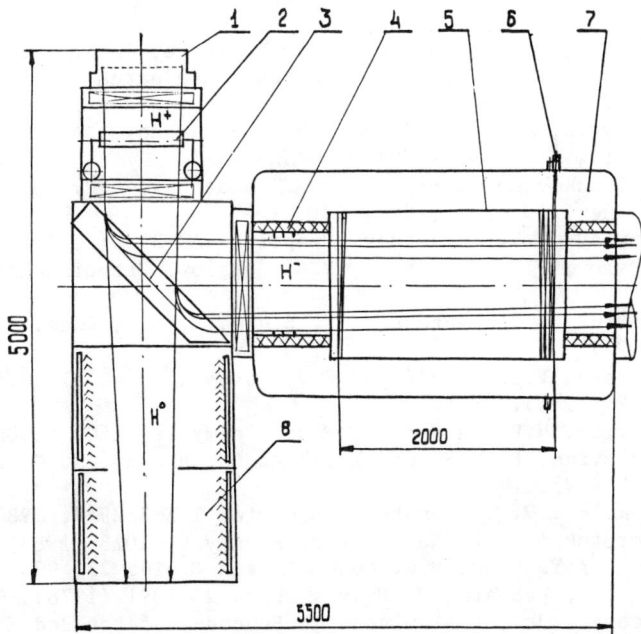

Figure 7. System design of a neutral injector based on negative ions produced by charge-exchange:
1 - positive ion source; 2 - charge-exchange target; 3 - magnet; 4 - accelerating system; 5 - photon target; 6 - laser radiation inlet; 7 - chamber filled with SF_6 gas for collecting the unused neutral beam.

The difficulties we came across in the acceleration were caused by the penetration of sodium into the acceleration system, by the ionization of sodium there, by the presence of almost 90% neutral component in the beam beyond the jet target, and by the diffusion of electrons from the target. The system design proposed in Figure 7 is free of these shortcomings. The negative ion beam deflection performed with a magnetic mirror (achromatic) allows the utilization of negative ions with all three energies. The separated neutral beam is directed to an inlet chamber, where it is easy to achieve differential pumping of gas. Molecular gas from the source is intercepted by the sodium charge-exchange jet target. Thus, the gas load in the high-vacuum part of the injector, where the transport acceleration, and stripping of negative ions occur, can be reduced to a level determined by the outgassing and by beam losses at limiters. It is convenient to use a photon target for stripping. The circuit design with a magnetic bend allows the use of a shield against the neutron flux from the reactor.

REFERENCES

1. A.I. Krylov, V.V. Kuznetsov, N.N. Semashko, In Proceed. of the 8th Symp. on Engin. Problems of Fusion Research (San Francisco, 1979), p. 853.
2. V.M. Kulygin, A.A. Panasenkov, N.N. Semashko, I.A. Chukhin, J. Tekhn. Fiz. $\underline{49}$, 168 (1979).
3. B.A. D'Yachkov, A.I. Krylov, V.V. Kuznetsov, N.N. Semashko, Atomn. Ener. $\underline{49}$, 246 (1980).
4. A.I. Krylov, V.V. Kuznetsov, Preprint IAE-3330/7, 1980.
5. A.I. Krylov, V.V. Kuznetsov, V.G. Sokolov, Preprint IAE-3566/7, 1982.
6. D.P. DeBruijn, J. Neuteboom, V. Sidis, J. Los, Chem. Phys. $\underline{85}$, 215 (1984).
7. E.A. Gileev, V.I. Davydenko, G.V. Roslyakov, et al., Fiz. Plasmy $\underline{11}$, 1502 (1985).
8. A.I. Krylov, V.V. Kuznetsov, Fiz. Plasmy $\underline{11}$, 1508 (1985).
9. M.D. Gabovich, L.S. Simonenko, I.A. Soloshenko, J. Tekhn. Fiz. $\underline{48}$, 1983 (1978).
10. V.G. Sokolov, V.V. Kuznetsov, Preprint IAE-3329/7, 1980.
11. V.P. Goretskij, A.P. Naida, Fiz. Plasmy $\underline{9}$, 1023 (1983).
12. P. Leung, A.Y. Wong, B.H. Quon, Plasma Fluids $\underline{23}$, 993 (1980).
13. G. Joyce, R. Habbard, J. Plasma Phys. $\underline{20}$, 391 (1978).
14. E.B. Hooper, Jr., P. Poulsen, In Proceed. of the 2nd Int. Symp. on the Prod. & Neutr. of Neg. Ions & Beams (Brookhaven, 1980), p. 247.

DISCUSSION

Alessi: In the accelerated beam, what was the ratio of H^- to electrons?

Semashko: The electron current is effectively suppressed, and the ratio I_e/I^- is lower than 0.2 when the "magnetic plasma grid" is used.

Forrester: What is the nature of the beam neutralization in the propagating H^- beam? Does it gather in Na^+ ions from the target to provide the neutralized plasma through which the H^- ions traverse? Or is space charge no problem at these levels of voltage and current? I think it is a problem if there isn't space-charge neutralization. You must have some plasma, some ions neutralizing the space charge.

Semashko: Beam space-charge neutralization in the region between the sodium jet and accelerator is mainly produced by sodium ions. The H^- beam potential is positive (about +1V).

Hellblom: You quoted a plasma potential in the collision chamber of 4V, and 1V potential in the beam drift.

Semashko: In the sodium cell, the plasma potential is about 4V. Outside of this cell, in the region between the cell and the first electrode of the accelerating system, approximately +1V.

Hellblom: This potential of 4V in the collision chamber, how did it scale with the incident positive ion beam? With a very large positive beam, could you increase this potential to larger values than +4V?

Semashko: Plasma potential in the jet is 3-4V and slightly increases with the beam current.

Peterson: What is the optimum target thickness for your beams?

Semashko: It is up to 5×10^{15} thickness.

Tsai: What is the additional divergence due to scattering in the target plasma?

Semashko: Our measurements show that increasing of beam divergence angle in the jet is lower than 0.1°.

Forrester: You must have a multi-aperture accelerating structure between the target and the beam. Can you describe the nature of the potentials on those?

Semashko: The accelerating structure is multi-slot for the ion source. The distance between the ion source and the target is approximately 1 meter. The additional acceleration system is one aperture.

Ehlers: 35 cm by 5 cm.

EXTRACTION OF H⁻ IONS FROM A DC CUSP SOURCE

D.H. Yuan, R. Baartman, K.R. Kendall, M. McDonald, D.R. Mosscrop
and P.W. Schmor
TRIUMF, 4004 Wesbrook Mall, Vancouver, B.C., Canada V6T 2A3

ABSTRACT

Recent investigations have concentrated on the parameters affecting the quality of the H⁻ beam from a newly developed dc volume multicusp ion source with a single-stage-post-acceleration (SSPA) extraction system for injection into the TRIUMF cyclotron. The experimental results indicate that the beam emittance is nearly independent of the arc current in the range of 10 A to 40 A. Also described are modifications to the extraction system which have significantly reduced the electron contamination in the H⁻ beam.

1. INTRODUCTION

Volume production of H⁻ ions has become the preferred mechanism for achieving intense, bright beams of H⁻ ions, partly as a result of the cross-section calculations by Wadehra[1] for collisions between cold electrons and vibrationally excited H_2 molecules. Leung et al.[2] and K.R. Kendall et al.[3] have extracted dc H⁻ beams from cusp sources which use a magnetic filter to divide the source into two chambers. Initially, a source chamber uses "hot" electrons ($\gtrsim 100$ eV) to generate vibrationally excited H_2 molecules. This is followed by an extraction chamber where H⁻ ions are formed by dissociative attachment of thermal electrons to these highly vibrationally excited hydrogen molecules and extracted. These sources are characterized by their good beam qualities, i.e. low emittance, high brightness, stable output, low beam noise and relatively long filament lifetimes.

The beam emittance and brightness have been measured with the dc H⁻ volume cusp source built at TRIUMF both for various discharge conditions as well as for different extraction parameters. These results permit us to select the portion of the beam that best matches the acceptance of the TRIUMF cyclotron. The extracted electron current has been minimized with the aid of a computer program which was used to optimize the magnetic field configuration across the extraction lens.

2. SOURCE SET-UP

An outline drawing of the cusp source and the SSPA extraction system is shown in Fig. 1. The extraction system is an axially symmetric four-electrode structure designed to produce a 25 keV beam. The long collimator of the fourth electrode serves to permit differential pumping and reduces gas stripping of the extracted beam while still allowing the source to be run at the optimum pressure. The source is a cylindrical full-line cusp source with 10 rows of 3.2 kG SmCo magnets located axially on the outside of an all-copper water-cooled 20 cm diameter × 26 cm deep plasma chamber. This plasma chamber is divided into two regions by a strong magnetic filter which

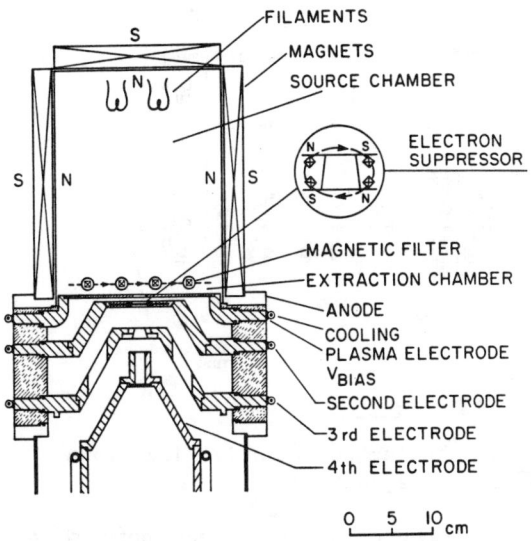

Fig. 1. A cross-sectional schematic of the TRIUMF H⁻ cusp ion source and single-stage post-accelerator.

creates a magnetic field transverse to the beam axis with $\int B \cdot d\ell \approx 0.2$ kG-cm. Two filaments of tungsten wire (3 mm ϕ) are mounted on the back face extending approximately 10 cm into the plasma chamber. The filament electrodes are shielded with quartz tubes to protect them from plasma breakdown and from being hit by the positive ions in the plasma. The beam-forming electrode is insulated from the cusp body and has a 6.5 mm diameter aperture at the centre. Hence, it can be biased separately from the anode. The value of this bias potential together with the hydrogen gas pressure inside significantly influence the extracted electron current. Under certain conditions the ratio of e⁻ to H⁻ current extracted from the source could be as low as 1. These electrons are trapped by a second electrode (biased at ~4 kV) in which two pairs of small magnets (with cross sections of 3 mm × 5 mm) produce ≈100 G (peak) magnetic field transverse to the beam axis.

3. THE EXPERIMENTAL RESULTS

A. H⁻ and Electron Currents

Since the H⁻ and electron currents are separated by the magnetic field in the extraction system, it is possible to determine the dependence of both currents with respect to the discharge parameters. The weak magnetic field in the extraction system acts primarily as a H⁻-e⁻ separator. The electron current was deduced from the high-voltage drain current while the H⁻ current was measured by a Faraday cup placed downstream in the fourth diagnostic box shown in Fig. 2.

The dependence of the e⁻ and H⁻ currents on the bias potential of the first electrode is shown in Fig. 3. At a lower range of bias potentials, a plateau of H⁻ current occurs for each particular value of H_2 gas pressure inside the source (a H_2 gas flow of 10 cc/min is equivalent to a source pressure of ~5 mTorr) as shown in Fig. 3(a). The plateau shifts towards a smaller bias potential as the source pressure increases, suggesting that the plasma potential in the extractionn chamber decreases as the pressure increases. The normalized beam brightness as a function of the bias for a H_2 gas flow of 10 cc/min is also shown in Fig. 3(a). A plateau in the beam brightness occurs when the H⁻ current is on its plateau, indicating that the H⁻ temperature in the extraction chamber is independent of the bias. The

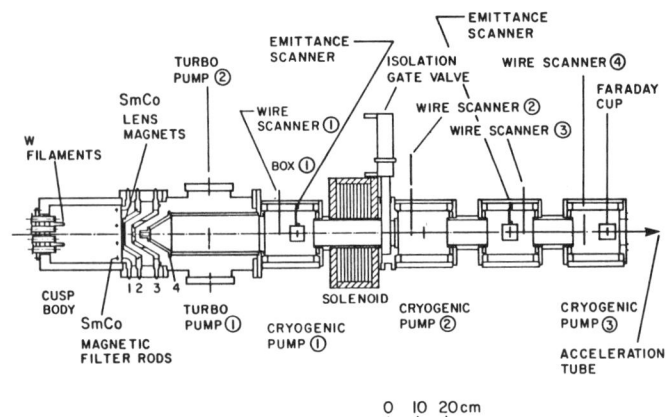

Fig. 2. A schematic diagram showing the relative locations of the cusp source and H⁻ beam diagnostics

brightness decreases at higher values of the bias due both to the drop of H⁻ current and an increase in the emittance resulting from spherical aberrations of the extraction system. The H⁻ current at higher bias potentials decreases when the bias becomes equal to or higher than the plasma potential in the source chamber. In this case, the amount of positive ions (mainly protons) and cold electrons that penetrate through the magnetic filter into the extraction chamber is substantially reduced.[2] The electron current decreases by one order of magnitude for a ~4 V change in the bias while the H⁻ current remains relatively constant [Fig. 3(b)]. Similar results have been

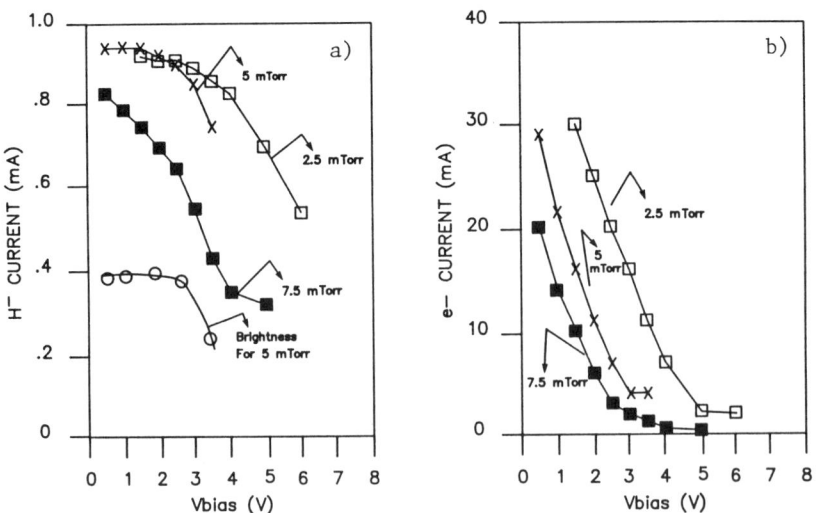

Fig. 3. H⁻ current (a), relative brightness (a) and electron current (b) as a function of the bias potential on the beam forming electrode for the specified source gas pressures (6.5 mm ϕ extraction hole, Arc = 20 A at 140 V).

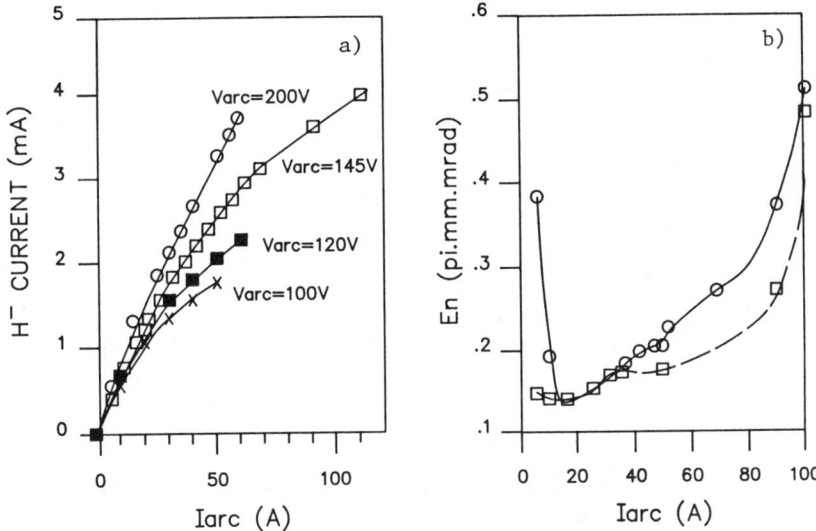

Fig. 4. (a) H⁻ current as a function of the arc discharge current for various discharge voltages (6.5 mm φ extraction hole). (b) Normalized emittance (90% contour level) vs arc current for 145 V. Solid line has a 25 keV fixed beam energy. For the dashed line the beam energy is an optimized variable. (Note offset zero)

obtained by Holmes[4] in a cusp source using a virtual filter. The H⁻ current is a relatively weak function of H_2 pressure in the source. A maximum current was obtained at a pressure of 5 mTorr. The e⁻ current which goes through the extraction system, however, strongly depends on the pressure. For less e⁻ loading, higher pressures are preferable.

B. <u>H⁻ Emittance</u>

Figure 4 shows the extracted H⁻ current and its normalized emittance at the 90% level as a function of the discharge current for a constant arc voltage. The H⁻ current initially increases linearly with arc current. At higher arc powers the H⁻ current exhibits a tendency to saturate. Higher arc voltages appear to delay the onset of this saturation.

The beam emittance appears not to depend on the discharge current (over a range of 10-40 A) when the beam energy is used as an optimizing parameter [see dash line in Fig. 4(b)]. The rapid emittance increase observed at high arc currents (above 40 A) is due to non-linear space charge effects in the first gap of extraction system and due to lens aberrations. These non-linear effects can be observed by comparing the measured emittances shown in Figs. 5 and 6. The emittances of Fig. 5 are for a H⁻ current of 0.8 mA and 15 keV beam energy whereas the emittances of Fig. 6 are for a current of 3.2 mA and 25 keV beam energy. The normalized emittances at the 90%

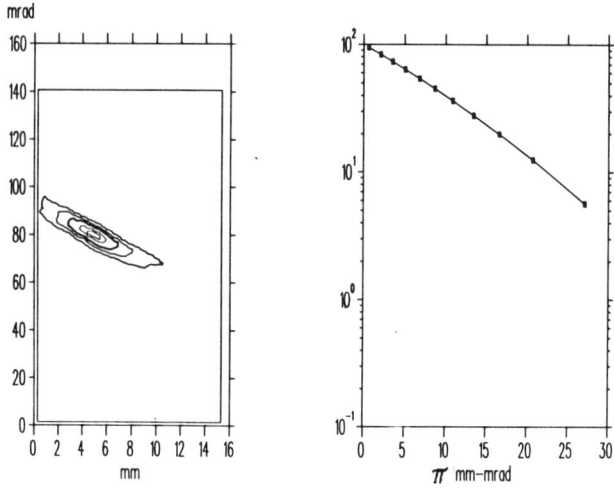

Fig. 5. Emittance contours for a 0.8 mA H⁻ 15 keV beam. The contour levels are 6%, 16%, 46%, 72% and 94%. The right graph plots the fraction of beam outside the contour level vs the unnormalized emittance. In this semi-log plot a Gaussian distribution yields a straight line.

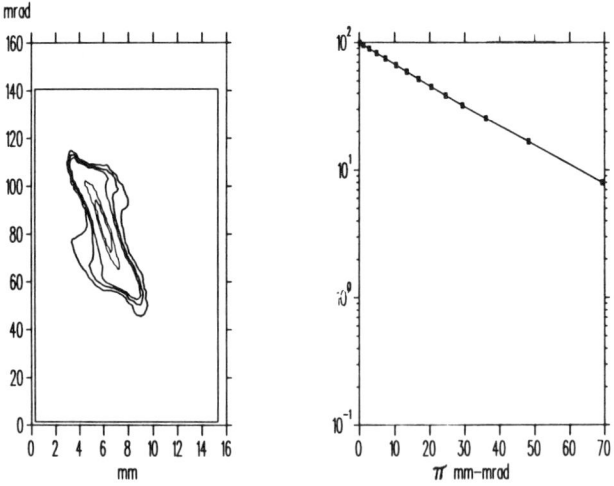

Fig. 6. Emittance contours for a 3 mA H⁻ 25 keV beam. The contour levels are 10%, 25%, 68%, 83% and 92%.

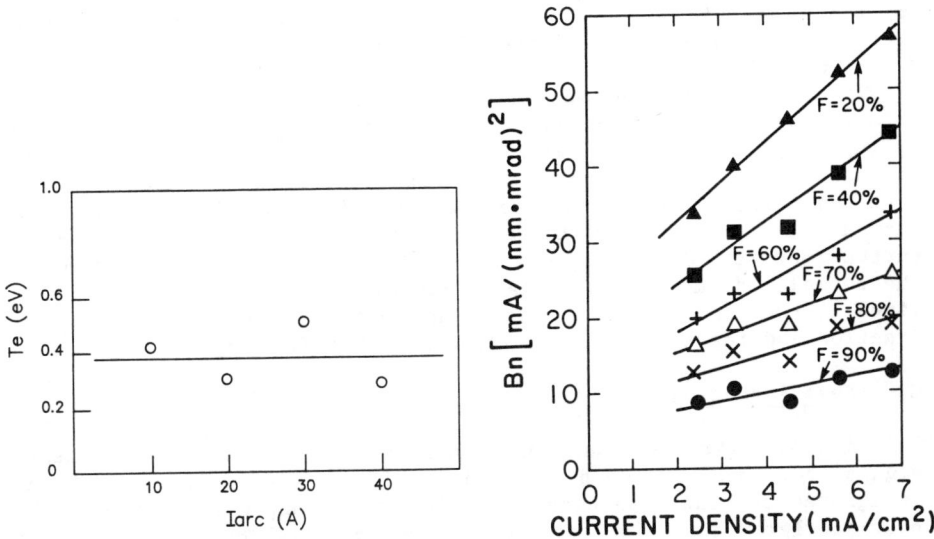

Fig. 7. Electron temperature deduced from measurements with a simple Langmuir probe at the extraction hole vs the arc current.

Fig. 8. The normalized beam brightness as a function of H⁻ current density at various (labelled) emittance contour levels.

level for the two currents are in the ratio 3.6. By comparison, at the 20% level the ratio is only 2.7.

An H⁻ temperature of 0.4-0.6 eV in the extraction chamber was calculated using the equation,

$$\varepsilon(63\%) = 1.05 \, r \left(\frac{kT}{m_0 c^2}\right)^{1/2},$$

where $\varepsilon(63\%)$ is the normalized emittance for a 0.63 fraction of the total beam, r is the radius of the emitter, kT is the H⁻ temperature and $m_0 c^2$ is the rest energy of the H⁻ ions. It has been assumed that the H⁻ ions have a Gaussian distribution of energies, a uniform spatial density at the plasma surface and there is no ion-optical aberration or non-linear space charge forces in the extraction and transport system.[5] Recently, Wadehra[6] calculated the average H⁻ ion temperature as a function of the e⁻ temperature. His results suggest that H⁻ ions are formed with an average of 0.1-0.4 eV for an e⁻ temperature of 0.5-3 eV. The e⁻ temperature was measured in our source with a simple Langmuir probe mounted at the extraction hole. An e⁻ temperature of ~0.4 eV was obtained which appears to be independent of the discharge current for arc currents less than 40 A. The measured data are shown in the graph of Fig. 7.

The normalized beam brightness as a function of H⁻ current density for various beam fractions is shown in Fig. 8. The brightness is defined as $B_n = 2If^2/\varepsilon_n^2$, where I is the total beam current as measured on the Faraday cup of Fig. 2 and f is the appropriate emittance fraction. A linear relation is observed for each beam fraction f^2. The slope of the line increases as the beam fraction decreases, which implies that in order to have a bright beam it is possible to restrict the acceptance and increase the arc power.

The normalized brightness increases slightly with current density (up to ~6 mA/cm^2). Above 10 mA/cm^2 the apparent brightness falls because the extraction geometry is not optimized for such high current densities. The maximum measured normalized brightness for a 81% beam fraction (a 90% emittance contour) was 14 mA/(mm-mrad)^2 at a current density of 5.6 mA/cm^2. For these data ε_n = 0.147π mm-mrad. The normalized rms emittance is 0.038π mm-mrad. Brightness is sometimes defined in terms of rms emittance without consideration of beam fraction and for this case the value is 250 mA/(mm-mrad)^2.

C. Beam Formation

A SSPA extraction and acceleration system is used to maintain high current density and beam quality. If the effect of space charge is ignored one can expect the beam divergence angle to vary with the post-acceleration ratio[7] [defined as the potential difference between the second and first electrodes divided by the potential difference between the third and second electrodes, i.e., $(V_{ext}-V_2)/V_2$]. The initial divergence of the beam after the extraction hole can be balanced by changing this ratio. Figure 9 shows our results. By adjusting V_2 (the potential on the second electrode), the beam diameter (equivalently the beam divergence), which is measured at a fixed distance downstream of the extraction system, can be minimized. As V_2 increases to ~3.5 kV not only does the beam size reach a minimum but so also does the emittance. A further increase in V_2 results in an increase to both beam divergence and emittance. Both the smaller beam divergence and the corresponding smaller emittance are due to a

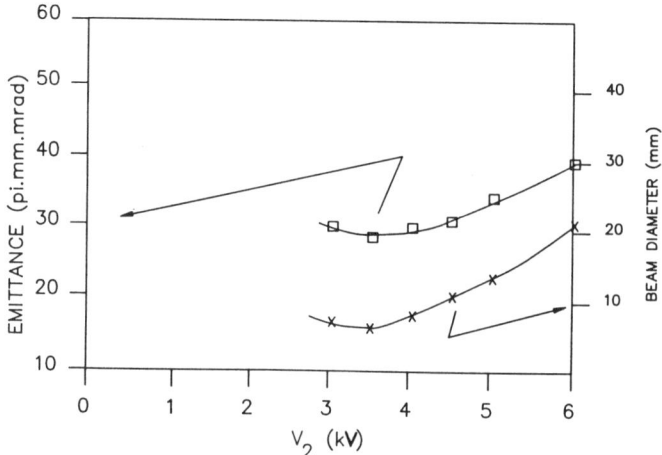

Fig. 9. The measured dependence of the beam diameter and the unnormalized emittance on the potential of the second electrode.

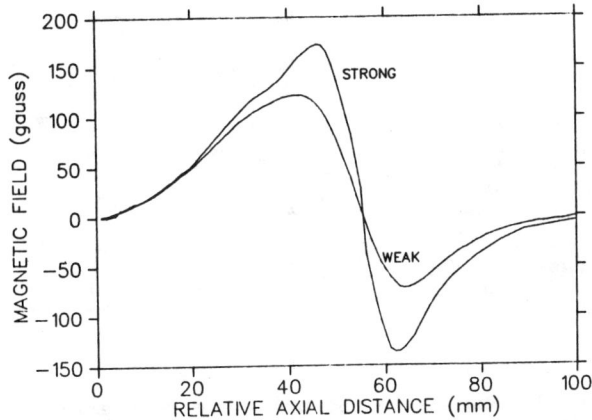

Fig. 10. Profiles of the transverse magnetic field along the beam axis of the extraction system for the strong field and weak field cases. The filter field is included.

reduction in the lens aberrations. The best beam performance is obtained when the ratio is set to about 5 for our extraction system.

The magnetic field in the extraction region is a sum of that created by the filter rods and the two pairs of magnets in the second electrode. It serves to sweep the simultaneously extracted e^- from the beam onto the second electrode while giving the heavier H^- ions only a small net displacement. The field should be as low as possible in order to minimize the distortion to the plasma surface at the extraction hole.[7] Two different magnetic fields have been used (Fig. 10). When the source is operated with the weak field, higher e^- drain current and e^- contamination in the H^- beam is observed (about 10 mA for a H^- current of 1 mA). A hole in the third electrode due to full energy e^- sputtering is observed after short periods of operation. With the stronger magnetic field the electron contamination in the full energy H^- beam is negligible. The e^- falls on the second electrode (about 1.5 mA for a H^- current of 1 mA) with a fraction of the extraction energy. The measured emittance degradation due to this field is less than 5%. The H^- beam has a displacement of the order of 0.1 mm and a deflection of ~3 mrad. A dipole magnet whose field and direction can be varied has been located after the final electrode in order to reduce this angular deflection. A computer program was used to calculate the trajectories of the charged particles in the extraction region. Figure 11 shows the results with weak and strong magnetic fields. It can be seen that the strong magnetic field deflects e^- onto the second electrode while e^- can leak through the weak field onto the third electrode and contaminate the H^- beam.

The bias voltage on the third electrode is intended to create a positive potential hill at the centre of the electrode which should reduce the backstreaming positive ions (mainly protons) from the beam generated plasma downstream of the extraction column (giving heat and X-rays which reduce the reliability of the system). A high bias on

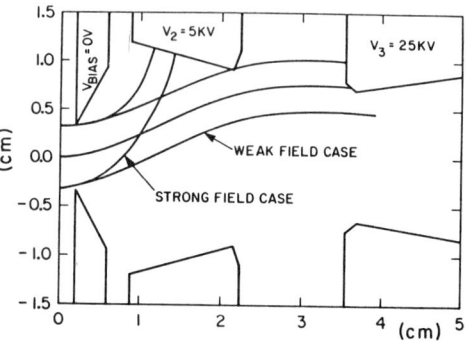

Fig. 11. The calculated electron trajectories in the extraction system for both the strong and the weak field cases. (Only the on-axis field was used.)

the third electrode results in secondary electron emission which also reduces the reliability of the systems. Beam diameters measured by the wire scanner after a drift distance downstream of the extraction system and the beam emittance as measured by the emittance scanner decrease as the bias on the third electrode, V_3, increases to ~10 V, then remains unchanged provided V_3 is lower than 200 V (Fig. 12). A further increase in V_3 results in an increase of emittance and a negative current flow through the V_3 power supply.

4. PRESENT STATUS

The cusp source has been installed into a recently constructed 300 kV terminal and the H⁻ beam has been injected into the TRIUMF cyclotron through a 50 m long transport line which contains only electrostatic optics. The source has already demonstrated advantages over the existing PIG source. With the cusp source a cw proton

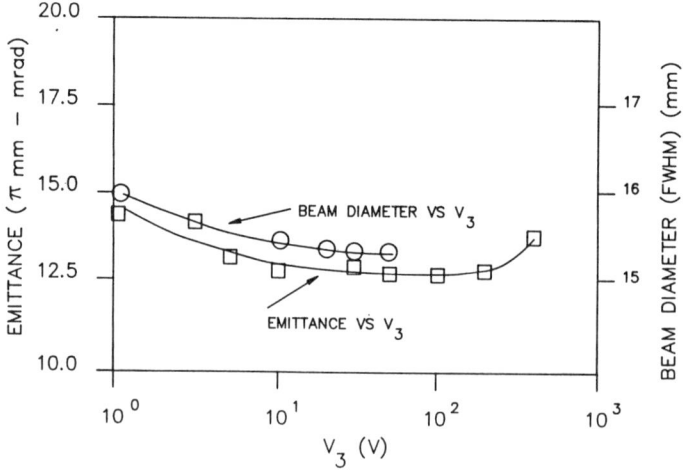

Fig. 12. The measured beam diameter and emittance as a function of the voltage on the third electrode. Note the suppressed zero. The voltage is here referenced to ground.

current of 120 µA was extracted from the cyclotron (which is equivalent to the PIG source) with less beam spill inside the cyclotron than is normal with the PIG source at these currents. It is felt that the cusp source is already capable of delivering four times more current into the cyclotron acceptance compared to the PIG source.

Toroids (single turn current transformers) in the injection line, used as non-intercepting current monitors, confirm the quiescent nature of the source plasma. In particular, for frequencies up to 1 MHz, no beam oscillations greater than 1% are apparent. The PIG source, by comparison, typically has ~10% intensity oscillations with a frequency around 200 kHz.

ACKNOWLEDGEMENTS

We would like to express our gratitude for the special efforts made by P. Chigmaroff with the electronics, H. Wyngaarden on the apparatus, M. Mouat and C. Kost in software.

REFERENCES

1. J.M. Wadehra, Appl. Phys. Lett. $\underline{35}$, 917 (1979).
2. K.N. Leung, K.W. Ehlers and M. Bacal, Rev. Sci. Instrum. $\underline{54}$, 57 (1983).
3. K.R. Kendall et al., Rev. Sci. Instrum. $\underline{57}$, 1277 (1986).
4. A.J.T. Holmes, G. Dammertz and T.S. Green, Rev. Sci. Instrum. $\underline{56}$, 1697 (1985).
5. Paul Allison, J.D. Sherman and H.V. Smith, 'Comparison of Measured Emittance of an H^- Ion Beam with a Simple Theory', LA-8808-MS (1981).
6. J.M. Wadehra, Phys. Rev. A 29, 106 (1984).
7. E. Thompson, Proc. of the 2nd Symposium on Ion Sources and Formation of Ion Beams, 1974, Berkeley.

OPERATION OF A DUDNIKOV TYPE PENNING SOURCE
WITH LaB$_6$ CATHODES*

K. N. Leung, G. J. DeVries, K. W. Ehlers,
L. T. Jackson, J. W. Stearns,
and M. D. Williams

Lawrence Berkeley Laboratory
University of California
Berkeley, CA 94720

M. G. McHarg, D. P. Ball, and W. T. Lewis

Air Force Weapons Laboratory
Kirtland Air Force Base
Albuquerque, New Mexico 87117-6008

and

P. W. Allison

Los Alamos National Laboratory
Los Alamos, New Mexico 87545

ABSTRACT

The Dudnikov type Penning source has been operated successfully with LaB$_6$ cathodes in a cesium-free discharge. It is found that the extracted H$^-$ current density is comparable to that of the cesium-mode operation and H$^-$ current density of 350 mA/cm^2 have been obtained for an arc current of 55 A. The H$^-$ yield is closely related to the source geometry and the applied magnetic field. Experimental results demonstrate that the majority of the H$^-$ ions extracted are formed by volume processes in this type of source operation.

INTRODUCTION

The reflex discharge originated by Maxwell[1] and Penning[2] has found important applications as ion sources for generating positive ion, high charge state ion, or negative ion beams. By operating with a hot cathode and a reflector, the Penning or PIG source was first studied and optimized by Ehlers[3] to generate steady-state beams of H$^-$ ions for cyclotron operation.

* This work is supported by the Air Force Weapons Lab. at Kirtland, the Air Force Office of Scientific Research and by the Director, Office of Energy Research, Office of Fusion Energy, Development and Technology Division, of the U.S. Dept. of Energy under Contract No. DE-AC03-76SF00098.

© American Institute of Physics 1987

Discovery of H^- enhancement by introducing <u>cesium</u> into the discharge has led Dudnikov to modify the original Penning source, by changing the dimensions and adding cesium vapor.[4] Since then, there have been extensive studies at Brookhaven and Los Alamos to optimize the source performance and the H^- yield. It has been demonstrated by Allison that high intensity H^- beams ($J^- > 2$ A/cm^2) can be extracted from the Dudnikov type Penning source if the proper amount of cesium is present.[5]

It is generally believed that H^- ions in the Dudnikov source are first formed on the cesiated cathode surface, either by desorption or backscattering.[6] Some of these ions undergo resonance charge exchange with the neutral hydrogen atoms near the emission slit.[7] The low energy H^- ions formed are then extracted from the source. However, a recent investigation shows that H^- ions formed by volume processes can play an important role in this type of Penning discharge.[8] These volume-produced H^- ions can also account for the high negative ion output current if sufficient plasma density is present in the discharge volume.

Instead of employing cesium, we have operated the Dudnikov source equipped with a LaB_6 cathode of similar low work-function. It is found that the extracted H^- current density in this cesium-free operation is comparable to that of the cesium-mode operation, and discharge current as high as 100 A can be obtained for short pulse durations. These results together with the measurements obtained from a "hybrid" production of H^- ions indicate that the majority of the H^- ions extracted from the Dudnikov source, when it is operated with LaB_6 cathodes, are formed directly by volume processes. Additional experimental investigation is required in order to understand the H^- production process when the Dudnikov source is operated with cesium.

I. EXPERIMENTAL SET-UP

A general description of the Dudnikov type Penning source has been discussed previously by Allison.[5] Figure 1 illustrates the different components of the Dudnikov source. The cathode, anode insert, extractor and emission slit are normally made of molybdenum while the anode housing is constructed from stainless steel. In this experiment, the original molybdenum cathode is replaced by LaB_6 material which can provide a work

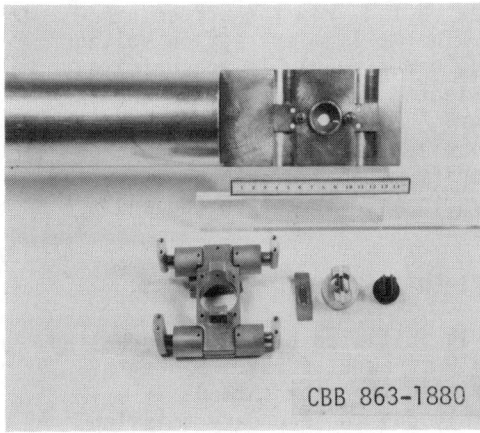

Fig 1. Different components of the Dudnikov type Penning source.

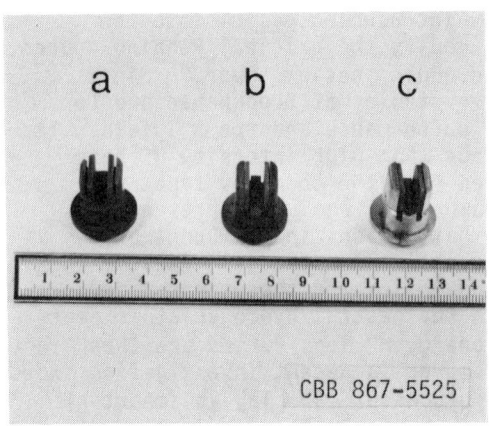

Fig. 2 The three different cathode arrangements: (a) a complete LaB_6 cathode; (b) a graphite cathode with two LaB_6 inserts; and (c) a molybdenum cathode with two LaB_6 inserts.

function as low as 2.3 eV.[9] Three different LaB_6 cathode arrangements have been tested. The complete cathode structure in Fig. 2(a) was fabricated from LaB_6. Only two planar LaB_6 inserts (13 mm x 12mm x 1.5mm) were installed on either a graphite or a molybdenum mounting in Figs. 2(b) and 2(c) respectively. Since LaB_6 is very reactive with refractory metals, it is separated from the molybdenum mounting by a rhenium foil.

The ion source is mounted on a water-cooled copper block. Hydrogen gas is introduced into the source chamber from three small holes located in the anode insert. The gas flow rate can be adjusted and is monitored by a digital mass flow meter. The magnetic field required for the source operation is generated by a pair of large electromagnets which can provide a uniform B-field as high as 7 KG. The ion source assembly is placed inside a 50 cm x 25 cm x 152 cm vacuum chamber which in turn is installed in the gap between the two magnet pole faces.

A 2 kV, 1 A power supply is used to start the discharge in a dc mode, and a 700 V, 400 A transistor pulser is used for pulsed operation. The ion source and its electronics are at high potential. H^- ion beams are extracted from the arc column through a 1-mm-diam aperture. The maximum extraction voltage available from the high voltage power supply is approximately 15 kV. The H^- ion beam is collected in a Faraday cup and is measured across a 1 kΩ resistor. Both the arc voltage and arc current are measured with an oscilloscope via fiber-optic analog telemetry. Due to the E x B drift motion, electrons extracted from the source drift side-ways in the extraction gap and are collected at the chamber wall.

II. EXPERIMENTAL RESULTS

A discharge in the source is initiated by first increasing the dc "keep-alive" power supply to about 500 V. Since the LaB_6 is cold, electrons are emitted from the cathode (approximately 20 - 50 mA) mainly due to secondary emission. As the arc current gradually increases to about 200 mA, the arc voltage decreases to 200 V. At this stage, the transistor pulser is switched on and the arc is operated in short pulses with a repetition rate of 5 - 20 Hz. The 200 V, 200 mA dc discharge is maintained during the time between pulses.

For new LaB_6 cathodes, the source requires some initial conditioning, therefore the pulse duration is kept below 100 μs so that source damage is minimized. Once conditioned, the pulse length can be gradually increased to several milli-seconds. The temperature of the LaB_6 cathode rises with the pulse length and the repetition rate. As the LaB_6 becomes sufficiently hot, electrons can be emitted thermionically.

The oscilloscope traces in Fig. 3 illustrate the arc current, the H^- beam current, and the arc voltage during a 700 μs pulse operation. The arc current and voltage and the H^- beam current stay constant during the time of the pulse. The noise level of the arc current is approximately 25% peak-to-peak.

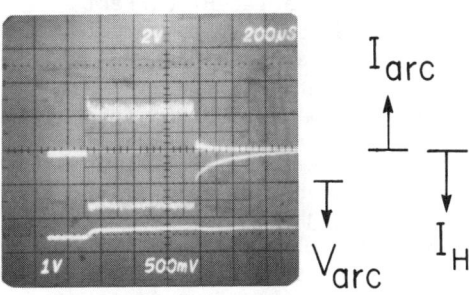

XBB 869-7148

Fig. 3 Oscilloscope traces showing the arc current, H^- beam current and arc voltage during a 700 μs pulse operation.

Occasionally, two discharge modes can appear in a single pulse. The oscilloscope trace in Fig. 4 shows that during the early part of the pulse, the arc current oscillates between two discharge modes; a high current and a low current mode. It then settles to the higher current mode in the later part of the pulse. This type of two mode operation eventually disappears as the LaB_6 cathode temperature increases or conditioning occurs. For arc current below 40 A, uniform pulses longer than 5 ms have been recorded. As the arc current increases, the pulse length is reduced in order to prevent possibility of damaging the LaB_6 cathode.

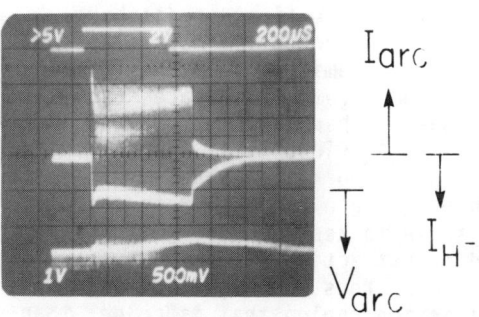

XBB 869-7147

Fig. 4 Oscilloscope traces showing the two discharge modes which occur during the early part of a 600 μs pulse operation.

Figure 5 shows a plot of the extracted H^- current density versus the arc current. The data points have been

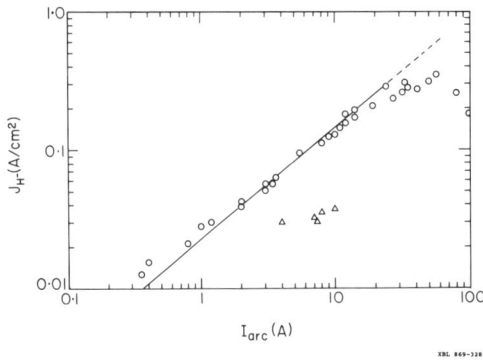

Fig. 5 Extracted H⁻ current density versus arc current The five (Δ) data points are obtained without the anode ribs in the source assembly.

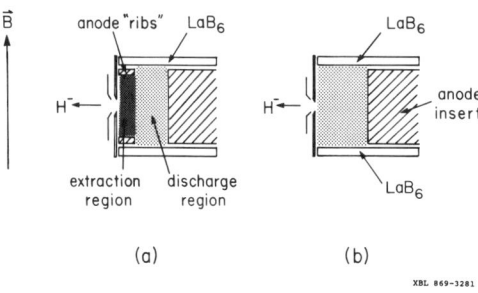

Fig. 6. Schematic diagram showing the discharge geometry (a) with and (b) without the ribs in the anode insert.

accumulated for source operations with all the three LaB_6 cathode arrangements shown in Fig. 2. The result demonstrates that there is essentially no difference in the H⁻ output between the three different LaB_6 cathode geometries, and a wide range of arc current can be easily obtained from them. Overall, the extracted H⁻ current densities are comparable to those obtained when the source is operated with cesium,[5] except that both the optimized gas flow rate (36 sccm) and the applied magnetic field (7 kG) are higher.

Figure 6(a) is a cross-sectional view of the Dudnikov Penning source showing the LaB_6 cathode, the anode insert with the "ribs", and the plasma column. The two anode ribs essentially divide the source chamber into two regions; the discharge and the extraction region. In the first or discharge region, primary electrons emitted from the two LaB_6 cathode surfaces oscillate back and forth along the field lines. They ionize or vibrationally excite the background gas, forming a dense hydrogen plasma. Both the anode ribs and the applied B-field serve to keep the energetic primary electrons out from the second or extraction region. However, both ions and plasma electrons can cross the magnetic field and they form a plasma in the extraction region that is colder than the plasma in the discharge region. It is very likely that the majority of the H⁻ ions extracted are formed in this part of the source by volume processes, either by electron collision with the H_2^+ or H_3^+ ions or by dissociative attachment of very low energy electrons to vibrationally excited H_2 molecules.

The effect on the H⁻ yield by operating the Dudnikov source **without** the anode ribs has been investigated. Figure 6(b) is a cross-sectional view of this particular source arrangement. With the shadowing effect of the anode ribs removed, primary electrons emitted from the LaB_6 cathode now exist throughout the entire chamber. The extracted H^- current density (as shown by the five (Δ) data points in Fig. 5) is reduced by a factor of 3 for the same source operating conditions. The reason for this drop in source efficiency is being investigated.

When the Dudnikov source was operated with cesium, the extracted H^- current saturated when the applied B-field reached about 2 - 3 kG.[5] The B-field dependence of the H^- current when the Dudnikov source is operated with the LaB_6 cathodes has also been studied. For a given arc current, the H^- current increases monotonically with the B-field. As the B-field is increased from 2 to 6 kG, the extracted H^- current (Fig. 7) increases by about a factor of two. It is observed that the increase in B-field is accompanied by an increase in the arc voltage in order to maintain a constant arc current. As a result, the arc power used for the discharge is increased, producing a higher plasma density. A stronger B-field can also provide better plasma confinement which in turn will contribute to the increase in the plasma density of the source. Since the H^- ion production rate is proportional to the density of the source plasma, the increase in B-field will result in a higher H^- output current.

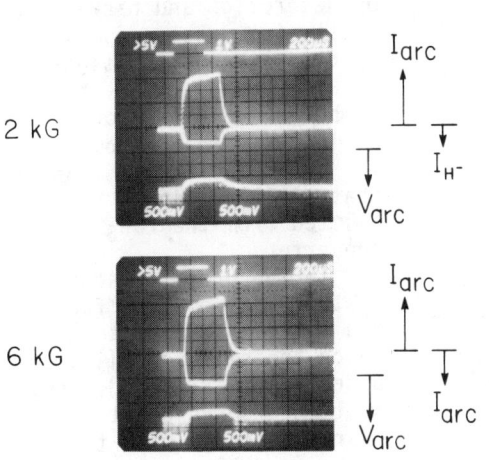

XBB 869-7150

Fig. 7 Oscilloscope traces showings the arc current, H^- current and arc voltage for two different applied B-field.

LaB_6 is a low work function material ($\phi_W \approx 2.3$ eV). If it is used as a converter in a surface production type negative ion source, the surface-produced H^- ions should be enhanced as in the case of a cesium-coated molybdenum surface ($\phi_W \approx 2$ eV). We have operated a filtered multicusp source[10] with a LaB_6 converter to generate both volume and surface produced H^- ions.[11] A mass spectrometer detects the H^- ions leaving the source and therefore can provide a comparison between the volume produced H^- ions extracted near the filter region (extraction voltage = 700 V) and the "self-extracted" H^- ions formed on the LaB_6 converter surface.

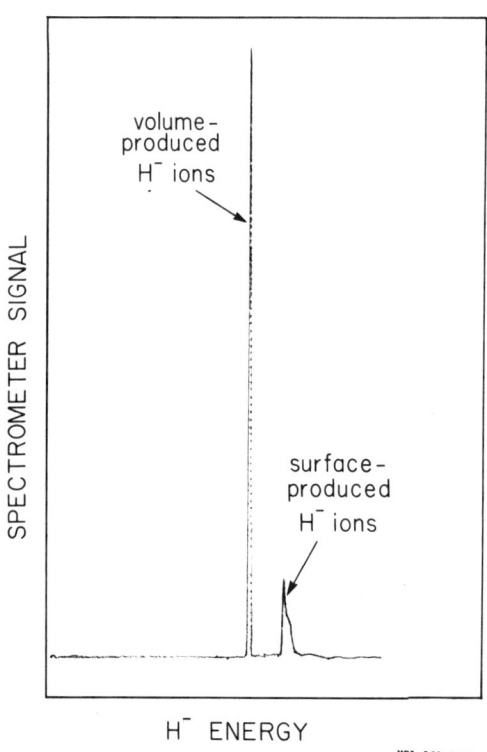

Fig. 8 Spectrometer output signal showing the extracted volume-produced H⁻ ions and the H⁻ ions formed on the LaB$_6$ converter surfaces.

In this test, a background hydrogen plasma is generated by dc discharge (80 V, 3 A) between the filament and the anode chamber wall. The converter is biased at -200 V with respect to the anode. Thus, positive hydrogen ions are impinging on the LaB$_6$ converter surface with an energy appropriate to this bias potential and H⁻ ions are formed both by desorption and back-scattering processes. The spectrometer output signal in Fig. 8 shows that the number of volume-produced H⁻ ions extracted from the source is larger than that of the H⁻ ions emitted from the LaB$_6$ converter.

In the Dudnikov source, fast H⁻ ions emitted from the cathode surface cannot reach the emission slit. In order to be extracted, they must undergo resonant charge exchange with the background H⁰ atoms, forming low energy H⁻ ions, which are then extracted from the source.[7] Since the result of this experiment shows that the H⁻ generated from a LaB$_6$ surface is lower than those formed directly by volume processes, it is unlikely that this two-step process (surface production followed by resonant charge exchange) can account for the high H⁻ ion current observed in the Dudnikov source when operated with the LaB$_6$ cathodes.

CONCLUSION

The results of this experiment demonstrate that H⁻ current densities as large as 350 mA/cm^2 can be obtained from a Dudnikov Penning source when it is operated with LaB$_6$ cathodes in a cesium-free hydrogen discharge. When the arc current is less than 40 A, pulse length of several milli-second have been achieved. For higher arc current, source operation is limited to short pulses of several hundred micro-seconds. In order to extend the pulse length and duty factor, source cooling must be improved; in addition, a feed-back circuit must be employed to stablize the arc current so that uniform H$^-$ current can be maintained.

When the Dudnikov source is operated with cesium, the life-time of the source is limited due to the formation of a deep hole on the molybdenum cathode, arising from positive ion sputtering. Further, cesium vapor migrating out of the source can cause voltage break-down problems in the accelerator column. In the pure hydrogen mode operation however, the life-time of the Dudnikov source is expected to be much improved because of the absence of massive Cs$^+$ ions in the discharge together with the lower LaB$_6$ sputtering rate. However, additional experiments must be performed in order to understand the H$^-$ production mechanism when the source is operated with cesium.

ACKNOWLEDGMENT

We would like to thank D. Moussa, C. Hauck and S. Wilde for technical assistance.

REFERENCES

1. L. R. Maxwell, Rev. Sci. Instrum. 2, 129 (1931).
2. F. M. Penning, Physica, 4, 71 (1937).
3. K. W. Ehlers, Nucl. Instr. Meth. 32, 309 (1965).
4. V. G. Dudnikov, Proc. IV All-Union Conf. on Charged Particle Accelerators, Moscow, 1974, Nauka 1975 Vol. 1, p.323.
5. P. W. Allison, IEEE Trans. on Nucl. Sci., NS-24, No. 3, 1594 (1977).
6. K. N. Leung and K. W. Ehlers, Proc. of the 3rd Int. Symp. on the Production and Neutralization of Negative Ions and Beams, Brookhaven, 1983. AIP conf. Proc. No. 111, p. 265.
7. G. E. Derevyankin and V. G. Dudnikov, Proc. of the 3rd Int. Symp. on the Production and Neutralization of Negative Ions and Beams, Brookhaven, 1983. AIP Conf. Proc. No. 111, p.376.
8. K. Jimbo, K. W. Ehlers, K. N. Leung, and R. V. Pyle, Nucl. Inst. Meth. A248, (1986) 282.
9. J. Pelletier and C. Pomot, Appl. Phys. Lett. 34, 249 (1979).
10. K. N. Leung, K. W. Ehlers and M. Bacal, Rev. Sci. Instrum. 54, 56 (1983).
11. K. N. Leung and K. W. Ehlers, J. Appl. Phys. 52, 3905 (1981).

DISCUSSION

Forrester: I want to make a comment about one of the motivations that is current through here. I can remember six years ago, when people were having difficulty with the use of cesium, making a suggestion that would solve one of the major problems associated with this use of cesium that has been totally ignored; I think it's just as valid now as it was then. The main problem that people encounter when they use cesium is that it makes voltage very hard to hold. And I've been in discussion with several people here who have talked about the same problem. One can solve the problem of holding voltage in the presence of cesium by either using insulators made of lead glass or having your insulators glazed with a lead glass glaze. It's a remarkable thing. The phenomena is probably due to the fact the lead oxide in the glass plus cesium probably forms a cesium oxide, plus free lead, and the free lead doesn't form a conducting coating. Whether that's the real explanation or not, I'm not sure, but it really, really works. No comment about your experiment, but the motivation for getting rid of cesium is less than people think, for that reason.

Alessi: If this source is volume H^- production, then if you would put a plasma electrode in that you could bias the same way that people do in other volume sources, you might expect to see an improvement in the output; because right now it's a grounded electrode.

Leung: Yes. That is one of our plans in one of the experiments. What we'd like to do is change the bias and see what happens. We did do an experiment on Ken Ehler's Penning source, by operating the source with a dual anode so that you can bias one anode with respect to the other. And we did see a factor of two improvement in H^-. So it's interesting to check this source with a separate bias anode.

Hiskes: Two questions, Kao. When you're varying the discharge current and watching the negative ion extracted current rise, are you holding the gas pressure fixed, or are you constantly adjusting the gas pressure?

Leung: We did optimize the gas flow.

Hiskes: So you're optimizing at every point, every value of the discharge?

Leung: Yes. But, it's not a big variation, the optimum pressure. It won't change by from, say 25-35 sccm.

Hiskes: And when you break away from the linear rise, do you have any idea what's happening to the electron temperature in either of those chambers?

Leung: The source is so small, you couldn't even put something inside to measure it. So we don't know what the electron temperature is. Maybe Paul did measure it spectroscopically. You did present a paper on that yesterday, on the electron temperature.

Allison: The spectroscopy, of course, was done without the LaB_6, and there was no evident variation in electron temperature with our current from 1A to 500A discharge current.

Bacal: Why don't you have surface production on LaB_6?

Leung: It does. I'll show it to you, here. It's small.

Bacal: Why is it small? Because, actually, you showed us this

kind of picture three or six years ago in the multicusp surface
source, except you said the surface peak was very big compared to
the volume peak.
Leung: No, that was cesium on moly.
Bacal: I know. So why is cesium on moly good, and LaB_6 is not
good? The work function is too high?
Leung: No, there's something else. When you have cesium on
molybdenum, besides lowering the work function, there's some kind of
chemistry going on on the surface. For example, hydrogen may be
able to sit better on the surface. If you put in cesium, you will
see that the desorbed portion is pretty high. But in this spectrum,
you can see that the desorbed part is not high at all, it's very
low. And that shows that there's not much hydrogen sitting on the
surface for you to desorb. So when you have cesium on moly, maybe
the cesium is doing something to grab hold of the hydrogen, I don't
know.
Hershcovitch: I would like to make a comment to support Kao Leung's
assertion. We did some experiments with LaB_6 and LaB_6 doped with
cesium and sodium. And when we doped it with cesium, which was too
big an atom - it's doped by taking a lanthanum out and putting a
cesium in, so you end up with 10% cesium, instead of the lanthanum -
one was able to get a measurable yield of H^- from the LaB_6 doped
with cesium. When we put sodium - I think Jim did the experiment -
which was sitting well inside the lattice, apparently the results
were negative. We couldn't get any H^-, which means that somehow,
when the surface is too smooth, the H^0 cannot desorb on the surface.
Leung: Yes, Chuck Schmidt at Fermilab did a similar experiment by
putting cesium on LaB_6 and he didn't see any big increase in the H^-.
Seidl: The molybdenum, when it's bombarded with cesium ions at
100V, has a steady state work function of 1.6, while LaB_6 has something like 2.3. So it should make a very big difference for surface
production.
Tsai: Kao, could you tell me what is the width of the anode
ribs?
Leung: Well, these are two molybdenum ribs. They extend all the
way out, so there's no energetic electrons coming from the cathode
that can get into this area.
Tsai: About how wide?
Leung: About 1 mm wide. In fact, Paul Allison changed the width
of this, and changing the width of this is like changing the filter
width, and it's a very interesting experiment to see how the H^-
changes. We would like to do that experiment. But, again, you have
to make a lot of inserts to do that.
Tsai: When you increase the magnetic fields, the output increases? Do you ever see a maximum?
Leung: The maximum? Good question. At 10A, 20A, we didn't see a
maximum. At 30A, we started to see some rolling off. And we can
get up to only 7 kG. We would like to go up higher, but that's the
limit of the magnet, so that's one reason why Kirtland is getting
bigger magnets, so they can go up to, maybe, 10 kG or 12 kG and see
what happens. But we're limited by the B field, so we don't know
what happens.

ACCELERATED BEAM EXPERIMENTS WITH THE ORNL SITEX AND VITEX H$^-$/D$^-$ SOURCES*

W. K. Dagenhart, C. C. Tsai, W. L. Stirling, P. M. Ryan,
D. E. Schechter, J. H. Whealton
Oak Ridge National Laboratory, Oak Ridge, TN 37831

J. J. Donaghy
Washington and Lee University, Lexington, VA 24450

ABSTRACT

Beam parameters have been measured for both the Surface Ionization with Transverse Extraction (SITEX) and Volume Ionization with Transverse Extraction (VITEX) H$^-$/D$^-$ ion sources. Both sources use a reflex discharge to generate the main plasma. Beam energies up to 18 keV were used for pulse lengths up to several seconds. For SITEX, Faraday cup magnetically analyzed D$^-$ beam currents of 110 mA at extraction densities of 48 mA/cm^2 and at a source ion temperature of 4 eV have been measured. For the VITEX results, Faraday cup magnetically analyzed beam currents of up to 80 mA at extraction densities of 27 mA/cm^2 and at a source ion temperature of 0.5 eV have been measured. Virtually all extracted electrons were recovered at an energy of 10–30% of the accel beam energy, and there were none in the analyzed beam.

INTRODUCTION

The original SITEX and VITEX H$^-$/D$^-$ ion sources were configured for fusion applications to provide eventual scaling to >10-A beams at energies above 200 keV for multisecond pulse lengths.[1-9] This hardware and these concepts are now being adapted to produce high-brightness, negative-ion beams for radio-frequency quadrupole (RFQ) acceleration. The goal here is to produce much smaller beams at 100 keV for further RFQ acceleration and for pulse lengths ranging from the submillisecond level through many seconds to steady state. The present sources are at the proof-of-principle stage and are evolving rapidly. The SITEX results presented here are from December 1985 to May 1986, after which the effort concentrated solely on the VITEX source. Since that time the hardware and results have been evolving rapidly. The VITEX results included in this paper are from May 1986 to October 1986.

The main goals of the SITEX experiments reported in this paper were (1) to use an emittance scanner on the accelerated beam to determine the source plasma generator ion temperature and (2) to determine the beam emittance. The VITEX experiments were aimed at developing higher beam current and beam current density and measuring the source ion temperature with an emittance scanner. Secondary goals of both sets of experiments were the development

---

*Research sponsored by the Department of Defense under Interagency Agreement DOE No. 40-1442-84 and Army No. W31RPD-53-A180 with the Office of Fusion Energy, U.S. Department of Energy, under Contract No. DE-AC05-84OR21400 with Martin Marietta Energy Systems, Inc.

© American Institute of Physics 1987

of plasma generators[10] and accelerators that would operate for short pulses, for long pulses, and steady state. Other goals were to improve gas efficiency, beam brightness, and arc efficiency; to reduce beam emittance and extracted electron current; to provide low-energy collection of electrons that are extracted; and to develop reliable systems with faster startup and with good endurance.

EXPERIMENTAL APPARATUS

Both VITEX and SITEX used a reflex discharge with a hot cathode electron emitter on one end of the discharge and a biased electron reflector on the other end. A second version of these sources replaced the reflector with a hot cathode electron emitter that essentially doubled the amount of arc power that could be put into the arc plasma. A biased electrode was mounted behind the discharge, as shown in Fig. 1.

Gas was independently fed to both cathode cavities and the anode cavity. Figure 1 shows the hookup of the independently controlled two-filament, two-arc discharge and one converter supply. The figure also shows that single gap extraction was used. For SITEX, cesium was fed from a temperature-controlled oven; secondary control was provided by a manually adjusted high-temperature series valve. The ion source was operated in an adjustable uniform magnetic field. Vacuum was provided by two oil diffusion pumps with a delivered speed

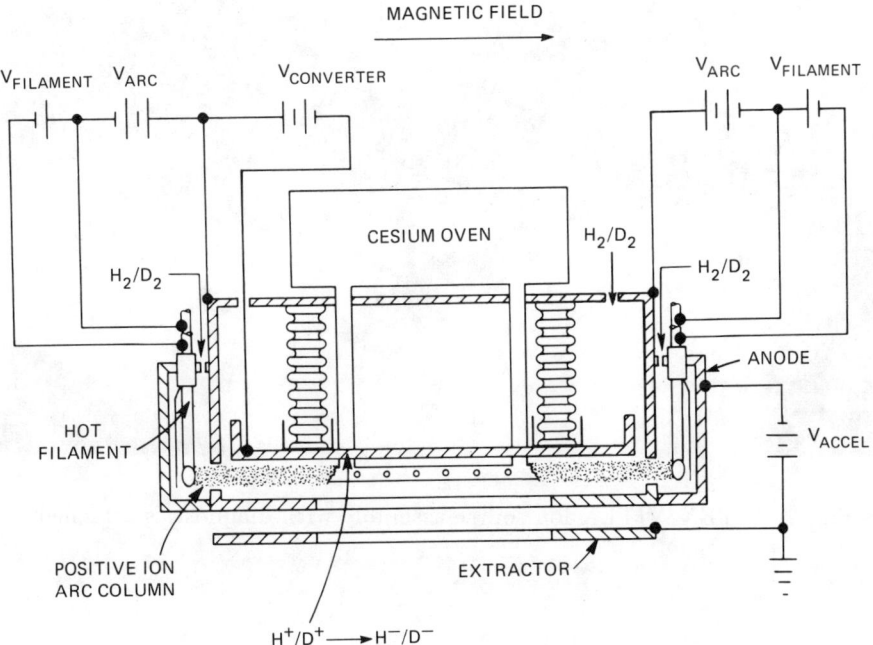

Fig. 1. Top view of plasma ion source, showing power supply connections.

of ~2000 L/s at the source. Faraday cups and electron recovery electrodes were operated in the magnetic field with secondary electron shields to provide complete current accountability. Two LANL-type emittance scanners[11] were used to measure the x- and y-plane emittances transverse to the beam propagation direction. Figure 2 shows the source and the diagnostic structure. The arc chamber was machined from graphite and was then plasma sprayed with molybdenum. The arc collimating slot was also machined from graphite and plasma sprayed. For SITEX operation, the graphite structure ran at high temperatures, controlled so as to control the location of cesium in the arc chamber. Plasma grids were machined from molybdenum.

Electrons extracted with the H^-/D^- beam were separated from it immediately after extraction by the $\mathbf{E} \times \mathbf{B}$ forces and cycloided up the front of the

Fig. 2. SITEX/VITEX ion source assembly with diagnostics attached.

plasma electrode to a recovery electrode biased positively with respect to the plasma generator. The recovery voltage needed to collect virtually all electrons varied between 10% and 30% of the first gap potential difference. Electrodes were used to measure the small number of electrons that were not recovered, and these proved to be negligible. All of the SITEX and some of the earlier VITEX experiments used an optics system that was optimized for beam focusing only in the direction perpendicular to the long slit. Since the magnetic field was uniform and in the direction of the slit, we could also measure beam profiles in the direction of the slit without the effects of ion optics and could then calculate the ion temperature of the extracted ions. Most VITEX measurements used extraction slits for which the optics have not been calculated. Both the extraction gap and the electrode relative locations could be adjusted during all experiments. The emittance scanners were mounted to scan in two orthogonal directions, which were also orthogonal to the beam path after it had been deflected through 90° in the uniform magnetic field. All beam currents were obtained by measuring the current voltage drop across small resistors with oscilloscopes or with the computer system analog-to-digital converters. All of the emittance data presented were measured over a 30-ms period during the multisecond beam pulses. One emittance scan of beam angle versus beam current was taken for each of 15 to 30 mechanical slit positions as it scanned across the beam. Each mechanical position required one multisecond beam shot with data taken for 30 ms at the same point in time during each pulse.

SITEX AND VITEX EXPERIMENTAL RESULTS

Table I lists the SITEX parameters achieved with cesium operation during the production of D^- beams. Generally, pulses lasting 3 to 10 s were run every 60 s. The repetition rate was mostly determined by our data acquisition system, and in order to eliminate repetition rate as a variable it was held constant during these experiments. Some modes of operation gave flat beam current pulses after about 1 s both for VITEX and SITEX. Square short pulses with a sharp rise time generally had to be achieved with source parameters different from those required for long, flat pulses. Figure 3 shows a set of SITEX beam waveforms in which the beam pulse came to equilibrium in about 0.25 s. Figure 3(a) shows an acceleration voltage of ≈ 11 kV and an acceleration power supply drain current of 175 mA. The Faraday cup was located about 3 cm from the extraction grid in a uniform magnetic field and measured 100 mA of H^- current. Beam interception by the extractor grid was ≈ 70 mA, leaving about 5 mA for unrecovered electrons that received full energy. These electrons were magnetically removed from the beam. Figure 3(b) shows the electron recovery current of ≈ 100 mA at a recovery energy of ≈ 4 keV. A shaped focused converter as specified from the ORNL optics code was used and had a current of ≈ 8 A and voltage of -130 V with respect to the anode. Notice that the electron recovery current doubled for about 100 ms when the arc and converter were pulsed off. Figure 3(c) shows a fairly constant arc pulse at 90 V and 6 A. The highest parameters have so far not been achieved with flat beam pulses but show a single maximum somewhere during the pulse.

Figure 4 shows a similar set of beam traces for VITEX with an H^- beam. Here the system reached steady state after 1.5 s. Note that the reflector (reflex

Table I. Source experimental status

Parameters (simultaneous)	VITEX status (H⁻)			SITEX status (D⁻)
	Long Pulse	High current density	High current	
H^-/D^- current, mA	19	25	80	110
H^-/D^- extraction current density, mA/cm²	6.3	125	27	48
Pulse length, s	5.5	0.500	0.100	10
Voltage, kV	16	15.8	16.4	10
Brightness, mA/cm²·eV	13	310	67.5	12
Source pressure, mTorr	9	192	—	10
Gas efficiency (atom⁻¹/atom)%	0.09	0.10	—	2
Electron control, % × IV	857	219	232	20
Arc efficiency (arc power/H⁻ current), kW/A	140	286	41	10
Reliability Turn-on, %	100	100	100	100
Full pulse, %	>95	>95	>95	>90
Electrode power loading, W/cm²	<2500	<2500	<2500	<2500

discharge electrode opposite the hot emitter) voltage [Fig. 4(c)] is about −170 V with respect to the anode. The ratio of electron recovery current to beam current is much higher for VITEX than for SITEX. The current accountability,

$$I_{\text{drain}} = I_{\text{Faraday cup}} + I_{\text{extractor}} + I_{\text{electrons to ground}} ,$$

gives about 13 mA of unrecovered electron current (electrons to ground) out of 1213 mA of electrons extracted with the H⁻ beam. This H⁻ current resulted mainly from the use of a high anode gas flow with no cathode or reflector gas and was restricted by the low (2000-L/s) pumping speed of the test facility, which resulted in pressures of $\approx 10^{-3}$ Torr and hence much charge-exchange loss of the beam. Figure 5 shows an intermediate beam current achieved by pulsing the gas but limited by pumping. The Faraday cup signal was actually ≈38 mA when corrected for beam that passed through a small hole in the Faraday cup. Electrons accelerated to ground are the sum of the side shield current and the top plate current. The 1/8 SLP (μA) is the current that passed through the

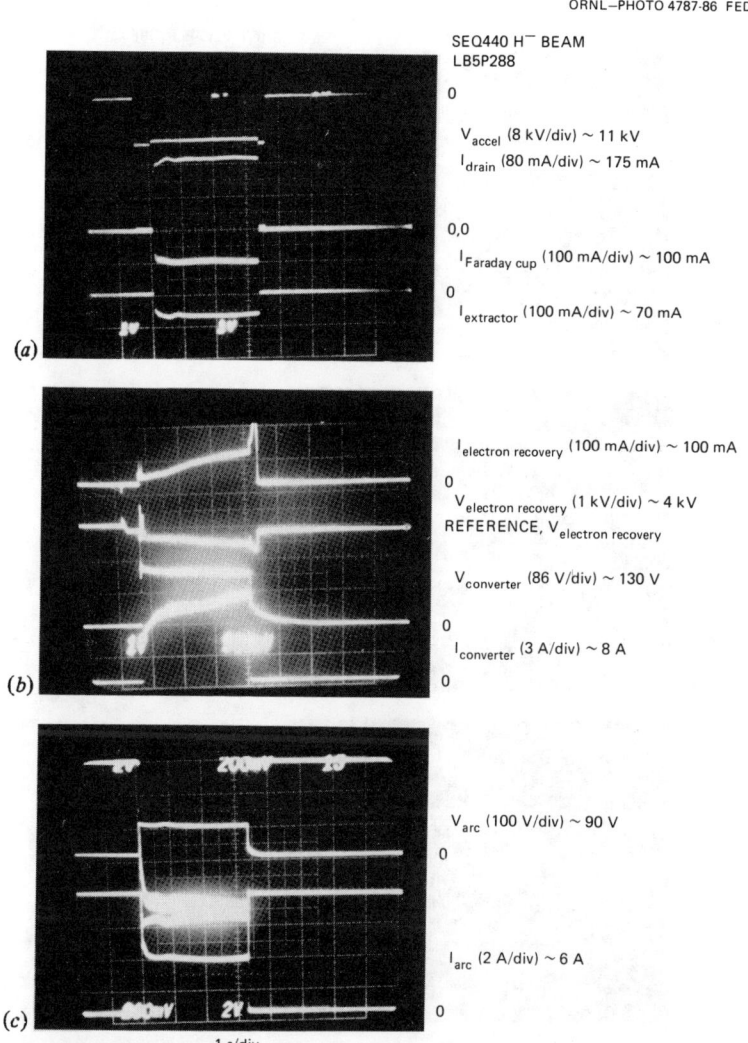

Fig. 3. SITEX beam waveforms with the source adjusted for a flat pulse.

Faraday cup and was not charge exchanged in the 23-cm flight path to the emittance scanner.

In Table I the high electrode power loading was on the graphite arc collimating slot. An intermediate arc electrode has been successfully tested and removed most of this power from the collimating slot, but it was not used for any of the data reported in this paper. It will be used when conversion to an all-metal VITEX source occurs.

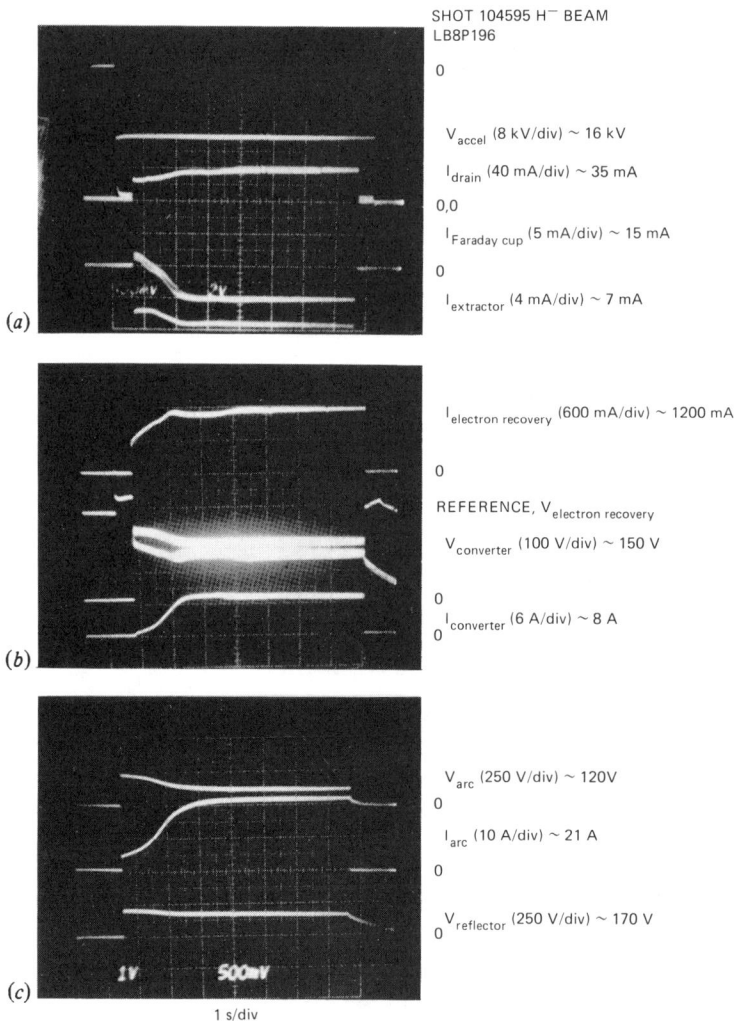

Fig. 4. VITEX beam waveforms with the source adjusted for a flat pulse.

ION TEMPERATURE AND EMITTANCE DATA

Figure 6 shows the layout of the emittance scanners, which are of the LANL type employing parallel plate electrostatic scanning. The emittance measurement is performed after 90° of beam deflection in a homogeneous magnetic field. Emittance measurements are made both parallel and perpendicular to the magnetic field. Figure 7 shows the beam current density distribution from an emittance scan parallel to the magnetic field.

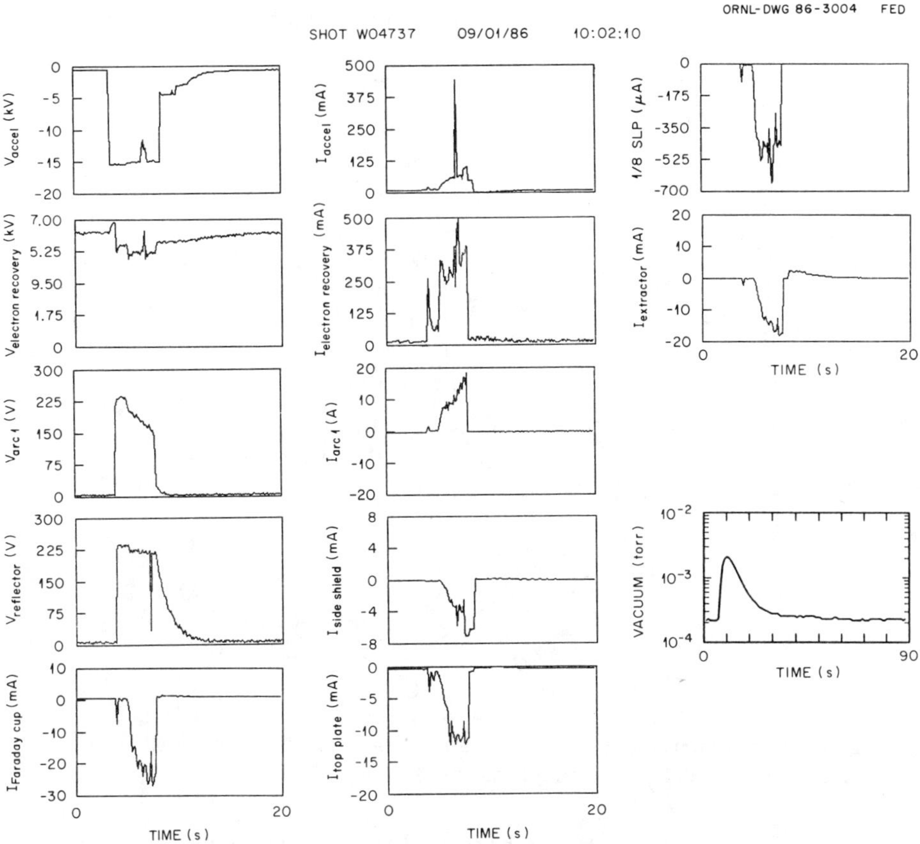

Fig. 5. VITEX beam waveforms using pulsed gas.

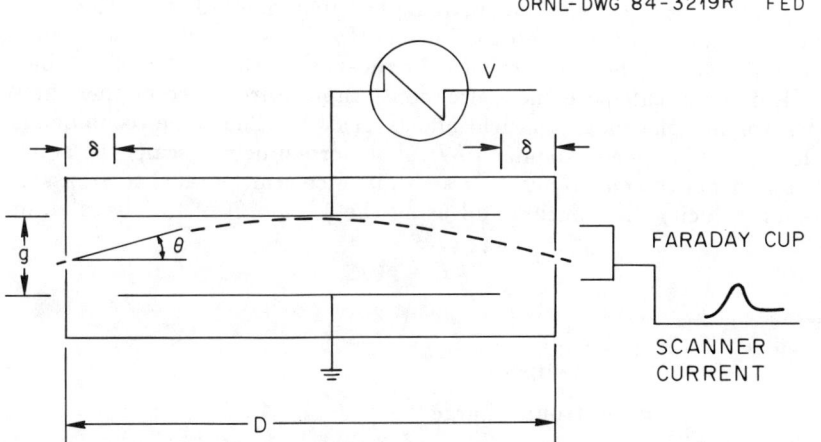

Fig. 6. Emittance scanner layout using parallel plate deflection.

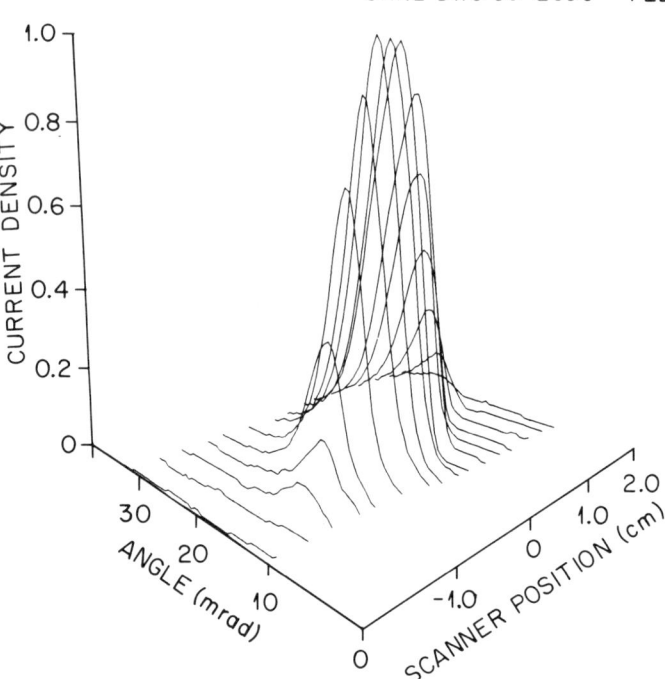

Fig. 7. Beam current density after 90° of beam deflection, derived from an emittance scan parallel to the magnetic field.

The orthogonal coordinates x,y,z are used to describe the beam at the emittance scanner, where y is parallel to the magnetic field B, x is perpendicular to B, and z is the direction of beam propagation. An emittance scan parallel to the magnetic field is done using a slit that is long enough in the x direction to integrate over $I(x,x')$. The slit is moved incrementally over the full extent of the beam in y so that the distribution $I(y,y')$ is measured. The source ion exit slit is long in the y direction. By measuring the center of the beam for $I(y,y')$, one can calculate the source ion temperature, since neither the source accelerator nor the magnetic field affects $I(y,y')$. The same technique is used for the emittance perpendicular to B. For perpendicular scans the accelerator focusing affects the emittance. For an emittance scan parallel to B, assuming a Gaussian velocity distribution and using the $1/e$ point of the distribution, one gets

$$kT = eV\phi_e^2$$

where

kT = ion temperature

e = electronic charge

V = beam acceleration voltage

ϕ_e = angular width of the beam at the $1/e$ point.

Table II gives a summary of the ion temperature and emittance data for SITEX and VITEX. Ion temperatures are computed from the emittance measurements parallel to the magnetic field using the width of the angular distribution. Ion temperatures for VITEX at 0.5 eV are about a factor of 10 lower than those for SITEX. These ion temperature measurements using an emittance scanner are in reasonable agreement with those from other surface and volume ion sources.

SUMMARY

Ion temperatures in the source have been established at 0.5 eV for VITEX and at 4 eV for SITEX by use of an emittance scanner on the analyzed beam. Further improvements in the extracted current density for SITEX are expected with the increased arc power that will be available through the use of a triode arc discharge. Better control of the arc discharge is expected to result in faster turn-on and constant discharge characteristics.

The output H$^-$ current and current density for VITEX have risen steadily in experiments during the last few months, which represent most of the work with this type of source. Future improvements in output current and current density are expected through use of the triode arc to permit higher arc power, geometry optimization, and biased plasma grid.

By acceptance of this article, the publisher or recipient acknowledges the U.S. Government's right to retain a nonexclusive, royalty-free license in and to any copyright covering the article.

Table II. Summary of typical ion temperature and emittance data

	SITEX		VITEX	
	Perpendicular measurement[a]	Parallel measurement[b]	Perpendicular measurement[a]	Parallel measurement[a]
Source slit dimensions $x \times y$, cm^2	0.21 × 0.16	0.21 × 0.16	0.1 × 2.0	0.1 × 2.0
Normalized rms emittance, $\pi \cdot$cm$\cdot$mrad	0.010 ± 0.005	0.003 ± 0.001	0.013	0.017
Normalized emittance (63% beam fraction)	0.010 ± 0.005	0.005 ± 0.002	—	—
Ion temperature at extraction, eV	6 ± 2	4 ± 2	—	0.5 ± 0.1
Current density j_{H-}, mA/cm^2	48	48	60	60
Beam current I_{H-} included in emittance, mA	—	—	8.3	8.3

[a]Entire beam emittance in this direction.
[b]Partial beam emittance in this direction. Useful for determining ion temperature without effects of beam optics.

REFERENCES

1. W. K. Dagenhart, W. L. Stirling, H. H. Haselton, G. G. Kelley, J. Kim, C. C. Tsai, and J. H. Whealton, in *Proceedings of the Second International Symposium on the Production and Neutralization of Negative Ions and Beams*, Report No. BNL-51304 (Brookhaven National Laboratory, Upton, N.Y., 1980).
2. W. K. Dagenhart, W. L. Stirling, and J. Kim, *Negative Ion Beam Generation with the ORNL SITEX Source*, Report No. ORNL/TM-7895 (Oak Ridge National Laboratory, Oak Ridge, Tenn., May 1982).
3. W. K. Dagenhart, W. L. Gardner, G. G. Kelley, W. L. Stirling, and J. H. Whealton, in *Proceedings of the Third Neutral Beam Heating Workshop*, Report No. CONF-811018 (Oak Ridge National Laboratory, Oak Ridge, Tenn., 1982).
4. W. L. Gardner, W. K. Dagenhart, G. G. Kelley, W. L. Stirling, and J. H. Whealton, in *Proceedings of the Third Neutral Beam Heating Workshop*, Report No. CONF-811018 (Oak Ridge National Laboratory, Oak Ridge, Tenn., 1982).
5. W. K. Dagenhart, W. L. Gardner, W. L. Stirling, and J. H. Whealton, Nucl. Technol./Fusion **4**, 1430 (1983).
6. W. K. Dagenhart, W. L. Stirling, G. M. Banic, G. C. Barber, N. S. Ponte, and J. H. Whealton, "Short-pulse operation with the SITEX negative ion source," *Proc. Third Int. Symp. on Production and Neutralization of Negative Ions and Beams*: AIP Conf. Proc. **111**, 353 (1984).
7. W. L. Stirling, W. K. Dagenhart, J. H. Whealton, and J. J. Donaghy, "Normalized emittance of SITEX negative ion source," *Proc. Third Int. Symp. on Production and Neutralization of Negative Ions and Beams*: AIP Conf. Proc. **111**, 450 (1984).
8. J. H. Whealton, R. J. Raridon, R. W. McGaffey, D. H. McCollough, W. L. Stirling, and W. K. Dagenhart, "2-D accelerator design for SITEX negative ion source," *Proc. Third Int. Symp. on Production and Neutralization of Negative Ions and Beams*: AIP Conf. Proc. **111**, 524 (1984).
9. W. L. Stirling, W. K. Dagenhart, and C. C. Tsai, *Volume H^- Generation with the SITEX Source in the VITEX Reflex Mode Without Cesium*, Report No. ORNL/TM-9753 (Oak Ridge National Laboratory, Oak Ridge, Tenn., 1986.)
10. C. C. Tsai, W. K. Dagenhart, W. L. Stirling, G. C. Barber, J. Donaghy, H. H. Haselton, P. M. Ryan, D. E. Schechter, and J. H. Whealton, "Discharge characteristics of a plasma generator for SITEX and VITEX ion sources," these proceedings.
11. P. W. Allison, D. B. Holtkamp, and J. D. Sherman, "An emittance scanner for intense low-energy ion beams," IEEE Trans. Nucl. Sci. **NS-30**, 2204 (1983).

THE CUSP H⁻ ION SOURCE AT KEK

Y. Mori, A. Takagi, K. Ikegami and S. Fukumoto

KEK, National Laboratory for High Energy Physics
Oho-machi, Tsukuba-gun, Ibaraki-ken, 305, Japan

Abstract

A cusp H⁻ ion source has been operated at the 12 GeV proton synchrotron at KEK. The ion source is pulsed (200 μsec x 20 Hz) and provides 15 - 20 mA H⁻ beams to the linac. The long lifetime of the LaB_6 cathodes resulted in stable operation.

Introduction

In order to increase the beam intensity of the 500 MeV booster and the 12 GeV main ring at KEK, charge-exchanged multi-turn injection with an H⁻ ion beam at the entrance of the booster synchrotron has been used since June 1985. A new intensity record of 1.5×10^{12} ppp at the booster has been recorded.[1]

A surface-plasma type of cusp-field H⁻ ion source has been developed for this purpose since 1983. Previously, we tried to use a magnetron type of H⁻ ion source. However, it had a serious stability problem at the relatively high duty factory of $\simeq 0.5\%$. The cusp H⁻ ion source was originally developed at LAMPF[2], based on the H⁻ ion source developed at LBL for their fusion project.[3]

The H⁻ ion source at KEK was mounted directly into the accelerating column of the 750kV Cockroft-Walton preinjector and provided stable 15 - 20 mA beams with a normalized 90% emittance (ϵ = phase area x $\beta\gamma$) of 1.8 - 2.2 mm·mrad to the linac. The beam extracted from the source was very quiet without any significant plasma oscillation. This stable operation was mainly the result of the long lifetime of the hot cathodes. Multicrystalline lanthanum hexaboride (LaB_6) filaments, which have been developed for cathodes of the ion source since 1984, operate at a relatively low temperature, 1450°C. Even after continuous operation for 2600 hours, only a slight reduction in the thickness was observed. We found that LaB_6 filaments showed very good performance when used as the cathode of the cusp H⁻ ion source.

Operation of the Ion Source

The setup of the cusp H⁻ ion source is shown schematically in Fig. 1. The ion source consists of a cylindrical plasma chamber, a molybdenum converter and two LaB_6 filaments. The operating characteristics of the ion source are summarized in Table 1.

© American Institute of Physics 1987

Table 1 OPERATING PARAMETERS OF THE H⁻ ION SOURCE

Arc Current	25 – 30 A
Arc Voltage	100 – 150 V
Filament Current	130 A
Filament Voltage	2 V
Filament Temperature	1450°C
H_2 Gas Flowing Rate	3 – 5 atm·cc/min
Converter Voltage	– 500 V
Cs Reservoir Temperature	155°C
Pulse Width	250 μsec
Repetition Rate	20 Hz
Cesium Consumption Rate	< 1mg/hr.

The beam shape extracted from the ion source is shown in Fig. 2. As can be seen from this figure, the H⁻ ion beam is stable and quiet. However, the beam intensity is rather sensitive to the temperature of the cesium reservoir. It was very important to control the temperature within a few degrees centigrade for stable operation. This differs from the case of using tungsten filaments and probably is due to the rather low working temperature of the LaB_6 filaments. The condition of the cesium coating on the converter is affected by the temperature at the surface of the converter, which is heated by radiation from the filaments. The low temperature also helps to reduce the cesium consumption rate, which is surprisingly low compared with the case of using tungsten filaments.

The emittance of the beam at the entrance of the linac is shown in Fig. 3. The normalized 90% emittance (phase space area x βγ) is about 1.8 – 2.1 mm·mrad.

The hydrogen flowing rate is 3 – 5 atm·cc/min and the gas pressure in the ion source is about 8×10^{-4} Torr. The ion source is installed directly to the accelerating column and is evacuated together with the beam-transport line by four turbomolecular pumps (2 x 650 l/s + 1500 l/s + 1000 l/s). The pressure in the column is about 1×10^{-5} Torr and

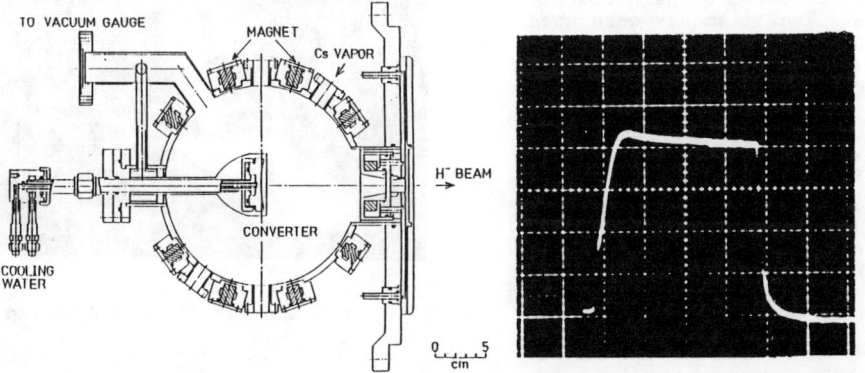

Fig. 1 Schematic setup of cusp H⁻ ion source

H: 50 μsec/div   V: 5 mA/div
Fig. 2 Beam shape of H⁻ ion beam.

15 - 20% of the H⁻ beam is lost by change stripping during passage through the beam line.

The ion source is installed directly to the acceleration column. This had led to worries that cesium contamination would cause frequent sparking in the column. In actual operation, the sparking rate was about o.5 ~ 1 times per one hour. This is not so serious at the moment, although high voltage pre-conditioning of the acceleration column is necessary for stable operation.

LaB_6 Cathode

One of the problems with an ion source using hot cathodes is the lifetime of the filament. For this type of cusp-field ion source, tungsten filaments have been commonly used so far, although there are problems with long period operation. K. Leung et al. used directly heated LaB_6 filaments for a PIG H⁻ ion source and a multicusp H⁺ ion source[3]. They found that LaB_6 filaments operated well at high arc currents. The work function of LaB_6 is about 2.66 eV, making possible operation of the filaments at 1450°C - 1500°C. However, it was unclear whether LaB_6 filaments could be used because a large amount of cesium is present in the plasma. Since lanthanum hexaboride reacts with active species, for example oxygen and chlorine, it was predicted that the filament might be damaged by cesium in the ion source.

In order to clarify these points, we tested the LaB_6 filament. The shape and dimension of the commercially available filaments used in the experiments are shown in Fig. 4.

A thin rhenium metal plate was placed between the filament and the molybdenum supporting electrode to avoid the poisoning effect of boron at high temperature. Two filaments were used at the same time. Each one was heated directly by its own power supply so as to control the temperature separately. The temperature of the filaments was first raised to 1650°C for activation, which was very important to get high arc current at lower temperature. Sometimes the beam was noisy at the beginning of the discharge, but after operating 40 - 50 hours the beam became

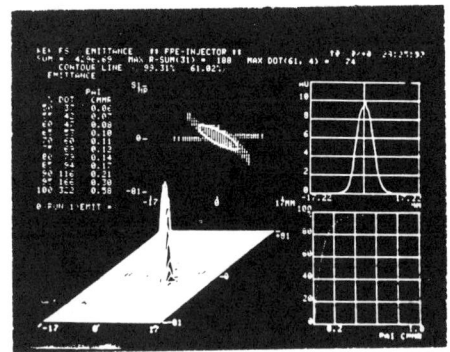

Fig. 3 Emittance of 750 KeV H⁻ Beam.

very quiet. The noise was probably due to impurities in the extracted beam, for example, O^- ions.

After 1000 hours of operation, the efficiency of electron emission from the filaments was rather improved and an arc current of 25 - 30 A was easily obtained at a temperature of 1400°C. However, near 2000 hours the arc voltage began to increase gradually and it was necessary to increase the temperature of the filaments to 1450 - 1500°C. No serious problems have occurred up to 2500 operating hours, which is almost half of the total operating time of the machine in one year.

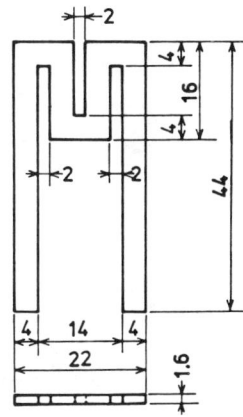

Fig. 4 Shape of LaB_6 filament.

When we opened the ion source after long period operation, we found a grey coloured coating on the filament, especially on the surface facing the converter. By SEM examination the coating was found to be molybdenum, which was sputtered from the converter, as shown in Fig. 5. The reduction in the efficiency of the electron emission after 2000 hours operation was probably caused by this coating, not by cesium.

In order to examine the sputtering characteristics of LaB_6 filaments, we have made experiments with a small cusp ion source[4]. A single LaB_6 filament was placed in the discharge chamber and argon was used as the discharge gas. The discharge was in DC mode and the cathode material loss was measured as a function of the time elapsed. The experimental conditions are summarized in Table 2.

The material loss after 148 hours was about 30 mg/cm^2, as shown in Fig. 6. This loss was mainly due to sputtering. The evaporation loss which has been reported[5] is two orders of magnitude lower than this value, as shown in Fig. 6 with a broken line.

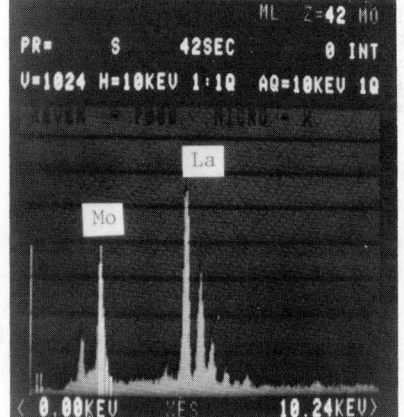

Fig. 5 Surface analysis by SEM.

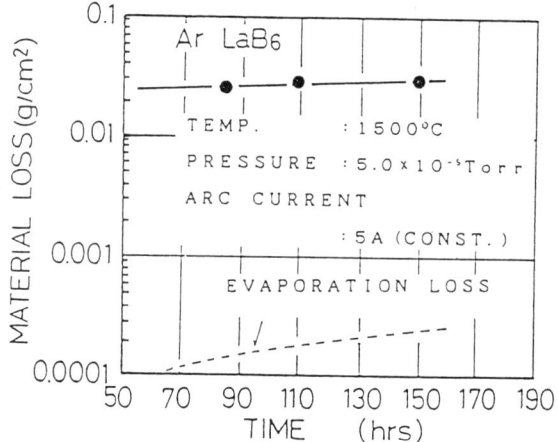

Fig. 6 Material loss of LaB_6 filament.

Table 2 DC DISCHARGE CONDITIONS

Arc Current	5 A
Arc Voltage	45 V
Filament Temperature	1500°C
Gas Pressure	5×10^{-5} Torr

Conclusion

A surface-plasma type of cusp H^- ion source has been operated at the KEK 12 GeV synchrotron since June 1985 and has delivered stable beams of 15 - 20 mA to the linac. The stable operation is mainly due to the use of LaB_6 filaments. The lifetime of the filament exceeds 2500 hours and the relatively low temperature also helps to reduce the consumption rate of cesium. The sputtering rate of this type of filament was measured with a DC argon discharge, and a loss of 30 mg/cm^2 was observed after a continuous discharge of 148 hours at an arc current of 5 A.

The authors would like to express their appreciation to Dr. Hagiwara and Mr. Yamada of DENKA Co. for their collaboration in the development of the LaB_6 filaments. They are also indebted to Profs. Kihara and Nishikawa for their continuous encouragement.

References

1) S. Fukumoto et al., Proc. of Linear Accelerator Conf., 1986 SLAC.
2) R.L. York et al., AIP Conf. Proc. No. 111, 410 (1984).
3) K.N. Leung et al., Rev. Sci. Inst., 55, 1064 (1984).
4) S. Yamada et al., ISIAT'86, Tokyo, 153 (1986).
5) J.M. Lafferty, J. Appl. Phys., 22, 229 (1951).

DISCUSSION

York: You said you found the moly on the LaB_6 filament. After the 2600 hours, was the effect still there?
Mori: Yes.
York: But it didn't seem to effect the arc efficiency at all?
Mori: Oh, I didn't tell you. Of course, after 1000, 1500 hours, the arc voltage increased gradually. So, probably, the efficiency of the electron emission gradually decreased by such a coating.
York: Did raising the arc voltage eliminate the moly coating at all?
Mori: I don't think so. Probably not.
Stirling: Concerning the low temperature emitters, have you considered using the lanthanum-doped moly instead of the LaB_6? It's comparable, I think it works easily, machines easily.
Mori: That's interesting.
Leung: I've tried it.
Stirling: You did? We find that it works very well.

A Surface Conversion Source of H⁻ with Hot Walls and Variable Converter Temperature

O.F. Hagena, P.R.W. Henkes

Kernforschungszentrum Karlsruhe, D-7500 Karlsruhe 1, Federal Republic of Germany

ABSTRACT

A surface conversion source has been constructed, whose walls are hot (ca. 900 K) during operation to keep adsorption of Cs low. The temperature of the converter can be varied between 300 and 670 K. From the response to temperature changes it is concluded that deactivation of Cs absorbed at the converter plays a major role. The best performance with respect to Cs economy is obtained by keeping the converter close to the upper limit of its temperature range. A self extracted, mass analyzed H⁻ current of 214 mA was measured from a converter of 120 cm² surface area.

1. INTRODUCTION

Surface conversion sources of the bucket type using self extraction, as introduced by Ehlers and Leung /1/, usually employ cooled walls that serve as anode and as vacuum vessel at the same time. In these sources cesium, which is introduced in the discharge, is not only deposited on the converter surface but also on the surfaces of the wall. This leads to a large inventory of Cs in the source, which poses a potential hazard to an extraction and acceleration system when it leaks out of the source, e.g. by surface migration through the extraction aperture. In order to reduce the inventory of Cs we proposed a design of a surface conversion source with hot walls /2/, which was intended to deliver 250 mA of H⁻ for a 400 keV acceleration test. First results presented in /3/ showed the advantage of an elevated temperature of the converter. Improvements of performance by using warm liners in sources with cooled walls were reported by Piosczyk et al. and Kwan et al. /4, 5/.

2. DESIGN

Fig. 1 shows a partial cross section of our source. The whole source is placed into vacuum. Its anode (volume 26 l) is fabricated from Mo-sheet and is cooled by radiation only. The magnetic cusp field is produced by Co-Sm magnets canned in stainless steel tubes and cooled by Frigen. The tubes also serve as supports for iron plates - not shown in Fig. 1. - that are used to shape the magnetic field and to reflect heat radiation. The size, position and orientation of the magnets was chosen as to minimize the field in centre of the source where the converter is situated. The field at the extraction slit is adjusted to produce no net deflection of the H^- emitted by the converter. Fig. 2 shows a contour plot of the magnetic field.

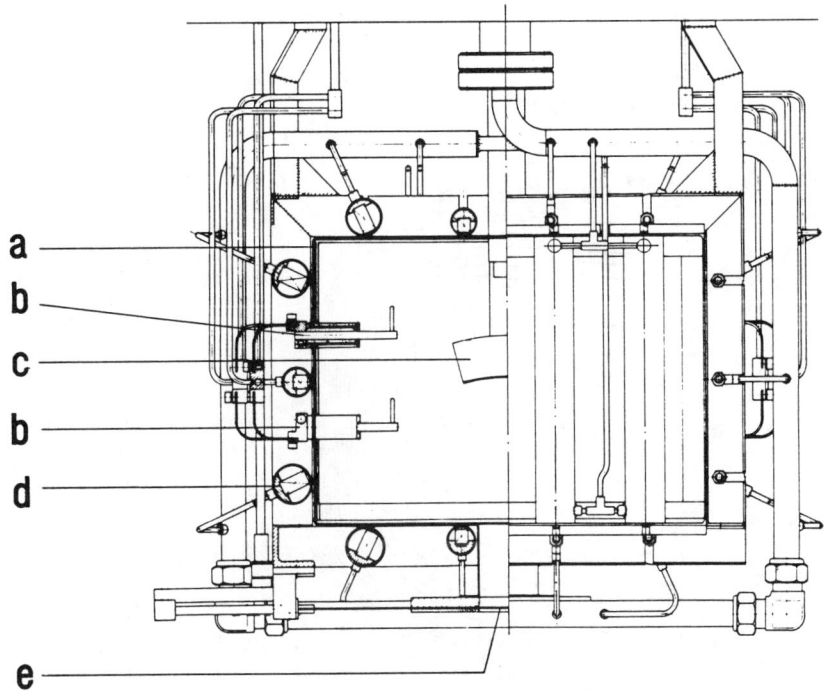

Fig. 1. Partial cross section of the ion source
    a) Molybdenum anode 47 x 28 x 20 cm^3
    b) Tungsten filaments, three per row
    c) Converter
    d) Canned Sm-Co magnets cooled by Frigen R 113
    e) Extraction slit with shutter

Only the lower row of filaments shown in Fig. 1 have been installed. To increase their lifetime they are heated by ac current. Each of the six filaments is connected to one phase of a six-phase transformer.

Fig. 3 shows a cross section of the converter. It is fabricated from OFHC copper. The front surface is detonation plated with molybdenum. This and its all-welded construction allows in principle operating temperatures up to ca. 1000 K. The front surface measures 20 × 6 cm² and is cylindrically curved to focus the H⁻ on the extraction slit. The back and sides are insulated by applying a coat of alumina by plasma spraying. The insulation is shielded from deposition of cesium and evaporated tungsten by molybdenum sheet that is on floating potential. The converter was initially cooled by water, later by thermo-oil.

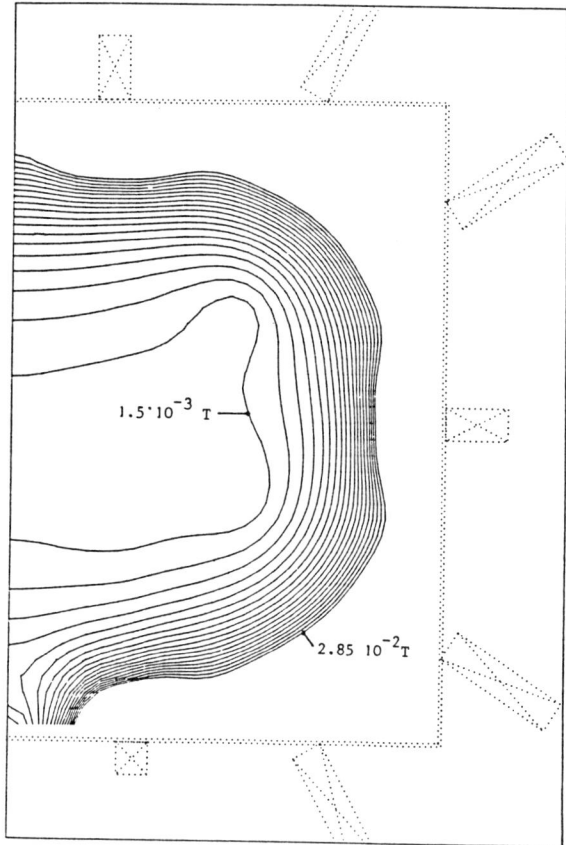

Fig. 2. Computed contour plot of the magnetic field. The increment between two adjacent contours is 1.5 mT.

Two tubes of 24 mm ⌀ penetrate the top of the source behind the converter and connect via valves to a Cs oven and a Cs condenser resp. The Cs influx is controlled by the temperature of the oven. Before switching off the discharge the extraction slit can be closed by a shutter. To remove the Cs from the converter its cooling may be turned off to let it run hot. The Cs vapor from the discharge volume may then be removed by the water cooled condenser.

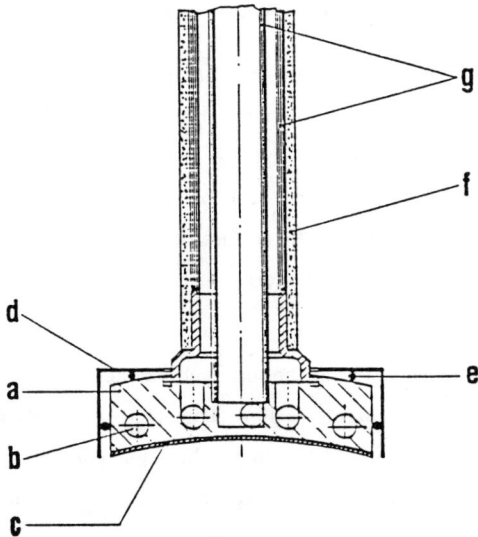

Fig. 3. Cross section of the Converter
 a) OFHC-copper
 b) Cooling ducts
 c) Front surface of molybdenum, 6 x 20 cm^2
 d) Molybdenum shield
 e) Alumina spacers
 f) Alumina tube (Molybdenum shield around lower part not shown)
 g) Concentric stainless steel tubes (Coolant ducts)

The following diagnostics were employed:

1. A stationary flat Langmuir probe, placed at the side of and somewhat behind the converter, monitored the discharge.

2. A movable flat Langmuir probe could be introduced through the extraction slit to determine the plasma properties in front of the converter.

3. A gridded ion collection plate measured the self extracted current escaping the source. The potential of the grid was kept at -10...-15 V with respect to the earthed anode to reflect plasma electrons. The collection plate was kept at +10 V to reflect positive ions.

4. The negative ion current was mass analyzed by a small, movable mass spectrometer with a mass range from 0.3...23 amu /2/, which was later replaced by another that could measure electron and ion currents at the same time /6/, thus giving reliable ratios of H$^-$/electron currents.

3. RESULTS WITHOUT CESIUM

The source was operated with a discharge voltage of 80 V and a current of up to $I_a = 50$ A, where it was limited by frequent mode flipping. At 40 A the stationary mean temperature of the anode exceeded 860 K. The converter was usually biased -120 V with respect to anode potential. The largest negative currents at the ion collector were found at fairly low hydrogen pressure. Therefore the neutral density in the source was usually kept at $1.4 \cdot 10^{19}$ m^{-3}, corresponding to a pressure of 0.42 mTorr at room temperature.

After thorough discharge cleaning the total self extracted negative current was as low as 0.14 mA at a discharge current of 40 A. The electron density determined by the Langmuir probes was of the order of 10^{17} m^{-3}. The probe characteristics display a bi-Gaussian distribution with electron temperatures of ca. 3.5 eV and 29 eV resp. The ratio of negative/positive saturation current is only around 11, suggesting that the plasma is disturbed by the probe at positive potential, and consequently, the undisturbed plasma density may be larger.

4. RESULTS WITH CESIUM

4.1 COLD CONVERTER

At $I_a = 40$ A and full water flow the temperature of the converter was $T_c \approx 300$ K. First only small amounts of Cs were admitted to the discharge by keeping the temperature of the oven low (ca. 410 K) and opening its valve only for 1 - 2 min, ("low Cs" mode). From the vapor pressure and the conductance of the tube the rate of influx is estimated to be lower than 5 mg/min. During the Cs "puff" the total negative current collected rose to a maximum of $I_- = 194$ mA, the mass analyzed current of H$^-$ increased by a factor of 300. The amount of cesium in the plasma, however, was so low that it did not influence the plasma parameters. The electron temperature did not change, the negative saturation current rose only slightly, which in part was due to a slightly larger discharge current and to the electrons emitted by the converter, whose current increased from 3.3 A to 6 A.

After I_- reached its maximum (and the oven was closed again) it decayed at first with a time constant of ≈ 12 min. After dropping to 60 - 80 mA it did not change much over periods of the order of one hour.

When the cooling water of the converter was shut off it was observed that the mass analyzed negative ion current as well as I_- first increased with rising T_c before dropping off again. At $T_c = 800$ K the discharge had to be cut off. Some preliminary tests on the effect of the converter temperature, using a compressed air - water mixture as coolant at $T_c > 373$ K, yielded the following information:

a) When after the injection of Cs the negative current with <u>cold</u> converter has dropped in time from its maximum to 40 % and then T_c is raised to 620 K, I_- as well as the mass analyzed current of H^- recover to almost the same level as at their maximum.

b) Conversely, cooling down the converter from an elevated to room temperature always results in a drop of I_-, sometimes as much as 50 %.

4.2 HOT CONVERTER

The results of the previous section shows that it is desirable to control the temperature of the converter over a wide range. Therefore, the cooling water was replaced by thermo-oil. After this the lowest temperature attainable was $T_c = 410$ K at $I_a = 40$ A. At the same time the the mass analyzer was replaced by the new one.

Fig. 4 shows an example of the influence of the converter temperature on the negative current for the case of "low Cs". With T_c rising from 420 K to 640 K I_- increases, after a transient decrease, from 130 mA to 190 mA. This procedure is reversible and reproducible. The currents of the individual species - H^-, electrons, and impurity ions - measured during the break of the curve in Fig. 4, all change by the same factor as the total current. After repeated injection of Cs with the temperature of the oven increased to 420 K[*] (estimated Cs-flux 12 mg/min), that is in a state, where the source may be "over cesiated", T_c no longer affects the negative ion current.

---

[*] This is still less than the temperature of any other part of the ion source.

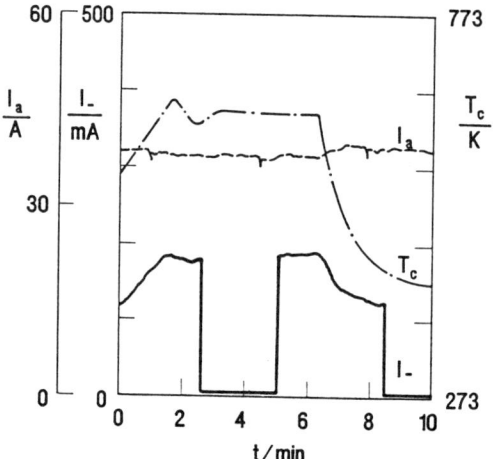

Fig. 4. The dependence on the temperature of the converter of the self extracted total negative current.

The temperature of the walls proved to be equally important. Since they are cooled by radiation their temperature depends on the discharge current. Fig. 5 shows two examples of the change of I_- with that of I_a. One observes an immediate change due to the change of plasma density and a slow one which obviously corresponds to the change of the wall temperature. Even in the case of "high Cs" quoted above raising the wall temperature results in a increase of the negative current.

The long-time behavior was tested with $I_a = 33$ A and at a converter temperature of 570 K. Typically the negative current rises during the admission period of Cs (1-2 min) to a maximum of 220 - 240 mA. It then decays with an initial time constant of 23 - 27 min. This decay is arrested at about 100 - 120 mA, and I_- remains virtually unchanged over a period of 85 minutes, when the experiment was terminated.

The composition of the negative current was determined at various phases and was found to be fairly constant: 75 % H-, 22 % electrons, and 3 % impurity ions.

a)

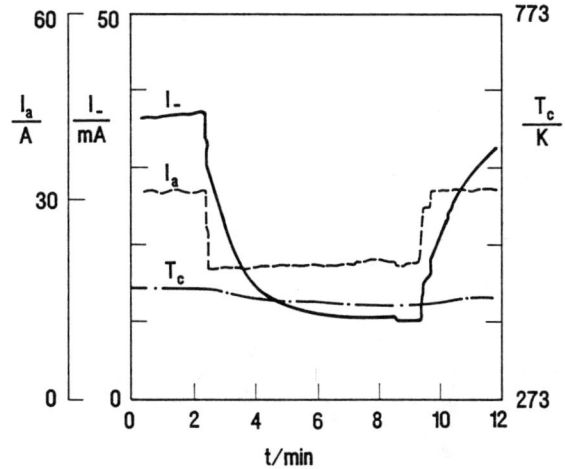

b)

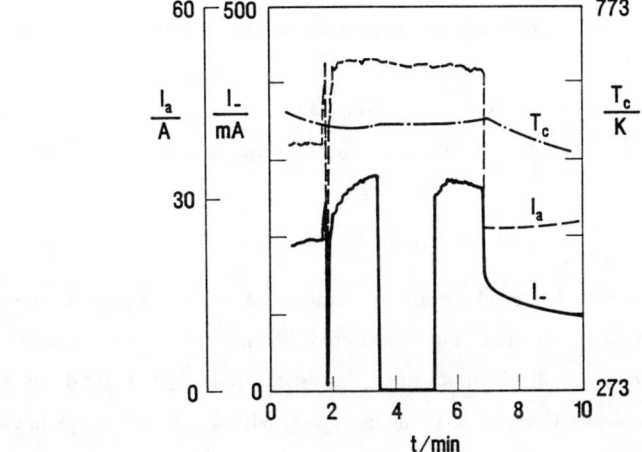

Fig. 5. The dependence on the wall temperature of the selfextrated total negative current.
a) With very little Cs. (Shortly after the start of the discharge before admitting Cs).
b) After repeated injection of Cs. The short disruption at $t \approx 2$ min was due to a mode flip.

Fig. 6 shows a beam profile obtained by scanning the mass spectrometer across the ribbon beam 36 mm below the exit of the source. Its full width at half maximum is 29 mm.

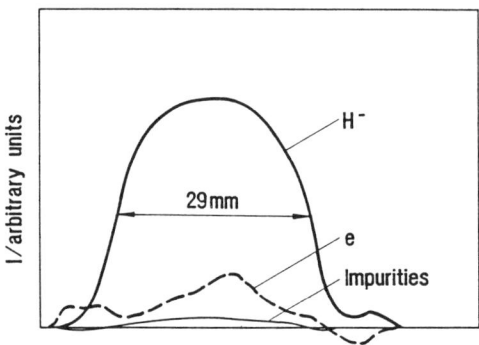

Fig. 6. Beam profile taken with the mass analyzer

The highest negative current we observed was 285 mA at a discharge current of 51 A (Fig. 5b). This corresponds to a current density of 1.8 mA/cm^2 of H$^-$ at the converter surface.

5. DISCUSSION

Our results show clearly the advantage of running the walls of the ion source hot. In contrast to sources with hot liners there is no cold spot where Cs could accumulate, as for instance at the cold wall behind a liner. As a consequence only small amounts of Cs are needed to operate the source. If one is content with about half the peak current at the time of the admission of Cs, an initial inventory of less than 50 mg of Cs is sufficient to run the source for hours. Despite the high temperature of the walls they still adsorb Cs. This is well known from the dependence of the adsorption of Cs on tungsten in the Langmuir-Taylor detector. Because of the high ratio of wall/converter surface, even a small coverage of the wall may be sufficient to supply Cs to the converter when the wall temperature is raised. On the other hand, if the coverage of the walls is less than a monolayer, it may pump Cs off the converter even if its temperature is $< T_c$.

The dependence of the converter temperature of the H⁻ current suggests that other effects than the degree of coverage may control the production of H⁻. The initial decay of the H⁻ current after injection of Cs can not be explained in terms of under- or oversupply of Cs alone for the following reasons:

a. Sputtering of Cs could result in an increasing Cs-deficiency. However, in that case raising T_c should not counteract but accelerate the decay of H⁻ current.

b. After closing the valve of the oven the converter may continue to collect Cs from the discharge and the walls and thus become overcesiated. In this case, however, one would expect the H⁻ current to recover when Cs gets lost in time - which was never observed - and to decrease with rising wall temperature. Furthermore, raising T_c should have the strongest effect after to much Cs has been injected, when it produces no effect at all.

One effect that may explain our findings is a partial deactivation of Cs by adsorption of hydrogen as observed by Geerlings at al /7/. Raising T_c will desorb hydrogen and tends to restore the activity of the converter surface. With large amounts of Cs in the source there may be sufficient recirculation of Cs from the discharge volume and the walls to the converter resulting in a fairly fresh surface. Thus T_c will have no effect. Deactivation of Cs would also explain that with rising temperature of the anode the H⁻ current <u>always</u> increases, even after injection of much Cs.

REFERENCES

/1/   K.N. Leung, K.W. Ehlers, Rev. Sci. Instr., 53 (1982) 803.

/2/   O.F. Hagena, P.R.W. Henkes, R. Klingelhöfer, B. Krevet, H.O. Moser, Proc. of the 3rd Int. Symp. on Production and Neutralization of Negative Ions and Beams, Brookhaven 1983, AIP conf. Proc. No. 111, p. 96.

/3/   O.F. Hagena, P.R.W. Henkes, Proceedings of the 2nd Europ. Workshop on the Production and Application of Light Negative Ions, Palaiseau, March 5- 7, 1986, M Bacal and C. Mouttet editors, p. 171.

/4/   B. Piosczyk, G. Dammertz, Rev. Sci, Instr., <u>57</u>, 1986, 840.

/5/ J.W. Kwan, G.D. Ackerman, O.A. Anderson, C.F. Chan, W.S. Cooper, G.J. deVries, A.F. Lietzke, L. Soroka, W.F. Steele, Rev. Sci. Instr. 57, 1986, 831.

/6/ P.R.W. Henkes, to be published.

/7/ J.J.C. Geerlings, R. Roding, P.W. Amersfoort, J. Los, H.J. Hopman, Proceedings of the 2nd Europ. Workshop on the Production and Application of Light Negative Ions, Palaiseau, March 5-7, 1986, M Bacal and C. Mouttet, editors, p. 179.

DISCUSSION

York: Have you ever scanned the converter voltage under those conditions to see if that effected the current?

Henkes: No, we have made practically no measurements with the converter voltage. We didn't have the time for this. We also have not made many changes in the pressure of the source. We usually run the source at very low pressure, only at 0.5 mTorr at room temperature. And the only thing we found is that it seems to be that the lower the pressure is, the higher the negative current is. But the converter voltage was not varied.

York: Have you ever tried transferring cesium at a constant rate at a lower temperature with the valve open?

Henkes: No, that was not possible with our system. We have no good control, we have no needle valve or something to control the cesium. The only way to control the cesium flux is by changing the temperature of the oven, and that is too sluggish. When we tried to do this, too much cesium was getting into the source. The total time we can run the source with just a few mg of cesium in the source is hours and hours. The last experiment we had, the source was running for over 6 hours, without any drop of cesium. If you're satisfied with a current of 1/2 the peak current - that is, when we let in the cesium, the maximum current we got was about 280 mA - if you're satisfied with 1/2 of the value, you can just keep on, without adding cesium, for hours and hours.

Allison: I assume this is a dc operated source. Is that correct?

Henkes: Yes.

Allison: When you said you had over-cesiated conditions and temperature did not play an important role, was the performance the same as the peak with the various temperature effects?

Henkes: Except for the converter temperature, it was. I didn't say it is really over-cesiated; it behaves like that. But it may just mean that the effect of the converter temperature is not important, because if you have enough cesium in the source, then you will always have a recirculation of cesium from the anode to the converter, which is always providing a fresh surface, and will work against the deactivation by hydrogen.

SURFACE-PLASMA SOURCE OF H⁻ IONS

G.E. Derevyankin and V.G. Dudnikov
Institute for Nuclear Physics, Siberian Branch of the
Academy of Sciences of the USSR,
Novosibirsk, USSR

ABSTRACT

A surface-plasma source of H⁻ ions with a semiplanotron geometry of the discharge chamber was investigated. The discharge was localized around the emission slit of 0.5×10 mm^2, and a cylindrical cathode surface serves for geometrical focussing of H⁻ ions. The H⁻ beam current was up to 0.15A, normalized brightness to 10^8A/cm^{2}rad^2, pulse length 200 μs and no cooling was required for operation up to 50 Hz. The source is very efficient in producing H⁻ ions and has a simple construction. Extraction of ions and primary beam formation have been studied for the geometrical focussing of ions in the emission slit. Reasons for a reduction of the beam brightness during the transport are considered as well.

INTRODUCTION

Charge-exchange injection of protons into a synchrotron was one of the first applications of surface-plasma sources (SPS) of H⁻ ions. Experiments with charge-exchange injection into large synchrotrons, using surface-plasma sources, have been reviewed in Ref. 1. Requirements imposed on sources used on large accelerators are very strict concerning the beam intensity and quality, reliability, and an easy operation and maintenance. To satisfy these requirements, different variants of surface-plasma sources with a planotron geometry were developed at the very beginning of studies of surface-plasma processes for production of negative ions in gas discharges.[2,3]

At about the same time, Penning sources were proposed for accelerator applications, having H⁻ ion beams with a substantially higher brightness. Formation of beams with the highest possible brightness was treated in Ref. 4. The search for new variants of surface-plasma sources, optimized for particular applications, has been described in Ref. 5.

In this paper, characteristics of a SPS with a semiplanotron geometry will be described, having an improved efficiency of H⁻ ion generation, a high pulsed and average beam intensity, a high brightness and a simple construction. The discharge takes place in the vicinity of the emission slit, without a closed ExB drift, and the H⁻ ion flux is geometrically focussed toward the emission slit, as described first in Ref. 6.

CONSTRUCTION OF THE SOURCE AND ITS CHARACTERISTICS

Figure 1 shows schematically the construction of the source. The cathode (3) of the discharge chamber is the emitter of H⁻ ions; it is enclosed in the anode (2). The active H⁻ ion emitting surface

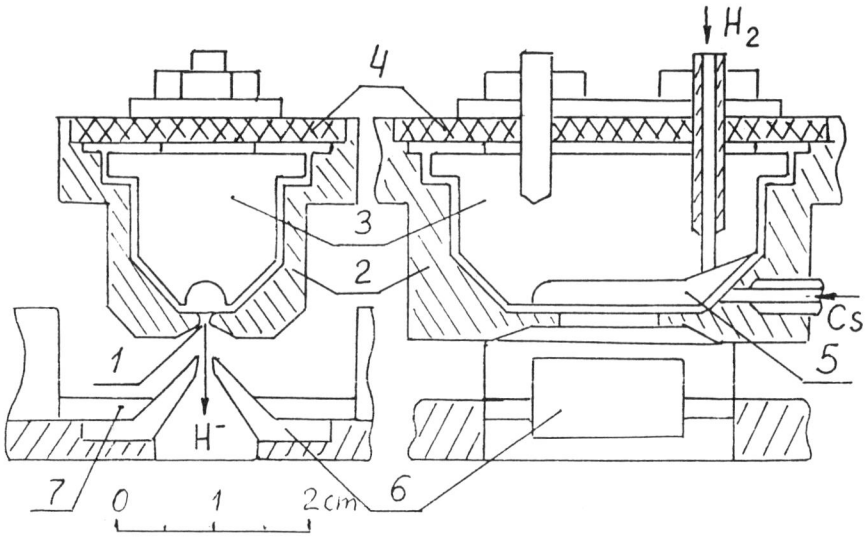

Figure 1. SOURCE CROSS SECTIONS. 1 - emission slit (0.5×10 cm^2); 2 - anode; 3 - cathode; 4 - insulator; 5 - cathode groove; 6 - extraction electrode; 7 - mild steel inserts.

of the cathode has the shape of a cylindrical groove with a 2.5 mm radius. The depth of the groove was varied from 1 mm, which corresponded to its curvature, to 2.5 mm. The groove is parallel to the anode emission slit (1). The center of curvature of the cathode groove was in the plane of the emission slit. Negative ions receive, in the thin cathode voltage fall, an initial velocity directed toward the emission slit. In this way, the ion beam consisted of ions from the total emitting surface of the cathode (geometrical focussing). The area of the emission slit (0.5×10 mm^2) and the geometry of the extraction gap were the same as in the Penning source.[7] The outside surface of the anode, acting as a focussing electrode, formed, with the beam boundary, a Pierce angle of 67.5°.

The discharge gap was in the magnetic field, perpendicular to the emission slit. The discharge is limited to the gap between the active part of the cathode surface and the anode cover. Hydrogen is injected into the cathode cavity through a channel in the cathode by means of an electromagnetic valve,[8] while cesium is injected through an opening in the anode from a heated container.

Plasma, produced by the discharge in the cathode cavity, is drifting perpendicularly to the magnetic field along the cathode groove, maintaining the discharge. For that reason the direction of the magnetic field should agree with the direction of the drift.

A pulse forming network generates the voltage for the discharge gap. The line impedance of 12Ω is substantially higher than the impedance of the discharge gap while operating. This high line impedance allows a wide range of discharge currents while keeping, at the same time, the voltage between the limits required for the initiation and maintenance of the discharge. Source electrodes are conditioned before the initial operation by an arc discharge. At a current of 30-40A this procedure takes less than 10 minutes and proceeds in a quiescent mode.

The extraction voltage for negative ions is a square negative voltage pulse with an amplitude up to 25 kV applied to the body of the source and with respect to the grounded extraction electrode (6). The amplitude of the pulse is stabilized to 10^{-3} accuracy.

Mild steel inserts (7), mounted on the extractor, serve to produce a convex magnetic field in the extraction gap. As a result, secondary electrons drift rapidly out of the beam along magnetic field lines and arrive at the extractor. In this way, conditions are such as to prevent the multiplication of electrons in the extraction gap, and the electrical strength of the gap is greatly increased. On the contrary, if the magnetic field is uniform, an electron trap may appear in the gap, where they oscillate and multiply, which leads to a breakdown. In the latter case, the maximum voltage is 10-12 kV.

The source cathode and the extractor are made out of molybdenum and the rest of metal parts out of stainless steel.

The cathode is mounted in the chamber by using a ceramic plate (4). Good results were achieved with the ceramic 22 HS. Contact lines of the anode and the cathode on the ceramic are shielded by deep recesses. The gap between the anode and the cathode is about 0.5 mm. Such a construction prevents sputtering of the ceramic and eliminates surface breakdowns.

Typical source operating parameters are:

discharge voltage	100 V
discharge current	100 A
extraction voltage	20-25 kV
H^- beam current	100 mA
extraction gap current	to 0.2A
pulse length	250 µs
repetition rate	to 50 Hz

Average gas and cesium consumptions per pulse were 2-3 Torr-cm^3 and 2×10^{15} atoms, respectively. In such a mode of operation there was no need for cooling. The design of the source (accessibility of the cathode) makes the forced cooling of the cathode easy to apply, as this electrode receives about 70% of the source input power. It should be noted that forced cooling of the cathode in a planotron is quite difficult.

A stable discharge is possible in the source at low magnetic fields (about 500G). Under such circumstances it is much easier to form an ion beam having a low fluctuation level. In this way it was possible to compare the brightness of the beams from this source and from the Penning source under similar conditions.

Figure 2 shows emission characteristics of the source for different geometries of the emission slit by keeping the area constant.

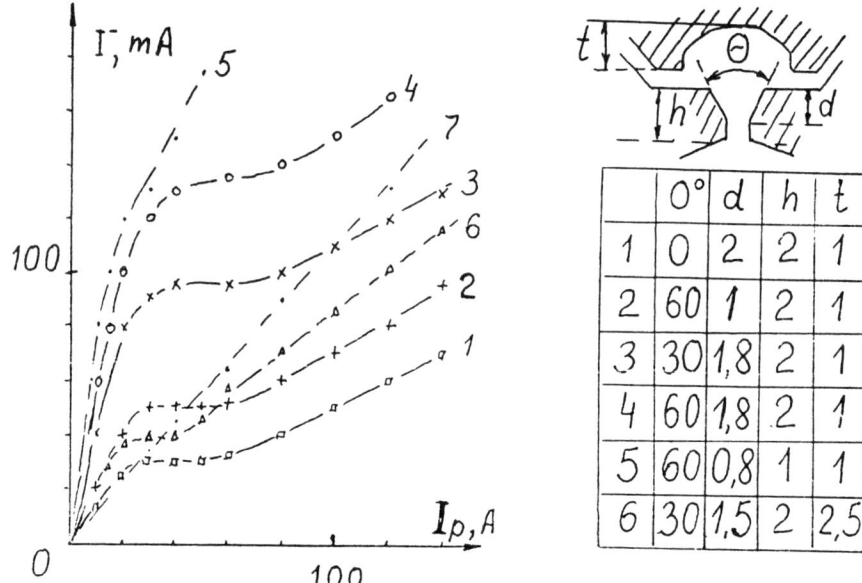

Figure 2. Dependence of the beam current (I^-) on arc current (I_p), for different geometries of the emission slit. Curve 7 shows $I^-(I_p)$ for the Penning source.

Two parameters were varied: angle θ (opening toward the cathode) and the depth of the emission slit h. A maximum output of H^- ions was obtained for the smallest depth of the cathode groove (t ≈ 1 mm). For comparison, curve 7 represents the characteristics of a Penning source.[7]

It is evident that the output of negative ions rapidly increases when the depth of the emission slit is reduced from 2 to 1 mm and the angle θ increased from 0 to 60°. A further increase of the angle θ does not result in any appreciable increase in the ion output because ions from almost the total emitting cathode area are collected within 60°.

Curves on Figure 2 show three characteristic regions, distinguished by the magnitude and behavior of the efficiency of ion generation processes. Over the initial part of the curves, for small discharge currents, the plasma density is not high (≈ $10^{13} cm^{-3}$). Negative ion flux, emitted from the cathode, suffers a negligible attenuation in the plasma and reaches, almost without losses, the emission slit to be extracted and formed into the beam. Therefore, in this part the beam current rises linearly with the discharge current.

As the discharge current increases so does the plasma density, and ion destruction becomes more and more evident. As a result, the ion beam current does not increase beyond a discharge current of 30A and it may even decrease because the H⁻ ion emission from the cathode grows linearly with the plasma density, while the destruction depends exponentially on the plasma density. Linear increase of the intensity over the third region is a witness for the appearance of new processes for H⁻ ion formation: ion emission from the walls of the emission slit under bombardment by fast atoms, and resonant charge exchange between fast negative ions and atoms in the vicinity of the emission slit (when increasing the discharge current it is necessary to increase somewhat the hydrogen flow). An increase of the H⁻ ion output for wider angles θ and for small slit depths h is related to a better transmission of primary ions emitted from the cathode (for small discharge currents) and to a deeper penetration of the extraction field through the slit leading to a more efficient collection of slow ions into the beam.

EXTRACTION OF IONS AND PRIMARY BEAM FORMATION

Two groups of ions are extracted through the slit: fast ions emitted at the cathode and accelerated in the cathode sheath, and slow ions produced mainly by the resonant charge exchange in the vicinity of the extraction slit.

Over the initial part of the curves on Figure 2, for currents up to 20A, the production rate of the first process is dominant. The maximum ratio of the rates for the two processes can be estimated from the slopes of the curves 5 and 7; the latter corresponds to a Penning source where the beam is formed basically by the second process.[4] It is evident that the initial ratio is equal to seven. In this case the source operates as with a solid emitter. Therefore, the plasma boundary in the vicinity of the extraction slit is basically determined by the space charge of electrons and not by the space charge of negative ions. For negative ions, the plasma boundary does not represent an emission surface because they reach it with substantial initial velocities, corresponding to the energy they gain in the cathode sheath (~ 100 eV). It means that the strength of the electric field at the plasma boundary (i.e., inside a transition layer, which is determined by the smaller of the two parameters, Debye length or Larmor radius of electrons) is not equal to zero. As the strength of the electric field inside the plasma can be considered to be zero, the plasma boundary represents, for the flux of negative ions, a thin, strong lens. The character of the lens will depend on the relative position of the centers of curvature of the cathode emitting surface and of the refracting plasma boundary. The lens will be divergent if its center of curvature is far, and convergent if it is close.

For small values of plasma densities the boundary shifts deeper into the emission slit. If the slit is deep enough and has a small angle of opening to the plasma, the resulting lens will be convergent and very nonlinear. The angular scattering of the ion beam while passing through the lens will be limited by the geometry of the emission slit and the extraction gap.

This model is in agreement with the emittance measurements. Beam formation and measurements of its ion-optical characteristics have been described in detail in Ref. 10. In this mode of operation, the normalized emittance perpendicular to the emission slit (along the magnetic field) is $E_x \simeq 10^{-4}$ cm-rad, which corresponds to a transverse energy spread of $\Delta W_x \simeq 10^3$ eV. In the other direction, along the slit, the emittance is $E_y \simeq (6 - 8) \times 10^{-5}$ cm-rad, corresponding to a mean transverse energy spread of ions arriving to the emission slit of $\Delta W_y \simeq 10$ to 15 eV. This is very close to the results in Ref. 12 for a magnetron source, where the energy spread was estimated to be $\Delta W_y \simeq 20$ eV.

Overall, these results are in a satisfactory agreement with the latest measurements of the energy and angular distributions of negative ions leaving the surface of a cesiated emitter in a plasma.[11] From the latter it was estimated that $\Delta W_\perp \simeq 10$ eV.

As the plasma density increases (with the discharge current), the contribution of "cathode" negative ions to the extracted ion beam decreases because of large charge-exchange and destruction losses in the plasma. But the contribution of slow negative ions produced by the charge exchange increases because the density of thermalized hydrogen atoms in the vicinity of the extraction slit increases. Space charge of slow negative ions affects more and more the formation of the plasma boundary. Judging by the shape of the curves around the discharge current of 50A, this effect becomes dominant and the plasma boundary becomes an emitting surface. In this region the source operates as a plasma emitter, similar to a Penning source. The appearance of maxima in curves in Figure 3 (current density vs. discharge current) agrees with such a transition into a plasma emitter mode.

Plasma ion emitters operate in a regime where an equilibrium exists between the ion flux density from the plasma into its boundary, and the ion

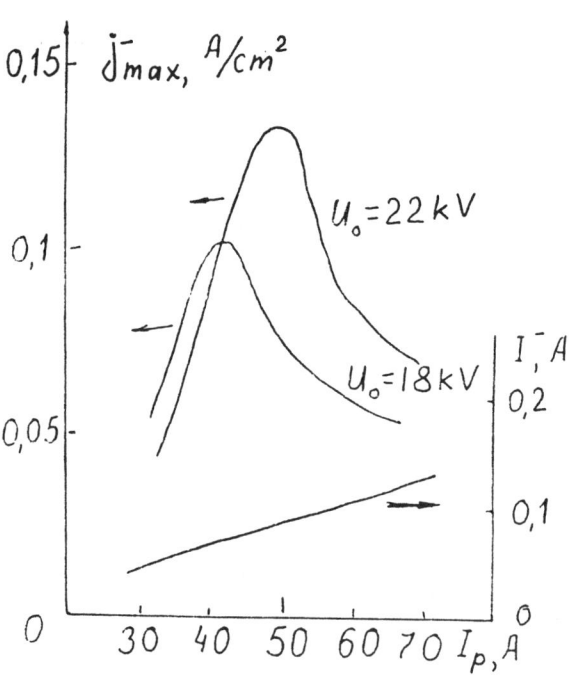

Figure 3. Dependence of the maximum current density in the beam center (j^-_{max}) and the total beam current (I^-) on the arc current (I_p), for two values of the extraction voltage (U_o).

flux density in the space-charge layer, determined by the "3/2" law

$$j^- = \frac{en_1 \langle v_1 \rangle}{4} = A \frac{U_o^{3/2}}{d^2} \qquad (1)$$

where e is the charge, n_1 and $\langle v_1 \rangle$ are ion density and ion mean velocity in the plasma, respectively, U_o and d are the voltage drop and the thickness of the space-charge layer (or extraction gap), respectively. For a fixed voltage and geometry of the extraction gap, the maximum beam current density is achieved for an optimum position and shape of the plasma boundary or, therefore, for an optimum plasma density (discharge current).

It should be noted that the best match of the extraction gap to the space-charge conditions is observed only when the beam intensity fluctuation level is very small (< 1%) and magnetic field weak (500 Gauss).

For the selected geometry of the extraction gap, an optimum ion extraction requires that a plasma boundary, either flat or slightly concave (to compensate for a divergent effect of the slit in the extractor), be formed at the exit of the extraction slit. This is possible only if the magnetic field is weak enough. A strong magnetic field would press the plasma deeper into the emission slit (away from the exit) and formation of an optimum plasma boundary is not possible. Leveling off or appearance of a minimum in beam current characteristics (Fig. 2) is related to a reduction of the plasma diffusion rates across the magnetic field in a narrow channel. It is evident that this behavior is more pronounced if the slit is deeper or the angular aperture of the emission slit smaller.

Transition of the source into the plasma emitter regime, where the extraction of slow negative ions is dominant, corresponds to an improvement of the brightness of the beam. Still, they are not better than the best results obtained for a Penning source.[4] It is clear that there is still a substantial contribution of fast cathode ions in the beam. It was possible to approach the best results only when the depth of the cathode groove was increased from 1 to 2.5 mm. A beam of 80 mA, with a fluctuation level < 0.5%, had emittances, $E_x \simeq 6 \times 10^{-6}$ cm-rad and $E_y \simeq 2 \times 10^{-5}$ cm-rad. Figure 4 shows phase diagrams of beam in this regime of operation.

As it was found in Ref. 4 as well, an increase in the ordered beam divergence that follows a change of the extraction voltage U_o away from the optimum value for a particular emission current density (Fig. 3), causes a substantial increase of E_x and ΔW_x. A pronounced beam spreading due to an ordered diverging should result in a "cooling" - reduction of the local transverse energy spread to a very low level ($\simeq 10^{-3}$eV). But fluctuating electric fields in the beam prevent the achievement of such low transverse "temperatures", with the result that the emittance is very much increased. This is why a minimization of transverse dimensions of the beam and minimization of the fluctuation level are so important to prevent an emittance growth during beam transport.

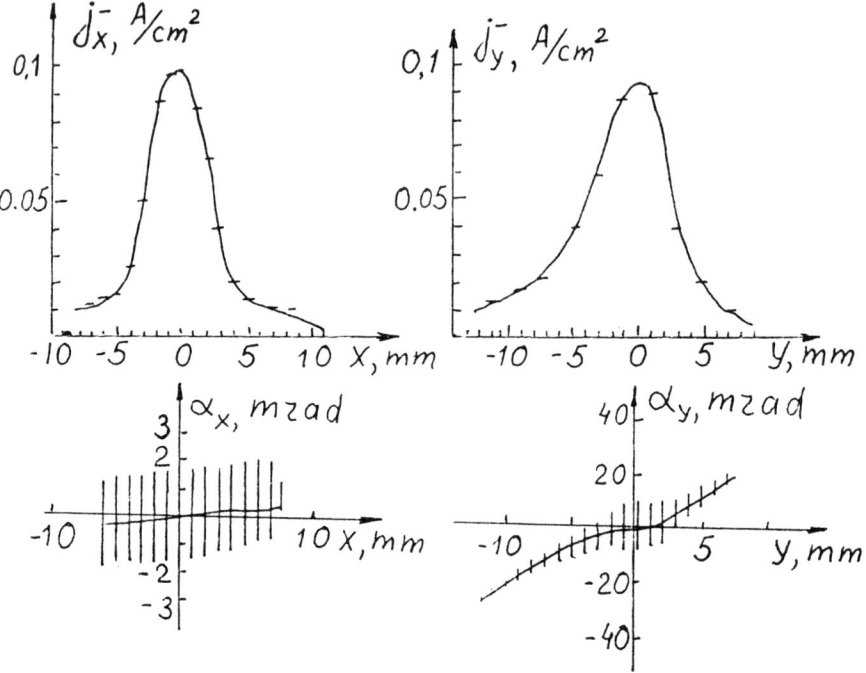

Figure 4. Phase diagrams of the beam, at a low fluctuation level, for a source with a groove depth of 2.5 mm.

$I^- = 80$ mA $\quad E_x = 6 \times 10^{-6}$ cm-rad
$U_o = 23$ kV $\quad E_y = 2 \times 10^{-5}$ cm-rad
$\Delta W_x = 27$ eV $\quad B = 10^8$ A/cm^{2}rad^2
$\Delta W_y = 0.75$ eV

REFERENCES

1. Ankenbrandt, C. Charge-Exchange Injection Systems. Proc. 11th Intern. Conf. on High Energy Accel., Geneva, 1980, p. 260.
2. Belchenko, Yu. I., Dimov, G.I., Dudnikov, V.G. Emission of Intense Fluxes of Negative Ions from a Surface, Bombarded by Fast Particles from the Discharge. Izvestiya AN SSSR, Seriya Fiz. 1973, 37, No. 12, p. 2573.
3. Belchenko, Yu. I., Dimov, G.I., Dudnikov, V.G. Surface-Plasma Sources of Negative Ions. Proc. Sec. Symp. on Ion Sources and Formation of Ion Beams, Berkeley, 1974 (LBL Rep. No. 3399) p. VIII-1.
4. Derevyankin, G.E., Dudnikov, V.G. Production of High Brightness H$^-$ Beams in Surface Plasma Sources, Production and Neutralization of Negative Ions and Beams (3rd Int. Symp., Brookhaven, 1983) AIP Conf. Proc. 111, 376 (1984).
5. Production and Neutralization of Negative Ions and Beams, 3rd Intern. Symp. (Brookhaven), AIP Conf. Proc. No. 111, (1983).
6. Belchenko, Yu. I., Dudnikov, V.G. Surface-Plasma Source without a Closed ExB Electron Drift. Inst. Nucl. Phys. SO AN SSSR, Preprint 78-95, Novosibirsk, 1978.
7. Dimov, G.I., Derevyankin, G.E., Dudnikov, V.G. 100 mA Negative Hydrogen Ion Source for Accelerators, IEEE Trans. Nucl. Sci., NS-24, No. 3, 1545 (1977).
8. Derevyankin, G.E., Dudnikov, V.G., Zhuravlev, P.A. Electromagnetic Valve for Pulsed Gas Injection Into Vacuum. Prib. Tech. Eksp. No. 5, 1975, p. 168.
9. Belchenko, Yu. I., Dudnikov, V.G. Surface-Plasma Source with an Increased H$^-$ Production Efficiency. Inst. Nucl. Phys. SO AN SSSR, Preprint 80-34, Novosibirsk 1980.
10. Derevyankin, G.E., Dudnikov, V.G. Formation of H$^-$ Ion Beams in Surface-Plasma Sources for Use in Accelerators. Inst. Nucl. Phys. SO AN SSSR, Preprint 79-17, Novosibirsk 1979.
11. Wada, M., Pyle, R.V., Stearns, J.W. Work Function Dependence of Surface Produced H$^-$ in the Presence of Plasma, Production and Neutralization of Negative Ions and Beams (3rd Intern. Symp., Brookhaven 1983), AIP Conf. Proc. 111, 247 (1983).
12. Smith, H.V., Allison, P.W. H$^-$ Beam Emittance Measurements for Penning and Asymmetric Grooved Magnetron Surface-Plasma Sources, Rev. Sci. Instrum. 53, No. 4, 405 (1983).

HIGH BRIGHTNESS ION SOURCE: THE PENNING RINGATRON

J. H. Whealton, W. L. Stirling, P. M. Ryan, M. A. Akerman, W. R. Becraft,
W. L. Gardner, H. H. Haselton, K. E. Rothe, M.A. Bell, R. J. Raridon,
D. E. Schechter, L. A. Berry, P. L. Goranson,
J. T. Greer, and W. K. Dagenhart
Oak Ridge National Laboratory
Oak Ridge, Tennessee 37831

A negative ion source is proposed which is a variant of the "hollow cathode" duoplasmatron (HCD)[1] (shown in Fig. 1) and the planar Penning source of Ehlers[2] and Stirling[3] (shown in Figs. 2 and 3). A basic description of the source is shown in Fig. 4. The top view of a transverse extraction[2,3] source is shown in Fig. 4, where a excited neutral from the primary Penning discharge, D1, undergoes dissociative attachment in the secondary Penning discharge, D2 (which has a low electron temperature[6]). The first issue relates to ions extracted across a magnetic field. Only the high temperature ions get extracted from a significant depth into the discharge; the low temperature ions deep in the discharge are left to circle around aimlessly until they undergo collision, which probably also results in electron detachment. To wit, extraction across a magnetic field preferentially selects the high temperature ions.[7] In order to avoid such selection, in the context of a Penning discharge one might wish to extract parallel to the magnetic field.[1,4,8] Furthermore, the geometry which may lead to enhanced production density is also shown in Fig. 4; this ring geometry has been shown to increase negative ion current a significant factor over a more simple geometry.[4,8]

It is expected that there is a rather sharp fall off a short distance away from the primary Penning discharge, D1, where negative ion net production is maximized[4,5] (illustrated in Fig. 5). Therefore a hollow discharge like Ref. 1, illustrated in Fig. 1, is not optimum since the distance from the axis to the discharge varies and so does not remain on the peak of net production of negative ions. Furthermore, the negative ions so generated have to go through the gauntlet of the tail end of the Penning discharge, D1, on the way out (see Fig. 1). Nevertheless, an impressive current density of 224 mA/cm^2 has been achieved for the extraction of H$^-$ ions.[1c]

Both of the difficulties mentioned above are eliminated by use of the cylindrical configuration (ringatron) of Fig. 4. Namely, the distance from the axis to the discharge, D1, is kept constant at the optimum value and the negative ions once generated do not have to run the gauntlet of D1 on the way out (illustrated in Fig. 1b).

---

*Research sponsored inpart by the Department of Defense under Interagency Agreement DOE No. 40-1442-84 and Army No. W31RPD-53-A180 with the Office of Fusion Energy, U.S. Department of Energy, under Contract No. DE-AC05-84OR21400 with Martin Marietta Energy Systems, Inc., and in part by Energy Systems IRAD support.

In the ringatron the percent ionization fraction of D1 is also expected to be smaller since the uniform-B Penning discharge operates at a lower density ($\sim 5 \times 10^{13}$ cm^{-3}) than that of the constricted duoplasmatron discharge ($\sim 5 \times 10^{14}$ cm^{-3}). The diminshed ionization fraction of the ringatron may increase the negative ion density by maximizing the excited H_2 population.

Another possible advantage of the ringatron over the duoplasmation is the pulse length of negative ion production. The hollow cathode duoplasmation[11] as well as some of the higher current density off axis extraction[12] sources runs for less than a millisecond. On the other hand, the transverse extraction Penning sources generally run for much longer lengths of time ($\sim$ sec). This is due in part to the lower arc density of D1 in the transverse extraction sources. It appears that the feature of long pulse length will obtain for the ringatron negative ion source.

Yet another advantage of the ringatron over the HCD is the lack of coherent oscillation[13] in the discharge with electrostatic end confinement[9] compared with magnetic mirror confinement as in the duoplasmatron.[8] Therefore, the emittance of the ion beam is expected to be less with a ringatron than with a duoplasmatron. One possible source of ion acoustic waves in the duoplasmatron that would not be present in the electrostatic end Penning discharge (as in the ringatron) is the flagrant violation of the Bohm sheath instability at the magnetic mirror end. In this case the anode need not be biased as negative as it would be in a purely electrostatic trap.

Extraction of electrons along the field lines remains an issue; presumably they can be cheated of some of their just deserts by suitable orientation of electric and magnetic fields.

The above comparisons were made between the ringatron configuration and the duoplasmation; now a comparison will be made between the ringatron and the planar Penning discharge configuration.[2,3,9] The ringatron has a possible current density enhancement over the planar transverse extraction Penning source[2,3,9] due to both the increased negative ion density from the wraparound ion column and the fact that deep extraction of cold negative ions is possible due to parallel extraction.

An embodiment is shown below in Fig. 6: filament 1, anode 2, reflector 3, extraction electrode 4, water cooling channels 5, gas feeds 6, potential controller and shield 7, and solenoid 8.

Since the current density achieved with a HDC is 230 mA/cm^2 [1c] and the current density obtained with the planar discharge[10] is 108 mA/cm^2, improvements over this represent significant current densities. The addition of cesium would probably increase the current density output to greater than 1300 mA/cm^2.[1c] An argument can be made that the current density achievable in the ringatron configuration is at least $5 \times 108 = 540$ mA/cm^2 (without cesium and 6 times more with cesium); that argument is as follows. The difference between the duoplasmatron and plasma Penning discharge current density is possibly the trade-off between depth of extraction in the transverse extraction Penning discharge and the gauntlet of D1 in the duoplasmatron (illustrated in Fig. 1). The amplification achieved in the hollow configuration is (~ 5) for the duoplasmatron. Therefore, at least the same amplification should occur in the ringatron case; furthermore, since the depth of extraction and the absence of the gauntlet of D1 effects were neglected, the extractable current density should further increase.

Another negative ion source, denoted as the reverse tandem plasmatron, considered is also a variant of the HCD[1] and the ringatron. It is proposed to ameliorate some of the potential problems of the ringatron. Figure 7 shows a simplified embodiment indicating the filament, 1, anode, 2, reflector, 3, magnet poles 4N, 4S, plasma electrode, 5, extraction electrode, 6, and water, 7, and gas, 8, lines. The magnet is arranged to produce the B field indicated along which lies the inhomogeneous Penning discharge (as in a plasmatron) - designed to excite neutrals. The low electron temperature part of the tandem discharge,[5] interior to the inhomogeneous Penning discharge, P2, is confined electrostatically at both ends like the transverse extraction planar Penning negative ion source.[2,3] This confinement of the P2 plasma is greater than in the ringatron because of the magnetic field configuration (a configuration which also helps to quickly deter potentially escaping electrons). The configuration appears somewhat similar to the HCD[1] but the negative ions are able to be extracted through a low T_e plasma instead of being forced to go across the intense plasmatron discharge[1]. This is reversed from the HCD. Much of the potential gain of the ringation over the plasma Penning source[2,3,9] is preserved in this design except for the available depth of extraction. In addition, the outer discharge density may by higher than the cylindrical ringatron since the discharge is constricted; a concomitant effect may be the stability of said discharge and a higher ionization fraction.

An example of self-consistent extraction sheath-optics calculation is shown in Fig. 8.

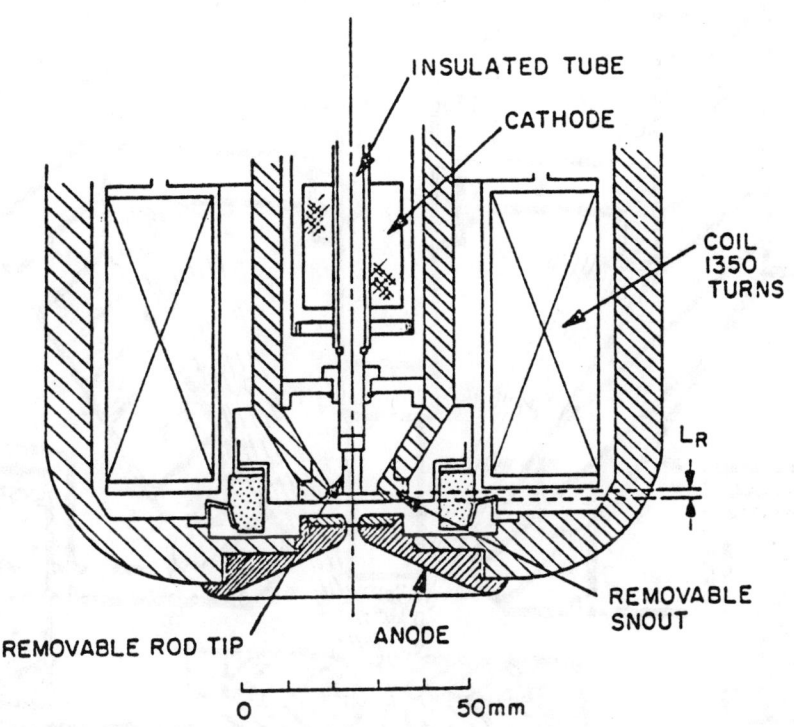

Cross section of the HDD source.

From K2-76 of Ref. 8

Fig. 1A

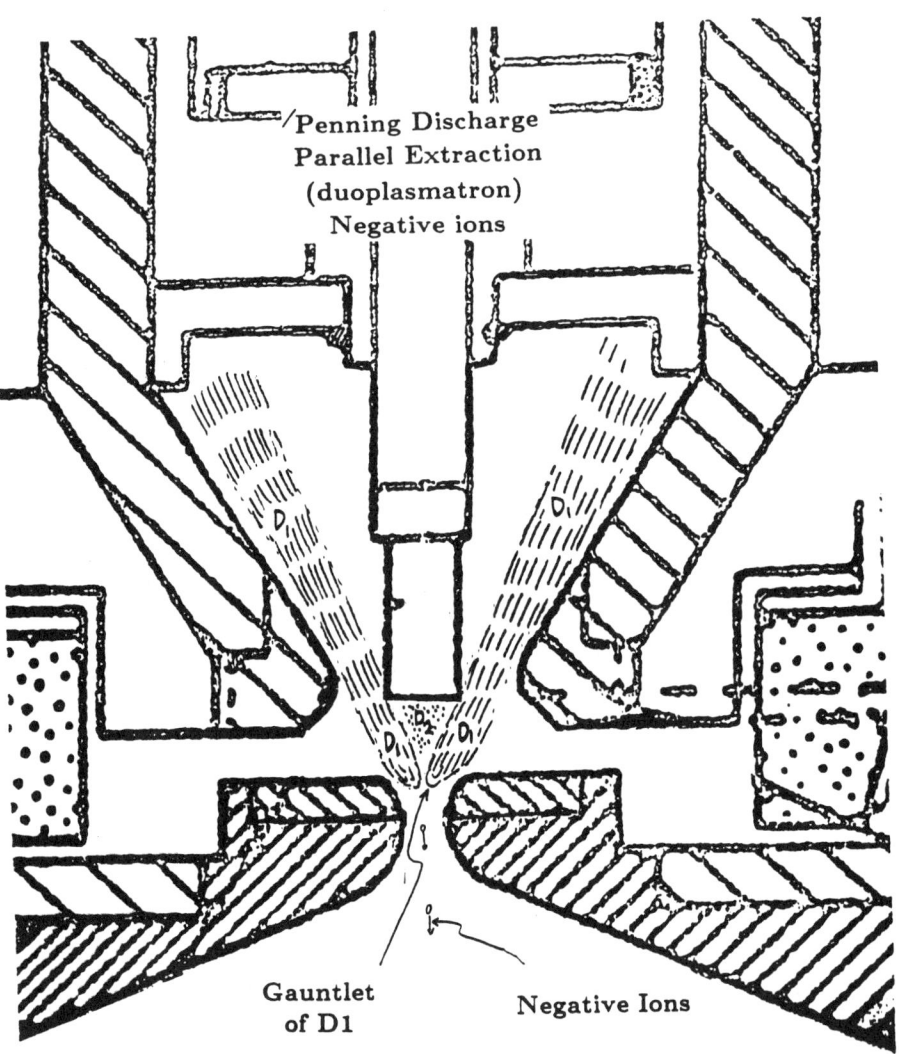

Fig. 1B

Penning Discharge
Parallel Extraction
(duoplasmatron)
Negative ions

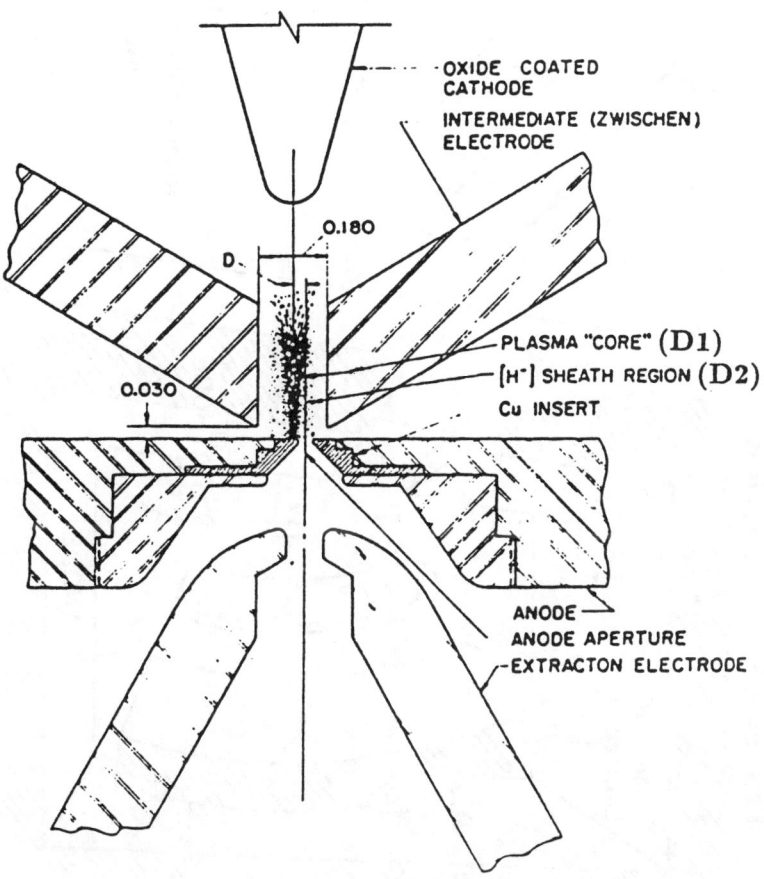

Critical region of the duoplasmatron.

From L-65 of Ref. 8

Fig. 1C

410

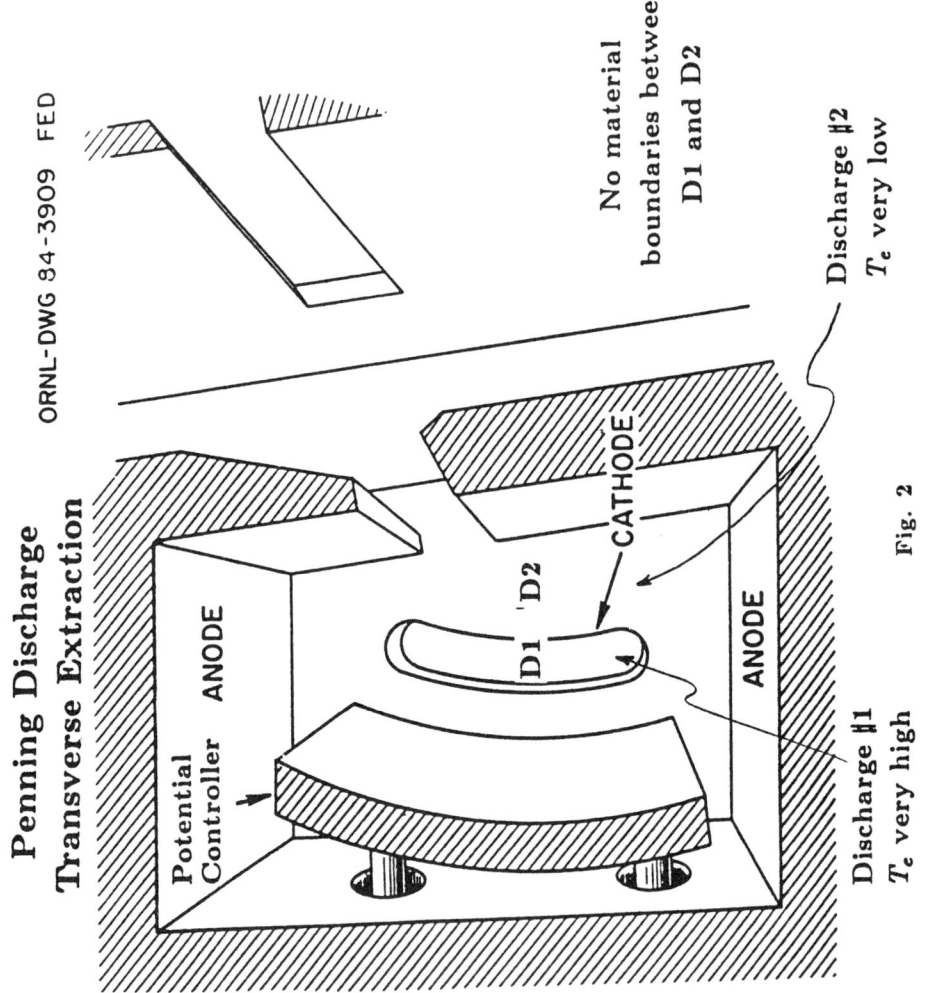

Fig. 2

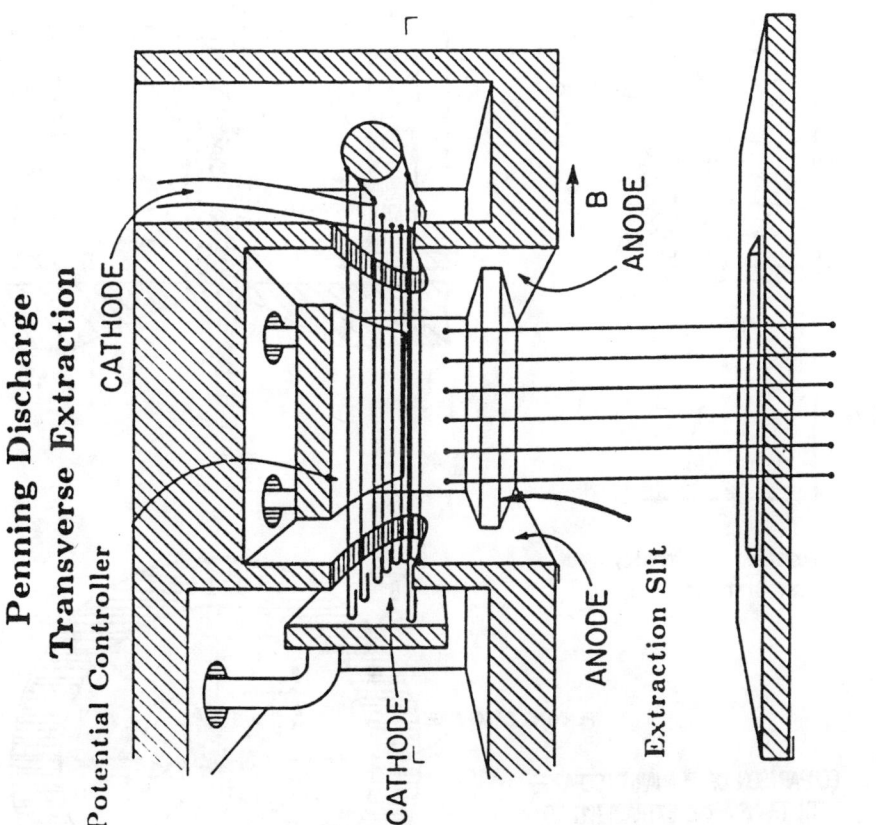

Fig. 3

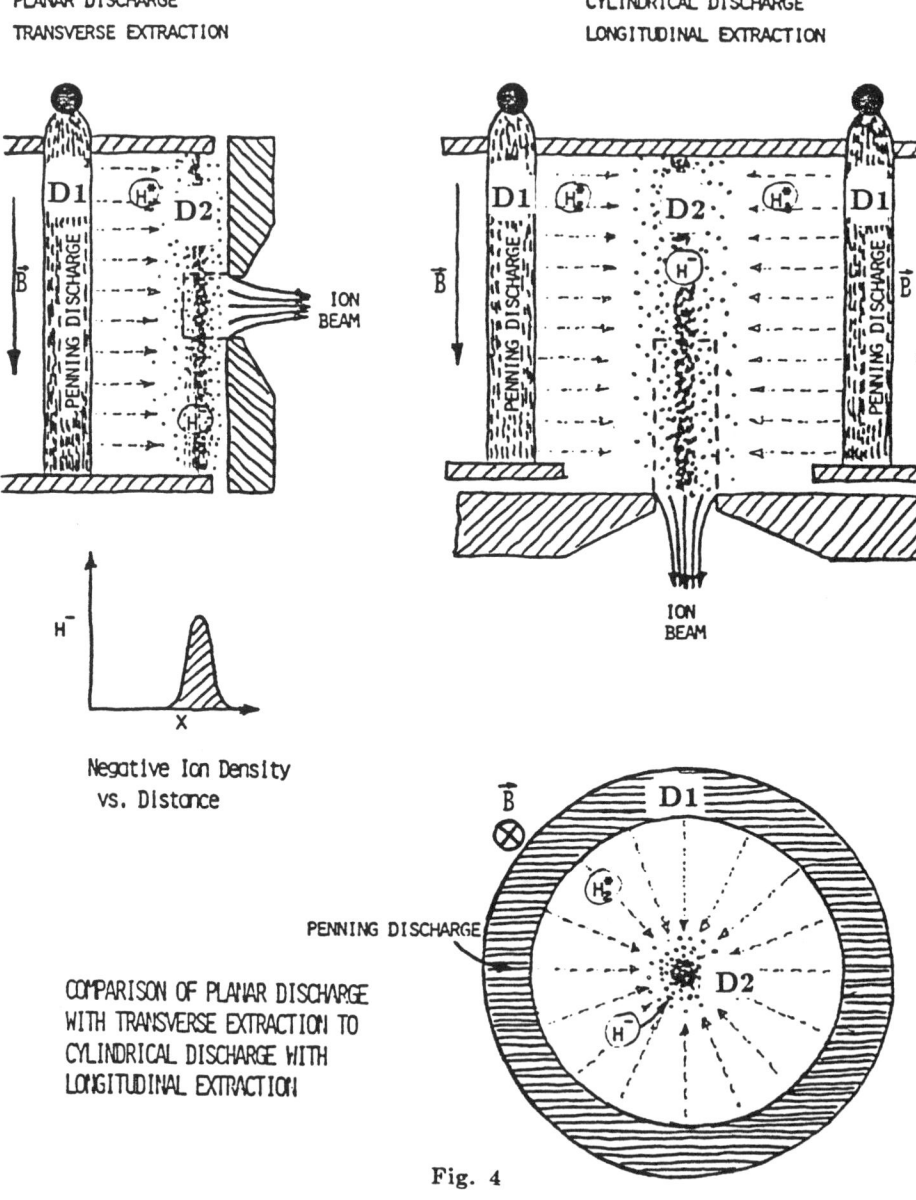

Fig. 4

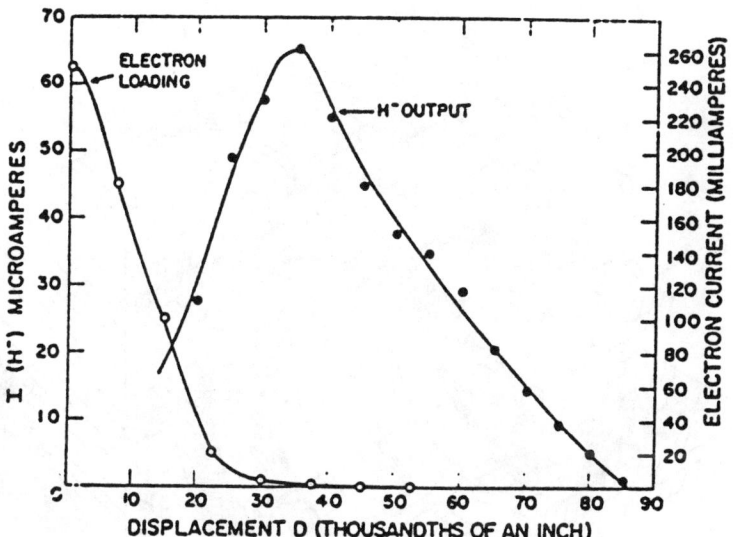

H⁻ output and electron loading as a function of radial displacement.

From L-65 of Ref. 8

Fig. 5

THE RINGATRON
(a new ORNL negative ion source concept)

ORNL-DWG 86-2440A2 FED

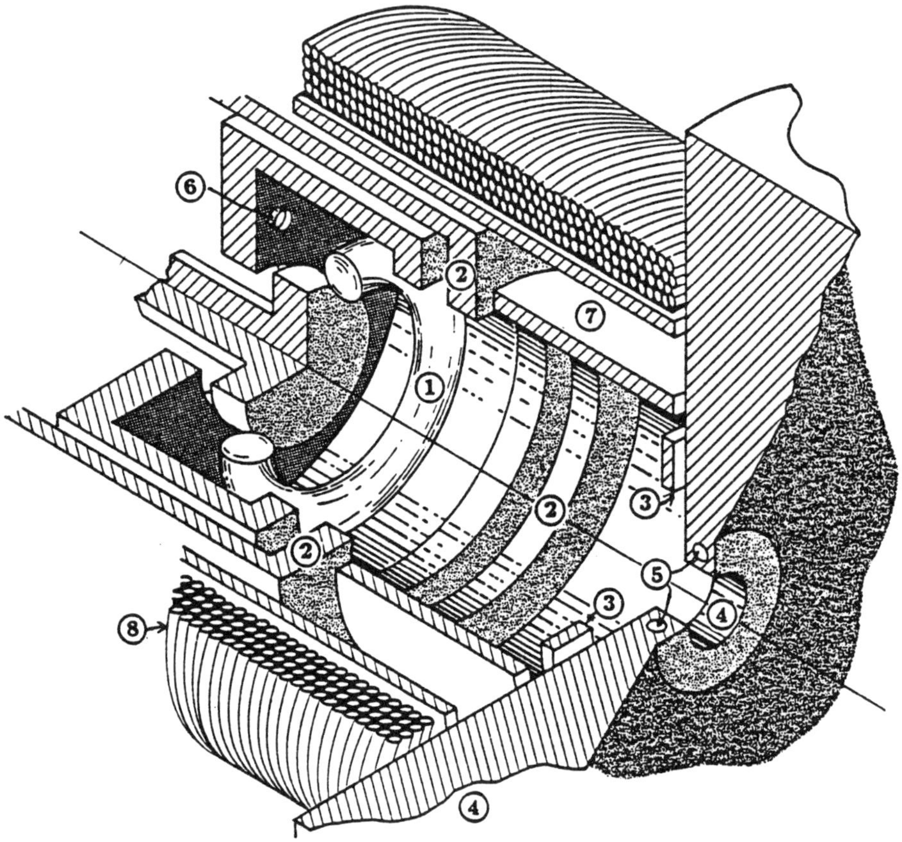

Fig. 6

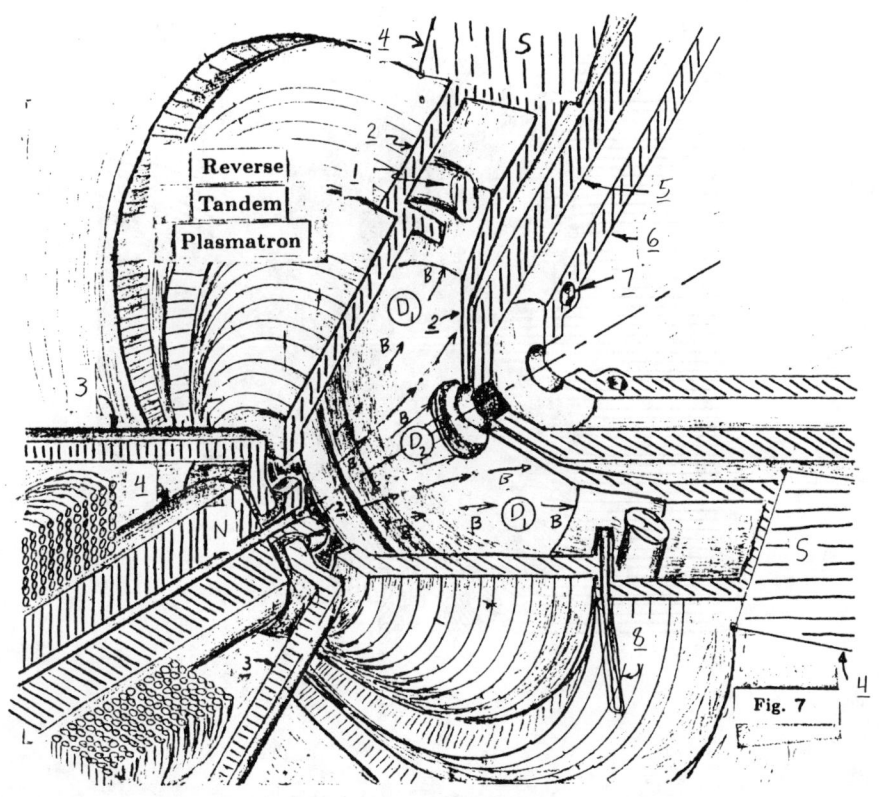

Fig. 7

Ringatron

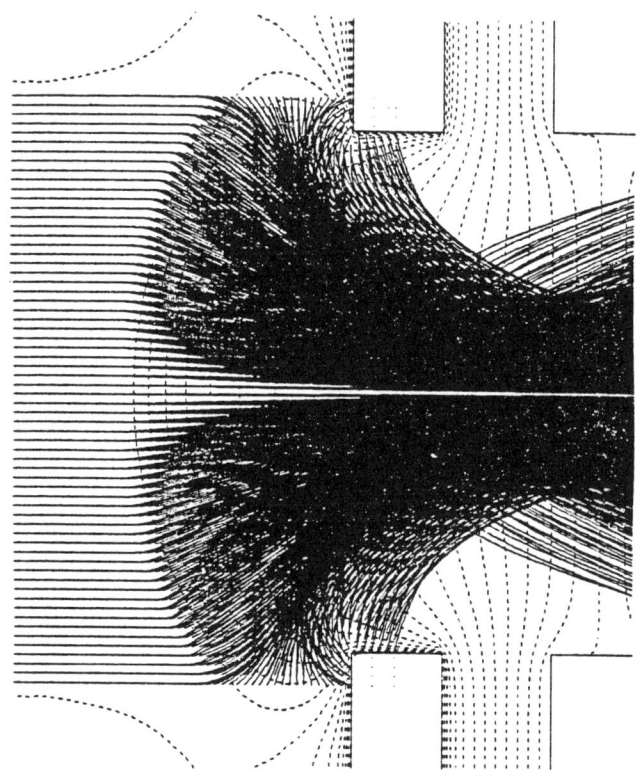

Penning Discharge
Parallel Extraction

Fig. 8

REFERENCES

1. V. P. Golubev, G. A. Nalivaiko, and S. G. Tsepakin, Proc. Proton LINAC Conf., LANL (1972) p. 356; T. Slüyters and K. Prelec, Nuc. Instrum. Meth. **113**, 299 (1973). The development of this concept is indicated in Fig. 21. M. Kobayashi, K. Prelec, and T. Slüyters, Rev. Sci. Instrum. **47**, 1425 (1976).

2. K. W. Ehlers, B. F. Gavin, and E. L. Hubbard, Nucl. Instrum. & Meth. **22**, 87 (1963). The development of this concept is shown in Fig. 21.

3. W. L. Stirling, W. K. Dagenhart, and C. C. Tsai, ORNL-TM/9753 (1986).

4. G. P. Lawrence et al., Nucl. Instrum. Meth. **32**, 357 (1965).

5. A. M. Karo, J. R. Hiskes, and R. J. Hardy, J. Vac. Sci. Technol. **3**, 1222 (1985); J. R. Hiskes, A. M. Karo, and P. A. Willmann, J. Appl. Phys. **58**, 1759 (1985); J. R. Hiskes, A. M. Karo, and P. A. Willmann, J. Vac. Sci. Technol. **3**, 1229 (1985); C. Garse, M. Capitelli, J. Bretagne, and M. Bacal, Chem. Phys. **93**, 1 (1985); O. Fukumasa, S. Saedi, J. Phys. **D18**, L21 (1985); M. Pealat, J-P.E. Taran, M. Bacal, and F. Hillion, J. Chem. Phys. **82**, 4943 (1985); A. M. Bruneteau and M. Bacal, J. Appl. Phys. **57**, 4342 (1985); J. R. Hiskes and A. M. Karo, J. Appl. Phys. **56**, 1927 (1984); K. N. Leung and M. Bacal, Rev. Sci. Instrum. **55**, 338 (1984); J. R. Hiskes, A. M. Karo, M. Bacal, A. M. Bruneteau, and W. G. Graham, J. Appl. Phys. **53**, 3469 (1982); M. Bacal, Physica Scripta T2, 467 (1982); M. Bacal, A. M. Bruneteau, W. G. Graham, G. W. Hamilton, and M. Nachevan, J. Appl. Phys. **52**, 1247 (1981); J. R. Hiskes, J. Appl. Phys. **51**, 4592 (1980); M. Bacal and G. W. Hamilton, Phys. Rev. Lett. **42**, 1538 (1979); M. Bacal, G. W. Hamilton, A. M. Bruneteau, and H. J. Doucet, Rev. Sci. Instrum. **50**, 719 (1979); E. Nicolopoulou, M. Bacal, and H. J. Doucet, J. Physique **38**, 1399 (1977).

6. C. C. Tsai et al., private communication (1985).

7. J. H. Whealton, Negative Ion Source Workshop, LANL, (1985).

8. K. Wiesemann, K. Prelec, and T. Sluyters, Jour. Appl. Phys. **48**, 2668 (1977) W3-77; B. Bouchy-Lorentz, M. Remy, and C. Muller, Inter. J. Mass Spec. and Ion Phys. **22**, 147 (1976) B-76 ; M. Kobayashi, K. Prelec, and T. Sluyters, Rev. Sci. Instrum. **47**, 1425 (1976) K2-76; C. Muller, B. Lorentz, and M. Remy, Int. J. Mass Spec. and Ion Phys., **18**, 33 (1975) M2-75; W. Aberth, R. Schmitzer, F. Engeaser, and M. Aubar, Ion Source Conf. LBL (1974) A2-74 ; T. Sluyters and K. Prelec, Nuc. Instrum. Meth. **113**, 299 (1973) S-73;A. Cesati and F. Christofori, Energia Nucleare **19**, 36 (1972) C2-72; M. A. Abroyan, G. A. Nalivaiko, and S.G. Tsepakin, Sov. Phys. -Tech. Phys. **17**, 690 (1972) A-72; A. Cesati and F. Christofori, Energia Nucleare **19**, 36 (1972) C2-72; V. P. Bolubev, G. A. Nahvaiko, and S. G. Tsepakin, Proc. Proton LINAC, Los Alamos (1972) G-72; M. Dubarry, These, Orsay (1972) D-72; M. A. Abroyan, V. P. Golubev, V. L. Komarov, G. V. Tarvid, G. M. Tokarev, and S. G. Tsepakin, Particle Accelerators **2**, 133 (1971) A-71;N. Wells and P. R. Hauley, Ion Source Conf. (BNL, 1971) W2-71; V. L. Komarov, Z. G. Solnishkova, G. V. Tarvid, and S. G. Tsepakin, Ion Source Conf.,Saclay (1969) K-69; M. Dubarry and G. Gautherin, Ion Source Conf., Saclay (1969) B-69;V. P. Golubev, V. L. Komarov, and S. G. Tsepakin, Ion Source Conf., Saclay

(1969), G-69; M. A. Abroyan and G. M. Tokarev, Ion Source Conf. (1969) Saclay (A-69); W. Aberth and J. R. Peterson, Rev. Sci. Instrum. **38**, 745 (1967) A2-67; K. Bethge and G. Rau, Nuc. Instrum. Meth. **39**, 157 (1966) B-66; L. E. Collins and R. H. Gobbett, Nuc. Instrum. Meth. **35**, 277 (1965) L-65; G. P. Lawrence, R. K. Beauchamps, and J. L. McKibben, Nuc. Instrum. Meth. **32**, 357 (1965) L-65; A. B. Wittbower, P. H. Rose, R. P. Bastide, and N. B. Brooks, Rev. Sci. Instrum **35**, 1 (1964) W-64; A. B. Wittbower, R. P. Bastide, M. B. Brooks, and P. H. Rose, Phys. Lett **3**, 336 (1963), W-63; C. D. Moak, H. E. Banta, J. N. Thurston, J. W. Johnson, and R. F. King, Rev. Sci. Instrum. **30**, 694 (1959) M-59; (seminal work is indicated in Fig. 21).

9. W. L. Stirling, W. K. Dagenhart, and C. C. Tsai, ORNL/TM-9753 (1986) S3-86; K. Jumbo, K. W. Ehlers, K. N. Leung, and R. V. Pyle, Nuc. Instrum. Meth. **A248**, 282 (1986) J-86; M. D. Gabovich, Y. N. Kozyrev, A. P. Naida, L. S. Simaveuko, and I. A. Solosheuko, Sov. Tech. Phys. Lett. **4**, 153 (1978) G2-78; M. Baribaud, S. Bliman, J. M. Dolique, and F. Zadwarny, Ion Source Conf., Saclay (1969) B3-69; K. W. Ehlers, Nuc. Instrum. Meth. **32**, 309 (1965) K-65; K. W. Ehlers, B. F. Gavin, and E. L. Hubbard, Nuc. Instrum. Meth. **22**, 87 (1963) E-63; K. W. Ehlers, Nucl. Instrum. Meth. **18**, 571 (1962) E-62; E. L. Hubbard et al., Rev. Sci. Instrum. **32**, 621 (1961) H-61; R. J. Jones and A. Zucker, Rev. Sci. Instrum. **25**, 562 (1959) Z-54; C. E. Anderson, K. W. Ehlers, Rev. Sci. Instrum. **27**, 809 (1956) A3-56; R. S. Livingston, R. J. Jones, Rev. Sci. Instrum. **25** 552 (1954) L2-54; M. S. Livingston, M. G. Holloway, and C. P. Baker, Rev. Sci. Instrum. **10**, 63 (1939) L2-39; F. M. Penning, Physica **4**, 71 (1937) P2-37; (seminal work is indicated in Fig. 21).

10. C. C. Tsai et al., private communication (1986).

11. G-72, S-73, and K2-76 of Ref. 8.

12. A-72 of Ref. 8.

13. M. Kobayashi and A. Takagi, Ion Source Conf. (1974) LBL.

A CIRCULAR APERTURE MAGNETRON FOR INJECTION INTO AN RFQ*

James G. Alessi
AGS Department, Brookhaven National Laboratory
Associated Universities, Inc., Upton, New York, 11973

ABSTRACT

A magnetron with a circular anode aperture and a spherical dimple in the cathode has been operated. With this configuration, a normalized emittance (90%) of 0.1πcm-mrad has been measured in both planes for an H^- current of > 50 mA. Other than this symmetric emittance, the source performance is the same as with the typical anode slit and grooved cathode.

INTRODUCTION

The AGS will be replacing the 750 keV Cockcroft-Walton (C-W) preinjector with an RFQ accelerator for H^- injection into the 200 MeV linac. For this, we need a 50 mA, 35 keV H^- source having an emittance which can be properly matched to the RFQ acceptance. It is preferable to have the x- and y-emittances identical, and then transport the beam to the RFQ using magnetic solenoids to preserve the symmetry. The AGS routinely operates with a magnetron surface-plasma source. These sources normally run with a slit aperture, and have emittances which are considerably different in the two planes. While modifications to operate with a circular aperture were straightforward, there was some question as to the effectiveness of cathode focussing from a dimple. Electron loading from the circular aperture was also an uncertainty since the slit, with the narrow dimension parallel to the magnetic field, might be more effective in reducing the extracted electron current. The results of operation with the circular aperture, and the comparison with operation for the slit aperture, will be given in the following sections.

EXPERIMENTAL SET-UP

For all tests, the magnetron source was the same as is used for normal AGS operation.[1] Figure 1 shows a schematic of the source test stand. Following the source, the beam travels through a 90° gradient bending magnet. This magnet provides beam focussing, and reduces the gas load and possibility of cesium contamination of the 750 kV accelerating column when installed in the C-W. Horizontal and vertical emittances can be taken with slit and collector devices 8.3 cm beyond the exit pole of the magnet. A Faraday cup can be inserted approximately 1 cm from the extractor. A second Faraday cup, 15 cm from the exit of the magnet reads the transported current. The test box is pumped by a 1500 ℓ/sec cryopump.

---

*Work performed under the auspices of the U.S. Department of Energy.

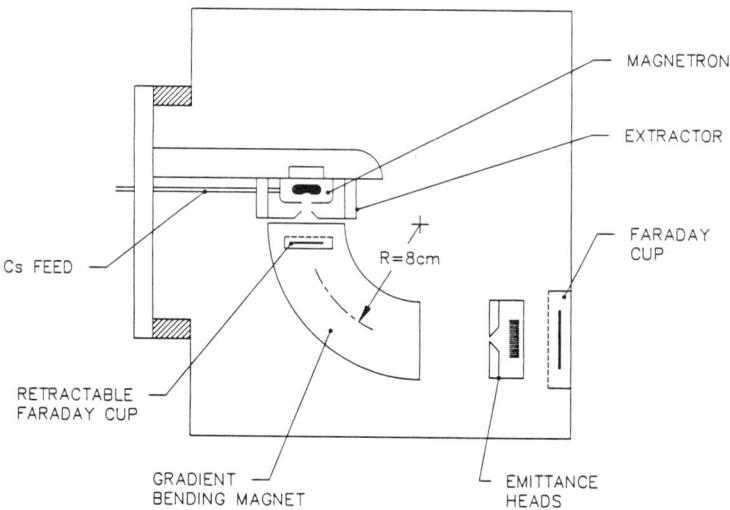

Figure 1. Schematic of the source test stand.

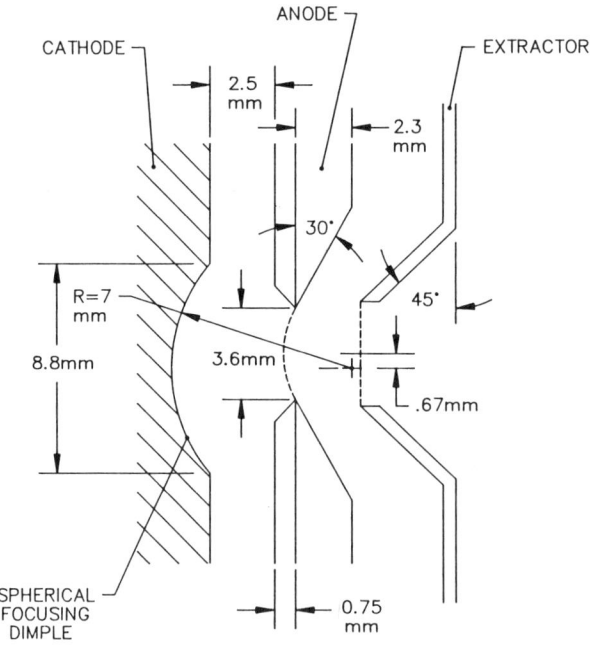

Figure 2. Geometry of the circular aperture magnetron.

The normal slit source has a 1 mm × 1 cm anode slit, and a 4 mm wide cathode focussing groove. The extractor plate opening is 2 mm, with a 2 mm extractor gap. A schematic of the circular aperture geometry used for the present experiments is shown in Figure 2. The cathode dimple was offset in one plane to compensate for the deflection of H$^-$ ions by the source magnetic field while travelling between the cathode and anode. The anode aperture diameter was chosen so as to have the same area as the aperture in the slit source (0.1 cm^2). In this way, gas loading, H$^-$ output vs. discharge current, electron loading, etc. could be directly compared between the two.

RESULTS

Figure 3 shows an emittance measured with the normal slit source and a gradient of n = 1.35 in the 90° magnet. This is the magnet normally used with the C-W. The normalized emittances for a 45 mA (450 mA/cm^2), 17.75 keV beam were 0.049 and 0.177πcm-mrad for 90% of the beam. Although the measured emittances are somewhat larger than some previous measurements,[1] the ratio of horizontal (x; = narrow slit direction) and vertical (y; = wide slit direction) emittances are the same as for the previous measurements, and the emittance orientations are reproduced. The orientations are reasonable for the injection into quadrupole focussing elements, as is presently done in the C-W. The emittances in the two planes are not in the ratio that one might expect from the source geometry, and the differences are normally attributed to some coupling of the emittances in the two planes by the magnetic field, aberrations in the extraction optics in the narrow slit direction, etc.

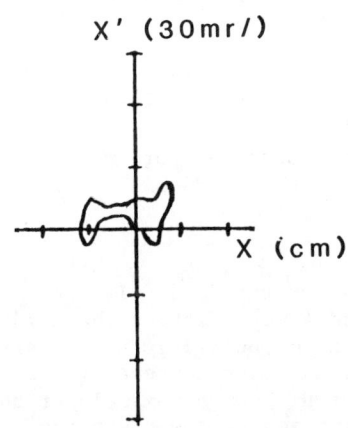

E(x) = 0.049πcm-mr
(normalized, 90%)

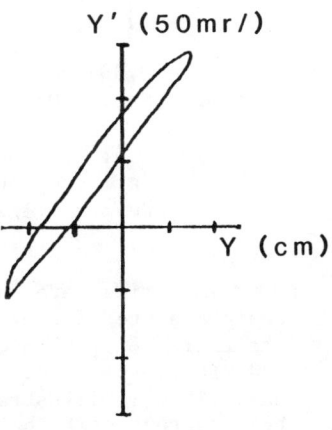

E(y) = 0.177πcm-mr
(normalized, 90%)

Figure 3. Emittances measured after the bending magnet for the normal magnetron used with the Cockcroft-Walton. The source has a grooved cathode and a 1 mm × 1 cm anode aperture. I(H$^-$) = 45 mA; V = 17.75 kV; n = 1.35

The source having the circular geometry was then installed. The magnet poles were replaced with ones having a gradient of n = 1/2, to preserve the expected symmetry of the beam. For simplicity, first tests were made with an 88% transparent tungsten grid over the anode aperture (to prevent any plasma boundary effects) and over the extractor aperture (to eliminate aperture defocussing). Emittances taken under these conditions for a 36 mA, 20 keV beam are shown in Figure 4. The normalized (90%) emittances were 0.094 and 0.101π cm-mrad. There was some scraping of the beam on the magnet poles in the x-direction, which probably accounts for the slightly smaller emittance in that plane. Other than this beam scraping, and some misalignment, the emittances are symmetric. Essentially the same emittances as those shown were still obtained at a beam current of 70 mA.

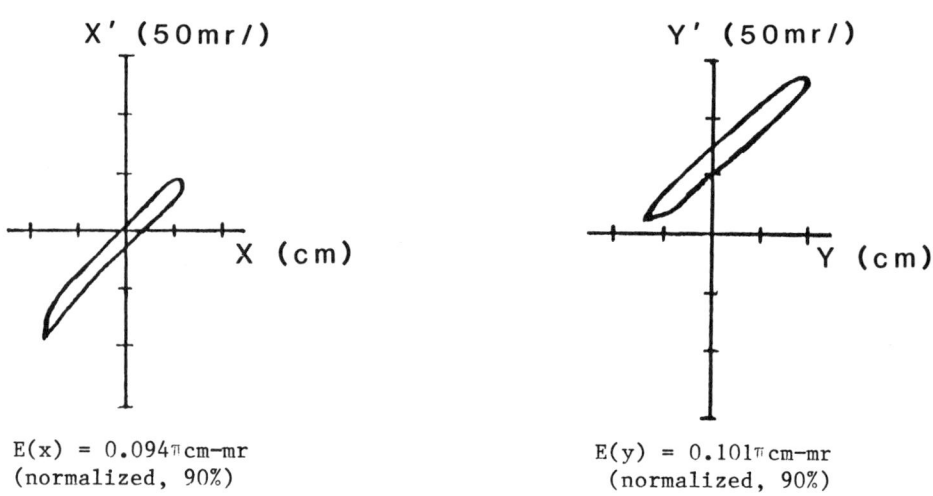

Figure 4. The emittances from the circular aperture source (grids on anode and extractor). $I(H^-)$ = 36 mA; V = 20 keV; n = 1/2.

The grids were then removed, and beam current and emittance again measured for extractor gaps of 3 and 4 mm. Without the grids the extractor perveance was lower, and a beam could only be focussed and transported around the bending magnet at lower currents (< 25 mA). This is illustrated in Figure 5, which shows an example of the beam current on both Faraday cups as a function of the discharge current. The emittances taken at the current giving optimum transmission were the same as the emittances with the gridded extractor (Fig. 4). Since these measurements were taken at ≤ 20 keV, the perveance of the ungridded extractor should be high enough to transport the desired 50 mA at the final extraction voltage of 35 keV.

The total extractor power supply loading was measured, and the difference between the power supply loading and the current measured

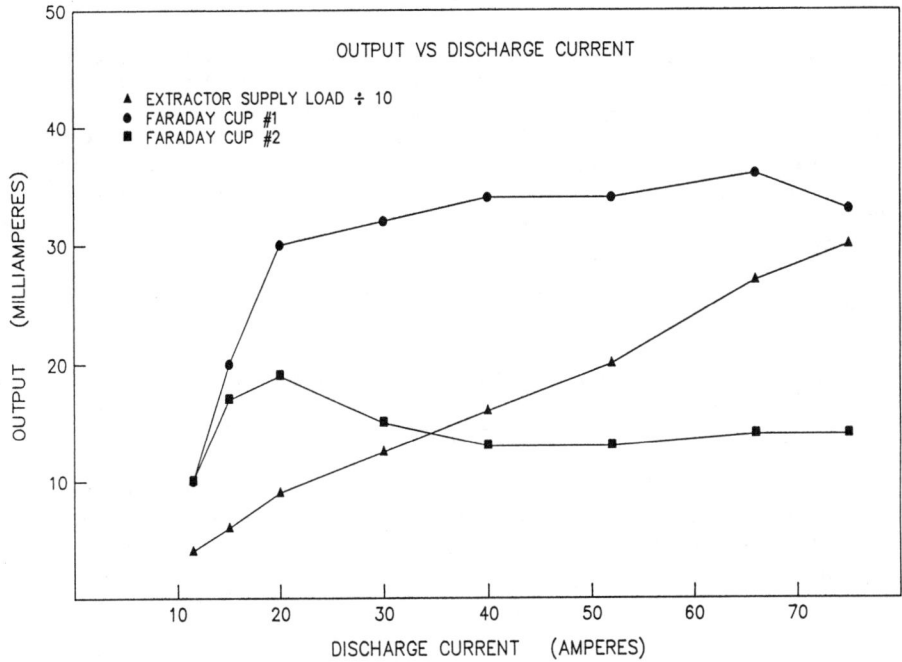

Figure 5. Measured currents vs. discharge for the circular aperture, without the extractor grid (n=1/2 magnet).

on the first Faraday cup was assumed to be electrons. Typically, the ratio of electrons to H$^-$ current was 1.5-2.0, and there was no significant difference in this ratio among the slit source, the circular aperture with grids, or the ungridded circular source when below the perveance limit. When operating without the extractor grid, and comparing operation with and without a grid over the anode aperture, the H$^-$ current, electron loading, and emittance were the same in both cases. This indicates that under the present operating conditions plasma boundary effects are unimportant.

When the slit source was tested with the n=1/2 magnet, one was able to get essentially 100% transmission around the magnet (this is not the case with the slit and n=1.35 magnet). The emittance measured for a 55 mA, 18.5 keV beam was 0.065 × 0.105π cm-mrad (normalized, 90%). This is different than the emittance given at the beginning of this section for the slit source and n=1.35 magnet.

SUMMARY

We have obtained, with the circular aperture magnetron, a symmetric emittance well suited for injection into an RFQ. Our plans are to next study the beam transport, looking for any emittance growth from beam instabilities in the 2 m transport line. We will then make emittance measurements without the 90° bending magnet to see if operation in that simpler configuration is a possibility. Finally, there will be a more careful design of the extractor for 35 keV operation.

ACKNOWLEDGEMENTS

I would like to give special thanks to Tom Russo for his work on the preparation and operation of the ion source test stand.

REFERENCES

1. R.L. Witkover, Proceedings of the Third Inter. Symp. on the Prod. and Neutral. of Neg. Ions and Beams (Brookhaven, 1983), AIP Conf. Proc. No. 111, p. 398.

OPERATION OF THE FERMILAB H⁻ MAGNETRON SOURCE

C. W. Schmidt and C. D. Curtis
Fermi National Accelerator Laboratory*
P.O. Box 500, Batavia, IL. 60510

ABSTRACT

Installation of a grooved cathode and addition of a resistive heater around the source body have greatly improved the performance of the magnetron source. Lifetimes of nine months with stable pulsed beams of 50 mA have been achieved. Improvements and operation of the magnetron source at Fermilab will be presented. Recent tests with lanthanum hexaboride as the cathode will be discussed.

INTRODUCTION

The Fermilab linac and booster were converted to H⁻ operation in early 1978. The conversion was made to provide charge-exchange injection into the booster accelerator and to accommodate the cancer therapy facility and electron cooling ring on a time-sharing basis. Since this time two 750-keV Cockcroft-Walton preaccelerators, each using an H⁻ magnetron source, have been used to provide a beam to the linac. The H⁻ magnetron source as initially constructed and operated at Fermilab was described in 1979 and 1980[1,2]. It is an outgrowth of sources originally developed at Novosibirsk and studied at Brookhaven by the Neutral Beam Group. Following this early experience several other laboratories have used this source so that considerable experience and several improvements have occurred. Operation and improvements of the source at other laboratories have been reported recently[3,4]. These improvements have led to a dramatic increase in the source lifetime, stability and performance.

ION SOURCE

In 1981 a problem developed with one of the two Fermilab preaccelerators which prevented its normal operation after 8 to 12 hours from start-up. Normally in a good source, once cesium has entered, the source starts in a low plasma-current (few amps), high plasma-voltage (> 200 V) mode. After several hours the current rises to 140 Amps while the voltage decreases to 140-150 Volts as the cesium reaches optimum condition, the cathode becomes hot (400-500 °C) and the source hydrogen pressure is adjusted (decreased). A good source operates 1-2 months at this level producing 40-50 mA of H⁻ ions after 12-24 hours of conditioning and careful adjustment of the source parameters. In the problem system the high current, low voltage mode would occur but soon after revert to a high voltage

---
*Operated by Universities Research Association, Inc., under contract with the U. S. Department of Energy.

condition with a very unstable plasma. Maintaining even unstable operation required high gas pressure and resulted in very erratic low-current H⁻-ion beams making the source unusable. Fortunately during this period sources continued to work well in the second preaccelerator system. For over a year the problem in the 'bad' system persisted even though magnetron assemblies from the good, bad and test bench systems were interchanged.

The cause of the source failures was speculated to be a contaminant in the vacuum system which poisoned the source surfaces after several hours of operation. Investigations using optical spectra analysis of the plasma emissions, surface analysis of the source and system, and residual gas analysis produced no satisfying results. The vacuum system was thoroughly dismantled and cleaned but still the failure persisted.

During this period, Witkover, at Brookhaven experienced a similar problem[5], possibly due to different causes, and at the suggestion of Sluyters[6] grooved the source cathode as a possible solution. The grooved cathode not only solved the problem but gave superior performance over earlier source operation. Meanwhile at Argonne, Stipp also used a grooved cathode to achieve improved source performance[7]. Following these successes a grooved cathode was installed in the Fermilab problem source. In addition to the grooved cathode a resistive heater was placed around the source body (fig. 1). Both Brookhaven and Argonne maintained the temperature necessary for proper cesium condition by increasing the arc duty factor to compensate for the lower arc current used with the grooved cathode. At Brookhaven the pulse length was increased while at Argonne the repetition rate was increased to meet the needs of each facility. At Fermilab the resistive heater proved very useful in maintaining the source temperature without having to change the duty factor. With the independent heater the source can be started more quickly, the source temperature can be optimized independently of the arc parameters, the plasma condition can be changed to give different ion currents without significantly changing the source operation, and the lifetime can be maximized by keeping a low duty factor.

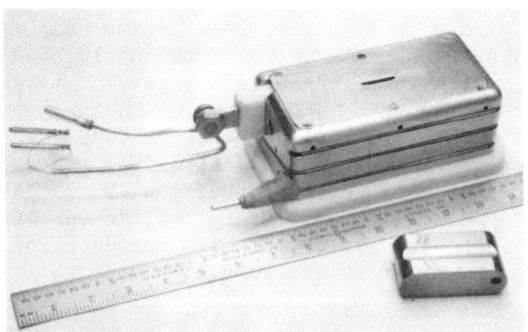

Fig. 1. H⁻ magnetron source showing resistive heating element around the source body (anode). A grooved cathode from within the source is in the foreground. The cathode electrical and thermocouple connections are on the left side and the anode beam aperture is on top.

The heated and grooved cathode source is now used successfully in both preaccelerator systems. The parameters for this source are given in table I and a 54-mA H⁻-beam pulse from the column at 750 keV is shown in figure 2. This source operated smoothly for nine months. In the last month the current decreased slowly from 54 to 45 mA. At this level the source was shut down although it was still running well except for the low output. Cleaning the source interior of sputtered metal and replacing the eroded cathode restored the source to normal output. The previous source in the other preaccelerator ran eight months before showing a slight decrease in beam. It was then shut down, cleaned and replaced. It was restarted, achieved normal operation and ran for about a week before being turned off. The source remained off, in standby, for eight months until the other source showed a decrease. It came back to normal operation in about twelve hours and is presently in operational service.

Table I. Ion source parameters.

Repetition rate	15	Hz
Arc width	85	μsec
Arc voltage	140–150	V
Arc current	35	A
Source magnetic field	1–1.5	kG
Cathode temperature	370–430	°C
Anode temperature	250–300	°C
Cesium boiler temperature	130–140	°C
Cesium valve and feed-tube temperature	> 250	°C
Source chamber pressure	3×10^{-5}	Torr
Extraction voltage	18	kV

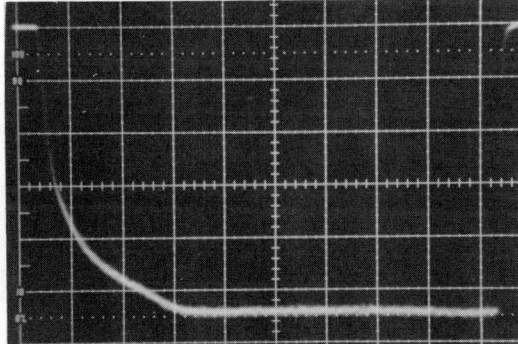

Fig. 2. Typical H⁻ beam pulse from the 750-kV accelerating column.

Vert: 10 mA/Div.,
Horiz: 10 μsec/Div.

The notable changes in the source parameters which have led to the improved performance and lifetime are the lower arc current (150 down to 35 Amps) which has reduced sputtering and erosion of the cathode, and the lower cesium boiler temperature which results in a lower consumption and deposit of cesium outside the source.

The beam pulse is stable to a few milliamperes for many months with noise variations being less than a few percent. Careful comparison between the grooved and previously flat cathode sources has shown little if any change in the emittance following the 750-kV column. The normalized emittance for 90% of a 50-mA H⁻ beam from the preaccelerator at 750 keV, as measured by a slit scanner, is:

$E_{nh} = 1.0 \; \pi$ mm-mrad, $E_{nv} = 1.5 \; \pi$ mm-mrad.

LaB$_6$ STUDIES

The good stability and relatively low work function (2.67 V) of lanthanum hexaboride makes it attractive to test as the cathode material in an H⁻ surface-plasma source. K.N. Leung and others at Berkeley have used LaB$_6$ as the cathodes in a Dudnikov-type Penning source[8] and following this work LaB$_6$ is being tried in our magnetron source.

For this test a cathode identical to the normal grooved-molybdenum cathode was machined from LaB$_6$, placed in a clean source without cesium and installed on the test bench. The system was vacuum pumped overnight after which the source heater was turned on to further outgas the source, primarily the cathode. As the cathode became hot ($\sim$100 °C) considerable outgassing occurred due primarily to water vapor and hydrocarbons. The hydrocarbons may have come from the oils used in machining the cathode or from cleaning solvents. Continued heating of the source overnight reduced the background gases to near normal levels for the test bench.

As usual the source was difficult to start in the hydrogen only mode. After several hours a stable discharge was obtained with very high source pressure and arc voltage. The first attempt at extraction gave an H⁻-beam current of 1 mA which increased to 2 mA over the next two hours before turning off. For the next running period the source was on continuously for four days. For best operation the source pressure was $\sim$1.5 times a normal source operating pressure, the arc voltage (205 V) and arc current (143 A) were extremely high (1.5 and 4 times respectively as compared to a normal source) and produced a cathode temperature of 663 °C. For all this the H⁻-beam current was only 8 mA and very noisy as compared to a smooth 50-60 mA current for a cesiated-molybdenum grooved-cathode source. During this run the cathode was heated to 800 °C but temperatures above 700 °C gave slightly lower H⁻-beam current.

Two additional attempts at using LaB$_6$ were tried. The first introduced argon gas into the source in addition to the hydrogen to hopefully lower the arc voltage. With argon the 8 mA in the hydrogen mode increased initially to 12 mA but within fifteen minutes decreased to 6.5 mA. With the argon off the current returned to 8 mA in the next hour but with a much lower arc voltage (167 V versus 200 V earlier). Several attempts with argon gave similar results and the source was left to run overnight without argon. By morning the H⁻-beam current was 9.5 mA with an arc voltage of 165 V and remained stabled through the day. To achieve this current still required a high arc current (183 A) and cathode temperature, and a relatively high source pressure. Further attempts with argon helped to lower the arc voltage to 155 Volts but gave no improvement in the H⁻ output.

For the second attempt the cesium supply was reconnected. Running the LaB$_6$ cathode with cesium gave a maximum H⁻-beam current of 20 mA but again with a high arc current (167 A) and gas pressure. Operating the source at low arc current consistent with grooved-cathode operation gave only 5-6 mA of H⁻ beam.

Some concern has arisen over the initial hydrocarbon outgassing which may have contaminated the cathode. Perhaps the LaB$_6$ cathode never became activated and cleaner conditions or higher temperatures are necessary. As a test a second LaB$_6$ cathode is being made without using oil or cleaning solvents and will be tried. Achieving higher temperatures would require a major redesign of the source. Initial operation with argon appears useful in conditioning the cathode.

In summary, operation of the H⁻ magnetron source with a lanthanum hexaboride cathode does not appear useful. In all attempts the H⁻ beam was low and noisy, the source pressure, and hence gas consumption, was high and the plasma conditions were extreme creating higher than normal temperatures as compared to a cesiated-molybdenum grooved-cathode source.

Another interesting test made with the LaB$_6$ cathode source was to reverse the extraction voltage and accelerate protons. In this mode a 25 mA beam of protons was obtained at a modest arc current of 100 Amperes. The reason for this test was to investigate a simple filament-free low-duty-factor pulsed proton source that could be closely coupled to an RFQ linac. This test was encouraging but further study and design would be necessary to make this type of source useful in this mode.

REFERENCES

1. C. W. Schmidt and C. D. Curtis, IEEE Trans. Nucl. Sci., NS-26, 3, June 1979, p. 4120.
2. C. W. Schmidt, 2nd. Symp. on the Production and Neutralization of Negative Hydrogen Ions and Beams, Brookhaven, Oct. 1980, BNL 51304, p. 189.
3. R. L. Witkover, et al., Proc. 1984 Linear Accel. Conf., (1984), (GSI-84-11), p. 223.
4. V. Stipp and A. DeWitt, IEEE Trans. Nucl. Sci., NS-32, 5, Oct. 1985, p. 1754.
5. R. L. Witkover, D. S. Barton and R. K. Reece, IEEE Trans. Nucl. Sci., NS-30, 4, Aug. 1983, p. 3010.
6. J.G. Alessi and Th. Sluyters, Rev. Sci. Instrum., 51 (12), Dec. 1980, p. 1630.
7. V. Stipp, A. DeWitt and J. Madsen, IEEE Trans. Nucl. Sci., NS-30, 4, Aug. 1983, p. 2743.
8. K. N. Leung, et al., This symposium.

H^- PRODUCTION AND BEAM FORMATION

SURFACE PRODUCTION OF NEGATIVE HYDROGEN IONS BY HYDROGEN AND CESIUM ION BOMBARDMENT

M. Seidl, W. E. Carr, J. L. Lopes
S. T. Melnychuk, and G. S. Tompa
Department of Physics and Engineering Physics
Stevens Institute of Technology, Hoboken, NJ 07030

ABSTRACT

Metal targets have been bombarded with cesium and hydrogen ions in the energy range 100 to 1000 eV. Angular and energy distributions and the yield of the H^- ions have been measured as function of energy and ion mix. Synergistic effects due to simultaneous hydrogen and cesium ion bombardment produce a higher H^- yield than the sum of the individual processes. Cesium coverage on metal surfaces due to cesium ion bombardment has been studied. High coverages corresponding to a workfunction of 1.6 eV can be obtained at 100 eV energy when the target mass is smaller than the cesium mass.

1. INTRODUCTION

In surface conversion sources[1], negative hydrogen ions are produced on the surface of a metal target placed into a hydrogen-cesium plasma and biased negatively with respect to the plasma. The surface of the target is bombarded by positive hydrogen and cesium ions typically in the 100 to 300 eV energy range. The target is also exposed to a flux of hydrogen molecules and atoms in ground and excited states. Surface production of negative hydrogen ions may be due to several processes, such as backscattering of hydrogen ions or atoms from the surface, or by sputtering of adsorbed or implanted hydrogen by cesium or hydrogen ion bombardment.[2]

The first section of this paper is a short account of our studies[3-5] on sputtering adsorbed hydrogen from a molybdenum target bombarded with cesium ions. In the second section we discuss our experiments on H^- ion production by simultaneous bombardment of a ruthenium target with cesium and hydrogen ions. Preliminary results indicate that H^- ion production due to combined bombardment with hydrogen and cesium ions is considerably larger than the sum of the two individual processes. In order to investigate this "target chemistry" in more detail, we have started to study low energy ion bombardment of metal and semiconductor surfaces. The first part of this program, involving cesium ion bombardment of metal surfaces, is presented in Section 4 of this paper. This work explores the mechanism of cesium coverage due to Cs^+ ion bombardment. In particular, it shows why molybdenum is a better target for surface conversion sources than tungsten.

© American Institute of Physics 1987

2. PRODUCTION OF H$^-$ IONS BY CESIUM BOMBARDMENT

In these experiments a metal target (molybdenum or ruthenium) is bombarded with Cs$^+$ ions. The target is also exposed to cesium vapor and hydrogen gas. Optimum cesium coverage for minimum work function is obtained by adjusting the ratio of the cesium ion to atom fluxes. Hydrogen coverage is provided by chemisorption of hydrogen on the target. Cesium ion bombardment is the only available process for desorption of the hydrogen from the surface of the target.

The latest experimental set-up[5] is shown in Fig. 1. The apparatus consists of a planar diode, rotating Faraday cup, and a rotating magnetic sector mass spectrometer.

The cathode of the diode is the molybdenum target. The anode is a fine tungsten mesh placed 2.0 mm from the cathode. During operation the mesh is heated to about 1000°C by passing current through it.

Cesium vapor is produced in a small oven and directed into a cesium manifold by means of a feeder tube. The cesium manifold is heated to 300°C and provides a uniform flux of cesium vapor to the diode region. Some of the cesium is surface ionized at the hot tungsten mesh which provides a source of Cs$^+$ ions. These ions are accelerated onto the negatively biased cathode. The specific perveance of the geometry is large enough to provide a space-charge-limited Cs$^+$ current of 100 μA/cm^2 at 100 V, almost two orders of magnitude larger than the residual water vapor flux. The lens effect of the mesh adds an intrinsic angular spread of 7.5 mrad to the negative ions accelerated by the mesh. This is about 10 X less than the typical spread due to sputtering. In order to monitor the uniformity of the Cs$^+$ ion current density over the cathode area, the Cs$^+$ ion current density is measured by two positive ion cups facing two 0.635-mm-diam holes drilled 1.19 cm apart in the molybdenum alignment plate. Typically, the two positive ion cup currents differ at most by 10%, which indicates that the neutral and ion fluxes to the Mo cathode are fairly uniform.

The basic operation of the experiment is as follows: Hydrogen and cesium are co-adsorbed on the molybdenum target surface which is continuously bombarded by Cs$^+$ ions. The sputtered H$^-$, Mo$^-$ ions, and electrons are accelerated back across the diode gap, partially attenuated by the mesh and collected by the cesium manifold. A small sample of the sputtered beam passes through a 0.635-mm-diam aperture. All particles emitted in a cone of 14° half angle are accepted by the detection system.

The Faraday cup measures the total current due to all (H$^-$, Mo$^-$, and e$^-$) negative particles. When the Faraday cup is rotated out of the beam path, the mass spectrometer is used to measure the angular dependence of each species.

A typical set of angular distributions is shown in Fig. 2.

Measurements indicate that the H$^-$ ion angular distribution is approximately Gaussian, $f(\theta) = \exp - (\theta/\theta_o)^2$, where θ_o varies slightly with Cs$^+$ bombarding energy. For U^o = 250 eV, θ_o = 4.5°, and for Cs$^+$ energy of 1000 eV, θ_o = 3.2°.

The exit angle θ of a negative ion is related to its parallel energy E at the surface of the target by the equation $\tan^2\theta \approx E/U$ where U is the accelerating voltage applied between the anode and cathode. Using this relationship, the angular distribution can be converted into parallel energy distribution. Fig. 3 shows that the H$^-$ ion parallel energy distribution is approximately Maxwellian, $f(E)_+ = \exp-(E/T)$. The ion temperature T depends on the bombarding Cs$^-$ energy U, ranging from T/U = 0.37% (for U = 750 eV) to T/U = 0.60% (for U = 250 eV). The ion temperature increases when the hydrogen or cesium coverage is incomplete.

Measurements show that most H$^-$ ions leave the target surface with an initial energy of 1.0 to 1.5% of the Cs$^+$ energy. All H$^-$ ions have an initial energy less than 3% of the Cs$^+$ energy. This is consistent with the notion that a few binary elastic collisions are sufficient to describe the desorption process. The simplest model consists of only two collisions; an incoming Cs$^+$ ion collides with a hydrogen atom that in turn is reflected from a molybdenum atom. The energy transferred from the Cs$^+$ ion to the hydrogen atom is given by the equation

$$E_2/E_o = 4M_1M_2 \cos^2\theta/(M_1 + M_2)^2,$$

where E_o and M_1 are the Cs$^+$ ion energy and mass, respectively; E_2 and M_2 are the hydrogen energy and mass, respectively; θ is the recoil angle (angle between the Cs$^+$ ion velocity and the velocity of the hydrogen atom). The maximum energy transfer, occurring for head-on collisions, is E_2/E_o = 0.03. The hydrogen-molybdenum collision does not change the hydrogen energy.

The negative ion yield is defined as the number of negative ions sputtered per incident cesium ion, $Y_{ion} = J_{ion}/J_{Cs}$, where J_{ion} is the current density of the sputtered ions leaving the target and J_{Cs} is the incident current density. The yield has a maximum value when the work function of the cathode reaches a minimum (about 1.6 eV for molybdenum covered with 2/3 monolayer of cesium). The yield also depends on hydrogen coverage of the target. Fig. 4 shows that the H$^-$ yield reaches a saturation when the hydrogen pressure reaches about 10^{-4} torr.

The optimum H$^-$, Mo$^-$, and e$^-$ yields are shown in Fig. 5 as function of Cs$^+$ ion bombarding energy. The yield of H$^-$ ions has a maximum value of 0.41 at a Cs$^+$ energy of 750 eV. Under these conditions most of the H$^-$ ions have an initial energy of about 10 eV. The same value of H$^-$ energy has been obtained for maximum H$^-$ production in backscattering experiments[9,10]. Fig. 5 shows that the H$^-$ yield is smaller than 10^{-3} for Cs$^+$ energies less than 120 eV. The maximum energy a hydrogen atom can obtain in a collision with a 120 eV Cs ion is 3.6 eV. Since the binding energy of hydrogen on molybdenum is about 2.7 eV the initial energy of the H⁻ ion can be at most 0.9 eV. At this energy the survival probability of the H⁻ ion is low.

Surface conversion sources are usually operated in the energy range of 100 to 200 eV. Fig. 5 shows that at these energies the H⁻ yield is very small. It follows that sputtering of <u>adsorbed</u> hydrogen by cesium ion bombardment is not the dominant process in these sources. Hydrogen ion bombardment has to be an important process either on its own merit or in combination with cesium bombardment. Preliminary experiments along this line are described in the next section.

3. PRODUCTION OF H⁻ IONS BY HYDROGEN AND CESIUM BOMBARDMENT

In these experiments the target is bombarded simultaneously with cesium and hydrogen ions. Cesium ions are again made by surface ionization while the hydrogen ions are produced in a planar magnetron discharge.

The apparatus, shown in Fig. 6, consists of the planar diode where the H⁻ ions are produced and of a diagnostic section which analyzes a small sample of the H⁻ beam. The cathode of the diode is a molybdenum plate with a center hole into which various targets can be inserted. The anode is again a heated tungsten mesh placed 0.5 cm from the cathode. Cesium vapor introduced into the diode region through slits in the cesium manifold is surface ionized at the hot mesh. The Cs^+ ions are accelerated to the cathode by a voltage of 100 to 500 V applied between cathode and anode.

Hydrogen ions are produced in a plasma slab filling the cathode-anode gap. The plasma is sustained by a magnetron discharge. A uniform magnetic field of 270 G, parallel to the cathode, is generated by two sammarium-cobalt magnets (not shown). The ExB drift is from the hot filament (cathode of the discharge) to the discharge anode which is connected with the anode of the diode (hot mesh). The plasma potential is close to the anode potential so that the kinetic energy of the hydrogen ions hitting the target is close to eU where U is the anode-cathode voltage of the diode. The target is bombarded with an unknown mix of H^+, H_2^+ and H_3^+ ions. The discharge typically operates at 60 V, 0.02 A and 1 torr pressure. This high pressure is only in the diode region, the rest of the chamber has a pressure much lower (10^{-6}-10^{-4}) due to differential pumping. The hydrogen ion current density at the target can be varied up to 0.5 mA/cm^2.

The positive hydrogen and cesium ion current densities bombarding the target are monitored by two pairs of Faraday cups collecting ions passing through two holes drilled 1.19 cm apart in the cathode plate. A magnetic field of 3.3 kG deflects the hydrogen ions into the hydrogen cups. The cesium ions follow essentially straight trajectories and are collected in the cesium ion cups.

The diagnostics system is identical to that in the previous experiment (Section 2) except for a retarding grid analyzer which measures perpendicular energy distributions of the H^- ions.

In this experiment the ion bombardment of the target can be varied from pure cesium ion bombardment to almost pure hydrogen ion bombardment. Optimum cesium coverage of the target is obtained by controlling the target temperature, cesium vapor pressure and the ion bombardment flux.

Several interesting phenomena can be observed when the target is exposed to mixed hydrogen and cesium ion bombardment. The bombardment energy threshold of 130 eV, typical for cesium ion bombardment, is no longer present. This can be attributed to hydrogen sputtering and backscattering. Indeed, the energy spectra of the H^- ions indicate the existence of these processes. Fig. 7a shows that the distribution of H^- ions in parallel energies can be represented by three exponential components. The part with the steepest slope corresponds to cesium sputtering, the intermediate slope is due to hydrogen sputtering and the smallest slope is due to hydrogen backscattering. The distribution of H^- ions in perpendicular energies, Fig. 7b, shows a similar pattern.

However, yield measurements indicate that H^- production is not a simple superposition of the three isolated processes of cesium sputtering, hydrogen sputtering and backscattering. Rather, the combined synergetic action of hydrogen and cesium gives higher yields than expected from the superposition of these processes taken individually. This point is illustrated in Table I which presents measured H^- ion yields for several cesium and hydrogen current densities. In this table $J_T = J_{Cs} + J_H$ is the sum of cesium and hydrogen current densities hitting the target, H^- yield is defined as the ratio J_H^-/J_T where J_H^- is the H^- current density integrated over all angles.

The brightness of an H^- beamlet along the axis is defined by the quantity $B = I/(S \times \Delta\Omega)$ where I is the H^- current emitted from a target area of 3.2×10^{-3} cm^2 in a solid angle of $\Delta\Omega = 5.9 \times 10^{-5}$ steradians. If H^- ion production was due to the superposition of cesium and hydrogen bombardment, the H^- yield would have the form

$$J_H^-/J_T = \gamma_{Cs} J_{Cs}/J_T + \gamma_H J_H/J_T$$

where γ_{Cs} and γ_H would be constant for a given energy. Table I shows that this is not the case. At 200 eV bombarding energy the H^- yields are small for pure cesium as well as for predominantly hydrogen bombardment. The yield increases to a high value of 0.31 for a 0.17/0.83 cesium/hydrogen mix. This yield is about 5 times higher than the yield in a surface conversion source operating at a similar voltage[11]. The specific brightness B/J_T follows a similar trend.

U (Volts)	200				125
J_{Cs} (mA/cm^2)	0.03	0.04	0.06	0.13	0.35
J_H (mA/cm^2)	0.00	0.76	0.23	0.63	2.34
J_T (mA/cm^2)	0.03	0.80	0.29	0.76	2.69
J_{Cs}/J_T	1.00	0.05	0.20	0.17	0.13
J_H/J_T	0.00	0.95	0.80	0.83	0.87
H$^-$ Yield	0.02	0.03		0.31	0.18
B/J_T (ster^{-1})	0.95	0.57	2.8	4.9	1.32

Table I. H- ion yield and H- beam brightness for several cesium and hydrogen bombardment current densities.

These and similar observations of synergetic reactions in surface production of H$^-$ ions by combined cesium and hydrogen bombardment bring up new questions: What is the reaction mechanism? How can it be used for improving H$^-$ yield? In order to answer these questions we have initiated some fundamental studies in ion surface interactions. The first part of this program is described in the next section.

4. WORKFUNCTION REDUCTION OF METAL SURFACES BY CESIUM ION BOMBARDMENT

In these experiments the formation of composite surfaces due to cesium ion bombardment of some metals has been studied. The results have direct relevance to surface conversion sources. The mean free path for cesium ionization by electron impact is smaller than the plasma dimensions in these sources. Consequently, cesium coverage of the converter surface is due to Cs^+ ion bombardment, not vapor deposition[12]. It has been suggested that the minimum workfunction surface may not be achieved by cesium ion bombardment at typical converter voltages.[12,13]

The apparatus used for measurements is shown in Fig. 8.[11] It consists of a UHV system, with a base pressure of 5×10^{-11} Torr, containing 4 experimental stations; a cesium ion gun, an electron gun for the work function measurement, an Auger electron spectrometer (AES) for detection of surface atoms, and an argon ion gun for sputter cleaning and depth profiling. A target, 3 mm in diameter, is mounted on a carousel, which allows rotation to the stations. The experiment is connected to an Apple computer, so that control and data acquisition are provided for the cesium gun, the work function station, and the AES.

The target is cleaned by argon sputtering, using typically 0.1 microamps at 5 KeV. The beam is rastered to cover the entire target. The Auger electron spectrum shows no observable contaminants for approximately 20 minutes after cleaning, and all of the measurements are obtained within this interval. In order to assure that the target remains clean each measurement is preceded by sputter cleaning.

The cesium ion beam is produced by a gun constructed in house. The source is a heated cesium mordenite pellet which thermionically emits cesium ions, mounted in a Pierce gun. This source eliminates the contamination that is a problem with sources using cesium vapor. The maximum current is a few microamps, of which approximately one microamp strikes the target. The beam is deflected before striking the target so that neutral cesium does not have a line of sight to the target. Beam energy and total dose to the target are computer controlled.

Work function shifts are measured using the retarding field diode method[14]. In this method an electron beam is directed toward a biased target and the bias voltage is adjusted to reflect part of the beam. The shift in bias voltage is equal to the work function shift of the surface.

Figs. 9-11 show the dependence of the work function shift on cesium ion dosage for several bombarding energies. Fig. 9 refers to a polycrystalline tungsten target, Fig. 10 to polycrystalline molybdenum, and Fig. 11 to polycrystalline beryllium. In all cases the initial work function shift is zero (clean metal). The shift decreases smoothly with increasing dosage until it reaches a steady state value dependent upon the Cs^+ ion energy. The total ion dose needed to reach steady state work function is less than 10^{16} ions/cm^2 in all cases. The steady state work function shift is a function of incident ion energy and target mass. The steady state work function (absolutely calibrated by photo-emission) is plotted in Fig. 12 as function of incident cesium energy for all three targets.

For tungsten (atomic mass 183.8) the work function reaches a minimum of 1.6 eV at incident energy of 40 eV. Vapor deposition experiments[15] give the same minimum work function for a cesium coverage of about 0.6. The steady state coverage at incident energies smaller than 40 eV is larger than the optimum coverage. This can be seen in the 30 eV curve in Fig. 9. The work function rises sharply with increasing bombarding energy reaching 2.9 eV at 100 eV and approaching a plateau of 3.4 eV. This is in general agreement with predictions of van Amersfoort, et al.[13]

In the case of molybdenum (atomic mass 95.6) the work function reaches the minimum of 1.6 eV at an incident energy of 100 eV. Comparison with vapor deposition experiments[15] again indicates optimum cesium coverage. At lower incident energies the coverage is larger than optimum and at higher energies the coverage decreases. This was confirmed by Auger electron spectroscopy. The increase of work function with energy is considerably slower than for tungsten.

For beryllium (atomic mass 9.0) there is a minimum work function for all energies. The minimum is almost energy independent (1.6 eV to 1.7 eV) only the required dose changes with energy as seen in Fig. 11. The steady state work function is 2 eV, constant in the entire energy range from 8 eV to 600 eV (Fig. 12). Auger electron spectroscopy confirms that cesium coverage in steady state is larger than optimum for all energies.

The fundamental processes responsible for cesium coverage by ion bombardment are cesium ion implantation and surface erosion due to sputtering by cesium ions. Starting with a clean surface the cesium concentration evolves smoothly to a steady state. In the beginning there are few cesium atoms on the surface because incident atoms are either backscattered or implanted. After target material is sputtered away the surface concentration of cesium increases due to exposure of previously implanted atoms. These are also sputtered, and the surface film stabilizes when the net cesium flux to the target is zero. Additional exposure to the cesium beam does not change the film. The threshold energy for composite film formation is a result of the zero net flux condition, since at low enough energy the sputtering is insufficient to achieve it and the beam forms an overlayer. A simple theoretical model[8] provides the following formula for the fractional steady state concentration c of cesium atoms on the surface:

$$c = \ln(1 + (1-\beta)/\gamma)$$

where β is the backscattering coefficient for cesium ions and γ is the sputter yield of the target material. This formula applies as long as c < 1. One can understand qualitatively Fig. 12 by noting that both β and γ increase with bombarding energy and target mass. When the target atomic mass is smaller than 133, which is the atomic mass of cesium, the reflection coefficient β is practically zero. If in addition the target material has also a large sublimation energy, the sputter yield γ is also small, resulting in a large cesium concentration.

It has been known for some time that molybdenum is a better converter material than tungsten[16]. Our results indicate that the reason is the high work function of tungsten exposed to cesium bombardment in the energy range of 100 to 200 eV where most converter sources operate. On the other hand, the work function of molybdenum reaches the minimum value of 1.6 eV at the operating energy of most surface conversion sources. The mass of beryllium is so low that the steady state coverage of cesium exceeds the optimum value.

Production of layered composite surfaces by means of multi-ion bombardment of some semiconductor materials may lead to surfaces providing improved H$^-$ yield. This is the next step in our research.

ACKNOWLEDGMENTS

This work was supported by the Air Force Office of Scientific Research, Department of Energy, and State of New Jersey Commission on Science and Technology.

REFERENCES

1. Production and Neutralization of Negative Ions and Beams, edited by K. Prelec (Brookhaven National Laboratories, NY, American Institute of Physics Proceedings, No. 111, 1984), pp. 331-457.

2. Production and Neutralization of Negative Ions and Beams, edited by K. Prelec (Brookhaven National Laboratories, NY, American Institute of Physics Proceedings, No. 111, 1984), pp. 171-288.

3. M. Seidl and A. Pargellis, Phys. Rev. B $\underline{26}$, 1 (1982).

4. J. A. Greer and M. Seidl, Production and Neutralization of Negative Ions and Beams, edited by K. Prelec (Brookhaven National Laboratories, NY, American Institute of Physics Proceedings 111, 1984), p 220.

5. J. L. Lopes, J. A. Greer, and M. Seidl, J. Appl. Phys. $\underline{60}$, 17 (1986).

6. G. S. Tompa, W. E. Carr, and M. Seidl, Appl. Phys. Lett. $\underline{48}$, 1048 (1986).

7. G. S. Tompa, W. E. Carr, and M. Seidl. To be published in Appl. Phys. Lett.

8. W. Carr, M. Seidl, G. Tompa, and A. Souzis. To be

published in J. Vac. Sci. Technol.

9. P. J. M. van Bommel, J. J. C. Geerlings, J. N. M. van Wunnik, P. Messmann, E. H. A. Granneman, and J. Los, J. Appl. Phys. 54, 5676 (1983).

10. J. J. C. Geerlings, P. W. van Amersfoort, L. F. Tz. Kwakman, E. H. A. Granneman, and J. Los, Surf. Sci. 157, 151 (1985).

11. K. N. Leung and K. W. Ehlers, Rev. Sci. Instrum. 53, 803 (1982).

12. K. W. Ehlers and K. N. Leung, Production and Neutralization of Negative Ions and Beams, (K. Prelec, Editor), American Institute of Physics Proceedings, No. 111 (1984) pp.227-236.

13. P. W. van Amersfoort, Ying Chun Tong and E. H. A. Granneman, J. Appl. Phys. 58, 2317 (1985).

14. A. G. Knapp, Surf. Sci. 34, 289 (1973).

15. L. W. Swanson and R. W. Strayer, J. Chem. Phys. 48, 2421 (1968).

16. K. N. Leung and K. W. Ehlers, Production and Neutralization of Negative Ions and Beams, (K. Prelec, Editor), American Institute of Physics Proceedings, No. 111 (1984) p. 265.

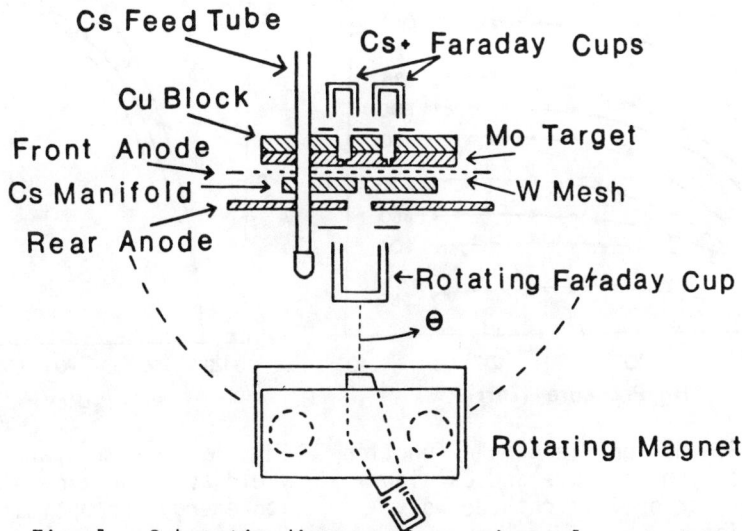

Fig. 1. Schematic diagram of experimental apparatus.

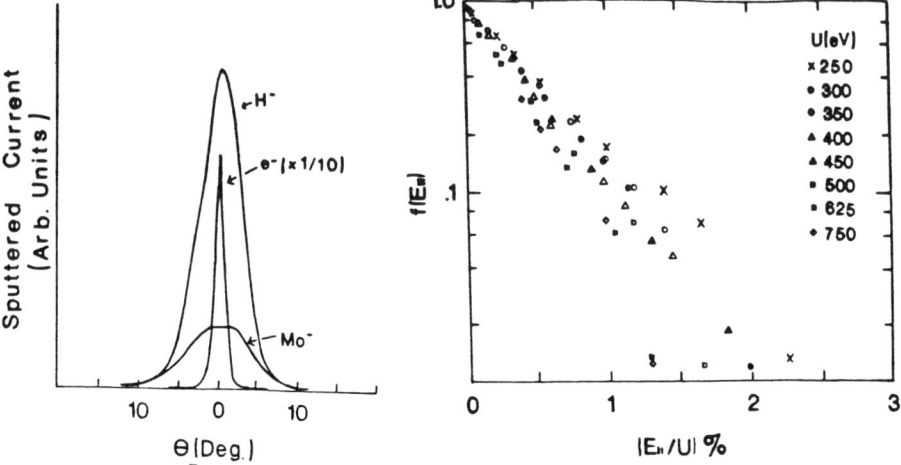

Fig. 2. H^-, Mo^-, and e^- angular distribution for a Cs^+ ion energy of 450 eV. Hydrogen pressure is 2.4×10^{-4} Torr.

Fig. 3. H^- ion parallel energy distribution for various Cs^+ ion energies for optimum work function.

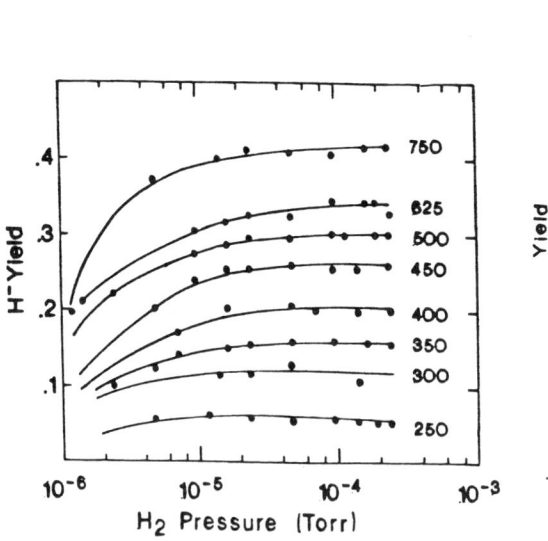

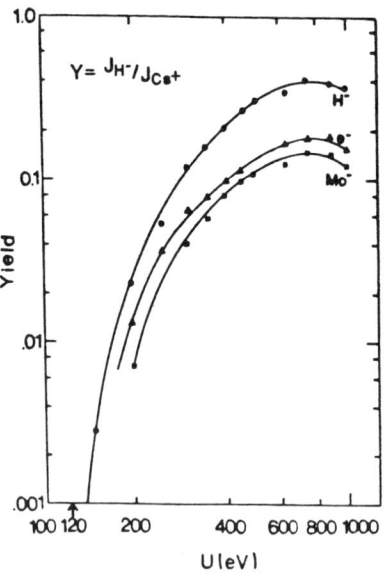

Fig. 4. H^- ion yield as a function of hydrogen pressure and Cs^+ ion energy for optimum cathode work function.

Fig. 5. H^-, Mo^-, and e^- yield as a function of Cs^+ ion energy for optimum work function.

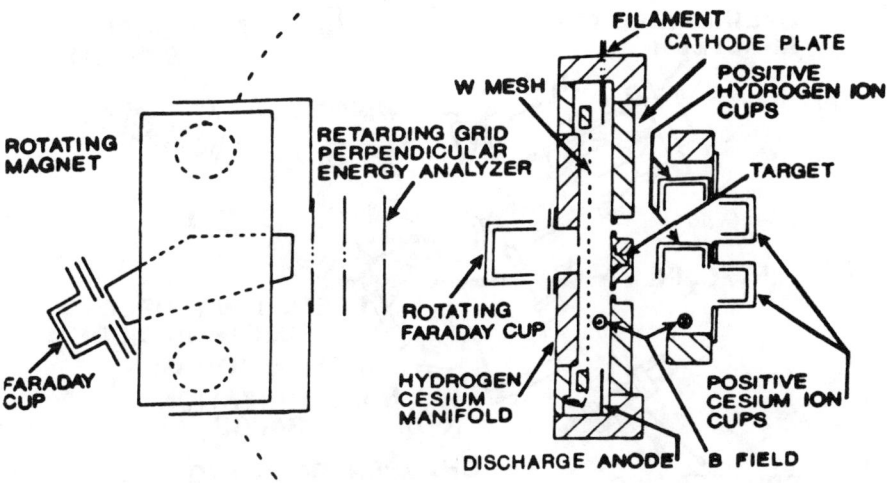

Fig. 6. Apparatus for H⁻ ion production by hydrogen and cesium bombardment.

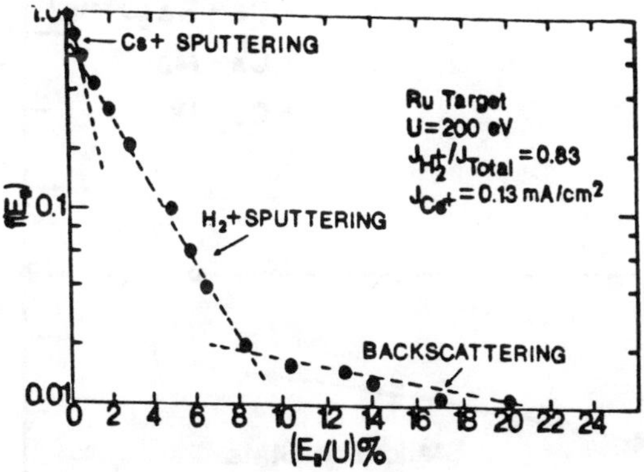

Fig. 7a. Distribution of H⁻ ions in parallel energies.

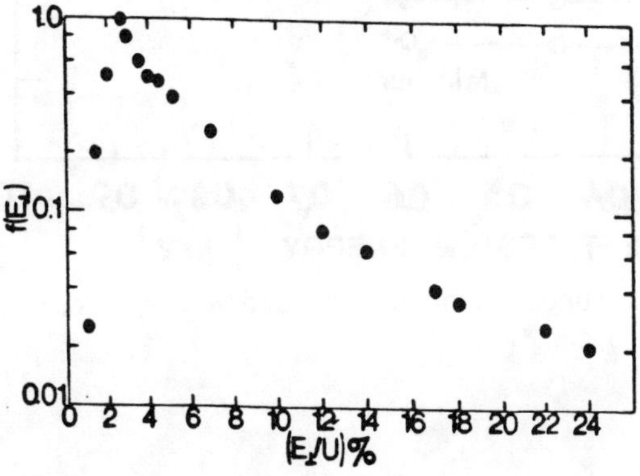

Fig. 7b. Distribution of H⁻ ions in perpendicular energies.

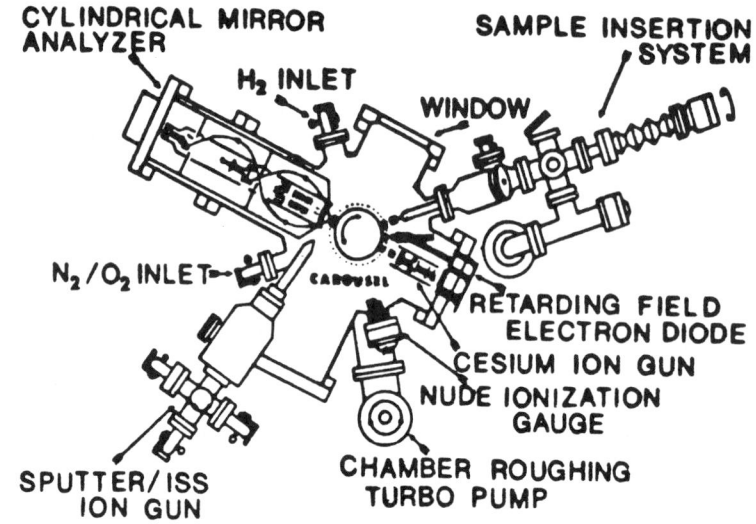

Fig. 8. UHV system for surface studies.

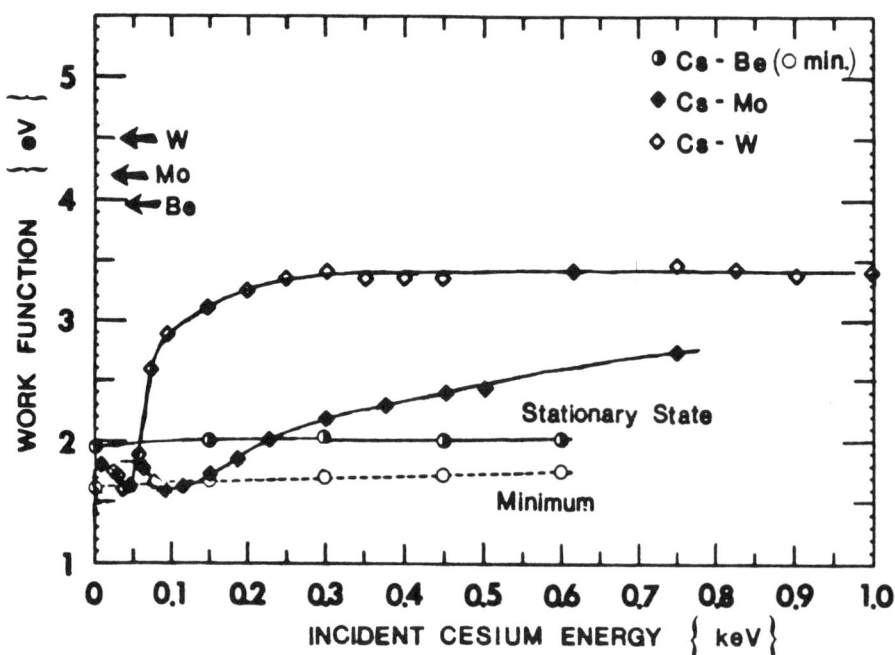

Fig. 12. Steady state work function of Be, Mo and W targets.

445

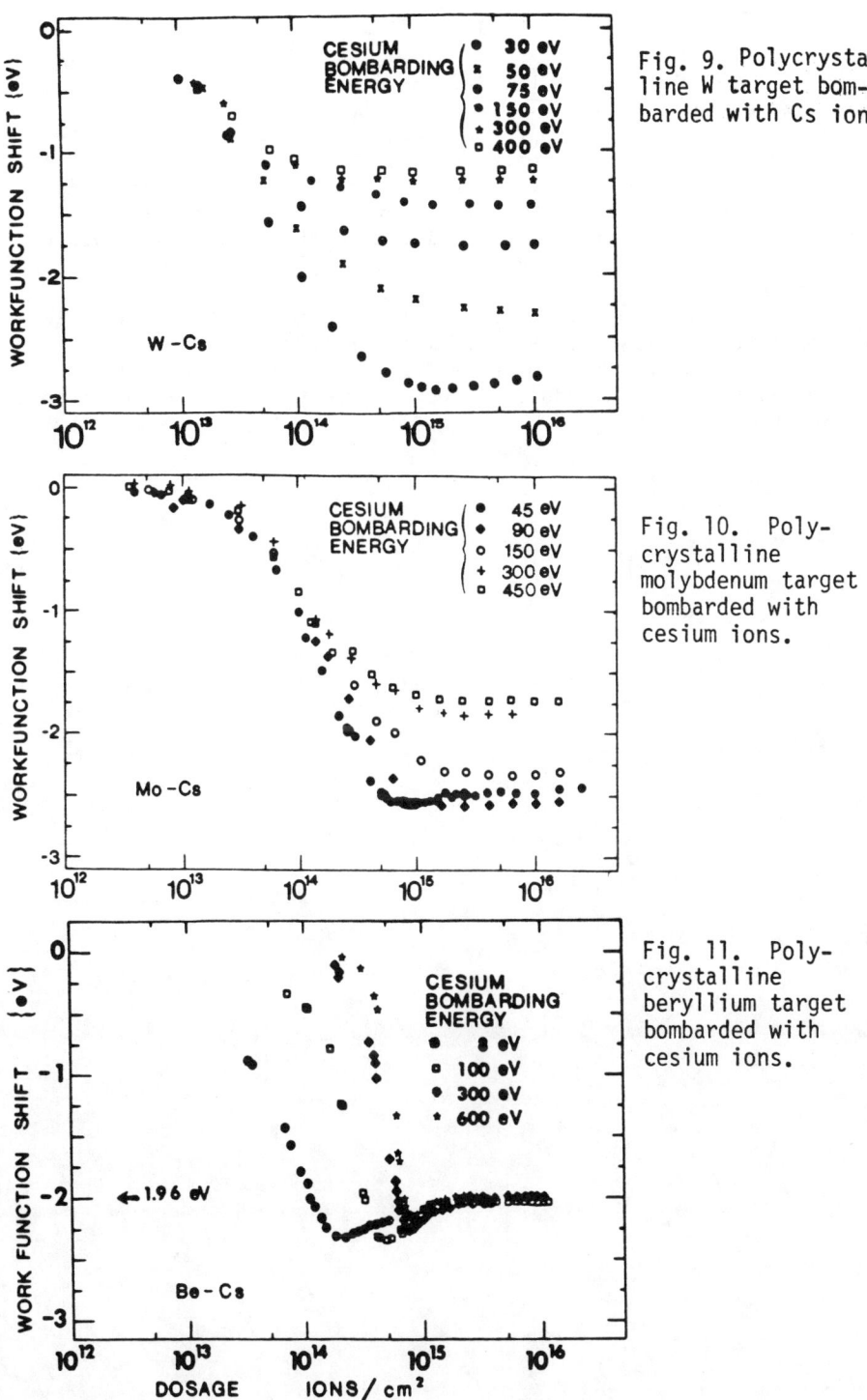

Fig. 9. Polycrystalline W target bombarded with Cs ions.

Fig. 10. Polycrystalline molybdenum target bombarded with cesium ions.

Fig. 11. Polycrystalline beryllium target bombarded with cesium ions.

DISCUSSION

York: You showed that beryllium had a lower work function. Did you take any data on the H^- yield from beryllium?

Seidl: Not yet. I didn't show you, but we also looked at the co-adsorption of hydrogen with cesium. That means we bombarded the surfaces in the presence of hydrogen. And, in general, there is a slight improvement of the work function due to the hydrogen by, on moly, like 0.2 eV or so. But we didn't do that yet with beryllium either, and of course we didn't do that beryllium target in a source condition.

Leung: You showed a threshold energy for cesium ions. What is the threshold energy for hydrogen ions hitting?

Seidl: Well, we didn't measure it. All I can say is that from this binary collision model, you would expect that it would be something like 1 eV.

DISCUSSION:
Relative Importance of Reflection and Desorption
in Surface H⁻ Sources
K. Prelec, Moderator

Prelec: This is a discussion about the relative importance of reflection and desorption in surface H⁻ sources. We thought of asking for short comments from the audience on any subject that is related to the title. We know that there were theories developed and experiments done both on backscattering and on sputtering, or desorption of H⁻ ions. On the other hand, we know something about the processes in or characteristics of sources. We have studied the light coming from the source, we know the H⁻ yield, we know the H⁻ beam current, we know the emittance from which we can deduce the energy spectra; sometimes we measure the energy spectra directly. Very often those two approaches do not agree completely, and this is why we thought we should talk about how to relate what we know from elementary measurements and theories, and what we really see in ion sources. That will be one area that we could touch upon. The second area could be on how to design a source to maximize the yield, to optimize the gas and power efficiencies, and to reduce the energy spectra of H⁻ ions. The floor is open to anybody who would like to comment on this.

Allison: Let me ask Milos Seidl a question. You showed, Milos, that the sputter energy of H⁻ produced by cesium bombardment followed a simple momentum equation. Does that also hold true for the production by H_2^+ bombardment?

Seidl: Well, for combined hydrogen-cesium bombardment, I showed a distribution where you could identify three separate regions. These lines were not calculated, they were just to guide the eye.

Allison: Well okay then, does the sputter energy then follow from the....?

Seidl: Well, I should say that it's not in contradiction to that model, because once you include the sputtered particles, there's always a broad distribution. So, the sputtering of metals, or things like that have been explored. But from the distribution of absorbates, that has really just started. Some people in Germany are doing, in Munich, the sputtering of absorbates, absorbed hydrogen on surfaces. It's a broad distribution. It's all done by computer simulation.

York: A comment on the conclusion that light targets are getting better concentrations of cesium on the surface. In a source such as a cusp field surface conversion source, the high sputter yield of H⁻ ions is mutually exclusive with the life time of the converter. We tried surfaces like titanium, niobium and molybdenum. Although we got higher yields with the niobium converter, its lifetime was about 8 hours. The lifetime of the titanium converter was about an hour. In a production source, such as we run at LAMPF, moly is about the only surface that can run for 2 weeks without significant deterioration due to long-term exposure to the plasma. Even the moly surface, after about 2 weeks, starts to erode from the ion bombardment. Even if a source may work better for short times with lighter, high-sputter yield surfaces, I think it would be very difficult to make a production source out of such material. The next material we're going to try is vanadium, which has a low-sputter yield, which would be harder to get a good concentration of cesium on, but....

Seidl: That is good, a low-sputter yield is good.

York: A low-sputter yield is good? I thought you said that a high-sputter yield got a higher concentration of cesium on the surface.

Seidl: No, no. A low-sputter yield is good.

York: Okay. Well, we've tried titanium. We observed a hydrogen embrittlement of that material and it cracked and spalled and deteriorated very quickly. Niobium was a little slower in that respect, but we got about twice the brightness out of a niobium converter, even though it didn't last very long. And our next material is going to be vanadium. I would love to try beryllium, but certain health restrictions make it very difficult to try it.

Seidl: Well, I understand what you are saying, because we've tried these materials too. What really happens is that they have a high solubility for hydrogen and form hydrides which swell and crack the material, so that anything that produces hydrides is not a good material. You must have a low solubility of hydrogen in that material, and, in beryllium, that is the case.

York: What would you think about vanadium?

Seidl: Vanadium will be a disaster too.

York: Well, I've had the converter for about a year, and I've been reluctant to try it after I'd tried the others. Is there any other materials close to beryllium that you'd recommend?

Seidl: Well, I want to make a contribution here, and I want to talk about that. It seems to me that if people put in so much basic physics work into the surface physics and chemistry as they did into the atomic physics in connection with volume sources, then this field would be in a totally different position. Let me just shortly say what everybody knows, but I would still like to emphasize. In a metal, there is a high work function. You reduce it by the dipole layer of the absorbed cesium, but you still end up with something like 1.5 eV at best. And the affinity is 0.75 eV, so that there has to be a resonant process of tunneling of electrons into empty states, and that produces this probability that depends on the velocity. Because of that, you need a high velocity ion leaving the surface, and since ions don't leave only perpendicular to the surface, these sources have a higher energy spread than the volume sources. If one could reduce the work function even more, then, of course, one would improve the yield at low energies; especially if one made the work function smaller or equal to the electron affinity, then, of course, every hydrogen particle which leaves the surface would leave as a negative ion, regardless of its speed. In this way, one could produce a very low energy spread in these beams. Now, it doesn't seem that in metals one can do that, at least not with cesium coating alone. But it's known that if you add oxygen to the system, you can lower the work function to 1 eV or less. In fact, Marthe Bacal did some work on that many years ago.

Leung: But you make a lot of O^-.

Seidl: Right. That's the problem, that you also make O^- ions. The question is whether one can tolerate it. Now, the basic problem in these sources is that whatever complex surface you make, you have to make it by ion bombardment, because that surface will be exposed to ion bombardment. You cannot prepare a surface and then put it into the source, because it would last about a second. So that's the basic question: Can you do it by ion bombardment? And our work indicates that one might, at least, try to do more complex surfaces by bombarding the surface with different kinds of ions. And what one could do if one could really produce such surfaces is illustrated in the case of negative electron affinity surfaces of p-type semiconductors which are, basically, heavily doped semiconductors, so that the Fermi level is close to the valence band edge. And if you now cover this with cesium, reducing the work function by the dipole layer, then you can obtain cases where the electron affinity – that means the difference between the conduction band and the vacuum level – is zero or negative. That's why they are called negative electron affinity. And, in that case, these surfaces may be good because there are no states for the electron to tunnel from the hydrogen ion because this is a forbidden band. So, in that case, the probability of producing negative ions might increase. By the way, that one can make these surfaces in semiconductors is known.

Hershcovitch: I think this was tried in Japan. They diffused deuterium gas through a semiconductor material and they got D^- ions.

Seidl: That was a different thing. They used a hot electron injection. They didn't use a negative electron affinity surface.

Hershcovitch: I remember distinctly that it did not involve any electrons, except the semiconductor itself.

Graham: Milos, isn't there a major problem in semiconductors, that the electronic states that you're talking about are very susceptible to the quality of the surface, if you get dislocations or if you get defects in the surface....?

Seidl: Sure, that has to be explored, right.

Graham: Isn't that one of the big problems in all these semiconductors? The present way of producing structures through dry etching is that we start to distort the semiconductor surface. You're trying to etch through ion bombardment and they're very unstable in an environment where they're being bombarded with ions.

Seidl: Sure. There are some problems that have to be explored. Let me finish this first, and then I can talk about the other. They also produced a cold cathode based on the negative electron affinity surface, which is an np junction. And, in this case, when it is forward biased, they could emit room-temperature electrons – now this is the difference, it's not hot electrons, but room-temperature electrons – by just forward biasing this pn junction and reported currents are like 2A/cm^2 dc and 200A/cm^2 pulsed. And the temperature was about 0.1 eV, not exactly room temperature, because they are heated up in this surface layer. If one could produce these surfaces, then the biasing could be simple because they are in contact with the plasma, so that the potential between the plasma and the base would produce an electric field in the p-type layer which would produce this effect. Now, there are all kinds of problems, right. But, silicon for example, can be made an NEA surface, but only at the certain orientation of a single crystal. On the other hand, gallium arsenide, or things like that, can be made an NEA surface just with cesium coating, with no oxygen coating, and the orientation of the crystals is not important. That would indicate that one could even get some effect in disordered structures. After all, amorphous silicon, which is just silicon with about 20-30% of hydrogen, is a semiconductor, and works like a single-crystal semiconductor. So, there are possibilities that amorphous materials could work in a similar way. But, I agree, there are many, many problems to solve.

<u>Azria</u>: I would like to make a comment about negative electron affinity surfaces. I worked on gallium arsenide sources, and the main problem is that when the vacuum goes from 10^{-10} to 10^{-9}, there is almost no more the negative electron affinity situation. That means you need, in order to get negative electron affinity in sources with gallium arsenide, a very very good vacuum, 10^{-10} or below. When you increase the vacuum from 10^{-10} to 10^{-9}, the work function increases by 0.5 eV. And the lifetime of the source at 10^{-10} is a few hours, and at 10^{-9}, a few minutes.

<u>Leung</u>: Let us go back to the topic for discussion – which process is more important, reflection or desorption? I think we still don't understand how the hydrogen is absorbed on the surface, and I think more work should be done on that. We worry about the low work functions, but I think we neglect how the hydrogen is absorbed on the surface. Speaking from the point of ion source development, there are two processes for forming the H$^-$ on surfaces. Here is the spectrum of the H$^-$ from a molybdenum converter. The converter is biased at, say, 200 or 300V. There is one group coming out with a couple of eV from the converter, and then there is another group where you can actually see three peaks. This group is formed by backscattering, and the theory of this has been worked out by John Hiskes. Now, you can see that the second group has a very large energy spread. You can have the spread as large as twice the converter voltage. The reason is, when you have a full energy H$^+$ ion hitting the surface, after it backscatters and forms a negative ion, it has an energy twice the converter voltage. When an H$_2^+$ hits the surface, it decomposes into two particles, and when they bounce back, they each have half of the energy. So you have H$^-$ ions with 1+1/2 the converter voltage. And with H$_3^+$ ions, which may decompose into three particles, the maximum energy is 1+1/3 of the converter energy. Now

this is taken without cesium and you can see that this backscattered group is substantially larger than the desorbed one, if you integrate over all the angles. But when you put in cesium, and if the coverage is right, the hydrogen flow is right, you can have mainly desorbed H^-. And this is what we want for ion sources for accelerator design. Not the backscattered group. We should try to optimize the desorbed H^- by understanding how the H is being absorbed on the surface and, at the same time, get a low work function surface. Just by getting a low work function surface, for example, LaB_6, you can see that the desorbed portion is not high. We even tried some barium oxide cathode, made by Spectra-Mat, which has also a very low work function and no cesium. The desorbed yield was very low. So I think more work should be done to understand how the hydrogen is being absorbed on the surface, in addition to studies of low work function surfaces.

Hiskes: I'd like to comment on Milos' data from this morning, which I thought was interesting. In the past, we wondered why moly converters were better than tungsten converters, since both gave about the same minimum work function. But Milos showed, this morning, that for tungsten the minimum occurred below 50V, whereas for moly it occurred between 100 and 200 eV, and I thought that explained, for the first time, why moly was preferable to tungsten.

Forrester: I'm hung up on a very elementary point. In the Saha-Langmuir equation for surface ionization, one would have an (A-B)/kT. And the temperature comes in because of the energy of the electron distribution. In your formula for the ionization at the surface, you have an (A-B) over the velocity of the escaping ion, and I'm just puzzled how that comes in, and where's the difference. I had thought, somehow, the bombardment would create a thermal spike which would raise the temperature that goes in there. But you have it related just to the velocity of the escaping ion and not to the energy distribution of the electrons.

Seidl: Well, I didn't create this formula. Many people have worked on that. This simplified version is by Lang from IBM. The reason is simple, because this is not equilibrium. In equilibrium, you would get zero negative hydrogen ions from the surface.

Forrester: In the thermal spike model, one assumes you have a sort of local equilibrium.

Seidl: Right, but you don't have any thermal spike.

Roberts: I don't know if I understand it either, but when he first put it up there, I had the same confusion you did. And I figured that it would be 1/kT if the surface was being bombarded by thermal particles.

Forrester: No, the high energy particles create a thermal spike. And I have found in trying to fit some data that most data fits when you put in an elevated temperature of the order of 0.5 eV, like 6000°. In a thermal spike model, one gets a reasonably good fit. I somehow failed to be aware of Lang's work with a velocity.

Allison: Isn't that just a tunneling probability?

Seidl: Yes. That spike model - people have applied at much higher energies, like tens of kilovolts of bombarding energy. And even there, this is considered now kind of doubtful, because there is probably no thermal equilibrium in a real sense in that little spot. So, I don't think that thermal equilibrium can be used in this case. But I would like to answer the question of whether, when you make an NEA surface, it disappears when the vacuum is not 10^{-11}. Well, this is very much related to how you make these surfaces. If you say that they will be made by ion bombardment on a continuous basis, and you will get a steady state that is determined by this ion bombardment, then there is no question of this sort, because you don't make the surface first. You sort of continuously make it.

Azria: It gets poisoned by carbon. I don't know where the carbon comes from.

Seidl: Well, that depends very much on the ratio of the good particles to the bad particles. If the good particles have a much higher flux than the bad particles, the bad particles are sputtered off before they can do damage.

Azria: We were not able to get a negative electron affinity surface in 10^{-8} vacuum, even under continuous bombardment of cesium. And we were checking that by shooting a Krypton laser infrared line and trying to see any electron current. And in bad conditions, as I said, it was impossible to detect a single nanoampere.

Anderson: I'm curious whether people think there's any prospect for obtaining surfaces where one could convert a polarized atomic beam of hydrogen into H^- ions in the time short compared to the relaxation time on the surface, which might be on the order of a few microseconds?

Hershcovitch: Something was done by Haglund at Los Alamos, but not with hydrogen.

York: I'd still like a recommendation of a non-toxic converter surface that we could try that would be better then moly.

Seidl: Well, you'll want to try aluminum, silicon, or anything light that does not desorb hydrogen.

York: But in a hot plasma, aluminum is not going to last very long.

Prelec: On this note, we shall finish the short discussion.

PRODUCTION OF INTENSE H$^-$-ION BEAMS IN HIGH-POWER PULSED DIODES

Henri J. DOUCET

Laboratoire de Physique des Milieux Ionisés
Laboratoire du CNRS
Ecole Polytechnique, 91128 PALAISEAU Cedex (France)

ABSTRACT

This paper presents a short review of the published work done in the USA, USSR, and France on the production of intense H$^-$-ion beams in pulsed, high-power magnetically insulated lines and diodes. The production of H$^-$ ions is discussed and the application of repetitive, pulsed intense H$^-$-ion beams to magnetic and inertial-confinement fusion is considered.

INTRODUCTION

During the last few decades, the intensity of negative-ion beams produced in laboratories has changed dramatically. Some 30 years ago, following discussions on the possible existence of H$^-$, the first beams obtained were in the range of pA-nA. However, during the last 20 years, both steady-state and long-pulse H$^-$-ion beams of up to several amperes have been widely produced and studied by double-charge-exchange and surface and volume techniques. To achieve much higher-current beams requires new techniques.

The existence of high-density plasmas in pulsed, magnetically insulated high-power lines and diodes makes such devices good candidates for the production of intense (kA-range) pulsed beams of negative ions at large energies (a few MeV). To our knowledge, the production of intense H$^-$-ion beams in a high-power diode was first mentioned by the group at IRVINE, University of California[1]; however, the first detailed reported of such experiments was presented by the group at Sandia National Laboratory who made studies of H$^-$-ion production in magnetically-insulated transmission lines[2-4]. This important work gave rise to the hope of producing intense beams of H$^-$ and D$^-$ in high-power diodes. Following this approach, the group at the Lebedev Institut in Moscow reported[5-9] a series of experiments in which H$^-$-ion beams with currents of 5 kA and a few hundred kV were produced. Independent efforts at the Ecole Polytechnique in France[10-11], at Tomsk in USSR[12], and at Irvine, U.C., in USA,[13] to reproduce and understand the FIAN results, have had only limited success. Larger efforts, focused on understanding both the prepulse effect in the FIAN experiments and the plasma-production processes before the extraction of the intense H$^-$-ion beam, are presently being made.

© American Institute of Physics 1987

In the present paper, we first give a short presentation of the main results and experiments going on in each group. We then discuss in some detail both the plasma production and the extraction of H⁻-ion beams. Finally we give a short overview of some possible applications of intense, repetitive H⁻- and D⁻-ion beams in both magnetic and inertial-confinement fusion.

PRODUCTION OF H⁻-ION BEAMS IN MAGNETICALLY INSULATED TRANSMISSION LINES

In MITL, the high-power transmission is due to the fact that the electrons are prevented from crossing the interelectrode gap by an azimuthal magnetic field produced by the large current flowing in the inner conductor of the line. Since negative ions, flowing almost freely across the vacuum gap between the cathode and the anode, can introduce significant current losses, thereby affecting the power transport in the MITL, H⁻-ion production has been studied extensively at Sandia[3-5] in magnetically insulated transmission lines. Sandia's negative-ion experiment is shown in Fig.1.

Several ion species have been identified: H^-, C^-, and C^{-2}. The extracted current densities were generally in the range of a few A/cm^2, but were sometimes as high as 30 A/cm^2. No clear-cut dependance on voltage was found, however the current density seems to scale as the third power of the total current I in the transmission line.[4]

In MITL, a cathode plasma is produced at early times,[3] possibly by explosive emission due to the large electric fields which can exceed 2.5×10^7 V/m. The first electrons are expected to cross freely the cathode-anode gap until the total current in the line produces an electric field large enough to reflect them back and permit magnetic insulation to be achieved. These early electrons will strike the anode, thereby emitting radiation which will participate in the next phase of gas desorption and plasma formation on the cathode. This plasma formation was identified in the Sandia work,[3] and possible mechanisms of H^-[14] and other negative-ion formation in volume were examined, but the authors were more in favor of a surface production, since continuous negative-ion sources were under study at that time[15-16] on surfaces.. We presently believe, as discussed in reference 10, that the density of the cathode plasma is sufficiently high that any H⁻ produced on the cathode surface would be destroyed by electron collisions, thereby preventing any such ion from crossing the cathode plasma and escaping across the vacuum gap.

Diagnostics used in this first carefull and remarkable study included time-resolved measurements made with a time-of-flight spectrometer, and time-integrated measurements of ion species and ion energies made with a Thompson Parabola. Figure 2 shows the current measured with the time-of-flight spectrometer, as a function of time, while Fig.3 shows the results of the determination of the negative-ion species by use of the Thomson Parabola.

PRODUCTION OF H⁻-IONS IN HIGH-POWER DIODES

The plasma parameters, field values, and geometries in an MITL and in a high-power pulsed diode with a magnetic field are quite similar, except that the magnetic field in a diode will generally be an externally applied field, whereas in an MITL it is a self-produced azimuthal field. Thus, one can expect to be able to produce and extract large current densities in a pulsed, high-power diode.

Presently the most detailed, published experiment is probably the work of the Lebedev Institute[7-9]. That experiment used a coaxial geometry, as shown in Fig.4. The negative ions are thought to be produced in the plasma close to the central cathode and then extracted radially by electric fields applied between the cathode and the cylindrical anode grid surrounding the cathode. The extracted ions were collected on graphite targets and were identified from the curvature of their trajectories (see Fig.5) as being H^+ and H^-. Time-integrated measurements of the number of collected ions were made by nuclear activation, using the reaction $^{12}C(p,\gamma)^{13}N(\beta^+)^{13}C$, which has a threshold of 457 keV and a half life of 9.96 min. The cathode voltage shape always consists of two successive pulses : a main negative pulse, followed by a positive pulse, the relative amplitudes of which depend on the diode mismatch. For this reason, the existence of the threshold allows a separation of the effects of the protons and H^- ions in a single shot, since the diode-voltage amplitude can be chosen to be larger than the threshold, but small enough in order that the positive pulse can be smaller than the threshold.

The main results of these experiments is that total currents of H^- ions of up to 5 kA have been measured, with current densities of up to 200 A/cm². Large current densities, generally in the range, 10-50 A/cm², have been obtained, also, with metallic cathodes on which hollow dielectric cylinders were fitted, or with dielectric cathodes (polyethylene, teflon, polymethylmethacrylate).

WHERE AND WHEN ARE THE H⁻ IONS PRODUCED ?

The Lebedev experiment was based on a simple theoretical model which assumed that the ratio, $\lambda = n_-/n_o$, of the negative ion density to the electron density was large enough to permit the extraction of large current densities of H^- ions. As the λ values measured in low-density plasmas at Ecole Polytechnique were found sometimes to be very large[14], values of a few 10^{-2} were used in the model. For a typical diode of 500-700 keV, 10-20 kA, 100 ns, with a magnetic field of 15 kG, the model predicts a current density of 15 A/cm² of H^-, in reasonable agreement with the experiment. The following parameters were typical : a plasma density of 2×10^{15} cm⁻³, $\lambda = 5\times10^{-2}$, a cathode-plasma layer of 1mm, and a neutral gas density of

10^{16}-10^{18} cm^{-3}. These parameters seem to be quite realistic for the cathode plasma some time after its production, that time corresponding to the time required for the dense (10^{19} cm^{-3}) low-temperature (a few eV) plasma to expand and recombine.

An attempt was made, in a collaborative effort between the Ecole Polytechnique and the Lebedev Institute, to reproduce the above results with a Physics International PULSERAD 110 A machine. It was found that an essential difference between the FIAN and French experiments was the importance of the prepulse voltage. Figure 6 shows the voltage pulse of the ERG generator used for the negative-ion experiment at the Lebedev Institute. In the experiment at the Ecole Polytechnique, the prepulse was usually supressed by a prepulse switch. For the French experiment, the prepulse switch (usually working with SF_6 under pressure) was connected to a vacuum pump to permit the voltage to be transmitted, even for small voltages. Despite this arrangement, the prepulse in the PULSERAD machine was always much smaller in duration and amplitude than the prepulse in the ERG machine, and no significant quantities of high-energy negative ions were ever detected, no matter what diagnostics were used : charge collectors with thin foils to separate between fast electrons and H$^-$ ions, time-of-flight spectrometer, nuclear activation. Also, in the FIAN experiment, it was found that no negative ions were ever observed without a prepulse. The importance of the prepulse was also confirmed by the group at Irvine, who found, using an APEX machine, that H$^-$ ions were never found without a large prepulse[13].

The importance of the prepulse focused attention on the cathode-plasma formation. An analysis of electron detachment by electron collisions lead to the conclusion that H$^-$ ions can be extracted only from moderate-density plasmas[10]. More-direct evidence of the plasma-formation effect was investigated at several places, using different techniques. At the Lebedev Institut, laser[9] irradiation of a conical cathode (see Fig.7) was used to demonstrate that an energy density as low as 1J/cm^2 leads to H$^-$ production in a high-power diode, if a delay of the order of few tens of nanoseconds is provided between the 20-ns laser pulse used to illuminate the cathode and the application of the high voltage in the diode. A laser pulse having 5 J of energy, with a delay of 30-60 ns, led to the production of 10^{13} H$^-$/pulse. Increasing the laser energy to 35 J, and using of a delay in the range of 50-250 ns, led to 10^{14} H$^-$/pulse. The very modest value of the required fluence led to the idea that, perhaps, the laser light does not directly produce an appreciable plasma density but, instead, produces enough gas desorption to permit a large plasma density to be produced in a very short time after the application of the large diode voltage. At the Ecole Polytechnique, another approach to produce the cathode plasma uses a flashboard to produce an intense, short pulse of VUV light[17]. Such a VUV source is presently routinely used to produce anode plasmas in high-power diodes used for the production of intense sources of positive ions for inertial fusion research[18]. At Irvine, another approach[13] is to add a prepulse generator which prepares the plasma before the application of the main voltage pulse. The formation of H⁻ in the expanding and recombining cathode plasma of a high-power diode is under investigation at the Lebedev Institut and at the Ecole Polytechnique.

APPLICATIONS TO MAGNETIC AND INERTIAL-CONFINEMENT FUSION

The magnitudes of the high current densities of negative ions obtained in the early experiments gave rise to the idea that such negative-ion beams could be used in magnetic-confinement fusion. Sandia proposed to produce Li⁻ ions from an MITL.[4] Their analysis showed that 50 MW of neutral-beam power could be produced in repetitive operation at 100 Hz, at low cost ($1 per Watt). But one difficulty, besides an incomplete understanding of the H⁻-ion production, itself, is the use of an MITL as the source of the high power. It is well-known, in fact, that an MITL works only if a low-impedance load is used at the end of the line (typically lower than, or of the order of, 60% of the characteristic impedance of the line, for the range of voltages of interest here). Therefore, a large fraction of the energy must be deposited into this load in order to permit magnetic insulation of the line, thereby producing an efficiency of the system which will be too low, perhaps, for any practical application at very-high-power levels. However, the same high-power technology can be used in a diode. In this case, the application is simpler, since the diode itself is a high-impedance device, so that the technological constraints are less severe. Also, the over-all efficiency can be much higher, since the magnetic insulation is produced by an external magnetic field. Such a system has been proposed by both the Lebedev Institute[7] and the Ecole Polytechnique[10-11].

Some other applications of negative-ion beams in magnetic fusion have also been suggested[19-20], which would involve using either single or repetitive intense pulses of negative ions.

The success in producing very intense beams of positive ions has given rise to the hope that intense beams of negative ions can be produced, also. Such beams would be of interest in the effort to produce large surface power densities for use in inertial-confinement fusion research. As the photodetachment of negative ions has been demonstrated to be very efficient, and does not require very high-power lasers, the use of intense negative-ion beams to produce intense neutral beams focused on a small target has been suggested.[ ] Simple analysis indicates that the highest surface power densities which are presently obtained using very large generators (a few TW/cm²), could be obtained from negative ions with a modest machine[17].

CONCLUSIONS

Pulsed, high-power magnetically insulated transmission lines and diodes can be used to produce intense beams of H⁻ and other light negative ions. Even if, presently, the mechanism responsible for the negative-ion formation is not precisely known, the importance of plasma formation before the beam extraction has been demonstrated. Currents as high as 5 kA, with current densities reaching 200 A/cm² have been obtained, but not yet extracted as directed beams. Present progress in the repetitive operation of pulsed diodes makes it possible to consider the application of pulsed negative-ion beams produced from high-power diodes to magnetic and inertial-confinement fusion.

AKNOWLEDGMENTS

It is a pleasure to acknowledge stimulating discussions with Dr. Martha Bacal, Dr. Jean-Max Buzzi, Dr. Stavros Moustaïsis, Pr. Norman Rostoker, Dr. Regan Stinnett and Dr. Vitali Papaditchev concerning the production of negative ions in high-power magnetically insulated lines and diodes.

REFERENCES

1. A. Fisher and N. Rostoker, Bull. Am. Phys. Soc., 21, 1097 (1976).

2. J.P. Van Devender, S.W. Stinnett and R.J. Andersen, Negative Io Losses In Magnetically Insulated Vacuum Gaps, Appl.Phys. Lett. 38, 229 (1981).

3. R.W. Stinnett and Tim Stanley, Negative Ion Formation In Magnetically Insulated Magnetic Lines, J. Appl.Phys. 53, 3819 (1982)

4. R.W. Stinnett and M.T. Buttram, A Magnetically Insulated Negative Ion Source For Neutral Beam Heating, Journ. Fus. Energy, 4, 253 (1983)

5. R.W. Stinnett, M.A. Palmer et al., Small Gap Experiments in Magnetically Insulated Transmission Lines. R.IEEE Trans. on Plasma Sci., PS-11, 216 (1983)

6. A.A. Agafonov, A.A. Kolomenski et al., Generation of Intense Fluxes of Negative Ions Zh. Eksp. Teor. Fiz. 84, 2040 (1983) ; Sov. Phys. J.E.T.P. 57, 1188 (1983)

7. A.A. Kolomensky, A.N. Lebedev et al., Generation and acceleration of Multicharge and negative Ions in High-Current Diodes and by Means of Collective Effects, Proc.of the 5th Int.Conf. on High-Power Particle Beams. San Francisco (1983). (BEAMS 83)

8. A.A. Kolomensky, I.I. Logachev et al., Cathode Plasma Formation and H^- Generation With 5-kA Current In A Magnetically Insulated Diode, Proc. of the 2nd European Workshop On The Production and Application of Light Negative Ions, Ecole Polytechnique, Palaiseau, France (1986)

9. A.A. Kolomensky, I.I. Logachev et al., Generation and Diagnostics of Pulsed Beams of Negative Hydrogen Ions, 6th Int. Conf. on High-Power Particle Beams, Kobe (1986). (BEAMS86)

10. S. Moustaïsis, H.J. Doucet et al., Towards MV, kA Beams of H^- from a Pulsed, Magnetically Insulated Diode, Proc. of the 2nd European Workshop On The Production and Application of Light Negative Ions, Ecole Polytechnique, Palaiseau, France (1986)

11. S. Moustaïsis, H.J. Doucet, et al., Towards A Magnetically Insulated Pulsed H^- Beam Source, 6th Int. Conf. on High-Power Particle Beams, Kobe (1986) (BEAMS 86)

12. V.M. Bitstritsky, Ya.E. Matvienko et al., Investigation of H^- Generation In The High Power Diode On The 10GW Power Level, 6th Int. Conf. on High-Power Particle Beams, Kobe (1986) (BEAMS 86)

13. A. Fisher, H. Lindenbaum and N. Rostoker, Intense Pulsed Sources of Negative Ions, 6th Int. Conf. on High-Power Particle Beams, Kobe (1986) (BEAMS 86)

14. M. Bacal, Volume Generation of H^- Ions in Plasmas Physica Scripta $\underline{T2/2}$, 467 (1982)

15. Yu. Belchenko, G.I. Dimov et al., Proc. of The Symp. On The Production and Neutralization Of Negative Ions and Beams, BNL Report No 50727, 79 (1977)

16. J.R. Hiskes and J.R. Schneider, Proc. of The Symp. On The Production and Neutralization Of Negative Ions and Beams, BNL Report No 51304, 15 (1980)

17. H.J. Doucet, J.M. Buzzi et al., Basic Plasma Physics Rresearch in Pulsed-Power Technology at Ecole Polytechnique. Proc. 2nd Int. Topical Symposium on ICF Research by High-Power Particle Beams, Nagaoka, Niigata (Japan), June 16-18, 1986.

18. J.R. Woodworth and P.F. McKay, Surface Discharges as Intense Photon Sources In The Extreme Ultraviolet, J.Appl.Phys. 58(9), 3364 (1985)

19. L.R. Grisham, D.E. Post and D.R. Mikkelsen, Multi-MeV Li Beam As A Diagnostic For Fast Confined Alpha Particles, Nucl. Tech/Fusion, 3, 121 (1983).

20. D.E. Post, Particle Diagnostics For Magnetic Fusion Exp. Atomic and Molecular Physics of Controlled Thermonuclear Fusion, Ed. C.J. Joachim and D.E. Post, NATO ASI Serie B : Physics, 101, 539 (1982).

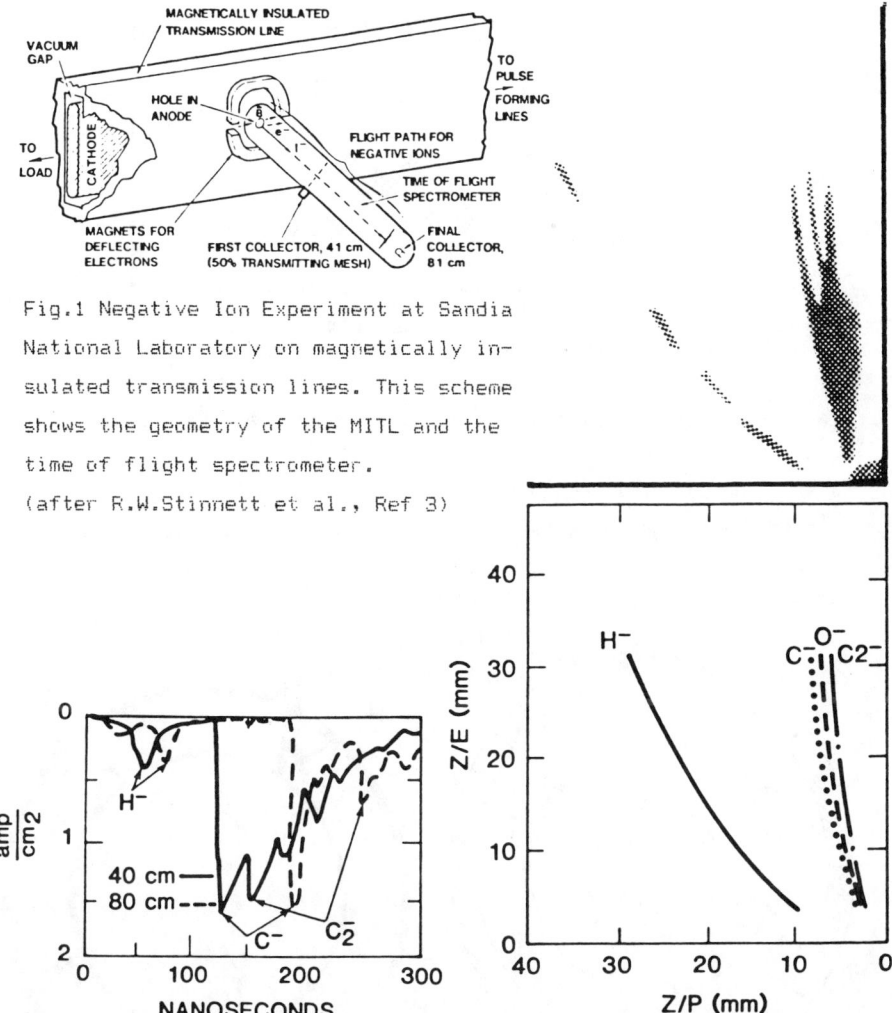

Fig.1 Negative Ion Experiment at Sandia National Laboratory on magnetically insulated transmission lines. This scheme shows the geometry of the MITL and the time of flight spectrometer.
(after R.W.Stinnett et al., Ref 3)

Fig.2 Negative ion species and ion energy determination from the time-of-flight spectrometer. (after R.W.Stinnett et al, Ref 3)

Fig.3 Experimental negative-ion parabolas together with computer simulation of Thompson parabola curves for H^-, C^-, O^- and C_2^- ions with enrgies from 200 keV to 2 MeV (after R.W.Stinnett et al, Ref 3).

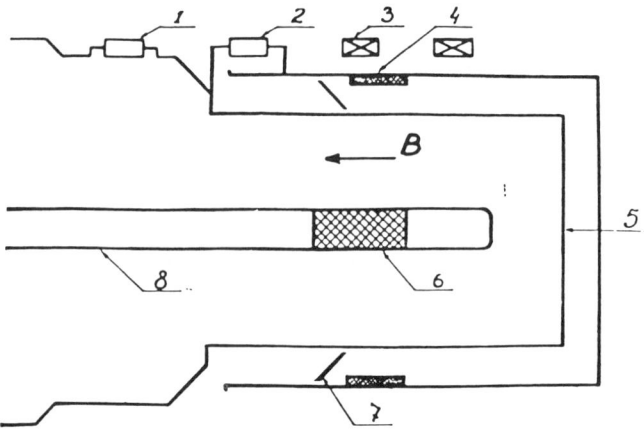

Fig.4 Negative-ion experiment in coaxial geometry at the Lebedev Institute. Scheme of the experiment: 1-resistive shunt to measure the total diode current; 2-shunt to measure the transverse current; 3- solenoid; 4- graphite target; 5- anode; 6- dielectric cathode; 7- Al-covered mylar foil; 8- cathode rod.
(after A.A.Kolomensky et al., Ref 8)

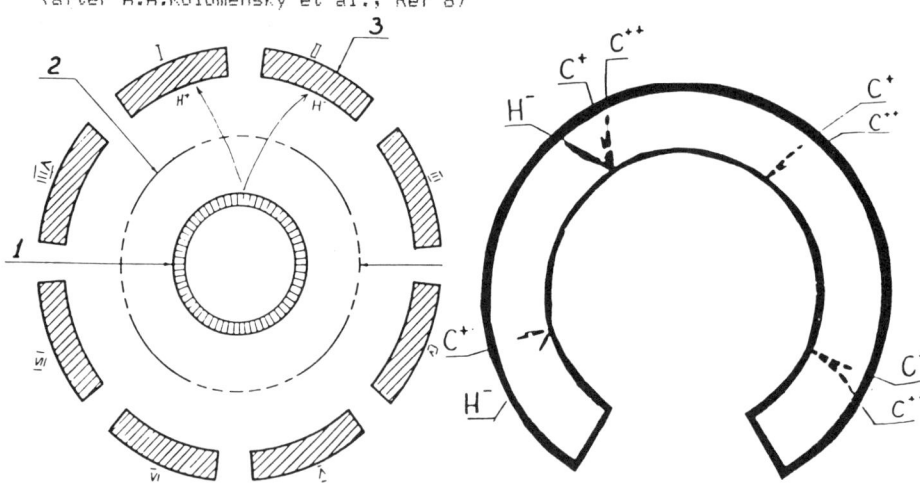

Fig.5
a) Experiment for determining the sign of the ions: 1- dielectric cathode; 2- anode; 3- segmented graphite target.

b) Ion track on the aluminized mylar foil

(after A.A.Kolomensky et al. Ref 8).

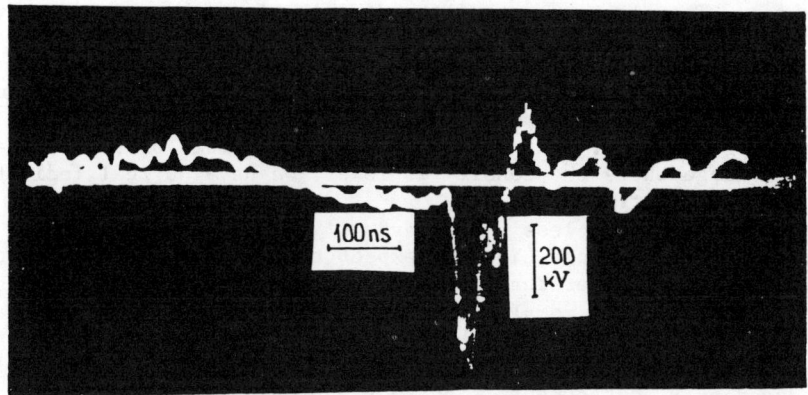

Fig.6 Voltage waveform in the diode of the ERG generator used for the ion experiment at the Lebedev institute.
(after A.A.Kolomensky et al., Ref 8)

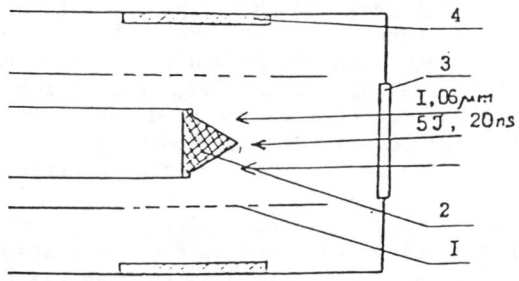

Fig.7 Scheme of the experiment in cathode plasma formation by laser:
1- anode; 2- cathode; 3- glass window; 4- graphite target.
(after A.A.Kolomensky et al. Ref 9).

DISCUSSION

Hiskes: That latter part of the distribution, where the density has fallen below 10^{14}, do you have any idea what the electron temperature is in that time?

Doucet: About the electron temperature, we know that the plasma temperature is in the range of 4-5 eV. We know that from the plasma expansion velocity. But this is for the main bulk of the plasma. In the range where the density is lower, we have no real indication at the present time.

Hiskes: It may be less than 4 ev?

Doucet: Yes, and no. You can expect that the high energy electrons will go farther than the low energy electrons. But, on the other hand, it will completely depend on the kinetics of the plasma because it's unlikely that the particles will start from the cathode and continue. Most of the electrons produced there will probably be produced by H^- detachment. So you are probably in a better situation than I am to tell me what the electron temperature could be. I mean, it will probably be the result of the plasma kinetics more than any special technology involved in the high power business.

Ehlers: What would be your practical estimate for a reliable, long-term duty factor, as equipment exists today?

Doucet: You mean for the high power technology? At the present time, you can say that using gas switches it is difficult to go any higher than a few hundred Hertz to 1 kHz, just for the problem of gas flow. But, you see, the currents are not very large, so you can turn to magnetic switches and there the repetition rate can be very high. Magnetic switches have been proven to work up to something like 100 MHz, so much higher than you can do for the rest. So the limit will not be the pulse technology problem for the switches. It will be mostly the rest of it. The need for starting from new conditions in the high-powered diodes, probably.

Ehlers: Do you have to replace the cathode periodically, or recondition it?

Doucet: We don't work in a repetitive system, but we can make several hundred shots a day. And we can say that we use mostly aluminum plastic-coated cathodes, and you can use them for a very long time.

Bacal: I would like to suggest that in the case of this, short pulses of extraction voltage, the atomic physics of negative ions shouldn't be considered in the same way as you consider in a steady state. It may be that all the negative ions formed in a very thin layer, just one mean free path away from the surface, are the important ones. The fact that we only deal with extraction in an extremely short time, may imply that we can extract all the negative ions before they have any destructive collisions. So some possibilities are here which we do not have when we have to consider steady state.

Doucet: That's right. In modeling, for example, it will be very important, not only to take care of the equilibrium distribution function that you can find, but the time dependence and the local dependence, as density gradient, are probably very important.

SOME COMMENTS ON EMITTANCE OF H⁻ ION BEAMS*

Paul Allison, AT-2, MS-H818
Los Alamos National Laboratory, Los Alamos, NM 87544

ABSTRACT

Some properties of emittance, emittance distributions, and measurement techniques are reviewed. In comparing the results of measurements with several different types of H⁻ sources with each other and with emittance formulae, it is concluded that the emittance of surface-type sources is dominated by the effective ion temperature. Other effects, such as ion-optical distortions, may account for the emittance of volume-type sources.

INTRODUCTION

The concept of beam emittance and the study of phase-space distributions in accelerator development began to be important as soon as the focusing properties of early accelerators were understood and increased intensities were demanded by users. Careful studies of ion-beam emittance started in the 1960s, corresponding to the availability of small computers, which could be used to automate what was otherwise a tedious set of measurements, perhaps requiring hours to complete. Recent high-power accelerators, for example FMIT,[1] and proposed future devices put stringent requirements on emittance that are not fully understood at present, but success will demand better control of the beams and their emittance distributions. The reasons for this are that the accelerator itself can be damaged by beam loss, and for some applications the beam deliverable to the target depends directly on its emittance.

CONCEPTS

An excellent and comprehensive review[2] of emittance and brightness has been written by LeJeune and Aubert, and the present discussion will follow much of their notation. The multiplicity of definitions used by researchers partly reflects the fact that no single emittance value is adequate to describe all requirements. For high-power dc beams, it may be necessary to know the emittance for 99.9% of the beam, but for low-power pulsed beams only the central core may be important. Most measurements reduce the four-dimensional distribution $\rho_4(xx'yy')$ to the two-dimensional distribution $\rho_2(xx')$, for example, and the normalized two-dimensional emittance ϵ is given by $\beta\gamma A/\pi$,

---

* Work performed under the auspices of the U.S. Dept. of Energy and supported by the U.S. Army Strategic Defense Command.

where A is the area described by ρ_2 and $\beta\gamma$ is the relativistic velocity factor. For careful study of the emittance of cylindrically symmetric beams, an rr' measurement across the beam diameter can yield more information about the ion-optical effects.[3] The normalization changes the phase space from xx' coordinates to those of xẋ/c, so that ϵ is generally conserved according to Liouville's theorem[4] as the beam is transformed by acceleration and focusing. Unfortunately, the phase-space distribution may become highly distorted while maintaining the same area, a process known as filamentation,[2] and the measured or effective emittance may increase in disregard of the theorem. Also, it is necessary to give the fraction of current within ϵ to give it meaning; thus for example, ϵ_{90} means the emittance containing 90% of the beam integrated over all emittance in the other plane.

The concept of rms emittance was introduced first by Agritellis[5] and later developed by Lapostolle[6] and Sacherer[7] in the derivation of first-order envelope equations for beam transport of arbitrary distributions of particles. From their statistical analysis, they identified a quantity closely related to ϵ, which (following Ref. 2) is called the rms emittance $\bar{\epsilon}$,

$$\bar{\epsilon} = 4\beta\gamma\sqrt{[\overline{x^2}\ \overline{x'^2} - (\overline{xx'})^2]} \qquad (1)$$

where the moments are weighted by the intensities ρ_2 in the distribution. This appears to be the most commonly used definition of rms emittance; however, the factor of 4 is omitted by many authors. The notation $\bar{\epsilon} = \epsilon_{4\mathrm{rms}} = 4\epsilon_{\mathrm{rms}}$ is suggested. The rms emittance, Eq. (1), has several desirable characteristics:

1. $\bar{\epsilon}$ enters directly the first-order beam transport equations, and hence it is the "correct" emittance to use for this purpose;
2. it is approximately equal to ϵ_{90} for simple distributions, but it is larger if distortions, such as filamentation, are present,
3. it is conserved under the influences of linear forces but grows otherwise; and
4. as a weighted measure, there is no question of the beam fraction involved.

The emittance may also be described[8] by the minimum-area ellipse that will enclose all or a given fraction of the beam. This may be useful in situations for which unusual halos or distortions are present, and it has been implemented in the beam extraction code SNOW[9] to supplement the rms calculations to help compare different beam extraction designs.

For beams extracted from a plasma, there are at least three sources of emittance: (1) ion temperature, (2) extraction aberrations, and (3) fluctuations in time of the plasma. If ions have a Maxwellian energy distribution with temperature kT across a spatially uniform beam and only (1) is important, then $\rho_4 = C(x,y)e^{-(\dot{x}^2+\dot{y}^2)M/2kT}$ where C is a constant within the beam boundaries and[10,11]

$$\bar{\epsilon} = 2r\sqrt{(kT/Mc^2)}\ for\ an\ emitter\ of\ diameter\ 2r \qquad (2)$$

and
$$\bar{\epsilon} = 4r\sqrt{(kT/3Mc^2)} \text{ for a slit of width } 2r \ . \qquad (3)$$

The fraction of beam within emittance ϵ for the latter case is

$$F = erf(\pi\epsilon/\sqrt{6}\bar{\epsilon}) \qquad (4)$$

with a slightly different result for the aperture.[11] Common experience indicates that real beam distributions are more typically "Gaussian," which is defined by a beam fraction F and intensity T given by

$$F = 1 - e^{-2\epsilon/\bar{\epsilon}} \qquad (5)$$

and

$$T = e^{-2\epsilon/\bar{\epsilon}} \qquad (6)$$

where T is ρ_2 normalized to a maximum of unity. This distribution is significantly different from that for the "Maxwellian" beam. A measure of $\bar{\epsilon}$ is often obtained by plotting ϵ vs $\ln[1/(1-F)]$ (Fig. 1).

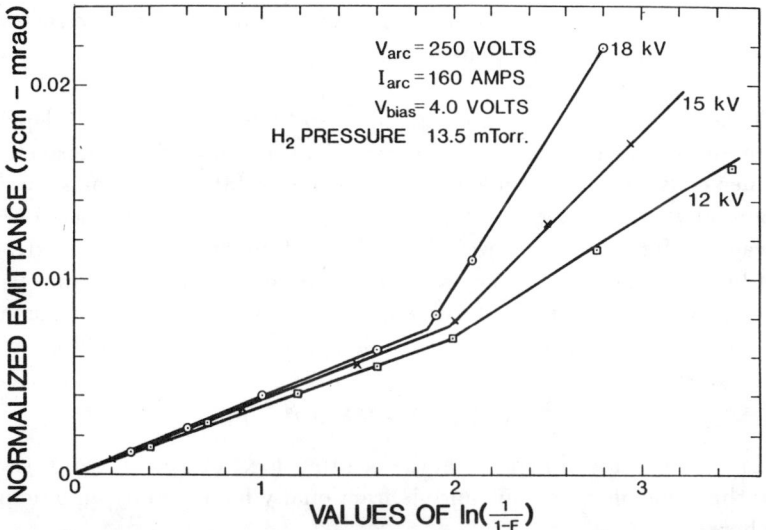

Fig. 1. Emittance ϵ vs $\ln(1/(1-F))$ for a 2-mA H$^-$ beam, ref. 36, showing quasi-Gaussian dependence for most of the beam, ion-optical effects at very high beam fractions.

The phase-space plot calculated by SNOW for ions with zero temperature extracted from a slit is shown in Fig. 2. The minimum area ellipse enclosing all particles is seen to be larger than the rms ellipse. Obviously, ϵ is zero but $\bar{\epsilon}$ is not, and an effective kT can be calculated from it, using Eq. (2) or (3) to benchmark the quality of the design. A given design will have a constant laboratory emittance $\bar{\epsilon}_L = \bar{\epsilon}/\beta\gamma$ at constant perveance if the effects of

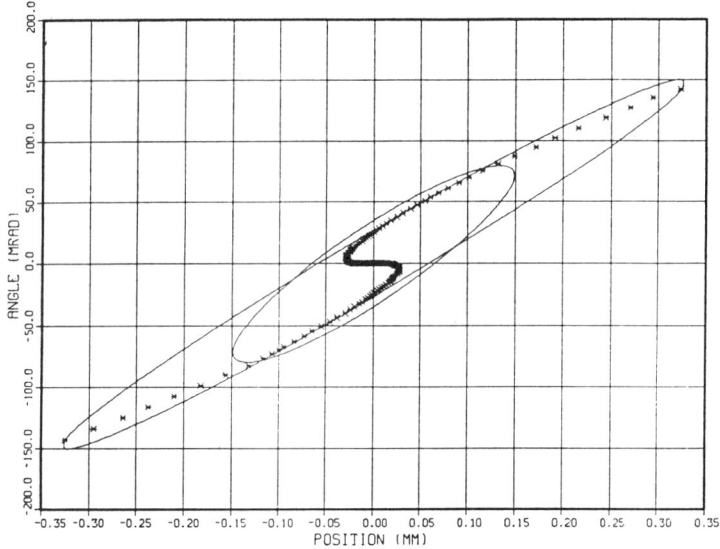

Fig. 2. Emittance diagram for 130 mA/cm beam from slit extractor, 20-kV, zero ion temperature.

ion and electron temperature are negligible, so the effective kT will depend on beam energy. In this paper, all ion temperatures quoted are those effective values consistent with the emittance. The emittance is the measured and nominally conserved quantity, of course, and kT is neither; therefore, it is not always useful to parametrize a beam by its temperature. If the extracted current fluctuates, $\bar{\epsilon}$ can be calculated from the moments of distributions superposed in time. A simpler alternate method uses the concept of mismatch factor,[12] in which the beam phase space is approximated by an ellipse with the $\gamma\alpha\beta$ parameters

$$\epsilon_L = (\gamma x^2 + 2\alpha x x' + \beta x'^2). \quad (7)$$

If time variations in space charge or other forces result in two variations of $\gamma\alpha\beta$, then the effective emittance is fractionally increased by an amount MM, where

$$MM = \sqrt{[1 + \Delta/2 + \sqrt{(\Delta + \Delta^2/4)}]} - 1 \quad (8)$$

and

$$\Delta = \beta_1 \gamma_2 + \beta_2 \gamma_1 - 2\alpha_1 \alpha_2 - 2 \quad (9)$$

with the subscripts 1 and 2 referring to the different ellipse orientations. The mismatch factor is the fractional increase in ellipse area needed to enclose two ellipses of equal area but of different orientation. For the extraction geometry shown in Fig. 2, MM $\simeq$ 0.2 for $\delta i/i = 0.1$. Although not an exact quantitative prediction, the mismatch factor can be used to gauge sensitivity to fluctuations of many sorts. This method was used to estimate the increase in emittance in a transport channel under conditions of fluctuations in the space-charge neutralization.[13]

MEASUREMENT TECHNIQUES

Perhaps the simplest type of ϵ-scanner is the pepper pot,[2] which can be used to obtain the four-dimensional phase-space density $\rho_4(xx'yy')$. This is often very valuable information, but the difficulty of automating the measurement has limited interest in its use. Possibly future demands will require automated measurements of ρ_4, but at present the two-dimensional distribution $\rho_2\,(xx')$ is measured by integration over yy' and vice versa. Many different types of scanners have been developed.[2] Two devices that are widely used are the slit and collector scanner (SCS[14]) (Fig. 3) and the electric-sweep scanner

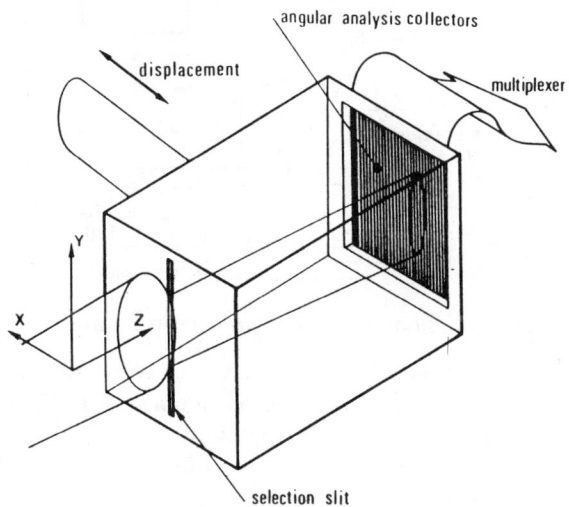

Fig. 3. Schematic of slit and collector emittance scanner, from Ref. 2, Fig. 12.

(ESS[15]) (Fig. 4). Both devices measure ρ_2 as a function of angle x' at one position of the beam and then are moved across the beam to complete the measurements as a function of x. A comparison of these two devices shows the following:
1. Both take measurements over a period of time (many seconds) necessary to move the scanner across the beam, hence some averaging is done.
2. The SCS can take the x' distribution over a short time interval, <1 μs, whereas the ESS requires 10-100 μs. Thus, time dependence at one position is better observed with the SCS.
3. The ESS is limited in use to beam energies $\phi \lesssim 1$ MeV to maintain reasonable scan voltage ($V_s \simeq 2x'^2_m \phi$ for maximum scan angle x'_m), whereas the SCS can be used at much higher energies.

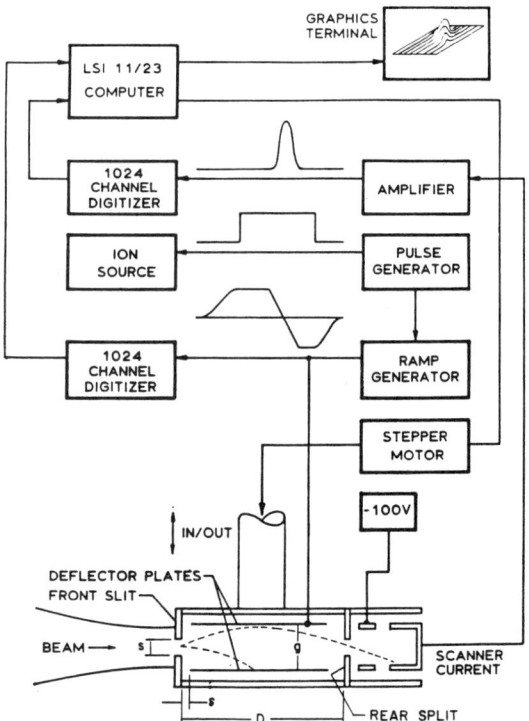

Fig. 4. Schematic of electric-sweep scanner.

4. The SCS, using many parallel collectors, usually requires longer beamline space to get the same angular resolution as the ESS, which has an inherently finer collector, its analyzing slit. This results generally in better angular resolution for the ESS, for example ±1/4 mrad for a 5-cm-long version at Los Alamos.
5. As a result of 4, the ESS is particularly useful for high perveance beams because the maximum drift D allowable before space-charge forces on the ribbon beam passing through the slit double its width is[15]

$$D^2 < 2^{5/2} \epsilon_o A / P_e (m/e)^{1/2} \qquad (10)$$

where A is the beam cross section and P_e is the electron equivalent perveance $P_e = \sqrt{\mu} I/\phi^{3/2}$, μ being the ratio of beam particle mass M to the electron mass m.
6. The SCS requires more electronics, but that for the ESS is more complex. On the whole, the ESS is perhaps somewhat simpler.

A tomographic method[16] (TM) has been used to provide a nonintercep-tive measurement, and it has also been used with intercepting profile scanners.[17] A comparison of measurements made with an ESS and the TM using three wires to measure the beam profile at three locations was made at TRIUMF,[18] and reasonable agreement was found, with the TM result being $\simeq 30\%$ higher than that for the ESS. Good agreement was found at Los Alamos with a 100-kV beam in comparing measurements with an ESS and an SCS.

Accuracy of emittance measurements is rarely discussed, particularly regarding the possibility that the device itself may alter the beam phase space by changing the space-charge forces in the beam by secondary emission of electrons or other beam-plasma effects. Guyard and Weiss[12] have investigated the effect of angular bin size and concluded that the data must cover at least five bins to reach 10% accuracy. We investigated[19] this by simulating a Gaussian beam, Eq. (5), for which the distribution ρ_2 had an elliptical boundary given by Eq. (7) and was sampled with a slit of zero width at the center of the bin size Δx, but was averaged in angle over the bin width $\Delta x'$. Normally the slit width is very small compared with the bin size, but the angular resolution equals the bin size. The parameters $\gamma \alpha \beta$ were chosen so that the intensity contour $T = 0.001$ just fit into the simulated scan region of half-widths x_m by x'_m (Fig. 5), with the number of bins being given by $N_x = 2x_m/\Delta x$ and $N_{x'} = 2x'_m/\Delta x'$. With these assumptions it was found that the rms emittance calculated by our data analysis program REANE for 100% of the beam fraction exceeded the simulated value by a fractional amount δ as shown in the table below. The quantity $N_s = \sqrt{(0.5 \ell n 1/T)}\, \bar{\epsilon} N_{x'}/x_m x_m'$ is the number

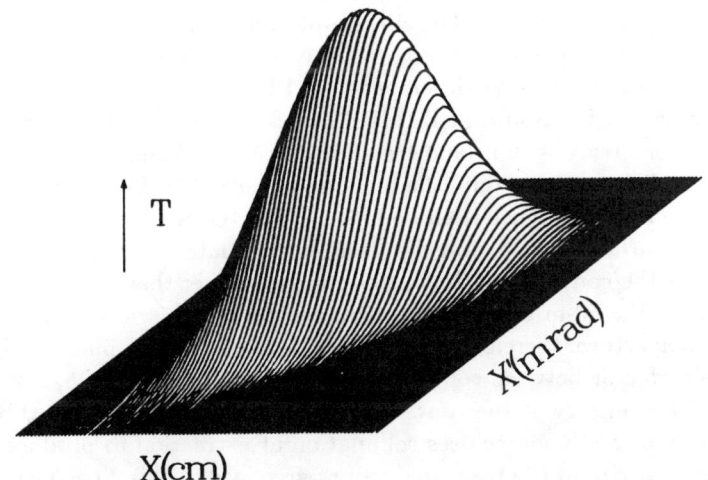

Fig. 5. Isometric plot of simulated Gaussian emittance distribution, T vs x and x'.

of angle bins that the data covers at the center of the distribution for the $\epsilon = \bar{\epsilon}$ contour. The effect of minimum sensitivity T_m, equivalent to the noise-to-signal ratio, is also shown for several cases.

TABLE I. Error in Calculated Emittance for Various Bin Sizes and T_m

δ	# x bins	N_s	T_m
-0.007	$\simeq\infty(100)$	$\simeq\infty(19)$	$\simeq 0(0.001)$
0.05	"	3.1	"
0.10	"	2.3	"
-0.01	8	$\simeq\infty(19)$	"
0.03	6	"	"
-0.28	5	"	"
-0.11	$\simeq\infty(100)$	"	0.03
-0.26	"	"	0.10

Although the rms emittance for such a distribution can therefore be determined accurately with rather coarse bins in Δx and $\Delta x'$, more bins with finer resolution are required for practical distributions.

RESULTS OF MEASUREMENTS

It is interesting to compare some H$^-$ sources to try to deduce the factors controlling the emittance. The surface-plasma source with independent converter was first proposed and tested[20] at Novosibirsk in a Penning version, and a cusped-field type source developed[21] at LBL has been widely used, similar versions having been built at LAMPF[22] and at KfK.[23] A Penning version of different geometry was built and tested at ORNL.[24] A full set of emittance measurements has not been published for the latter (SITEX), but sputter-ion temperatures of 5-10 eV have been deduced.[25] For the KfK source, a spherical converter surface is used, and the beam is collimated to prevent extraction of ions at the converter with sputter energies larger than 1.5 eV. The emittance $\bar{\epsilon}$ for the 50-mA beam was estimated to be 0.16 π·cm·mrad for a 13-mm-diam aperture; therefore, kT = 14 eV at the emission plane. With a compression factor between converter and emission aperture of 8/1.3, the weighted sputter energy at the converter is 0.4 eV, about that allowed by collimation. The LAMPF source uses collimation at extraction to produce a predictable emittance, and the measured values are close to this. The geometry of the extraction and converter are shown in Fig. 6. At the converter center, it can be seen that the maximum allowable transverse energy $\phi_t = (r/d)^2 \phi_c$ for transmitted ions, where it is assumed that they leave the converter with energy ϕ_c and after drift d are collimated by the emission aperture of radius r. If the sputter energy $\phi \gg \phi_t$, then the converter of radius R may be considered as an emitter with uniform density ρ_4 for $x^2 + y^2 \leq R^2$ and for $\dot{x}^2 + \dot{y}^2 \leq 2 e\phi_c (r/d)^2/M$, defining the coordinate system at the converter to be curved. (The result of the calculation is not changed if the usual coordinate system is

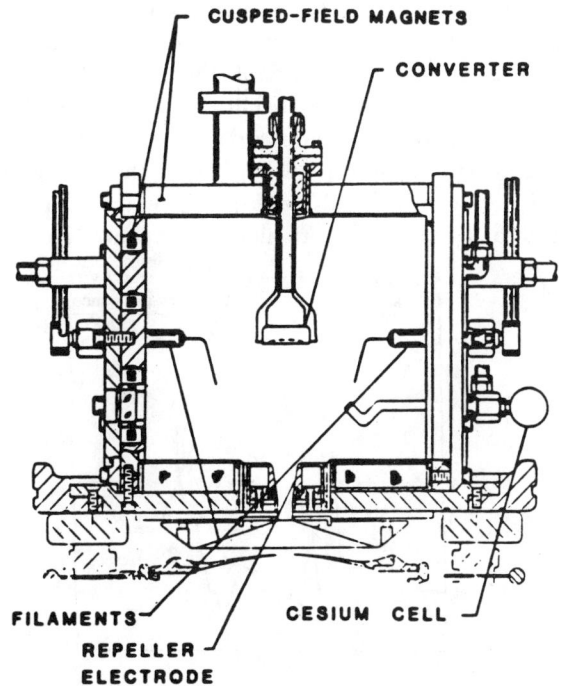

Fig. 6. Geometry of LAMPF converter H⁻ ion source.

used and the focusing is taken into account.) From these assumptions it follows that

$$\rho_2 = \sqrt{(1 - x^2/R^2)(1 - M\dot{x}^2(d/r)^2/2e\phi_c)} \tag{11}$$

and from Eq. (1)

$$\bar{\epsilon} = (rR/d)\sqrt{(2e\phi_c/Mc^2)} \,. \tag{12}$$

The emittance calculated from Eq. (12) with r = 0.5 cm, R = 1.905 cm, d = 12.6 cm agrees very well with the data and the prediction presented,[22] being 0.055 π·cm·mrad for the 20-mA beam vs the measured 0.052 with ϕ_c = 250 V. Then kT = 2.7 eV at the emission aperture and 0.19 eV at the converter, about half the maximum allowed by collimation, but in close agreement with the allowable weighted average. Thus, it seems likely that the converter fully illuminates the emitter and justifies the use of Eq. (11). The emittance is fully determined geometrically, with no further increase in extraction and transport having been observed except in turn-on transients and after a long beam transport. The scaling of $\bar{\epsilon}$ with ϕ_c over a limited range was shown to agree with the scaling of Eq. (12).

The source[26] used at LBL has a cylindrical converter to produce a dc beam current of 1.25 A, and the emittance measurements for the two planes provide an interesting comparison. The beam from the converter is collimated in the focusing plane by the extraction aperture set to a degree varying across the converter surface, with the maximum allowable $\phi_t = 2.0$ eV for $\phi_c = -130$ V, r = 15 mm, d = 120 mm. The emittance in the focusing plane measured 10-cm from the extractor and that calculated with the WOLF code[27] assuming large sputter energy are compared in Fig. 7, and these are seen to be in very good agreement. The conclusion is that the sputter energy is in fact $\gg 2.0$ eV, and the beam aberrations from the extraction fields are small but well-accounted for by the code. At higher perveance, ion-optical

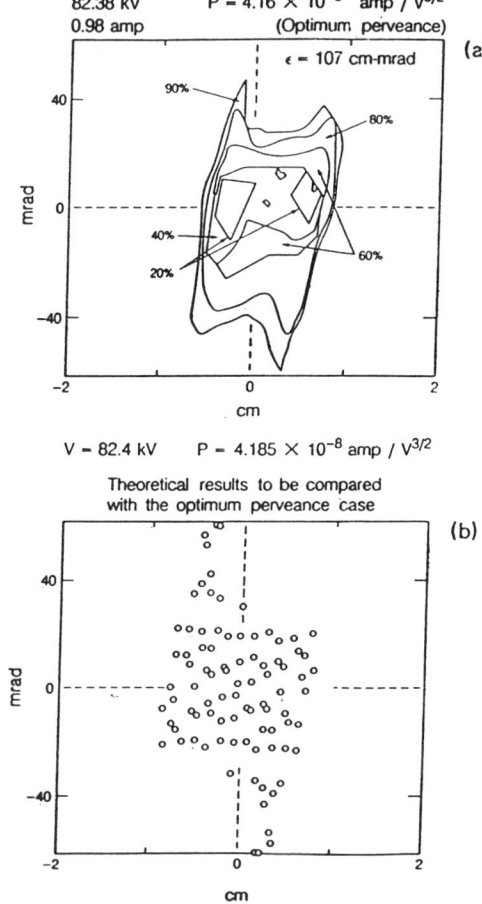

Fig. 7. Comparison of measured and calculated emittance in focusing plane cylindrical converter LBL source.[26]

effects cause the emittance in this plane to increase. The converter is 35 cm long and the emission plane is 25 cm long; therefore, the maximum acceptable sputter energy is about 130 V and it may be assumed that $\phi_t = \phi_s$ in the long, unfocused plane. The emittance plot (Fig. 8) then gives the energy distribution of converter-produced particles where $\phi_s = x'^2 \phi_b$, and ϕ_b

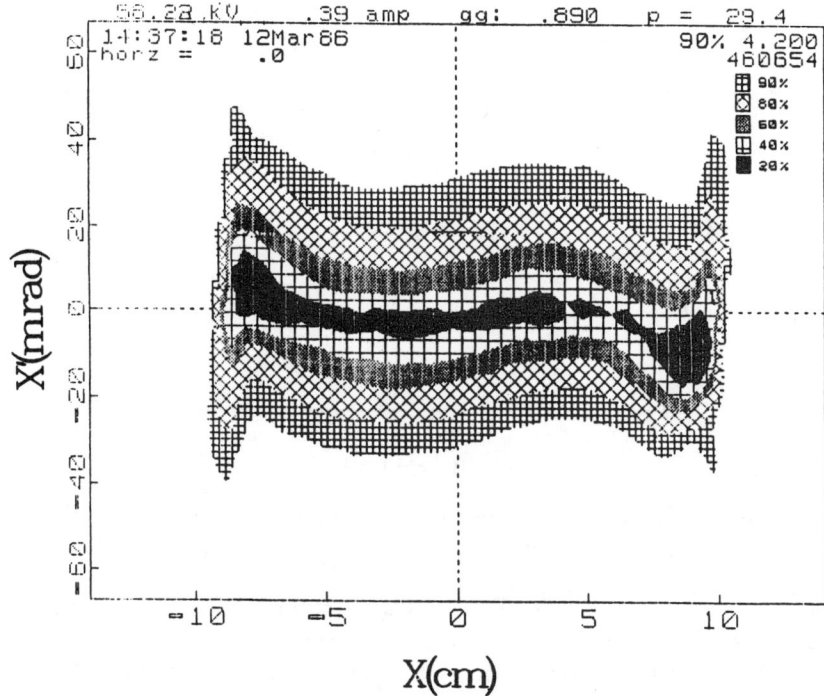

Fig. 8. Emittance plot in non focusing plane, showing beam fraction contours, for LBL source, courtesy of J. H. Kwan, LBL

= 56 kV. The plot in Fig. 9 of $\sqrt{\phi_s}$ vs $\ell n[1/(1-F)]$ shows that the emittance is nearly Gaussian, and the average sputter energy is $\simeq 15$ eV. This large energy of some of the ions suggests a substantial fraction of reflected particles from the converter. At lower converter voltages, the energy spread decreases, but it does not seem to increase at higher voltages as might be expected.

Grooved magnetrons are used to produce beam currents of about 50 mA at both FNAL[28] and BNL,[29] with 1- by 10-mm slits, and the values of ϵ_{90} reported along the 10-mm slit are nearly equal at 0.15 and 0.14 π·cm·mrad, respectively. Assuming that the beams have Gaussian distributions, $\bar{\epsilon} = 0.126$, from which kT = 11 eV according to Eq. (3). Since these sources directly extract the cathode-produced ions and since there is no collimation along this plane, this would be the sputter energy if there were no emittance growth in extraction and transport. Both sources have good quiescence, so that emittance growth caused by current fluctuations should be small. Both beams are transported considerable distance to the ϵ-scanner and are measured at 750-keV energy. Emittances in the narrow plane are $\bar{\epsilon} = 0.090$ and 0.030. In both cases, the maximum sputter energy that can be transmitted is about 16 eV, so collimation is minimal. Using Eq. (3), kT at the 1.4-mm-wide converter for $\bar{\epsilon} = 0.03$ is 33 eV; therefore, emittance in this plane is probably

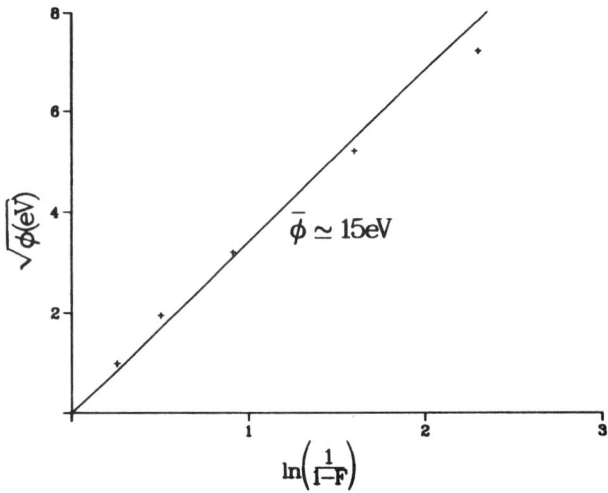

Fig. 9. Square root of sputter energy vs $\ell n(1/(1-F))$ for data of Fig. 8

determined by ion-optical effects. The emittance of a semiplanotron[30] at Novosibirsk is given as 0.02 by 0.006 (no beam fraction specified) for a 10- by 0.5-mm slit, and the authors give the energy spread at emission as 0.2 and 7 eV. This seems surprisingly low for a converter source and is in sharp contrast with the other results; however, the authors assume that the ions are essentially volume-produced in this mode of operation. The Penning source[31] used at Los Alamos produces a beam current of about 140 mA from a 7- by 0.8-mm slit, and $\bar{\epsilon} = 0.068$ by 0.024, implying kT = 6.6 and 63 eV. The emittance distributions in both planes usually closely follow the Gaussian model. The rms of a given beam fraction, $\bar{\epsilon}(F)$, is easily calculated in the experimental data analysis and is a more severe measure than total emittance. Using Eqs. (5) and (7), it is possible to show that for this distribution,

$$\bar{\epsilon}(F)/\bar{\epsilon} = 1 + (1 - F)ln(1 - F)/F. \tag{13}$$

The experimental value of $\bar{\epsilon}(F)$ is plotted against the right-hand side of Eq. (13) in Fig. 10 for the large plane, with similar results in the narrow plane. Since the ions are ultimately volume produced in this source, substantially smaller energies would be expected. In the narrow plane, the relatively large emittance may be due to a current nonuniformity of $\simeq 30\%$, which causes the ion optical properties to vary along the slit and to mix phase spaces as discussed in connection with Fig. 2. If the ions are created by resonant charge exchange of fast cathode ions with H atoms, then they should have the temperature of the atoms, which was found[32] to be 1-2 eV. This is closer to the temperature of 2-3 eV inferred by collimating the beam to eliminate the end-of-slot effects. Also charged particles produced in a transverse magnetic field

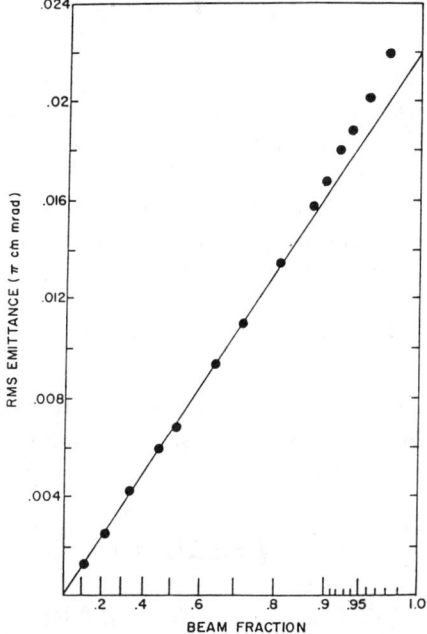

Fig. 10. Plot of $\bar{\epsilon}(F)$ vs $1+(1-F)\ln(1-F)/F$ for LANL Penning source.[31]

are dispersed by it, which in the absence of collisions increases the effective temperature.[33] It seems likely that, for this source, ion-optical effects play a dominant role in determining emittance.

The ion energies in volume sources are expected[34] to be a fraction of an eV, and therefore one may expect ion-optical effects to be much more important. The TRIUMF source[18] has been carefully studied with a four-element extraction system. There is obvious evidence of ion-optical aberrations, particularly at the highest current density, but the emittance distribution at lower density appears to be Gaussian with $\bar{\epsilon} = 0.0156$ π·cm·mrad, which for r = 3.25 mm corresponds to kT = 0.5 eV, about as expected. The beam current is about 4 mA dc. Measurements on an LBL[35] source at Los Alamos[36] with somewhat smaller apertures have shown that kT $\leq$ 1.5 eV for pulsed currents up to 10 mA, but it is clear that extraction effects still dominate the emittance because it varies strongly with extraction voltage. Preliminary measurements for D^- operation with this source at the same voltage as for H^- show that the emittance is lower by about 1.7 than for H^-, whereas $\sqrt{2}$ would be expected for either the same ion temperature or extraction aberrations at the same extraction voltage.

CONCLUSIONS

The emittance of beams from converter H^- sources is dominated by the ion energy at emission, and therefore this is the limiting factor in obtaining bright beams. The sputter energy is $\gtrsim 10$ eV for these sources. For the volume-type source, emittance measurements indicate that $kT \lesssim 0.5$ eV, and the extraction system may be the limiting factor. Fluctuation and nonuniformity of current may be important factors and for many cases must be $\leq 1\%$ to maintain emittance at the lowest level.

ACKNOWLEDGMENTS

For helpful discussions and permission to share their data, I am indebted to colleagues at the following institutions: BNL, Culham Labs, FNAL, LBL, Los Alamos, ORNL, and TRIUMF.

REFERENCES

1. W. D. Cornelius, "CW Operation of the FMIT RFQ Accelerator," IEEE Trans. Nucl. Sci. **32** (5), 3139 (1985).
2. Claude Lejeune and Jean Aubert, "Emittance and Brightness: Definitions and Measurements," Adv. in Electronics and Electron Physics, Supplement 13A, Applied Charged Particle Optics, A. Septier, ed., Part A, Academic Press 129 (1980).
3. Roderich Keller, GSI, Darmstadt, Germany, private communication, (1986).
4. Ref. 1, 167.
5. C. Agritellis, R. Chasman, and Th. J. M. Sluyters, IEEE Trans. Nuc. Sci. **16** (3), 221 (1969).
6. Pierre Lapostolle, "Possible Emittance Increase through Filamentation due to Space Charge in Continuous Beams," IEEE Trans. Nuc. Sci. **18** (3), 1101 (1971).
7. F. J. Sacherer (1971), "RMS Envelope Equations with Space Charge," IEEE Trans. Nuc. Sci. **18** (3), 1105 (1971).
8. Roderich Keller, Joseph D. Sherman, Paul Allison, "Use of a Minimum-Ellips Criterion in the Study of Ion-Beam Extraction Systems," IEEE Trans. Nuc. Sci. **32** (5), 2579 (1985).
9. J. E. Boers, "SNOW-A Digital Computer Program for the Simulation of Ion Beam Devices," Sandia National Laboratory report SAND79-1027 (1979).
10. J. D. Lawson "The Physics of Charged Particle Beams," (Oxford Press, Oxford, 175 (1977).
11. Paul Allison, Joseph D. Sherman, H. Vernon Smith, Jr., "Comparison of Measured Emittance of an H^- Ion Beam with a Simple Theory," Los Alamos National Laboratory report LA-8808-MS, (1981).

12. J. Guyard and M. Weiss, "Use of Beam Emittance Measurements in Matching Problems," Proc. 1976 Linear Accel. Conf., Atomic Energy of Canada, Limited report AECL-5677, 254 (1976).
13. Paul Allison, "Emittance Growth Caused by Current Variation in a Beam-Transport Channel," IEEE Trans. Nuc. Sci., **32** (5), 2556 (1985).
14. Robert W. Goodwin, Edward R. Gray, Glenn M. Lee, and Michael F. Shea, "Beam Diagnostics for the Fermi National Laboratory 200-MeV Linac," Proc. 1970 Proton Linear Accel. Conf., Batavia, Illinois, Fermi National Laboratory report LCO-058, 107 (1970).
15. Paul W. Allison, Joseph D. Sherman, and David B. Holtkamp, "An Emittance Scanner for Intense Low-Energy Ion Beams," IEEE Trans. Nuc. Sci., **30** (4), 2204 (1983).
16. D. D. Chamberlin, J. S. Hollabaugh, and C. J. Stump, Jr., "Images to Photodiode Beam-Profile Imaging System," IEEE Trans. Nuc. Sci., **30** (4), 2201 (1983).
17. O. R. Sander, G. N. Minerbo, R. A. Jameson, and D. D. Chamberlin, Proc. 1979 Linear Accelerator Conference, Montauk, New York, September 1979, Brookhaven National Laboratory report 51134, 314 (1979).
18. K. R. Kendall, M. McDonald, D. R. Mosscrop, P. W. Schmor, and D. Yuan, "Measurements on a DC Volume H^- Multicusp Ion Source for TRIUMF," RSI **(7)**, 1277 (1986).
19. Paul Allison, "Test of Emittance Analysis Program with Simulated Data," Los Alamos National Laboratory report AT-2 Tech. Note AT-2-ATS-23, (1986).
20. Yu. I. Bel'chenko, G. I. Dimov, V. G. Dudnikov, "Physical Principles of the Surface Plasma Method for Producing Beams of Negative Ions," Proc. Symp. on Production and Neutralization of Negative Hydrogen Ions and Beams, September 1977, Brookhaven National Laboratory report BNL 50727, 79 (1977).
21. K. W. Ehlers and K. N. Leung, RSI **51** 721 (1980).
22. Ralph R. Stevens, Jr., Rob L. York, John R. McConnell, Robert Kandarian, "Status of the new High Intensity H^- Injector at LAMPF," Proc. 1984 Linac Conf., Gesellschaft für Schwerionenforschung, May 7-11, 1984, Dormstadt report GSI-84-11, 226 (1984).
23. B. Piosczyk and G. Dammertz, "Continuous Operated H^- Surface Plasma Source,"RSI 57 (5), 840-846 (1986).
24. W. K. Dagenhart, W. L. Stirling, H. H. Haselton, G. G. Kelley, J. Kim, C. C. Tsai, and J. H. Whealton, "Modified Calutron Negative Ion Source Operation and Future Plans," Proc Second Int. Symp. on the Production and Neutralization of Negative Hydrogen Ions and Beams, Brookhaven, Brookhaven National Laboratory report BNL 51304, 217 (1980).
25. J. H. Whealton, ORNL, private communication, (1986).

26. J. H. Kwan, G. D. Ackerman, O. A. Anderson, C. F. Chan, W. S. Cooper, G. J. deVries, A. F. Lietzke, L. Soroka, and W. F. Steele, "One-Ampere, 80-kV, Long Pulse H^- Source and Accelerator," RSI **57**(5), 831 (1986).
27. W. S. Cooper, K. Halbach, and S. B. Magyary, Proc. 2nd Symp. on Ion Sources and Formation of Ion Beams, Berkeley, California, (1974).
28. C. D. Curtis, C. W. Owen, and C. W. Schmidt, "Factors Affecting H^- Performance in the Fermilab Linac," Proc. 1986 Linac Conf, Stanford Linear Accelerator Center, Stanford, California, June 2-6, 1986, to be published.
29. R. L. Witkover, Proc. Third Int. Symp. on the Production and Neutralization of Negative Ions and Beams, 1983, AIP Conf. Proc. No. 111, 398 (1983).
30. G. E. Derevyankin and V. G. Dudnikov, preprint IYaF 86-20, Novosibirsk (1986).
31. Paul Allison, H. Vernon Smith, Jr., and Joseph D. Sherman, "H^- Ion Source Research at Los Alamos," 171, Ref. 24, p 171.
32. Roderich Keller and H. Vernon Smith, Jr., "Spectroscopic Measurements on and H^- Ion Source Discharge," Ref. 13, p. 1736.
33. H. Vernon Smith, Jr., Paul Allison, and Joseph D. Sherman, "The 4X Source," Ref. 13, p. 1797.
34. This has been discussed by several authors, for example, J. M. Wadehra, "Negative Ion Production via Dissociative Attachment to H_2," Ref. 29, p. 47.
35. K. N. Leung, K. W. Ehlers, R. V. Pyle, "Optimization of H^- Production in a Magnetically Filtered Multicusp Source," RSI **56**, 364 (1985).
36. R. R. Stevens, Jr., R. L. York, K. N. Leung, and K. W. Ehlers, "Operation of a Magnetically Filtered Multicusp Volume Source," these proceedings.

DISCUSSION

<u>York</u>: Just a comment on the discrepancy between the TRIUMF 0.5 eV value and the 1.5 eV value. Under certain conditions, where we control the electron ratio using the plasma electrode, we approach ion temperatures of 0.5 eV at approximately the same currents that TRIUMF saw. So we believe that the problem is associated with the high flux of electrons that you extract with the H^- ions, and it's not reflective of the actual ion temperature in the source.

<u>Allison</u>: Right. Both the measurements that Ralph Stevens has made and the measurements that Culham has made show that rather clearly. So it has not been demonstrated that it's less than 0.5 eV, but the higher values can be taken with a grain of electrons, I guess.

<u>Witkover</u>: A comment about the slit and collector method being usable up to and above 100 MeV. At Brookhaven, we have been using it since about 1972, at 200 MeV, by the expedient of a partially degrading slit to wide-angle scatter the beam. A 1 cm-thick slit is all that's needed and the background is so thin that it's immeasureable. It makes an easy-to-do measurement. The other comment is, at lower energies, at extraction energies, for the H^- sources at least, the need for a suppressor grid across the collector array is very important. Otherwise, you can actually get the distributions split in half with an absence of any signal from the center line of the distribution.

<u>Allison</u>: Right. I think similar effects have been seen elsewhere on that.

REVIEW OF COMPUTER MODELING OF NEGATIVE ION BEAM FORMATION*

J. H. Whealton
Oak Ridge National Laboratory
Oak Ridge, Tennessee

Negative ion beam formation from volume-produced negative ion sources is an important and difficult issue. It is significantly different from positive ion beam formation; for positive ions the formation plasma can be thought to consist of positive ions and relatively fast and confined (read equilibrium) electrons. The ions see a downhill run to the accelerator column (presheath) and become extracted; the electrons want no part of the accelerator and stay away. A natural sheath occurs (illustrated in Fig. 1) as predicted by the collisionless Vlasov-Poisson (V-P) plasma equations:

$$\nabla^2 \phi(\mathbf{x}) = \int f(\mathbf{v},\mathbf{x}) d\mathbf{v} - \exp[-\Phi(\mathbf{x})] \tag{1}$$

$$(\mathbf{v} \cdot \nabla_x) f(\mathbf{v},\mathbf{x}) + \{[\nabla \phi(\mathbf{x}) - \mathbf{v} \times \mathbf{B}] \cdot \nabla_v\} f(\mathbf{v},\mathbf{x}) = f_0(\mathbf{v},0) \tag{2}$$

usually considered in two or three spatial dimensions. Appendix I describes the numerical difficulties associated with solving the V-P equations in 2 or 3D and discusses the state of the art in negative ion extraction modeling. This formulation appears to be adequate for describing the extraction optics for positive ions.

For negative ions, however, the formation plasma consists of three species: positive ions, electrons, and negative ions; equations for the ion dynamics are:

$$\nabla^2 \phi(\mathbf{x}) = \int f_-(\mathbf{v}_-,\mathbf{x}) d\mathbf{v}_- - \int f_+(\mathbf{v}_+,\mathbf{x}) d\mathbf{v}_+ + \int f_e(\mathbf{v}_e,\mathbf{x}) d\mathbf{v}_e \tag{3}$$

$$(\mathbf{v}_- \cdot \nabla_x) f_-(\mathbf{v}_-,\mathbf{x}) + \{[\nabla \Phi(\mathbf{x}) - \mathbf{v}_- \times \mathbf{B}] \cdot \nabla_v\} f_-(\mathbf{v}_-,\mathbf{x}) = f_{0-}(\mathbf{v}_-,\mathbf{x}) \tag{4}$$

*Research sponsored in part by the Department of Defense under Interagency Agreement DOE No. 40-1442-84 and Army No. W31RPD-53-A180 with the Office of Fusion Energy, U.S. Department of Energy, under contract DE-AC05-84OR21400 with Martin Marietta Energy Systems, Inc., and in part by Energy Systems IRAD support.

"The submitted manuscript has been authored by a contractor of the U.S. Government under contract No. DE-AC05-84OR21400. Accordingly, the U.S. Government retains a nonexclusive, royalty-free license to publish or reproduce the published form of this contribution, or allow others to do so, for U.S. Government purposes."

© American Institute of Physics 1987

$$(v_e \cdot \nabla_x)f_e(v_e,x) + \{[\nabla\Phi(x) - v_e \times B] \cdot \nabla_{v_e}\}f_e(v_e, x) = f_{0e}(v_e, x) \quad (5)$$

$$(v_+ \cdot \nabla_x)f_+(v_+,x) - \{[\nabla\Phi(x) - v_+ \times B] \cdot \nabla_{v_+}\}f_+(v_+, x) = f_{0_+}(v_+, x) \quad (6)$$

which differ significantly from Eqs. (1) and (2), which they replace. We call this a "triple Vlasov" model in celebration of the fact that there are three Vlasov equations, Eqs. (4) and (6), as opposed to the "single Vlasov" dynamical system represented by Eqs. (1) and (2).

In this case the electrons represented by Eq. (5) are no longer repelled by the acceleration fields and, in the absence of a confusing medium (such as a transverse component of magnetic field), will rocket out of the source causing destruction at the target and depletion in the source. The confined species in this picture, positive ions, represented by Eq. (6), is slow and less able to respond in the relatively rapid fashion expected of the electrons. While no attempts have been made, to our knowledge, to solve the full nonlinear multidimensional Eqs. (3)-(6), an attempt[1] has been made to dispatch a meaningful variant:

$$\nabla^2\phi(x) = \int f_-(v_-,x)dv_- - \int f_+(v_+,x)dv_+ - \exp[-\phi(x)] \quad (7)$$

coupled with Eqs. (4) and (6) ("double Vlasov" opposing). Another meaningful variant of Eqs. (3)-(6) is:

$$\nabla^2\phi(x) = \int f_-(v_-,x)dv_- + \int f_e(v_e,x)dv_e - \exp[-\phi(x)] \quad (8)$$

coupled with Eqs. (4) and (5). This analysis, more successful than the more difficult Eqs. (4), (6), and (7), is described in Refs. 2-4. One might be tempted to imagine that the successful negative ion sources that we hear about, Ref. 5-7, would have it that the negative ions, once formed, see a downhill run in analogue with positive ions. But the situation is different; negative ion production is maximized for a positive plasma potential.[8,9] One interpretation of this is that electron confinement is the most important parameter even if a concomitant effect is a possible potential hill, or saddle, for negative ion extraction. Another difference between positive and negative ion extraction is that fast electrons want to rocket out of the source into the accelerator; the retarded species is the positive ions in the plasma which can scarcely be thought of as an equilibrium species, as are the electrons in a positive ion source. Positive ions are only inertially confined: the furthest thing from equilibrium. If one wanted to treat accurately the positive ions in the plasma and assume the electrons will prevent significant space charge imbalance, then Eqs. (4), (6), and (7) would describe the dynamics in this approximation. Electron motion in the sheath is neglected. If one wanted to treat the electron sheath motion and assume that the effect of the positive ions is neutralized by a component of the electron distribution function, then Eqs. (4), (5), and (8) represent this approximation. If one is not too concerned with either the electrons in the sheath or the specific positive ion motion in the plasma, then the more tractable Eqs. (1) and (2) would appear to be an adequate representation. An alternative rationalization for the use of Eqs. (1) and (2) is as follows: suppose the electrons, before being sucked into the accelerator, noticed a potential imbalance between the negative ions and positive ions, then these electrons would move to correct this imbalance, locally, in their customarily swift manner (read equilibrium). For the appropriate sign of the source terms in the above equations, the equilibrium terms can be made to correspond to the rapid deployment of electrons away from, or absence of electrons toward, the local potential (or space charge) imbalance.

One small detail that needs to be attended to for the modeling of volume-produced negative ion sources is the source term for the negative ions. For positive ions the initial data for the ions are at a plane, $Z = 0$. This is justified in part because (a) the ions have a downhill run to the accelerator and (b) most of the actual generation of the positive ions is probably done to the left of the left-hand boundary (e.g., Fig. 1). In volume-produced negative ion sources, for reasons which will be described later, neither of these conditions appears to be true. Therefore, the ion source term in Eq. (2), $f_0(v,0)$, should be replaced by an actual "volume" source, $f_0(v,x)$. This is done for all the examples considered below relating to volume-produced negative ion sources. Solving Eqs. (1) and (2) with the above modification will enable many phenomena to be studied with respect to volume-produced negative ion extraction: to wit, the effect of saddle equipotential contours on ion optics, extraction across a magnetic field, and the advantages and disadvantages of deliberate field penetration.

In particular, the issue of a potential hill or saddle in the extraction region is of significance; experimental data suggest that increasing the extraction opening actually increases[10] the current density. This unexpected result suggests that field penetration from the accelerator (nulling the hill or opening up the saddle) is more important than the expected consequence of diminishing the local negative ion production rate (by decreasing the local neutral density, which happens because of the larger exit hole).

A specific indication of saddle point extraction effects is presented. A sheath from a positively biased plasma extracting negative ions is illustrated in Fig. 2; this is in contrast to a negatively biased plasma extracting negative ions illustrated in Fig. 1. A significant conclusion that may be drawn from Fig. 2 is that even though the extraction electrode is negatively biased with respect to the plasma, the accelerator fields penetrate the hole, allowing a downhill run for most available negative ions; in fact, the wall may even reflect some extra ions into the sheath. A transverse magnetic field of consequence may cause more significant effects, as the experiments of W. L. Stirling and colleagues[9] suggest.

Small perturbations of the extraction electrode cause enormous variations in current density; this effect may also have been observed by Allison, Smith, and colleagues.[11] A reliable modeling of these extraction processes appears warranted since not only is an improvement not only in optics possible but also an improvement in extractable negative ion current density.

As a specific example the extracted ions as a function of electrode bias and negative ion density are shown in Fig. 3. Specifically at low negative bias and low negative ion density, more negative ions get out than would be expected from geometrical considerations due to field penetration. (There is an axial magnetic field[12] present for this ringatron[13] configuration). However, for higher negative bias the reflection of negative ions by the potential barrier dominates and the ion current extracted diminishes. For this configuration a significant reduction of negative ion extraction occurs only for very high negative bias voltage (100 V) and so is not an issue. Other configurations are shown in Fig. 4 where the thickness of the electrode is varied. When the electrode is thick (high α), only a few volts cause significant diminishment of negative ion extraction.

We shall deal first with extraction across a strong magnetic field; negative ions sources of this kind are numerous[14] and are usually Penning-type discharges. A basic issue is illustrated in Fig. 5 which shows how the effective emittance of the volume-generated beam increases for increased magnetic field strength. The transverse magnetic field preferentially selects the high temperature ions for extraction.[3] For example, in the configuration of Fig. 6 if we consider how the negative ion current depends on initial speed and magnetic field strength, a result like that illustrated in Fig. 7 is obtained. Electrode configurations which will enable more ions to be extracted are possible.[15] Since the higher ion extraction configuration also allows more field penetration, they should also be relatively effective in sucking out reflected ions from the negatively biased wall (see Fig. 2, but note that the transverse magnetic field is zero).

Another variant which requires extraction across a magnetic field is the Dudnikov source.[17] Extensive optimization of the accelerator design was undertaken;[16] in this case the magnetic field is transverse to the slot as well as to the beam (in the case of Figs. 6 and 7 the magnetic field is parallel to the slot). However, the same issues of extraction inhibition in general, and for low energy ion filtration in particular, obtain.

It was later discovered[11] that the analysis reported in Ref. 16 made an unrealistic assumption that the plasma density was uniform. It turns out that there is a significant nonuniformity (factor of two along the slit), which has a drastic effect on ion optics. It is possible that the nonuniform density is due in part to the space charge of the electrons which are moving along the slit and building on one end of the slit. This effect, unlike the opposing double Vlasov attempts of Ref. 1, is amenable to a full resolution with a numerically stable double Vlasov treatment.

Surface production negative ion sources have seen significant development;[18,19] they have interesting attributes with respect to volume negative ion sources. Apparently there is an increase in yield possible with specially coated converter surfaces as opposed to volume production; the ions ejected from the converter usually fall down an electrostatic sheath of 100-300 V. If one assumes that the sheath electric fields were exactly normal to the surface, then the modeling of the ion optics is enormously simplified. This simplification has been taken advantage of in the modeling of such negative ion extraction optics. A question arises as to how close to normal to the surface the electric field is, especially with a configuration in which the converter surface is not parallel to the ambient magnetic field.[15]

Specific surface production ion sources are considered. The LBL "field-free self-extraction" negative ion source is considered in Refs. 3, 15, and 18; the adaption by Stevens and co-workers at LANL to a LAMPF injector[20] is discussed in Refs. 3 and 15. Embodiments by Stirling and co-workers[19] are considered in Refs. 3, 15, 21, 22, and 23. The temperature of ions ejected from the converter surface is found[18,20] to be higher (4-8 eV) than that of volume-produced ion sources. This may limit the applicability of such sources in some applications.

The above analyses include only steady-state effects; however, there are important phenomena which warrant a time-dependent analysis. Among these are transients and plasma instabilities caused by ion acoustic waves. Specifically the dynamical systems which we will consider are:

$$\nabla^2 \phi(\mathbf{x},t) = \int f_-(\mathbf{v}_-, \mathbf{x}, t) d\mathbf{v} - \exp[-\phi(\mathbf{x},t) T_e] \tag{9}$$

$$\frac{\partial f_-(\mathbf{v}_-,\mathbf{x},t)}{\partial t} + (v_- \bullet \nabla_x) f_-(\mathbf{v}_-,\mathbf{x},t) + [\nabla \Phi(\mathbf{x},t) \bullet \nabla_{v_-}] f_-(\mathbf{v}_-,\mathbf{x},t) = f_o(\mathbf{v}_-,\mathbf{x},t) \tag{10}$$

where T_e is the "electron" temperature. Some details of the analysis are given in Appendix II.

As an example of transient ion motion, a plasma approaching an electrostatic accelerator is studied. A value of T_e is chosen which is very low so that the Bohm sheath stability criterion is met. The evolution of the plasma is also described in Appendix III. A steady-state solution to the same problem, by solving Eqs. (1) and (2), is also considered in Appendix III, and the long time limit of Eqs. (9) and (10) is directly compared. If the "electron" temperature, T_e, is now increased and the above problem is reconsidered, an ion acoustic instability is excited; this is also shown in Appendix III. The inevitable excitation of ion acoustic plasma instabilities is relevant to either an ion source or an LEBT for an intense ion source (requiring a plasma for space charge quasi-neutralization). The effect of the waves diminishes for low values of T_e; this can be accomplished by sucking all the high energy electrons out of the plasma. In addition, if significant positive ion generation can be accomplished in the sheath then the waves may not be excited at all.

A logical procedure for adequately modeling the extraction of negative ions from a volume-produced negative ion source appears to be as follows (in order of difficulty). Apply existing positive ion plasma extraction analysis to positive space potential plasma neglecting negative ions. Consider, self-consistently, another Vlasov equation for the electrons (still retaining the Boltzman term). Consider the magnetic fields invariably placed at the extraction aperture (to cheat the electrons out of their just deserts). Include realistic volume equilibrium distribution from atomic physics considerations, and analyze all the edges of the plasma, including sources and sinks, to determine the probable nonlinear saturation amplitude of ion acoustic waves. Consider a hydrodynamics level analysis of the positive ion motion. Hopefully, meaningful predictive and corrective power can be obtained before all these steps are taken and experimental iteration appears warranted to most quickly expedite a reliable model.

In conclusion, we have examined the nature, pedigree, and limitation of the algorithm available to study negative ion extraction and acceleration. Several extraction issues have been resolved for at least some specific situations. In particular the effects of field penetration, negatively biased extraction electrodes, and transverse magnetic field on beam emittance and extractability have been delineated for several situations of interest. In addition, two issues with respect to beam transport through a plasma have been addressed. As specific examples, seven negative ion sources have been considered at a new level of detail.

ACKNOWLEDGMENTS

This review is the result of a long-standing collaboration with R. J. Raridon, M. A. Bell, K. E. Rothe, D. W. Wooten, R. W. McGaffey, J. W. Wooten, and J. C. Whitson, of the Computing and Telecommunications Division of Martin Marietta Energy Systems, Inc., and P. M. Ryan, G. L. Chen, W. L. Stirling, W. R. Becraft, and H. H. Haselton of the Oak Ridge National Laboratory.

REFERENCES

1. G. L. Chen, J. H. Whealton, and R. W. McGaffey, unpublished.
2. J. H. Whealton and R. W. McGaffey, Nucl. Instrum. Methods **203**, 377 (1982).
3. J. H. Whealton, Negative Ion Source Workshop, LANL (1985).
4. J. H. Whealton, Bull. Am. Phys. Soc. **26**, 1017 (1981); **27**, 1136 (1982); R. J. Raridon, J. H. Whealton, J. W. Wooten, R. W. McGaffey, D. E. Wooten, and C. C. Tsai, J. Appl. Phys. **57**, 819 (1985).
5. A. J. T. Holmes, G. Dammertz, and T. S. Green, Rev. Sci. Instrum. **56**, 1697 (1985); A. J. T. Holmes, T. S. Green, M. Inman, A. R. Wallser, and N. Hampton, Proc. Varenna-Grenoble Int. Symp. on Heating in Toroidal Plasma, Brussels (1982) p. 95; A. J. T. Holmes, G.

Dammertz, T. S. Green, and A. R. Walker, Proc. Int. Ion Engineering Congress (1983), p. 71; M. P. S. Nightingale, A. J. T. Holmes, J. D. Johnson, Rev. Sci. Instrum., in press; A. J. T. Holmes, Ion Source Workshop (1985), Los Alamos; A. J. T. Holmes, M. P. S. Nightingale and T. S. Green, Negative Ion Source Conf., Palaiseau (1986).

6. K. N. Leung, Ion Source Workshop, Los Alamos (1985); K. N. Leung, K. W. Ehlers, and R. V. Pyle, Rev. Sci. Instrum. **56**, 2097 (1985); R. L. York, R. R. Stevens, Jr., K. N. Leung, and K. W. Ehlers, Rev. Sci. Instrum. **55**, 681 (1984); K. N. Leung, K. W. Ehlers, and R. V. Pyle, Rev. Sci. Instrum. **56**, 364 (1985); Appl. Phys. Lett. **47**, 227 (1985); K. N. Leung, K. W. Ehlers, and M. Bacal, Rev. Sci. Instrum. **54**, 56 (1983).

7. M. Bacal and F. Hillion, Rev. Sci. Instrum. **56**, 2274, (1985); M. Bacal, F. Hillion, and M. Nachman, Rev. Sci. Instrum. **56**, 649 (1985).

8. K. Jimbo, K. W. Ehlers, K. N. Leung, and R. V. Pyle, Nucl. Instrum. Methods **A248**, 282 (1986).

9. W. L. Stirling, private communication (1986).

10. T. Green, private communication (1986); see also Ref. 5.

11. R. Keller, H. V. Smith, Jr., and P. Allison, Bull. Am. Phys. Soc. **30**, 1608 (1985).

12. **See Fig. 8 for the history of parallel extraction from a Penning source.** K. Wiesemann, K. Prelec, and T. Sluyters, Appl. Phys. **48**, 2668 (1977) W3-77; B. Bouchy-Lorentz, M. Remy, and C. Muller, Int. J. Mass Spectrosc. Ion Phys. **22**, 147 (1976) B-76; M. Kobayashi, K. Prelec, and T. Sluyters, Rev. Sci. Instrum. **47**, 1425 (1976) K2-76; C. Müller, B. Lorentz, and M. Remy, Int. J. Mass Spectrosc. Ion Phys., **18**, 33 (1975) M2-75; W. Aberth, R. Schmitzer, F. Engeaser, and M. Aubar, Ion Source Conf., LBL (1974) A2-74; T. Sluyters and K. Prelec, Nucl. Instrum. Methods **113**, 299 (1973) S-73;A. Cesati and F. Christofori, Energia Nucleare **19**, 36 (1972) C2-72; M. A. Abroyan, G. A. Nalivaiko, and S.G. Tsepakin, Sov. Phys. Tech. Phys. **17**, 690 (1972) C2-72; V. P. Golubev, G. A. Nahvaiko, and S. G. Tsepakin, Proc. Proton LINAC, Los Alamos (1972) G-72; M. Dubarry, These, Orsay (1972) D-72; M. A. Abroyan, V. P. Golubev, V. L. Komarov, G. V. Tarvid, G. M. Tokarev, and S. G. Tsepakin, Part. Accel. **2**, 133 (1971) A-71;N. Wells and P. R. Hauley, Ion Source Conf., BNL (1971) W2-71; V. L. Komarov, Z. G. Solnishkova, G. V. Tarvid, and S. G. Tsepakin, Ion Source Conf., Saclay (1969) K-69; M. Dubarry and G. Gautherin, Ion Source Conf., Saclay (1969) B-69; V. P. Golubev, V. L. Komarov, and S. G. Tsepakin, Ion Source Conf., Saclay (1969) G-69; M. A. Abroyan and G. M. Tokarev, Ion Source Conf. (1969) Saclay (A-69); W. Aberth and J. R. Peterson, Rev. Sci. Instrum. **38**, 745 (1967) A2-67; K. Bethge and G. Rau, Nucl. Instrum. Methods **39**, 157 (1966) B-66; L. E. Coilins and R. H. Gobbett, Nucl. Instrum. Methods **35**, 277 (1965) L-65; G. P. Lawrence, R. K. Beauchamps, and J. L. McKibben, Nuc. Instrum. Meth. **32**, 357 (1965) L-65; A. B. Wittbower, P. H. Rose, R. P. Bastide, and N. B. Brooks, Rev. Sci. Instrum. **35**, 1 (1964) W-64; A. B. Wittbower, R. P. Bastide, M. B. Brooks, and P. H. Rose, Phys. Lett. **3**, 336 (1963) W-63; C. D. Moak, H. E. Banta, J. N. Thurston, J. W. Johnson, and R. F. King, Rev. Sci. Instrum. **30**, 694 (1959) M-59.

13. J. H. Whealton et al., these proceedings.

14. **See Fig. 8 for the history of transverse extraction from a Penning source.** W. L. Stirling, W. K. Dagenhart, and C. C. Tsai, ORNL/TM-

9753 (1986) S3-86; K. Jimbo, K. W. Ehlers, K. N. Leung, and R. V. Pyle, Nucl. Instrum. Methods. **A248**, 282 (1986) J-86; M. D. Gabovich, Y. N. Kozyrev, A. P. Naida, L. S. Simaveuko, and I. A. Solosheuko, Sov. Tech. Phys. Lett. **4**, 153 (1978) G2-78; M. Baribaud, S. Bliman, J. M. Dolique, and F. Zadwarny, Ion Source Conf., Saclay (1969) B3-69; K. W. Ehlers, Nucl. Instrum. Methods. **32**, 309 (1965) E-65; K. W. Ehlers, B. F. Gavin, and E. L. Hubbard, Nucl. Instrum. Methods. **22**, 87 (1963) E-63; K. W. Ehlers, Nucl. Instrum. Methods. **18**, 571 (1962) E-62; E. L. Hubbard et al., Rev. Sci. Instrum. **32**, 621 (1961) H-61; R. J. Jones and A. Zucker, Rev. Sci. Instrum. **25**, 562 (1954) Z-54; C. E. Anderson and K. W. Ehlers, Rev. Sci. Instrum. **27**, 809 (1956) A3-56; R. S. Livingston and R. J. Jones, Rev. Sci. Instrum. **25**, 552 (1954) L2-54; M. S. Livingston, M. G. Holloway, and C. P. Baker, Rev. Sci. Instrum. **10**, 63 (1939) L2-39; F. M. Penning, Physica **4**, 71 (1937) P2-37.
15. J. H. Whealton et al., unpublished (1986).
16. M. A. Bell, J. H. Whealton, R. J. Raridon, R. W. McGaffey, and D. E. Wooten, Bull. Am. Phys. Soc. **30**, 1609 (1985).
17. H. V. Smith, Jr. P. Allison, and J. D. Sherman, IEEE Trans. Nucl. Sci. **NS32**, 1797 (1985); J. D. Sherman, P. Allison, and H. V. Smith, Jr., Neg. Ion Source Conf., Upton (1980); P. Allison and J. Sherman, Neg. Ion Source Workshop, Los Alamos (1985); P. W. Allison, IEEE Trans. Nucl. Sci. **NS24** (3), 1594 (1977); V. G. Dudnikov, IV U.S.S.R. Conf. Part. Accel. (1977).
18. K. W. Ehlers and K. N. Leung, Rev. Sci. Instrum. **51**, 721 (1980); Appl. Phys. Lett. **38**, 287 (1981); K. N. Leung and K. W. Ehlers, J. Appl. Phys. **52**, 3905 (1981); Rev. Sci. Instrum. **53** (1982).
19. W. K. Dagenhart, W. L. Stirling, and J. Kim, ORNL/TM-7895 (1982); W. K. Dagenhart et al., Neg. Ion Source Conf., Upton (1980); W. K. Dagenhart, W. L. Stirling, C. C. Tsai, and D. E. Schechter, Bull. Am. Phys. Soc. **30**, 1609 (1985); W. K. Dagenhart, W. L. Gardner, W. L. Stirling, and J. H. Whealton, Nucl. Technol./Fusion **4**, 1430 (1983); W. L. Stirling, W. K. Dagenhart, J. H. Whealton, and J. J. Donaghy, Neg. Ion Source Conf., Upton (1983): AIP Conf. Proc. **111**, 450; W. K. Dagenhart, W. L. Stirling, G. M. Banic, G. C. Barber, N. S. Ponte, and J. H. Whealton, ibid., 353.
20. R. L. York and R. R. Stevens, IEEE Trans. Nucl. Sci. **NS30**, 2705 (1983); Proc. Ion Source Conf., Upton (1983); R. L. York, R. R. Stevens, R. A. DeHaven, J. R. McConnell, E. P. Chamberlain, and R. Kandarian, 8th Conf. on Application of Accelerators in Research and Industry, Denton (1984).
21. J. H. Whealton et al., Ion Source Conf., Upton (1983), AIP Conf. Proc. **111**.
22. R. W. McGaffey et al., Ion Source Conf., Upton (1983), AIP Conf. Proc. **111**.
23. J. H. Whealton, R. J. Raridon, M. A. Bell, S. Y. Ohr, W. L. Stirling, D.E. Wooten. and R. W. McGaffey, Bull. Am. Phys. Soc. **30**, 1632 (1985).
24. B. Piosczyk and G. Dammertz, Rev. Sci. Instrum. **57**, 840 (1986); A. J. T. Holmes, M. P. S. Nightingale, and T. S. Green, Neg. Ion Conf. Palaiseau (1976), p. 215; A. J. T. Holmes and M. P. S. Nightingale, Rev. Sci. Instrum., in press; L. R. Elias and G. Ramian, IEEE Trans. Nucl. Sci. **NS32**, 1732 (1985); D. L. Friesel, T. Ellison, and W. P. Jones, IEEE Trans. Nucl. Sci. **NS32**, 2421 (1985); A. Kadish, W. Peter, and M. E. Jones, IEEE Trans. Nucl. Sci. **NS32**, 2576 (1985); R. True, IEEE

Trans. Nucl. Sci. **NS32**, 2611 (1985); A. I. Warwick, D. Vanecek, and O. Fredriksson, IEEE Trans. Nucl. Sci. **NS32**, 3196 (1985); A. J. T. Holmes, Negative Ion Source Workshop, LANL (1985); J. Ishikawa,Y. Takeiri, and T. Takagi, Rev. Sci. Instrum. **55** (1984); J. Ishikawa, Y. Takeiri, H. Tsuji, T. Taya, and T. Takagi, Nucl. Instrum. Methods **34**, 186 (1984); A. J. T. Holmes, E. Thompson, and F. Watters, *J. Phys. E.* **14**, 856 (1981); P. Raimbault, M. Fumelli, amd R. Becherer, 2nd Joint Grenoble Plasma Heating Conference (1980); W. B. Herrmannsfeldt, SLAC-226 (1979); Y. Ohara, *J. Appl. Phys.* **49**, 4711 (1978); Y. Ohara et al., *Jpn. J. Appl. Phys.* **17**, 423 (1978); J. Hauser and F. Tanzer, GK5S 77/E/52, ICPIG, Berlin (1977); H. R. Kaufman, *AIAA* **15**, 1025 (1977); T. Green, IEEE Trans. Nucl. Sci. **NS23**, 918 (1976); Y. Ohara, J. N. Inoue, H. Nihei, K. Asai, and T. Uchida, *Jpn. J. Appl. Phys.* **15**, 135 (1976); **15**, 1343 (1976); W. W. Hicks et al., Nucl. Instrum. Methods **139**, 25 (1976); D. Dirmikis, Thesis, Univ. Sheffield (1975);T. Sugawara and Y. Ohara, *Jpn. J. Appl. Phys.* **14**, 1029 (1975); V. S. Boldasov et al, *Sov. Phys. Dokl.* **19**, 652 (1975); K. W. Ehlers, W. R. Baker, K. H. Berkner, W. S. Cooper, W. B. Kunkel, R. V. Pyle, and J. W. Stearns, *J. Vac. Sci. Technol.* **10**, 922 (1973); B. I. Valkov, A. G. Sveshnikov, and N. N. Semashko, *Sov. Phys. Dokl.* **16**, 1040 (1972); W. S. Cooper, K. H. Berkner, and R. V. Pyle, *Nucl. Fusion* **12**, 263 (1972);K. H. Berkner et al., UCRL-72880 (1971); G. R. Nudd and K. Amboss, *AIAA* **8**, 649 (1970); J. L. Harrison, *J. Appl. Phys.* **39**, 3827 (1968); W. C. Lathem, *J. Spacecraft* **5**, 735 (1968); D. G. Bates, CLM-R53 (1966); C. D. Bogart and E. A. Richley, NASA TND-3394 (1966);P. T. Kirstein and J. S. Hornsby, *IEEE Trans. Elect. Dev.* **11**, 196 (1964); J. Hyman et al., *AIAA* **2**, 1739 (1964); V. Hamza, NASA TND-1711 (1963);P. T. Kirstein and J. S. Hornsby, CERN 63-16 (1963); V. Hamza, NASA Lewis TND-1711 (1963); V. Hamza and E. A. Richley, NASA TND-1323 (1962); TND-1665 (1963).

25. H. Nishimura, T. Sato, and S. Isobe, 9th Symp. Engr. Problems in Fusion Research (1981) Chicago; M. Mori, T. Nakamura, N. Inoue, and T. Uchida, *Jpn. J. Appl. Phys.* **19**, 1377 (1980); O. A. Anderson, Proc. 2nd Int. Symp. on the Production and Neutralization of Negative H_2 Ion Beam, BNL, p. 355 (1980); K. H. Berkner, W. S. Cooper, K. W. Ehlers, and R. V. Pyle, IEEE Engr. Prob. Fusion Res., Knoxville, TN (1977); K. W. Ehlers et al., LBL-4471 (1976); K. Halbach, LBL-4444 (1975); E. F. Jaeger and J. C. Whitson, ORNL/TM-4990 (1975); W. S. Cooper, K. Halbach, and S. B. Maygary, 2nd Symp. on Ion Sources, Berkeley (1974).
26. J. H. Whealton, E. J. Jaeger, and J. C. Whitson, *J. Comput. Phys.* **27**, 32 (1978).
27. J. C. Whitson, J. Smith, and J. H. Whealton, *J. Comput. Phys.* **28**, 408 (1978); J. H. Whealton and J. C. Whitson, *Part. Accel.* **10**, 235 (1980).
28. J. H. Whealton, *J. Comput. Phys.* **40**, 491 (1981); J. H. Whealton, *Nucl. Instrum. Methods* **189**, 55 (1981); J. H. Whealton, *IEEE Trans. Nucl. Sci.* **NS28**, 1358 (1981).
29. J. W. Wooten, J. H. Whealton, D. A. McCollough, R. W. McGaffey, J. E. Akin, and L. J. Drooks, *J. Comput. Phys.* **43**, 95 (1981).
30. J. H. Whealton, R. W. McGaffey, and P. M. Meszaros, *J. Comput. Phys.* **63**, 20 (1986).

31. L. R. Grisham, C. C. Tsai, J. H. Whealton, and W. L. Stirling, *Rev. Sci. Instrum.* **48**, 1037 (1977); J. Kim, J. H. Whealton, and G. Schilling, *J. Appl. Phys.* **49**, 517 (1978); J. H. Whealton, L. R. Grisham, C. C. Tsai, and W. L. Stirling, *J. Appl. Phys.* **49**, 3091 (1978); J. H. Whealton et al., *Appl. Phys. Lett.* **33**, 278 (1978); M. M. Menon et al., *Rev. Sci. Instrum.* **51**, 1163 (1980); C. N. Meixner et al., *Rev. Sci. Instrum.* **52**, 1625 (1981); M. M. Menon et al., *J. Appl. Phys.* **58**, 3356 (1985); W. K. Dagenhart et al., AIP Conf. Proc. **111**, 353 (1984); W. L. Stirling et al., AIP Conf. Proc. **111**, 450 (1984).

32. J. H. Whealton and J. C. Whitson, *J. Appl. Phys.* **50**, 3964 (1979); J. H. Whealton, R. W. McGaffey, and W. L. Stirling, *J. Appl. Phys.* **52**, 3787 (1981); J. H. Whealton, J. W. Wooten, and R. W. McGaffey, *J. Appl. Phys.* **53**, 2806 (1982); J. H. Whealton and R. W. McGaffey, *Nucl. Instrum. Methods* **203**, 377 (1982); R. J. Raridon, J. H. Whealton, J. W. Wooten, R. W. McGaffey, D. E. Wooten and C. C. Tsai, *J. Appl. Phys.* **57**, 819 (1985); J. H. Whealton, *J. Appl. Phys.* **53**, 2811 (1982); J. H. Whealton, R. J. Raridon, R. W. McGaffey, D. H. McCollough, W. L. Stirling, and W. K. Dagenhart, *Proceedings of the of 3rd International Negative Ion Conference* (Brookhaven National Laboratory) AIP Conf. Proc. **111**, 524 ff (1984); R. W. McGaffey, P. S. Meszaros, J. H. Whealton, R. J. Raridon, and D. H. McCollough, ibid. 533; J. H. Whealton, M. A. Bell, and R. J. Raridon, Second European Workshop on Production and Application of Light Negative Ions, Palaiseau (1986); M. A. Bell et al., *Bull. Am. Phys. Soc.* **30**, 1609 (1985); M. A. Bell et al., *Bull. Am. Phys. Soc.* **29**, 1369 (1984); J. H. Whealton, *Bull. Am. Phys. Soc.* **25**, 983 (1980).

33. H. R. Kaufman, J. J. Cuomo, and J. M. E. Harper, *J. Vac. Sci. Tech.* **21**, 725 (1982); M. Fumelli et al., 3rd Neutral Beam Heating Conf., ORNL (1981) J8; P. Rainbault, 3rd Neutral Beam Heating Conf., ORNL (1981), H6; J. F. Bonnal et al., 3rd Neutral Beam Heating Conf., ORNL (1981), H5; P. Rainbault, 3rd Neutral Beam Heating Conf., ORNL (1981), H3; A. J. T. Holmes, E. Thompson, and F. Watters, *J. Phys. E.* **14**, 856 (1981); M. Mori, T. Nakamura, N. Inove, and T. Uchida, *Jpn. J. Appl. Phys.* **19**, 1377 (1980); H. J. Hopman, BNL Neg. Ion Conf. (1980) p. 233; Y. Ohara, *J. Appl. Phys.* **49**, 4711 (1978); H. R. Kaufman, *AIAA J.* **15**, 1025 (1977); K. Asai et al., *Jpn J. Appl. Phys.* **15**, 1343 (1976); Y. Ohara et al., *Jpn. J. Appl. Phys.* **15**, 135 (1976); T. Sugawara and Y. Ohara, *Jpn. J. Appl. Phys.* **14**, 1029 (1975); H. R. Kaufman, *Electron Phys.* **36**, 265 (1974); R. Geller, C. Jacquot, P. Serrnet, Ion Source Conf., LBL (1974), p. II-5-1; R. L. Poeschel and H. F. King, Ion Source Conf., LBL (1974), p. II-4-1; J. E. Boers, J. R. Freeman, and J. W. Poulsey, Ion Source Conf., LBL (1974), p. II-2-1; W. S. Cooper, K. H. Berkner, and R. V. Pyle, *Nucl. Fusion* **12**, 1263 (1972).

34. R. Keller, J. D. Sherman, and P. Allison, IEEE **NS32**, 2579 (1985); O. A. Anderson, at al, IEEE **NS32**, 2509 (1985); R. Keller, F. Nohmayer, P. Spadtke, and H. M. Schouenberg, *Vacuum* **34**, 31 (1984); J. Aubert, These, Univ. de Paris - Sud, Centre D'Orsay (1984), p. 111, 138; M. R. Shubaly and M. S. de Jong, IEEE **NS30**, 1339 (1983); H. V. Smith, Jr., P. Allison, and T. Sherman, AIP Conf. Proc. **111**, 460 (1983); O. A. Anderson, *Nuc. Tech./Fusion* **4**, 1418 (1983); R. L. York and R. R. Stevens, Jr., IEEE Trans. Nucl. Sci. **NS30**, 2705 (1983); P. W. Allison, Ion Source Workshop, LANL (1982); M. R. Shubaly, R. A. Judd, and R. W. Hamm, IEEE Trans. Nucl. Sci. **NS-28**, 2655 (1981); R.

W. Hamm, D. W. Mueller, R. G. Sturges, unpublished (1981); G. D. Alton, Nucl. Instrum. Methods **189**, 15 (1981); M. R. Shubaly, 3rd Neutral Beam Heating Conf., ORNL (1981), B1; J. H. Whealton, R. W. McGaffey, and E. F. Jaeger, *Appl. Phys. Lett.* **36**, 91 (1980); J. E. Boers, SAND 79-1027 (1980); O. A. Anderson, 2nd Int. Symp. Prod. Neut. of Negative Hydrogen Ion Beam, BNL, p. 355 (1980); J. H. Whealton and J. C. Whitson, *J. Appl. Phys.* **50**, 3964 (1979); J. Kim, J. H. Whealton, and G. Schulling, *J. Appl. Phys.* **49**, 517 (1978); J. H. Whealton, L. R. Grisham, C. C. Tsai, and W. L. Stirling, *J. Appl. Phys.* **49**, 3091 (1978).

35. P. Spadtke, IEEE **NS32**, 2465 (1985); K. Ota, N. Inoue, H. Nikei, J. Morikawa, S. Ishida, and T. Uchida, *Jpn. J. Appl. Phys.* **23**, 1241 (1984); J. Aubert, Thesis, Univ. Paris - Sud Centre D'Orsay, p. 138 (1984); H. V. Smith, Jr., P. Allison, and T. Sherman, AIP Conf Proc. **111**, p. 460 (1983); M. R. Shubaly and M. S. de Jung, IEEE **NS30**, 1399 (1983); M. R. Shubaly, R. A. Judd, and R. W. Hamm, IEEE **NS28**, 2655 (1981); J. C. Whitson, J. Smith, and J. H. Whealton, *J. Comput. Phys.* **28**, 408 (1978); J. Hauser and F. Tanzer, GKSS 78/E/24, IEEE Conf. Plasma Science, Monterey, CA (1978); J. H. Whealton and C. C. Tsai, *Rev. Sci. Instrum.* **49**, 495 (1978).
36. G. D. Smith, *Numerical Solution of Partial Difference Equations: Finite Difference Methods*, 2nd ed., Oxford Univ. Press (1978).
37. E. F. Jaeger and J. C. Whitson, ORNL/TM-4990, 1975.
38. J. M. Ortega and W. D. Rheinboldt, *Iterative Solution of Nonlinear Equations in Several Variables*, Academic Press, New York (1970).
39. F. F. Chen, Introduction to Plasma Physics and Controlled Fusion, Plenum (1984), p. 292.

APPENDIX I

Computational Issues Relating to Steady-State Ion Extraction From a Plasma

The most elementary self-consistent optics analyses are the so-called "electron codes"[24] used in modeling electron guns. They are also used, with significant limitations, to model positive ions extracted from a plasma. They are also used to model negative ion extraction with the dubious space-charge-limited assumption (used for positive ion modeling boundary data) replaced by ad hoc boundary data. A significant advance was made by Cooper and Halbach in 1974[25] when some of the self-consistent collisionless sheath could be analyzed for the first time in more than one dimension. Later techniques have evolved to analyze the full sheath and presheath without significant limitation.[26-30] These analyses have come to be known as "positive ion codes," have seen considerable experimental confirmation,[31] and are used as design tools.[32]

As an example consider steady-state ion extraction from a plasma. A configuration which represents what may be the most difficult convergence issue is indicated in Fig. 1; the region to the left of the extraction sheath is dominated by the two nonlinear terms on the right-hand side of Eq. (1). In addition, a large stagnant region occurs in the pre-sheath region near the electrode. The details of the sheath structure are very important in determining the aberrations of the ion beam. Other attempts to describe the ion extraction optics cannot duplicate the kind of result shown in Fig. 1; they consider (1) no sheath region,[24] (2) a small sheath region with no pre-sheath,[25] (3) geometries with no stagnant region,[33] or (4) artificially high pre-sheath ion temperature or directed velocities.[34] In addition, some results have significant noise even with either zero ion temperature or a non-stochastic model of ion temperature.[35] For example, see Fig. 9 (negative ions from a volume produced negative ion source, which have no noticeable ion stochasticity if considered in detail). An enlargement of the actual injector made for MFTF long-pulse application[4] is shown in Fig. 10; it produced 30A, 80 kV positive ion beams with a divergence of 0.3 deg, a performance predicted by the design. The positive ion codes are probably capable of modeling surface production ion sources because of the relatively moderate role of the extraction plasma.

APPENDIX II

Details of the Time-Dependent Vlasov-Poisson Analysis

The subject analysis can best be understood with reference to Fig. 11, which shows the path of the calculation. First, the Poisson equation is considered. For this first pass, the source terms are set equal to zero and a Laplace equation is solved by SOR, finite difference, and boundary interpolation within a cell, using a Gauss-Seidel implicit method.[36] For the attributes A1: resource utilization and A2: accuracy, iteration reduces memory requirements (A1), and boundary interpolation contributes to the accuracy per cell (A2). Generally, individual convergence of the solutions is not warranted on each pass (contributes to A1), since the iteration procedure lends itself to incomplete convergence of the intermediate solutions. As noted before, the finite difference method compared with the finite element method has in our experience reduced A1 by a factor of 20 for the Poisson solution (ref. 29 vs ref. 30) for the same accuracy. Boundary conditions for arbitrarily shaped metal surfaces can be specified as time-dependent Dirichlet or ramped Dirichlet conditions (contributes to A2). Neumann boundary conditions can also be specified.

Second, the Vlasov equation is solved for an arbitrary initial condition using the solution to the Laplace equation above for a time step Δt. The technique is described in refs. 28 and 30 where significant advances in A1 and A2 are reported. Reference 28 speeds up the Vlasov solver by a factor of 10 (contributes to A1) from that in refs. 26 or 27 while at the same time improving the accuracy by over a factor of 10 (contributes to A2). Reference 30 decreases resource utilization (A1) over ref. 29 by a factor of 400 with the same accuracy. The trivial relationship between the coordinates inside an element and the global elements for the uniform Cartesian grid used in this algorithm allows a factor of 20 (of the 400) savings in the Vlasov solver (A1) over that employed in the irregular elements of ref. 29. As mentioned in ref. 28, the Vlasov solver is made self-regulating in accuracy; trajectory refinement is undertaken only in those places that need it (A2).

Third, charge deposition is done in three dimensions by interpolation over the grid and is "exact" in the sense that as the 3-D grid is made more fine and the number of trajectories is increased, a result as accurate as desired can be obtained (A2). Notice that nowhere is any paraxial-like assumption made, and the fields "to all orders" are directly calculated (attribute A3, nonlinear effects). Therefore, aberrations (to all orders) are also directly computed. Other nonlinear optics effects (A3) computed include space charge "to all orders" caused by nonuniform beam density and/or boundaries. (Boundaries cause nonlinear space charge forces also because they alter the delicate dependence of ϕ on r required to keep it linear.)

Fourth, the beam charge and the exponential plasma term (A3) are taken as inhomogeneous terms to the Laplace equation solved in step 1 above. Now the two inhomogeneous terms are, in many cases, large, of opposite sign, extremely nonlinear, and three-dimensional. This is the cause of numerical difficulties that were first surmounted (in 2-D steady state) in ref. 26. The technique used, accelerated under-relaxation, improved the prior art[37] by a factor of 1000 (A1) in the beam perveance of interest and by a greater factor for higher perveance. Another factor of 10 increase in speed (A1) was achieved, while at the same time the accuracy was increased by more than a factor of 100 (A2) in ref. 27. This technique was extended to three dimensions in refs. 29 and 30. Essentially the best technique we have found is to use an unconverged Newton SOR outside its established range of validity.[38]

Fifth, the time is moved back by Δt, the ions are moved back to their phase space positions a time Δt ago, and the Vlasov equation is re-solved with the new fields computed from the Poisson equation solution of step 4. The trajectories are different from those computed in step 2 because of the presence of the space-charge terms (steps 3 and 4).

Sixth, since the trajectories of step 5 are different from those of step 2, steps 3, 4, and 5 are repeated (Vlasov-Poisson iteration) until no change obtains. This completes the convergence procedure (A2), and it is time to proceed to the next time step. However, one should note the implication of the iteration consisting of steps 5 and 6.

Seventh, the time is advanced by Δt and steps 2 through 6 are repeated. This performs the beam evolution through the device under consideration.

The attributes A1 through A3 provide orbit accuracies of up to 10^{-8} radians in speedy calculations with significant nonlinearities. Six items contributing to a decrease in resource utilization (A1) total about 2×10^9 in the product of memory saved and CPU time (however, the accounting procedure leading to this figure is somewhat ambiguous). Five items contributing to increased accuracy (A2) make an improvement of about 10^6 for a significantly nonlinear problem.

APPENDIX III

Ion Extraction From a Moving Plasma

Consider ions extracted from an approaching plasma into an electrostatic accelerator. The accelerator is shown in Fig. 12, with the potential contours shown by dashed lines. An advancing plasma is shown in the sequence of Fig. 12. Only the ions of the approaching plasma can be seen. The Boltzmann distribution of electrons can only be imagined. In the time increments represented by Fig. 12, the extension of the plasma forms the well-known[39] sheath which approaches a steady state. Validation of the asymptotic long-time limit of the time-dependent solution by comparison to a direct and validated steady-state solution is obtained. In Fig. 13, the same situation as Fig. 12 is shown, but the plasma electron temperature is higher. On the scale of the Bohm speed, v_B, Fig. 12 represents $v_i = 8.85\ v_B$, and Fig. 13 represents $v_i = 1.25\ v_B$. As one approaches the Bohm sheath instability criteria[39] noticeable waves appear in the plasma causing visible deflections of the ions. Emittance growth of an ion beam traversing such a plasma would result.

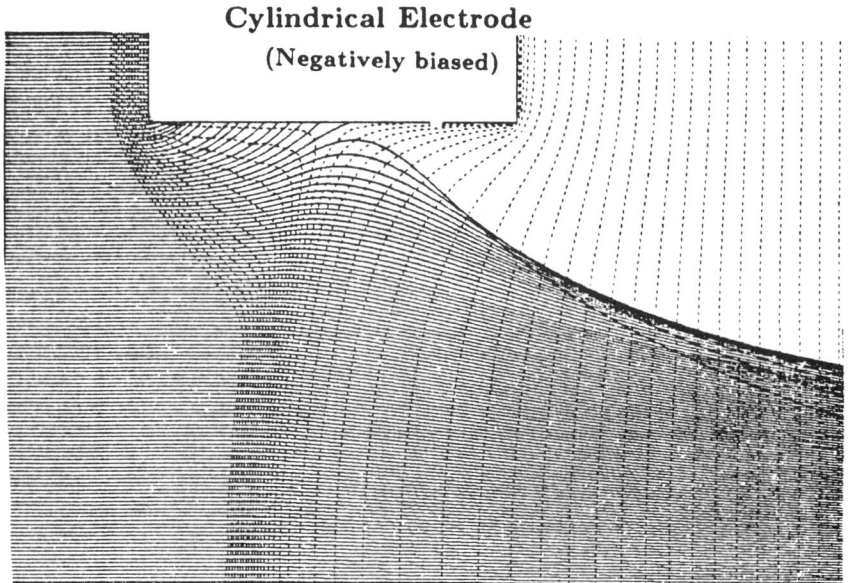

Fig. 1. Positive ion extraction with plasma sheath ion accelerator, showing monotonic downhill run. ----: electrostatic equipotential contours (solution to Poisson equation). ——: ion trajectories (solution to Vlasov equation).

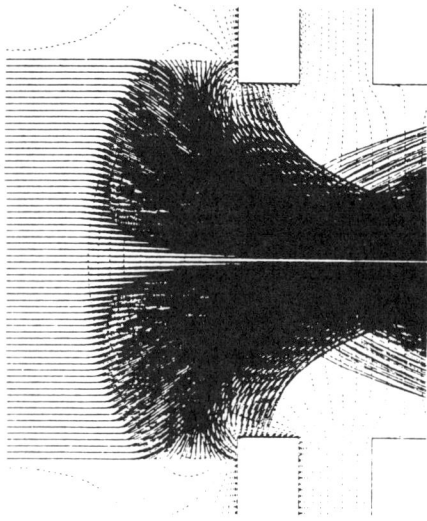

Fig. 2. Penning discharge parallel extraction, ringatron configuration.

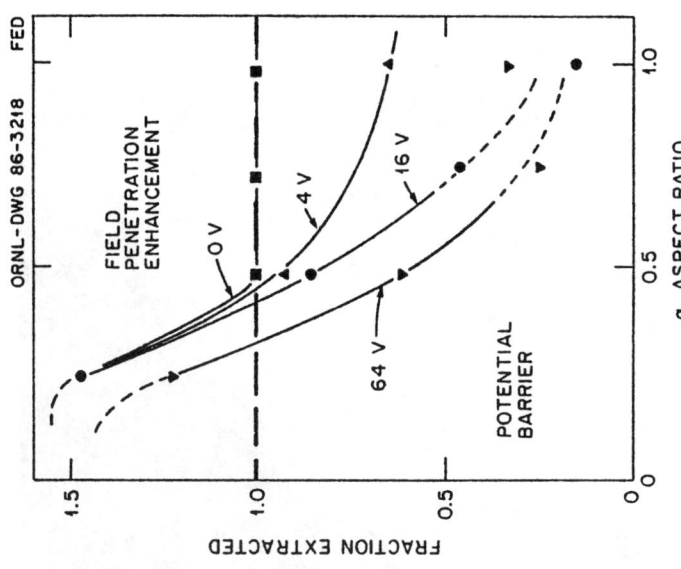

Fig. 4. Penning discharge parallel extraction, ringatron configuration.

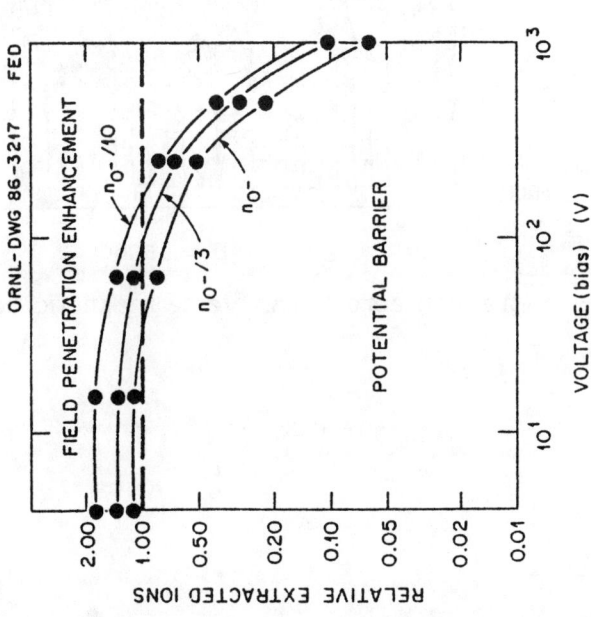

Fig. 3. Penning discharge parallel extraction, ringatron configuration.

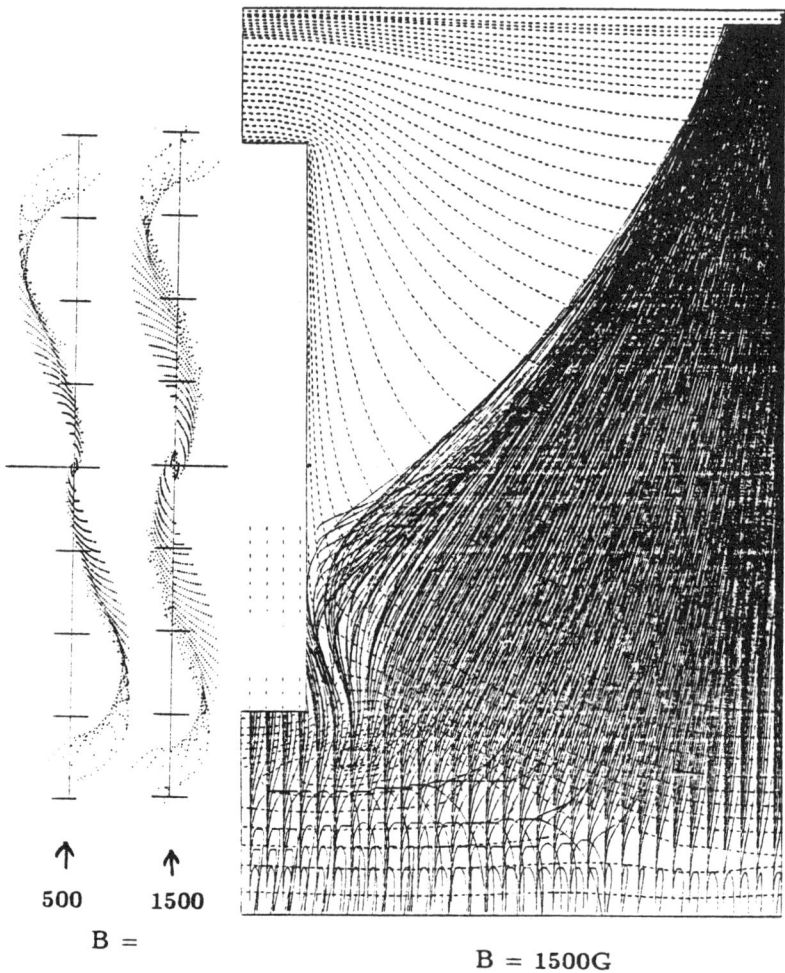

Fig. 5. Extraction across transverse magnetic field.

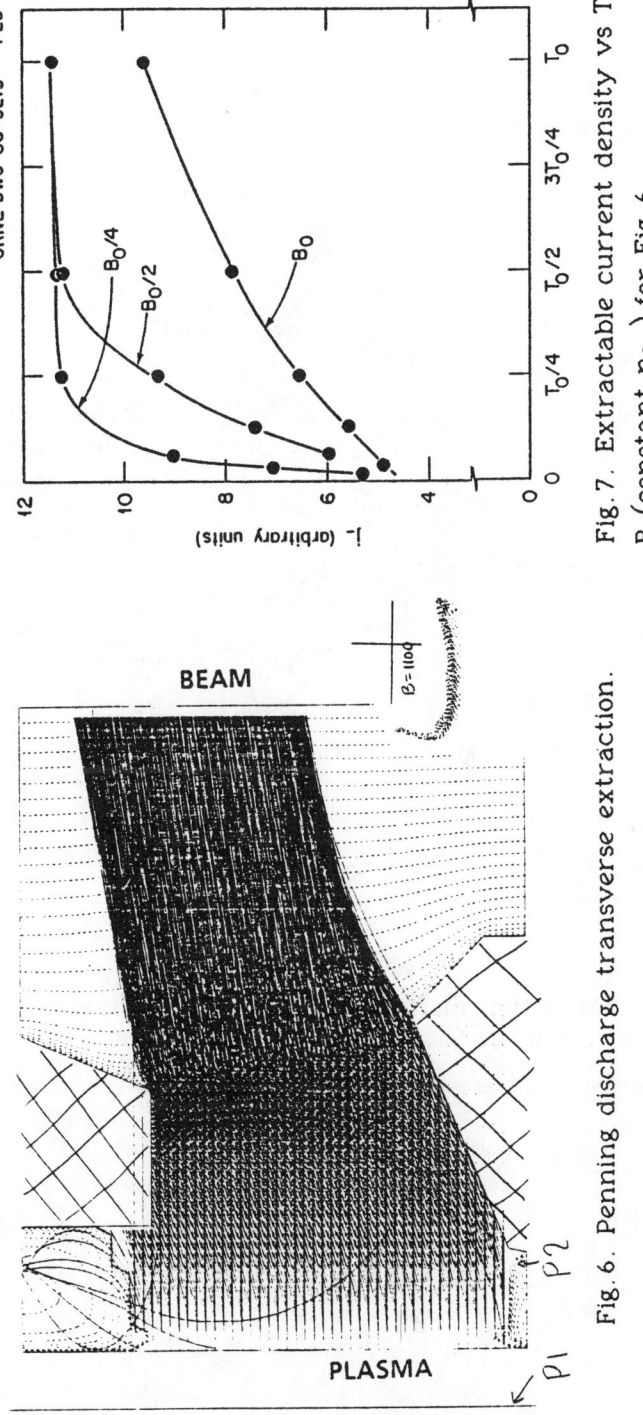

Fig. 6. Penning discharge transverse extraction.

Fig. 7. Extractable current density vs T, B (constant n_{0-}) for Fig. 6.

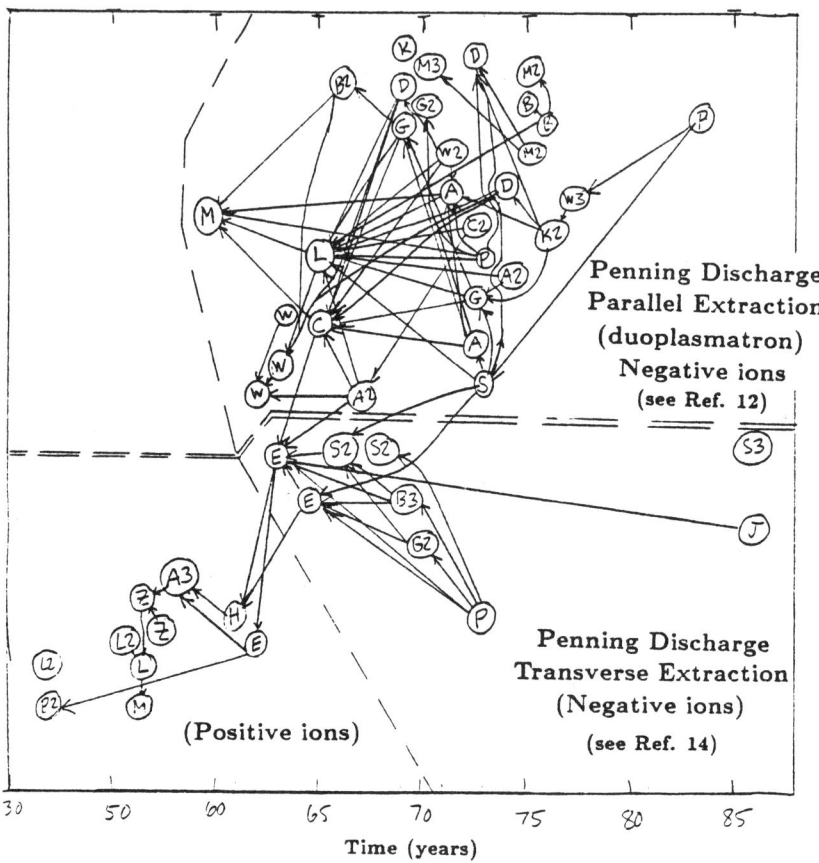

Fig. 8. Intellectual history of negative ion extraction from Penning discharge (no Cs, just H⁻) since 1930.

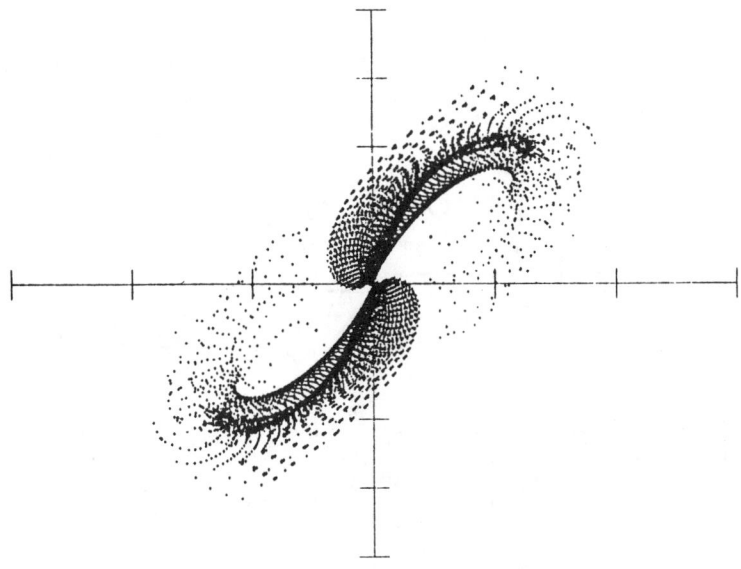

Fig. 9.

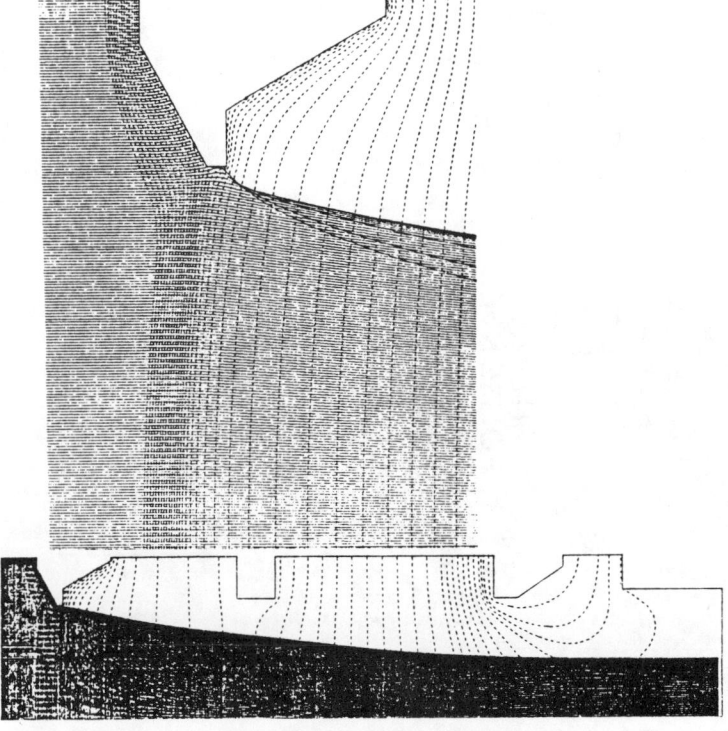

Fig. 10. MFTF injector

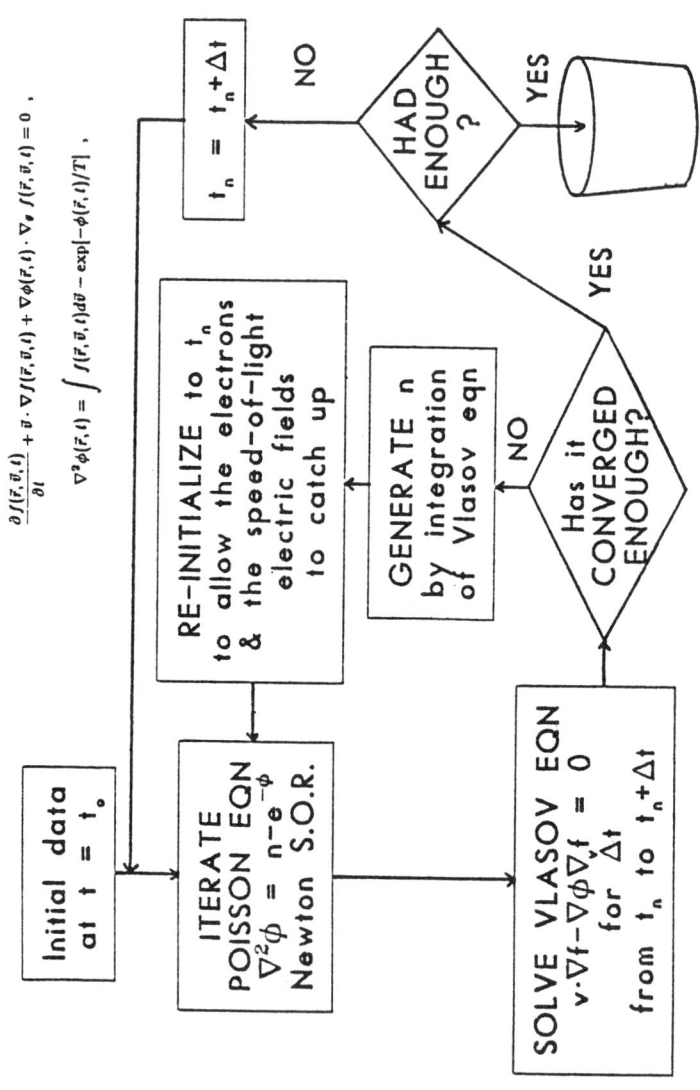

Fig. 11. Time-dependent algorithm.

503

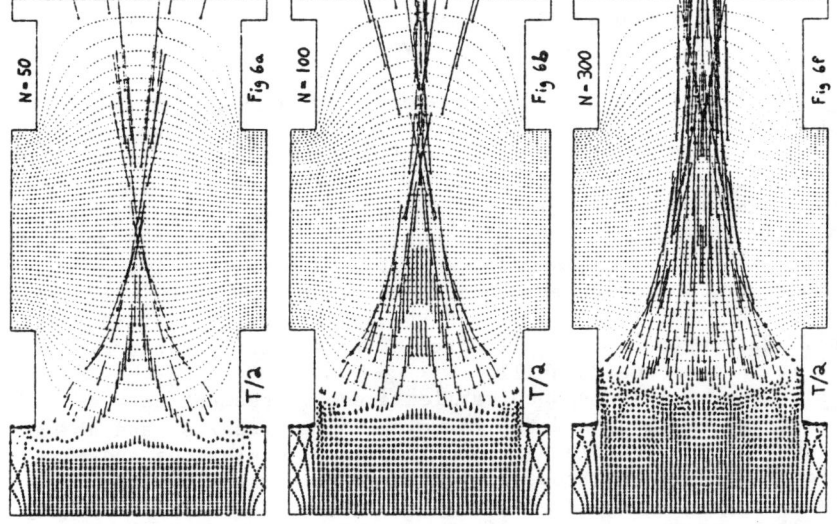

Fig. 13. Ion acoustic waves in plasma.

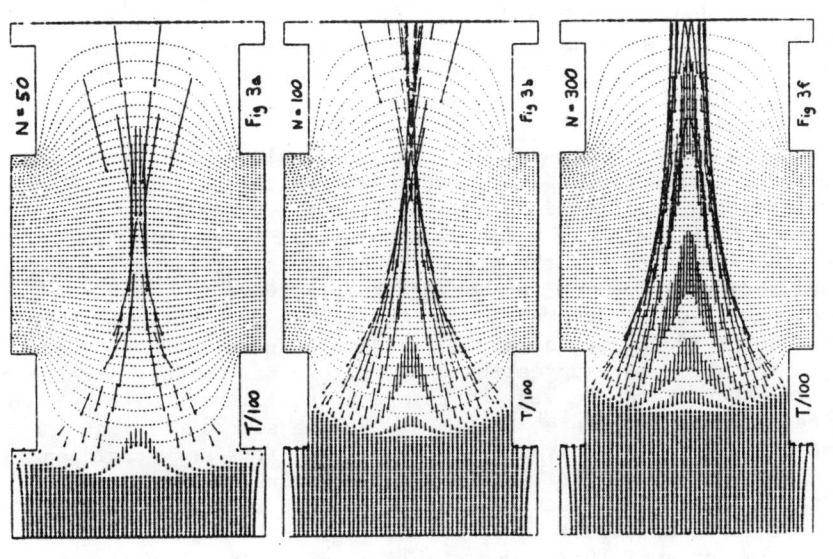

Fig. 12. Ion extraction from plasma.

DISCUSSION

Kwan: Would you describe the potential profile of your negative ion simulation from the plasma to beyond the plasma boundary, on the center axis as well as off axis.

Whealton: For virtually all the cases where negative ions came out, for the center axis it is a monotone downhill. Of course, off of the geometric center, in the case of field penetration, it would be a little bit downhill and then, as you got toward the wall, it would be uphill and the ions would turn around. For some place in between, near the edge of the hole, but still visible to the accelerator, it would either be monotone downhill with maybe a large sideways force or, in the cases where there was a very thick electrode or very high bias, it would perhaps be uphill enough to turn them around. But certainly, in all cases of interest, near the axis region it is monotone downhill.

Alessi: Paul Allison showed earlier a good agreement between the Wolf code calculations and the Berkeley source results, and at LAMPF also, they get reasonable agreement between calculations and measured emittances. But those sources are low current density, they have weak magnetic fields, and they're dc sources. For the Magnetron and the Penning sources, which are both high current density, pulsed, and have a strong magnetic field, do you know of any case where one could get a good agreement between a calculation and measured emittances?

Whealton: In the case of the Oak Ridge source we haven't established that yet because in the transport region between the source and the emittance scanner, there is probably gas focussing going on, which is going to have a significant effect on the beam optics. What we have done is proposed electrode designs, and they have shown increased current density extracted, so there's been a little bit of success in that region. But actual validation of the optics for the Penning discharge at Oak Ridge has not been done because of the beam transport issue.

Holmes: John, two things on what you just said. First, you don't include the destruction of negative ions from the plasma. You're almost treating negative ions in a similar way to positive ions. You have a drift velocity which is accelerated by the electric field, the ions approach the sound speed and then they're accelerated and form a beam. Now, as we were hearing yesterday and the day before, negative ion destruction in the plasma is a very big phenomenon. So to ignore it, as you're describing here, weakens the model. You were also suggesting that the plasma boundary is highly flexible, that it can be very concave and it will vary with current density. Yet the evidence, so far, in volume sources is that the plasma boundary is exactly the contrary; that it's very rigid and very flat, and it doesn't respond at all varying the arc current.

Whealton: About the first question, I indicated in the pictures that this is an idealistic generation, just uniform generation, over a rectangle, or a pillbox. One should couple this to an analysis or a higher dynamic model of the ion generation, so that one would account for the depletion mechanism. Numerically, this is a rather straightforward thing to do, and that should be done to model accurately the production in negative ion sources. However, I'm not sure about the second comment, where you said that you presented evidence, or you have interpreted the sheath as being rather inflexible. We're doing a rather sophisticated self-consistent analysis of the sheath, and we find that it is not inflexible. There may be a lot of interpretations for given experimental data. One experimental confirmation of the flexibility of the sheath is the fact that for razor blade type electrodes at low current densities, there are a lot more ions extracted than for other kinds of geometry, indicating that the field penetration or flexibility of the sheath is really a phenomenon to be considered.

Holmes: I can't agree with that, John, because we are using, at the moment, what you would term "razor blade edged" electrodes, i.e. very small, narrow angle in the metal. And what we see through the experiment, which is the final determinant of models, is that the plasma boundary is rigid, so that suggests the model is wrong.

Whealton: Well, I agree, the experiment is the final determinant of the model, and what we see in our experiment is enhanced production as one changes the shape of the electrode, indicating that the sheath is flexible. The experimental result is enhanced ion extraction current density. That is the experimental result, not a model result.

Holmes: How do you know?

Whealton: We measured the current.

Holmes: Where? In the source?

Whealton: No, we measured it in the emittance scanner.

Holmes: That's the beam, though. How do you know the current has increased over what it was in the source?

Bacal: I support John in saying that the plasma boundary is not rigid. We didn't have time to discuss that, but evidence is that extracted negative ion current is going up with extraction voltage very much. We've studied that very carefully and the only explanation for that is that the sheath is increasing when you apply higher and higher voltage. And I think you showed us the results which confirms that...remember the figures which were exactly the same in your and our case? So I don't know why the interpretation is different. However, I had a question for John. In the figures you showed for different current densities, the walls were always negative. Actually, the plasma electrode is usually positive, or no different. So what happens when the plasma electrode is positive?

Whealton: My understanding is that the plasma electrode is biased positively, with respect to the floating potential; however negatively, with respect to the plasma.

Bacal: No. Since the beginning, when it is not biased, it's negative with respect to plasma, but when you bias it positive, it comes to a moment when there is no potential difference between the plasma and the plasma electrode. And that's more or less an optimum situation.

Whealton: That's the easiest to model of all, because in that sense the similarity...that's the way it is for positive ion sources.

Bacal: So what would be the trend when you have no difference?

Let's say they are at the same potential.

Whealton: Well, you still have the effect of field penetration, which would dominate at the lower current densities.

Lietzke: What is the mesh size that you have to use relative to the local Debye length to avoid numerical artifacts? And the second question is: How much energy do you see the negative ions picking up before they reach the non-neutral location? The so-called plasma boundary.

Whealton: For the sheaths that you saw there, the mesh size is on the order of a Debye length. Or, in some cases we have found that we could actually have the mesh size...the distance between nodes larger than Debye length, and numerical experiments indicate that, in some cases, that is not a fatal flaw. But whenever that happens, we test to see that there is no anomalous mesh point variation. And some of the examples that you saw there, where there was significant field penetration, the ions gained a lot of energy before they got to the extraction aperture.

Lietzke: That's not my question. How much energy do they gain before they get to the non-neutral location where the positive ions no longer exist?

Whealton: That depends on the situation. We just varied them all over the map. We can follow any model we choose, between the left-hand boundary and the center of the plasma. In analog with the positive ion model we can use the results of Self in the one-dimensional case in the middle of the plasma to the left-hand boundary or...we have our choice.

TRANSVERSE-FIELD FOCUSSING BEAM TRANSPORT EXPERIMENT*

J. W. Kwan, G. D. Ackerman, O. A. Anderson, C. F. Chan,
W. S. Cooper, L. Soroka, and W. F. Steele
LBL-22387
Lawrence Berkeley Laboratory
University of California
Berkeley, CA 94720

ABSTRACT

The Transverse-Field Focussing (TFF) beam transport and accelerator system developed at LBL is useful for negative-ion-based neutral beam injection due to its unique differential pumping and neutron shielding properties. We have tested the first module of our TFF system transporting H$^-$ beams up to 80 keV beam energy. The testing addressed the most crucial physics and engineering issues involved in the principles of a TFF system including beam compression and differential gas pumping. At optimum perveance, the present design will transport 4 A/m of H$^-$ beam at 80 keV beam energy.

INTRODUCTION

In a conventional neutral beam injection system, the neutrons produced by the reacting plasma in a fusion reactor can travel back into the neutral beam system causing radiation damage and contamination to the beamline components. The Transverse Field Focussing (TFF) beam transport and accelerator system developed at LBL can solve this problem by transporting the ion beam through several bends in a neutron shielding region before converting the beam to neutrals, thus protecting the ion source and the accelerator from the energetic neutrons.[1] An additional advantage in the TFF beam transport system is that differential gas pumping can be applied to quickly pump away the hydrogen gas in the beamline. This unique differential pumping characteristic is important in reducing the loss by premature electron detachment of a H$^-$ beam by the background hydrogen gas.[2] Substantial beam losses can significantly affect the overall performance of a neutral beam injector.

The TFF system makes use of the focussing ability of the transverse field between a pair of concentric cylindrical electrode plates.[3] Charged particles with the correct initial velocity will be able to travel along the channel in between the electrodes in such a way that the electrostatic force is balanced by the centrifugal force. Hence the TFF system is inherently a 2-dimensional system which transports ribbon-shaped beams. The third dimension (in the y-axis) is usually limited by the length of the negative ion source being used. The transverse field can also compress the beam width. Typically the beam width is about 1.5 cm which corresponds to a current density of 27 mA/cm^2.

The original design of our TFF system is supposed to transport more than 4A of H⁻ per meter of source length.[4]

EXPERIMENTAL SET-UP

Figure 1 shows a plan view of the experimental set-up. The negative hydrogen ion beam, produced by a surface-conversion H⁻ source, is accelerated by a conventional single-slot accelerator (called the pre-accelerator) to an initial energy as required by the TFF system. Details of the design and testing of the ion source and the pre-accelerator have been reported elsewhere.[5] The ion source together with the pre-accelerator have been tested to produce 1 A of 80 keV H⁻ beam for 30 sec (the ion source is

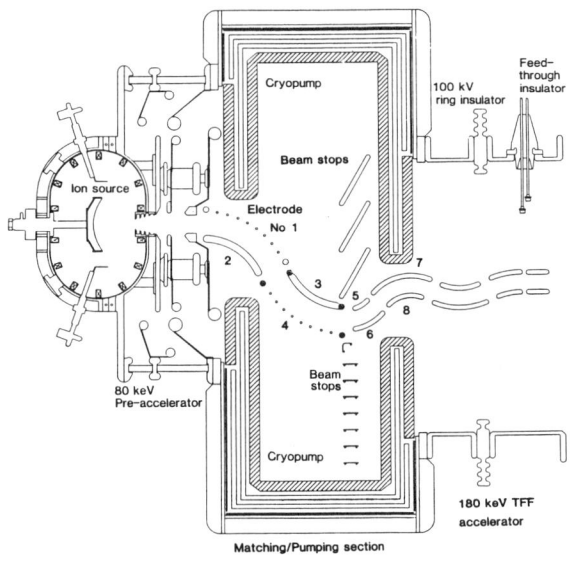

Fig. 1. Plan view of the TFF beam transport experimental set-up.

capable of producing more than 1.25 A of H⁻). Typical beam data taken at the exit of the pre-accelerator in absence of the TFF beam transport section are summarized in Table 1.

Both the surface-conversion H⁻ source and the pre-accelerator have problems related to beam uniformity along the slot length. The first problem is that it is difficult to maintain a spatially uniform H⁻ production in a large surface-conversion source when neither the cesium coverage on the converter surface nor the plasma density in front of the converter are very uniform.[6] The variation in H⁻ intensity can be greater than 20%. This effect is more pronounced at the edge than in the middle part of the converter.

The second problem lies in focussing of the ion beam by the ends of the aperture slot in the pre-accelerator. This end-of-slot effect is basically a three dimensional beam aberration problem which happens where the two dimensional planar symmetry fails to hold at the slot ends. The ions near the ends of the slot are deflected toward the center thus producing a nonuniform current density profile. The effect can be minimized, but not completely eliminated, by reducing the Pierce angle on the ends of the beam-forming electrode. Fig. 2 compares the typical H⁻ current density profiles along the slot length (after integrating the beam intensity across the short direction) before and after making the end corrections. This problem becomes less important as the slot length becomes much longer than the slot width.

Table 1. Typical beam data taken at the exit of the pre-accelerator in absence of the TFF beam transport system.

Beam current	up to 1 A (25 cm slot length)
Beam energy	up to 80 keV
Optimum perveance	4.4×10^{-8} A/V$^{3/2}$
Uniformity along slot	± 20%
Electron fraction	12%
Pulse length	30 sec
Normalized x-emittance	0.45π mrad-cm (90% contour)
Beam temperature	11.2 eV

Immediately following the pre-accelerator is the first section of the TFF beam transport system. This section is called the matching and pumping (M/P) section because besides transporting the ion beam it also serves the dual purposes of compressing the beam width to match the requirement of any subsequent TFF accelerator and pumping away the gas from the source rapidly.[2] It consists of two sets of curved electrodes (electrode number 1 through 4 in Fig. 1), which will transport the beam through two linked 70 degree bends. The electrodes with the larger radius are made up of parallel tubes so that there are large openings for vacuum pumping. The smaller radius electrodes are solid plates which act as partitions to separate the TFF chamber into two halves thus providing an effective differential pumping arrangement. All the electrodes are water-cooled. Vacuum pumping is done by ultra-thin cryopanels (developed at LBL) placed along the sidewall of the chamber.[7] The cryopanels are protected from any neutralized H⁻ beam particles by the beam stops as shown in Fig. 1. Also shown in the figure are the electrodes for the 180-keV TFF accelerator. This part of the TFF system will not be tested due to budget limitations despite the fact that many parts of the accelerator are already built. In testing the TFF M/P section without the 180-keV TFF accelerator, electrodes number 5 and 6 are replaced with a pair of flat plates used as steering electrodes.

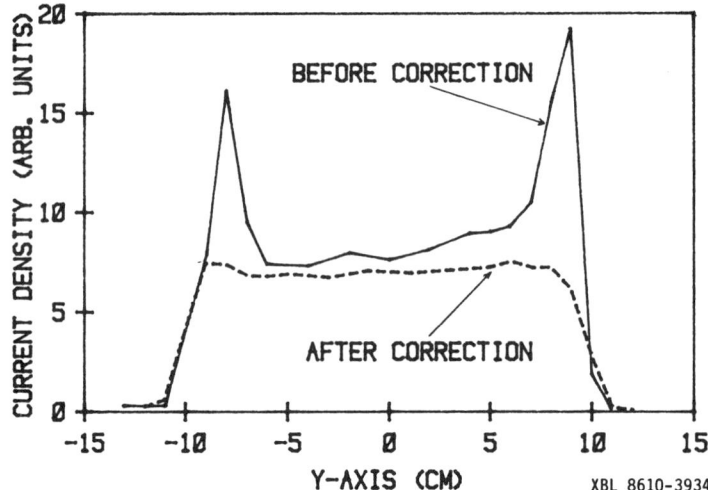

Fig. 2. Typical H⁻ current density profiles along the slot length before and after the end corrections.

Like all other negative ion sources, the surface-conversion source produces electrons as well as H⁻ ions. The electrons are accelerated together with the H⁻ ions through the pre-accelerator. At the exit of the pre-accelerator we find that the electron fraction in the beam is about 10%. This is an upper limit of the amount of electrons leaving the ion source because included in the 10% are electrons being generated inside the pre-accelerator due to electron detachment from the H⁻ ions and also secondary electrons emitted from the gradient electrode but not collected by the suppressor. Ideally, none of the electrons can get through the TFF M/P section. Any electron in the beam will be removed from the ions by a transverse magnetic field at the second bend of the TFF M/P section; the electrons will be collected on electrode number 3.

The ion beam leaving the TFF will be diagnosed by two emittance scanners and a pair of small-aperture (1.5 mm diameter) mass analysers which are capable of distinguishing heavy ions from electrons.[5] Both the emittance scanners and the mass analysers are located at 40.6 cm downstream of the TFF M/P section exit. Results of the x-emittance measurement can be compared with predictions from our beam optics code (WOLF). The y-emittance will provide useful information on the temperature of the H⁻ ions coming from a Cesium surface-conversion source. The mass analysers measure the ion beam intensity and the electron fraction which in turn gives an estimation of the total beam loss during acceleration and transport.

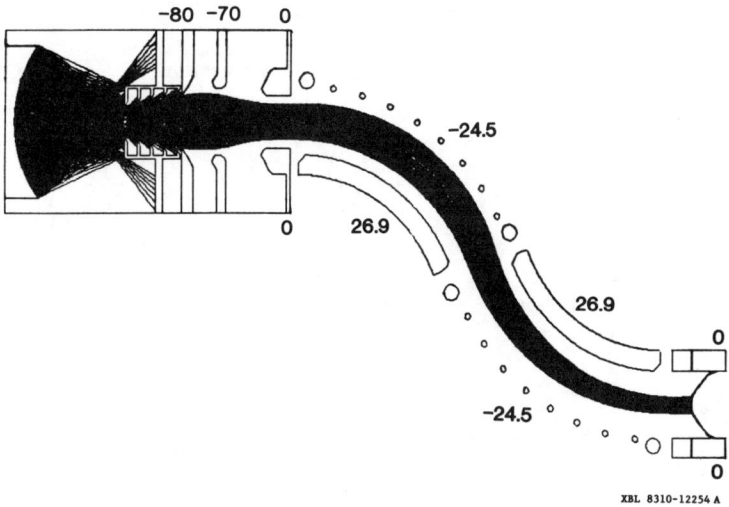

Fig. 3. Ion trajectories obtained from WOLF code calculation, the voltage bias on the electrodes are indicated in the figure.

TEST RESULTS

One 80-kV power supply is required to operate the pre-accelerator and two separate 30-kV power supplies to operate the TFF M/P section. The voltage on the gradient electrode is obtained from the 80-kV power supply using a resistive divider. All three power supplies are voltage regulated to within 0.3% of the operating voltages. A deviation of 1% in the accelerating voltage can amount to 14 mrad deflection in the final beam steering. Likewise, a 1% change in the TFF electrode voltage will produce a deflection of 8 mrad. The design values of the voltage on each electrode (according to WOLF code) are depicted in Fig. 3; these numbers are very close to the ones we have used in the experiment to properly steer the beam through the TFF. In running beams of energy lower than 80keV, one simply scales down the voltages on all the electrodes by the same factor and reduces the beam current according to the "three-halfs" law in order to keep the perveance constant. Also shown in Fig. 3 are the ion trajectories computed with the space-charge effect taken into consideration.

We have operated the TFF M/P section with the pre-accelerator to transport H$^-$ beam currents up to 640 mA and energies up to 80 keV. The beam pulse length was limited by breakdown in the TFF section. It appeared that some insulators failed to hold voltage after accumulating charged particles during a beam pulse. The pulse length is less at higher beam currents and energies. At 48 kV, we were able to transport 320 mA of H$^-$ beam for over 430 msec, but at 80 kV the pulse length was only

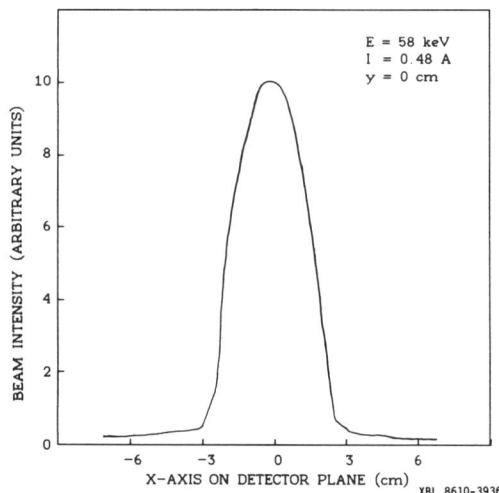

Fig. 4 H⁻ ion beam intensity profile obtained at 40 cm downstream of the TFF exit.

40 msec long. Nevertheless 40 msec was long enough to perform beam diagnostics and to demonstrate that the beam was properly steered through the TFF transport system without running into the electrodes. No damage to the TFF electrodes or the supporting insulators could be found after the experiment. We think that the breakdown problem can be solved by adding shields to protect any exposed insulator surfaces, but the experiment had to be terminated before this problem could be solved.

Fig. 4 shows a typical intensity profile of the H⁻ ribbon beam in the short direction of the slot (the x-direction), near the vertical center (y=0), measured with the mass analyser. The location of the peak is a quick reference of the amount of beam steering. The amount of electrons measured in the beam is so small that it can be ignored. The mass analyser was scanned up to 6.5 cm on each side of the centerline in order to make sure that there is no satellite beam missed by the emittance scanner.

The TFF M/P section was designed with a carefully shaped contour at the top and bottom ends of the electrodes in order to produce electrostatic barriers to confine the beam to within 25 cm. Unfortunately the length of the beam was reduced from 25 cm to 20 cm when we added shields to the beam forming electrode in order to correct the end-of-slot focussing effect in the pre-accelerator. Hence, without changing the location of the electrostatic barriers, the beam coming out of the pre-accelerator can expand from 20 cm up to the 25 cm limit. At the beam edge, the H⁻ ion velocity will become diverging as a result of the space charge expansion. The TFF beam edge confinement scheme has therefore never been fully tested although we have not detected any beam beyond the 25 cm limit (see the y-emittance diagram in Fig. 7).

The current density nonuniformity has a significant impact on the beam optics because it is impossible to maintain optimum perveance throughout the entire slot length. A condition where the center of the beam is at optimum perveance would imply that the beam edge is always "underdense". Thus the correct numerical value of the optimum perveance in this condition is equal to $I_c/(V_{acc})^{3/2}$ where I_c is the product of 20 cm times the current density at the center and V_{acc} is the final beam energy. Experimentally, we found the optimum perveance to be about 3.6×10^{-8} $A/V^{3/2}$, which corresponds to 4 A/m of H^- beam at 80 keV beam energy as expected.

The total detectable beam current (I_t), obtained by integrating the mass analyser signal in both x and y directions, can be compared with the electrical current (I_d) entering the ion source from the 80-kV power supply. The total beam loss fraction is given as

$$f = 1 - I_t / (I_d - I_e)$$

where I_e is the electron current leaving the ion source which is approximately equal to 10% of I_d or less. Naturally, the amount of beam loss depends significantly on the beamline pressure. At 1 mTorr of gas pressure in the source (gas flow = 130 SCCM) and with cryopumping, f is about equal to 15%. Without cryopumping, f is in the order of 33%. It is obvious why we want a pumping section in the beamline before the beam can be accelerated to higher energy.

It is interesting to observe the electrical currents drawn to the TFF electrodes. Fig. 5 shows the variation of these currents with gas flow for a 42 keV beam without cryopumping. The linear dependence of the electrode currents agrees with the model that the origin of these currents is secondary electron emission resulting from the impact of neutrals (created by electron detachment of the H^- ions) on the negative TFF electrodes. Any direct beam interception by the electrodes would produce an intercept on the graph in Fig. 5 indicating a finite amount of electrode currents at zero gas flow. This is completely in agreement with our experience in beam steering. A well steered beam would produce an intercept near the origin like the one shown in Fig. 5, and a badly steered beam would have an intercept far away from the origin.

Projectional emittance of the ion beam at the exit of the TFF section was measured with two double-slit emittance scanners, one in each plane. The entrance slit of the scanner is 3 cm long and 0.01 cm wide. The resultant emittance diagrams are shown in Fig. 6 and Fig. 7. First, we will discuss the x-emittance. According to the emittance diagram, there is no sign of significant beam aberration introduced by the TFF transport system. The normalized emittance value, associated with the area included in

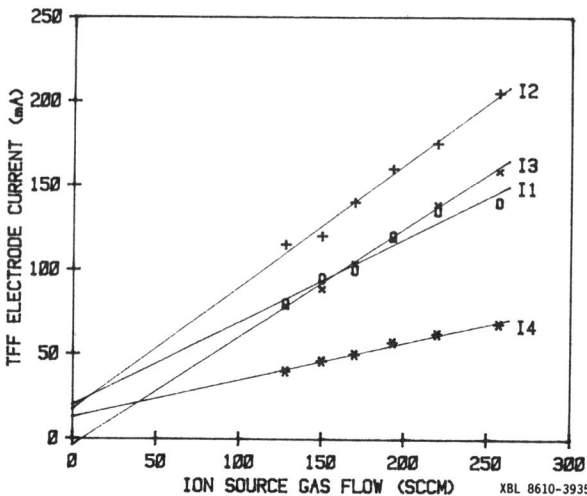

Fig. 5 TFF electrode currents as a function of ion source gas flow

the 90% beam fraction contour, for the beam transported through the TFF M/P section agreed with the value that we found earlier when we were operating the pre-accelerator alone to within 15%. Thus there is little emittance growth occurred in the TFF beam transport section.

The emittance diagram also shows that the width of the beam at the exit plane of the TFF system is about 1.5 cm provided that we ignore the "wings" of the beam which contribute to only a small percentage of the total beam current. This beam width is in excellent agreement with the prediction from WOLF code simulation (see Fig. 3). The -0.4 cm offset of beam center in the emittance diagram is considered to be an alignment error.

Now we will examine the y-emittance data. Since the system is basically a slot geometry, we would expect the y-emittance to look like a stack of rectangles. This is more or less what happens at the center portion of the slot. However the beam edges are diverging due to the end-of-slot effect discussed earlier.

Since there is no focussing effect in the y-direction (except at the very ends), the angular distribution of the H$^-$ ion intensity measured by the emittance scanner near y = 0 cm is the actual angular distribution when the ions leave the converter. A typical profile of the angular distribution of a 40-keV beam is shown in Fig. 8. The half width of the distribution at the 1/e intensity level is 16.75 mrad. This corresponds to a transverse energy of 11.2 eV. This ion temperature is almost as high as the one obtained by A. Lietzke[6] using a different method of measurement. It is also interesting to note the maximum transverse energy. The profile shows that H$^-$ ions were detected up to an

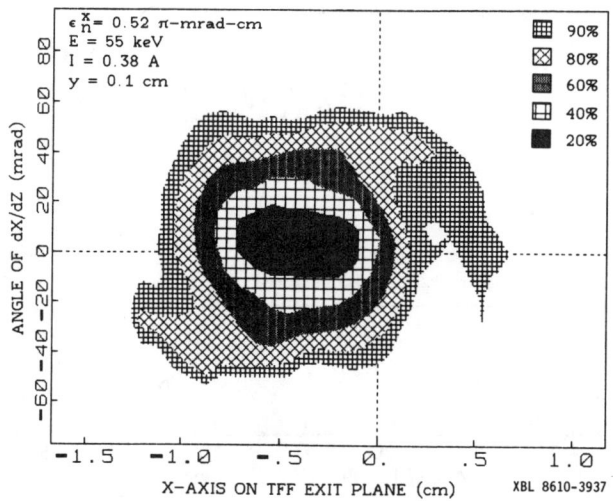

Fig. 6. x-emittance diagram

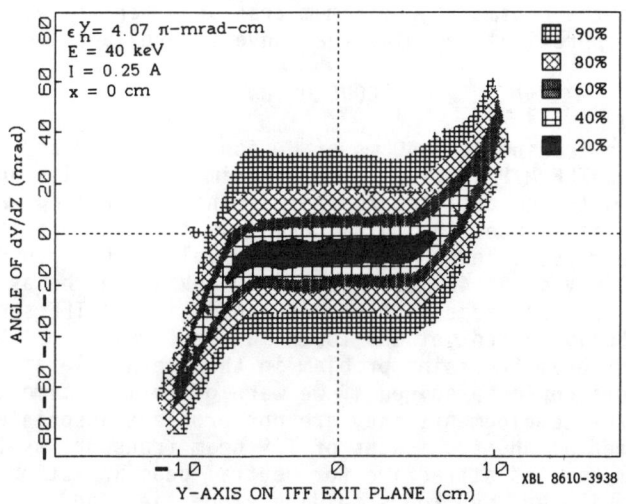

Fig. 7. y-emittance diagram

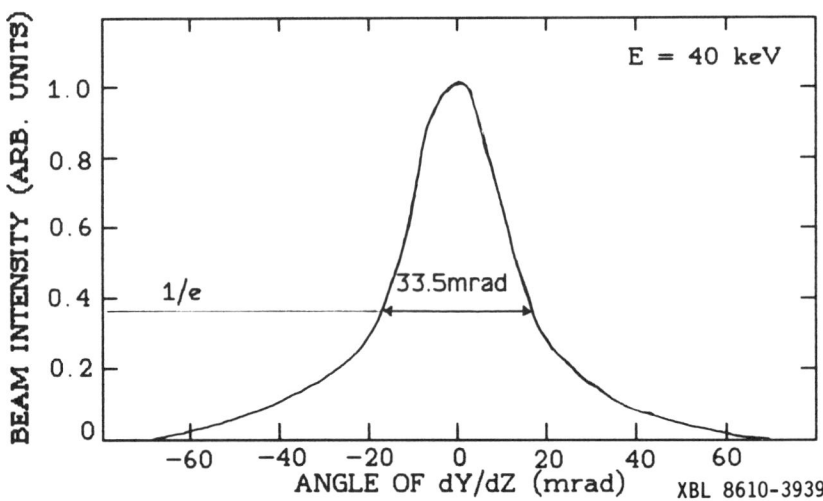

Fig. 8. Angular distribution of H⁻ intensity in the unfocussed direction

angle of 70 mrad which corresponds to a transverse energy of 196 eV. Since the converter voltage was approximately -100 V with respect to the plasma, the maximum transverse energy was therefore approximately twice the converter voltage.

CONCLUSION

We have transported 640 mA of H⁻ ion beam at 80 keV energy through the TFF M/P section. The experimental results agreed very well with the design parameters. The cryopumping was very effective; it reduced the total beam loss to less than 15%.

The beam pulse length is limited by voltage breakdown in the TFF--at 80 keV it is 40 msec, but at 40 keV it can be as long as 0.5 sec. The beam edge confinement scheme in the TFF section has not been fully tested yet due to an indirect result of the end-of-slot beam focussing problem in the pre-accelerator. These two problems could be solved if we were given more time to continue the development; they are not problems associated with the fundamental physics issues of TFF beam transport system. The TFF system remains attractive for neutral beam injection in view of its pumping and neutron shielding properties, and its demonstrated ability to transport a beam with little loss and with little emittance growth.

* This work is supported by US DOE contract no. DE-AC03-76SF00098.

REFERENCE

1. W. S. Cooper, Proc. of the 3rd International Symposium on the Production and Neutralization of Negative Ions and Beams (Brookhaven, New York), p605 (1983).
2. O. A. Anderson, C. F. Chan, W. S. Cooper, J. W. Kwan, C. A. Matuk, H. M. Owren, J. A. Paterson, P. Purgalis and L. Soroka, IEEE Trans. on Nuclear Sci., NS-32 (5), p3509, (1985).
3. O. A. Anderson, Proc. of the 3rd International Symposium on the Production and Neutralization of Negative Ions and Beams (Brookhaven, New York), p473 (1983).
4. W. S. Cooper, O. A. Anderson, J. W. Kwan, and W. F. Steele, Proc. 11th Symposium on Fusion Engineering, Austin, Texas, p110, (1985).
5. J. W. Kwan, G. D. Ackerman, O. A. Anderson, C. F. Chan, W. S. Cooper, G. J. deVries, A. F. Lietzke, L. Soroka, and W. F. Steele, Rev. Sci. Instrum., 57 (5), p831, (1986).
6. A.F. Lietzke, K.W. Ehlers and K.N. Leung, Proc. of the 3rd International Symposium on the Production and Neutralization of Negative Ions and Beams (Brookhaven, New York), p344 (1983).
7. P. Purgalis, G. Ackerman, J. W. Kwan, J. A. Paterson, and A. H. Wandesforde, Proc. 11th Symposium on Fusion Engineering, Austin, Texas, p497, (1985).

PANEL: TRANSPORT OF NEGATIVE ION BEAMS

PANEL SESSION: TRANSPORT OF NEGATIVE ION BEAMS

P. Allison (Moderator), A.J.T. Holmes, A.F. Lietzke,
J.H. Whealton, L. Wright

Allison: Andrew Holmes, from Culham Lab, probably has the most extensive experimental and theoretical experience on neutralization of beams of any of us, at least in the quiescent type beam transport. At Berkeley, it's been proposed to transport negative beams with an electrostatic quadrupole system, and Al Lietzke, from Berkeley, represents that interest. John Whealton, from Oak Ridge, has been studying theoretical beam transport instabilities, and Larry Wright, from Mission Research Corporation, is a Los Alamos consultant, and he's studied the results of our experiments and also the Russian experiments, and has done modeling of that. The panel discussions will center on space-charge effects in H^- beam transport, and each panel member will make a brief statement. In space-charge compensation of H^- beams in transport, there have been several "normal" and several "abnormal" situations for which instabilities have been observed. What is the expectation and experience in normal compensation and what is the expectation and experience in abnormal instabilities? And will unneutralized transport lead to emittance growth, due to whatever you prefer for emittance growth? I have two propositions that I'll throw out. First, under normal transport conditions, negative ion space-charge compensation in many situations has been shown to be better than 100% effective; so some slight focussing of negative beams occurs in transport. Second, under abnormal conditions, compensation is less than 100% effective and instabilities develop, leading to emittance growth. Now, you may wish to challenge these propositions or add your own conclusions to them. The panel does not include all of the expertise that is here in this room, and I've asked several of you to give your own experiences in transport, at Brookhaven, TRIUMF, and Fermilab. Now, let me show a list of some experiments in beam transport of negative ions that we have to draw from (Table I). Those experiments in which instabilities were observed are marked. Instabilities were not observed in the rest of them. So it's a little hard to sort out what parameters may be involved. It strikes me that, in the cases that instabilities have been observed, they're in high perveance situations; but perhaps that's not the only case. In most situations where it's fairly certain that there are no instabilities, the perveance is much lower. For example, the Los Alamos measurements, in one circumstance, have electron equivalent perveance of 1.5 μperv, and in another situation at the LAMPF accelerator, the perveance is .001. In the Russian experiments, the perveance was generally rather large, 1 to 2. Now, I would like to ask each of the other panel members to give you his view of the situation.
Lietzke: Well, I don't represent myself as an expert. I came on behalf of Oscar Anderson, who is our local expert. I am not prepared to talk about what we're planning to do, because that has not been approved by the funding agency. What I will talk about,

WHOM	i_-(mA)	j_{ext}(mA/cm^2)	ϕ(kV)	P_e(μperv)	L(cm)	τ(μs)	Gases
LANL							
SAS	100	2000	20	1.5	60*	3-60	H_2,Xe $\simeq$
	140	3000	100	0.19	80	15-30	2×10^{-5}T
4X	75	330	20	1.1	9	3-35 35*	H_2,Xe
LAMPF	20	25	80	0.038	150	35	H_2, 5×10^{-6}
			750	0.0013	$\simeq$1200	65	H_2, $\simeq10^{-5}$
TRIUMF	1		25	0.001			
FNAL	50	500	750	0.003			
BNL	50	500	750	0.003			
Culham	135	25	100	0.18	400		
Gabovitch, Dzhabbarov, Naida 1979							
	80	50*	15	1.9	50*	4-500	H_2,Ar
Dzabbarov and Naida 1980							
	100	50*	14	2.6	38*	450	H_2, 3×10^{-6}
						5-450	H_2,Ar $\simeq10^{-4}$
Goretskii and Naida 1985							
	30		14	0.8	70*	5-450	H_2,Ar
*Instabilities Observed							

TABLE I

however, is some of the issues involved, especially in terms of what we have already done, which was funded by the Department of Energy. What we have at LBL is the so-called TFF: Transverse Field Focussing System. That was tested on test stand 3A by Joe Kwan. The principal people involved in the calculations were Oscar Anderson, Ludmila Soroka, and Chung Fai Chan. The experiment consists of a converter source for negative ion production, a three gap accelerator, and one section of a transport for pumping and matching. The latter has two purposes: to get the gas out of the way, so that you can handle it for succeeding stages which were not tested or built, and to match the beam going into subsequent stages. This has been described in the literature. The matching is important in electrostatic optics to minimize emittance growth. We have simulated the particle dynamics for a strip beam using 3000 particles for 8 oscillations through an electrostatic system. So the beam goes through various oscillations in size, then settles down. It started out too big, in this simulation, and this is the size that was appropriate for the channel. And as the beam oscillates in size and rotates in phase space, the emittance eventually grows and settles down to a new value, a problem that can occur if you mismatch. Another problem that comes in has to do with the fact that is especially important in volume negative ion production. At the high pressures involved, you can lose quite a bit of beam. A Monte Carlo calculation of the pumping in that section shows that the pressure's fairly good. Approximately 90% of the ions survive. We are now beginning to look in terms of the quadrupole for round beams, using the expertise of the heavy ion fusion people at Berkeley, who are doing this routinely at the present time. We take a round beam that's going into an electrostatic transport section where now you're using quadrupoles, so in the x-direction, let's say, the beam increases in size, in the y-direction the beam decreases in size, because of the alternating squeezing and rarefaction of the beam. And it eventually settles down to something that you can then transport a long distance.

Whealton: We just got started in the area of beam transport. We are starting an analysis of beams going through plasmas, using very sophisticated algorithms. We've identified a lot of issues, but we haven't really related them, specifically, to individual situations. Experimental results from the Oak Ridge ion source tend to suggest that there is a significant gas focussing effect going on, and so we're certainly aware that a beam going through plasmas is a big issue.

Holmes: Several people have already mentioned gas focussing for negative ion beams. I'd just like to expand on that slightly. When you have a negative ion beam passing through a gas, two processes can occur. Negative ions can be stripped, which can produce, in addition to the fast neutral, a slow electron, or those negative ions can simply ionize the gas. The bottom line is that you produce an excess of electrons compared with positive ions. So you might think, at first glance, that you have a problem because you're producing even more negative charge than you had in the first place. However, these slow positive ions are created with essentially no energy. They stay in the beam potential well - which is initially negative, so it would trap them - until they fill it up, and then they can escape. So, you can write down a particle balance for positive ions, for thermal electrons, and also for the electron-ion energy balance. And, incidentally, in the last category we have to include the electron energy input from the stripping process, because the electron which is stripped off a negative ion has exactly the same velocity as the original beam ion itself, and that can be quite high. Even for a 100 kV beam, you're talking about energies in the order of 20 eV. That's a process which doesn't occur in positive ion beams. You can then put these three balance equations into either Poisson's equation, if you have a big computer, or into a plasma neutrality equation, if you've got a pocket calculator, and you can then solve the equation to derive the space-charge field. Several attempts have been made at this; Paul, himself has made one, and also Lemons, at Los Alamos, and also the Russians have looked at these problems. I have developed a simple model, based on the idea of plasma neutrality. This model predicts that the ratio of the space potential to beam energy is a single monotonic function of the product of gas density and beam radius. The space potential is positive relative to the outside world and saturates at a high gas density and radius product. This saturation potential, in this particular case, is a value slightly in excess of 10^{-4} of the beam energy. The positive potential causes a space-charge focussing action on the beam because the space-charge field is directed towards the axis for negative ions, and so the beam pinches. Unfortunately, as the beam pinches the field becomes even more intense, and so the process could cause the beam to collapse to zero radius. Fortunately, however, the beam has a finite emittance, which opposes this process. Our result for our 14 mm aperture beam shows the beam radius, at half width maximum, decreasing with the pressure in the drift space, which in fact is a representation of the gas flow from the source, for a fixed pumping speed. Now, as you can see, when we're running low pressures in the discharge, or low pressures in the drift space, the beam appears to be divergent, but once you get beyond the relatively low pressure, of 4×10^{-5} mbar, the beam actually becomes convergent. This is at a distance of 1.5 meters from the accelerator. The only way we can explain this is that we have a space potential which is positive, so the space-charge field acts as a lens which causes the beam to contract until it saturates at the value about half the initial radius of the beam. This agrees with the picture created by the simple model which says that when you get to high pressures, the potential saturates. The beam divergence just simply flattens out because now the space-charge field is not changing with increasing pressure. It does seem, from a beam optics point of view, we're having qualitative agreement with the early work by the Russians, which suggested that positive plasma potentials exist in negative ion beams.

Wright: I'll give a brief summary of what I consider preliminary results on computer simulations of low-energy beam transport. We're comparing to the two experiments that were operating in the unstable regime: the Russian experiment of Gabovich, Dzhabbarov, et al., who had a 14 kV beam and a fill gas of argon in the beam transport, and the work at Los Alamos by Paul Allison and others, which had a 20 kV beam propagating in xenon. Most of our calculations have been simulations of the Russian experiment. What are the phenomena that have been observed in these experiments? Well, it's a strong function of pressure. At the low pressure (10^{-5} Torr), you see emittance growth, instability; at the higher pressure, it's stable, but there's beam loss due to the collisions in the ionization of the beam. At low pressure, the potential oscillates at a frequency of about 1 MHz, whereas at the higher pressure it's stable and there's a difference in the sign of the potential; it's negative for the low pressure, assuming azimuthal symmetry, and it's positive for the higher pressure. Our goal was to do some simulations and see if we could see these effects, and then see if there was actually a strong quantitative comparison. In the calculations we're taking a 3-D electrostatic code called IVORY, but right now we're using it in the 2-D mode, radius and length. We're looking at the role that either an ion-ion, or an ion-acoustic instability could play in giving rise to the observed phenomena. The parameters used in the simulation are an H$^-$ beam of 15 kV, with a current density of 50 mA/cm^2, and a radius of the beam of 0.8 cm. The channel is Ar$^+$, the length of the channel is 45 cm, and radius of 0.8 cm, all located in a tube of radius 8 cm. The model includes the following: H$^-$ stripping on argon; H$^-$ coming in and ionizing the argon; and we have the option for putting in other species, although at this point we're not doing that. We assume a square profile for the radial distribution of the beam, and it's monoenergetic. The Ar$^+$ is created cold; we can modify that later if desired. For the electrons created by ionization, we have an energy distribution which is forward-biased and has an average energy of about 5 eV, and the electron stripping is also forward-directed, monoenergetic, about 8 eV, with a slight perpendicular velocity. We then, essentially, solve Maxwell's equations, putting in particles, letting them move, and solving for the electrostatic potentials. They range anywhere from zero to a couple hundred volts. We initialize the calculation by starting with a channel that's 95% neutralized with Ar$^+$, and from that point on, we allow the channel to evolve through the reactions and through the electromagnetic distributions. The potential variation on axis at the beam entrance was calculated (MRC report to LANL) as a function of time out to 13.6 µs. The potential has considerable high-frequency noise, which is an artifact of the calculation and is to be ignored. At low pressure (10^{-5} Torr) there are major low-frequency oscillations which do not damp, and the average potential is negative. At high pressure (10^{-4} Torr) the potential reaches an approximately steady value of +10V. At intermediate pressure (3×10^{-5} Torr) the oscillations appear to damp, but more slowly. Well, what are the conclusions and how does it compare with the experiments? We think there's good agreement with the Soviet experiment. They saw a transition in the potential at about 3×10^{-5} Torr, and that's roughly the pressure for which we saw a change in the sign of potential. We saw oscillations at low pressure at about the plasma frequency. It's hard to tell from the computer simulations, because the plasma frequency is changing as a function of time, but it was around 850 kHz whereas, experimentally, they observed oscillations around 1 MHz. Also, it's been observed that if you shorten the propagation length, you'll reduce the amplitude of oscillations, and this is confirmed by our calculations. And there is a strong variation in the number of electrons, as you'd expect, because of the negative potential. The variation in the electron density between the high-pressure case and the low-pressure case is essentially a factor of 35, and our feeling is that at the high pressures, where you have 35 times as many electrons, they're damping out the oscillations and you get the larger population density of electrons, of course, because you have that positive potential.

<u>Allison</u>: Let me ask Robbie York to comment on the LAMPF experience

on beam transport.

<u>York</u>: We don't see any instabilities in the transport of the H$^-$ beam after it reaches equilibrium. But I've recently taken some data that show some instabilities in the initial transport of the pulsed H$^-$ beam. We think we understand the emittance of the converter source that we run at LAMPF. It fits the admittance of the source quite well. The measured emittance is exactly what you would calculate from the admittance of the source. But, while we were running the source, and tuning it into the proton storage ring, the tuners came back to me and claimed that the emittance, lo and behold, was not what we had measured so many times before, but was actually different from that. Now, just a little background: when we run at LAMPF, we run an 800 μs pulse, and the emittance measurements were taken approximately 150 μs into the pulse. That is, 150 μs after the current reaches the flat value. It has about a 50 μs rise time, and I've never measured the emittance in that region. But it turned out that they were tuning in a region for which they were using a very short pulse, which was approximately 50 μs into the flat region of the pulse. I measured the emittance at different times into the pulse. At 150 μs it is normal. At 100 μs a wing starts to form on the phase space, and the area rose to 8.3 from 6. The trend continues at earlier times, and at 25 μs the emittance is so large that the angular divergence exceeds the capabilities of the slit collector and the total current collected drops from 160,000 counts to 25,000. Most of the beam has vanished, and I believe it's because the beam is so poorly space-charge neutralized, that it cannot transport from the slit to the collector, even though it has reached its maximum current. Now, Ralph Stevens and I have known about this character of the beam, and we attributed it to some type of equilibrium that the source had to reach. That is, the plasma inside the source gave us an instability in the first part of the beam. In our transport, we not only have a source, we have a transport in which we can deflect the beam prior to the emittance gear. So, using that, I deflected the beam away for the first 400 μs, and then deflected it onto the emittance gear. Now, if I measure the emittance right after this, I get poor emittance. In fact, if I go from 150 μs out, I get exactly the same emittance for the whole length of the beam pulse. The emittance of the deflected beam had exactly the same characteristics as did the first part of the pulse. And that is the instability that I'm seeing. The change in emittance has nothing to do with the source, but it has to do with transporting a beam in the vacuum. Now, I will admit to one other possible explanation: is it some sort of equilibrium that the emittance gear itself has to reach? That is a possibility I cannot rule out. I'm going to explore this as a function of gas pressure, and what I would predict is that at higher gas pressures, this effect occurs faster. All these measurements were done at 80 kV. At Fermilab, they have seen much faster response at higher energies, but there is still some variation in the emittance of the beam in the initial part of their pulses. We're going to explore it at 80 kV as a function of gas pressure, and compare 80 kV and 750 kV, to see how this phenomenon characterizes itself.

Lietzke: The background gas is hydrogen?
York: Yes. The vacuum for these measurements is approximately 3×10^{-6}. When we change the gas pressure, it will be hydrogen.
Allison: It's worth noting that that gives you about a $1/n\sigma v$ time constant of order 50 µs, so it's certainly not inconsistent that the emittance changes are due to neutralization transients.
Leung: Could you put xenon in the transport?
York: Not very easily. The problem is we have ion pumps in the transport line, and if I put xenon in the transport I'll have to replace them. But we are going to try hydrogen gas, and if I see the effect I might try argon or something a little heavier to see if it changes the neutralization time.
Allison: Let me ask Chuck Schmidt to talk about beam transport at Fermilab.
Schmidt: I am showing two recent emittance measurements in our 750 kV transport line. The first one is the one we were seeing on one of our study periods, and there are very large distortions in the plot with the transport set up as it's normally running. By chance, one of the ion pumps gurgitated, and we saw a momentary increase in beam current down the linac. At the end of that period, we wanted to see what increasing the pressure would do. So we throttled the ion pumps back and got the pressure up to about 1×10^{-5}. The distortions greatly decreased at the higher pressure. So we really are seeing nonlinear space-charge distortions in this beam and are certainly neutralizing the beam here. If we go much higher in pressure, we start seeing significant stripping of the beam; we have about 50 mA and 750 kV. We see this effect in both transport lines; one line is about 10 meters long, the other is about 4 meters long. We are increasing the gas pressure, only at one end of the line, near the column, the accelerating end of that line, so the pressure distribution on the line is very non-uniform, and the pressure is probably mostly changed near the source. We suspect we could get rid of a lot of this effect by re-tuning the line. The first time we really saw it, it was much more distorted than this. The phase space folded over into a big "S" shape. It just looked like a large beam; we almost couldn't see the filaments in it any longer. By re-tuning the line and then by adding gas, we did reduce the extent of the wings. We probably could do more tuning on the line to even improve this a little more. It may say that we're focussing the beam too tightly somewhere in the line. The pulse that we use down this line is only 30 µs long, and we're sampling it about in the middle of that pulse. So whatever is occurring, is occurring on the order of 10 µs. Whether it is stabilized here, I don't know. We really haven't done enough study with pressure, or looking at the beam pulse as a function of time to really say anything about that.
Witkover: At Brookhaven we also have had a similar experience. In fact, when we were testing the ion source in our test stand, we came across identical phase-space distributions with the same wings. And we found that as we scanned at 10 µs intervals from the start of the beam, we were able to not only see the wings vary, but also the beam rotate in the emittance plane. We were quite concerned about this. There was speculation that it was from lack of space-charge

neutralization of the beam. To test this out, we did scans at different extraction voltages and found some reduction of the effect, but a major change occurred when we started bleeding gas into the transport channel at 20 kV. By doing that, we were able to completely eliminate the rotation down to our shortest measurement time of 10 μs. The results are in the '81 Particle Accelerator Conference. The rotation was stopped, and was the same rotation that occurred after about 100 μs; it was the same orientation, and the wings that were present initially were no longer there. So our conclusion was, at that time, that this was a neutralization phenomenon, and it was possible to avoid it by bleeding gas into the channel. The normal pressures at which we worked were in the high 10^{-6} range. We had to bleed up to several times 10^{-5} range to eliminate the effect. We saw, also, similar results in our 760 kV transport. We have two Cockcroft-Waltons that we use to inject beam into the AGS. One of them is off-line, and the beam must go through a 10 meter transport to the main line to our linac, and that 10 meter line is weakly focussed and pumped with 4 ion pumps along the line. The thing that was puzzling us was a long rise time, typically of the order of 50 μs, until it reached saturation, for a pulse width of about 400 μs. As a very quick test, I successively turned off vacuum pumps in that line and found that the rise time gradually decreased until it was within 10 μs, or so. I reached a point at which the trailing edge of the beam started going down. We interpreted that as, in fact, the stripping. I believe these results were in the '83 Particle Accelerator Conference. The pressure ranges, although they were not uniform through this 10 meters, ran from an initial value of several times 10^{-7} to about 10^{-6}, when we did see the stripping occurring. This is still an issue, and we certainly would be interested in eliminating the problem.

Schmor: I shall make a few comments that are very preliminary. I have not yet satisfied myself that we've eliminated all of the possible systematic errors. Since moving our source from the test stand into the high-voltage terminal, we have observed that the apparent emittance that we are measuring decreases as we increase the focussing field of the solenoid. We were concerned that something unusual was happening; we therefore built a second emittance scanner, and this has just come into service this past week, so we only have about a week's worth of results. The emittances before and after the solenoid are equal to within 5% if the beam size after the solenoid is small enough. However, if the solenoid fields are weak or off, the emittance growth is 30-50%. These are currents of only 1 mA, at an energy of 25 keV. The vacuum in this region is low 10^{-7}. As I say, these results are very preliminary.

Leung: Paul, I wanted to ask you if you built an energy analyzer to measure the ion energy falling down from the center of the beam?

Allison: We've made preliminary measurements, with an energy analyzer, of the ion and electron energies coming out of the beam, yes.

Leung: Have you seen any 10 eV or 100 eV positive ions coming out?

Allison: We've seen ion energies of 5 eV to 10 eV, both with emissive probe and with the energy analyzer. In one of the Russian papers, however, energies of up to 50 eV were seen. That seems to be much higher than one would expect from any of the quiescent theories.

Leung: But when you run it with low gas pressure, suppose you have a negative potential. At that point, can you see that no positive ions are coming out?

Allison: That is right. We measured the total ion current falling out radially from a 20 kV beam, versus gas pressure, and we found a threshold pressure. Below about $8 \times 10^{11}/cm^3$ of xenon you see no ions whatsoever. And at that pressure, correspondingly, you have very high energy electrons, 20-30V. So, I think this is rather similar to the Russian measurements. It shows that there's an onset, there's a critical pressure before you really get complete neutralization.

Holmes: The fact that you see positive ions coming out of the beam means that you must have a positive plasma potential in order to expel them.

Allison: That's correct.

Holmes: And you say about 5 eV, more or less?

Allison: Roughly, yes.

Holmes: That would fit reasonably well with my model.

Allison: If one takes your old model for positive beams, Andrew, and converts that for negative ions, that's also about the same potential that you would calculate. We saw that approximate value both with the energy analyzer and also with the hot probe and, for the most part, that's in agreement with Russian measurements.

Alessi: At Brookhaven, we're building a 35 kV transport line between a magnetron source and an RFQ. We're wondering how concerned we have to be about emittance growth due to beam instabilities as a function of the length of the transport line. As it stands, it's several meters long. Secondly, we would like to put an electrostatic chopper in the 35 kV line, which would be chopping the beam at a 2.5 MHz rate. I think, rather than being a problem of the turn-on of space-charge neutralization, this is more like producing a noisy beam. I'd be interested in anybody's comments on what that might do.

Ehlers: It doesn't sound to me like any of the problems you're running into are the least bit unusual. By that, I mean, the same thing has occurred in positive ion sources for a long time. The only difference is, that with positive ions, you've got to neutralize with electrons, and there's always that problem of the formation time, versus their escape time. In the case of positive ions, the escape time is fast, because the electrons can get away; hence, if you have very short oscillations in the beams, electrons can get away and you have to re-neutralize again in phase with all of these little jitters in the beam. With positive ions as your neutralizing agent, and particularly heavy ones like xenon, it takes a little longer for them to escape, so you can get by with a little more murder, if you will. But there's really two problems. You've got the problem of making the ion, be it electron or positive ion, in the case of negative ions, and you've got the problem of it getting away. So, if you put electrostatic plates in there, you're accelerating the escape of these ions. For example, when we built the heavy ion source for the HILAC, where we were using nitrogen, say 4 times ionized, space-charge forces were so great that when the beam came out of that ion source it diverged in line with B, to where it blew up to 6 inches up and down in line with the slit, within an inch of the extractor. This meant we had to find some way of providing electrons to do the neutralizing, and we couldn't use gas, because if you added gas, you've lost all your high charge states. So we had to make the electrons by putting in some aluminum plates so that the ions, as they blew up, would breed secondary electrons in the region where the beam was running, inside that magnetic field. This made the time for neutralization much shorter, very, very short, compared to the on-time. Otherwise, it took almost the whole on-time to form enough electrons to really neutralize the beam. You could conceivably consider making positive ions by some other means, other than by making them off the background gas. Or, you could also look at the possibility of trying to trap them so they can't escape. Your magnetic field, for example, is a classic example because it prevents the escape. I used an electrostatic quadrupole, at one time, looking at the heavy ion beam, trying to focus it to present it to the Cockcroft-Walton. It took 15 kV on a lens designed to operate at 300V before I could get the thing focussed back to where it was originally; and the reason was that the electrostatic lens ate up all the neutralizing particles, and then it really blew up. So, two things: prevent the escape of the particle, and find some new way of breeding them - the neutralizing particles.

Allison: Well, I agree, at least in part. I think there are two different things going on here. For example, the Fermilab experience was with a 30 µs beam pulse; that is comparable to the time of formation of the positive ions. In other cases, however, you have instabilities where the beam pulse is, say, a millisecond long and formation times are a few microseconds. I think it's a different regime of operation.

Holmes: We've touched on two interesting points. I think everybody is saying that if we add more gas in the beam line, we can quench the instabilities and get better beam transport. The only problem is if you have too much gas in the beam line, you'll strip the negative ion beam. This is particularly tough, I think, at lower energies. You're working at 35 kV, Jim. The stripping cross-section is quite large at that energy; in fact, it's almost at its maximum value. This gets easier as you push up the energy, of course, because you have more room to maneuver. You can have a higher pressure in the beam line for a given loss of negative ions. The only thing I'd like to propose is to exploit the beam radius effect. That if you have a really large diameter beam in the transport channel, you can work at a lower pressure, because the space-charge neutralization phenomenon scales not as gas density alone, but gas density times beam radius.

Allison: It'd be interesting to know if this has any relevance to the effect that Paul Schmor sees.

Holmes: Yes, I was thinking about that, actually, because you get up to almost 6-7 cm diameter, and you're working at very low pressures. The huge diameter of the beam allows you to work at such low pressures.

Schmor: Our apparent emittance, however, increases with larger beam diameter.

York: Since Fermilab and Brookhaven and LAMPF have all observed effects related to the time to reach neutralization in the transport line, can you give me an explanation of what that should look like as a function of beam energy? It bothers me that at 80 kV it seems to take longer than it does at 750. And the cross-sections for neutralization are actually higher at 80 kV than they are at 750.

Lietzke: But the space charge is higher at the lower energy. You have to neutralize more.

York: Yes, maybe it's the effect of the space charge itself that's larger.

Tsai: Since all the questions relate to the space-charge effect, I'd like to know if people are thinking of creating a low density plasma for beam neutralization. A low density plasma, 10^{-6} or 10^{-7} pressure range. And the beam passing through automatically can space charge neutralize. Use beam particles to create the plasma.

Allison: My answer is that very low pressures are where the instabilities are observed. Gabovich has derived a critical pressure, roughly 10^{-5} Torr and, experimentally, the disappearance of the instabilities does seem to occur at about that pressure range. It may be just fortuitous.

Moses: Jim, I can relate one experience back at Livermore when we were working electron beams and tried to chop those beams. As observed, most of the instabilities started at the head and work their way back. When you start chopping, you're going to be introducing a large amount of perturbations at each one of those little sections, and I would caution you, that could be a problem.

Allison: An interesting point is that probably many of us are somewhat ignorant of the electron beam experience in this regard. One of the fundamental papers goes back to 1944 to Pierce in propagation of electron beams in vacuum tubes.

Lietzke: I'd like to make one comment on electrostatic transport. What one is trying to do physically is to, by imposing boundary conditions on the edge...on the average, the electrostatic field in the plasma is zero. So, in a case of a quadrupole, you have to average over space or average over time. The disadvantage, as has been pointed out by Ken Ehlers, is that the electric field sweeps out the natural neutralization that one could achieve with positive ions. So, in an electrostatic transport system, one has to take over that function of the ions by appropriately tuning the boundary conditions.

Holmes: I just have one comment to what Al has just said. I think in low energy beam transport systems, you have two options. One is to go for the highest degree, over-compensation of negative ion beams and use magnetic lenses to transport the beam to wherever you want to go; that avoids emittance growth problems, but you wind up with a Gaussian beam profile, because it's created from the thermal

distribution of electrons — the beam profile matches that, and you always have Gaussian profiles. The other option is to take all the charge out of the beam completely using electrostatic quadrupoles, and have a strong focussing channel. I don't know if this has ever been tried before.
Allison: Yes, the RFQ does that.
Holmes: Yes, that's true. But not directly from the ion source, where you have a gas plume coming out of the accelerator. And when you're into the RFQ, you have already established a pressure of 10^{-6} or thereabouts, and gas flow problems are not really a difficulty. Whether you can actually do this right up at the earth electrode of the accelerator, that's more difficult.
Whealton: We were designing such a LEBT, but I guess the principal operation depends upon the ion source having a relatively low current density. If the ion source current density is sufficiently high, there is probably no choice but to have a plasma in there to help you along.
Lietzke: Yes, the voltages get too high.
Holmes: Yes, I agree with that, John.
Lietzke: Does anyone see any kind of a beam plasma instability with this neutralizing plasma?
Allison: Yes. The Russians' experiments are quite well documented on that point. We've also had considerable experience with that ourselves, so I think there's no question that you do have instabilities in negative beam transport. I'm not aware, myself, of comparable instabilities in positive beam transport. If there are none, I take that as a sign that the accumulation of electrons, which is presumed to quench the instabilities of negative beam, is what's working in positive beams to prevent the instabilities in the first place.
Roberts: I didn't understand what you said. Don't positive beams...?
Allison: Well, with positive beams, you always accumulate electrons, at high pressure and low pressure, both. For negative beams, at very low pressures, you do not accumulate electrons. The accumulation of electrons is what quenches the instabilities. You have to get electron densities greater than 10^{-3} to 10^{-4} that of the beam to quench the instabilities, because otherwise, you have ion-ion, very low-frequency, instabilities.
York: If it is the accumulation of particles and, depending on gas pressures, it causes quenching of instabilities, why, when we run such a long pulse, from 150 µs out, don't we see any instabilities at all?
Allison: Well, that's a good question. Your perveance is very low, so it may depend on perveance. Also, your time constant in the LAMPF transport is about 60 µs, so you may well have accumulated significant electrons by then. In Pierce's original paper, the condition for stability for the electron beam transport was perveance-related.
Roberts: Let me ask a question to see if I can understand what you said. If I make electrons, the electrons are light, they can leave, and not only cancel the potential, but make a potential well for

negative particles, because the positives are heavier. And so, because of thermal velocities, they can leave more, and when you balance currents to the walls you can have a potential well for the negative particles. So you can have not only space-charge neutralization (so the beam drifts), but it could conceivably go to a smaller diameter and actually pinch. But suppose you go the other way around now, and you're using positive beams, and you create electron-ion pairs. You can attract enough electrons, I think, to cancel that. But after that, they're thermally going to leave. I don't see how the positive beam ever pinches. It looks like it would just drift without expansion, other than what its divergence was.

Allison: For positive ion beams, the experience is that you never reach a state of over-neutralization, for exactly that reason.

Roberts: I don't see that the instability problem is the same there.

Allison: I don't think it is.

Roberts: Okay, I thought maybe you said it was.

Moses: Robbie York, have you just measured your emittance at one position, or have you measured it as a function of distance or time later in the beam?

York: We have measured as a function of time later in the beam, but not as a function of distance. This emittance scanner is in the H$^-$ dome and there's only one position. But from 150 µs out, the emittance is absolutely constant.

Moses: What I would suggest, if you have enough beam line, is you march back in beam line and you may see that instability grow toward the tail of your pulse.

York: Well, it's not easy. I can do it as a function of energy; couldn't I do the same thing by changing the energy?

Moses: No, no. You have to allow for the growth time, and that's what I would suggest. But if you don't have beam line, you can't do it.

York: I could probably do it at the ion source test stand. Not in the dome. But I've not observed this phenomenon at the ion source test stand - yet. We don't have a deflector there. As I said, we saw it in the rise time of the pulse, and we attributed it to the source, but after I saw it with the deflector also, I couldn't attribute it to the source anymore. I have a new deflector that I can add to the test stand, and I can probably do the study there.

Moses: You might try to move forward. If you can move forward, you might see it shortening up.

York: In the H$^-$ dome, you can't move at all.

Roberts: That's a good point. But, if it takes 100 µs from the front of the pulse, so you have 800 down to 150.

York: At 150, it was absolutely constant. At 100, I first saw the variation.

Allison: What Robbie observes is not necessarily an instability. I believe I'm right in assuming that your emittance scanner integration time is about 20 µs.

York: Yes.

Allison: And your neutralization time constant is of that same order, so there's a dynamic change in the neutralization during the integration time. If you could measure with an integration time of 1 µs, the results might be very different. Chuck Schmidt, what is your integration time on the emittance measurement?
Schmidt: They should be typically very short. I notice the amplifiers are kind of slow lately.
York: I did look at the emittance of the function of the integration time, and I went from 20 out to 100 or so, and there is no variation. But I didn't try anything smaller than 20.
Allison: It seems to me that your results may be explained by the changing neutralization during the measurement, rather than by an instability.
Lietzke: What are the frequencies observed in the transport instability, and do they depend upon the mass of the background gas?
Allison: In the Russian experiments they are observed and they roughly match the frequency appropriate to the background gas. But I don't think they investigated gases other than argon.
Lietzke: Even at high density?
Allison: Yes. The frequencies are typically around 1 MHz, and that matches the ion-ion frequency.
Holmes: What's your threshold, Paul, for the quenching of the instabilities - without xenon?
Allison: Without xenon, I don't know, because I've never used hydrogen to do that. In the Russian experiments, the threshold is found to be almost 10^{-3} with hydrogen.

PRODUCTION OF OTHER NEGATIVE IONS

FUNDAMENTAL PROCESSES IN LOW ENERGY COLLISIONS OF ALKALI ANIONS AND ATOMS

R.L. Champion, L.D. Doverspike, D.M. Scott, and Yicheng Wang
Department of Physics, College of William and Mary
Williamsburg, VA 23185, USA

I. INTRODUCTION

Negative ion sources often employ alkali metal atoms in one way or another in order to increase their yield. Models which describe equilibrium conditions or the nature of energy transport within these sources require, among other things, information about the two-body cross sections for various scattering channels which involve either alkali negative ions or alkali atoms. The purpose of this report is to provide a brief summary of recent experimental observations in which collisions of alkali anions (M$^-$) with various atoms and molecules and collisions of H$^-$ and D$^-$ with alkali atoms (M) have been investigated. The energy range of the experiments, $5 < E < 500$ eV, includes those kinetic energies often found in discharge-type ion sources. The specific experiments which will be discussed focus upon measurements of total cross sections for collisional electron detachment and charge transfer of negative ions.

II. EXPERIMENTAL METHODS

Experiments which involve alkali negative ion beams or alkali atomic beams have problems associated with the handling, etc., of alkali metals and it is perhaps of interest to discuss those relevant features of the two types of experiments, M$^-$ + X and H$^-$ + M.

The alkali negative ions, Cs$^-$, K$^-$, and Na$^-$ are produced in a discharge-type source whose design is based upon the observation of K$^-$ production by electrical discharges in potassium vapor at low pressure[1] (0.01 - 0.3 torr). In the present source,[2,3] the discharge is maintained in a pure alkali vapor; the source and oven temperatures at which the yield is maximum imply that the principal mechanism for M$^-$ production may be via dissociative attachment to alkali dimers. The oven which supplies the source is maintained at 230C, 250C and 300C for Cs, K, and Na, respectively; the source is maintained at 350C for all three alkali metals.

The performance of this source is adequate for these experiments: typical mass-analyzed currents produced are 0.11 nA of Na$^-$, 0.45 nA of K$^-$, and 0.5 nA of Cs$^-$. The energy width of these beams can be as low as 0.2 eV and is generally less than 1 eV.

It has been observed that this ion source produces anions with

masses corresponding to NaH⁻ and KH⁻. During the first several
hours of operation, the hydride ions are more abundant than the
desired alkali-metal anion, but the mass peak associated with the
hydrides diminishes with the time. These hydrides are thought to
result from reactions of the alkali-metals with water vapor
present in the source. In order to ensure that there are no
undesired ions in the primary beam, a 90° double-focusing section
magnet with a resolution of 1% has been employed for mass
analysis. The K⁻ and KH⁻ mass peaks are clearly resolved; the
observed contamination of the K⁻ primary beam by KH⁻ is less than
1%. The Na⁻ and Cs⁻ beams are similarly pure.

The beam of alkali anions is directed into a collision region
which contains a gas-phase target; σ_e and σ_{ct} are measured by
separating and trapping the detached electrons and the (relatively slow) charge transfer products.[4]

A gaseous target can not be employed for studies of alkali
atomic targets; rather a crossed beam apparatus has been constructed for that purpose. A schematic diagram of that apparatus
is given in Fig. 1. The negative ion beam is mass-selected and

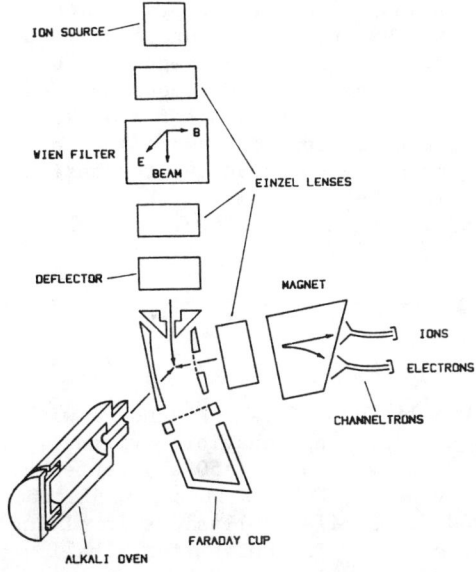

FIG. 1

Schematic diagram of
crossed-beam apparatus.
The alkali oven (or a
gas jet which is not
shown) can be rotated
into and out of the
collision region.

focused into the collision region which is within a one-sixth
section of a 127° cylindrical electrostatic energy analyzer. The
voltage across the two curved plates of the analyzer allows the
ion beam to pass resonantly through the analyzer section. The
ion beam intersects, midway of its path inside the analyzer, with
a neutral target beam. The electric field maintained across the
analyzer is used to extract the slow anions and electrons
produced in the collision region through a grid on the inner

plate. The extracted anions and electrons are then focused, separated by a magnetic field and detected with conventional particle multipliers.

The alkali oven, along with a separate gas nozzle identical in shape to the oven's exit cylinder, may be rotated into the collision region. This design enables one to monitor and calibrate the apparatus, before and during experiments with alkali vapors, using some previously studied reactants[5] such as $H^- + O_2$ and $H^- + Ar$. Liquid-nitrogen-cooled surfaces are positioned to trap the undesired alkali vapor. A vexing source of noise is related to the presence of alkali atoms on the surfaces within the collision region. Even with no ion beam in the collision region, some negatively-charged particles are observed to desorb from the alkali-coated surface and arrive at the two particle multipliers. The intensity of these particles increases dramatically as one increases the partial pressure of oxygen or water vapor in the vacuum chamber and, in fact, can easily saturate both particle multipliers. This problem prevents one from using $H^- + O_2$ to calibrate the apparatus during the alkali experiments. These negatively-charged particles which come from alkali-coated surfaces include both electrons and anions; their production mechanism remains unexplained at this time. Fortunately, in the present experimental environment, the extraneous signals caused by the alkali-coated surfaces decrease when the temperature of the oven increased; at the appropriate temperature to do the experiments, these extraneous signals are about one-eighth of the authentic signals. The background contribution due to these particles is measured by steering the ion beam away from the collision region and observing the resultant signals present with a zero-intensity negative ion beam.

III. Results

A. $M^- + X$

Measurements of $\sigma_e(t)$ for collisions of Cs^-, K^- and Na with rare gas (RG) targets reveal a surprising behavior: virtually no detachment is observed until relatively high (~50 eV) center-of-mass collision energies are reached.[2] This is in contradistinction to what has been observed for similar collisions involving another nsns' negative ion, viz., H^-. The onsets for alkali anion detachment are approximately equal to the thresholds for *excitation* observed in collisions of neutral alkali atoms with these same targets.[6] The similarity between the dynamics of the neutral system and that of the negative ion system, together with the observation (at greater energies) of detachment accompanied by excitation of the alkali parent, suggests that electron detachment in the present experiments is mediated by a two-electron process.

Measurements of $\sigma_e(E)$ for the Cs⁻+Ar and K⁻+Ar systems are shown as a function of relative collision energy in Fig. 2. Data

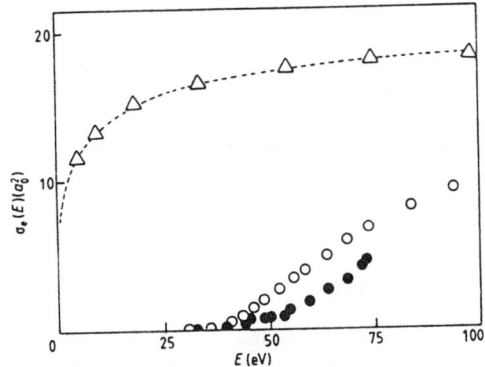

FIG. 2

Total cross sections for electron detachment by Argon; H⁻ (triangles), K⁻ (open circles) and Cs⁻ (solid circles).

for H⁻+Ar has been included for comparison. Note that the energetic threshold for detachment of H⁻ is on the order of a few eV, whereas the thresholds for the K⁻ and Cs⁻ projectiles are about 50 eV, approximately 100 times the electron affinity of the alkali atom. A similar result is observed for the Na⁻+ Ne system, where the apparent threshold for detachment is about 55 eV. These distinctively high thresholds are typical for the M⁻+RG systems where thresholds have been observed. The striking dissimilarity between $\sigma_e(E)$ for M⁻+Ar and that for H⁻+Ar at these low energies, may be a manifestation of electron correlation. It is well known that H⁻ is best described by a split shell (1s1s') configuration, implying that the correlation between the two electrons is primarily radial in nature. On the other hand, recent calculations[7] of M⁻ wave-functions suggest that angular correlation between the valence electrons is significant for the M⁻ ions. Perhaps it is this difference in character that leads to a difference in the threshold behavior of $\sigma_e(E)$.

An M⁻+RG correlation diagram (for the example of Na⁻+Ne) is proposed in Fig. 3. The diagram has been inferred in part from

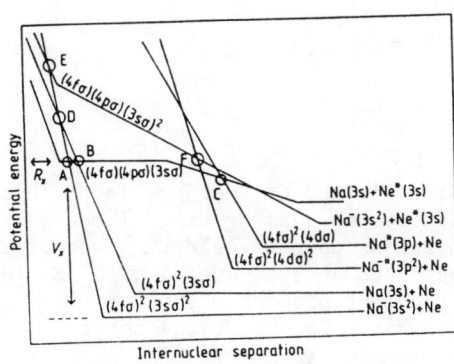

FIG. 3

A schematic representation of molecular configuration correlation diagram for Na + Ne + e.

calculations for the Na+He system[8] and also from the observation of a common threshold for detachment (of Na$^-$) and of excitation (of both Na and Ne). The incoming Na$^-$+Ne state remains below the $X^2\Sigma^+$ continuum of Na+Ne+e until the crossing with the continuum of Na+Ne*(3s)+e is reached at A. This crossing (with coordinates R_x, V_x) accounts for the onset of electron detachment at the collision energy $E = V_x$. A subsequent transition at B can result in ground state products. If one assumes that the probability for detachment at crossing A is unity, then the total cross section is given by

$$\sigma_e(E) = \pi R_x^2 (1 - V_x/E) \quad \text{for } E > V_x$$

$$= 0 \quad \text{for } E < V_x.$$

Estimates of R_x and V_x can be made by fitting this expression to the data. The recovered values for the Na$^-$+Ne system are V_x=65 eV and $R_x = 1.6 \, a_0$. The fact that the threshold for Na(3s)+Ne → Na(3P)+Ne is the same as that for detachment of Na$^-$ by Ne suggests that A is located near B. The remarkable agreement between the estimated position of A and the calculated position of B indicates that this picture has some validity.

Measurements of $\sigma_e(E)$ and the cross sections for charge transfer have also been completed for H_2, D_2, N_2, O_2, CO, CO_2, and CH_4 targets. The energetic thresholds for electron detachment are found to be about 5 eV for H_2 and D_2 and range from 3-10 eV for the other molecular targets.

The detachment channel is especially interesting for the CO_2 target. As can be seen in Fig. 4, significant structure is

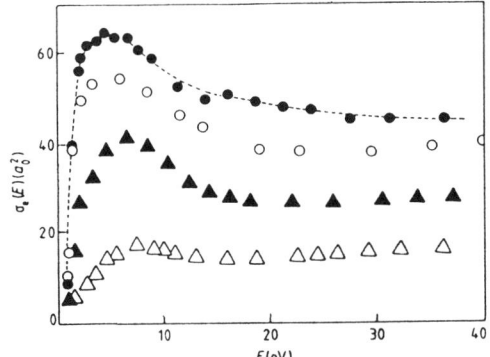

FIG. 4

Detachment cross sections for H$^-$(△), Na$^-$(▲), K$^-$(○), and Cs$^-$(●) + CO_2.

present in $\sigma_e(E)$ for the H$^-$, Na$^-$, K$^-$, and Cs$^-$ projectiles. This structure is peculiar to projectiles possessing two valence s-electrons; O$^-$, S$^-$, and halogen anions do not yield this structure for CO_2. The relatively strong dependence on energy of $\sigma_e(E)$ for E < 15 eV suggests that the detachment process may occur via a negative-ion resonance of the target. The CO_2^-($^2\Pi_u$) shape resonance, which is an important detachment mechanism in high energy H^-+CO_2 collisions,[9] has an endoergicity of about 4 eV. In contrast, the present measurements of $\sigma_e(E)$ indicate a substantial cross section (~10 Å^2 for K^-+CO_2) at 2 eV. Even after taking into account instrumental and doppler broadening inherent to the experiment, the observed threshold (~1.2 eV) is much lower than is necessary to access the $^2\Pi_u$ state. There is, however, a metastable state of CO_2^- ($^2A_1$) which lies at a lower energy.[10] This state (which has a lifetime of ~90 μs) is bent in its equilibrium geometry with a bond angle of ~135°, and its energy lies approximately 0.7 eV above that of the CO_2 ($^1\Sigma^+_g$) ground state. The endoergicity of charge transfer to this state is therefore about 1.2 eV (0.7 eV plus the electron affinity of the alkali), suggesting that for E < 15 eV, electron detachment is mediated by charge transfer to this metastable $^2A_1$ state of CO_2^-.

For the systems listed above, it is only for the O_2 target that charge transfer is observed to be the dominant electron removal mechanism. In fact, σ_{ct} is as large as 400 a_o^2 for Cs^-+O_2 at E≈4 eV.

B. $H^-(D^-)$ + M

The measured cross sections for charge transfer and electron detachment for collisions of H^- with Na are shown in Fig.5 as

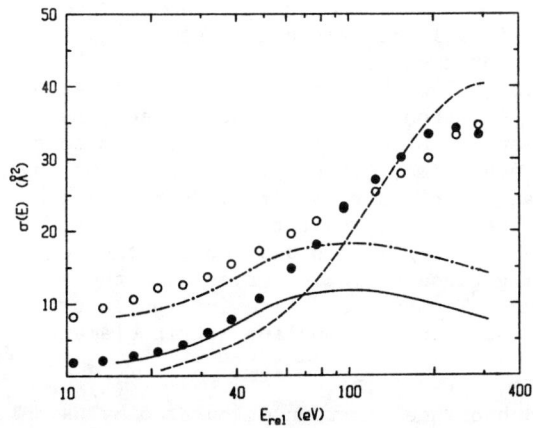

FIG. 5

Charge transfer cross section for H^- + Na - solid circles (exp.) and dashed line (eq. 4); detachment: open circles (exp.) and solid line (eq. 3).

functions of relative collision energy. Since the target thickness in the crossed-beam experiments could not be measured accurately, the absolute value of the cross sections reported here were not experimentally determined. We have chosen to normalize $\sigma_{ct}(E)$ to a calculation of Olson and Liu[11] at high energy in order to facilitate comparison. As may be seen in Fig. 5, the energy dependence of $\sigma_{ct}(E)$ agrees well with this calculation. However, the measured electron detachment cross sections seriously disagree with their prediction that $\sigma_e/\sigma_{ct} \leq 0.5$.

There are several possible reasons for this disagreement; they will be discussed below.

First, it is useful to refer to the potential curves for NaH and NaH⁻ in Fig.6. These curves are obtained directly from the

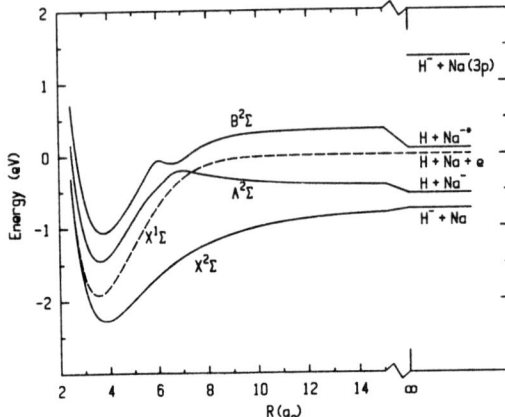

FIG. 6

Low-lying molecular states of NaH⁻ (solid lines) and ground state of NaH; from Ref. 11.

numerical results reported by Olson and Liu, adjusted to account for the correct electron affinity of hydrogen. As may be seen, the $X^2\Sigma$ state crosses into the $X^1\Sigma$ continuum around $R = 2.7 \, a_0$, agreeing with the calculation of Karo et al.[12] This crossing (which was ignored in Ref. 11) will clearly contribute to $\sigma_e(E)$ with a maximum cross section of about 6 Å².

Second, it is useful to review the two-state PSS calculation for the charge transfer and electron detachment cross sections. In the calculation, as presented in Ref. 11, it is presumed that the electron-loss of H⁻ is due primarily to a long-range coupling between the $X^2\Sigma$ and $A^2\Sigma$ states. In order to determine a diabatic coupling matrix element between these two states of NaH⁻, a point-charge-induced interaction $(\alpha/2R^4)$ is assumed for the diabatic curves. By comparing these diabatic curves with the calculated adiabatic $X^2\Sigma$ and $A^2\Sigma$ curves at large internuclear distance ($R = 10 - 20 \, a_0$), the diabatic coupling matrix element is found to be:

$$H_{12}(R) = 0.0274 \exp(-0.171R) \qquad (1)$$

Probability evolution on each channel is then calculated by using a two-state PSS method with straight-line trajectories. In that model, the amplitude for the particles to be in the upper $A^2\Sigma$ state is

$$C_2(b,v,x) = (i\hbar v)^{-1} \int_{-\infty}^{x} H_{12} \exp(i\phi) \, dx' \qquad (2)$$

with $\qquad \phi(b,v,x) = (\hbar v)^{-1} \int_{0}^{x} (H_{22} - H_{11}) \, dx'$

in first-order approximation ($C_1 = 1$), where H_{11} and H_{22} are the diabatic potential curves, C_1 and C_2 the amplitudes on states $X^2\Sigma$ and $A^2\Sigma$, b is the impact parameter, v the collision velocity and x is vt.

Although detachment directly from $X^2\Sigma$ was neglected, autodetachment from $A^2\Sigma$ was included. As may be seen in Fig.6, the $A^2\Sigma$ state crosses into the $X^1\Sigma$ continuum at $R_2 \approx 7.4\ a_0$. Thus, for trajectories with b < R_2, the $A^2\Sigma$ state may yield autodetachment when R < R_2. In fact, unit probability was assumed for electron ejection from the $A^2\Sigma$ state at this crossing. In other words, the long-range interaction between the target and the incoming projectile (t < 0) on a trajectory with b < R_2 yields an electron detachment cross section

$$\sigma_e(v) = \int_0^{R_2} |C_2(b,v,-(R_2^2 - b^2)^{1/2})|^2\ 2\pi b\ db \qquad (3)$$

For b > R_2, the coupling leads only to charge transfer. The resultant charge transfer cross section is

$$\sigma_{ct}(v) = \int_{R_2}^{\infty} |C_2(b,v,\infty)|^2\ 2\pi b\ db \qquad (4)$$

The contribution to σ_{ct} for outgoing trajectories (i.e., t > 0) on a trajectory with b < R_2 is not included in (4). The magnitude of this contribution is uncertain due to a lack of information about the coupling for R < R_2.

Numerical calculations using (3) and (4) have been performed and the calculated charge transfer cross sections duplicate those presented in Ref. 11 to within 5%. Yet, we get very different results for electron detachment, viz., $\sigma_e(E)$ is found to be larger than $\sigma_{ct}(E)$ for energies below 70 eV or so, as shown in Fig.5.

The detachment due to the crossing of the $X^2\Sigma$ and $X^1\Sigma$ states, combined with that predicted by (3), can well account for the measured electron detachment cross sections for low energy; this is shown by the dot-dash line in Fig.5. The gap between the dot-dash line and the measured electron detachment cross sections shows an energy dependence similar to the charge transfer cross sections. This strong energy dependence suggests that there exists another electron loss channel that is near resonant in nature.[13]

The exit channels for electron loss discussed so far include only H + Na⁻ and H + Na + e. The energy loss spectra of neutral H for collisions of H^-_* + Na, as measured by Tuan and Esaulov,[14] suggest that the Na^{-*} shape resonance[15,16] plays a role comparable to that for charge transfer to Na^-. Of course, Na^{-*} will autodetach and contribute to electron detachment. To the best of our knowledge, no theoretical prediction or explanation exists to account for a substantial production of Na^{-*} in slow collisions of H^- + Na.

To summarize, charge transfer in slow collisions of H^- with Na is due to long-range coupling between $X^2\Sigma$ and $A^2\Sigma$; a previous calculation[11] agrees well with our measurements. For the electron detachment, on the other hand, there are several responsible mechanisms: (1) autodetachment due to the crossing of $X^2\Sigma$ into $X^1\Sigma$; (2) charge transfer to $A^2\Sigma$ for $b < R_2$ and $t < 0$ and thereafter detachment due to the crossing of $A^2\Sigma$ into $X^1\Sigma$; (3) charge transfer to $A^2\Sigma$ in $R < R_2$ and thereafter transition to $B^2\Sigma$ or $H + Na^{-*}$ shape resonance state due to the avoided crossing between $A^2\Sigma$ and $B^2\Sigma$.

The measurements of σ_{ct} and σ_e for collisions of D^- with Na (and H^- + Na) display velocity-dependent isotope effects which are consistent with the theories discussed above; the two-state PSS method with straight-line trajectories inherently contains a velocity-dependent isotope effect for H^- and D^-.

The measurements of σ_{ct} and σ_e for collisions of H^- and D^- with K are shown in Fig.7 as functions of collision velocity.

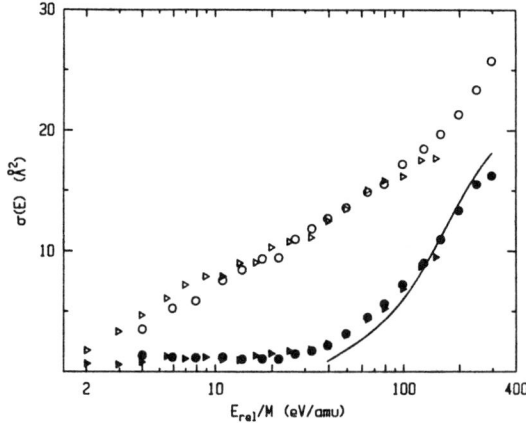

FIG. 7

Charge transfer (solid symbols) and electron detachment (open symbols) for H^- (circles) and D^- (triangles) + K. The solid line is from Ref. 11.

As for Na, we have chosen to normalize the measured σ_{ct} for H^- + K to Olson and Liu's calculation at high energy. As may be seen in the figure, the general structure of σ_{ct} and σ_e for K are very similar to those for Na. This feature suggests that the electron-loss mechanisms are the same for these two targets. The threshold of σ_{ct} for K is higher than that for Na and the overall cross sections for K are smaller than those for Na. These differences are due to the fact that K has a smaller electron affinity and a larger dipole polarizability, and consequently a larger energy gap between the $X^2\Sigma$ and $A^2\Sigma$ states of KH^-.

In summary, both charge transfer and electron detachment are significant electron-loss mechanisms in slow collisions of H^- and D^- with Na and K. Both processes exhibit a velocity-dependent isotope effect for H^- and D^-. $\sigma_{ct}(E)$ displays a strong energy dependence and high energetic thresholds (about 20 eV for H^- + Na and 40 eV for H^- + K), while $\sigma_e(E)$ displays weaker energy dependence and apparent zero energy thresholds.

Measurements for the reactants $H^-(D^-)$ + Cs are currently underway.

ACKNOWLEDGEMENTS

The work reported in this paper was supported in part by the U. S. Department of Energy, Office of Basic Energy Sciences, Division of Chemical Sciences.

REFERENCES

1. Yu. P. Korchevoi and V. N. Makarchuk, Ukr. Fiz. Zh. 24, 897 (1979).
2. D. Scott, M. S. Huq, R. L. Champion and L. D. Doverspike, Phys. Rev. A 32, 144 (1985).
3. D. Scott, M. S. Huq, R. L. Champion and L. D. Doverspike, Phys. Rev. A 33, 170 (1986).
4. M. S. Huq, L. D. Doverspike and R. L. Champion, Phys. Rev. A 27, 2831 (1983).
5. M. S. Huq, L. D. Doverspike and R. L. Champion, Phys. Rev. A 27, 785 (1983).
6. V. Kempter, B. Kubler and W. Mecklenbrauch, J. Phys. B 7, 2375 (1974); W. Mecklenbrauch, J. Schon, E. Speller and V. Kempter, J. Phys. B 10, 3271 (1977).
7. B. L. Christensen-Dalsgaard, J. Phys. B 18, L407 (1985); J. L.Krause and R. S. Berry, Comment. At. Mol. Phys. 18, 91 (1986).
8. C. Courbin-Gaussorques, P. Wahnon and M. Barat, J. Phys. B 12, 3047 (1979).
9. V. N. Tuan, V. Esaulov and J. P. Gauyacq, J. Phys. B 17, L133 (1984); V. A. Esaulov, J. P. Grouard, R. I. Hall, M. Landau, J. L. Montmagnon, F. Pichou and C. Schermann, J. Phys. B 17, 1855 (1984).
10. R. N. Compton, P. W. Reinhardt and C. D. Cooper, J. Chem. Phys. 63, 3821 (1975).
11. R. E. Olson and B. Liu, J. Chem. Phys. 73, 2817 (1980).
12. A. M. Karo, M. A. Gardner and J. R. Hiskes, J. Chem. Phys. 68, 1942 (1978).
13. J. P. Gauyacq, J. Phys. B 12, L387 (1979).

14. V. N. Tuan and V. A. Esaulov, Phys. Rev. A<u>32</u>, 883 (1985).
15. A. R. Johnston and P. D. Burrow, J. Phys. B <u>15</u>, L745 (1982).
16. A. L. Sinfailam and R. K. Nesbet, Phys. Rev. A <u>7</u>, 1987 (1973).

DISCUSSION

Peterson: Do those first calculations on H^- + Na include the crossing into the second continuum?

Champion: Yes.

Peterson: After it makes the first transition?

Champion: No, if the charge transfer occurs, and the impact parameter is less than the outer crossing, then it is assumed with unit probability that electron detachment occurs.

Peterson: But, in fact, it crosses over into....

Champion: Well, if detachment has occurred, it doesn't matter. I can't have detachment twice.

Peterson: No, I think it's just the charge transfer part. Because that $A^2\Sigma$ crosses over that at R_x as well.

Champion: Not if the impact parameter is larger than R_x. That's how the integral is carried out for charge transfer. Only for impact parameters larger than R_x.

Peterson: I see. That explains the difference too.

Champion: That's exactly the same as the calculation of Olson and Liu. The electron detachment cross section is not the same. The charge transfer is.

Michels: Roy, way back in the beginning you had a potential diagram for rare gas alkalis. What was the separation scale? Because the excited alkali rare gasses should all be bound. Their Rydberg to an ion core is bound. Like the second curve down, is that way up on the repulsive wall, or....? What's the R scale?

Champion: It's highly schematic, obviously. The crossing coordinates here are 60 eV and 1.6 a_o, a very tiny distance, indeed.

York: Is it possible to extend the experimental data to energies like 5 kV or so? Do you have that capability?

Champion: Not at the moment.

York: Do you have any plans of acquiring that capability?

Champion: Not at the moment.

DISSOCIATIVE ATTACHMENT TO LITHIUM DIMERS

J.M.Wadehra
Department of Physics and Astronomy, Wayne State University
Detroit, Michigan 48202

ABSTRACT

The cross sections and the rates of production of negative ions of atomic lithium by the process of dissociative electron attachment to lithium dimers are obtained by using the resonant scattering theory. Both the cross sections as well as the rates of attachment are enhanced if the lithium molecule is initially vibrationally excited. General expressions for approximately obtaining the rates of electron attachment to any vibrational level of Li_2 are presented.

INTRODUCTION

One possible way of producing beams of light negative ions is by the process of dissociative electron attachment to neutral molecules. This process is understood[1] to proceed through the formation of an intermediate resonant anion state which is capable of dissociation as well as of electron autodetachment. These two possibilities lead, depending upon the particular molecule, to negative ion formation by electron attachment or to a vibrationally excited molecule, respectively. Schematically, for a molecule AB

$$AB(v_i) + e \rightarrow AB^- \begin{cases} \rightarrow A + B^- & \text{(Dissociative Attachment)} \\ \rightarrow AB(v_f) + e & \text{(Vibrational Excitation)}. \end{cases}$$

Previous theoretical[2] and experimental[3] studies have shown that in the case of molecular hydrogen the cross sections for dissociative attachment are strongly enhanced if the molecule H_2 is initially rovibrationally excited. In order to ascertain whether similar enhancement of attachment cross sections occurs for other molecules we have investigated the process of electron attachment to lithium dimers which are isovalent with the molecular hydrogen. It is found that analogous to molecular hydrogen the rate of electron attachment to lithium dimers by the process of dissociative attachment is enhanced if the dimers are initially vibrationally excited.

CALCULATIONS

The fact that both the lithium dimers and the hydrogen molecules are isovalent leads to similarities between the two molecules as far as the configurations of the electronic states are concerned[4]. For example, the lowest electronic states of the negative molecular ions with configurations $(1\sigma_g)^2 (1\sigma_u)^2 (2\sigma_g)^2 (2\sigma_u)$ for Li_2^- and $(1\sigma_g)^2 (1\sigma_u)$ for H_2^- have similar symmetry, namely, $^2\Sigma_u^+$. However, unlike hydrogen molecule, the lithium dimers possess a large polarizability

© American Institute of Physics 1987

and a weak bond strength which makes the ground state of Li_2^- a true bound state. In the case of H_2^-, on the other hand, the $^2\Sigma_u^+$ state is a true bound state only for internuclear separations R larger than 2.9 a.u. and an autodetaching state for smaller values of R. The first excited state with symmetry $^2\Sigma_g^+$ and configurations $(1\sigma_g)^2 (1\sigma_u)^2 (2\sigma_g) (2\sigma_u)^2$ for Li_2^- and $(1\sigma_g) (1\sigma_u)^2$ for H_2^- is a partly Feshbach and a partly shape resonance in nature for both molecular anions[4]. This resonance is the lowest-lying resonance of Li_2^- and because of its nature (namely Feshbach) the resonance is expected to have a small width and a long lifetime. This resonant state is essentially responsible for dissociative electron attachment to lithium dimers.

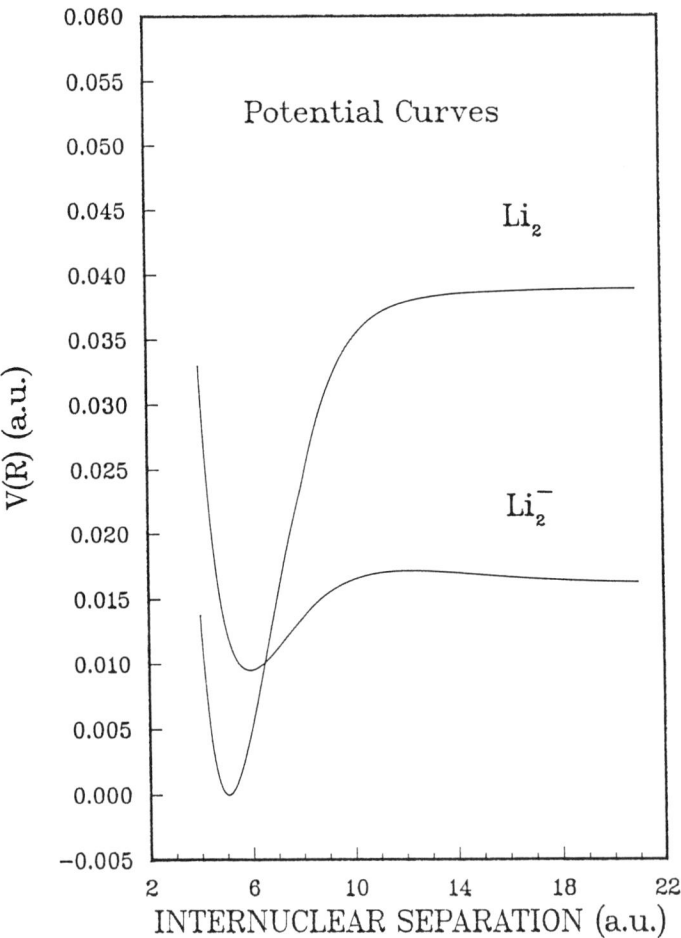

Figure 1. Potential curves of the ground X $^1\Sigma_g^+$ electronic state of Li_2 and the lowest A $^2\Sigma_g^+$ resonant state of Li_2^-.

Potential curves of the electronic states of Li_2 and Li_2^- relevant to the attachment process are shown in Figure 1. Due to its autodetaching nature the A $^2\Sigma_g^+$ electronic state of Li_2^- exhibits a complex potential energy curve whose real part along with the potential curve of the ground $^1\Sigma_g^+$ electronic state of the neutral Li_2 is shown in the Figure. The two curves cross at R = 6.5 a.u. so that only for internuclear separations smaller than 6.5 a.u. the A state is autodetaching. Detailed orbital optimized CI calculations[4] reveal that the X and the A states have their respective potential minimum at 5.1 a.u. and 5.9 a.u. The imaginary part of the complex potential energy curve of the A state of Li_2^- is related to the width of this resonant state. For internuclear separations smaller than 6.5 a.u.

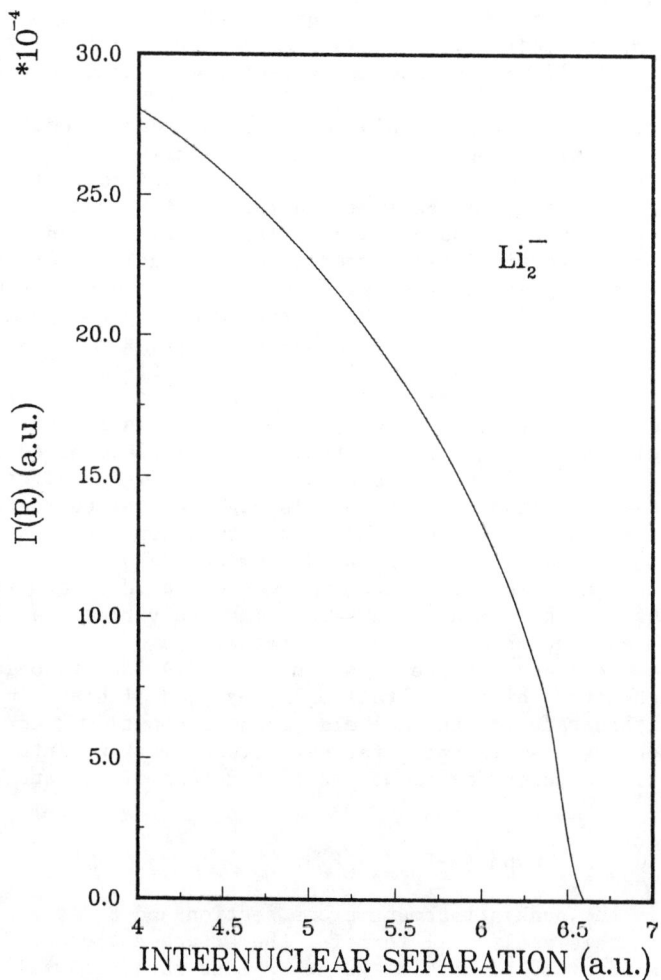

Figure 2. The width of the A $^2\Sigma_g^+$ resonant state of Li_2^- as a function of the internuclear separation R.

this resonant state can autodetach into Li_2 + e. In this autodetachment process [$^2\Sigma_g^+ \to {}^1\Sigma_g^+$ + e] the lowest contributing partial wave is an s-wave. Thus Wigner's threshold law for the width of this state implies $\Gamma(R) = c \cdot k(R)$, where $k(R)$ is the wave number of the electron emitted at internuclear separation R and c is a constant. To obtain this constant c the fully optimized orbital exponents of the CI wave functions were smoothly extrapolated from the variationally stable region (R $\gtrsim$ 6.5 a.u.) into the autodetaching region (R $\lesssim$ 6.5 a.u.) to obtain the matrix elements coupling the discrete and the continuum states. These matrix elements are related to the autodetachment width by Fermi's golden rule. This procedure yielded c = 0.0143 a.u. Thus the width of the A $^2\Sigma_g^+$ state of Li_2^-, which is primarily responsible for dissociative electron attachment to lithium dimers, as a function of internuclear separation is given, in atomic units, by $\Gamma(R) = 0.0143 \cdot k(R)$ and is shown in Figure 2. The small value of the width is characteristic of the Feshbach nature of the resonance.

The similarities between lithium dimers and hydrogen molecules suggest that theoretical approaches used successfully in the past[5] for investigating the cross sections and rates for dissociative electron attachment to H_2 can be employed for similar investigations for Li_2. Thus local width resonant scattering theory is used for obtaining the cross sections as a function of the incident electron energy and the corresponding rates as a function of the electron temperature T (or, equivalently, the average electron energy $\bar{E}$ = 3kT/2.) for dissociative electron attachment to Li_2, e + $Li_2 \to Li$ + Li^-. The contribution of only the $^2\Sigma_g^+$ resonance of Li_2^- is taken into account. Analogous to molecular hydrogen the cross sections for dissociative electron attachment to lithium dimers, as a function of the incident electron energy, show a rapid increase leading to a peak in the cross section, followed by a gradual decrease. The difference, however, is that the cross section peak is right at the energetic threshold in the case of hydrogen while the attachment cross section in the case of lithium peaks at an energy somewhat above the energetic threshold. This difference in behavior could be explained in terms of the Franck-Condon factors relating the vibrational levels of the neutral and the anion electronic states.

Recent measurements[6] of the rate constant k(T) for dissociative electron attachment to highly vibrationally excited lithium dimers indicate that this rate for thermal electrons is about 10^{-8} cm^3 sec^{-1}. In order to convert the present attachment cross sections into attachment rates the cross sections are fitted to a simple analytical form:

$$\sigma_{DA}(E) = \sigma_{peak} \cdot \exp\{-(E-E_{peak})/E_o\} \quad (1)$$

where σ_{peak} is the peak attachment cross section and E_o is a constant. Using this analytical form for the attachment cross sections it is possible to obtain the attachment rates as well in an analytical form if a Maxwellian distribution is assumed for the electron energies. The attachment rate as a function of the electron temperature is then given by

$$k(\bar{E}) = \left[\frac{27\bar{E}}{\pi m}\right]^{\frac{1}{2}} \sigma_{peak} \exp(-1.5 E_{peak}/\bar{E}) \left[1 + \frac{3E_{peak}}{2\bar{E}} + \frac{E_{peak}}{E_o}\right] \left[\frac{3}{2} + \frac{\bar{E}}{E_o}\right]^{-2}. \quad (2)$$

The average electron energy $\bar{E}$ is related to the electron temperature T by $\bar{E} = 3kT/2$.

RESULTS

It is observed that the cross sections as well as the rates of Li^- formation are enhanced if the molecule Li_2 is initially vibrationally excited. The factors by which the peak attachment cross sections are enhanced, on vibrationally exciting the lithium molecule initially, are summarized and compared with the corresponding factors for H_2 in the table I below.

TABLE I. Enhancement factors for electron attachment to Li_2.

Initial vibrational level, v, the molecule is in.	Factor by which the peak attachment cross section is enhanced over that for v=0.		
	Li_2	H_2 (Theory)	H_2 (Experiment)
1	7.4	32.5	30 ± 10
2	16.4	465	500 ± 175

The reason for enhancement of the peak attachment cross section is that, as the internal energy of the molecule is initially increased via vibrational excitation, the range of internuclear separations R over which electron capture occurs is increased due to an increased vibrational amplitude.

Finally, the rates of electron attachment to Li_2 (that is, the rates of production of Li^- beams) are calculated as a function of the electron temperature T using Eq.(2) and are shown in Figure 3. The rate is as low as 10^{-11} cm^3 sec^{-1} when the molecule is in its lowest vibrational level and the rate increases by almost an order of magnitude for each quantum of vibrational excitation of the molecular Li_2. It is thus plausible that the total attachment rate can approach the experimental value of 10^{-8} cm^3 sec^{-1} as the initial excitation of the molecule is raised to the v = 10 level. The enhancement of the attachment rate, which is a direct result of the enhancement of the attachment cross section, is expected for any distribution of the electron energies. It is to be noted that for the isovalent molecule H_2 as well the maximum predicted rate for electron attachment via the process of dissociative attachment is about 10^{-8}

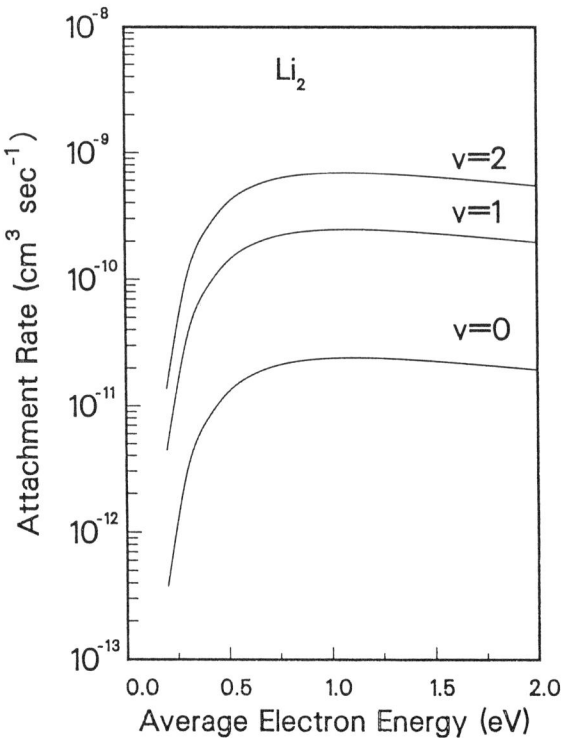

Figure 3. Rates of dissociative electron attachment to various vibrational levels of Li_2 as a function of the average electron energy $\bar{E}$ ($\bar{E} = 3kT/2$).

$cm^3\ sec^{-1}$.

FUTURE POSSIBILITIES

Analogous to molecular hydrogen, the lithium dimers exhibit an enhancement of the cross sections as well as of the rates of dissociative electron attachment if they are vibrationally excited. Detailed calculations have been carried out only for the lowest three vibrational levels of the dimers. In the immediate future these calculations will be extended to higher vibrational levels upto and including the endoergic regime. Experimental observations[6] of rates of electron attachment to lithium indicate that this rate can be as high as 10^{-8} cm^3 sec^{-1} when the molecule Li_2 is in the vibrational level v = 10.

In the case of molecular hydrogen it has been established that the rotational excitation of the molecule also aids in the enhancement of the electron attachment rate; however, the enhancement factor is larger for initial vibrational excitation than for initial

rotational excitation. Recent experimental observations[6] on the
electron attachment to lithium dimers, on the other hand, seem to
suggest that initial rotational excitation plays little role in
controlling the attachment to Li_2. A theoretical investigation of the
effect of initial rotational excitation on the rate of Li^- formation
via the process of dissociative electron attachment to Li_2 will be
made within the resonance scattering model.

Finally, the effect of higher resonances of Li_2^- (other than the
$A\ ^2\Sigma_g^+$ resonance) on the production of Li^- will be investigated since
the complex potential energy curves of these resonances are just
becoming available[4].

It is a pleasure to thank Dr. H. H. Michels for providing the
potential curves of Ref. 4 in numerical form. The support of the Air
Force Office of Scientific Research through Grant Number
AFOSR-84-0143 is gratefully acknowledged.

REFERENCES

1. For a recent review see, J. M. Wadehra in Nonequilibrium
   Vibrational Kinetics, ed. M. Capitelli (Springer-Verlag, 1986).
2. J. M. Wadehra and J. N. Bardsley, Phys. Rev. Lett. 41, 1795 (1978)
3. M. Allan and S. F. Wong, Phys. Rev. Lett. 41, 1791 (1978).
4. H. H. Michels, R. H. Hobbs and L. A. Wright, Chem. Phys. Lett.
   118, 67 (1985).
5. J. M. Wadehra, Phys. Rev. A29, 106 (1984).
6. M. W. McGeoch and R. E. Schlier, Phys. Rev. A33, 1708 (1986).

DISCUSSION

<u>Mitchell</u>: Could you explain to me why the rate coefficient vs.
temperature dependence for dissociative attachment is not the same
as for dissociative recombination? Why it comes up from zero,
whereas dissociative recombination continues upwards as you go to
zero temperature?
<u>Wadehra</u>: Dissociative recombination is, in principle, a very similar process to dissociative attachment, except you start off with a
molecular ion. Now, I'm not an expert on dissociative recombination, since I've not done calculations, obviously, for it. But the
process of attachment goes by forming this state in which you form a
resonance, which ultimately dissociates. In the case of dissociative recombination, my impression is that you go via formation of
Rydberg states, and Rydberg states of the neutral molecule.
<u>Mitchell</u>: No, normally you go directly through a resonance which is
very similar to dissociative attachment. The Rydberg state is just
a secondary mechanism, it's an indirect mechanism.
<u>Wadehra</u>: Indirect? But my impression was that that was the more
important channel.
<u>Mitchell</u>: No, that's not true.
<u>Wadehra</u>: Then, maybe Harvey Michels has done some calculations on
that....

Michels: In dissociative recombination, you have ion pair combination, which has a 1/e dependence. You don't have that for electron-neutral. So you have a big threshold effect. The cross-section goes to infinity as the energy goes to zero for dissociative recombination.

Mitchell: But I was wondering if you do get a match of the energy between the molecule and the resonance curve, why, for zero-energy electron, wouldn't you get a large cross-section there?

Michels: Well, it peaks at threshold for dissociative attachment.

Mitchell: Joe's curve showed them rising up from zero at zero-energy.

Michels: I think that's due just to the detailed nature of that system that has a slightly displaced resonant state.

Mitchell: So, if you were in the right vibrational level of the molecule....

Michels: Typically, in dissociative recombination, the states are more vertical.

Schlachter: Joe, you said that Li_2^- is a bound state?

Wadehra: The lowest electronic state of Li_2^- is a bound state.

Schlachter: It hasn't been observed, has it? Experimentally?

Wadehra: No.

Schlachter: But we should be able to find it in principle, yes?

Wadehra: Yes, I think so. Theoretical calculations indicate it's a bound state, but since the theoretical calculation is done in a variational manner, they provide an upper bound, meaning that the real energy is always lower than what you get, so you can be assured that it's a bound state. There's no question about that.

McGeoch: On the same question, is there any chance that we could observe Li_2^- by radiative stabilization from the resonance? What would be the probability of that happening?

Wadehra: Good question. I think that's what Harvey Michels will be talking about.

NEGATIVE ION FORMATION IN LITHIUM ATOM COLLISIONS

H. H. Michels
J. A. Montgomery, Jr.
United Technologies Research Center, East Hartford, CT 06108

ABSTRACT

The formation of Li⁻ by dissociative attachment in e + Li_2 collisions is characterized by a large cross section for electron attachment to highly vibrationally excited Li_2 molecules. However, the electronic structure of the Li_2^- system dictates that low energy electrons ($\lesssim 0.2$ eV) are needed to minimize loss processes. An alternate production route for Li⁻ is possible via direct ion-pair formation through collisions of highly excited (Rydberg) Li atoms. The mechanisms for ion-pair formation will be discussed in terms of the high-lying molecular Rydberg states of Li_2.

INTRODUCTION

The negative ions of light atoms are currently being studied for their possible application in gaseous discharges, fusion plasmas and gas lasers[1]. One source for the volume production of atomic anions is the process of dissociative electron attachment to molecules[2]. This process has been studied experimentally for low energy electron-hydrogen molecule collisions by Allen and Wong[3]. A parallel theoretical study has been reported by Wadehra and Bardsley[4]. More recently, McGeoch and Schlier[5] have examined dissociative attachment (DA) in electron-lithium molecule collisions and have found large DA rates for attachment to highly vibrationally excited Li_2 molecules. The effect of vibrational excitation on the DA rates has been studied by Hiskes[6] for e + H_2 collisions and by Wadehra and Michels[7] for the e + Li_2 system.

In the e + Li_2 system, it has been shown[7] that excitation of Li_2 to the A $^1\Sigma_u^+$ state, either by electron impact or by photon pumping, results in an enhanced vibrational distribution upon radiative decay of the A $^1\Sigma_u^+$ state to the ground X $^1\Sigma_g^+$ state of Li_2. The cross sections for vibrational excitation of Li_2 X $^1\Sigma_g^+$ via low energy electron collisional excitation of the A $^1\Sigma_u^+$ state exhibit a relatively flat vibrational distribution in the range $3 \leq v'' \leq 9$. This suggests that Li_2 molecules which are vibrationally excited to the DA threshold for Li⁻ formation should exhibit the largest attachment cross sections, a result that is indicated by the studies of McGeoch and Schlier[5].

---

*Supported in part by AFOSR under Contract F49620-85-C-0095.

An alternative mechanism for negative ion formation was suggested by Lee and Mahan[8], who proposed that ion-pair formation:

$$X^* + X \rightarrow X^+ + X^- \tag{1}$$

should occur for systems where the energetics for ion-pair formation are competitive with associative ionization:

$$X^* + X \rightarrow X_2^+ + e. \tag{2}$$

The alkali metals, including lithium, are therefore good candidates for such a study of ion-pair formation. Ciocca, et. al.[9] have recently reported on the formation of Na^- ions in highly excited (n=7-40) Rydberg atom collisions. The purpose of the present study is to examine the possible mechanisms for ion-pair formation in lithium atom collisions.

THEORETICAL CONSIDERATIONS

In order to determine the possible paths for Li excited state interactions, we have undertaken a series of <u>ab initio</u> calculations of the electronic states of Li_2 up to the $Li[^2S(3s)] + Li[^2S(3s)]$ dissociation limit. Since the formation of the ion-pair $Li^+[^1S] + Li^-[^1S]$ can occur only for $^1\Sigma_g^+$ and $^1\Sigma_u^+$ symmetries, we have restricted our studies to these two representations (see Table 1).

Theoretical potential curves for the $^1\Sigma_g^+$ and $^1\Sigma_u^+$ symmetries of Li_2 were obtained from valence configuration interaction (VCI) calculations. A 64 function Slater orbital basis, containing a (2s1p) optimized valence basis augmented by a (3s3p3d2f) Rydberg basis, was used in this study. The basis was transformed to $D_{\infty h}$ symmetry orbitals and CI calculations were performed in the space of all symmetry adapted configurations having the $1\sigma_g$ orbital doubly occupied. The resulting CI expansions contained 344 and 331 configurations for the $^1\Sigma_g$ and $^1\Sigma_u$ symmetries, respectively. As the interactions in this system are very long-ranged, calculations were performed over a range of internuclear separations from 5 to 50 bohrs.

CALCULATED RESULTS

The results of our theoretical calculations are shown in Fig. 1 and Fig. 2 for the $^1\Sigma_g^+$ and $^1\Sigma_u^+$ symmetries, respectively. The ionic curves which dissociate to $Li^+[^1S] + Li^-[^1S(2s^2)]$ are clearly evident in both Fig. 1 and Fig. 2, where they exhibit a nearly diabatic crossing behavior with the normal valence states of Li_2 at large internuclear separations, but show strong mixing at distances ⩽10Å. Ion-pair formation may occur for long-range interactions of the type

$$\text{Li}\,[^2S(2s)] + \text{Li}^{**}[^2S(n=5)] \rightarrow \text{Li}^+\,[^1S] + \text{Li}^-[^1S] \qquad (3)$$

but clearly there are few effective curve crossing channels leading to $\text{Li}^+ + \text{Li}^-$. The curve crossing mechanism suggested by Ciocca, et al[9], thus appears not to be an important mechanism for ion-pair production in the Li_2 system.

A second mechanism suggested by Ciocca et al[9] for alkali negative ion formation would involve excited state negative ion formation via

$$\text{Li}^*[^2P(2p)] + \text{Li}^{**}(n \geqslant 9) \rightarrow \text{Li}^+[^1S] + \text{Li}^-[^1P(2s2p)] \qquad (4)$$

or from electron attachment to a Rydberg state of Li^{**} followed by radiative stabilization to the ground $\text{Li}^-[^1S(2s^2)]$ state. Both of these processes can be ruled out for the Li_2 system since the autoionization lifetime of the $\text{Li}^-\,[^1P(2s2p)]$ state is known to be very short[10] relative to radiative stabilization and higher autoionizing states should exhibit even shorter lifetimes.

DISCUSSION

The analysis of our calculated results given above leaves us with radiative stabilization of excited state Li_2 interactions as the most probable mechanism for ion-pair formation in this system. Three separate types of interactions leading to ion-pair production can be identified:

$$\text{Li}[^2S(2s)] + \text{Li}^{**}(n>5) \rightarrow \text{Li}_2^{**} \rightarrow \text{Li}^+ + \text{Li}^- + h\nu \qquad (5)$$
$$(18.0\mu \geqslant \lambda \geqslant 2.0\mu)$$

$$\text{Li}^*[^2P(2p)] + \text{Li}^{**}(n \geqslant 3) \rightarrow \text{Li}_2^{**} \rightarrow \text{Li}^+ + \text{Li}^- + h\nu \qquad (6)$$
$$(2.8\mu \geqslant \lambda \geqslant 0.5\mu)$$

$$\text{Li}^{**}(n \geqslant 3) + \text{Li}^{**}(n \geqslant 3) \rightarrow \text{Li}_2^{**} \rightarrow \text{Li}^+ + \text{Li}^- + h\nu \qquad (7)$$
$$(\lambda \leqslant 0.6\mu)$$

The mechanism illustrated by Eq. (7) exhibits a quadratic dependence on the Rydberg atom concentration and, similar to the studies reported by Ciocca, et al[9] for Na, should be distinguishable from the reactions given by Eq. (5) and (6). The latter reactions should exhibit a linear dependence on reaction rate with Rydberg atom concentration. A preliminary analysis of the radiative decay of these high-lying Li_2 states shows a strong preferential coupling to the $\text{Li}^+[^1S] + \text{Li}^-[^1S(2s^2)]$ dissociating channel. This is as expected since the transition moments are very large for covalent Rydberg interactions radiating to a lower ionic molecular state. In addition, the range for these Rydberg atom interactions is predicted to be very large ($\geqslant 30\text{Å}$) where the radiative moments are increasing proportional to the internuclear separation.

An experiment similar to that reported by Ciocca et al[9] is suggested to test the radiative mechanism for Li^+ - Li^- ion-pair formation through Rydberg atom collisions. The two laser excitation mechanism:

$$Li[^2S(2s)] + h\nu \to Li^*[^2P(2p)] \quad \lambda = 671.0 \text{ nm} \quad (8)$$

$$Li^*[^2P(2p)] + h\nu' \to Li^{**}[^2S(3s)] \quad \lambda' = 812.8 \text{ nm} \quad (9)$$

should produce Rydberg Li atoms in either the $^2S(3s)$ or $^2D(3d)$ states, provided near-saturation is achieved in the first resonant line excitation. The subsequent reactions of these laser-produced Rydberg state Li atoms is illustrated in Table 2. Two photon excitation by the first laser, and excitation of the $^2S(3s)$ Rydberg state by the 671 nm laser source, are both below the ionization threshold for Li, thus eliminating any first order ionization process. The suggested experiment should look for Li^+ - Li^- ion-pair formation accompanied by excimer-like radiation at wavelengths < 600 nm. The ion pairs could be electrostatically extracted from the laser reaction region and mass analyzed to identify the Li^- signal.

As mentioned above, the reaction given by Eq. (7) should exhibit a quadratic dependence on the Rydberg state density and a very large cross section for ion-pair formation. Similar volume processes to that represented by Eqs. (5)-(7) are energetically possible for H^+ - H^- ion pair formation, provided that a useful mechanism for $n = 2$ production in H can be found.

REFERENCES

1. K. Prelec, ed., Proceedings of the Third International Symposium on the Production and Neutralization of Negative Ions and Beams, AIP Conf. Proc. 111 (AIP, New York, 1984).
2. M. Bacal, Physica Scripta, T2/2, 467 (1982).
3. M. Allan and S. F. Wong, Phys. Rev. Letters, 41, 1791 (1978).
4. J. M. Wadehra and J. N. Bardsley, Phys. Rev. Letters, 41, 1795, (1978); Phys. Rev. A20, 1398 (1979).
5. M. W. McGeoch and R. E. Schlier, in: Proceedings of the Third International Symposium on the Production and Neutralization of Negative Ions and Beams, AIP Conf. Proc. 111, ed. K. Prelec (AIP, New York, 1984), p. 291; Phys. Rev. A, 33, 1708 (1986).
6. J. R. Hiskes, J. Appl. Phys. 51, 4592 (1980).
7. J. M. Wadehra and H. H. Michels, Chem. Phys. Letters, 114, 380 (1985).
8. Y. T. Lee and B. H. Mahan, J. Chem. Phys. 42, 2893 (1965) and references therein.
9. M. Ciocca, M. Allegrini, E. Arimondo, C. E. Burkhardt, W. P. Garver and J. J. Leventhal, Phys. Rev. Letters, 56, 704 (1986).
10. Y. K. Bae and J. R. Peterson, Phys. Rev. A, 32, 1917 (1985).

TABLE I

LOW-LYING MOLECULAR STATES OF Li_2 AND THEIR DISSOCIATION LIMITS

Separated atoms	E(cm^{-1})*	Molecular states
Li + Li		
$^2S_g(2s) + {}^2S_g(2s)$	0.0	$^1\Sigma_g^+(1)$, $^3\Sigma_u^+(1)$
$^2S_g(2S) + {}^2P_u(2p)$	14903.0	$^1\Sigma_g^+(1)$, $^1\Sigma_u^+(1)$, $^3\Sigma_g^+(1)$, $^3\Sigma_u^+(1)$
		$^1\Pi_g(1)$, $^1\Pi_u(1)$, $^3\Pi_g(1)$, $^3\Pi_u(1)$
$^2S_g(2s) + {}^2S_g(3s)$	27206.0	$^1\Sigma_g^+(1)$, $^1\Sigma_u^+(1)$, $^3\Sigma_g^+(1)$, $^3\Sigma_u^+(1)$
$^2P_u(2p) + {}^2P_u(2p)$	29807.0	$^1\Sigma_g^+(2)$, $^1\Sigma_u^-(1)$, $^3\Sigma_u^+(2)$, $^3\Sigma_g^-(1)$
		$^1\Pi_g(1)$, $^1\Pi_u(1)$, $^3\Pi_g(1)$, $^3\Pi_u(1)$
		$^1\Delta_g(1)$, $^3\Delta_u(1)$,
$^2S_g(2s) + {}^2P_u(3p)$	30925.0	$^1\Sigma_g^+(1)$, $^1\Sigma_u^+(1)$, $^3\Sigma_g^+(1)$,
		$^3\Sigma_u^+(1)$, $^1\Pi_g(1)$, $^1\Pi_u(1)$,
		$^3\Pi_g(1)$, $^3\Pi_u(1)$
$^2S_g(2s) + {}^2D_g(3d)$	31283.0	$^1\Sigma_g^+(1)$, $^1\Sigma_u^+(1)$, $^3\Sigma_g^+(1)$,
		$^3\Sigma_u^+(1)$, $^1\Pi_g(1)$, $^1\Pi_u(1)$, $^3\Pi_g(1)$,
		$^3\Pi_u(1)$, $^1\Delta_g(1)$, $^1\Delta_u(1)$, $^3\Delta_g(1)$,
		$^3\Delta_u(1)$
$Li_2^+[^2\Sigma_u^+] + e$	33200.0	
$^2S_g(2s) + {}^2S_g(4s)$	35012.0	$^1\Sigma_g^+(1)$, $^1\Sigma_u^+(1)$, $^3\Sigma_g^+(1)$, $^3\Sigma_u^+(1)$
$^2S_g(2s) + {}^2P_u(4p)$	36469.0	$^1\Sigma_g^+(1)$, $^1\Sigma_u^+(1)$, $^3\Sigma_g^+(1)$,
		$^3\Sigma_u^+(1)$, $^1\Pi_g(1)$, $^1\Pi_u(1)$,
		$^3\Pi_g(1)$, $^3\Pi_u(1)$

*C. E. Moore, Atomic Energy Levels, 1, 9 (1949).

TABLE I (Continued)

$^2S_g(2s) + {}^2D_g(4d)$	36623.0	$^1\Sigma_g^+(1)$, $^1\Sigma_u^+(1)$, $^3\Sigma_g^+(1)$, $^3\Sigma_u^+(1)$, $^1\Pi_g(1)$, $^1\Pi_u(1)$, $^3\Pi_g(1)$, $^3\Pi_u(1)$, $^1\Delta_g(1)$, $^1\Delta_u(1)$, $^3\Delta_g(1)$, $^3\Delta_u(1)$
$^2S_g(2s) + {}^2F_u(4f)$	36630.0	$^1\Sigma_g^+(1)$, $^1\Sigma_u^+(1)$, $^3\Sigma_g^+(1)$, $^3\Sigma_u^+(1)$, $^1\Pi_g(1)$, $^1\Pi_u(1)$, $^3\Pi_g(1)$, $^3\Pi_u(1)$, $^1\Delta_g(1)$, $^1\Delta_u(1)$, $^3\Delta_g(1)$, $^3\Delta_u(1)$, $^1\Phi_g(1)$, $^1\Phi_u(1)$, $^3\Phi_g(1)$, $^3\Phi_u(1)$
$^2S_g(2s) + {}^2S_g(5s)$	38299.0	$^1\Sigma_g^+(1)$, $^1\Sigma_u^+(1)$, $^3\Sigma_g^+(1)$, $^3\Sigma_u^+(1)$
$Li^+[{}^1S_g(1s^2)] + Li^-[{}^1S_g(2s^2)]$	38474.0	$^1\Sigma_g^+(1)$, $^1\Sigma_u^+(1)$
$^2S_g(2s) + {}^2P_u(5p)$	39015.0	$^1\Sigma_g^+(1)$, $^1\Sigma_u^+(1)$, $^3\Sigma_g^+(1)$, $^3\Sigma_u^+(1)$, $^1\Pi_g(1)$, $^1\Pi_u(1)$, $^3\Pi_g(1)$, $^3\Pi_u(1)$
$^2S_g(2s) + {}^2D_g(5d)$	39094.0	$^1\Sigma_g^+(1)$, $^1\Sigma_u^+(1)$, $^3\Sigma_g^+(1)$, $^3\Sigma_u^+(1)$, $^1\Pi_g(1)$, $^1\Pi_u(1)$, $^3\Pi_g(1)$, $^3\Pi_u(1)$, $^1\Delta_g(1)$, $^1\Delta_u(1)$, $^3\Delta_g(1)$, $^3\Delta_u(1)$
$^2S_g(2s) + {}^2F_u(5f)$	39104.0	$^1\Sigma_g^+(1)$, $^1\Sigma_u^+(1)$, $^3\Sigma_g^+(1)$, $^3\Sigma_u^+(1)$, $^1\Pi_g(1)$, $^1\Pi_u(1)$, $^3\Pi_g(1)$, $^3\Pi_u(1)$, $^1\Delta_g(1)$, $^1\Delta_u(1)$, $^3\Delta_g(1)$, $^3\Delta_u(1)$, $^1\Phi_g(1)$, $^1\Phi_u(1)$, $^3\Phi_g(1)$, $^3\Phi_u(1)$
$^2S_g(2s)$ + Doublets $(5 < n \lesssim 9)$	39987.0– 42003.0	Singlet, triplet Σ, Π, Δ, (g,u)

TABLE I (Continued)

$^2P_u(2p) + {}^2S_g(3s)$	42110.0	$^1\Sigma_g^+(1)$, $^1\Sigma_u^+(1)$, $^3\Sigma_g^+(1)$, $^3\Sigma_u^+(1)$, $^1\Pi_g(1)$, $^1\Pi_u(1)$, $^3\Pi_g(1)$, $^3\Pi_u(1)$
$^2S_g(2s)$ + Doublet $(9 \lesssim n=\infty)$	42118.0–	Singlet, triplet Σ, Π, Δ (g,u)
$Li^+ [{}^1S_g(1s^2)] +$ $Li [{}^2S_g(2s)]$	43482.0	$^2\Sigma_g^+(1)$, $^2\Sigma_u^+(1)$
$^2P_u(2p) + {}^2P_u(3p)$	45829.0	$^1\Sigma_g^+(2)$, $^1\Sigma_u^+(2)$, $^3\Sigma_g^+(2)$, $^3\Sigma_u^+(2)$, $^1\Sigma_g^-(1)$, $^1\Sigma_u^-(1)$, $^3\Sigma_g^-(1)$, $^3\Sigma_u^-(1)$, $^1\Pi_g(2)$, $^1\Pi_u(2)$, $^3\Pi_g(2)$, $^3\Pi_u(2)$, $^1\Delta_g(1)$, $^1\Delta_u(1)$, $^3\Delta_g(1)$, $^3\Delta_u(1)$
$^2P_u(2p) + {}^2D_g(3d)$	46187.0	$^1\Sigma_g^+(2)$, $^1\Sigma_u^+(2)$, $^1\Sigma_g^-(1)$, $^1\Sigma_u^-(1)$, $^3\Sigma_g^+(2)$, $^3\Sigma_u^+(2)$, $^3\Sigma_g^-(1)$, $^3\Sigma_u^-(1)$, $^1\Pi_g(3)$, $^1\Pi_u(3)$, $^3\Pi_g(3)$, $^3\Pi_u(3)$, $^1\Delta_g(2)$, $^1\Delta_u(2)$, $^3\Delta_g(2)$, $^3\Delta_u(2)$, $^1\Phi_g(1)$, $^1\Phi_u(1)$, $^3\Phi_g(1)$, $^3\Phi_u(1)$
$^2P_u(2p)$ + Doublet $(n=4, 5)$	51088.0– 53782.0	Singlet, triplet Σ, Π, Δ, Φ, Γ (g,u)
$^2S_g(3s) + {}^2S_g(3s)$	54412.0	$^1\Sigma_g^+(1)$, $^3\Sigma_u^+(1)$
$Li^+ [{}^1S_g(1s^2)] +$ $Li^- [{}^1P_u(2s2p)]$	57920.0	$^1\Sigma_g^+(1)$, $^1\Sigma_u^+(1)$, $^1\Pi_g(1)$, $^1\Pi_u(1)$

TABLE II

LITHIUM ATOM INTERACTIONS

(1) Resonant line excitation

$$\text{Li}[^2S(2s)] + h\nu \rightarrow \text{Li}^*[^2P(2p)]\text{(saturated)} \quad \lambda_{2s-2p} = 671.0 \text{ nm}$$

(2) Rydberg state excitation

$$\text{Li}^*[^2P(2p)] + h\nu' \rightarrow \text{Li}^{**}[n \geq 3] \text{ Rydberg state excitation}$$
$$\lambda_{2p-3s} = 812.8 \text{ nm}$$

(3) Rydberg-ground state reaction

$$\text{Li}[^2S(2s)] + \text{Li}^{**}[n \geq 5] \rightarrow \text{Li}^+[^1S] + \text{Li}^-[^1S]$$

(4) Rydberg-excited ^2P state reaction

$$\text{Li}^*[^2P(2p)] + \text{Li}^{**}(n \geq 3) \rightarrow \text{Li}^+[^1S] + \text{Li}^-[^1S]$$

(5) Rydberg-Rydberg excited state reaction

$$\text{Li}^{**}(n \geq 3) + \text{Li}^{**}(n \geq 3) \rightarrow \text{Li}^+[^1S] + \text{Li}^-[^1S]$$

(6) Associative ionization reaction (loss of Li^{**})

$$\text{Li}[^2S(2s)] + \text{Li}^{**}(n \geq 4) \rightarrow \text{Li}_2^+[X^2\Sigma_g^+] + e$$

(7) Dissociative ionization reaction (loss of Li^{**})

$$\text{Li}[^2S(2s)] + \text{Li}^{**}(n \geq 4) \rightarrow \text{Li}_2^+[A^2\Sigma_u^+] + e \rightarrow \text{Li}^+[^1S] + \text{Li}[^2S] + e$$

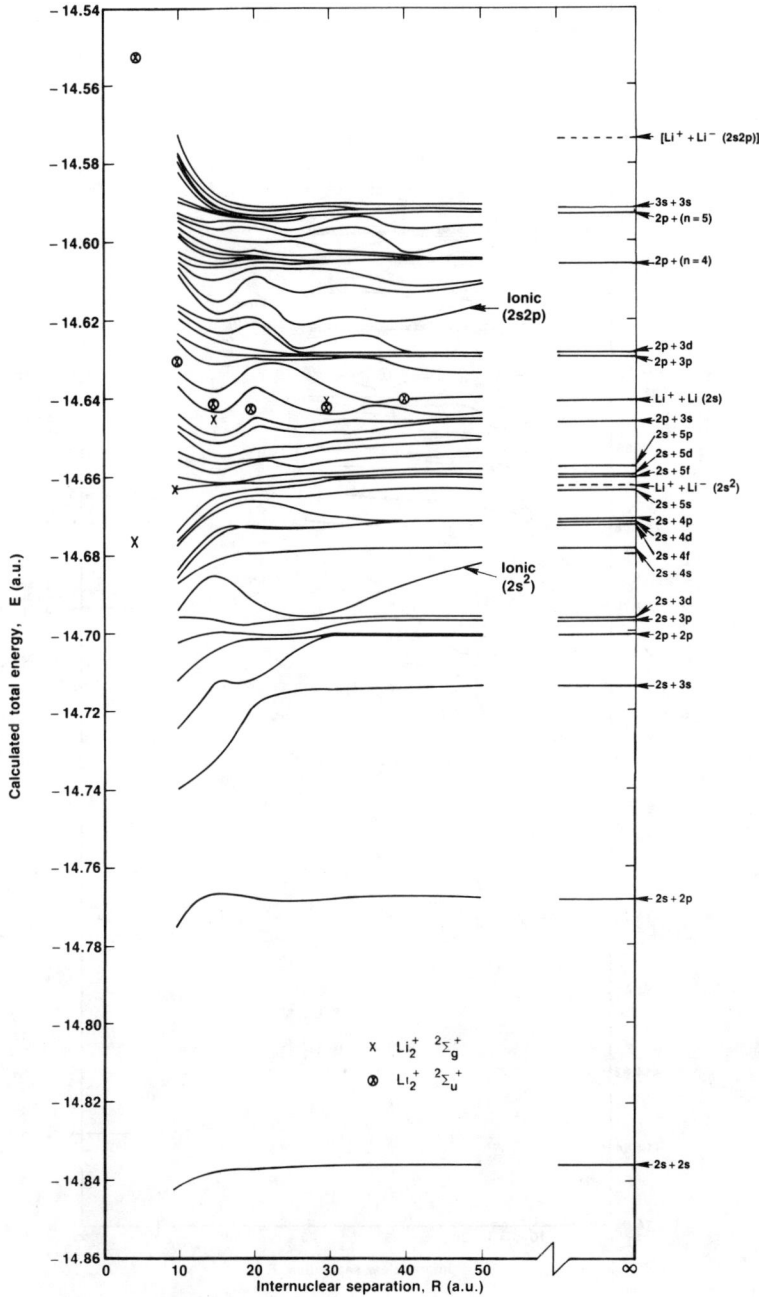

Fig. 1 Long-range behavior of excited $^1\Sigma_g^+$ states of Li$_2$.

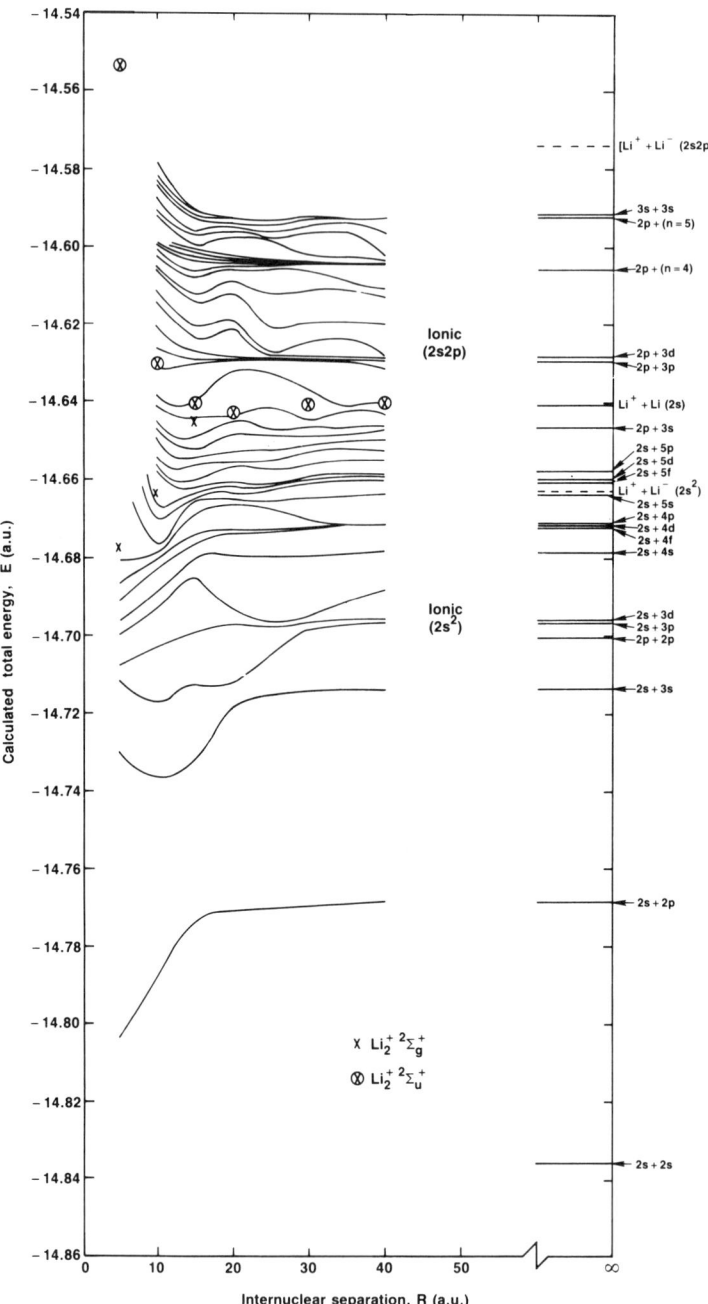

Fig. 2 Long-range behavior of excited $^1\Sigma_u^+$ states of Li_2.

DISCUSSION

Hiskes: What reaction would you propose for the formation of the ground state Li_2^-?
Michels: Li_2^-? A lot of people look for it and nobody's seen it.
Hiskes: It looks like a very hard one to get into.
Michels: Yes. There are two shape resonances. What you really want is a Feshbach state, lying just above where I did these calculations, in which the electron will get trapped in that state and then radiate back down to the ground state. The $^2\Sigma_g$ state, of course, is not connected. I've got to go a little higher, I think, to get the optimum state but, in terms of energies, it's just about ....well, the zero of the scale is Li_2, so - the density of states is getting quite high up in here - it's about a volt. Somewhere around a volt. There are states that will be connected strongly back down to the ground state. We're still working on it.
Mitchell: Harvey, the radiation from the Rydberg states, that normally takes quite a while to occur, like a nanosecond, or something like that. But these collisions are happening at very large...
Michels: These are happening at 50 Bohr radii. The oscillator strength is enormous because it goes like r.
Mitchell: I'm trying to think of it physically. If these atoms were moving together very rapidly, if the lithium was very hot, then they would get down into this curve.....
Michels: Then they would get down into where they're all mixed up, but if you have a thermal, like Malcolm's sources, like 800°, you envision that they would radiate long before they get into the region of mixing up.
Mitchell: So, if you were to, say, form the lithium vapor in a supersonic nozzle so that they're all going the same direction, and a very small energy spread in the beam, you should get a much larger cross-section, I would think. Because they're going to collide much slower with each other. Then that cross-section should go even higher, in fact.
Michels: Yes. The less thermal energy they have, the more you enhance this radiative process.
McGeoch: I've done some preliminary measurements on that process with lithium in n=5, and the cross-section that I observed for the 5d state plus the 2s state to give an ion pair is the order of $1 Å^2$, and for the 5s state, the rate was 40 times less than that. The reason is that it's below energy threshold for the process. Now, if I say that there is a very slow rate for the 5s state, then it seems that 5s + 5s, which is also possible in the system, cannot be too great a rate. And, from the densities, I would estimate that, too, is probably less than $10 Å^2$, for instance.
Michels: I guess I don't understand the argument, because the 5s-5s interaction is very much removed, energetically, from ground state 5s.
McGeoch: The question is, could it have a large cross-section as by the mechanism of stabilization?
Michels: Yes, in principle. You would predict at least two orders of magnitude improvement in the cross-section from two Rydberg

atoms, because they're two very large atoms, rather than one small on ground state.

McGeoch: What I've seen doesn't bear that out. I emphasize, a preliminary experiment was performed.

Wadehra: Do you have any information about the width of the $^2\Pi_u$ state order of magnitude?

Michels: No, we haven't analyzed it. It's not a Feshbach state. It's a shape resonance. It's very large.

DENSITY SCALING OF AN OPTICALLY PUMPED LITHIUM NEGATIVE ION SOURCE

M. W. McGeoch and R. E. Schlier
Avco Research Laboratory (Textron), Everett, MA. 02149

ABSTRACT

An experiment is described in which a high density of lithium negative ions (1 x 10^{10} cm^{-3}) is generated by dissociative attachment of electrons to optically pumped lithium molecules. During a three microsecond period up to 7% of electrons are attached. The possibilities for increased Li$^-$ density are explored.

INTRODUCTION

In prior work [1] a high rate constant (2 x 10^{-8} cm^3 sec^{-1}) was measured for the process

$$Li_2(v^*) + e^- \rightarrow Li + Li^- \qquad (1)$$

where $Li_2(v^*)$ is a vibrationally excited molecule and thermal electrons (0.1eV) are present. It was found that the dissociative attachment rate constant did not vary with vibrational or rotational state provided that the internal energy threshold ($v^* \geq 11$) was exceeded. Large cross sections for (1) were calculated by Michels and Wadehra [2], in agreement with experiment.

In both the prior work and the present experiment $Li_2(v^*)$ molecules were prepared by optical pumping via the $Li_2(A)$ or $Li_2(B)$ states:

$$Li_2(X,v=o) + h\nu \rightarrow Li_2(A,v=v_A) \qquad (2a)$$
$$Li_2(A,v=v_A) \rightarrow Li_2(X,v=v^*) + h\nu' \qquad (2b)$$

Typically half of the pumped molecules decay into a close group of vibrational states which can be selected to be at the correct internal energy (0.44eV). The pumping scheme is illustrated in Figure 1.

In the present experiment we have introduced broad band optical pumping which allows a much wider range of rotational states to be utilized, and hence produces a higher density of $Li_2(v^*)$.

EXPERIMENTAL

The lithium oven previously described [1] was used to generate a supersonic expansion with the following characteristics (for 850°C oven temperature):

© American Institute of Physics 1987

Li density	5×10^{13} cm^{-3}
Li$_2$ density	2×10^{12} cm^{-3}
Li$_2$ rotational temperature	60°K
Li$_2$ vibrational temperature	82°K
Beam velocity	3×10^5 cm sec^{-1}

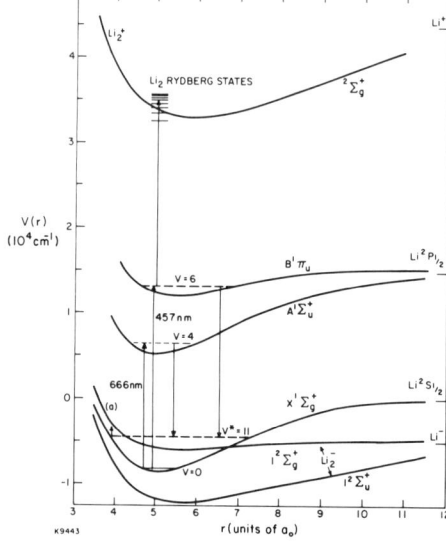

Fig. 1. Potential energy diagram for Li$_2$, Li$_2^-$ and Li$_2^+$, showing the Li$_2^-$ $1^2\Sigma_g^+$ attachment channel and routes for optical pumping to Li$_2$(v*).

Previous modelling of beam conditions [1] showed that the beam becomes collisionless as its density drops through the 'freezing' value of 6.7×10^{14} cm^{-3}, hence Li$_2$(v*) molecules are metastable once formed in the experimental region.

Optical pumping via process (2) requires a tunable source of red light (~650 nm). In previous experiments we employed a 2μJ, 0.25Å bandwidth pulse from a N$_2$-laser pumped dye laser (duration 5 nsec). In the present experiment we also used a 2 mJ, 15Å bandwidth (excimer laser pumped) dye laser (duration 20 nsec). This allowed us to optically pump a much greater fraction of Li$_2$ rotational states (spectrum shown in Figure 2).

Electrons were produced by the two-step ionization of Li atoms via

$$Li(^2S) + h\nu \rightarrow Li(^2P) \quad (3a)$$
$$(671 \text{ nm})$$

$$Li(^2P) + h\nu \rightarrow Li^+ + e^- \quad (3b)$$
$$(337 \text{ nm}) \quad (0.13 \text{eV})$$

where the Li resonance transition was pumped by a 2 μJ N$_2$-laser pumped dye laser and the ionization step was performed using a N$_2$-laser oscillator-amplifier combination.

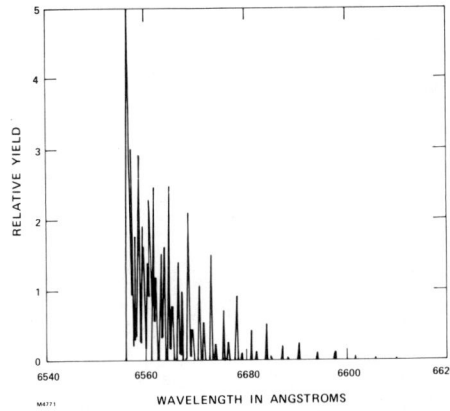

Fig. 2. Computed line strength on the $v_x=0$ to $v_A=5$ band assuming a rotational temperature of 75°K

Ions were extracted after a variable delay from optical pumping by the application of a ± 2 kV pulse to grids enclosing the interaction volume. Ions were detected by channeltron at the end of a 20 cm drift tube which contained a magnetic separator to remove electrons. Ion identification was by time-of-flight analysis. Total ion charge was also determined by a close-mounted Faraday collector cup.

The optically pumped volume was measured by magnified projection onto a viewing screen outside the vacuum chamber.

ATTACHMENT RESULT

With the 0.25Å laser pumping to $v_A=5$ (bandhead 655 nm) only 11% of the Li_2 rotational population was accessed, and the estimated $Li_2(v^*)$ density was 1×10^{11} cm^{-3}. The attachment of ~ 1% of electrons occurred after a 3 μsec delay. When this laser was replaced by the 15Å laser tuned to 656 nm, 57% of the Li_2 molecules were accessed and $Li_2(v^*)$ rose to 6×10^{11} cm^{-3}. The negative ion signal increased sevenfold to equal 7% of the positive ion (Li^+) signal (Figure 3).

There were background Li^- signals which could be associated with coincident pumping from $v_x=1$ to $v_A=5$ by the 671 nm photon (section (c) in Figure 3), and also with coincident production of electrons via the accidental three-step ionization of Li_2 (section (a) in Figure (3)), a point verified separately by monitoring the Li_2^+ signal.

Measurements of the total collected charge (1.8×10^{-10} coulombs) and the illuminated volume (0.8×10^{-2} cm^3) gave a density of ions or electrons of 1.4×10^{11} cm^{-3}. In conditions when 7% attachment occurred the Li^- density was 1.0×10^{10} cm^{-3}.

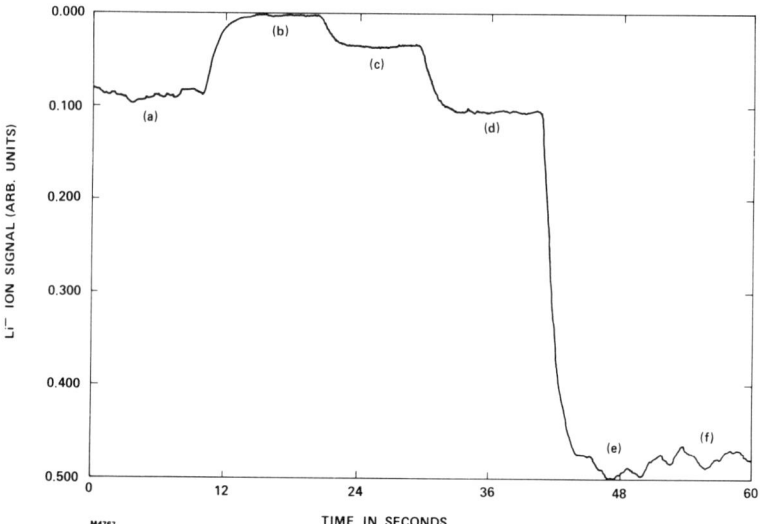

Fig. 3. Li⁻ signal for various optical pumping conditions:
(a) 656 nm, 15Å bandwidth; (b) no lasers, (c) 671 nm Li resonance laser + 337 nm ionizer only; (d) same as (c) plus molecular pumping at 655 nm in 0.25Å; (e) same as (c) plus molecular pumping at 656 nm in 15Å and 655 nm in 0.25Å; (f) same as (c) plus molecular pumping at 656 nm in 15Å bandwith

PROJECTION TO A CONTINUOUS SOURCE

At the velocity of particles in the Li beam a source current density of 1 mA cm^{-2} would require a production density of 2×10^{10} cm^{-3} Li⁻ ions, continually renewed at some position in the beam. The present experiment was pulsed, and it is of interest to consider what beam or laser parameters would be needed for a continuous source.

An increased Li_2 fraction is possible by operating at much higher Li oven temperatures. These conditions have been demonstrated by Schumacher et al with a molybdenum oven at 1260°C and 450 torr which gave a beam with approximately 50% Li_2 content. The increased $Li_2(v*)$ density that this would allow would be sufficient to generate 1×10^{11} cm^{-3} Li⁻ ions at approaching complete electron attachment.

The laser power requirement for molecular pumping is calculated to be

$$P(w) = 0.39 \, W_B \text{ Watts per cm}^{-1} \qquad (4)$$

where W_B(cm) is the width of the lithium beam transverse to the direction of optical pumping. For typical rotational temperatures a bandwidth of about 20 cm^{-1} is required to access a high percentage of rotational states (Figure 2). The laser power required for a 0.5 cm beam is thus only 4 Watts, within the reach of commercially available argon ion-laser pumped dye lasers.

A continuous plasma can be maintained by optical pumping of an alkali supersonic beam on the resonance transition [4]. This plasma is characterized by a low electron temperature (0.2eV) which allows the production of cold Li$^-$ ions by process (1). The laser power requirement for plasma production is comparable to that for molecular pumping.

With some refinements it is possible to project a Li$^-$ production temperature as low as 0.03eV by optically assisted dissociative attachment. The decay transition (2b) could be made to lase by the addition of an optical resonator, thereby channeling all the decays into a single terminal vibrational state (e.g. v*=10, energy defect for attachment = -0.018eV). If the Li(^{2}D) state is optically pumped to form a plasma, then an electron temperature as low as 0.07eV is sufficient to create a plasma density of 1×10^{12} cm^{-3}. The attachment energy being shared between products in process (1) could give a Li$^-$ production temperature of 0.03eV.

REFERENCES

1. M. W. McGeoch and R. E. Schlier, Phys. Rev. A33, 1708 (1986).
2. H. H. Michels and J. M. Wadehra, Bull. Am. Phys. Soc. 30, 129 (1985).
3. E. Schumacher, W. H. Gerber, H. P. Harri, M. Hofmann and E. Scholl, 'Metal Bonding and Interactions in High Temperature Systems', Section 8, p83, American Chemical Society Symposium 179, Washington, D. C., 1982.
4. J. L. LeGouet, J. L. Picque, F. Wiulleumier, J. M. Bizau, P. Dhez, P. Koch and D. L. Ederer, Phys. Rev. Lett. 48, 600 (1982).

DISCUSSION

<u>Anderson</u>: Why, in your excitation to the 5s and then looking for ions, do you believe that it's not excitation transferred to the 5d that's producing the ions?

<u>McGeoch</u>: The collision cross-section for that would also have to be very high to produce an effect. There's an energy defect to the 5d, which is even greater, so we wouldn't expect a large cross-section for that, at the temperatures in the beam. It effectively wouldn't occur.

<u>Wadehra</u>: In regard to your attachment, since you can selectively form various vibrational levels, have you tried to look for attachment to low vibrational levels?

<u>McGeoch</u>: Well, we don't see it. I have, yes. It's below our resolution. One problem is that our electrons are cool, so there's an energy defect if we go down to lower levels.

<u>Michels</u>: I'd like to point out that you're doing the experiment just right, with regard to low energy electrons. If you raise the electron energy, you're going to get into all these resonant states, which are all loss mechanisms. You were about 150 mV...?

<u>McGeoch</u>: Yes. And we would like to keep the energy even lower in real ion sources. One other thing I forgot to mention...in the last experiment on pair production, we also saw associative ionization rates which were very, very high. And we saw Li_2^+ signals, which were 1000 times greater than the Li^- signals, which would indicate that the associative ionization cross-section for the 5d or the 5s plus the 2p was extremely high, on the order of at least $1000 Å^2$.

<u>Michels</u>: If you probe only to n=3, you're below that threshold, though.

<u>McGeoch</u>: Correct. And we haven't done that yet.

PRODUCTION OF A COOLED, POLARIZED ATOMIC HYDROGEN BEAM

P.A. Schmelzbach, W. Grüebler, D. Singy and Z.W. Zhang
Institut für Mittelenergiephysik, Eidg. Techn. Hochschule,
CH-8093 Zürich, Switzerland

ABSTRACT

A new atomic beam apparatus using low velocity atoms (35 K) is presented. Beam optics design, recombination and beam forming problems are discussed. The use of this technique to improve the output of the ETH polarized ion source is shown. New possibilities in polarization state selection are indicated.

INTRODUCTION

At present, an increasing variety of methods for the production of polarized atoms and ions has been implemented or proposed. All methods have in common that at first an ensemble of polarized neutral atoms is prepared, which will either be used for targets or will be ionized selectively to positive or negative ions. The oldest method, i.e. the use of an atomic beam at thermal velocity, is still one of the most promising technique for both purpose.

A schematic diagram of the conventional atomic beam source is shown in Fig. 1. Hydrogen atoms are generated by dissociation of molecules in a rf or microwave discharge. An atomic beam of thermal velocity is formed by a nozzle and a skimmer. This beam is polarized in passing through an inhomogenous magnetic field (usually a sextupole field). This Stern Gerlach separation of magnetic substates of the hydrogen atoms delivers a fully electron polarized atomic beam. The application of rf transitions between hyperfine states of the atoms selects different nuclear polarization and provides rapid spin reversal. This nuclear polarized neutral beam can be used as a nearly 100 % polarized hydrogen jet target or can be compressed in a vessel. For polarized ions the atoms are converted to positive or negative ions in ionizers or charge exchange devices. This simple conventional scheme, which is shown in Fig. 1 by the solid line boxes is used in many operational polarized ion sources producing over 100 µA positive ions or about 5 µA negative ions. In the last 5 years substantial effort has been made to cool the atomic beam from room temperature to $T > 4$ K in order to obtain higher density and intensity of the atomic beam. Further improvement involves the single state selection by a second Stern-Gerlach separation and an additional rf transition to reverse the polarization direction. These new improvements, which are shown in Fig. 1 by the dashed line boxes, are now operational or are expected to be available in the very near future.

A large increase in ion beam intensity is expected if the atomic beam stage is redesigned in order to produce atomic beams at low temperature. The velocity dependence of the solid angle accepted by

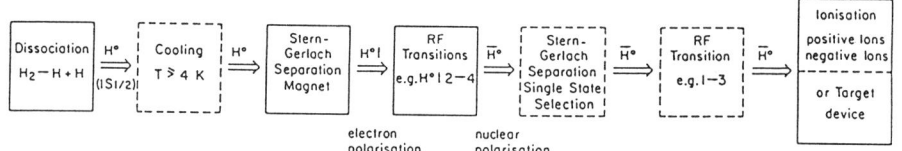

Fig. 1. Production of polarized atoms and ions by the atomic beam method.

a sextupole magnet together with the increased dwell time of the atoms in the ionizer suggest a $T^{-3/2}$ variation of the ionic output of the source. For example a reduction of the temperature from 300 K to 35 K results in a gain by a factor 25. The practical realisation of such large gains can, however, be hampered by several new problems. Constraints on the beam optics are imposed e.g. by the need for long drift spaces for rf transitions or by the geometry of the existing ionizers. Further, the question arises if a cooled beam can be produced without losses due to recombination of atoms in the cooling procedure and due to different conditions in the beam formation and scattering. In this paper we want to summarize results obtained at ETH during the development of a cooled atomic beam and to discuss the problems relevant to the design of such a device.

EXPERIMENTAL SET-UP

The experiments have been performed with the arrangement shown in Fig. 2. On the right hand side the upper part shows the diagnostic elements used to investigate the neutral beam, the lower part the ionizer for the production of the polarized ions [1]. Compression tubes allow beam intensity measurements and a quadrupole mass spectrometer is used for relative density determination.

The atoms generated in a room temperature dissociator are transported through a Pyrex and Teflon tubing to a copper accommodator, of which the end forms the nozzle. This cooling arrangement is shown in Fig. 3. Cooling is provided by a 10 W closed-cycle He refrigerator which allows investigations in a temperature range above 15 K. Numerous geometries of the accommodator have been investigated, the goal being to simultaneously optimize thermal accommodation, recombination and beam formation. Best results are obtained with a conical channel 20 mm long having an exit aperture of 3 mm diameter.

The loss due to recombination during the transportation of the atoms from the warm discharge to the cool accommodator is not critical, even at the high fluxes considered here. Different configurations have been used with equally good results. The Teflon can also be replaced by Macor [2]. The accommodator is made of good conductivity materials as e.g. copper to reduce temperature gradients since in the case of non negligible recombination, the cooling of several 10^{19} atoms s^{-1} in a DC mode may result in a thermal load in

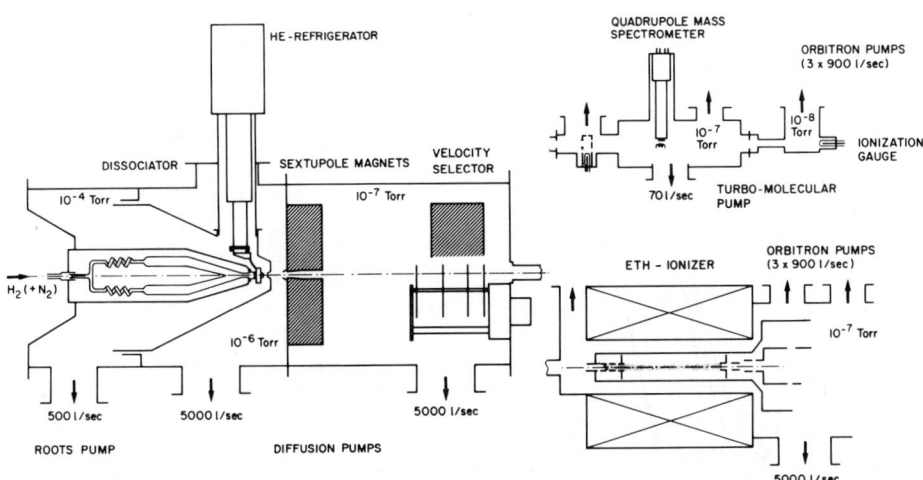

Fig. 2. Test bench for a cold atomic beam. The ETH ionizer is also shown.

the range of several Watt.

RECOMBINATION OF THE ATOMS

The nature of the accommodator surface is crucial for avoiding recombination of the atoms. The material of the accommodator, however, is not of primary importance. Oxydized or unoxydized Cu or Al, with and without Teflon coating show the same behaviour at low temperatures, i.e. a strong recombination below 50 K, leading to the conclusion that under practical conditions of gas and vacuum cleanness the role of frozen or adsorbed species dominates. A review of recombination phenomena in the temperature range of interest for a cold source is discussed in ref. 3. In this paper we showed that the behaviour is characterized by the recombination coefficient of H on H_2O. We also found that adding a small amount of nitrogen to the hydrogen gas allows to create and maintain a good recombination inhibiting surface in a range of a few degrees

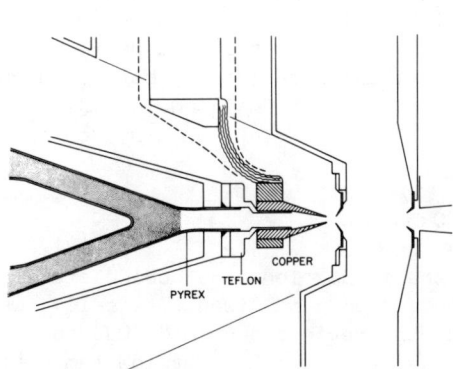

Fig. 3. Details of the cold beam production.

around 35 K. The observation of the thermal load of the accommodator indicates that under normal operating conditions less than 10 % of the atoms recombine. This method has been successfully applied up to hydrogen gas flow of 150 cm^3 min^{-1} into the dissociator. A typical evolution of the beam density as a function of the temperature is shown by the thick curve (1) in Fig. 4. A copper accommodator is used in this experiment and the focusing magnets are switched off. The broken line indicates the behaviour one expects if the recombination remains constant at the value at 150 K over the whole temperature range. It is obtained by correcting the T$^{-1/2}$ dependence for the loss due to the temperature dependent beam attenuation (beam forming, scattering) measured for this experiment. The most striking feature is the rapid drop observed below 50 K, indicating the dramatic increase of the recombination below this temperature.

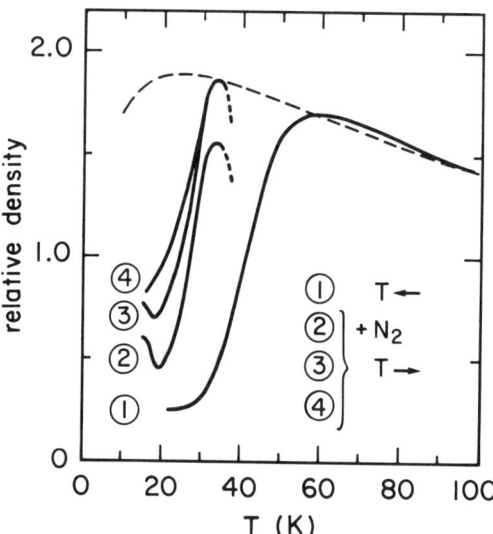

Fig. 4. Behaviour of the beam density for different amounts of N$_2$ added to the gas at 15 K.

The results obtained by adding different amounts of N$_2$ (estimated to be at most a few percents) to the gas and measuring the density of the atomic beam with increasing temperature is shown by the curves (2) to (4). Figure 5 a) shows a measurement performed with the focusing magnets turned on, illustrating the better matching of the 35 K beam with the optics of a system designed for v≲1000 m/s. In Fig. 5 b) the effect of the addition of N$_2$ at ~60 K by decreasing temperature is shown.

BEAM FORMATION

Further critical points in the production of a cooled atomic beam are the beam forming and the vacuum conditions. Basically, we use the same vacuum system as in our previous source [4]. Due to the increase of the relevant scattering cross sections at low temperature, beam formation and attenuation are expected to differ from the situation at room temperature. A typical behaviour of the beam intensity as a function of the total gas flow at different temperatures is shown in Fig. 6. The dominating feature is the strong decrease in beam intensity as the temperature is lowered. A closer inspection of these results show that this effect is not primarily due to the increased

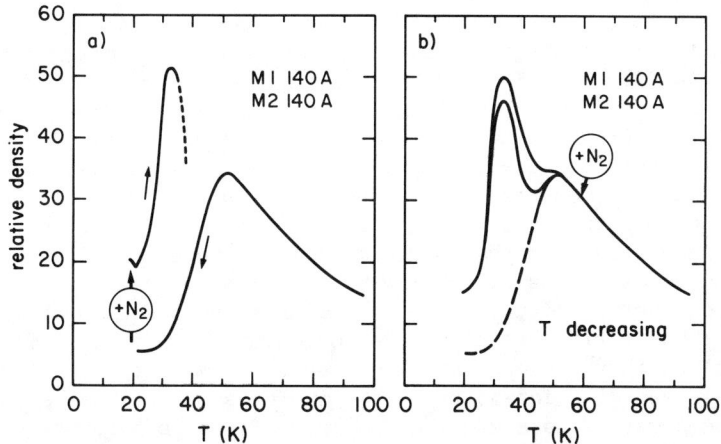

Fig. 5. Focussed beam density as a function of the temperature
a) N_2 added at ~15 K   b) N_2 added at ~60 K

beam scattering but to the mode of beam formation itself.

At room temperature and at 90 K, the observed behaviour can be described by a relation containing a term proportional to the total flow Q and an exponential part due to the beam attenuation in the residual gas. At low temperature the proportional term has to be replaced by a $Q^{1/2}$ dependence, which is typical for beam formation in the opaque mode [5]. This relation is also approximately valid in the case of "cloud" formation in front of the nozzle. The transition to hydrodynamical flow occurs above the practical range considered here. In contradiction to supersonic beam design considerations a small nozzle-skimmer distance (~4 mm) and a short and widely open skimmer ($\alpha = 90°$) give the best results, the diameters of the nozzle and skimmer being 3 and 4 mm respectively.

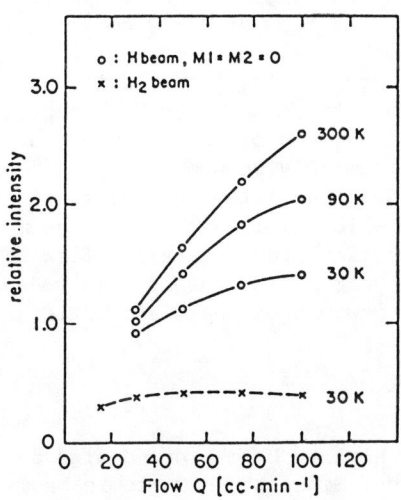

Fig. 6. Flow dependence of the beam intensity at various temperatures.

The performance of the vacuum system has been checked by varying the residual gas pressure or adding a 3500 ℓs^{-1}

diffusion pump at several locations. The increase of the pumping speed in the first stage (nozzle chamber) produces a maximum intensity gain of 20-30 %. For the second stage (skimmer-collimator chamber) a gain of 10 % is obtained. The variation of the intensity as a function of the distance between nozzle and skimmer follows an exponential dependence above ~3 mm (= 1 nozzle diameter). The behaviour of the measured curves and of their extrapolation to the origin confirms the precedent finding on the beam formation mode. As mentioned above, in the opaque or "cloud" mode, the intensity on axis depends on the total flow Q like $Q^{1/2}$. In addition, it depends on the velocity like $V^{1/2} \propto T^{1/4}$ and on the temperature dependent scattering cross section. Calculations with this model agree well with the experimentally observed temperature dependence. The decrease of the intensity of the (unfocussed) beam as a function of the temperature is a genuine property of the beam formation mode and only in a much smaller amount due to scattering by the residual gas. As a consequence of this fact, the amount of beam injected into the sextupole focussing system will be significantly smaller at 30 K than at room temperature and the situation cannot be simply improved by a better vacuum system.

The velocity distributions measured with a velocity selector (cf. Fig. 2) are quite narrow, approximately one half of a Maxwellian distribution. Velocities and shapes show a strong flux dependence. Favourable conditions in respect to the beam velocity (most probable velocity ~1000 ms^{-1}) are obtained at a fas flow of 30-40 cm^3 min^{-1}. This flow is about 4 times smaller than for standard operation of the room temperature source [4]. At higher gas flow the velocity increases and becomes too large for an optimum design of the magnet system. Investigation with molecular beams and different accommodator geometries suggest that this problem is mainly connected to the beam formation and not to an insufficient cooling of the particles. At present this is the dominating effect which limits the atomic beam output. Unfocused beam intensity saturation occurs at a gas flow of 80-100 cm^3 min^{-1} but in this case the velocity of the beam increases to more than 1500 ms^{-1} and the relative width of the distribution becomes about 15 % larger.

BEAM OPTICS

To take full advantage of a low velocity beam a new design of the sextupole magnet system is required. Since the new atomic beam is intended to upgrade an existing source a difficulty arises from the fact that some parameters like minimum total length, location and geometry of the ionizer are given by existing hardware and impose constraints on the choice of the optics. We have used the acceptance diagram technique to investigate the properties of the whole system (diaphragms, magnets, drift spaces, ionizer) [6]. The geometry of the existing ionizer sets an upper limit to the acceptance of the system. The situation is illustrated in Fig. 7, for a two magnet system.

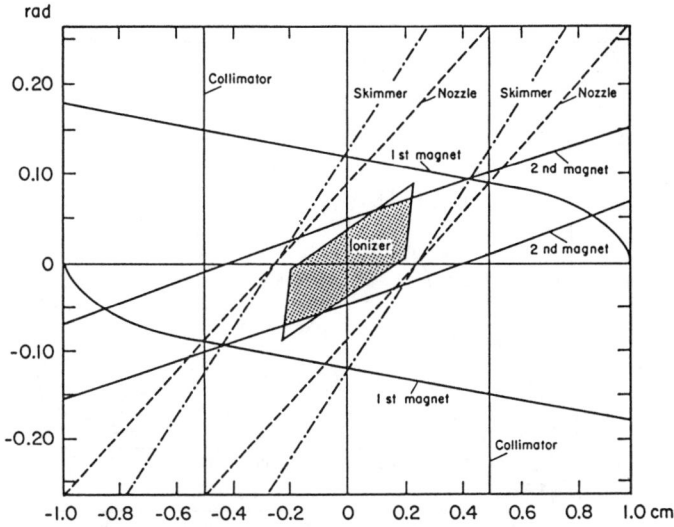

Fig. 7. Acceptance diagram for a two magnet system.

Here the acceptance diagrams of all components of the system have been transferred to the collimator position at the entrance of the first magnet, and the area of the overlap region (i.e. the acceptance area) of the whole system is shown. The aim of the design is then to find a beam velocity and a magnet configuration matching as close as possible this maximum acceptance. This condition is satisfied by the system shown in Fig. 2, with a velocity of the hydrogen atoms around 1000 m s^{-1}, corresponding to an accommodator temperature (T≅35 K) at which the recombination problem can be solved as discussed previously.

This configuration also provides the necessary free space for the rf-transition units. Two magnets are needed for achromatic focusing. The system is characterized by the shortness of the magnets (10 and 15 cm), large apertures (12 to 20 mm and 30 mm) and a large distance between the two magnets. The maximum fields on the pole tips are 1 Tesla and 0.7 Tesla. A typical dependence of the acceptance as a function of the velocity for a two magnet system is shown in Fig. 8. A criteria on the quality of a given configuration is obtained by folding these curves with the effective velocity distribution.

The efficiency of the beam transportation, defined as the ratio of the mean atomic density in the ionizer to the density at the skimmer, is calculated to be one order of magnitude larger than in the previous room temperature source [1,3]. The calculations have been performed with measured velocity distributions. On the test bench good agreement was found between calculated and measured properties of the magnet system. From the calculations, a strong dependence of the mean density as a function of the distance along the ionizer axis is expected. (Decrease by a factor 4 between 30 cm and 60 cm from the

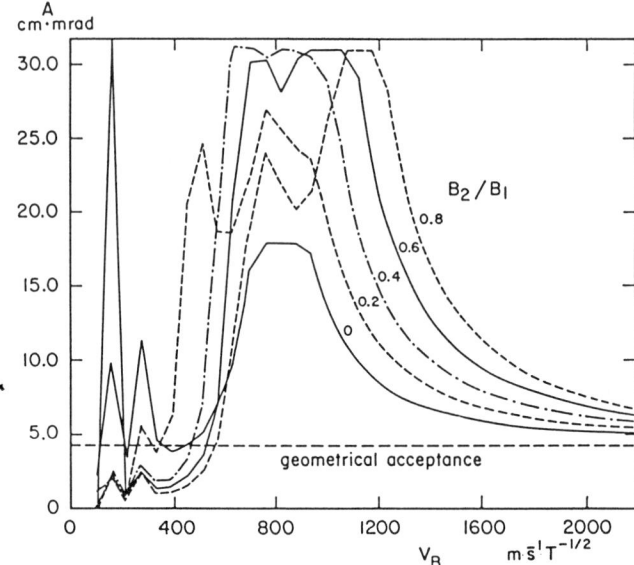

Fig. 8. Variation of the acceptance of a system equipped with two magnets as function of the reduced velocity $v_B = v \cdot B_1^{-1/2}$ and the ratio of the magnetic fields.

exit of the second magnet [6].)

PERFORMANCES

The preceeding discussion leads to the conclusion that the cold atomic beam technique results in a substantial gain due to the new optics but peculiarities of the beam generation require operation at a reduced gas flow. Further work is needed to solve this problem. Beams of 10^{17} atoms s^{-1} have been observed at the exit of the atomic beam apparatus with density on axis of $2 \cdot 10^{12}$ atoms cm^{-3}. The intensity is approximately 6 times higher than observed previously with the room temperature atomic beam [4] and the density has increased by a factor 15. These figures demonstrate that a significant improvement has been achieved in the atomic beam production. In such comparison, one should, however, keep in mind that the optical properties of the beams are different and that only atoms matching the phase space accepted by the device in which they are used are of interest to judge the performance of a system.

APPLICATIONS

The present apparatus has been specifically designed to improve the performance of a polarized ion source equipped with a long ionizer.

Recently the cooled atomic beam was installed at the ETH polarized ion source, replacing the atomic beam with the parameters reported in ref. 4. The results obtained on the test bench have been confirmed at the complete source system. Even if operated at a reduced gas flow, the new apparatus delivers a significantly higher ionic output. The positive ion beam intensity actually observed at the location of the double charge exchange device after acceleration of the beam to 5 keV is ~0.4 mA. The polarized negative beam intensity produced by double charge exchange in sodium vapour was measured after the acceleration to 60 keV and resulted in a negative beam output of 16 µA. Compared to the performances observed just before the installation of the cooled atomic beam, the intensity has increased by a factor 4. These results are obtained with a simplified version of the ionizer of ref. 1. The filament grid has been removed and the number adjustable electrode potentials has been reduced in order to simplify maintenance and operation of the device. Scaling to the previous system results in beam intensities of ~0.5 mA and 25 µA respectively. For a further increase of positive or negative ion beam intensities, which would reveal the whole benefit of a cooled atomic beam, an adapted ionizer with larger acceptance is necessary. A new electron bombardement ionizer or a ECR ionizer as proposed in ref. 7 may be suitable for this purpose and have, in addition, a higher efficiency. A combined approach to ideally match both stages should result in beams well in the mA range.

With an appropriate design of the optics a cooled atomic beam can also be used directly as target or to be compressed in a cell. As for a polarized source a great flexibility in the state selection is of very strong interest. For example the use of pure states make it possible to be very free in the choice of magnetic field strength at the ionizer or target and might be a must to avoid recombination in a low temperature compression cell. To illustrate the possibilities attainable with a two magnet system a deuterium beam is presented as example.

The production of polarized deuteron beams or targets by polarized atomic deuterium beams is particularly interesting since in this case high vector and/or tensor polarization can be obtained. While at present, in conventional operational cryogenic deuteron targets the tensor polarization p_{zz} is small ($p_{zz} \leq 0.08$) polarized atomic deuteron beams can reach p_{zz} values near -2.0. Further all different vector and tensor polarization components of spin -1 particles can be produced in an atomic beam target.

The required combination of transitions depends on the application. From a large variety of schemes a number of interesting combinations of Stern-Gerlach separations and rf transitions are shown in Table I. Mode (b) and (c) give a pure vector polarization but with different signs. This can be used to switch the sign of the polarization periodically. The modes (d) to (g) produce a combined vector and tensor polarization with $|p_z| = 1/3$ and $|p_{zz}| = 1$. Two of these modes can be combined to a measuring scheme such that either

Table I Production of polarized atomic deuterium beams

Mode	Ionizer or target B-field	Stern-Gerlach separation (S-G) and rf-transitions	Substates	Vector pol.p_z	Tensor pol.p_{zz}	relative intensity I	Figure of merit $I \cdot p_z^2$	Figure of merit $I \cdot p_{zz}^2$
a	strong	S-G	1 + 2 + 3	0	0	1	0	0
b	strong	S-G 1 → 4 (WF)	2 + 3 + 4	-2/3	0	1	4/9	0
c	strong	S-G 2 → 6, 3 → 5	1 + 5 + 6	+2/3	0	1	4/9	0
d	strong	S-G 3 → 5	1 + 2 + 5	+1/3	-1	1	1/9	1
e	strong	S-G 3 → 5 $\left.\begin{array}{l}1→4\\2→3\\5→6\end{array}\right\}$ (WF)	3 + 4 + 6	-1/3	+1	1	1/9	1
f	strong	S-G 2 → 6	1 + 3 + 6	+1/3	+1	1	1/9	1
g	strong	S-G 2 → 6 $\left.\begin{array}{l}1→4\\3→2\\6→5\end{array}\right\}$ (WF)	2 + 4 + 5	-1/3	-1	1	1/9	1
h	strong	S-G 1 → 4 (WF) S-G 3 → 5	2 + 5	0	-2	2/3	0	8/3
i	strong	S-G 1 → 4 (WF) S-G 2 → 6	3 + 6	0	+1	2/3	0	2/3
k	strong	S-G 3 → 5 S-G 2 → 6	1 + 6	+1	+1	2/3	2/3	2/3
ℓ	strong	S-G 3 → 5 $\left.\begin{array}{l}1→4\\2→3\end{array}\right\}$ (WF)	3 + 4	-1	+1	2/3	2/3	2/3
m	any	S-G 2 → 6 S-G 3 → 4	1 + 4	0	+1	2/3	0	2/3
n	any	S-G 3 → 5 2 → 6 S-G	1	+1	+1	1/3	1/3	1/3
o	any	S-G 3 → 5 2 → 6 S-G 1 → 4 (WF)	4	-1	+1	1/3	1/3	1/3

strong B-field: >>117 Gauss

the sign of the vector polarization or the tensor polarization or both signs are switched. The mode (h) with a pure tensor polarization mode (i) with $p_{zz}=+1$. In spite of the reduced intensities of mode (i) and (h) totally the highest figure of merit results from this combination, namely $I \cdot (\Delta p_{zz})^2 = 2/3 \cdot 9 = 6$. The modes (k) and (ℓ) give $p_{zz}=+1$ and $p_z=|1|$ with inverted sign of p_z. The pure states 1 and 4 allowing ionization in any field can be produced in the modes (m) to (o).

Since the required magnetic field in the ionizer or at the target is in any case relatively low, the sign of the vector polarization can also be easily reversed by changing the field direction in the ionizer or target region. In this case the sign of the tensor polarization is unchanged.

CONCLUSION

The production of a dense atomic beam at low temperature has been discussed. Despite the fact that an unfavourable temperature dependence of the beam forming reduces the possible gain an appreciable improvement of the output of a polarized ion source has been achieved. With the optics realisable for slow atoms the increased flexibility in the choice of polarization states becomes a unique feature of atomic beams for ion source or target use.

REFERENCES

1. P.A. Schmelzbach, W. Grüebler, V. König and B. Jenny, Nucl. Instr. Meth. 186, 655 (1984).
2. S. Jaccard, private communication.
3. P.A. Schmelzbach, Proc. Int. Workshop on Polarized Sources and Targets, Montana, 1986, Helv. Phys. Acta, 49, 539 (1986)
4. R. Risler, W. Grüebler, V. König and P.A. Schmelzbach, Nucl. Instr. Meth. 121, 42 (1974).
5. J.A. Giordmaine and T.C. Wang, J. Appl. Phys. 31, 463 (1959).
6. W.Z. Zhang, P.A. Schmelzbach, D. Singy and W. Grüebler, Nucl. Instr. Meth. A240, 229 (1985)
7. T.B. Clegg, V. König, P.A. Schmelzbach and W. Grüebler, Nucl. Instr. Meth. A238, 195 (1985)

DISCUSSION

Hershcovitch: There was a curve in which you showed the increase in the most probable velocity as a function of the throughput. Did the width of the velocity distribution also increase?

Schmelzbach: I show, in these pictures, an increase of the most probable velocity; but the width increased, also.

Hershcovitch: How much? In the same fashion?

Schmelzbach: Yes, also by 10 or 20 percent. But these are very strong effects. It is important to have a distribution which is quite sharp.

Stevens: You quoted that you had achieved currents of 16 µA. What was the polarization of the beam, then?

Schmelzbach: The measured polarization of the deuteron was about 90%; and the proton about 80-85%. This is the same polarization we had with our room temperature source. In fact, if you make a calculation, you have not only to calculate the focussed beam, you have also to calculate the unfocussed beam. We have a better ratio than before in the atomic beam. But in our source, the main origin of bad polarization is due to the conditions we choose to operate the ionizer stably. You have to make a compromise between maximum beam and low beam polarization. In fact, there are quite a few parameters. The best polarization we ever measured was something like 92 or 93% with deuterons.

POLARIZED H⁻ SOURCE DEVELOPMENT AT BNL*

A. Kponou, J.G. Alessi, A. Hershcovitch, T.O. Niinikoski
Brookhaven National Laboratory, Upton, New York 11973

ABSTRACT

A polarized H⁻ source has been in operation at BNL for several years. It produces 500 μs pulses of ~40 μA $\vec{H}^-$ with polarization of ~75% measured at 200 MeV. Normal operation is at 0.5 Hz, but the source has recently been operated at 5 Hz with encouraging results. Progress in developing a milliampere source is reported.

INTRODUCTION

Since the last Symposium in 1983[1], the AGS $\vec{H}^-$ source, PONI-I, has become operational and has provided a polarized proton beam which has been accelerated to 21 GeV/c in the AGS. Physics experiments have been done at ~15 GeV/c with ~50% polarization. The present peak output of PONI-I is still about 10^{-3} of the output of the AGS unpolarized magnetron source. A polarized H⁻ source with milliamperes of output would make possible new classes of spin physics experiments. Such a source is being developed at BNL. We shall review the operation of PONI-I and then report on the milliampere source development program.

PONI-I

At one end of PONI-I, shown schematically in Figure 1, H_2 is dissociated into H° in a conventional 20 MHz rf dissociator. As the atoms flow into the vacuum, they collide with the wall of the dissociator nozzle, which is maintained at ~100 K by a closed-cycle helium refrigerator, and are slowed down. This atomic beam is electron-spin polarized by sextupole magnets which focus atoms with $m_j = 1/2$, while defocussing the -1/2 component (the Stern-Gerlach effect). The focussed beam then passes through two rf transition units which, by adiabatic spin reversal, flip the electron spin of either m_I state in the focussed beam. The proton spin is also flipped via the hyperfine interaction. Thus, by energizing these units alternately, on a pulse-to-pulse basis, beam polarization is modulated between the two proton polarization states.

At the other end of the source, Cs⁺ ions are produced by surface ionization of neutral cesium atoms on a hot porous tungsten button. The ions are extracted with up to 50 kV and neutralized by charge exchange in a cesium vapor target.

The collinear pulsed beams of proton polarized, ground state, thermal hydrogen atoms and 50 keV cesium atoms, collide in a region of moderate magnetic field. $\vec{H}^-$ ions are produced by the reaction[2]:

$$Cs° + \vec{H}° \rightarrow Cs^+ + \vec{H}^-.$$

---

*Work performed under the auspices of the U.S. Department of Energy.
+On leave from CERN, Geneva, Switzerland.

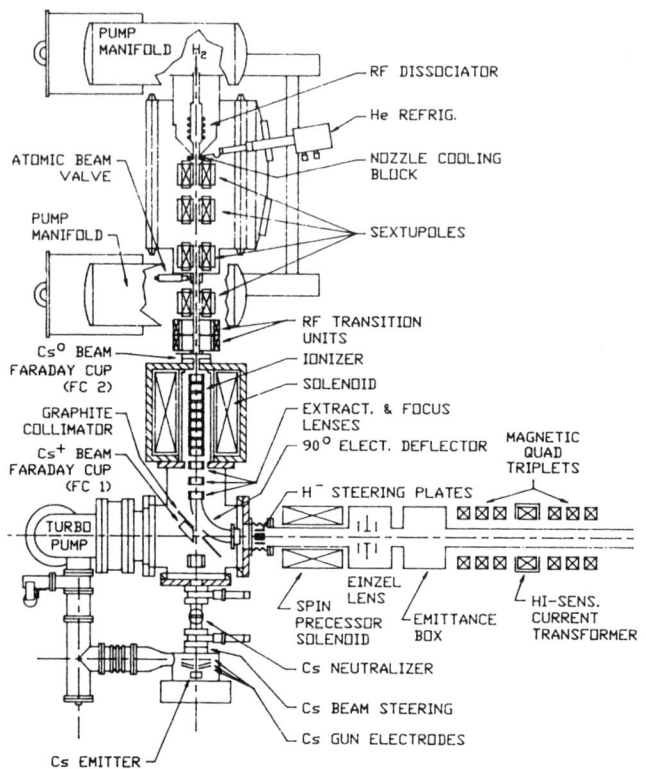

Fig. 1 Schematic diagram of PONI-I, the AGS polarized H⁻ source.

They are swept toward the extraction end of the ionizer by a weak electric field (~50 V over 30 cm), accelerated to 20 kV, and deflected by a 90° electrostatic deflector. A solenoid at the exit of the deflector precesses the spins into the vertical. The beam is then injected into the RFQ and accelerated to 760 kV before it is injected into the 200 MeV Linac.

PONI-I has operated very reliably since it went on-line. During the physics runs, the tungsten button is changed every two weeks and the electrodes are wiped clean of cesium once between button changes. Down-time for a button change is about six hours. The dissociator and neutralizer operate for months without maintenance.

Nearly all power supplies as well as the timing pulses are under computer control. This makes for both easier start-up and tuning the source, and accounts for its long-term stability.

A more detailed account of PONI-I can be found in Ref. 3.

Recently, the source was successfully operated at 5 Hz for several days. The purpose of this test was to find out if, in a contingency, PONI-I could supply beam to the AGS Booster, which is in its initial stages of construction and will operate at 7 Hz. The button and boiler of the cesium gun were operated about 100°C and 30°C higher respectively than at 0.5 Hz, and the gun was more stable. The same peak Cs° signal was obtained, but the pulse was about half as wide. A higher boiler temperature might have increased the pulse width. The neutralizer was operated d.c. for this test.

At the atomic hydrogen beam end, the indications were that the present pumping capacity would be inadequate at 7 Hz operation with the existing dissociator. We will replace the latter with a dissociator which was designed for the cold hydrogen beam source (see the next section) and is considerably more gas and rf power efficient.

HIGH INTENSITY SOURCE DEVELOPMENT PROGRAM

Development of the milliampere $\tilde{H}^-$ source is proceeding on three fronts:

1) Developing a very cold high-intensity atomic hydrogen beam
2) Developing a more efficient ionizer
3) Efficiently transporting the beam from the accomodator to the ionizer.

COLD HYDROGEN BEAM

The solid angle acceptance of the focussing (spin selection) lens depends inversely on the beam temperature T, and slower moving particles have a higher probability of being ionized (a $T^{-1/2}$ dependence). Thus, provided that there is no loss of flux with cooling, and the beam optics is optimized, one should ideally expect a $T^{-3/2}$ dependence of the source output with temperature.

The atoms are cooled by wall collisions in the channels through which they pass. This poses a problem because hydrogen very readily recombines on materials in the 100 to 30 K temperature range. To overcome this problem, the teflon and copper channels are separated by a 0.3 mm gap and they are maintained at 130 K and 6 K respectively. At 130 K, recombination of hydrogen on teflon is negligible. Below 20 K, molecular hydrogen freezes on the copper, and recombination of atomic hydrogen on it is also negligible. The flow regime of the beam in the channels dictated a very short gap in order not to lose beam collimation.

A schematic of the interior of the test stand which was constructed to try out these ideas is shown in Figure 2. A conventional rf dissociator produced the atomic hydrogen beam and the copper accomodator was cooled by a liquid helium cryostat. The time-of-flight technique was used to measure velocity distribution in the beam to determine how effectively the beam was cooled. Beam density was measured with a quadrupole mass analyzer.

We investigated the following:

1) Variation of beam density with dissociator orifice diameter.
2) Variation of beam density and velocity with accomodator temperature and diameter.
3) Beam focussing with a conventional sextupole and a permanent quadrupole magnet.

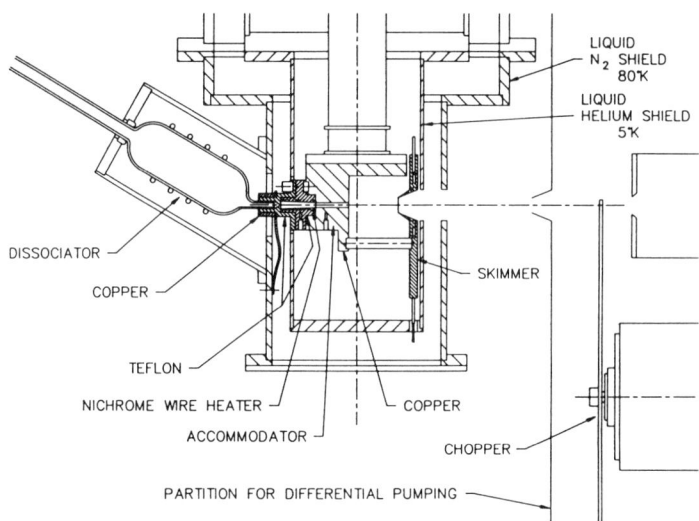

Fig. 2. Schematic of the interior of the cold hydrogen beam test-box. The skimmer plate is coated with activated charcoal and cooled to ~2.5 K to act as a very efficient cryopump.

The density increased as the orifice size was increased from 1 mm^2 to 9 mm^2, although by a smaller factor. The variation of beam density with accomodator temperature is shown in Figure 3, and velocity distribution measurements at the peaks are shown in Figure 4.

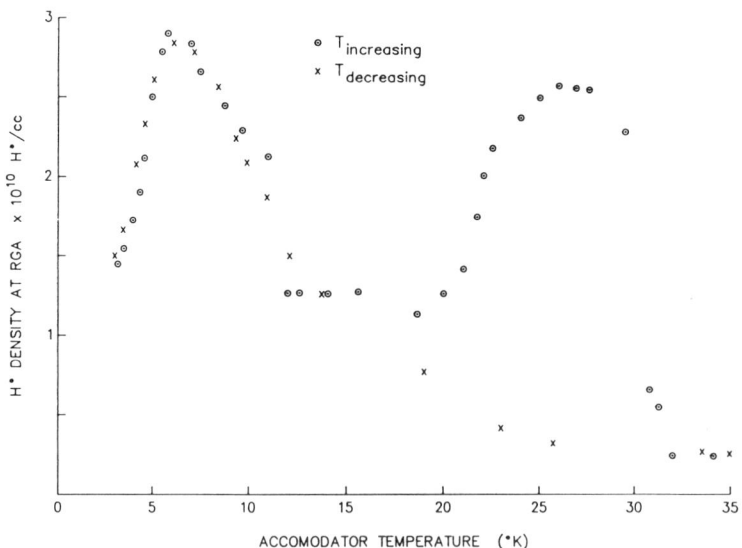

Fig. 3. Variation of peak hydrogen beam density accomodator temperature.

These results have been discussed in detail elsewhere, together with an attempt to give them a gas dynamical interpretation[4]. The much narrower distribution observed at 5.8 K means that the beam optics should be better at this temperature. Fluxes of 9.4×10^{18} and 1.1×10^{19} atoms/s/sr were obtained at accomodator temperatures of 5.8 K and 26 K respectively.

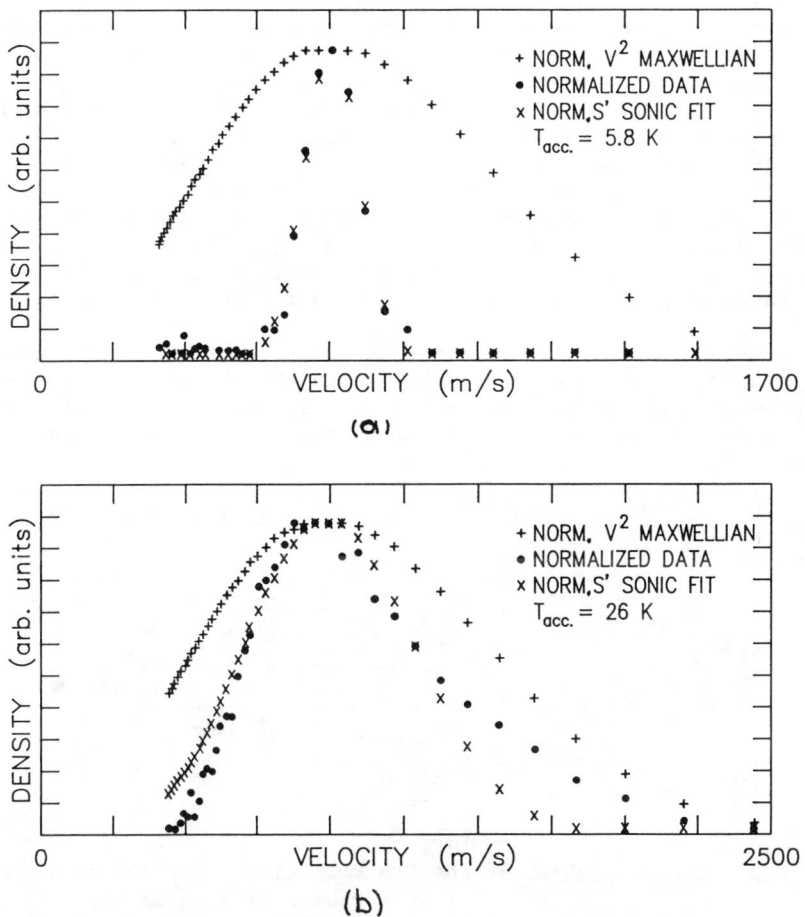

Fig. 4. Results of velocity distribution measurement with $T_{acc} \sim 6$ K (a) and 26 K (b).

Based on the results of these studies, a dissociator and accomodator for the prototype source have been designed and are being fabricated. The usual bulb in the dissociator tube, for example in Figure 2, has been eliminated in order to reduce the gas load expected with 7 Hz operation. The spiral, rather than helical, design of the rf coil makes for a more compact system. Bench tests indicate comparable H° output at 60% reduced gas load and 66% lower rf power consumption.

SUPERCONDUCTING SOLENOID

The acceptance solid angle of a sextupole magnet is $\Omega \sim 2\mu B_0/kT$ for a source at the magnet's entrance and a beam having a Maxwellian velocity distribution. For $B_0 \lesssim 1$ T and a most probable velocity of 680 m/s (when the accomodator temperature is 5.8 K), the calculated solid ange is $\Omega \lesssim 0.05$ sr. It is smaller still because, for practical reasons, the source is usually several centimeters away from the magnet entrance. It was decided to test a superconducting solenoid as the lens element since it can produce suitable field gradients over a large aperture[5]. The solenoid consists of three coils which are connected in series, with the current in the outer two counter to the current in the middle coil. The overall length of the solenoid is 10 cm and it has a useful i.d. of 9.4 cm. Field maps for the solenoid were generated with the program POISSON. A second program tracked H° atoms through the calculated fields. The result of a track-tracing calculation is shown in Figure 5. The box on the right represents the ionizer. We estimate the acceptance angle to be 0.1 sr.

Mounting the source very close to the solenoid was considered since pumping should not be a problem with the abundance of cryopanels between the accomodator and solenoid, but track-tracing showed that the beam could not be brought to a tight enough focus at the ionizer.

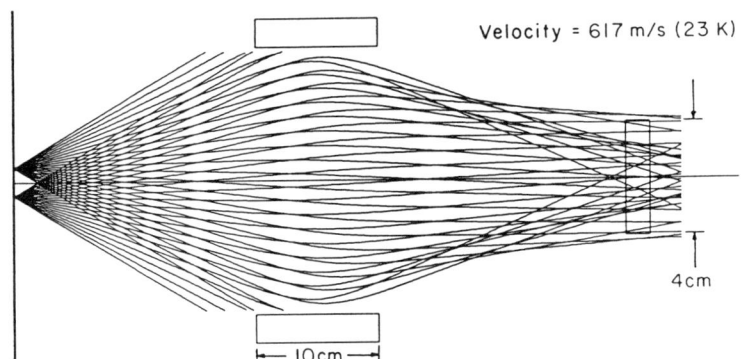

Fig. 5. Output of the track-tracing calculation for $B_0 \sim 5$ T at a radius of 5 cm in the mid-plane of the solenoid. A 2 cm dia. aperture at the ionizer (box on the right) intercepts about 50% of the focussed beam.

The calculated field of the solenoid at r = 4.7 cm in the median plane is 4.5 T for a current of 110 A. A calculation with POISSON using the actual ampere-turns, and a subsequent track-tracing calculation showed that the effect on the optics was negligible.

We estimate that, with the atomic beam charateristics measured on the test stand, we can obtain a flux of ~ 3×10^{17} atoms/s within a 1 cm radius at the ionizer, and an average density of 1.3×10^{12} atoms/cm^3.

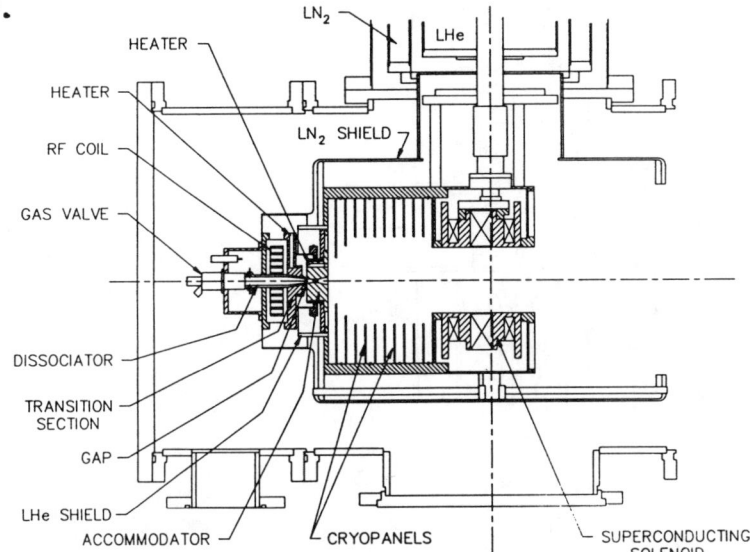

Fig. 6. Schematic of the cold hydrogen beam source mounted to the LH_e cryostat. It incorporates a new dissociator which is more gas and power efficient.

The solenoid will be mounted on the foot of the LHe cryostat and the dissociator will be supported as shown in Figure 6. Testing of this assembly should start soon.

RING MAGNETRON IONIZER

The ring magnetron ionizer[6] was proposed as a way to surmount the problem of space charge blow-up of the ionized beam inherent in the ionizing reaction:

$$\vec{H}^\circ + D^- \rightarrow \vec{H}^- + D^\circ.$$

This reaction has a cross section about five times greater than the cesium charge exchange reaction used in PONI-I. A schematic of the ionizer is shown in Figure 7. The molybdenum cathode is the cylinder with two grooves cut on its inside surface. The anode is a stainless steel cylinder with 0.05" wide slits, having a total length of 9 mm. D^- ions emerge through the slits as shown with ~200 eV energy (from the cathode voltage) and ionize the $\vec{H}^\circ$ beam passing through the magnetron. Up to 700 mA of self-extracted D^- has been measured on the axis. Enough D^+ ions drift into the reaction volume to provide space charge neutralization. The magnetic field also serves to preserve the polarization.

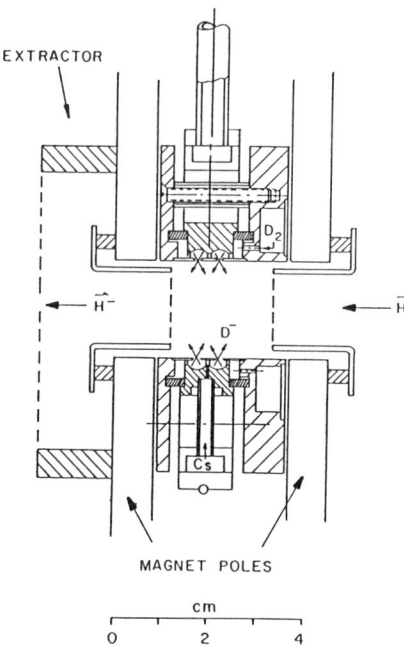

Fig. 7. Schematic of ring magnetron ionizer for the proposed AGS milliampere polarized H⁻ source.

To test the ionization, unpolarized H° was injected through the magnetron as shown in Figure 7. 500 µA of H⁻ coming from the resonant charge exchange was extracted at 2 - 3 kV. With an estimated H_o density in the center of the ionizer of 10^{12} atoms/cm^3, the ionization efficiency is ~ 4 better than for the Cs ionizer in PONI-I.

The ring magnetron has now been combined with a polarized atomic beam stage, to see if there is any depolarization during ionization. We are presently working on optimizing the extraction optics of the new system to first get back to the ionization efficiency observed on the test stand. Following the polarization measurements, the ionizer will be combined with the cold atomic beam and superconducting solenoid.

REFERENCES

1. Th. Sluyters, J. Alessi, A. Kponou, AIP Conf. Proc. No. 111, Am. Inst. Phys., New York, p. 736 (1984).
2. W. Haeberli, Nucl. Instr. and Meth. 62, p. 355 (1968).
3. J. Alessi, A. Kponou, Th. Sluyters, Helvitica Physica Acta 59, p. 563 (1986).
4. A. Hershcovitch, A. Kponou, T.O. Niinikoski, Helvitica Physica Acta 59, p. 527 (1986).
5. T.O. Niinikoski, S. Pentilla, J.-M. Rieubland, A. Rijllart, AIP Conf. Proc. No. 117, Am. Ins. Phys., New York, p. 139 (1984).
6. J.G. Alessi, Helvitica Physica Acta 59, p. 547 (1986).

OPTICAL PUMPING OF Na FOR USE IN POLARIZED H$^-$ ION SOURCES[*]

L. W. Anderson, D. R. Swenson, and D. Tupa
University of Wisconsin, Madison, WI 53706

ABSTRACT

Relaxation rates for optically pumped polarized Na atoms colliding with a variety of surfaces is reported. The relaxation rates have been studied as a function of the time of exposure to Na vapor, of the temperature, of the magnetic field and of the Na atom density. Silicone surfaces have relaxation times corresponding to more than one hundred wall collisions even after weeks of exposure to Na vapor at 225 C. Our measurements yield both the amplitude and the correlation time for the magnetic interaction that produces the relaxation. The correlation time increases with time of exposure to Na vapor. The correlation time is about 4×10^{-11} s for a fresh silicone surface and increases to about 1.5×10^{-10} s after four weeks exposure to Na. The amplitude is about 15 mT. For relaxation times corresponding to more than 100 wall collisions radiation trapping should not limit the optical pumping of a Na target until the Na density is 10^{19}-10^{20} atoms/m^3.

INTRODUCTION

Several optically pumped polarized H$^-$ ion sources have been built or are under construction.[1-4] Ion currents and nuclear polarizations as large as 65 µA and 0.65 respectively have been obtained using an optically pumped polarized H$^-$ ion source.[2] Because of the large H$^-$ ion currents and high nuclear polarization obtained the optically pumped polarized H$^-$ ion source appears promising for future use with high energy accelerators. The optically pumped polarized H$^-$ ion source was based on a proposal by Zavoiskii[5] with modifications suggested by Haeberli.[6] This polarized H$^-$ ion source has been analyzed in detail by Anderson.[7] Briefly the optically pumped ion sources described in Ref. 1 and 3 work as follows. A beam of H$^+$ ions is extracted from an ECR ion source that is in a large magnetic field. The H$^+$ ion beam is incident on a Na vapor target with an energy of about 5 keV. This target is in the same large magnetic field as the ECR ion source. The magnetic field and the Na target are colinear with the ion beam axis. The Na vapor target is electron spin polarized by optical pumping using a dye laser beam.[7,8,9] In this target some of the H$^+$ ions are neutralized by the reaction H$^+$ + $\vec{\text{Na}}^0$ → $\vec{\text{H}}^0$ + Na$^+$ where the arrow indicates that the polarized electron of a Na^0 target atom is transferred to the H^0 atom that is produced in the reaction. The fast H^0 atom emerges from the first target and enters a second alkali vapor target. The magnetic field at the second target is directed oppositely from the field in the first target. The fast electron spin polarized H^0 atoms pass suddenly through a region of near zero field between the two targets. This sudden passage through zero field transfers the electron spin polarization to the nuclear spin.[10] Some of the fast nuclear spin polarized H^0 atoms are converted to polarized H^- ions in the second alkali vapor target. The optically pumped ion source described in Ref. 2 is similar but works in a slightly different manner. The ion source of Ref. 4 is under construction.

Several improvements are possible in the optically pumped polarized H^- ion source. Probably the most important possible improvement in the ion source would be the use of a wall surface for the polarized Na target that does not result in depolarization at each wall collision. We have previously studied the spin relaxation of Na on a wide variety of wall surfaces.[11] Our previous measurements of Na relaxation times are presented in Table I.

Table I. Na vapor wall relaxation times. The relaxation time has been converted into the number of wall collisions before relaxation.

MATERIAL	NUMBER OF WALL COLLISIONS
stainless steel (250 C)	1
graphite (250 C)	1
anodized aluminum (250 C)	1
chrome plated copper (250 C)	1
teflon (200 C)	1
Kel-F (200 C)	1
Kapton (250 C)	1
Invex (225 C)	1
polyethylene (125 C)	15
fluorocarbon rubber (250 C)	
Viton	20
Fluorel	15
silicone rubber (225 C)	150
RTV 3145 (250 C)	200
"dry film" (275 C)	
on glass $(CH_3)SiCl_3$, $(CH_3)_2SiCl_2$	> 500
on glass $(CH_3)_2SiCl_2$	>1000
on copper $(CH_3)_2Si(OCH_3)_2$, $CH_3Si(OCH_3)_3$	>1000

As can be seen, both fluorocarbon rubber such as viton or fluorel and silicone surfaces such as dry film have relaxation rates that correspond to many wall collisions. We have made a detailed study of the relaxation of Na by a silicone surface. Relaxation rates have been studied as a function of the time of exposure to Na vapor, of the temperature, of the magnetic field, and of the Na atom density. The results of these measurements and some of their implications for the Na target used in the optically pumped polarized H^- ion source are discussed in this paper.

THE EXPERIMENTAL APPARATUS

The Na vapor target is a tube open to a vacuum system at both ends. The composition and diameter of the inner wall of the target are easily changed by sliding in liners. The Na target is inside a solenoid that produces a field up to 140mT. The target has a small hole at the center through which Na vapor enters from the Na oven. The temperature of the target and Na oven are measured using thermocouples. The Na density in the target is determined from the known vapor pressure of Na and the calculated conductances.

The Na vapor is optically pumped using a ring dye laser. The laser beam is split into a weak probe beam and a strong pump beam. The probe beam intensity is less than 10^{-7} W/cm^2. For such a weak probe beam the measured relaxation rates are observed to be independent of the probe intensity. The pump beam passes through an acousto-optic modulator (AOM) which allows the beam to be switched on and off in 10^{-6} sec. The intensity of the probe beam is monitored by a photomultiplier tube (PMT), and the signal is recorded as a function of time after the pump beam is shut off, using a transient digitizer. In order to study the relaxation of $\langle \vec{S} \cdot \vec{I} \rangle$, which is determined by the average state population difference between the two hyperfine levels, the laser is linearly polarized and the frequency is tuned so that the absorption is predominantly due to the $(F=2\ 3^2S_{1/2}) \rightarrow (3^2P_{1/2})$ transition. In the expression "$\langle \vec{S} \cdot \vec{I} \rangle$" $\vec{S}$ is the electron spin and $\vec{I}$ is the nuclear spin. In order to measure the relaxation of $\langle S_z \rangle$, the longitudinal electron spin polarization, the beams are circularly polarized by quarter wave plates and the laser frequency is tuned midway between the $(F=1\ 3^2S_{1/2}) \rightarrow (3^2P_{1/2})$ and the $(F=2\ 3^2S_{1/2}) \rightarrow (3^2P_{1/2})$ transitions so that the absorption from the two ground hyperfine sublevels is the same.

THEORY

When an atom from the vapor strikes the target wall, it to be adsorbed for a brief time. Typically an adsorbed atom in our experiment vibrates at a site where it is bound for a time on the order of 10^{-10} sec and then either jumps to another site or is desorbed. Bouchiat and Brossel[12] determined that the time-dependent magnetic fields generated by these motions cause relaxation by inducing transitions among the different Zeeman sublevels. The probability of inducing a particular transition depends on the frequency spectrum of the modulation. The frequency dependence of the moduation is taken as

$$j(\omega) = (1 + \omega^2 \tau_c^2)^{-1} \qquad (1)$$

where ω is the angular frequency and τ_c is the correlation time of the interaction. The relaxation rate for an isolated spin $\vec{S}$ in a magnetic field is simply

$$T_s^{-1} = C\ j(\omega_s) \qquad (2)$$

where ω_s is the Larmor frequency of the spin. The coefficient C, the amplitude of the relaxation, is given by:

$$C = \frac{2}{3} \frac{\tau_s}{\tau_s + \tau_v} \gamma_s^2 h^2 \tau_c \qquad (3)$$

where τ_s is the average time an atom remains on the surface at each wall collision, τ_v is the flight time between wall collisions, γ_s is the electron magnetic moment and h is the strength of the magnetic field producing the relaxation. The factor 2/3 is due to the fact that only the components of h perpendicular to the static magnetic field induce relaxation. The time the atom remains on the surface due to physical adsorption, τ_s, is given by:

$$\tau_s = \tau_o \exp(\frac{E}{kT}) \qquad (4)$$

where τ_o is the high temperature limit of τ_s (typically $\tau_o \sim 10^{-12}$ sec), E is the adsorption energy, and kT is the thermal energy. The time the atom remains on the surface τ_s, is much less than the flight time between wall collisions τ_v.

Hyperfine structure and external static magnetic fields make the relaxation more complicated. The evolution of the state populations is given by:

$$\frac{d}{dt} N_i = CR_{ij} N_j \qquad (5)$$

where the N_i is the population of state i and CR_{ij} is the transition probability between states i and j. Therefore R_{ij} is directly proportional to the square of the matrix element of the scalar product of the alkali magnetic moment and the modulated magnetic field between states i and j. Determining R_{ij} involves calculations of both the magnetic field dependent energy splittings of the states and the magnetic field dependence of the wave functions. Equation 5 is an eigenvalue problem. The solutions are weighted sums of exponentials, with the eigenvalues as the decay rates of the exponentials. Once the state populations are known other quantities such as $\langle S_z \rangle$ or $\langle \vec{S} \cdot \vec{I} \rangle$ can be calculated. Bouchiat and Brossel[12] have derived analytic expressions for the relaxation rates of the observables $\langle \vec{S} \cdot \vec{I} \rangle$ and $\langle S_z \rangle$ in a weak magnetic field for an atom with J = S = 1/2.[12] The time dependence of $\langle \vec{S} \cdot \vec{I} \rangle$ and $\langle S_z \rangle$ are given by

$$\langle \vec{S} \cdot \vec{I} \rangle = \langle \vec{S} \cdot \vec{I} \rangle_o \exp(-t/T_{wh}) \qquad (6)$$

where $\langle \vec{S} \cdot \vec{I} \rangle_o$ is the initial expectation value for $\langle \vec{S} \cdot \vec{I} \rangle$ and where T_{wh}^{-1} is the relaxation rate for $\langle \vec{S} \cdot \vec{I} \rangle$, and

$$\langle S_z \rangle = \left\{ \langle S_z \rangle_o - \frac{2}{(2I+1)^2 - 2} \langle I_z \rangle_o \right\} \exp(-t/T_e)$$

$$+ \frac{2}{(2I+1)^2 - 2} \langle I_z \rangle_o \exp(-t/T_n) \qquad (7)$$

where $\langle S_z \rangle_0$ is the initial expectation value for S_z, $\langle I_z \rangle_0$ is the initial expectation value for I_z, and T_e^{-1} and T_n^{-1} are the relaxation rates of the two exponentials for the relaxation of $\langle S_z \rangle$. Thus for low magnetic fields $\langle \vec{S} \cdot \vec{I} \rangle$ varies as a single exponential where as $\langle S_z \rangle$ varies as the sum of two exponentials. The relaxation rates are given by

$$T_{wh}^{-1} = Cj(\omega_{hfs}) \tag{8}$$

$$T_e^{-1} = C\left\{ \frac{j(\omega_f) - j(\omega_{hfs})}{(2I+1)^2} + j(\omega_{hfs}) \right\} \tag{9}$$

$$T_n^{-1} = \frac{C}{(2I+1)^2} \{ j(\omega_f) + j(\omega_{hfs}) \} \tag{10}$$

where $\omega_{hfs}/2\pi$ is the hyperfine frequency and $\omega_f/2\pi$ is the Zeeman transition frequency.

The loss of atoms in the polarized state also occurs through polarized atoms effusing from the target and being replaced by unpolarized atoms from the sodium reservoir. The rate at which this occurs T_p^{-1} is determined from the temperatures and conductances. Spin exchange collisions between Na atoms causes relaxation of $\langle \vec{S} \cdot \vec{I} \rangle$ at a rate T_{se}^{-1}.

ANALYSIS

Experimentally we monitor the transmission of the probe beam as a function of the time after the pump beam is shut off. We call the observed initial relaxation rate of the vapor T_{1e}^{-1}. In order to obtain a good signal to noise ratio, a large polarization is obtained by optical pumping. In order to obtain the true initial relaxation rate T_1^{-1} from T_{1e}^{-1} a correction is made to account for the changing optical thickness of the vapor.[13] With the absorption correction applied the relaxation rate is relatively independent of the percent absorption of the laser beam.

The relaxation rate for $\langle \vec{S} \cdot \vec{I} \rangle$ is given by $T_1^{-1} = T_{wh}^{-1} + T_p^{-1} + T_{se}^{-1}$. The initial relaxation rate for $\langle S_z \rangle$ is given by $T_1^{-1} = T_{wz}^{-1} + T_p^{-1}$ where T_{wz}^{-1} is the initial wall relaxation rate for $\langle S_z \rangle$. In the expression for T_1^{-1} for $\langle S_z \rangle$ we have ignored the effects of spin exchange since we find experimentally that these effects are small.

RELAXATION RATE VS TIME

Figure 1 shows the relaxation time T_{wh} for $\langle \vec{S} \cdot \vec{I} \rangle$ as a function of the time of exposure to Na vapor at 225°C for a copper liner coated with dryfilm. The main feature to be noted is that the

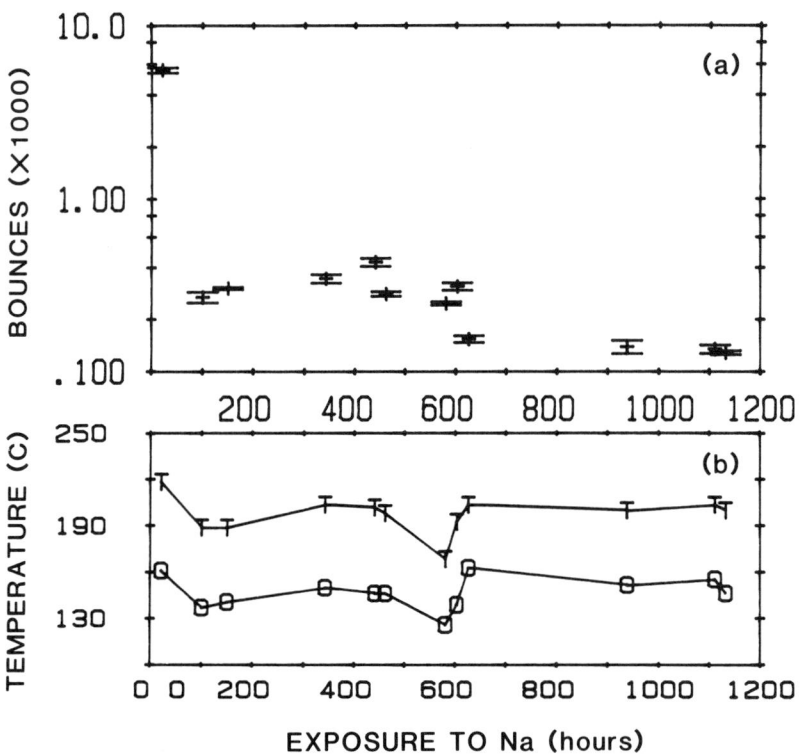

Fig. 1. (a) Relaxation time T_{wh} for $\langle \vec{S} \cdot \vec{I} \rangle$ as a function of time of exposure to Na at 225 C for dryfilm on copper. The relaxation time has been converted into the number of wall collisions before relaxation and has been corrected for spin exchange. During the time of the experiment the target temperature was kept at or below 225 C and the sodium density was kept at or below 10^{11} atoms/cm^3. (b) The temperature "T" of the target and the temperature "O" of the Na oven during the experiment.

relaxation time initially is longer than 1000 wall collisions and decreases rapidly to 400 wall collisions after which the rate of decrease is much smaller. Increasing the wall temperature causes the decrease to occur more rapidly. At 275 C the relaxation time is diminished to one bounce in one week, and at a temperature of 325 C the walls become completely depolarizing in one day. This is probably due to chemical reaction of the walls with the Na vapor.

The relaxation time for $\langle S_z \rangle$ decreases with time in a manner that is similar to the decrease in the relaxation time for $\langle \vec{S} \cdot \vec{I} \rangle$.

RELAXATION RATE VS WALL TEMPERATURE

The relaxation rates for $\langle S_z \rangle$ and $\langle \vec{S} \cdot \vec{I} \rangle$ increase with increasing wall temperature for a dry film surface. After a heating and cooling cycle the relaxation rate is greater than before the cycle. This indicates that the Na reacts chemically with the wall and the process is irreversible.

RELAXATION RATE AS A FUNCTION OF NA VAPOR DENSITY

We have measured the relaxation rate $T_1^{-1} = T_{wh}^{-1} + T_{se}^{-1} + T_p^{-1}$ for $\langle \vec{S} \cdot \vec{I} \rangle$ as a function of the average Na density $\bar{N}$ in the target. The relaxation rate T_{wz}^{-1} and $\langle \sigma_{se} v \rangle$, the thermal average of the Na-Na spin exchange cross section σ_{se} times the relative velocity v, are obtained by making a linear least-squares fit of $T_1^{-1} - T_p^{-1}$ as a function of $\bar{N}$ in order to obtain the slope, $\frac{\Delta(T_{se}^{-1})}{\Delta \bar{N}} = \langle \sigma_{se} v \rangle$, and the intercept, T_{wh}^{-1}. We obtain $\sigma_{se} = 1.6 \pm 0.5 \times 10^{-14} cm^2$ at a thermal velocity corresponding to 300 C. The error corresponds to a standard deviation in the data and does not include possible systematic errors such as may arise from uncertainties in the Na vapor pressure as a function of the temperature. Similar data for the relaxation of $\langle S_z \rangle$ shows that the relaxation of $\langle S_z \rangle$ is relatively independent of the density.

RELAXATION RATE AS A FUNCTION OF MAGNETIC FIELD STRENGTH

The measurement of the relaxation rate of $\langle S_z \rangle$, T_{wz}^{-1} as a function of the magnetic field allows the determination of τ_c. τ_c can also be determined either from the ratio T_n/T_e or from the ratio T_{wh}/T_{wzl}, where T_{wzl}^{-1} is the initial rate of relaxation rate of $\langle S_z \rangle$ in a low magnetic field. In order to obtain τ_c from our experimental data we must know how T_n, T_e, T_{wh}, and T_{wz} depend on τ_c. For T_n, T_e, and T_{wh} the dependence on τ_c is given by Eqns. 8-10. We determine the dependence of T_{wz} on τ_c and the magnetic field strength by solving Eqn. 5. In general the relaxation $\langle S_z \rangle$ is the sum of several exponentials. A theoretical determination of the relaxation rate T_{wz}^{-1} is made from Eqn. 5 by calculating the time dependence of $\langle S_z \rangle$ and performing a least-squares fit of a single exponential over the time interval $t=0$ to $t < T_{wz}/4$ as is done with the experimental data. The calculations are made assuming the initial condition that all the atoms are pumped into the state $M_I = 3/2$ $M_S = 1/2$ when pumped by σ^+ light (or the state $M_I = -3/2$ $M_S = -1/2$ when pumped by σ^- light).

We have measured the magnetic field dependence of T_{wz}^{-1} on different dryfilm surfaces that were exposed to Na vapor for variable amounts of time. These measurements show a variety of

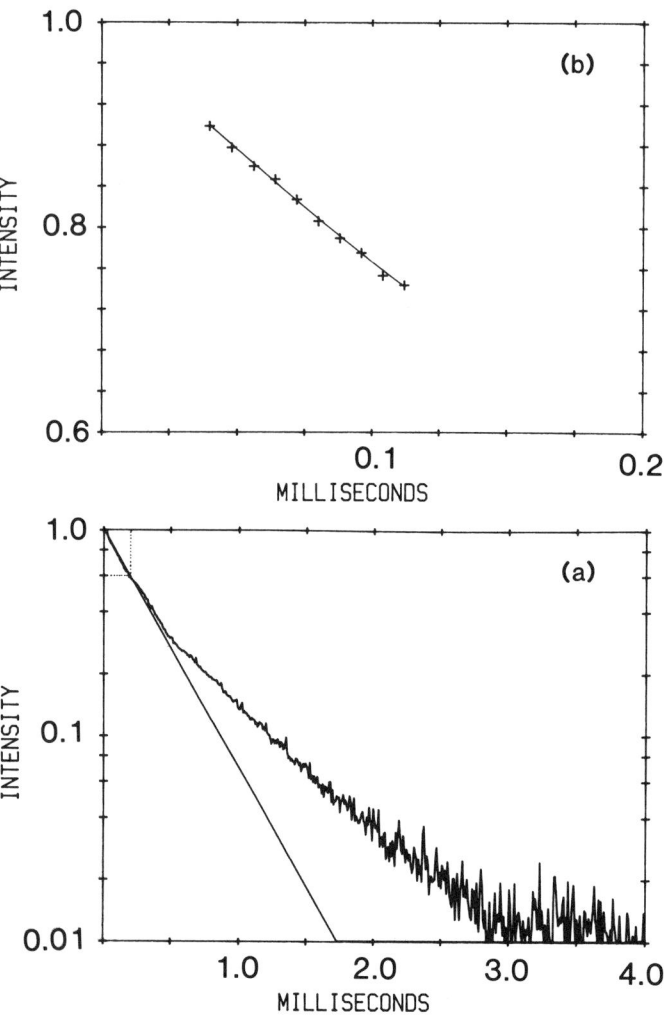

Fig. 2. (a) Semi-log plot of experimental data showing the relative transmission of the probe beam as a function of time for $\langle S_z \rangle$ pumping. The signal is determined by $\langle S_z \rangle$ which relaxes as the sum of two exponentials. The straight line is a least-squares fit of the data from which T_{1e}^{-1} and hence T_{wz}^{-1} is determined. (b) An expanded linear plot of the data in the dotted square showing the least-squares fit in detail. The surface had been exposed to Na vapor for four weeks.

effects. Our experience is that the relaxation of polarized Na colliding with a dryfilm surface is essentially independent of whether the dryfilm is on a glass, stainless steel, or copper substrate. In our experiment, with a fresh surface, the relaxation rate is nearly equal to the the escape rate T_p^{-1}, and we are not able to measure surface relaxation effects. As the relaxation rate of the surface increases it is possible to see the magnetic field dependence of the relaxation rate. We have measured T_1^{-1} for $\langle S_z \rangle$ using Zeeman pumping as a function of magnetic field for a glass liner treated with dryfilm, which was exposed to Na vapor for 2 days. A theoretical best fit to the data is obtained by solving Eqn. 5 as a function of magnetic field with a correlation time of $\tau_c = 4.0 \times 10^{-11}$ sec. At this short exposure time the relaxation is still dominated by T_p^{-1} and we are unable to measure T_{wh}/T_{wzl} or T_n/T_e accurately enough to obtain τ_c.

After longer exposures to Na vapor, as the relaxation rate becomes larger, it is possible to obtain τ_c by measuring the ratio T_{wh}/T_{wzl} the ratio T_n/T_e, or by measuring the magnetic field dependence of T_{wz}^{-1}. Figure 2 shows the relaxation of $\langle S_z \rangle$ in a low magnetic field for a copper surface coated with dryfilm after 4 weeks exposure to Na vapor. This figure shows that the relaxation is the sum of two exponentials. Also shown is a least-squares fit from which T_{1e}^{-1} and hence T_{wz}^{-1} is determined. A least-squares fit of the data is made to the sum of two exponentials in order to obtain the ratio T_n/T_e. We obtain $T_n/T_e = 4.8$ from which we determine $\tau_c = 1.3 \times 10^{-10}$ sec. Figure 3 shows corresponding data for the relaxation of $\langle \vec{S} \cdot \vec{I} \rangle$ which relaxes with a single exponential. For this surface we measure $T_{wh}/T_{wzl} = 1.0$ from which we obtain $\tau_c = 2.2 \times 10^{-10}$ sec. We have also measured T_1^{-1} for $\langle S_z \rangle$ as a function of the magnetic field for this 4 week old surface. An analysis of this data yields $\tau_c = 1.5 \times 10^{-10}$ s. The agreement between these three means of measuring the correlation time is good. The correlation time is longer by a factor of 3-6 than $\tau_c = 4 \times 10^{-11}$ sec measured on the fresher surface.

CONCLUSIONS FROM RELAXATION STUDIES

Our data is consistent with a single weak fluctuating magnetic interaction causing the relaxation. It appears that the surface is progressively damaged with time in such a way that the correlation time becomes longer and $\tau_s h^2$ becomes larger. We estimate that the adsorption time for Na on dryfilm is about 0.1 of the adsorption time measured by Bouchiat and Brossel for Rb on parafin. The estimated adsorption time is probably smaller than the time associated with an atom jumping from site to site on the surface. If the adsorption time is less than the jumping time then $\tau_c = \tau_s$ which is an upper limit on the correlation time. Under these conditions the amplitude of the interaction C would be proportional to the square of $\tau_c = \tau_s$, which is consistent with our data. If $\tau_c = \tau_s$ the measured values of C and τ_c yield a value of h = 15 mT for the strength of the interaction for both the data of Fig. 2 and the data of Fig. 3.

EFFECT OF RADIATION TRAPPING

During the past year we have carried out theoretical calculations on the limitations imposed by radiation trapping on the polarization that can be produced by optical pumping an alkali vapor in a high magnetic field.[14] Trapping occurs when the alkali vapor is sufficiently dense that multiple scattering of the light is important for one or more radiative decay branches of the vapor.[15]

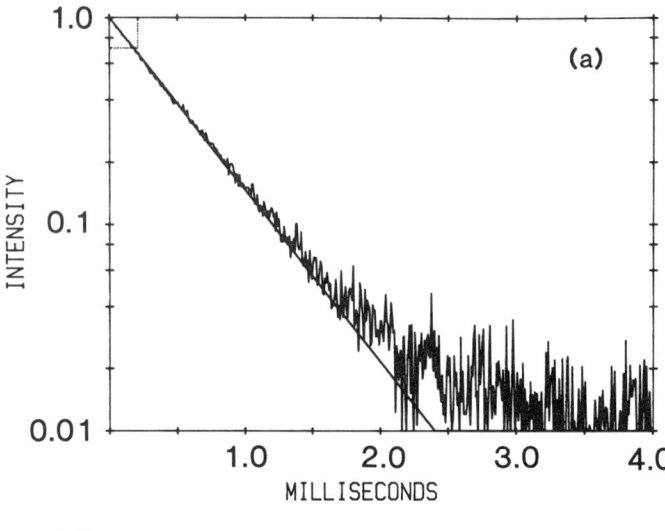

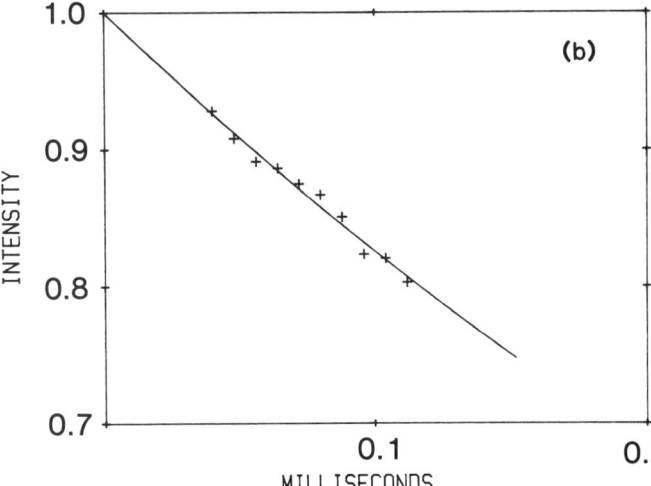

Fig. 3. (a) Semi-log plot of experimental data showing the relative transmission of the probe beam as a function of time for $\langle \vec{S} \cdot \vec{I} \rangle$ pumping. The signal is determined by $\langle \vec{S} \cdot \vec{I} \rangle$ which relaxes as a single exponential. The straight line is a least-squares fit of the data from which T_{1e}^{-1} and hence T_{wh}^{-1} is determined. (b) An expanded linear plot of the data in the dotted square showing the least-squares fit in detail. These data were obtained the same day as the data of Fig. 2.

We find that it is possible to obtain high polarization using optical pumping with a total optical intensity per Hz of 2.55×10^{-6} W/m^{2}Hz (which corresponds to a power of 1W distributed over a diameter of 0.015m and over a frequency spread about equal to the Doppler line width) even for densities of 10^{19}/m^3 if the relaxation time is 150 μs (which corresponds to 10 wall collisions) and higher densities are possible for longer relaxation times. For shorter relaxation times high polarization can not be obtained in a dense vapor even for infinite laser power. Thus a surface with a small relaxation rate such as silicone is essential for obtaining a high polarization in a dense optically pumped alkali vapor.

REFERENCES

1. Y. Mori, K. Ikegami, Z. Igarashi, A. Takagi, and S. Fukumoto, Polarized Proton Ion Sources, TRIUMF, G. Roy and P. Schmor, eds., 117 (AIP New York) p.123.
2. A. N. Zelenskii, S. A. Kokhanouskii, V. M. Lobashev, and V. G. Polushkin, JETP Lett. $\underline{42}$, 5 (1985); see also A. N. Zelenskii, S. A. Kokhanouskii, V. M. Lobashev, V. G. Polushkin, and K. N. Vishevskii, Helvetica Phys. Acta. $\underline{59}$, 681 (1986).
3. C. D. P. Levy, M. McDonald, P. W. Schmor, and J. Uegaki, Helvetica Phys. Acta $\underline{59}$, 674 (1986).
4. R. L. York, private communication.
5. E. K. Zavoiskii, Soviet Physics JETP $\underline{5}$, 338 (1957).
6. W. Haeberli, Proc. 2nd Intl. Symp. on Pol. Phenomena in Nucl. Reat. (P. Huber and H. Schopper, Eds.) Birkuauser Basel 1966, p. 64.
7. L. W. Anderson, Nucl. Instr. and Methods $\underline{167}$, 363 (1979); see also L. W. Anderson, IEEE Trans. on Nucl. Sci. $\underline{NS30}$, 1051 (1983).
8. M. S. Feld, M. M. Burns, T. V. Kuhl, P. G. Pappas, and D. E. Murnick, Optics Letters $\underline{5}$, 79 (1980); see also P. G. Pappas, M. M. Burns, D. D. Hinshelwood, M. S. Feld, and D. E. Murnick, Phys. Rev. A $\underline{21}$, 1955 (1980).
9. W. D. Cornelius, D. J. Taylor, and R. L. York, and E. A. Hinds, Phys. Rev. Letters $\underline{49}$, 870 (1982).
10. P. G. Sona, Energia Nucl. $\underline{14}$, 295 (1970).
11. D. R. Swenson and L. W. Anderson, Nucl. Instr. and Methods $\underline{B12}$, 157 (1985); see also D. R. Swenson and L. W. Anderson, Helv. Phys. acta $\underline{59}$, 652 (1986).
12. M. A. Bouchiat and J. Brossel, Phys. Rev. $\underline{147}$, 41 (1966).
13. H. M. Gibbs and R. J. Hull, Phys. Rev. $\underline{153}$, 132 (1967).
14. D. Tupa, L. W. Anderson, D. L. Huber, and J. E. Lawler, Phys. Rev. A $\underline{33}$, 1045 (1986); see also D. Tupa, L. W. Anderson, D. L. Huber and J. E. Lawler, Helvetica Phys. Acta $\underline{59}$, 657 (1986).
15. W. Happer, Rev. Mod. Phys. $\underline{44}$, 169 (1972).

Supported by Department of Energy Grant #DE-FG02-86ER 40266.

DISCUSSION

<u>Mori</u>: What is your time resolution of your measurement of the relaxation time?

<u>Anderson</u>: The system we have is a transient digitizer, and we can resolve about 10^{-6} seconds.

<u>Mori</u>: Did you see any fast relaxation time in your dense targets on sodium atoms in high density?

<u>Anderson</u>: Yes. In low field, you see a situation that looks something like this, with two relaxation times. If you go to a higher field, you see a relaxation that has got seven relaxation times in it. Needless to say, it's not possible to resolve all these, but some of them are pretty fast. What we have done, in this curve that I showed here of the relaxation as a function of B-field, we have taken the initial slope of that, which is the highest relaxation we see. So this is the fast part of the relaxation that I'm showing you here. There are slower components to it.

<u>Peterson</u>: From a novice - what are your definitions of a low field and a high field?

<u>Anderson</u>: Low field is a field that is appreciably below the field that decouples i and j in the ground state, and high field is a field that is appreciably above it. The field that corresponds to just having a field that's sort of exactly equal to the field that the electron puts on the nucleus, in the case of sodium, is on the order of 560G, 570G. I'm sorry, I don't remember the exact number. So low field is appreciably below that; high field is appreciably above that.

RECENT PROGESS ON THE OPTICALLY PUMPED POLARIZED H⁻ ION SOURCE AT KEK

Y. Mori, A. Takagi, K. Ikegami and S. Fukumoto
KEK, National Laboratory for High Energy Physics
Oho-machi, Tsukuba-gun, Ibaraki-ken, Japan, 305

A. Ueno
Department of Nuclear Engineering, Kyushu University
Fiukuoka-ken, Japan, 810

·C.D.P. Levy and P.W. Schmor
TRIUMF, 4004 Wesbrook Mall
Vancouver, B.C., Canada, V6T 2A3

ABSTRACT

Recent progress on the optically pumped polarized H⁻ ion source is described. Beam current of 50 µA with nuclear polarization of 56 ± 5% has been obtained with an intense pulsed dye laser. Experimental results for radiation trapping are also described.

INTRODUCTION

The optically pumped polarized ion source has been developed at KEK for the 12 GeV proton synchrotron[1]. The principle of this type of polarized ion source was proposed by Haeberli[2] and its actual feasibility was examined by Anderson[3]. This source is based on the electron-capture reactions of H⁺ ions with optically pumped electron polarized sodium atoms. In order to get a high beam current with high polarization, it is important to obtain a thick sodium target with high polarization. Until recently, the optically polarized H⁻ ion source at KEK used three cw dye lasers to produce ~ 50% polarization in sodium vapor of thickness ~ $1 \times 10^{14} \times$ atoms/cm². The source is pulsed at a repetition rate of 20 Hz with a pulse width of ~ 70 µs. If a pulsed dye laser were used at the ion source pulse rate, the very high peak intensity could possibly result in a sodium target polarization near 100% during the ion source pulse[4]. With that in mind, we built a flashlamp-pumped pulsed dye laser, which can produce two orders of magnitude more peak power than cw lasers[5]. So far, we have obtained 50µA H⁻ ion current with 56 ± 5% polarization from the ion source using this pulsed dye laser.

© American Institute of Physics 1987

Radiation trapping reduces the efficiency of optical pumping at high atomic densities. Recently, Tupa et al.[6] examined the effect of radiation trapping theoretically and proposed new rate equations for three-level optical pumping, having additional radiation trapping terms. In order to test their equations we made measurements with the pulsed dye laser, which was useful for these experiments because its high power saturated the polarization even at high atomic densities where radiation trapping occurred. The results of the experiments are in this paper.

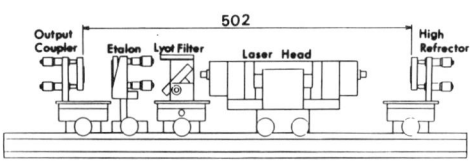

Fig.1. Setup of the pulsed dye laser.

STATUS AND OPERATION WITH THE PULSED DYE LASER

The pulsed dye laser uses two Xe flashlamps which pump R6G dissolved in ethanol flowing through a pyrex tube. The power supply uses a resonant charging circuit in the PFN. The wavelength of the laser light is tuned with a three-plate birefringent filter and a 1.1 mm thick coated etalon. The setup of the pulsed dye laser is shown in Fig. 1. The peak power output ranges from 300-500 W and the pulse width is 60 μsec. The laser bandwidth is about 28 GHz.

In order to measure the polarization of sodium atoms, a Faraday rotation technique[5,7,8] was used. A linearly polarized probe beam, passing through the sodium cell, underwent Faraday rotation in the presence of sodium vapor and also a further rotation if the sodium was polarized. The time-dependent Faraday rotation was detected as the variation in signal from a photomultiplier, which detected the probe light transmitted by a linear polarizer. Figure 2 shows the photomultiplier signal is flat-topped, indicating that the polarization is saturated (at ~ 100% in this example). The measured sodium polarization as a function of the sodium atomic density, at a

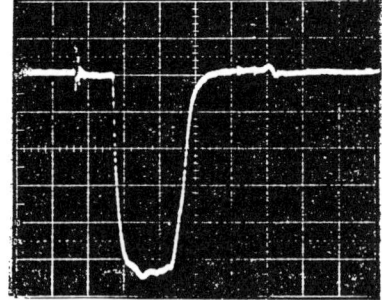

Fig.2. Photomultiplier signal of Faraday rotation. H: 50μs/div., V: 5mV/div.

peak laser power of 370 W, is shown in Fig. 3. Above a density of ~ 1 × 10^{13} atoms/cm^3 the polarization was usually still saturated at a level less than 100%. This effect is well explained by radiation trapping and is described later. When three cw dye lasers were used to optically pump the sodium, the polarization at an atomic density of 1 × 10^{12} atoms/cm^3 was about 90%, decreasing to 40% at 5 × 10^{12} atoms/cm^3. As can be seen from Fig. 3, a polarization of 90% at an atomic density of 5 × 10^{12} atoms/cm^3 was achieved using the pulsed laser. Figure 4 shows the polarized H$^-$ ion current as a function of the sodium cell temperature, measured at an energy of 750 keV (open circles) after acceleration by the Cockcroft preaccelerator and the 40 MeV linac respectively. The polarization measured at 40 MeV by p-C reactions is also shown. The H$^-$ ion beam current of 50 μA at a polarization of 56 ± 5% was obtained at a cell temperature of 230°C. Raising the cell temperature to 245°C increased the H$^-$ ion current to 180 μA, although the polarization decreased to 43 ± 5%.

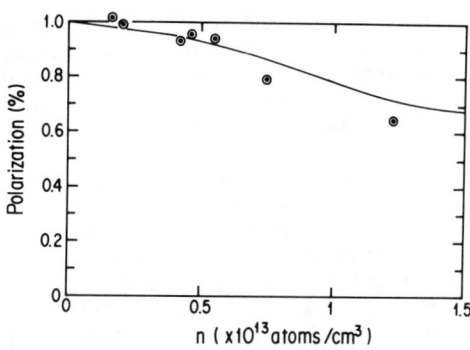

Fig.3. Measured (circles) and calculated (solid line) polarization of the optically pumped sodium vapour.

RADIATION TRAPPING

Recently, Tupa et al. examined theoretically the effect of radiation trapping in optically pumped sodium vapor[6]. According to their calculations, the polarization of optically pumped sodium atoms in thick targets is affected to a large degree by radiation trapping.

The polarization of the sodium atoms calculated according to Tupa et al. is shown in Fig. 3 as a solid line. As can be seen, the experimental results show good agreement with the calculated curve.

An interesting characteristic of radiation trapping appears in the polarization relaxation process. In a thick target, photons emitted by excited atoms are usually re-absorbed by ground

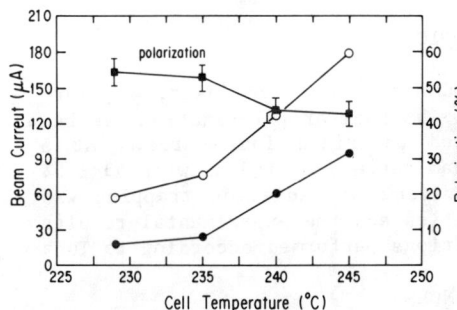

Fig.4. Beam intensities and polarizations of the accelerated H$^-$ beams. Open circles show the beam currents at 750 keV and closed ones the beam currents at 40MeV, respectively.

state atoms and are confined by the system. This leads to an increase in the proportion of excited atoms and affects ground state polarization relaxation. Therefore, at high densities two different polarization relaxation rates can be seen, one rapid rate due to excited state decay, and a slower rate due to wall collisions.

The calculated time-dependent polarization and the corresponding Faraday rotation signal derived from the polarization are shown in Fig. 5. In this calculation, we assumed a relaxation time due to wall collisions of 250 μs, a laser beam filling the cell volume, and a trapezoidel laser pulse shape. As can be seen, the calculated curves show two different polarization relaxation rates. In a preliminary experiment using silicon dry film coating on the sodium cell walls, we observed two different relaxation rates very similar to the calculated rates.

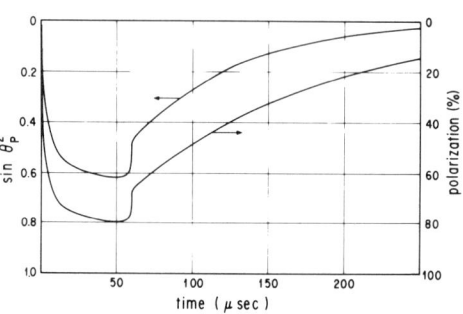

Fig.5. Calculated time-dependent polarization (lower) and the corresponding Faraday rotation signal (upper).

We have thus confirmed the validity of the assumptions used by Tupa et al. in their equations predicting the effects due to radiation trapping in alkali vapor cells. This will remove many uncertainties in the design of alkali vapour cells and in the polarization attainable.

CONCLUSION

Using an intense flashlamp-pumped pulsed dye laser, a polarized H⁻ ion beam of 50 μA with 56 ± 5% nuclear polarization has been obtained from the optically pumped polarized ion source. At a higher sodium atomic density, a beam current of 180 μA with 43 ± 5% polarization was achieved. The effect of radiation trapping was also measured at high atomic densities and the experimental results showed good agreement with calculations performed according to Tupa et al.

REFERENCES

1. Y. Mori et al., AIP, New York, 123 (1984).
2. W. Haeberli, Proc. 2nd. Int. Symp. Polarization Phenomena in Nuclear Reactions, 64 (1966).
3. L.W. Anderson, Nucl. Inst. Meth., 167, 363 (1979).
4. A.N. Zelenskii et al., J. Phys. Soc. Jpn. 55, 1064 (1986).
5. Y. Mori et al., KEK Report 86-2, 1986.
6. D. Tupa et al., Phys. Rev. A33, 1045 (1986).
7. W.D. Cornelius et al., Phys. Rev. Lett., 49, 870 (1982).
8. Y. Mori et al., Nucl. Instr. Meth., 220, 264 (1984).

DISCUSSION

Peterson: Could it be that the difference between your measured cross-sections and the theory has to do with the velocity of your atoms?

Mori: The theory says that the cross-section is gradually decreased as a function of the energy. So we averaged within the range of the thermal velocities.

Peterson: But do you know that your atoms are thermal?

Mori: Of course. But there are two predictions, especially the Dalgarno, that predict the thermal atom cross-section.

Anderson: I don't have a question, but I'll just comment on Jim Peterson's question. We make calculations of spin exchange cross-sections at higher energies, and in order to get as low as 3×10^{-15}, at least in our cross-sections, you'd have to be at least at 100 eV of energy, which these wouldn't even approach. I have to say that our cross-sections, when we extrapolate, really weren't good at these low energies. But nevertheless, when we extrapolate them back they do agree closely with the numbers that Olson had.

STATUS OF THE TRIUMF OPTICALLY PUMPED POLARIZED H⁻ ION SOURCE

M. Law, C.D.P. Levy, M. McDonald, P.W. Schmor and J. Uegaki
TRIUMF
4004 Wesbrook Mall, Vancouver, B.C., Canada V6T 2A3

ABSTRACT

The optically pumped polarized H⁻ ion source at TRIUMF is capable of producing ≈ 10 µA of H⁻ at a polarization of ≈ 65%, within a normalized emittance at the 60% contour level of 0.4 Π.mm.mrad for an ionizing field of 1.5 kG. Installation of the ion source on the TRIUMF cyclotron has begun and is scheduled to be completed by the end of 1987.

INTRODUCTION

TRIUMF presently uses a Lamb-shift polarized H⁻ source capable of providing ~1 µA of ~ 75% polarized beam on target. Since 1983 an optically pumped source has been under development at TRIUMF. This is expected to eventually produce intense d.c. H⁻ beams (~ 50 µA) at a polarization of $\geq$ 70%, with an emittance suitable for injection into the cyclotron. To date, the optically pumped source has been shown to produce ~10 µA of H⁻ at a polarization of ~65% within a normalized emittance of 0.4 Π.mm.mrad, at an ionizing field of 1.5 kG and work has begun on attaching the source to the cyclotron.

Optically pumped H⁻ ion sources have been successfully demonstrated in pulsed operation by Mori et al.[1] at KEK, using an electron cyclotron resonance (ECR) proton source, and by Zelenskii et al.[2] at INR. The TRIUMF source is the only existing d.c. optically pumped source.

The optical pumping technique as proposed by Anderson[3] is as follows. Circularly polarized dye laser light tuned to the sodium D_1 transition is used to electron polarize ground state sodium atoms in an optically thick vapour. An electron-spin polarized atomic hydrogen beam is produced by passing protons through the polarized sodium vapour, where charge exchange occurs. A diabatic field reversal technique, similar to that used in Lamb-shift sources, transfers the electron polarization to the nucleus. Charge exchange in a second unpolarized alkali vapour cell yields a nuclear polarized H⁻ beam. Figure 1 shows the layout of the TRIUMF source and the optics to match the 5 keV beam to a 300 kV acceleration column.

ECR PROTON SOURCE

A hydrogen plasma is produced in the multi-mode ECR cavity by up to 400 W absorbed cw microwave power at 28 GHz, from a Varian

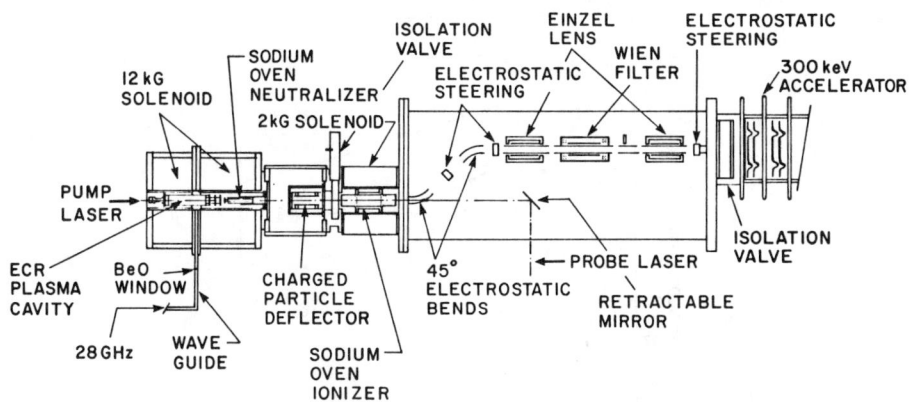

Fig. 1. Layout of the TRIUMF optically pumped polarized H⁻ ion source.

Extended Interaction Oscillator (Model VKQ-2H35F). The extraction electrodes and polarized sodium vapour cell are located in a ~ 12 kG axial magnetic field. The field has a mirror configuration with a minimum at 8 kG, where the microwave power is fed in radially to avoid resonance at 10 kG. Hydrogen gas is fed in through the same waveguide, and with a quartz liner in the ECR cavity the proton ratio, $[H^+/(H^+ + H_2^+ + H_3^+)]$, is greater than 0.75. The current density extracted at an energy of 5 keV from the water-cooled accel-decel multi-aperture molybdenum electrodes increases linearly with absorbed microwave power to a maximum of 120 mA/cm^2, limited by the available power. Earlier results showed that up to 300 mA/cm^2 could be extracted through a relatively small 2 mm diameter hole.

SODIUM POLARIZATION

The sodium in the neutralizing cell is polarized by circularly polarized light at up to 1 W at 5896 Å from a Coherent CR - 590 broadband dye laser, and the polarization is measured using a Faraday rotation technique.[4,5] The polarization depends on the rate of optical pumping by the laser light and the depolarization rate due to wall collisions. Figure 2 shows the improved polarization produced by narrowing the laser bandwidth from a nominal 30 GHz to ~ 6 GHz with an uncoated 0.5 mm thick intra-cavity etalon. The increase is due to the increased spectral power density of the laser light within the 3.0 GHz Doppler width of the sodium D_1 transition. The polarization is also improved by a Viton wall, instead of metal, since on average a sodium atom undergoes ~ 15 wall collisions in the former case[6,7] before depolarization. A more satisfactory "dry film" coating[5] is currently being tested at TRIUMF. The measurements in Fig. 2 were done using a cell 1.3 cm in diameter and 7 cm in length, and the ion beam quickly destroyed the Viton. A new cell 3.0 cm in diameter with 0.8 cm entrance and exit apertures has been

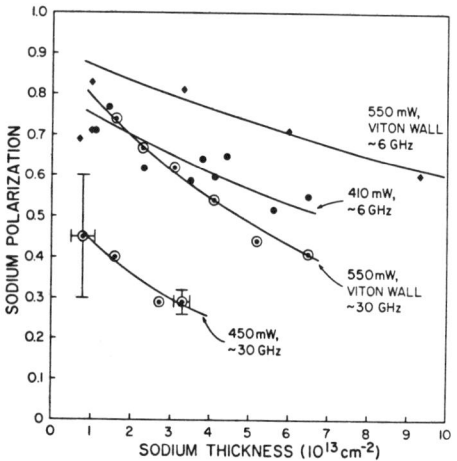

Fig. 2. Sodium polarization vs. thickness.

built, in order to protect the wall coating from the ion beam. Further experiments and calculations[8,9] show that a sodium polarization near 90% at an appropriate density of ~ 4×10^{12} atoms/cm^3 can be attained with a laser power of a few watts over the sodium absorption bandwidth, even with metallic walls. Zelenskii[2] has found similar high polarization levels at higher spectral power densities. Other work[2,10,11] shows that the transfer of polarization from the sodium to the hydrogen nucleus has an efficiency of ~ 65% at a field strength of 12-13 kG, compared with a calculated value of ~ 80%.[12] Therefore, a nuclear polarization of > 65% can be achieved in the TRIUMF source, assuming near 100% efficiency of polarization transfer from the hydrogen electron to the nucleus. Increasing the magnetic field in the neutralizer cell region will increase the polarization transfer between sodium and hydrogen.

BEAM CURRENT AND EMITTANCE

Using a multi-aperture electrode configuration, 8 mA of proton current was transmitted through the 8 mm neutralizer cell aperture, in the absence of sodium vapour. The proton current decreased to near zero as the sodium cell temperature was raised, as an increasing proportion of the beam was neutralized. Other results indicated a neutral beam current of 220 µA particle equivalent downstream of the ionizer cell, as shown by both secondary electron emission and calorimetry, at a neutralizer cell sodium thickness of 5×10^{13} atoms/cm^2. Given an equilibrium H$^-$ yield of 7% at 5 keV in a thick sodium target,[13] such a neutral current corresponds to an H$^-$ current of ~ 15 µA.

The emittance of the H$^-$ beam is determined mainly by the emittance growth as the beam leaves the 1.5 kG field of the ionizer

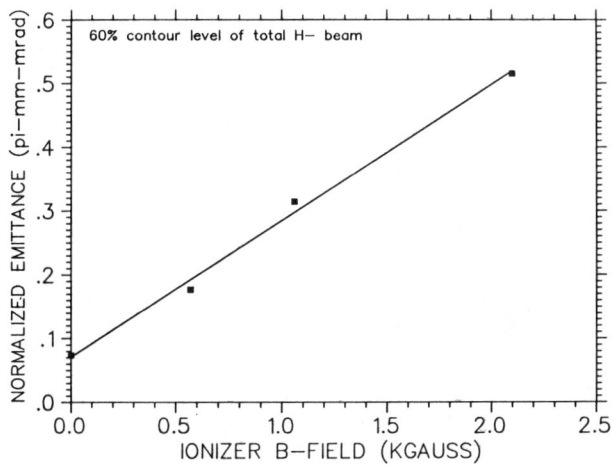

Fig. 3. H^- beam emittance vs. ionizer magnetic field.

cell. The H^- beam emittance was measured as a function of ionizer magnetic field using a slit scanner of a type developed at Los Alamos, and the results are shown in Fig. 3. As can be seen, the H^0 beam accepted by the ionizing cell has a normalized emittance of .07 Π.mm.mrad at the 60% contour level (the same as the H^- beam at zero field), and an emittance of 0.4 Π.mm.mrad after leaving a magnetic field of 1.5 kG.

We have also measured the H^+ beam normalized emittance and brightness within the neutralizer cell region at a magnetic field of 12 kG, for different aspect ratios S=r/d, where r is the radius of the circular electrode apertures, and d is the surface to surface separation between them. Figure 4b shows the optimum brightness, which is attained at an aspect ratio of 0.25 and a corresponding normalized emittance of 0.1 Π.mm.mrad. shown in Fig. 4a. The multi-aperture electrode configuration actually used has a central aperture of 2 x 8 mm, implying an effective aspect ratio of 0.33 given that d=3 mm. Increasing the beam brightness in future may necessitate reversing the direction of the optical pumping laser from its present upstream to downstream configuration.

FUTURE PLANS AND CONCLUSION

The TRIUMF source is capable of producing ~ 10 μA H^- ion beams polarized at ~ 65% within a normalized emittance of 0.4 Π.mm.mrad. at the 60% level, suitable for injection into the TRIUMF cyclotron. It is expected that suitable wall coatings on the neutralizer cell wall will make that current possible with modest laser power. The polarization can be increased by using superconducting coils around the neutralizer cell to increase the magnetic field to 20-25 kG. We are now designing steering and focusing optics for the H^- beam,

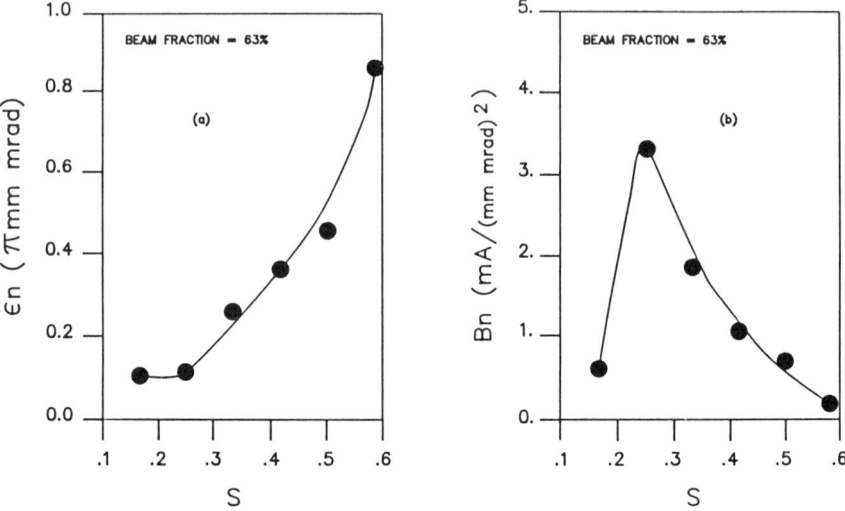

Fig. 4. The normalized emittance a) and brightness b) of the H^+ beam within the 12 kG neutralizer cell magnetic field region.

prior to scheduled installation of the source onto the cyclotron by the end of 1987.

ACKNOWLEDGEMENTS

We would like to thank Dr. Y. Mori for his considerable efforts on our behalf, and to Dr. R.R. Stevens, Jr., R.L. York, S. Kaplan, L.W. Anderson and W.D. Corelius for many helpful discussions. We also wish to thank Mr. H. Wyngaarden and Mr. P. Chigmaroff for their support.

REFERENCES

1. Y. Mori, K. Ikegami, Z. Igarashi, A. Takagi and S. Fukumoto, AIPCP 117, 123 (1984).
2. A.N. Zelenskii, S.A. Kokhanovskii, V.M. Lobashev and V.G. Polushkin, Nucl. Instrum. Methods, A245, 223 (1986).
3. L.W. Anderson, Nucl. Instrum. Methods, 167, 363 (1979).
4. W.D. Cornelius, D.J. Taylor, R.L. York and E.A. Hinds, Phys. Rev. Lett. 49, 870 (1982).
5. Y. Mori, K. Ikegami, A. Takagi, S. Fukumoto and W.D. Cornelius, Nucl. Instrum. Methods, 220, 264 (1984).
6. D.R. Swenson and L.W. Anderson, Helv. Phys. Acta 59, 652 (1986).
7. C.D.P. Levy, M. McDonald, P.W. Schmor and J. Uegaki, Helv. Phys. Acta, 59, 674 (1986).
8. Y. Mori, A. Takagi, K. Ikegami, S. Fukumoto, A. Ueno, C.D.P. Levy and P.W. Schmor, KEK Report 86-2, (1986).

9. D. Tupa, L.W. Anderson, D.L. Huber and J.E. Lawler, Helv. Phys. Acta, 59, 657 (1986).
10. Y. Mori, A. Takagi, K. Ikegami, S. Fukumoto and A. Ueno J. Phys. Soc. Jpn. 55, 453 (1986).
11. J. Uegaki, P.W. Schmor and C.D.P. Levy, Phys. Lett. A115, 216 (1986).
12. E.A. Hinds, W.D. Cornelius and R.L. York, Nucl. Instrum. Methods, 189, 599 (1981).
13. A.S. Schlachter, K.R. Stalder and J.W. Stearns, Phys. Rev. A22, 2494 (1980).

NEUTRALIZATION

NEUTRALIZER OPTIONS FOR HIGH ENERGY H⁻ BEAMS*

Joel H. Fink[†][1]
Lawrence Berkeley Laboratory
University of California
Berkeley, CA 94720

ABSTRACT

An energy spectrum of neutral beams is presented and the ranges over which various neutralizers give the most satisfactory performance is discussed.

INTRODUCTION

A neutralizer converts a negative ion beam into a neutral beam, but it also increases the beamline cost, weight and size while reducing its output power, efficiency and possibly the reliability of the entire system. In addition it scatters the newly formed neutrals, altering the beam current density distribution, causing the beam divergence to get larger and the brightness to go down.

In the following, the role of neutralizers for hydrogen ion beams is reviewed, and the problems encountered over a range of beam energies are discussed. Consideration is given to enhancing the goals of the neutral beam application, be they the highest neutral fraction, optimum overall efficiency or maximum beam brightness, etc.

Efficiency is undoubtedly the most critical parameter. No matter what beamline characteristics are considered important, a good design will provide them efficiently. Thus the maximum neutral fraction that can be obtained from a passive neutralizer, one which requires negligible additional power, is very important. The efficiency of a neutral beamline using such a neutralizer is proportional to the fraction of the incident ion beam that becomes neutral F^o i.e.;

$$n_o = n_{NEG} \cdot F^o \qquad [1]$$

where n_{NEG} is the efficiency of the ion beam from which the neutrals were formed.

For a driven neutralizer, in which the neutral fraction is a function of the power P_n needed to operate the neutralizer, the efficiency of a neutral beamline is more complex:

$$n_o = n_{NEG} \frac{F^o}{1 + (P_N - P_{ER})/P_{IN}} \qquad [2]$$

where P_{IN} is the power needed to form the the negative ion beam,

---

* This work was supported by the USASDC under Contract No. MIPR W31RPD-63-A087 through U.S. DOE Contract No. DE-AC03-76SF00098.
† On assignment from Negion, Inc., Hayward, CA 94542.

and P_{ER} is the energy recovered from the ions in the un-neutralized fraction of the beam.

In this case, there is an optimum beam line efficiency with respect to line density which is a function of the relationship between the neutral fraction and P_N. In general, this relationship is difficult to estimate because it depends upon features of the design of the neutralizer rather than fundamentals of physics. It is evident that the preferred neutral fraction will be less than the maximum.

There are applications where the divergence of the neutral beam θ_0 must be below a given maximum. As an approximation, for a neutral beam formed from an ion beam of divergence θ_{NEG}:

$$\theta_0^2 = \theta_{NEG}^2 + \theta_N^2 \qquad [3]$$

in which θ_N is the neutralizer's contribution to the divergence of the neutral beam. With non-relativistic beams, θ_N is roughly inversely proportional to the square root of the beam energy. Because θ_N is a function of the line density, or wavelength with photodetachment, of the neutralizer, it is related to the neutral fraction and the efficiency with which the neutral beamline operates. The upper limit to θ_0 might be attained at less than maximum efficiency and optimum neutral fraction.

A similar situation exists for beam brightness. On the assumption that the beam is gaussian and a peak brightness B_0 at a focus is of importance, it can be shown that:

$$B_0 = B_{NEG} \frac{F^0}{1 + (\theta_N/\theta_{NEG})^2} \qquad [4]$$

where B_{NEG} is the peak brightness of the ion beam. Once again a compromise is required to attain the highest beam brightness at the most favorable beamline efficiency.

Whereas the previous discussion was concerned with neutralizer options as they relate to the application of neutral beams, in the following options are considered in relation to the beam energy.

AT LESS THAN 50 keV

If one were to examine the energy spectrum of neutral hydrogen beams, he would find positive ions to be the preferred source of neutrals at low beam energies. Such neutral beamlines are more desirable because they are more efficient, support higher current densities and operate with higher gas efficiencies than negative ion sources. In addition the loss of positive ions at fractions of the beam line energy, resulting from charge exchange with the background gas in the accelerator, is considerably less than the corresponding loss of negative ions would be.

With respect to divergence and brightness, a comparison of the two types of neutral beamlines reflects the relative divergence and brightness of the ion beams. However, it is known that the increase in divergence caused by neutralizing negative ions is significantly less than that of positive ions.[2]

FROM 50 keV to ABOUT 80 keV

At higher energies than 50 keV, the choice of a positive ion beam as a source of neutrals is not as evident. With increasing beam energy, the efficiency of the beamline goes down. As shown in Fig. 1, the cross sections for electron attachment in a gas cell fall off, causing the optimum neutral fraction, equivalent to that of a positive ion beam passsing through a hydrogen neutralizer of infinite line density, to become less and less. See Fig. 2.

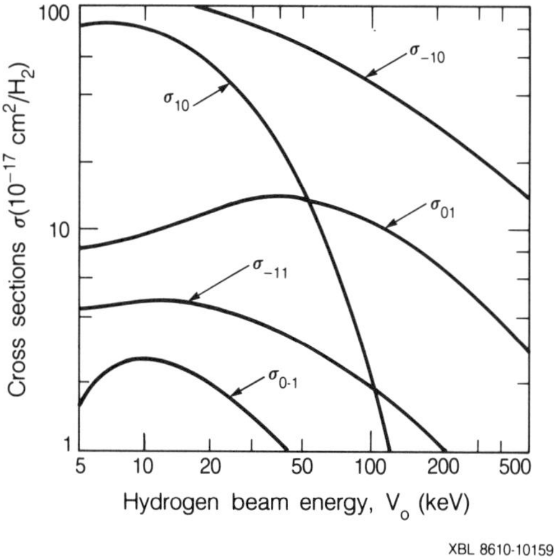

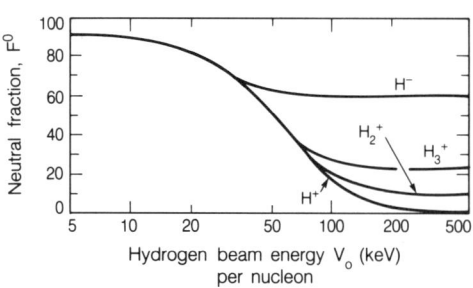

Fig. 1. Hydrogen cross sections in H_2.

Fig. 2. Maximum neutral fractions for hydrogen beams in hydrogen.

As the power consumed by the accelerator increases, the efficiency of the positive ion source becomes a smaller factor in the overall beamline performance. In addition, the decreasing neutral fraction cuts into the advantages of a higher emitted current density and the better gas efficiency a positive ion source provides. Problems arise if the application requires a mono-energetic beam of neutrals because molecular ions break up when neutralized, into neutrals of fractional beam energy. Of course the fraction of positive molecular ions extracted from the source can be minimized and those that remain can be removed before they are accelerated to any appreciable energy. In addition, the neutral beamline efficiency can be enhanced by means of energy recovery.[3] Nevertheless, there is an energy, of the order of 80 keV, above which the reduced neutral fraction makes even such measures ineffective.

On the other hand, the maximum neutral fraction of a negative hydrogen ion beam traveling through a hydrogen gas cell levels off at about 60%, and becomes relatively independent of additional increases in the beam energy. Whereas the efficiency of a negative ion beam, including the ion source, accelerator and miscellaneous services, is about 80% over a broad range of energies, the overall efficiency of a neutral beamline based on negative ions will not vary much with the beam energy and will be about 48%. Furthermore, there are no negative molecular ions to worry about.

It is also possible to take advantage of energy recovery with negative ion beams. Whereas the components of positive and negative ions are of the same order of magnitude in a beam of optimum neutral fraction, the positive ion component is considerably smaller in a neutralizer that is under dense, i.e., at less than optimum line density. Under these circumstances at maximum efficiency, the neutral fraction will be less than maximum. If the energy of both the positive and negative ions could be recovered the maximum beamline efficiency would be increased by almost 10%, while the neutral fraction remained at just about maximum.

As a result, neutral hydrogen beams formed out of negative ions of this energy range can be made competitive with beams formed of positive ions. Above 80 keV, the poor efficiency of neutral beams based upon positive ions make them impractical.

FROM 80 keV TO ABOUT 1 MeV

In this range of energy, the simple hydrogen gas cell becomes marginal and consideration must be given to alternative neutralizer designs. Problems arise from the increasing line density of the neutralizer needed to form a maximum fraction of neutrals. As the energy goes up, the neutralizer must either become longer or operate at a higher average pressure.

In practice there are limits to the neutralizer length. Depending upon the application, these could relate to its weight, but most likely to its cost. This includes not just the weight or cost of the neutralizer, but also its housing, containment structure, and magnetic shielding.

Problems originating from excessive neutralizer pressure result from the extra gas it introduces into the beamline. Because the beam losses are proportional to the background gas pressure, it is essential that the gas flow be minimal and whatever gas does get into the beamline is pumped away at an acceptable pressure. This is difficult to accomplish with a neutralizer of large diameter.

It should be recognized that the loss of a small fraction of the neutral beam, due to ionizing collisions with the background gas, is not as serious as the damage high energy ions can cause if they are neutralized before they have been properly aimed and focussed. To mitigate this, consideration must be given to more extensive pumping or, possibly, other types of neutralizers. Different gases, vapors, plasmas, gas or plasma jets or even photodetachment might be made to form acceptable neutralizers. A brief discussion of some of the choices follows.

Because the cross sections for electron attachment σ_{10}, σ_{0-1} and σ_{1-1}) drop off rapidly[4] with increasing beam energy V_0, the characteristics of neutralizers of high energy negative ion beams are functions of the electron detachment cross sections. At energies in excess of 100 keV, these cross sections (σ_{-10}, σ_{01} and σ_{-11}) are proportional to each other and they decrease, for non-relativistic beams, in accordance with the Born approximation[5] at a rate proportional to $v_0^{-1} \ln\{A\, V_0\}$ in which A is a function of the material of which the neutralizer is formed.

Compared to the other electron detachment cross sections, σ_{-11} is small. Thus only σ_{-10} and σ_{01} are needed to approximate the neutral fraction that corresponds to a gas cell of any line density.[6] The following approximation is acceptable because the cross-sections are only known to no better than ±15% and the neutral fractions to ±5%. Given the ratio of the cross sections:

$$r_{-10} = \sigma_{-10}/\sigma_{01} , \qquad [5]$$

the neutral fraction, as a function of the neutralizer line density, π is:

$$F^0 = \left(\frac{r_{-10}}{r_{-10} - 1}\right)(e^{-\sigma_{01}\pi} - e^{-r_{-10}\sigma_{01}\pi}) \qquad [6]$$

from which the **maximum** neutral fraction:

$$\text{MAX } F^0 = (r_{-10})^{(1/[1 - r_{-10}])} \qquad [7]$$

is found at the **optimum** line density:

$$\pi_0 = \left(\frac{1}{\sigma_{01}}\right)\frac{\ln\{r_{-10}\}}{[r_{-10} - 1]} \qquad [8]$$

Values of Max F^0 and the product ($\sigma_{01} \cdot \pi_0$) are shown in Fig. 3 as functions of the ratio r_{-10}. It is evident that the larger r_{-10}, the higher the maximum neutral fraction and the smaller the product ($\sigma_{01} \cdot \pi_0$) will be.

Because r_{-10} equals the ratio of two terms with the same energy dependence, it is independent of the beam energy. Therefore, neither Max F^0 or the product ($\sigma_{01} \cdot \pi_0$) are functions of the energy. On the other hand, the optimum line density π_0, being inversy proportional to σ_{01}, increases with the beam energy.

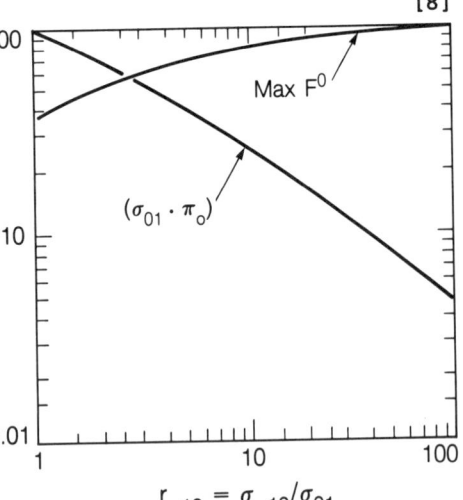

Fig. 3. F^0 and ($\sigma_{01} \cdot \pi_0$) as functions of r_{-10}.

To reduce the length of an optimum neutralizer, other gases with electron detachment cross sections greater than those of hydrogen could be used. For instance, an optimum argon neutralizer for a 200 keV hydrogen beam has a line density that is about 30% that of hydrogen.[7] Unfortunately however, the maximum neutral fraction obtained with other gases may not be as high as that of hydrogen. Indeed, the optimun neutral fraction obtained with argon is about 48% in contrast to 58% for hydrogen. Clearly, there is the possibility of a trade off between neutralizer length and beamline efficiency.

In general maximum neutral fractions observed with most elements range from below 50% to something over 60%. Furthermore[8], it appears that large molecules have lower optimum neutral fractions than might be expected from the sum of their atomic components. The few experimental studies that have been carried out on large molecules confirm this, frequently yielding maximum neutral fractions of less than 50%.

Measurements of the additional beam divergence obtained with different gas neutralizers[2] indicate a very weak dependence on Z. It is largely the result of inelastic collisions corresponding to electron detachment. With increasing beam energy it varies, in the non-relativistic range, inversely with the square root of the energy, V_o.

It is also possible to reduce the neutralizer length and minimize the gas flow by means of gas or plasma jets. This is discussed with regard to gases other than hydrogen in the 1977 proceedings of these symposiyms.[9] A cesium metal vapor jet, using a plug nozzle of the type first proposed for the formation of negative ions by double charge exchange[10] could be advantageous for these applications. The optimum line density of cesium is only 17% of that of hydrogen, while its maximum neutral fraction is about the same.[7] As for plasma jets, they have been tried[11] and found to be effective but not efficient in that all of the power used to ionize the plasma is lost in one pass of the jet across the beamline.

A high Q, multiply ionized plasma provides another approach.[12] With a plasma consisting of ions of density n_Q, primarily of charge state Q, and of neutral gas density n_o, the ionization fraction is:

$$f = \frac{n_Q}{[n_Q + n_o]} \quad [9]$$

The effective electron detachment cross sections of the plasma, being the sum of the weighted contributions of the ions, electrons and neutrals, are:

$$\sigma(Q,f)_{-1o} = f\sigma(Q)_{-1o} + Qf\sigma(e)_{-1o} + (1-f)\sigma_{-1o} \quad [10]$$

$$\sigma(Q,f)_{o1} = f\sigma(Q)_{o1} + Qf\sigma(e)_{o1} + (1-f)\sigma_{o1} \quad [11]$$

with the following approximations:

$$\sigma(Q)_{-1o} = Q^2\sigma(i)_{-1o} = Q^2\sigma(e)_{-1o} \quad [12]$$

$$\sigma(Q)_{o1} = Q^2\sigma(i)_{o1} = Q^2\sigma(e)_{o1} = Q^2\sigma_{o1} \qquad [13]$$

in which $\sigma(i)$ is the cross section of a singly ionized $Q = 1$, ion, the ratio of the plasma cross sections becomes:

$$r(Q,f)_{-1o} = \frac{(Q^2 + Q)f[\sigma(i)_{-1o}/\sigma_{-1o}] + (1 - f)r_{-1o}}{(Q^2 + Q - 1)f + 1} \qquad [14]$$

The maximum neutral fraction along with the product of the ionizing cross section, per Eq. 11, and the optimum line density of the ions in the plasma are the same as shown in Fig. 3, when $r(Q,f)_{-1o}$ is substituted for r_{-1o}, and $[\sigma(Q,f)_{o1} \cdot \pi(Q,f)_o]$ for $\sigma_{o1} \cdot \pi_o$. The optimum line density of the plasma in the neutralizer, i.e. the sum of the ion and neutral line densities, is then:

$$\pi(Q,f)_o = \frac{\ln\{r(Q,f)_{-1o}\}}{\sigma_{o1}[Q^2f + Qf - f + 1] \cdot [r(Q,f)_{-1o} - 1]} \qquad [15]$$

while that of the ions is:

$$\pi(Q)_o = f \cdot \pi(Q,f)_o \qquad [16]$$

and that of the neutrals is:

$$\pi(0)_o = [1 - f] \cdot \pi(Q,f)_o \qquad [17]$$

Figure 4 shows curves of the maximum neutral fraction, for $Q = 1$ and 5, versus the ionization fraction of the plasma, with the assumption that $\sigma(i)_{o1}/\sigma_{o1} = 25$ and $r_{-1o} = 3.5$. Figure 5 shows how the product of the ionizing cross section $\sigma(Q,f)_{o1}$ and 1) the optimum plasma line density $\pi(Q,f)_o$, 2) the corresponding ion line density in the plasma $\pi(Q)_o$ and 3) the corresponding neutral line density $\pi(0)_o$ vary with the plasma ionization fraction for ion charge states $Q = 1$ and 5.

When $Q = 1$, the above equations are applicable to plasma neutralizers with

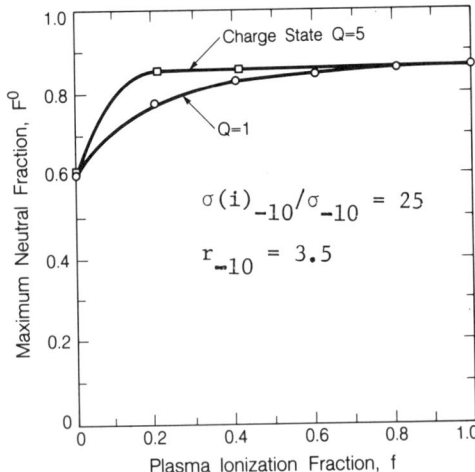

Fig. 4. The maximum neutral fraction as a function of the ionization fraction for plasmas with charge states of $Q=1$ and 5.

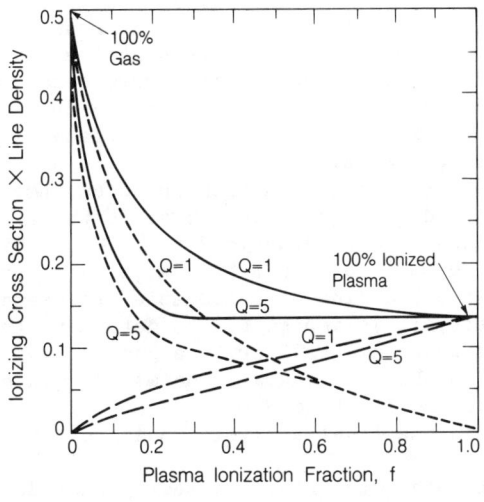

Fig. 5. The product of the ionizing cross section $\sigma(Q,f)_{01}$ and:
_____ The optimum plasma line density.
_ _ _ The ionline density.
\-\-\-\-\- The neutral line density.

singly charged ions. When $f = 0$, they describe a gas neutralizer and when $f = 1$, a fully ionized plasma. Both Figs. 4 and 5 show the advantages of a high Q plasma at relatively low ionization fractions. But the most significant advantage results from the cross section $\sigma(Q,f)_{01}$ which increases, and thereby reduces the optimum line densities by a factor of approximately Q^2.

The maximum neutral fraction obtained with any plasma neutralizer depends upon the gas of which it is formed. A fully ionized hydrogen plasma, for instance, will provide a maximum neutral fraction of approximately 85%.

Plasma neutralizers have been investigated in the past[13] and the high neutral fraction found at relatively low ionization fractions make them more efficient than they otherwise might be. However, it is important to keep the magnetic field that confines the plasma from interacting with the ion and increasing the beam divergence to an unacceptable level.[14]

It should be noted that there is some uncertainty about the effect of the plasma on the beam optics. Depending on whether it is slightly positive or negative, the plasma can act as a divergent or convergent lens to the un-neutralized ions traveling through the neutralizer. In addition, any plasma oscillations can cause a significant increase in the beam divergence.

To establish the efficiency of a neutral beamline equipped with a plasma neutralizer, it is necessary, per Eq. 2, to determine the ratio of the power needed to sustain the plasma in the neutralizer with respect to the power needed to form the negative ion beam. The intricacies of the various proposed neutralizer designs, make it difficult to estimate this ratio with any confidence. Obviously, the beam diameter, as well as the effectiveness of the radial confinement of the plasma are critical.

Photodetachment[15] has the potential of being the basis of a very desirable neutralizer. Without the introduction of additional gas, it could provide a high neutral fraction with a neutralizer of reasonable length and help form a neutral beamline of high efficiency. Most concepts call for an intense laser beam to be reflected back and forth across the path of the negative ions to be neutralized. The performance of such neutralizers depend upon the laser, its wavelength and efficiency, the mirrors, their reflectivity and ability to withstand high levels of radiation, and the design of the optical cavity.

Because the neutral fraction obtained from photodetachment is a function of power used to drive the optical system, the efficiency of a photoneutralized beamline is described by Eq. 2. While the neutral fraction can be made to approach 100%, it will probably be considerably less in a beamline of reasonable efficiency. As with plasma neutralizers, it is impossible to determine the power needed to operate a photo-neutralizer in a general way. It depends upon such things as the beam energy, the wavelength of the photons, the efficiency of the light source and the gain of the optical resonator which, in the final analysis, must be determined experimentally.

As for the increase in beam divergence caused by photodetachment, it is a consequence of the difference between the energy of the photon and the binding energy of the extra electron which forms the negative ion. Upon neutralization, this energy is shared between the newly released electron and the newly formed neutral. Because the photodetachment cross section and the beam divergence are functions of the wavelength, that wavelength which establishes the minimum divergence, the maximum neutral beamline efficiency and the maximum brightness are not the same. In fact, tunable lasers of wavelengths suitable to maximize the beamline efficiency are probably not available today.

Some of the lasers under consideration include a huge array of small cw Gallium Arsenide lasers with a wavelength of 0.84 μm, a vast chemical Iodine laser of 1.31 μm and a high power Neodymium-YAG laser of 1.06 μm.

The choice of cw lasers of wavelengths suitable for stripping H⁻ is limited but by means of Raman shifters, a range of wavelengths can be obtained. However, to tailor a photo-neutralizer to optimize the maximum the beamline efficiency, neutral beam brightness, etc. will require the development of efficient free-electron-lasers.[16]

FROM 1.0 TO 25 MeV

Until now, it has been assumed that the negative ion beams were accelerated in a simple gridded structure. But intense electric fields are required to form high energy beams in this manner. At high voltages, the electrodes are subject to occasional breakdown, making the accelerator unreliable. Furthermore, high voltage terminals exposed to the atmosphere must be constructed with very large radii to inhibit corona, while the stray capacitance about the system has to be limited to prevent the stored energy from sustaining a catastrophic arc at breakdown. While it is possible to contain the electrodes in oil or a pressurized gas such as SF_6, the container increases the capacitance of the system, possibly making matters worse. Thus at some energy ranging from 0.75 to roughly 2 MeV, depending upon the skill of the designer, rf accelerators will have to be used to form the negative ion beams from which neutrals are formed.

With rf accelerators, the efficiency of the negative ion beamline is decreased by a factor related to the efficiency with which the wall plug is converted to rf power, roughly 60%. As a consequence, at energies in excess of 2 MeV, the neutral beamline efficiency is expected to be less than 30%.

With regards to neutralizers, the optimum line densities continue to increase as the beam energy goes up, making this a particularly difficult region. Unless constraints on the neutralizer length are greatly relaxed, gas neutralizers are most probably unacceptable. Gas and vapor jets remain a possibility if they can be well confined and, depending upon the beam diameter and the beamline efficiency requirements, either a high Q plasma or a photo-neutralizer might be considered.

ABOVE 25 MeV

At 25 MeV, the optimum line density is so large that the neutralizers can be formed out of solid foils or possibly liquid films of several mg/cm^2. Because the optimum line density of any neutralizer is of the order of 1 to 2 mean-free-paths, irrespective of the beam energy, and the collisions per mg/cm^2 is nearly the same for all elements at any given energy[17], the mg/cm^2 for all optimized gas, vapor, liquid and foil neutralizers is about the same. Thus, the higher the beam energy, the greater the optimum line density and the more weight there is in an optimum foil or film to provide strength.

Preliminary experiments were conducted on a thin liquid sheet, produced by spraying high-purity Fomblin oil on a rapidly rotating disk, to determine its effectiveness as a neutralizer.[18] The concept of a liquid sheet or film is attractive because it is self replenishing and self healing while under bombardment. The results were encouraging and it is likely that the maximum neutral fraction will be of the order of gas or foil neutralizers.

As for foils, in particular carbon foils, present technology[19] is such that large diameter assemblies of great strength can be formed at line densities of more than 1 to 2 mg/cm^2. The assemblies consist of thin foils mounted on a strong, fine meshed grid of about 90% transparency, made of carbon fibers of up to 100 microns in diameter. The maximum neutral fraction, averaged over the grids obtained with such foils is expected to be about 55%. Because the required structural strength of the grid, i.e., the diameter of the carbon fibers, is independent of the beam energy, the grid is heated more at lower beam energies at which their stopping power is greater. Calculations indicate that the carbon fibers should be able to withstand the resulting thermal stress. It is expected that hydrogen beams of 25 MeV will have little affect on the foil, but the useful life of a carbon foil could be reduced by atomic oxygen erosion. In some environments special precautions may be required.

To operate at higher efficiencies, it will be necessary to use a driven neutralizer based upon either a high Q plasma or photodetachment. To be efficient, the power needed to operate either of these neutralizers must be considerably less than that of the negative ion beam. However, as can be seen from Eq. 2, the beamline efficiency cannot be better than that with which the negative ion beam was formed.

At 50 to 100 MeV, the beam velocity becomes an appreciable fraction of that of light and relativistic effects start to become important. Figure 6 shows the ratio of velocities, $\beta = v/c$ as a function of the kinetic energy of the beam. Specific proton accelerators are noted in the figure as items of general interest. Although I do not intend to discuss relativity in any detail, there are three effects which should be mentioned here. At relativistic energies: 1. The electron detachment cross sections increase with the beam energy at a slower rate, making the optimum neutralizer line densities somewhat larger than they otherwise might be.

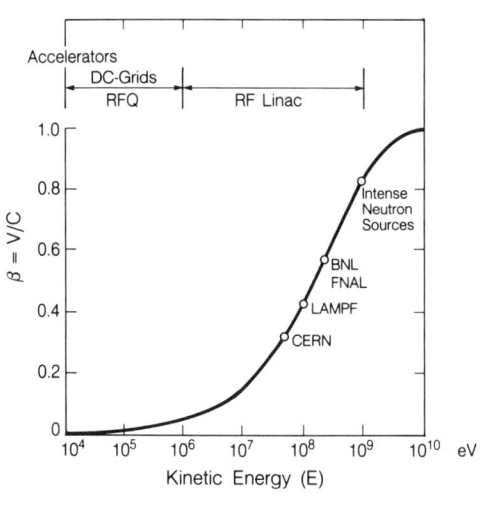

Fig. 6. β as a function of the kinitic energy of a hydrogen beam.

2. The added beam divergence resulting from neutralization decreases at a slower rate with increasing beam energy.
3. The wavelength of the photons seen by the moving ions is less than that in the rest frame of the laser, while the photon line density appears to be greater. Figure 7 shows the laser wave length as seen in the laboratory in contrast to the non-relativistic photodetachment cross section as seen in the frame of the moving ions.

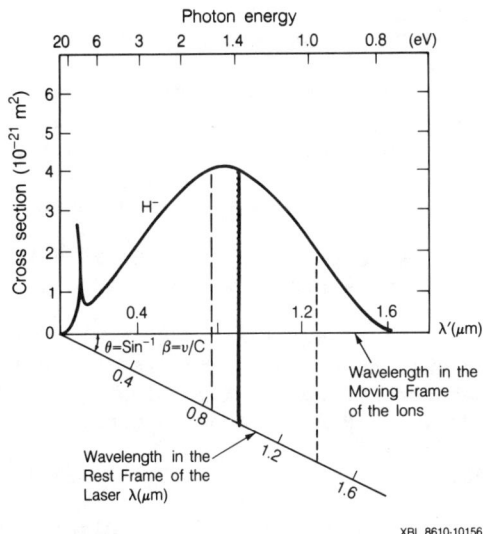

Fig. 7. Photo detachment cross section in the frame of the moving ions.
---- Galium Aresnide Laser.
——— Neodymium-YAG Laser.
----- Atomic Iodine Laser.

CONCLUSIONS

Figure 8 shows a spectrum of beam energies, indicating the approximate range over which the various types of neutralizers might be useful.

Obviously there are many options for neutralizers of high energy negative hydrogen beams. Future delopment should make photodetachment a more viable prospect and there probably are a few concepts that have not been thought of yet. But before the most desirable neutralizer can be selected, the critical items of the neutral beam specification, such as the efficiency, divergence, or brightness, etc. must be identified.

REFERENCES

1. Part of the work was done while under contract to GA Technologies Inc. La Jolla, CA 92138.
2. B.A. D'yachov, V.I. Zinenko, G.V. Kazantesv, ZhTF 47, 416 (1977).
3. W.L. Barr, R.W. Moir Moir, G.W. Hamilton, J. of Fusion Energy, 2, 131 (1982).

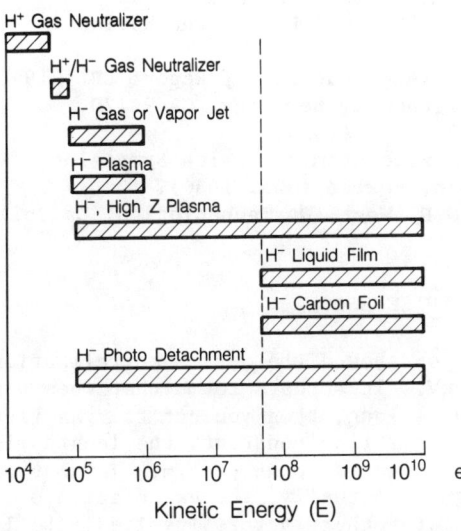

Fig. 8. Approx. energy range of various neutralizers.

4. D.R. Sweetman, et al., UKAEA Research Group Report CLM-R 112, Culhan Laboratory, Abingdin Berks., (July 1971).
5. E.W. McDaniel, "Collision Phenomena in Ionized Gases," John Wiley and Sons, 322 (1964).
6. G.I. Dimov, A.A. Ivanov, G.V. Roslyakov, ZhTF 47, 1881 (1977).
7. L.W. Anderson, C.J. Anderson, R.J. Girnius, A.M. Howard, Proceedings of the Second International Symposium on the Production and Neutralization of Negative Hydrogen Ions and Beams, 285 (Oct. 6-10, 1980), Brookhaven National Laboratory, Upton, NY.
8. G.H. Gillespie, Physical Dynamics Report #PDL-82-22-9-TM (1982) Physical Dynamics, Inc., P.O. Box 1883, La Jolla, CA 92038.
9. C.K. Lam, Proceedings of the Production and Neutralization of Negative Hydrogen Ions and Beams, 195 (Sept. 26-30, 1977) Brookhaven National Laboartory, Upton, N.Y. p195.
10. E.B. Hooper, Jr., P. Poulsen, P.A. Pincosy, Journal of Applied Physics, 52, 7027 (1981).
11. G.I. Dimov, A.A. Ivanov, G.V. Roslyakov, ZhTF, 47, 1881 (1977).
12. A.S. Schlachter, K.N. Leung, J.W. Stearns, R.E. Olson, These proceedings, (1986).
13. A.A. Ivanov, G.V. Roslyakov, ZhTF 50, 2300 (1980).
  A.I. Herscovitch, B.M. Johnson, V.J. Kovarik, M. Meron, K.W. Jones, K. Plrelec, L.R. Grisham, Rev. Sci. Instrum. 55, 1744 (1984).
14. M.W. Grossman, Proceedings on the Production and Neutralization of Negative Hydrogen Ions and Beams, 189 (Sept. 26-30, 1977) Brookhaven National Laboratory, Upton, N.Y. p189.
15. J.H. Fink, AIP Conference Proceedings, Number 111, Production and Neutralization of Negative Ions and Beams, 3rd International Symposium, 547 (1983) Brookhaven National Laboratory. p547 (1983).
16. A.M. Sessler, Lawrence Berkeley Laboratory Report LBL-21948 (1968), University of California, Berkeley, CA 94720.
17. H.A. Bethe, Phys. Rev. 89, 1256 (1953).
18. A.S. Schlachter, 2nd US-Mexico Atomic-Physics Symposium, Two Electron Phenomena, Coyoco, Mexico (Jan. 1986).
19. Private communication: W.P. West, GA Technologies, La Jolla, CA 93138.

DISCUSSION

<u>Henkes</u>: In the beginning, you showed the JAERI neutralization line, which was 30m for 250 kV. It appears to me that when you go to 500 kV, and make it twice as long, then you cut the gas flow by factor of 2. If you keep the gas flow constant, the length will be only multiplied by $\sqrt{2}$, and that would be about 42m. Is that right?
<u>Fink</u>: I think you're right. Actually, the calculation of that was a bit icky because there are other factors involved. So I ducked it and took the simple worst case. You're right.

A HIGH-CHARGE-STATE PLASMA NEUTRALIZER FOR AN ENERGETIC H⁻ BEAM*

A.S. Schlachter, K.N. Leung, and J.W. Stearns

Lawrence Berkeley Laboratory
University of California
Berkeley, CA 94720

and

R. E. Olson

University of Missouri, Rolla
Rolla, MO 65401

LBL-22333

ABSTRACT

A high-charge-state plasma neutralizer for a beam of energetic H⁻ ions offers the potential of high optimum neutralization efficiency (~85%) relative to a gas target (50-60%), and considerably reduced target thickness. We have calculated cross sections for charge-changing interactions of fast H⁻ and H⁰ in collision with highly charged ions using a semiclassical model for H⁻, and the Classical-Trajectory Monte Carlo method plus Born calculations, to obtain correct asymptotic cross sections in the high-energy limit. Charge-state fractions as a function of plasma line density, and f_0^{max}, the maximum H⁰ fraction, are calculated using these cross sections; we find that $f_0^{max} \simeq 85\%$ for ion charge states in the range 1+ to 10+, and that target ion line density for f_0^{max} decreases approximately as the square of the plasma ion charge state. The maximum neutral fraction is also high for a partially ionized plasma. We have built a small multicusp plasma generator to use as a plasma neutralizer; preliminary results show that the plasma contains argon ions with an average charge state between 2+ and 3+ for a steady-state discharge.

INTRODUCTION

The use of an accelerated beam of H⁻ to provide an intense source of neutral H⁰ requires the efficient removal of one electron from the H⁻ ion. A gas cell or a foil will produce a maximum H⁰ yield of 50-60% at high energies. Laser photodetachment could, in principle, provide a very high H⁰ yield, if a laser of sufficient power existed.

A dense plasma can be used as a neutralizer for an energetic H⁻ beam; cross-section data indicate that the conversion efficiency will be about 85%. A hydrogen plasma neutralizer has

---

* Supported by U.S DOE Contract DE-AC03-76SF00098 and USASDC MIPR W31RPD-63-A087.

been discussed by Grossman and by Berkner et al.,[1] and neutralizers for fast negative ions have been discussed by Schlachter.[2] A plasma neutralizer for fast negative ions[3] has been studied experimentally by Dimov and co-workers and by Hershkovitch et al, and a recent publication in the Soviet literature by D'yachkov et al.[4] has discussed apparent focusing of H^0 observed in the neutralization of 180-keV H^- in a Cs^+ plasma.

The purpose of our theoretical work is to calculate cross sections for single-electron-detachment

$$H^- + X^{q+} \rightarrow H^0 \quad \ldots \quad \sigma_{-0} \quad (1)$$

and double-electron detachment

$$H^- + X^{q+} \rightarrow H^+ \quad \ldots \quad \sigma_{-+} \quad (2)$$

for collisions of H^- with multiply charged ions. Ion charge states from q = 1+ to 10+ and H^- energies from 250 keV/amu to 10 MeV/amu are considered. These cross sections are then used to predict H^0 yield and target thickness requirements for neutralizing H^- in partially and fully ionized plasmas.

THEORETICAL METHOD

The theoretical method used at the lower energies ($E \leq 1$ MeV/amu) was the four-body Classical-Trajectory Monte Carlo (CTMC) method. A program code was developed for this application which included the Coulomb interactions between the proton, its two electrons, and a fully-stripped multiply charged ion. The work led to the understanding of the collision mechanisms so that extension could be made to higher energies using a modified Born-approximation method.

The principal difficulty in implementing the CTMC method is the development of a suitable classical model for the H^- ion. It is well known that a classical two-electron atom is unstable and autoionizes unless constraints are placed upon the initial orbits for its electrons. It is crucial, however, that two electrons be explicitly included in the calculations in order to determine an accurate ratio of the single to double electron removal reactions. Statistical models based on one-electron Hartree-Fock calculations overestimate the importance of the double-electron-removal reaction. Such comparisons have been clearly demonstrated for collisions of multiply charged ions with helium.[5] The direct application of the independent-electron model or the Born approximation predicts the σ_{-0} cross section for (1) should scale as $\sim q^2/E$, while the double-electron-removal cross section (2) should scale as $\sim q^4/E^2$. The E^{-2} dependence for double-electron removal is not confirmed in experiments[6] using electrons as projectiles. Furthermore, the q^4 dependence leads to serious questions regarding the wisdom of using multiply charged ions in a plasma neutralizer since a large σ_{-+} cross section significantly degrades its efficiency.

Our application of the CTMC method is based on semiclassical quantization rules given by Einstein, Brillouin, and Keller.[7] If one employs a self-consistent-field model for the H⁻ atom where each electron experiences an averaged potential due to the nucleus and the other electron, one can directly show that half-integer quantum numbers require the planes of the electron orbits to be offset 120° from one another and that the eccentricity of the orbits are equal to 0.866.

Within the context of a self-consistent-field model, the orbits of each electron are hydrogenic. Recalling the quantum-mechanical two-term spatial wavefunction for the H⁻ atom:

$$\Psi = N [e^{-\alpha r_1} e^{-\beta r_2} + e^{-\alpha r_2} e^{-\beta r_1}], \tag{3}$$

it is possible to relate classically the parameters α and β to the effective charge experienced by each electron. Using Slater's rules, $\alpha = \beta = 0.6875$, and the total electronic energy calculated for H⁻ is -0.4727 a.u. (i.e. not bound!). The experimental value is -0.5277 a.u. Moreover, such values for α and β classically require each electron to be bound by 0.2363 a.u. ($\alpha^2/2$ and $\beta^2/2$), which is unacceptable since the first and second ionization potentials are 0.0277 a.u. and 0.500 a.u., respectively. Since classical electrons are distinguishable, it is possible to use effective charges of $\alpha = 0.2354$ and $\beta = 1.0000$ for the two electrons. With these values the experimental ionization potentials are reproduced, and quantum mechanical calculations using the split-shell wavefunction given by Eq. (3) yield an electronic energy of -0.5125 a.u., 72% closer to the experimental value than the single-parameter result.

A further consideration is that the classical calculations allow us to consider directly the effects of "shake-off" or "shake-up" in the collision. "Shake-off" is the phenomenon whereby the sudden removal of one of the electrons leaves the other electron in a configuration that is not an eigenfunction of the remaining ion or atom, resulting in further ionization. In "shake-up", this electron also has a finite probability of being found in an excited state.[8] The excited-state production must be considered in the application to a plasma neutralizer. The excited atoms or ions will have a greatly enhanced possibility of ionization in subsequent collisions since their geometrical cross sections increase as n^4, where n is the principle quantum number. Our calculations include the effects of both the direct impact ionization and "shake-up" and shake-off." For the application to modeling a plasma neutralizer we have assumed that any single-ionization events that leave the neutral H⁰ in an excited level will be further ionized and should be classified as double ionization of the H⁻ projectile.

A model for the probability of double ionization is the product of the probability of removing the outer, loosely bound electron P_0 times the probability of removing the more tightly bound inner electron P_I:

$$P_{2-ion} = P_0 P_I . \tag{4}$$

The cross section is

$$\sigma_{-+} = 2\pi \int P_0 P_I \, b db . \tag{5}$$

However, since the limits of integration of Eq. (5) are restricted to small impact parameters by P_I where P_0 is approximately constant, we can rewrite Eq. (5) as

$$\sigma_{-+} = 2\pi P_0 \int P_I \, b db . \tag{6}$$

One can immediately see that Eq. (6) is simply

$$\sigma_{-+} = P_0 \sigma_{0+} , \tag{7}$$

where σ_{0+} is the well-known single electron-removal cross section from neutral H^0:

$$H^0 + X^{q+} \to H^+ \ldots \tag{8}$$

Equation (7) must be further expanded upon to include the "shake-up" and shake-off" processes:

$$\sigma_{-+} = P_0 \sigma_{0+} + (1 - P_0) S \sigma_{0+} . \tag{9}$$

The second term in Eq. (9) includes the possibility of ionizing the inner electron and <u>not</u> ionizing the outer electron, hence, the outer electron has a probability S of not finding itself in the ground state of the H^0 atom. The factor S equals 0.675. Note, a similar term exists for the "shake-up" of the inner electron but it is negligible. Equation (9) can be rewritten as:

$$\sigma_{-+} = (0.675 + 0.325 \, P_0) \sigma_{0+} . \tag{10}$$

CTMC values for P_0 are given in Table I for charge states $q = 1, 2, 5,$ and 10 and energies of 250 keV/amu, 1 MeV/amu, and 10 MeV/amu. The overall factor in Eq. (10) that relates the σ_{-+} and σ_{0+} cross sections is given in Table II. It is quite evident that "shake-up" is a very important process at the higher energies and cannot be ignored.

The CTMC calculations were performed at 250 keV/amu and 1 MeV/amu. Above 1 MeV/amu, the CTMC method loses validity due to the neglect of quantum tunneling. Correspondingly, above 1 MeV/amu, the Born approximation provides accurate cross sections. At 1 MeV/amu, the CTMC and Born-approximation cross sections for σ_{-0} are found to agree to within 20%, and the σ_{0+} values to within 30%. The Born cross sections of Kim and Inokuti[9] were used in our studies. Tabulated values for the calculated cross sections are given in Table III.

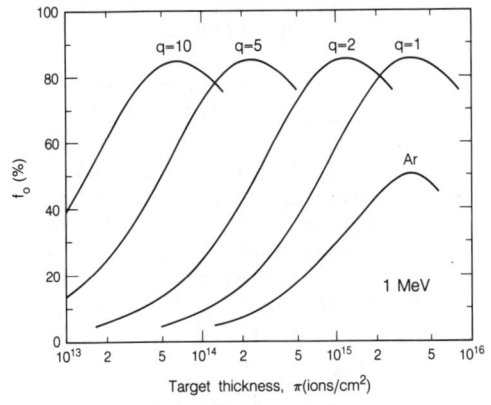

Fig. 1 Calculated H^0 yield as a function of target ion (atom) line density for 1-MeV H^- in an Ar gas neutralizer and in a plasma neutralizer with ion charge state 1+ to 10 +.

RESULTS

Charge-state fractions as a function of plasma line density for H^- incident on a plasma with ions in charge-state q and q electrons have been calculated by integrating a set of three first-order linear differential equations with effective cross sections obtained by adding q times the cross section for an electron collision to the cross section for collision with an ion. Examples are shown in Fig. 1 for 1-MeV H^- in Ar gas and in plasmas with charge states 1, 2, 5 and 10+. Figure 2 shows maximum neutral fraction f_0^{max} and π^{opt}, the target thickness to obtain f_0^{max}, as a function of plasma charge-state q for 1-MeV H^-; figure 3 shows the energy dependence of π^{opt} and f_0^{max} for H^- in Ar and in plasmas with charge states 2 and 10.

For the case of an energetic H^- beam, where electron capture collisions can be neglected, it is easy to calculate f_0^{max} and π^{opt}:

$$f_0^{max} = \left(\frac{\sigma_{-0}}{\sigma_{-0} + \sigma_{-+}} \right) \left(\frac{\sigma_{0+}}{\sigma_{-0} + \sigma_{-+}} \right)^\gamma \quad (11)$$

where $\gamma = \dfrac{\sigma_{0+}}{\sigma_{-0} + \sigma_{-+} - \sigma_{0+}}$

$$\pi^{opt} = \frac{1}{\sigma_{-0} + \sigma_{-+} - \sigma_{0+}} \ln \left(\frac{\sigma_{-0} + \sigma_{-+}}{\sigma_{0+}} \right) . \quad (12)$$

The maximum neutral fraction f_0^{max} ranges from 84% to 86% for the cases studied. The line density required for maximum neutralization fraction decreases with increasing charge state of the plasma and increases with increasing beam energy.

The effect of partial ionization of the gas in a plasma neutralizer is shown in Fig. 4 for 1-MeV H^- in a mixture of Ar gas and a plasma with charge-state 5+; f_0^{max} and π^{opt} are shown as a function of the degree of ionization of the plasma (0 meaning pure Ar gas, 100% meaning pure q=5+). A high value of f_0^{max} is obtained even for a relatively low degree of ionization.

PLASMA-NEUTRALIZER DEVELOPMENT

The permanent-magnet-generated multicusp plasma device is capable of producing large volumes of uniform and high density plasma. Since the dipole fields confine the energetic electrons very effectively, discharge efficiency and ion charge-state distributions in the plasma can be optimized by using a "mixture" of primary electron energies. In addition, electro-static confinement of ions can be achieved by operation with a negative plasma potential or by using positively biased electrodes located between the line-cusps.

We have constructed and operated a small multicusp plasma generator containing high-charge-state ions for use as a plasma-neutralizer target. The plasma chamber

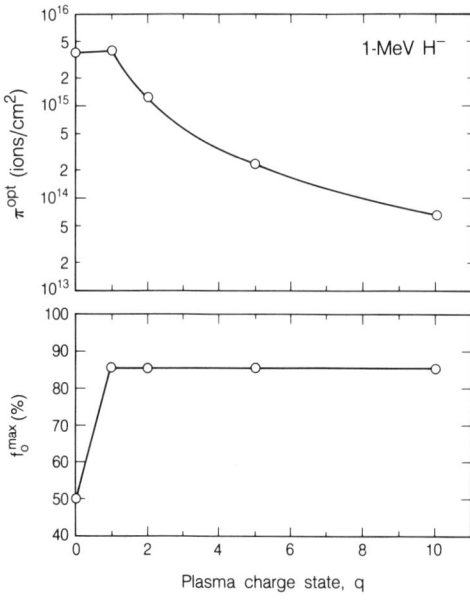

Fig. 2 Charge-state dependence of maximum H^0 fraction f_0^{max} and of target thickness to achieve F_0^{max}, π^{opt}, for 1-MeV H^- in an Ar-gas neutralizer ($q=0$) and in a plasma neutralizer with charge state q.

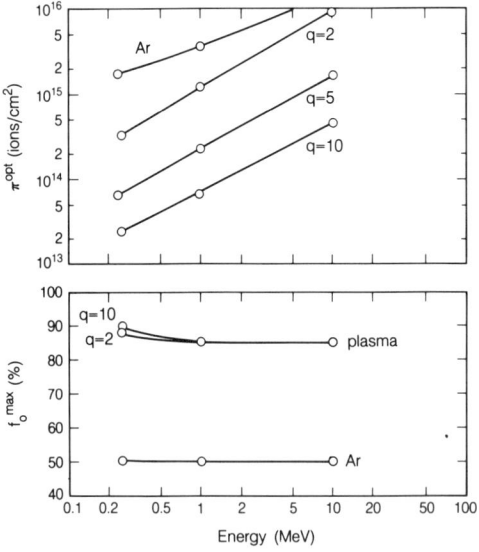

Fig. 3 Energy dependence of f_0^{max} and π^{opt} for H^- in an Ar-gas neutralizer and in a plasma neutralizer.

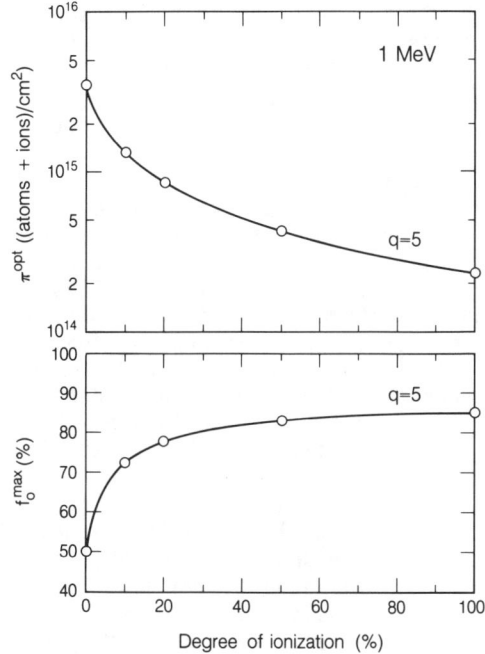

(15-cm-diam by 20-cm long) is made of copper and is surrounded externally by eight columns of samarium-cobalt magnets which form a longitudinal line-cusp configuration for primary-electron and plasma confinement. These magnet columns are connected at one of the end flanges by eight rows of magnets which converge to the axis of the chamber. In this configuration, a small field-free area is generated at the center of the flange. In order to determine charge-state distribution in the plasma, positive ions are

Fig. 4 Dependence of f_0^{max} and π^{opt} on degree of ionization of a plasma. 0 refers to pure Ar gas.

extracted from the plasma through a small aperture (1.5 mm-diam) located in this field-free region. The extracted beam is then analyzed by a compact magnetic-deflection mass spectrometer located just outside the plasma chamber.

Table IV summarizes the plasma parameters for operation with an argon plasma. The spectrometer output signal shows that the plasma is capable of producing argon ions with charge states as

Fig. 5 Photograph of plasma-neutralizer module.

high as 7+, with an average charge state between 2+ and 3+ for a steady-state discharge of 250 V, 25 A. In order to reduce the operating discharge voltage, we are now investigating the use of other gases, for example, xenon. Initial results demonstrate that electron densities as high as 4×10^{12} cm^{-3} can be achieved with arc power of 150 V, 20 A in a low-pressure xenon discharge.

To achieve the proper line density for the neutralization of energetic H$^-$ beams, a larger and longer modular-type multicusp plasma generator has been designed and fabricated. This neutralizer (25-cm-diam by 25-cm long) is bigger and therefore has a larger-field-free volume than the prototype neutralizer. Figures 5 and 6 show the appearance and the schematic of this new plasma-neutralizer module. The large number of ports around the chamber can provide easy access for the tungsten filaments or LaB$_6$ cathodes which are needed to generate a steady-state high-density plasma. With this new plasma neutralizer, we plan to measure the magnetic-field distribution, the neutralization efficiency, the dependence of H^0 yield on plasma line density, and the angular divergence, for energetic H$^-$ incident on this multiply charged plasma.

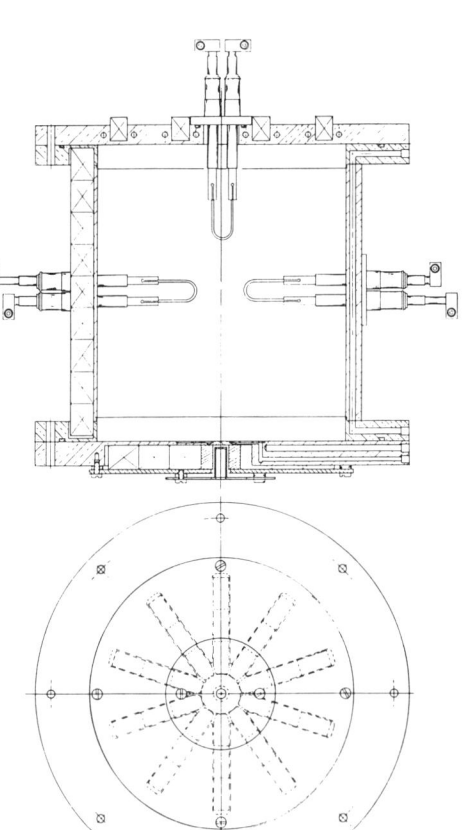

XBL 8610-4060

CONCLUSION

A theoretical method was developed to calculate the single- and double-electron-removal cross sections needed to model a plasma neutralizer. The maximum neutral H^0 fraction in a plasma neutralizer is ~ 85% and independent of energy for high-energy H$^-$. We are presently beginning to test a plasma target as a neutralizer for energetic H$^-$.

Fig. 6 Schematic diagram of plasma-neutralizer module, configured for extraction of ions for charge-state-distribution measurement.

Table 1. Probability P_0 of removing the outer electron on H^- by direct impact ionization in a collision with a fully-stripped multiply charged ion.

q	250 keV/amu	1 MeV/amu	10 MeV/amu
1	0.260	0.105	0.017
2	0.565	0.266	0.052
5	0.906	0.614	0.191
10	0.981	0.744	0.388

Table II. Calculated ratio of σ_{-+}/σ_{0+} which includes the possibility of "shake-up" and "shake-off."

q	250 keV/amu	1 MeV/amu	10 MeV/amu
1	0.760	0.709	0.681
2	0.859	0.761	0.692
5	0.969	0.875	0.737
10	0.994	0.917	0.801

Table III. Calculated cross section in cm^2 for reactions 1, 2, and 8, for H^- colliding with an ion in charge-state q.

q	σ_{-0}	σ_{0+}	σ_{-+}
		250 keV/amu	
1	1.9(-15)	5.5(-17)	4.2(-17)
2	6.8(-15)	2.1(-16)	1.8(-16)
5	4.0(-14)	1.1(-15)	1.1(-15)
10	1.4(-13)	3.1(-15)	3.1(-15)
		1 MeV/amu	
q			
1	4.6(-16)	1.8(-17)	1.3(-17)
2	1.9(-15)	7.4(-17)	5.6(-17)
5	1.2(-14)	4.6(-16)	4.0(-16)
10	4.6(-14)	1.8(-15)	1.7(-15)
		10 MeV/amu	
q			
1	6.2(-17)	2.6(-18)	1.8(-18)
2	2.5(-16)	1.0(-17)	6.9(-18)
5	1.6(-15)	6.5(-17)	4.8(-17)
10	6.2(-15)	2.6(-16)	2.1(-16)

Table IV. Plasma parameters for plasma-neutralizer operation with Ar.

Discharge power:	250 V, 25 A
Electron density n_e:	6×10^{11} cm^{-3}
Gas pressure:	5.3×10^{-5} torr
Neutral density n_0:	1.75×10^{12} cm^{-3}
n_e/n_0:	0.34
Electron temperature T_e:	3.5 eV
Average charge state:	between 2 and 3

REFERENCES

1. M. W. Grossman, "Plasma Neutralizer for H$^-$ Beams," in Proceedings of the Symposium on the Production and Neutralization of Negative Hydrogen Ions and Beams, Brookhaven, 1977, BNL 50727, p. 189; K. H. Berkner, R. V. Pyle, S. E. Savas, and K. R. Stalder, "Plasma Neutralizers for H$^-$ and D$^-$ Beams," Proceedings of the Second International Symposium on the Production and Neutralization of Negative Hydrogen Ions and Beams, Brookhaven, 1980. BNL 51304, p. 291.
2. A. S. Schlachter, "Neutralization of a Fast Negative-Ion Beam," Invited paper presented at Second U.S.-Mexico Atomic Physics Symposium (Two-Electron Phenomena), Cocoyoc, Mexico, 1986 (to be published).
3. G. I. Dimov and G. V. Roslyakov, Nucl. Fusion 15, 551 (1975); G. I. Dimov, A. A. Ivanov, and G. V. Roslyakov, Sov. Phys. Tech. Phys. 22, 1091 (1977); A. A. Ivanov and G. V. Roslyakov, Sov. Phys. Tech. Phys. 25, 1346 (1980); G. I. Dimov, A. A. Ivanov, and G. V. Roslyakov, Sov. J. Plasma Phys. 6, 513 (1980); A. L. Hershcovitch, B. M. Johnson, V. J. Kovarik, M. Meron, K. W. Jones, K. Prelec, and L. R. Grisham, Rev. Sci. Instrum. 55, 1744 (1984).
4. B.A. D'yachkov, V. E. Meshkov, and G. V. Kazantsev, Sov. Tech. Phys. Lett. 11, 299 (1985).
5. S. J. Pfeifer and R. E. Olson, Phys. Lett. 92A, 175 (1982).
6. B. Peart, D. S. Walton, and K. T. Dolder, J. Phys. B4, 88 (1971).
7. J. B. Keller, Ann. Phys. (NY), 4, 180 (1958).
8. J. A. Stone and T. J. Morgan, Phys. Rev. A 31, 3612
9. Y-K Kim and M. Inokuti, Phys. Rev. 3, 665 (1974).

DISCUSSION

Stewart: Have you measured the fractional ionization in your plasma stripper?
Leung: At low arc power, there is 20-30% ionization.
Prelec: The photos of the neutralizer that you showed seem to indicate that the ends were closed completely. Those values for the plasma densities - do they refer to a closed system or a system that is open-ended, that should be the only one usable for plasma neutralizers.
Leung: It's a closed structure.
Mitchell: Fred, what's going to happen when you put an amp of H^- through a plasma neutralizer? Will that differ in focussing from, say, a microamp from an accelerator?
Schlachter: We don't know.
Mitchell: Any feeling for that?
Schlachter: Well, presumably, any high intensity beam that goes through any kind of a neutralizer will create a plasma anyway. But we're starting with a plasma, so it depends upon the density, upon the diameter of the beam and, since there can be plasma effects, there can be nonlinear effects that could depend upon, for example, the currents. So I can't really answer it except that, depending upon the application, sometimes you want a very intense dense beam, and sometimes you want a big, old, spread out one.
Nightingale: You very correctly showed the advantage of generating very high q values. Would you care to comment on the power that you're going to need in order to generate those high densities in a real fusion device?
Schlachter: For any application, it depends upon the diameter, since the power is a function of the volume, and it depends upon the density that's required, which means it depends upon the energy of the beam. I think the right person to address these questions to is Joel Fink who discussed, in one of his transparencies, the questions of system efficiency. I personally have not addressed questions of system efficiency.
Anderson: Is it understood why a beam is focussed when it goes through one of these plasma neutralizers? I didn't understand that point.
Schlachter: I didn't explain why because I didn't understand it either, from the paper. They speculated that there's a radial electric field in the plasma which causes the focussing. One of the first things that we've proposed to do is, in fact, repeat the measurement under similar conditions. That means at a few hundred keV to first see the observation - I presume it's correct, since I think D'yachkov does good work - and then to try and understand it, to see its dependence upon the various parameters and to see its energy dependence. But the only explanation I have right now is their explanation, which is a radial electric field in the plasma. It's not an atomic physics effect. It's not an effect of binary collisions. That gives you positive angles.
Roberts: May I ask a question about that? If you shoot the beam into a gas, like in a LEBT, it can create its own plasma, expel

electrons, and gas focus. In a plasma, the electrons are already there. The Coulomb repulsion kicks the light electrons out much faster than these H^- ions. Do you think that gas focussing could occur much quicker in a plasma?

Schlachter: Do you mean quicker in time?

Roberts: In distance. So that the focussing could just be gas-like focussing.

Schlachter: If I understood their explanation, it is that the positive ions in the plasma have low mobility, and so that they're there and the electrons leave.

MEASUREMENT OF PLASMA PRODUCTION AND NEUTRALIZATION IN GAS NEUTRALIZERS

D. Maor, M. Meron, B. M. Johnson, K. W. Jones,
A. Agagu, and B.-L. Hu
Brookhaven National Laboratory, Upton, NY 11973

Several previous talks have emphasized the need of experimental data for the designing of gas neutralizers. We have started a project aimed at measuring all relevant cross sections for the charge exchange of H^-, H^0, and H^+ projectiles, as well as the cross sections for the production of ions in the target. The expected results of these latter measurements are shown schematically in Table I. Each square in the table represents a full charge state distribution of recoiling ions in coincidence with the colliding hydrogen ion of given incoming and given outgoing charge state. The diagonal elements represent cross sections for production of recoiling ions with different charge states in collisions where the projectile does not undergo a charge exchange.

Table I.

out \ in	-1	0	1
-1	$\sigma_{-1,-1}(q_T)$	$\sigma_{0,-1}(q_T)$	$\sigma_{1,-1}(q_T)$
0	$\sigma_{-1,0}(q_T)$	$\sigma_{0,0}(q_T)$	$\sigma_{1,0}(q_T)$
1	$\sigma_{-1,1}(q_T)$	$\sigma_{0,1}(q_T)$	$\sigma_{1,1}(q_T)$

9 RECOIL CHARGE-STATE DISTRIBUTION

The results of these experiments will give important information on the following questions:

1. Absolute values and energy dependence of stripping cross cross sections for H^- and H^0 are sorely needed for the accurate design of the gas neutralizer.

---
*This work was performed under the sponsorship of the United States Army SDC, Contract No. MIPR W31RPD-63-A169, and the Fundamental Interactions Branch, Division of Chemical Sciences, Office of Basic Energy Sciences, United States Department of Energy, Contract DE-AC02-76CH00016.

2. The total charge produced in the target and its distribution is a critical input for the calculation of the plasma density and the determination of its effect on beam divergence.

3. Since the experiments will be performed on many gases, differences in both neutralization efficiency and plasma production can be determined providing an important input to the final choice of the neutralizer gas.

4. The cross sections for ion production without projectile charge exchange for H^- and H^0 will give indirect but important information on the contribution of such collisions to beam divergence, a factor which, due to lack of information, has been disregarded in theoretical calculations.

The experiments will make use of two unique BNL facilities. The first stage of experiments will be performed at the MP6 Tandem Van de Graaff accelerator. This accelerator is equipped with a source at the high energy terminal and can be run at negative voltage. As a result, negative ions (including H^-) can be produced and accelerated to energies of 2-10 MeV. In the next stage, the experiments will be performed at the 200-MeV BNL Linac.

The location of the H^- source and the beam path are shown in Figure 1 which is a schematic of the BNL Double-MP Tandem facility. Figure 2 shows a schematic of the experimental beam line. Gas cell 1 is for the production of positive ions or the reduction of the H^- beam intensity. Gas cell 2, in conjunction with the first deflector, results in the production of a neutral beam. The position of the first deflector and the antiscatter collimator have been interchanged in the actual experiment. Also, the first deflector deflects in the horizontal plane while the second one deflects in the vertical plane.

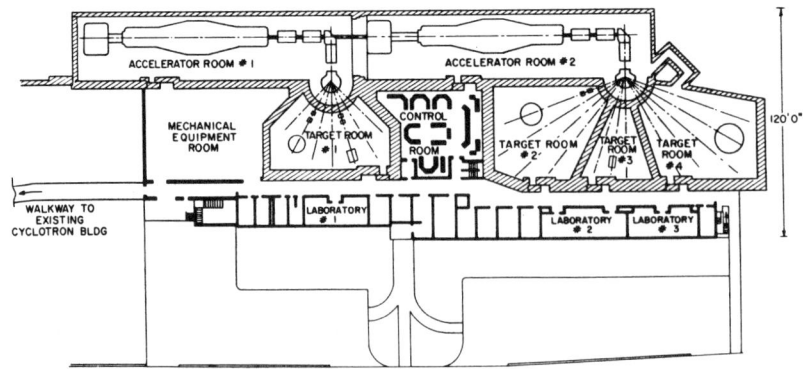

Figure 1.

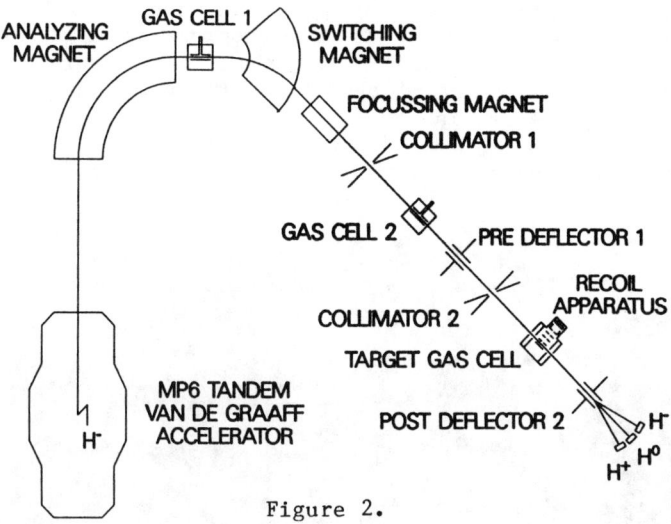

Figure 2.

Figure 3 shows in more detail the target area, including the recoil time-of-flight (RTOF) spectrometer. The electrons and ions produced in the gas are extracted by two grids, designated + and −. The electrons impinge directly on a microchannel plate detector (MCP) while the ions first pass through a specially designed drift space. Thus, the time of arrival at the MCP uniquely determines their charge state. The recoiling ions are detected in coincidence with the electrons and the different charge states of the emerging projectiles.

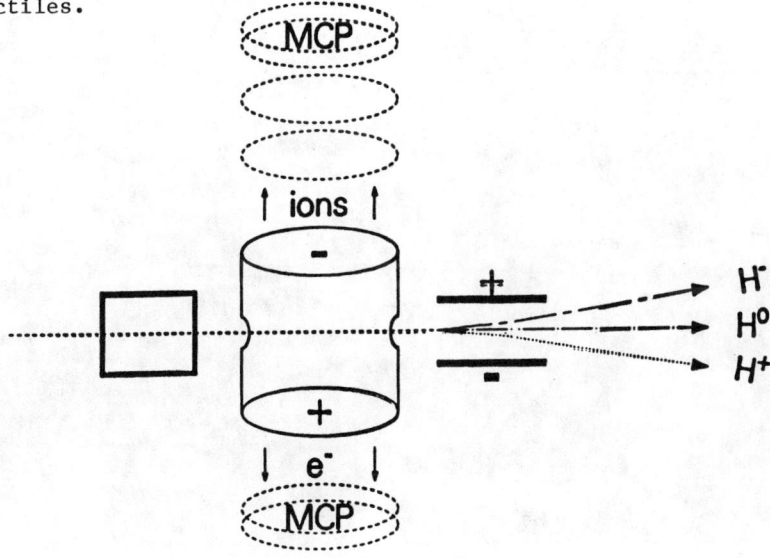

Figure 3.

Figures 4 and 5 are photographs of the beam line during a preliminary run. Figure 6 is a photograph of a phosphor screen situated in the beam path (the viewer in figures 4 and 5). Operating both the pre- and postdeflectors and choosing an appropriate pressure in the neutralizer gas cell, all three components H^+, H^0, and H^- could be obtained in comparable intensities and separated both horizontally and vertically.

Figure 4.

Figure 5.

647

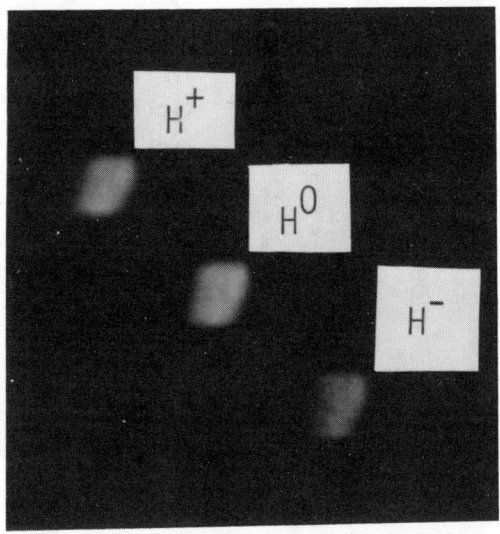

Figure 6.

Figures 7 and 8 display the available experimental cross sections for the one-electron and two-electron stripping of H⁻. The circles represent our preliminary results at 2.5, 5.6, and 8.0 MeV. The two additional data points for N_2 and Ar target gases are from two different sources in the literature, one at 10 MeV and the other at 14.6 MeV. Although there seems to be some systematic problem in our preliminary data, the data points seem to follow the 1/E dependence predicted by the Born approximation for stripping cross sections in this energy region, as shown in the figures for the N_2 target.

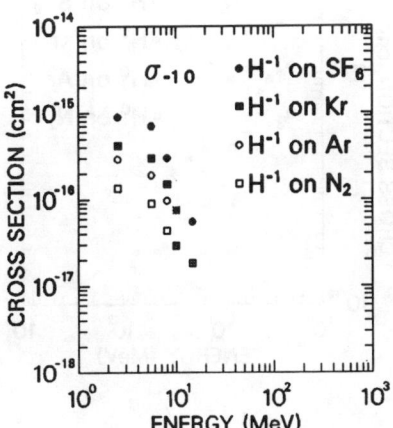

Figure 7.

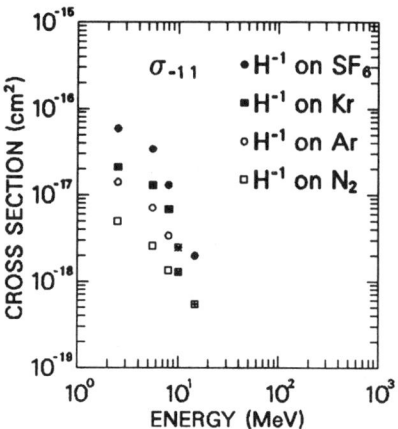

Figure 8.

Figure 9 displays the stripping cross section of the H^0 projectile. Here some more data points are available in the literature for N_2 and Ar targets. The circles represent our preliminary results. Again, it can be seen that the N_2 data points seem to follow the $1/E$ line, to the available accuracy. However, surprisingly, the Ar data would be in much better agreement with a $1/E^{0.8}$ line. Considering the quality of the available data, this may be a spurious occurrence, but clearly more measurements are needed.

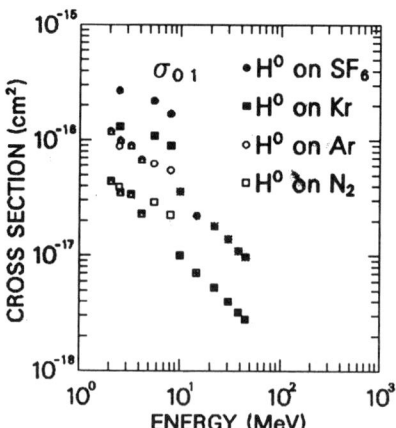

Figure 9.

Figure 10 displays the neutralization efficiency as a function of projectile energy. The neutralization efficiency is defined as the maximum neutral fraction obtainable, using a target of optimum thickness, and is calculated from the previous cross section. The data points represent our results (circles) and other results for N_2 and Ar, and the dotted line represents data from literature for H_2 target gas. The dependence of the efficiency on both target gas and projectile energy is very weak. Still, the efficiency seems to decrease slightly with the increasing atomic number of the target gas. Since the use of very light targets, such as H_2, does not appear to be practical, we have decided to include in our future studies gases with relatively large molecules of low-Z atoms, such as hydrocarbons. Also, when trying to extrapolate to a projectile energy of 200 MeV, one would tend to draw a slightly declining curve, arriving at an efficiency value of about 45%, which is considerably lower than the presently accepted 57%. Again, this might be just an artifact due to the scarcity of available data, and further measurements are necessary.

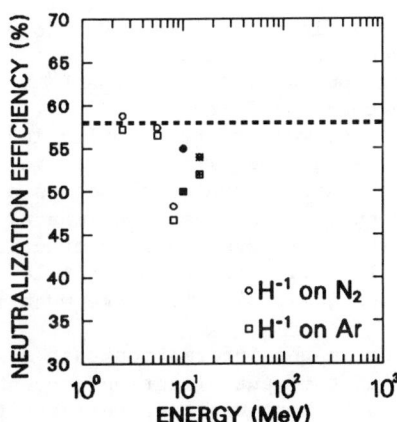

Figure 10.

After performing the proposed experiments on the Tandem Van de graaff accelerator, the following steps are planned:

1. Linac experiment. The whole apparatus for the detection of the recoils can be transferred without changes from the tandem to the 200-MeV BNL Linac. The charge separation and detection of the projectiles at the Linac are also feasible. Thus, since the experiment can be performed at several energies, interpolation and extrapolation to any relevant energy is possible.

2. _Ejected electron spectroscopy._ The average kinetic energy of the ejected electrons and their distribution is the crucial factor in determining the expected charge imbalance in the target plasma. The electron spectrum can readily be measured in both the tandem and Linac experiments by adding an electron spectrometer to the available setup.

3. _Charge production in thick targets._ All previous measurements are to be performed on thin gas targets in order to measure accurate cross sections. However, additional target ions can be produced in thick targets by the electrons ejected in primary collisions. Also, the energy distribution of the electrons will change in a thick target due to their frequent collisions. It is thus proposed to develop a thick gas target where these effects can be measured.

DISCUSSION

Anderson: I have two questions. One is: How do you get negative ions out of a tandem accelerator? And the second one is: Do you have any information on what atomic level the H^0 is in and what this cross-section you have corresponds to?

Meron: On the first one, we get the negative ions because one of the two tandems, MP-6, is equipped with a negative terminal source. It was done so that both tandems could be used as a single multiple tandem for high energy nuclear physics experiments. So the tandem can be operated at negative voltage and give practically any negative beam energy up to 10 MeV. Now, about what level the H^0 was – at the moment, we have no idea. It will take much more experimental refinement to find it.

Roberts: Did you notice in the information that Joel Fink showed from George Gillespie, of the neutralization percent as a function of the atomic number, that there was some periodicity in that chart? The more tightly bound things, noble gases and all, show less neutralization efficiency than the very lightly bound atoms. It may be that, in the case of CH_4, the molecular electronic orbitals behave somewhat like the more lightly bound ones. I'm not sure, but it wouldn't be inconsistent with the data.

Meron: We'll consider this possibility, it's quite possible. In fact, the reason we started with the CH_4 target was in an attempt to get something similar to hydrogen targets without the problems involved with pumping hydrogen away. But it seems to be a possibility with the CH_4. That would explain the slightly higher neutralization efficiency. Still, it's not an indication that the situation will stay like this once we go to high enough energy. One would expect that getting to high enough energy, a few hundred MeV, the binding energy of electrons in the target won't be very important.

CHARACTERISTICS OF AN RF PLASMA NEUTRALIZER*

J. R. Trow and K. G. Moses
JAYCOR, Plasma Technology Division, Torrance, CA 90503

ABSTRACT

Measurements of the electron density and temperature of a high-power, rf-driven, magnetic multicusp contained plasma are presented. The plasma chamber is a shortened version of an rf plasma neutralizer design for slab H$^-$ beams; hence its ends have no cusp fields, but instead contain large openings to emulate the plasma lost through beam apertures. Data is presented from operation with hydrogen in the 1-10 mTorr range and rf power up to 60 kW. Peak electron densities of 1.2×10^{13} cm^{-3} and maximum degrees of ionization of 12% have been recorded. This density is sufficient for an H$^-$ plasma neutralizer; but for a significant enhancement of the neutralization efficiency over that of a gas cell, the degree of ionization must be increased. Based on these results, estimates of the power requirements of a practical rf plasma neutralizer are presented.

INTRODUCTION

A highly ionized, low temperature plasma is a more efficient neutralizing agent for negative ions than are neutral media such as gases or foils. The theory for plasma neutralizers was presented at the second of these symposia by Berkner,[1] and has been reviewed at this meeting by Fink,[2] so only the results relevant to this work will be mentioned here. These are twofold: first, the maximum obtainable neutral fraction from an H$^-$ beam with a fully ionized plasma neutralizer is ~ 85%, about 20% more than from un-ionized gas; and second, the necessary target line density for optimum yield is only a fraction (~ 0.4) of what is necessary for an un-ionized neutralizer. The performance of a neutralizer rapidly changes with the degree of ionization until around 30%, at which point nearly all of the increase in the maximum neutral yield and most of the decrease in the optimum target thickness have been obtained. The increase in neutral yield is an obvious improvement; but the decrease in the required target thickness is also advantageous, since it means less gas loading for a fixed neutralizer length, or a shorter neutralizer for a fixed pumping capacity.

We have undertaken an experimental program to establish the operational characteristics of an rf-driven plasma neutralizer. In particular, this data will be used to design a neutralizer compatible with the slab negative ion beam of the Lawrence

---

*This work performed under U. S. Department of Energy Contract No. DE-AC03-84ER80153.

Berkeley Laboratory Transverse Field Focus Accelerator[3] (3 x 25 cm, 1 A, 200-500 keV). The conceptual design parameters for a slab beam neutralizer cell are presented in Table I. The plasma is produced by rf fields from an inductive antenna immersed in the discharge, a technique which has also been used successfully for producing plasmas in positive ion sources.[4,5] In this neutralizer design, the negative ion beam passes through the center of the antenna loop. Charged particle losses transverse to the beam path, especially ionizing electrons, are retarded by a strong, permanent magnet, multicusp arrangement. The information necessary to design such a plasma neutralizer is the dependence of electron density on rf power and the optimum antenna design including the matching circuitry. The rest of this paper reports on the design data and operational characteristics of a plasma neutralizer determined by Langmuir probes in experiments on a small test plasma cell.

Table I. Conceptual design for plasma neutralizer.

Parameter	Appropriate Value
Cross sectional area	15 x 35 cm^2
Length	100 cm
Target line density (plasma and gas)	3 x 10^{15} cm^{-2}
Electron density	1 x 10^{13} cm^{-3}
Neutral gas density	2.5 x 10^{13} cm^{-3}
Degree of ionization	20-30%
Electron temperature	low (to reduce loss)
Transverse profile	uniform
Longitudinal profile	not important

EXPERIMENTAL APPARATUS

The test plasma cell (Fig. 1) has the same configuration as the final neutralizer cell, except it is shorter and does not have the large cooling capacity required for long pulse operation. The

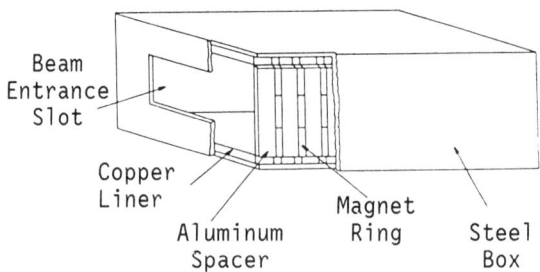

Fig. 1. Cutaway of 1/3-scale Plasma Neutralizer Cell, showing layered wall construction.

multicusp magnetic field is generated by 12 rectangular rings, each composed of 20 Nd-Fe-B magnets (B_{rem} = 12 kG, 7.5 x 50 mm pole faces, 7.5 mm thick) arranged so that all the magnets in any one of the rings present the same pole inward, and the polarity alternates from ring to ring. The magnet array is mounted inside a mild steel box which serves as a flux return, enhancing the field inside, while eliminating stray magnetic fields outside. The magnets are separated from the discharge by 0.7 mm thick copper liner.

A two-dimensional field calculation simulating the field in the plane containing the beam axis and the short dimension of the box (Fig. 2) shows that a cusp field strength exceeding 4 kG is achieved, while a vacuum field of less than 20 G is obtained throughout the region occupied by the beam. The ends of the box have no cusp-producing magnets but instead contain large open slots (5 x 25 cm), an arrangement which emulates the worst-case conditions for plasma loss from a neutralizer.

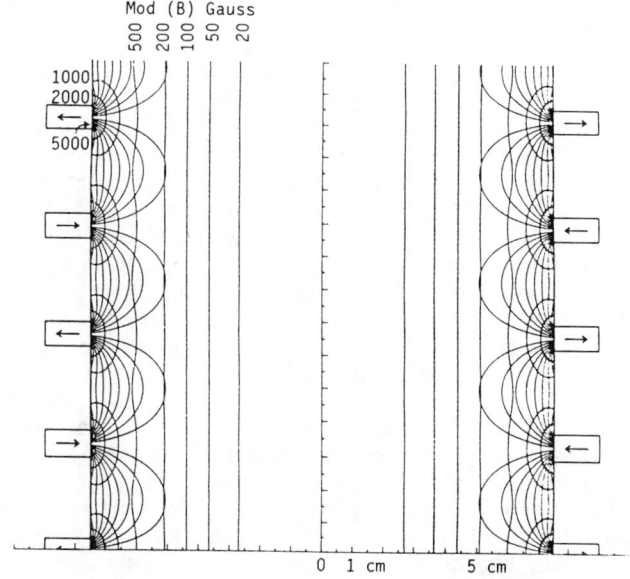

Fig. 2. 2-D magnetic field with the same narrow dimension (15 cm) but infinite in the other. The steel box is allowed for by making the magnets twice as high.

The rf power circuit consists of an oscillator, two amplifier stages, and an antenna circuit. The first amplifier stage is a broad band, high gain, pulsed, solid state amplifier capable of delivering 20 kW into a 50 Ω load. The final amplifier has a tunable operating frequency, a low gain, a low output Q, and a high pulse output power (see Table II); and it is also designed

Table II. High power rf amplifier characteristics

Nominal maximum power	80 kW
Pulse length	1 msec
Maximum duty cycle	< 0.1%
Operating frequency band	0.4-4 MHz
Gain	~ 10-20
Output Q	3-4
Bandwidth $\delta\omega/\omega_0$	~ .3

for optimum performance with a 50 Ω resistive load. The low Q, low gain design permits the power amplifier to tolerate the large change in the effective load impedance encountered during rf breakdown without the power level oscillating or being seriously reduced. The rf voltage, current, and gross rf power delivered to the antenna circuit are monitored at the output of the amplifier. In this paper, no attempt has been made to distinguish between antenna circuit losses and net power delivered to the plasma. The envelopes of the rf output signals are used as an indication of the effective impedance and the relative phase between antenna current and voltage. The antenna matching circuit is adjusted to provide an effective 50 Ω resistive load with the antenna immersed in plasma. No readjustment was necessary for any setting over the range of parameters used. The operation thus far has been with a single-turn, rectangular loop antenna, mounted at the midplane of the chamber (Fig. 3). This antenna is formed from 6.5 mm dia copper tubing with a thin layer of teflon for high voltage insulation and an outer protective jacket of braided alumina. The plasma cell is operated at low gas pressures, which makes it necessary to provide some free electrons to initiate the discharge. The electrons are produced by a small, hot tungsten filament (~ 1-10 mA emission current) mounted just beyond one end slot. This location is convenient for the present test configuration; but in a practical neutralizer, the filament would be located inside the cell.

The gas is introduced through two apertures located along the axis of the cell floor. The fill pressure is controlled by adjusting the voltage and duration of the pulse applied to a piezoelectric valve in the gas feed line. Calibration of fill pressure is accomplished by a separate series of measurements in the absence of rf power using a movable milliTorr ion gauge (Fig. 3). The gas valve pulse is 150 msec in duration and terminates approximately 1 msec before the initiation of the rf pulse. The triggering sequence chosen yields a uniform, central fill-gas pressure-profile, which is relatively constant with time during the interval corresponding to the rf discharge.

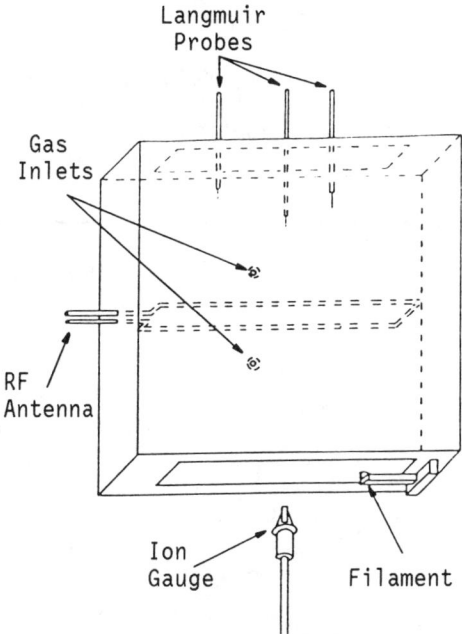

Fig. 3. Schematic of 1/3-scale Plasma Neutralizer Cell.

Two cylindrical Langmuir probes (0.51 mm dia x 5.1 mm long) which can be moved parallel to the beam direction through the chamber are used to measure the local plasma density and electron temperatures: one is on the central axis of the cell, and the other is located 8.3 cm to one side. The probes are repositioned between shots. A common, gated triangular voltage pulsetrain (~ 25 kHz) is applied to the probes. The triangular voltage is channeled to each probe by a separate, high current, floating amplifier which is grounded by a 1 Ω, 0.1% resistor across which the probe current signal is measured. The 1 Ω resistor is part of a series, parallel RLC filter circuit that yields a minimum current signal at the rf frequency, but is not resonant at that frequency. The probe characteristic is uneffected by the filter since the L and C have no effect at frequencies below 500 kHz. The current and voltage signals from a single simultaneous 20 μsec sweep of each probe during the time of maximum plasma density are displayed by oscilloscopes and photographed. The plasma density is determined from the ion saturation current using a Laframboise[6,7] fit parameter. Since the two probes yield essentially the same results only the data from the center probe is presented.

RESULTS

The electron density and temperature, derived from Langmuir probe measurements, are presented in Figs. 4 and 5 as functions of axial position, the rf power applied to the antenna, and the hydrogen gas pressure in the cell. The data in these figures was

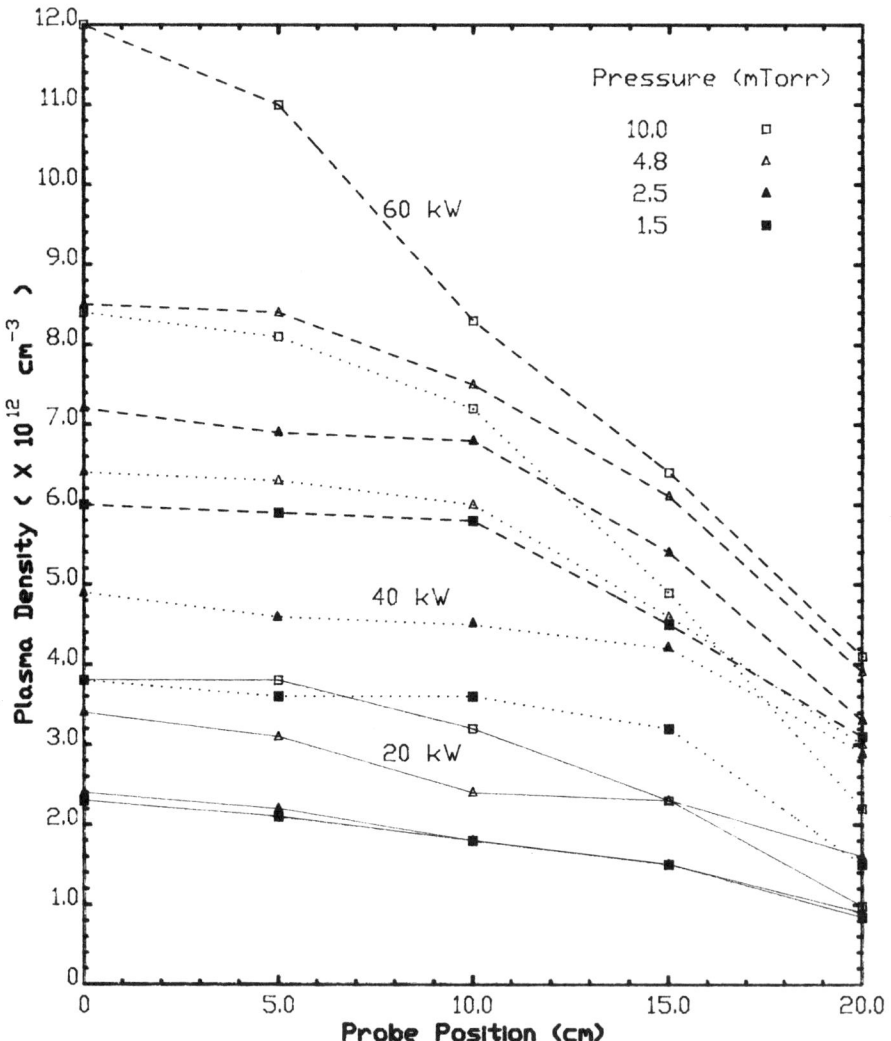

Fig. 4. Electron density as a function of position. The chamber center is 0 cm, and the edge is 20. There is some overlap with respect to power.

taken at five positions spanning the region from the plane of the antenna to the beam slot in the end of the cell. The position x = 0, in the figures refers to the position of the antenna (the midplane of the cell). Four different fill pressures spanning the range from 1 to 10 mTorr and three different rf power levels of approximately 20, 40, and 60 kW were used for each probe position.

In general, the trends found in the data of Fig. 4 are as expected; the plasma density increases with applied rf power and operating gas pressure. There is a noticeable overlap of plasma density for the three rf power levels with the density at 10 mTorr exceeding that at 1 mTorr at the next higher power. The plasma

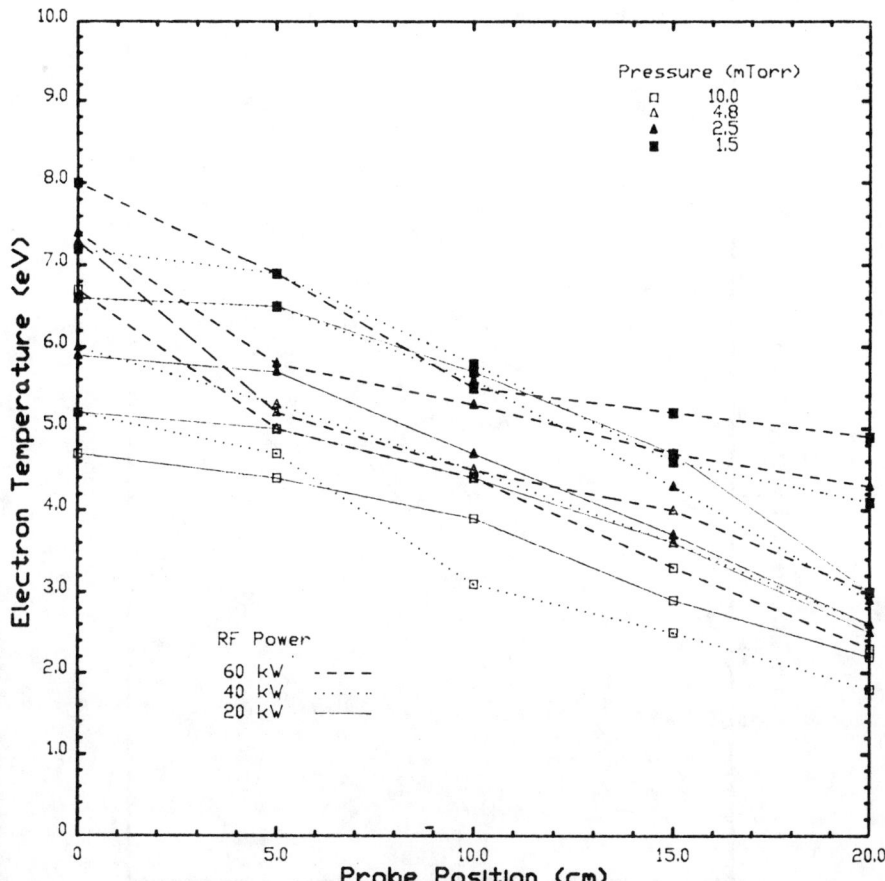

Fig. 5. Electron temperature as a function of position. For each pressure higher power yielded slightly higher temperatures.

density falls off more steeply with axial distance from the antenna as the gas pressure is raised, which is consistent with the concomitant decrease in the mean free path for electron-neutral collisions. Likewise, the data presented in Fig. 5 is also reasonable; higher electron temperatures result from applying higher rf power or operating at lower gas pressure. The range of electron temperatures produced in the rf discharge is fairly narrow, with the temperature decreasing almost uniformly with distance from the antenna. The central density (x = 0 position) is plotted in Fig. 6 as a function of the gross rf power applied

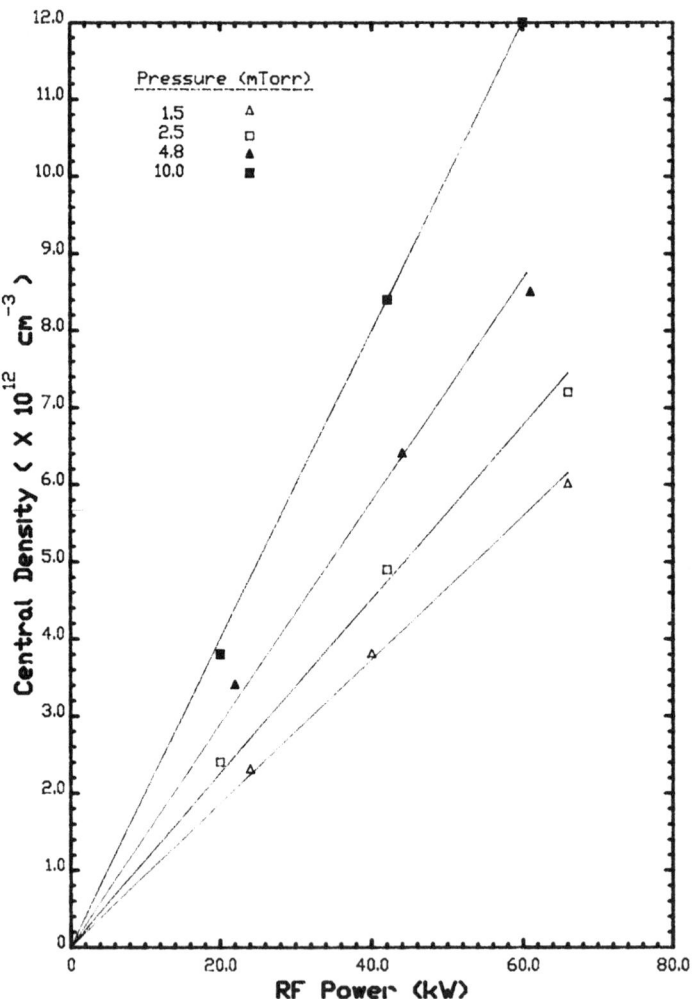

Fig. 6. Electron density at the center as a function of rf power applied to the antenna circuit.

to the antenna at four different gas pressures. These plots show a linear dependence of the central plasma density with applied rf power without any indication of saturation suggesting that higher plasma densities can be obtained at roughly the same efficiency with the application of more rf power.

SCALING ESTIMATES

We estimate the effective ionization energy, ε_{eff}, by

$$\varepsilon_{eff} = \frac{\text{Power} * \text{Loss Time}}{\iiint n_e(x) \, d^3x} , \quad (1)$$

where $n_e(x)$ is the plasma density at the position x. The loss time, τ_{loss}, is estimated by

$$\tau_{loss} \sim \frac{\ell}{c_s} = \ell \sqrt{\frac{m_i}{2T_e}} . \quad (2)$$

Where c_s is the sound speed, ℓ is the cell length, m_i is the average ion mass, and T_e is the electron temperature. This is a reasonable means to estimate τ_{loss} since only the transverse plasma losses are inhibited by the multicusp field, but the plasma flows freely in the axial direction and hence out through the end slots. Loss times calculated using $m_i = 2$, $\ell = 39$, and T_e from the probe data are consistent with the observed decay time of the ion saturation current signal from a probe with fixed bias. The volume integral of electron density in Eq. (1) is approximated by the product of the cross sectional area and twice the numerical integral of the electron density versus axial position. The average electron density $\bar{n}_e$ is defined as the value of this numerical integral divided by half the cell length. Under the conditions discussed above, and using the dimensions of the test cell (15 x 35 x 39 cm), ε_{eff} can be expressed as

$$\varepsilon_{eff} \sim 12 \frac{\text{Power(kwatts)}}{\sqrt{T_e} \, \bar{n}_e \, (\div 10^{12})} . \quad (3)$$

Using this expression yields ε_{eff} = 64 eV ± 10 eV for almost all data points; the exceptions being ε_{eff} = 84 eV and 97 eV evaluated from data taken with 20 kW of rf power at 2.5 and 1.5 mTorr, respectively.

The value of 64 eV for ε_{eff} is used to estimate the power necessary to run a full-sized neutralizer. The average plasma density can be given in terms of the required target line density π, the degree of ionization χ, and the neutralizer length ℓ. Since

$$\pi_{total} = (\bar{n}_o + \bar{n}_e + \bar{n}_i)\ell , \qquad (4)$$

and $\bar{n}_e = \bar{n}_i$, then

$$\bar{n}_e = \frac{\chi \pi}{(1 + \chi)\ell} . \qquad (5)$$

Rearranging (3) and substituting $\bar{n}_e$ from (5) yields

$$\text{Power(kwatts)} \sim \frac{64 \chi \pi \sqrt{T_e}}{12 \times 10^{12} (1 + \chi)\ell} . \qquad (6)$$

As shown in Fig. 5, T_e increases slowly with power, so $T_e \cong 10$ eV will be assumed. For a 500 keV H$^-$ beam the required neutralizer parameters are $\pi = 3 \times 10^{15}$ cm^{-2}, $\ell = 100$ cm, and $\chi = 0.3$. The required power calculated by using Eq. (6) is 120 kW.

We can also estimate the power required for a multi-MeV H$^-$ beam plasma neutralizer requiring $\pi \sim 10^{17}$ cm^{-2}, however, the value of 64 eV may not be as accurate for such large values of ℓ since the transverse plasma loss will no longer be negligible in comparison to the axial losses. Assuming $\ell = 1000$ cm, $T_e = 10$ eV, and $\chi = 0.3$, Eq. (6) gives 400 kW rf power required for a neutralizer with a cross sectional area of 525 cm^2.

SUMMARY

An approximately 1/3-length test plasma cell has been operated with up to 66 kW of applied rf power producing electron densities in excess of 10^{13} cm^{-3}. The electron density is a linear function of power with no apparent saturation, up to the maximum applied power. The electron temperatures were found to be $\sim$ 6-8 eV. The maximum degree of ionization obtained so far has been about 12%. A plasma neutralizer with an electron density of the order of 10^{13} cm^{-3} is sufficient for use in a 500 keV H$^-$ beamline; however, the degree of ionization must be increased to approach optimum neutralizer performance.

The areas to be emphasized in subsequent research are (1) increasing the degree of ionization to between 20 and 30%; (2) exploring the effect of various antenna geometries; (3) determining if there is a frequency dependence; and (4) verifying the density scaling with cell length.

REFERENCES

1. K. H. Berkner, R. V. Pyle, S. E. Savas, and K. E. Stalder, "Plasma Neutralizers for H^- or D^- Beams," Second International Symposium on the Production and Neutralization of Negative Ions and Beams, BNL 51304 (Brookhaven National Laboratory, 6-10 Oct. 1980).
2. J. Fink, "Neutralizer Options for High Energy H^- Beams," these proceedings.
3. O. A. Anderson, C. F. Chan, W. S. Cooper, J. W. Kwan, C. A. Matuk, H. M. Owren, J. A. Patterson, P. Purgalis, and L. Soroka, "Overview and Status of the Transverse Field Focus Accelerator," 1985 Particle Accelerator Conference, Vancouver, B.C., 13 May 1985, also LBL-19593 (Lawrence Berkeley Laboratory, 1985).
4. W. F. DiVergilio, H. Goede, and V. V. Fosnight, "Development of Radio Frequency Induction Plasma Generators for Neutral Beams," Rev. Sci. Instrum. 57 (7) 1254 (1986).
5. H. Goede, W. F. DiVergilio, V. V. Fosnight, M. C. Vella, K. W. Ehlers, D. Kippenhan, P. A. Pincosy, and R. V. Pyle, "Radio Frequency Induction Plasma Generator 80-kV Test Stand Operation," Ibid, p. 1261.
6. J. G. Laframboise, "Theory of Spherical and Cylindrical Langmuir Probes in a Collisionless Maxwellian Plasma at Rest," UTIAS Rep. No. 100 (University of Toronto, Institute of Aerospace Studies, 1966).
7. P. M. Chung, L. Talbot, and K. J. Touryan, <u>Electric Probes in Stationary and Flowing Plasmas: Theory and Application</u> (Springer Verlag, 1975).

DISCUSSION

<u>Lietzke</u>: Having operated RF antennas and filaments in the same bucket source, I have yet to determine that an RF antenna will really last longer. Do you have any idea, from your experience, how long an antenna lasts?

<u>Trow</u>: I've got no idea. We've run short pulse, and I've seen no deterioration over the full time that we've operated with the antenna.

<u>Moses</u>: Our antennas are 1/4" copper tubing, and it would take a long time, I think, to wear that out.

<u>Lietzke</u>: But it's probably the insulation on the antenna that deteriorates with time. I ran mine steady state and I couldn't get it to last longer than a 1.5 mm filament...not to say that can't be improved. With the technology that we inherited from TRW, we couldn't do better than that. The second question has to do with end effects in the magnetic field. If you take a periodic row of magnets, you will have, basically, a filter at the exit of that system. What kind of care do you have to take in the plasma neutralizer?

<u>Trow</u>: On the question of end effects, I haven't been so concerned with the actual beam constraints. However, in the construction of the box - this end is also an iron plate, which serves as a flux return - and, if you remember from the plots, the 20G contours is actually below the lip of the plate, so it should serve....

<u>Lietzke</u>: It should be very wide open, then, in your case. Now, the other half of the question is: Can you tolerate small amounts of confining magnetic field?

<u>Trow</u>: As far as how much of the field can be tolerated, I'm not really in a position to answer that, but it's more for people who do the actual beam optics. I've been looking at the requirements to produce the plasma cell.

<u>Moses</u>: I'm not sure I understand your question.

<u>Lietzke</u>: One can design magnetic buckets that have filters at the output. In that particular situation, if you didn't have that iron frame around the outside, there would be a cusp-type magnetic filter on the outside with, basically, losses only along the lines down the middle, which would improve your efficiency somewhat. But if you sent a big beam through there, that would produce some deflections of the top and the bottom of the beam in different directions.

<u>Moses</u>: Right. As you may recall, we were building this, really, for testing at Berkeley, and we wanted to avoid any magnetic fields in the region of the beam.

<u>Forrester</u>: You didn't couple into any magnetic field mode or anything of the kind. You were coupling into a field-free region, just into the resistive plasma, am I right?

<u>Trow</u>: The antenna sat in a region that was less than about 30 or 40G.

OSCILLATIONS IN PHOTODETACHMENT CROSS SECTIONS FOR IONS IN MAGNETIC FIELDS[1]

Oakley H. Crawford
Health and Safety Research Division, Oak Ridge National Laboratory,
Oak Ridge, Tennessee 37831-6123

ABSTRACT

The photodetachment cross section of a charged particle bound in a short range potential is an oscillating function of frequency of incident light, in the presence of a magnetic field. The theory of this effect is described, and calculated cross sections are shown. For photodetachment of electrons from negative atomic ions, this constitutes an additional magnetic field effect beyond the ones associated with fine and hyperfine structure of the ion and atom and with spin of the detached electron.

INTRODUCTION

Electron detachment by laser in a magnetic field is of interest as a possible method for neutralizing negative ion beams. For example, Hershcovitch and Hinds[1] have proposed a laser photodetachment scheme to generate a spin-polarized proton beam, taking advantage of the final-state interaction of the atom with a magnetic field. Also, we suggest that a magnetic field might improve the efficiency of photodetachment, due to a cross section enhancement arising from the interaction between the freed electron and the field. The theory of the effects on the photodetachment process of this electron-field interaction will be described here. A novel experimental approach to measurement of these effects is discussed in the following paper.[2]

Magnetic-field-dependent structure in the photodetachment cross section of a negative ion was first observed by Blumberg, Jopson, and Larson.[3] Performing high resolution photodetachment spectroscopy on sulfur negative ions in an ion trap, they discovered oscillations in the cross section versus laser frequency, with a period proportional to the magnetic field strength. Blumberg and coworkers[3,4] attributed these features to detachment into a series of Landau states of the electron in the magnetic field.

A charged particle in a magnetic field has an infinite set of quantized energy levels (Landau levels) for motion perpendicular to

---

[1]Research sponsored jointly by the U.S. Air Force Office of Scientific Research, under Interagency Agreement DOE No. 40-1262-82 and the Office of Health and Environmental Research, U.S. Department of Energy, under contract DE-AC05-84OR21400 with Martin Marietta Energy Systems, Inc.

the field (while motion parallel to the field is unquantized).[5] The spacing between these levels is $\hbar\omega_H$, where $\omega_H = 2\pi v_H = eH/mc$, v_H being the cyclotron frequency and H the magnetic field.

As a consequence, there is an infinite series of thresholds, separated by energy $\hbar\omega_H$, for detachment of an electron in a magnetic field. The theory of Blumberg et al. predicts that the cross section σ_n for photodetachment into the nth Landau state is infinite at the threshold frequency, v_n and decreases as $(v-v_n)^{-\frac{1}{2}}$ for laser frequency $v > v_n$. The total photodetachment cross section, obtained by summing the σ_n, thus consists of a series of infinite spikes. However, the result of convoluting over a distribution of Doppler shifts (to account for thermal motion) is finite. Using as free parameters the scale of the cross sections, the origin of the frequency scale, and the ion temperature, a good fit[4] to experiment was achieved.

Subsequent work has shown the importance of incorporating the final state interaction,[6-10] i.e., between the electron and the atom into the theory. It has been proven[7,9] that the cross section remains finite, given this interaction.

Recently, Gurvich and Zil'bermints[9] presented a detailed theory for photodetachment of a charged particle bound to a small-radius center in a magnetic field. Although motivated by experiments on D^- and A^+ centers in semiconductors, this theory is applicable to the problem of interest here, as well. The treatment of Gurvich and Zil'bermints has been extended[11] to handle arbitrary values of angular momentum and arbitrary laser polarization. Below, theoretical results for some cases of interest will be displayed and discussed.

THEORY

Consider a particle of charge $-e$ bound to a potential of radius r_o, beyond which its value is negligible. We choose initial angular momentum and laser polarization to be such that dipole selection rules (ignoring the field) would determine unique values,[12] ℓ and m, of the particle's final angular momentum quantum numbers, where m is[13] the component in the direction of the magnetic field. The charged particle of interest is an electron, but its spin does not enter into the theory of the effect under consideration here. The interaction of the spin with the magnetic field simply shifts the cross section in frequency.

Following Gurvich and Zil'bermints,[9] we assume

$$kr_o \ll 1, \quad |\beta_\ell| \ll 1, \qquad (1)$$

where

$$\beta_\ell = -2^{-\frac{1}{2}}(2\ell+1)!!\, a_\ell\, a_H^{-2\ell-1}. \tag{2}$$

where k is the wavenumber of the electron and a_ℓ the scattering length, both in the final state, and $a_H = \omega_H^{-\frac{1}{2}}$ is the cyclotron radius of the electron. The second condition in (1) is satisfied when the scattering length exists, thus excluding a zero-energy bound state, and when the magnetic field is correspondingly not too large.

Now, consider the case $|m| = \ell$. This holds, for example, for detachment into an s wave or a $p_{\pm 1}$ wave.[12] One finds that the cross section σ^H (knm) for photodetachment into the Landau level with quantum numbers nm is given approximately by

$$\sigma^H(knm) = \alpha(2\ell+1)!!\, 2^{-\frac{1}{2}}\, a_H^{-2\ell-1}\, |M|^2 \binom{\ell+n}{n}\left[\frac{\omega-\omega_{mn}}{\omega_H}\right]^{-\frac{1}{2}},\quad |m|=\ell \tag{3}$$

where

$$M = \left[1 - i\beta_\ell \sum_{n'}\binom{\ell+n'}{n'}\left[\frac{\omega-\omega_{mn'}}{\omega_H}\right]^{-\frac{1}{2}}\right]^{*-1} \tag{4}$$

where ω_{mn} is the threshold frequency times 2π,

$$\omega_{mn} = A + (n + \tfrac{1}{2}|m| + \tfrac{1}{2}m + \tfrac{1}{2})\,\omega_H \tag{5}$$

A being the (field-free) binding energy in the initial state. The sum in Eq. (4) goes from n = 0 to the first value of n for which $\omega - \omega_{mn} < 0$. The constant α in Eq. (3) is defined as the proportionality factor between the wavenumber and the cross section $\sigma(k\ell)$ in the absence of a magnetic field,

$$\sigma(k\ell) = \alpha k^{2\ell+1}. \tag{6}$$

These equations are equivalent to the results of Gurvich and Zil'bermints.[9] When written in the above form, they remain valid for two- or multi-photon processes, provided intermediate states are not affected by the magnetic field.

Fig. 1 shows a calculation for detachment into an s state, taking $a_\ell = \alpha = 1$. This result is relevant to photodetachment of S^-. Of course, the latter would be modeled by a superposition[3,4] of curves such as shown here, of differing magnitudes and threshold

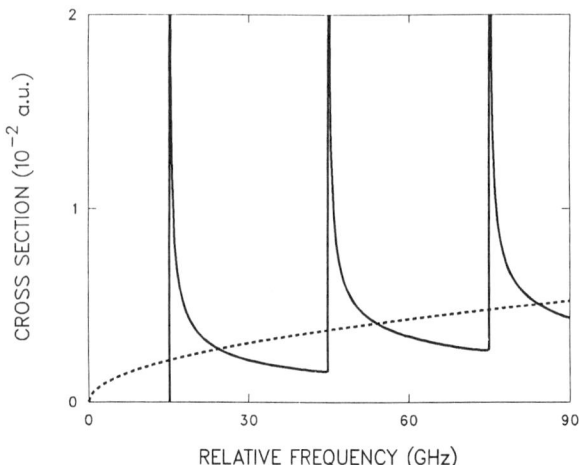

FIG. 1. Cross sections for photodetachment into an s state in a 1.07 T magnetic field (solid curve) and in zero field (dashed curve), as functions of relative frequency of incident light. The frequency here and in Figs. 3 and 4 is measured relative to the threshold for H = 0 photodetachment.

frequencies due to the presence of fine structure in both the ion and the atom, and the spin of the detached electron. The solid curve is calculated for a magnetic field of 1.07 T (where v_H is 30 GHz or $\omega_H = 4.56 \times 10^{-6}$ a.u.) and the dashed curve is for zero field.

The solid curve in Fig. 1 consists of a series of sharp spikes appearing identical to the k^{-1} singularities predicted by Blumberg et al.[3] This is also the behavior predicted by Eq. (3) when M = 1. From Eq. (4) it follows that M is identically unity when the potential vanishes, in agreement with the original theory.[3,4] Also, $M \approx 1$ when, for all n, $|\omega - \omega_{\ell n}|/\omega_H \gg \beta_\ell^2$. Here, $a_H = 468$ a.u., and $\beta_\ell = -1.51 \times 10^{-3}$ a.u., so the major departures from k^{-1} behavior occur within only narrow regions, defined by $|\omega - \omega_{\ell n}| \lesssim 2.3 \times 10^{-6} \omega_H = 68$ KHz, around each threshold.

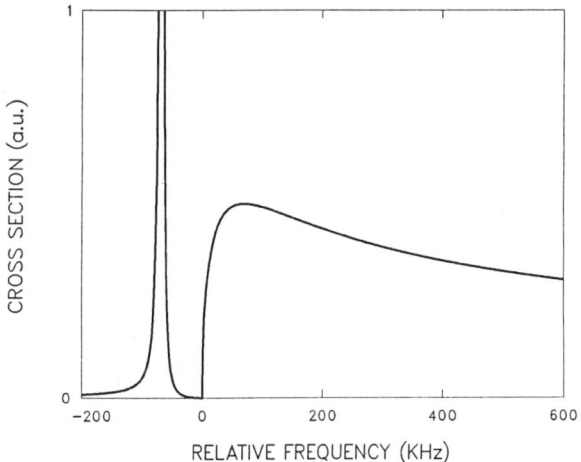

FIG. 2. Cross section for photodetachment into an s state in a 1.07 T magnetic field, as a function of relative frequency of incident light. The frequency is measured relative to the n = 1 threshold.

The details of σ near the n = 1 threshold are shown in Fig. 2. This is part of the cross section curve displayed in Fig. 1, but now on a greatly expanded frequency scale. Here, zero relative frequency denotes the threshold. We see peaks on both sides of threshold, with maxima at $\omega - \omega_{\ell n} = \pm \beta_o \omega_H = \pm$ 68 KHz.

The above features of the cross section have the following explanation. The Hamiltonian in the n = 1 channel has, whenever $\beta_\ell > 0$, a bound state, of (very small) binding energy $\beta_\ell \omega_H$, in its spectrum, when coupling to the n = 0 channel is ignored. That state manifests itself in two ways. First, because of the channel coupling, a resonance appears in the n = 0 channel (at the same total energy), which shows up in Fig. 2 as the very sharp peak below the n = 1 threshold. Second, the n = 1 cross section becomes very large just above its threshold, just as do elastic s-wave cross sections when a barely bound state exists.

The nature of the resonances below thresholds is clarified by considering that, in the final state wave functions, the x and y coordinates (perpendicular to H) play the role of internal degrees of freedom. The problem becomes a one-dimensional multichannel one. (This representation is central to some of the previous work, and is implicit in the above discussion of coupled channels.) The features below thresholds are then identified as core-excited (Feshbach) resonances. These resonances occur below all but the lowest

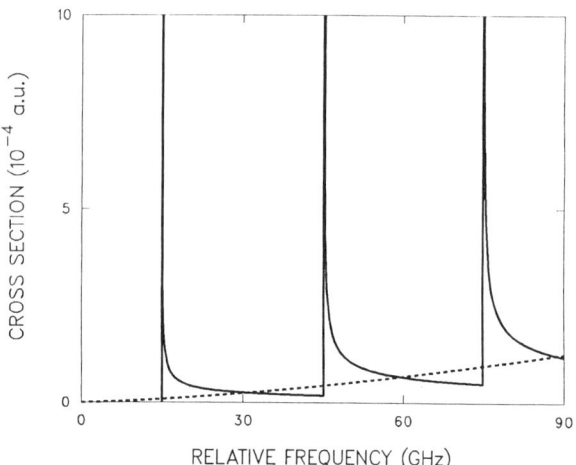

FIG. 3. Cross sections for photodetachment into an $(l,m) = (1,-1)$ state in a 1.07 T magnetic field (solid curve) and in zero field (dashed curve), as functions of frequency of incident light.

threshold when $\beta_l > 0$, but are missing altogether when $\beta_l < 0$. On the other hand, the structures above the thresholds are virtually independent of the sign of β_l. This is a familiar feature of single-channel s-wave scattering in the usual 3D situation. What is essential in the case ilustrated by Figs. 1 and 2, however, is the even symmetry of the final-state wave function with respect to reflection in the xy plane (guaranteed whenever $|m| = l$), rather than the rotational symmetry at small r.

Fig. 3 gives the cross section for photodetachment into an $(l,m) = (1, -1)$ state, calculated from Eq. (3). (As above, l has a definite meaning only at small r.) An example is H^- neutralization by a circularly polarized light beam propagating parallel to H. To strengthen this connection, the constant α was given the value[14] 871 a.u. appropriate to H^-. However, a problem arises in assigning the p-wave scattering length a_1 for e - H scattering. Because the potential falls off only as r^{-4} at large r, due to the polarizability, the p-wave scattering length cannot be defined. For the calculation, a value of unity was chosen for this quantity. The figure resembles the one, Fig. 1, for detachment into an s wave, in having a series of what appear (at GHz resolution) to be sharp spikes, separated by the cyclotron frequency. There are some differences, however. The zero-field cross section (dashed line)

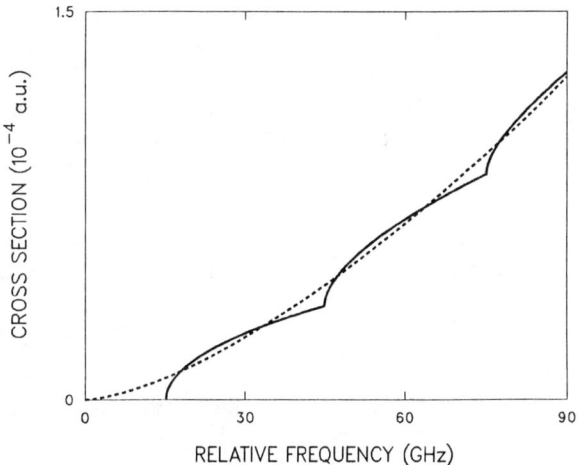

FIG. 4. Cross sections for photodetachment into an $(l,m) = (1,0)$ state in a 1.07 T magnetic field (solid curve) and in zero field (dashed curve), as functions of frequency of incident light.

has a different shape. Still, σ^H, which oscillates about σ, has approximately the same value, when averaged over an ω_H interval, as σ. The other difference, apparent from Eqs. (2) – (4), is that the near-threshold maxima are located much closer to thresholds when $l = 1$.

Now consider the case where the final states are characterized by $|m| = l - 1$. Here, the final-state wave functions have odd symmetry with respect to reflection in the xy plane. We find[11] that the photodetachment cross section for $(l,m) = (1,0)$ is given by

$$\sigma^H(kn0) = 3\sqrt{2}\ \alpha\ \omega_H^{3/2} \left[\frac{\omega - \omega_{mn}}{\omega_H} \right]^{\frac{1}{2}}, \quad l = 1 \qquad (7)$$

We have assumed that neither bound nor virtual states of odd symmetry occur very close to threshold when $H = 0$. They will not be created when the field is applied. Therefore the large features seen above where $|m| = l$ would not be expected here, and Eq. (7) shows that such is the case. A plot of σ^H for photodetachment from H^- in a 1.07 T field by a laser polarized in the z direction ($\alpha = 871$ a.u., $m = 0$) is shown in Fig. 4. One can see that although σ^H increases monotonically, staying close to σ, the thresholds are clearly evident.

Detailed comparison with existing experimental data is not possible because of broadening of spectroscopic features, due mainly to thermal motion. Nonetheless, Larson and Stoneman were able to find an indication, in their recent S^- experiment,[10] that the peaks are shifted upward in frequency compared with the locations of the Landau level thresholds, in qualitative agreement with a proposal by Clark.[7] These shifts are orders of magnitude greater than expected from present theory (cf. Fig. 2). (Indeed, at the level of resolution achieved so far, predictions of the theory discussed here for the $m = \pm \ell$ case, where the large oscillations occur, are experimentally indistinguishable from the results of the theory of Blumberg and coworkers.) Furthermore, Larson and Stoneman find that the data taken with π polarization tend to fit the model better than those taken with σ polarization. These findings suggest that a closer examination of the theoretical treatment of transitions between fine structure components would be useful.

CONCLUSIONS

Experiments at higher resolution are needed[2] to provide critical tests of the theory of photodetachment in a magnetic field. In the meantime, the theory requires further development. There is already a suggestion that it be reexamined, particularly where transitions between fine structure components of ion and atom are involved.[10] Furthermore, the theory requires extension to cover cases where the scattering length in the exit channel is undefined, as in photodetachment from H^-.

REFERENCES

1. A. I. Hershcovitch and E. A. Hinds, Proc. 3rd Int. Symp. on Production and Neutralization of Negative Ions and Beams, K. Prelec, Editor, Brookhaven National Laboratory, AIP Conf. Proc. No. 111 (1983) p. 763; A. Hershcovitch, Nucl. Instr. Methods in Phys. Research A243, 271 (1986).
2. H. F. Krause, "A Novel High-Resolution Experimental Approach for Studying Laser Photodetachment in a Magnetic Field," this conference.
3. W. A. M. Blumberg, R. M. Jopson, and D. J. Larson, Phys. Rev. Lett. 40, 1320 (1978).
4. W. A. M. Blumberg, W. M. Itano, and D. J. Larson, Phys. Rev. B19, 139 (1979).
5. L. D. Landau and E. M. Lifshitz, Quantum Mechanics: Non-Relativistic Theory, 3rd Ed. (Addison-Wesley, Reading Mass., 1977), p. 456.
6. D. J. Larson and R. Stoneman, J. Phys. Supp. 11, 43, C2-285 (1982).
7. C. W. Clark, Phys. Rev. A28, 83 (1983).

8. W. P. Reinhardt, J. Phys. B: At. Mol. Phys. **16**, L635 (1983).
9. Yu. A. Gurvich and A. S. Zil'bermints, Sov. Phys. JETP **58**, 754 (1983).
10. D. J. Larson and R. Stoneman, Phys. Rev. **A31**, 2210 (1985).
11. O. H. Crawford, to be published.
12. Although the angular momentum is not well defined in the field, the assumptions made allow us to treat the wave function at small r as though it had a definite value of l.
13. Atomic units, a.u. (e = m_e = $\hbar$ = 1), are used except where otherwise stated.
14. A. L. Stewart, J. Phys. **B11**, 3851 (1978).

DISCUSSION

<u>Nightingale</u>: You've shown the enhancement near threshold; you said it's an order of magnitude or more compared to the underlying continuum near threshold. How does it compare with the continuum at much shorter wavelengths near the peak?

<u>Crawford</u>: That depends on things like what the field is and what the scattering length is. In some of these calculations it's probably about the same order of magnitude; it might be larger. Certainly, if you resolve that little Feshbach resonance it would be larger. But that's a good point you're making because we are looking very close to threshold where the underlying cross-section is quite small.

<u>Hershcovitch</u>: You flashed a viewgraph about the π polarized laser on H^-. Could you elaborate on the big spikes you showed, like in sulphur?

<u>Crawford</u>: Say this is the magnetic field direction. I said it's π^- polarization because I wanted a definite m quantum number. But the important thing was the final state had a symmetry, even with respect to reflection in this plane, and so we still get the effect. In the other curve, where we detached with the σ polarized laser, you're going into a final state that has a symmetry that does not give you the effect. So the laser polarization makes a difference.

<u>K. Smith</u>: These huge cross-sections you have for photodetachment... do you think that they're only in the presence of a magnetic field, or do they exist also when the magnetic field isn't there?

<u>Crawford</u>: Oh, no, they require the magnetic field. The important thing is the electron has to go into that state, which only occurs in the magnetic field.

<u>K. Smith</u>: I seem to recall there were some calculations done, maybe 10 or 15 years ago, on photodetachment of things like S^- and Cl^-, which certainly weren't done in the presence of a magnetic field, and did show those huge spikes.

<u>Crawford</u>: I suspect you're talking about, maybe near some excited state threshold. In the case of S^-...

<u>K. Smith</u>: Local ionizing resonances.

<u>Crawford</u>: Absolutely. Those resonances do not occur in a region that we're looking at. Not at these photon energies.

Roberts: Is the main difference there between a bound-bound transition and a bound-free? And with the magnetic fields letting you do a bound-bound?

Crawford: That's pretty darn close. Because these dashed lines for the states, they're kind of like bound states. They're not really bound, but they're quasi-bound. It's just as though the transition is going up to those states that are extremely close to the continuum. That's a good way to look at it. And it causes the oscillator strength to clump up right around those...

Roberts: And you can't have those without the magnetic field?

Crawford: Absolutely.

A NOVEL HIGH-RESOLUTION EXPERIMENTAL APPROACH FOR STUDYING LASER PHOTODETACHMENT IN A MAGNETIC FIELD

H. F. Krause
Oak Ridge National Laboratory
Oak Ridge, TN 37831 USA

ABSTRACT

The need for new high-resolution photodetachment data is discussed. An $\vec{E}\times\vec{B}$ photoneutralizer concept that is capable of achieving the resolution required for spectroscopic studies is described. Feasibility of the concept has been evaluated in Monte Carlo trajectory studies. The calculations indicate that spectral resolutions in the range 300-30 MHz are achievable in a five-parameter coincidence experiment. These bandwidths are 20 to 200 times narrower than have been achieved heretofore in experiments that used the Penning trap technique. Stray electric fields in the $\vec{E}\times\vec{B}$ photoneutralizer also are much smaller than those in a trap (< 1/20). The improved apparatus resolution should allow Landau resonances to be studied at magnetic field strengths well below those used previously.

I. INTRODUCTION

Interest in the laser photodetachment process $(A^- + h\nu \rightarrow A + e)$ in the presence of a magnetic field began with an experimental investigation of the S^- ion.[1,2] In this investigation and in subsequent studies where the same technique was employed, a narrow-band CW tunable dye laser illuminated low-density ions contained in a Penning ion trap. The photodetachment yield was observed to increase periodically near threshold whenever the laser was tuned to a quantized state of the free electron. More specifically, the extrema occurred whenever different Landau levels in the continuum, $E = (n+1/2)hf$, were addressed by the laser; where n = the quantum number, h = Planck's constant, and f = the classical electron cyclotron frequency (f = 28.18 GHz/T). Although the overall spectral resolution was low (7 GHz), several Landau levels were resolved at magnetic field strengths close to 1 Tesla (T) and the extrema scaled correctly with the magnetic field strength. The resolution was limited by the temperature of the trapped negative ions (Doppler broadening), the existence of motional electric fields (Stark broadening), and other factors. Absolute photodetachment cross sections were not reported.

My interest in a new investigation arose from the following considerations. In a recent theoretical study, Crawford[3] found that the cross section for S^- is enhanced by orders of magnitude in the narrow frequency band (1 MHz) surrounding each Landau level and that the shape near resonance will depend sensitively on the polarization of the laser beam (e.g., whether the electric vector for linearly polarized light is parallel or perpendicular to the magnetic field). Zeeman structure in the Landau resonances[2,3] also should be observable in an experiment performed at much higher resolution. Theories could not be fully tested without the design of an improved experimental configuration capable of high to ultra-high frequency resolution (300-30 MHz).

The opportunity to exploit inherent spin alignment provided an additional incentive to perform an experiment at much higher resolution. Neutrals produced in a magnetic field will be electron spin aligned and the polarization can be controlled by accurate tuning of the laser to particular fine-structure levels. In addition, photodetachment of a negative ion having nuclear spin (e.g., H$^-$) would yield a neutral beam that is both electron and nuclear spin polarized. For ions of this type, each Landau resonance would exhibit hyperfine structure that should be resolvable at a much higher resolution than was used heretofore (x 20). Each hyperfine state corresponds to different combinations of spin alignment for the neutral atom (nuclear and electron spin) and the ejected electron (6 unique frequencies for H$^-$).[4,5] The high resolution apparatus could also function as a low-energy-polarized neutral beam source or perhaps produce an alignment-depleted, negative-ion beam that could be further accelerated.

Whether the spin alignment opportunities are exploited, a better understanding of the Landau resonances could lead to a very substantial increase in the efficiency of a negative-ion photoneutralizer (considering the cross section predictions) if the spacing of Landau resonances were tuned to match the mode structure of the photodetachment laser. The tuning could be accomplished by varying the magnetic field and/or the laser cavity design.

II. The HIGH-RESOLUTION TECHNIQUE

Consideration of the physical processes that limited spectral resolution in previous experiments and the potential applications (noted above) suggested the $\vec{E} \times \vec{B}$ photoneutralizer concept discussed here. The basic idea shown in Fig. 1 consists of (1) photodetaching a negative-ion beam (of finite angular spread) inside an $\vec{E} \times \vec{B}$ device at a position where the ion beam is focused (minimum spatial extent) and (2) detecting the neutrals on a high resolution two-dimensional position-sensitive detector (TDPSD).

The $\vec{E} \times \vec{B}$ photoneutralizer offers unique features that simplify the study of Landau resonances and the interpretation of the results. When the electrostatic force is adjusted to balance the Lorentz force produced by the applied magnetic field of interest (at the known projectile speed), the ion beam will pass through the photoneutralizer without angular deflection. The force balance condition, $v = E/B$ (rationalized MKS units), can be maintained at every point along the trajectory, including the entrance and exit regions of the filter where the electrostatic and magnetic fields are

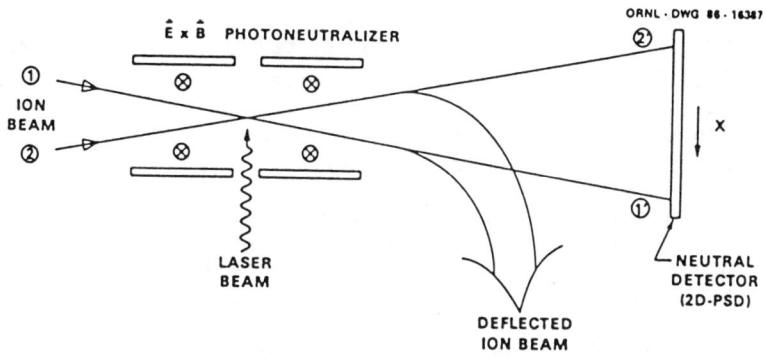

Fig. 1. Conceptual layout of the experiment.

changing, by the use of electrostatic and magnetic shims.[6] Proper shimming of the filter also can virtually eliminate cylindrical lensing effects of the $\vec{E} \times \vec{B}$ photoneutralizer device. Since it is possible to pass an ion beam through the photoneutralizer with no deflection, a simple linear beam-line geometry is possible. This layout also leads to a well-defined interaction region for the ion and laser beams (in a perpendicular or inclined arrangement) required in absolute cross section measurements and it would simplify the interfacing of this apparatus to another, should it be used in beam source applications.

Another attribute of the $\vec{E} \times \vec{B}$ photoneutralizer arises from the fact that crossed E and B fields adjusted for balance in the laboratory frame transform into the projectile frame as a slightly diminished but pure magnetic field.[7] Electric fields are particularly undesirable in a high resolution experiment because the field gives rise to Stark splitting. The magnetic field also must be very homogenous. While exact cancellation of the applied electrostatic field by the motional electric field occurs only for ions moving at velocity v in the $\vec{E} \times \vec{B}$ direction, a perturbation analysis indicates that the spurious uncancelled electric field in the projectile frame will be much smaller (< 1/20 worst case at 1 T) than in a trap environment[1,2] at the same magnetic field. Non-cancellation is being limited to less than 5×10^{-4} of the applied electric field in the apparatus being built. Magnetic field non-homogeneity in the projectile frame is not expected to affect the measurements. The analysis considered non-cancellation due to beam velocity spread, deviation of beam rays from the $\vec{E} \times \vec{B}$ direction, and electric and magnetic field homogeneity in the detachment region.[8] Cancellation of the motional electric field will simplify the comparison of experimental results with theory since the inclusion of electrostatic field effects in the theoretical calculations is much more difficult. Of course, the $\vec{E}\times\vec{B}$ photoneutralizer can be operated out of balance to investigate the influence of small electric fields on the Landau state spectroscopy.

The use of a fixed-frequency laser having a narrow bandwidth with a high-resolution TDPSD to detect the neutrals provides the means of spectrally dispersing the photodetachment yield from each Landau resonance. It also provides a Doppler self-tuning capability. The self-tuning feature (1) avoids a tedious search for widely spaced but narrow resonances using a laser of very narrow linewidth, (2) simplifies the laser setup in an actual experiment, (3) provides a frequency stability monitor for the entire experiment, and (4) suggests methods of achieving goals in application areas. Each of these areas will be explored once we examine a gedanken experiment using the layout shown in Fig. 1. Although the arguments are discussed specifically for the layout shown in Fig. 1, the concepts apply equally well if a merged beam (collinear) geometry were assumed.

Let us consider a cylindrically symmetric ion beam having a Gaussian angular distribution at the detachment zone (see Fig. 1). We shall assume that the beam energy and angular width have been adjusted so that the full Doppler tuning range of the beam (e.g., +5 GHz and -5 GHZ corresponding to rays 11' and 22', respectively) is 10 GHz and that the TDPSD is located to detect all neutrals born in the beam. The X-direction of the TDPSD also is assumed to be antiparallel to the laser-beam propagation direction. Considering the first-order Doppler shift, neutrals detected at different angles in the X-direction must have arisen from ions that had sampled a different laser frequency. In fact, the frequency dispersion in the X-direction is about 20 MHz/channel for a 512 x 512 2D histogram. If the magnetic field in the photoneutralizer were set to 0.05 T, where Landau resonances are separated by 1.409 GHz, several resonances would be fanned out across the face of the PSD in the X-direction. Although the neutrals arising from each resonance have a banana shaped image in a Y vs. X contour plot, we need to discuss only the projection of the two-dimensional (TD) neutral particle distribution onto the X-axis (parallel to the laser beam direction). Thus, we present the X-position-dependent neutral yield calculation plotted in Fig. 2 that illustrates the situation discussed above.

The calculations shown in Fig. 2 were obtained by the Monte-Carlo trajectory method. Although several assumptions have been made in order to simplify the discussion, the computer simulation includes a fair degree of realism that directly applies to an experiment involving an S^- ion beam at an assumed energy of 2 keV and angular divergence of 1.0° (FWHM ;TD Gaussian). Other realistic parameters were selected for the simulation that included a TD Gaussian angular width for the laser beam (0.05° FWHM), Gaussian laser resolution (100 MHz FWHM), finite diameter for the ion and laser beams (0.1 cm) and the spatial resolution of the TDPSD (0.05 cm). The routine selected ion and laser beam trajectories from within the assumed

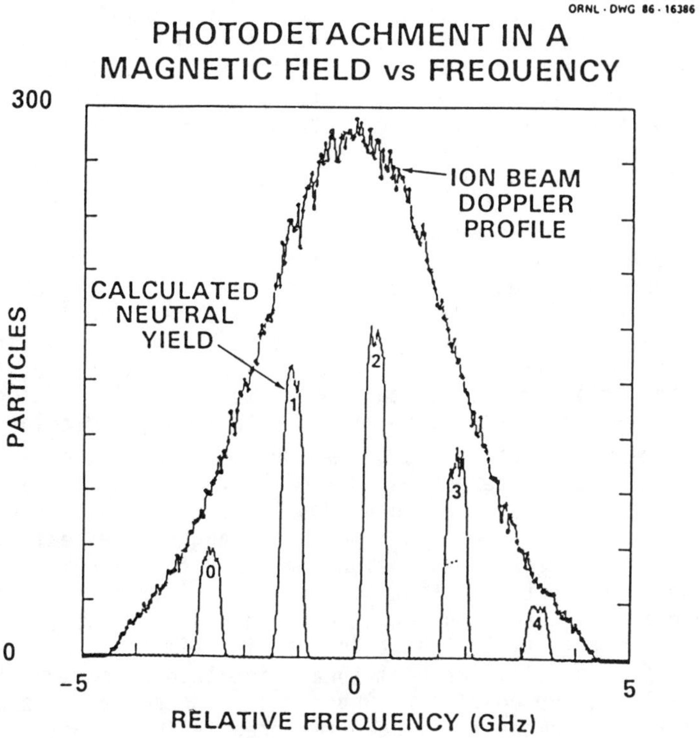

Fig. 2. Calculated angular distribution for five Landau resonances (n = 0-4) that illustrates the spectral resolving power of a perpendicular beam arrangement at a magnetic field of 0.05 T. The ion-beam Doppler profile shown with its associated neutral yield are projections of two dimensional angular distributions in the the X-direction of the TDPSD. The relative frequency scale (proportional to the beam deflection) is normalized to the Doppler shifted laser frequency for ions traveling at zero degrees.

beam constraints, Doppler shifted the frequency required to photodetach through the Landau levels in the projectile frame (n=0-4 levels in this example) into the selected photon direction, and calculated the photodetachment probability associated with this particular ion beam trajectory assuming a Gaussian laser linewidth. Gaussian distributions for the beam trajectories were generated using importance sampling methods. The photodetachment probability and the trajectory count were TD histogrammed separately into bins equivalent to the spatial resolution of the TDPSD. For the purpose of evaluating spectral resolution, the frequency dependence of the photodetachment cross section was assumed to be a delta function at the electron cycloton frequency (equal in magnitude for each Landau resonance) and fine structure was ignored. In the calculations, the product of laser power and cross section were adjusted to give a peak photodetachment probability of 0.60 (see Fig. 2).

Since the technique will be used for studying photodetachment spectroscopy in a magnetic field, the potential for achieving high resolution was evaluated. The spectral width in the X-direction of each Landau level in an actual experiment is the convolution of the Landau resonance shape with the laser bandwidth, the ion beam Doppler profile and other factors that affect resolution. The width of each spatially resolved neutral peak (350 MHz) shown in Fig. 2 is dominated by the assumed spatial width of the ion beam, since the Landau resonance width was assumed to be a delta function and the assumed laser bandwidth was 100 MHz (notice that the line shape is trapezoidal within Monte-Carlo statistics). The beam waist cannot be reduced much below 1 mm without difficulty; therefore, the ratio of the spectral width of a peak to the total Doppler tuning range, $\Delta f(FWHM)/f_{DTR}$, must always be of order 1/25 (perpendicular or parallel beams imaged onto a 25 mm TDPSD). Hence, the spectral width of a line can be decreased somewhat below that obtained in Fig. 2 by using a narrower total Doppler tuning range. The tuning range is adjusted in the experiment by changing the beam energy and angular divergence and moving the TDPSD to a longitudinal location consistent with that tuning. The ultimate resolution for a perpendicular beams configuration is probably about 150 MHz, limited primarily by the short interaction time of the laser with the ions. In a collinear beam (beams inclined at several degrees) configuration, the Doppler tuning range is decreased by about a factor of 10, and the residence time greatly increases; therefore, a bandwidth of order 30 MHz should be achievable if a single-frequency laser is used. Even with the spectral bandwidth shown in Fig. 2, the resolution is about a factor of 20 higher than was achieved in the ion trap measurements.[1,2] Velocity spread in the beam does not significantly affect the spectral line width in a perpendicular arrangement, it only changes the shape of the Doppler envelope.

In order to examine the ramifications of self-tuning, we discuss the behavior of the spectrally resolved Landau resonances shown in Fig. 2 when experimental parameters are varied. If the frequency of the laser is held constant while the magnetic field is increased, each of the Landau resonances would appear on the TDPSD at more positive angles (move to the right in Fig. 2) and the frequency separation between the levels would also increase proportionally with the magnetic field. The Doppler tuning range would have to be increased in order to still observe the same five Landau states shown in Fig. 2. If the laser frequency were decreased while the magnetic field is held constant, each Landau resonance would appear at more negative angles on the TDPSD (i.e., move to the left in Fig. 2). Since the TDPSD can image a broad Doppler tuning range, we see that the laser frequency has to be adjusted only to the approximate range of interest (within about 10 GHz in the example), not to the exact frequency of a Landau resonance. The absolute frequency of any observed resonance can be calculated by measuring (1) the angle where neutrals appear on the TDPSD and (2) the laser frequency, while transforming the relativistic Doppler shift between the laboratory and projectile frames. If the exact frequency of a Landau resonance is known, the apparatus can be used similarly as a wavemeter. If the magnetic field, the laser frequency, and/or the beam steering were to wander in an actual experiment, the TDPSD signals can be used to diagnose the origin of the instability. Thus, we see that the TDPSD is an important tool for performing a practical high-resolution spectroscopic study.

The self-tuning feature also could be used to advantage if the $\vec{E} \times \vec{B}$ photoneutralizer were used to produce electron and nuclear spin aligned neutral species such as H atoms. The frequency spacing of the Landau levels used in the simulation is close to the spacing of hyperfine states for the hydrogen system.[5] In this context, the lines shown in Fig. 2 would be different hyperfine levels of the same Landau state. From the preceding discussion it is clear that if the photodetachment laser were tuned precisely to address a particular spin state (one of the peaks shown), the spin selected neutral beam position could be 'steered' horizontally (left- right in Fig. 1) by adjusting the laser frequency. If the photodetachment laser bandwidth were large or the transitions to different Landau levels were power broadened in a high laser power application, long slits parallel to the Y-axis could be used to pass only one spin state. The total Doppler tuning range also could be adjusted (energy and beam divergence) to preferentially enhance a particular spin state produced in the axial direction. We see that the number of possibilities is substantial for an $\vec{E} \times \vec{B}$ photoneutralizer in an arrangement such as the ones considered.

Loss of signal is a major concern in any experiment performed at high resolution; hence, a few comments in this area are appropriate. Although a small fraction of the ions in the beam Doppler profile leads to photodetachment via a given Landau resonance (i.e., are removed from the ion beam as neutrals), the predicted large enhancement in the photodetachment cross section at resonance can more than make up for the shortfall in terms of total count rate. The small accessible ion fraction in the Doppler profile is not severe at low magnetic fields (e.g., 15% in Fig. 2) but it could become substantial in the high field range (1 T) particularly if a large Doppler scanning range were selected. Trade-offs between high resolution and count rate are clearly necessary at the highest field strengths. The use of a broader bandwidth laser in a perpendicular beam-beam configuration can be used to advantage without sacrificing resolution if the experimental spectral resolution is beam width limited as in Fig. 2. The fractional yield can be enhanced greatly in a nearly collinear beam configuration since the total Doppler tuning range here is greatly reduced. The fractional yield can also be increased without deleterious effects by the use of a more powerful laser provided that the increased power does not significantly perturb the spectroscopic measurements (e.g., power broadening) or exceed the maximum count rate at which the TDPSD signals can provide interpretable X-Y positions (20 KHz). (Power broadening effects were not included in the simulation shown in Fig. 2.)

III. THE EXPERIMENTAL ARRANGEMENT

The high-resolution apparatus that is being set up is in the configuration of Fig. 1. Negative ions (e.g., S$^-$; no nuclear spin) are accelerated to about 2 keV and then momentum analyzed before injection into an $\vec{E} \times \vec{B}$ photoneutralizer. The laser beam from a nitrogen-pumped pulsed tunable dye laser, outfitted with a Fabry-Perot interferometer to obtain a bandwidth below 100 MHz, intersects the focal point of the ion beam inside the photoneutralizer (propagation direction parallel to the electrostatic field). The dye laser cavity is outfitted with a temperature stabilized etalon to greatly narrow the linewidth (800 MHz) and to achieve long-term frequency stability. The laser linewidth is further narrowed by the insertion of a Fabry-Perot interferometer to obtain a Fourier transform limited bandwidth below 100 MHz. The linearly polarized laser beam (adjusted to be parallel or perpendicular to the magnetic field) is intercepted on a power meter located outside the vacuum system that is maintained at ultra high vacuum. The ion beam is deflected into a Faraday cup after leaving the photoneutralizer. Neutrals are detected downstream on a high-resolution TDPSD. The use of a TD rotating wire beam monitor together with a neutral beam mask, between the photoneutralizer and the TDPSD, facilitates the adjustment of the ion beam divergence and location before a photodetachment experiment. An NMR probe is used to accurately measure the magnetic field. The apparatus is controlled by very stable power supplies (better than 1 part in 10000).

Absolute cross sections for the spectrally dispersed resonances will be measured using different orientations of the laser-beam electric vector (different Zeeman transitions). The five-parameter data (position X and Y; time-of-flight, TOF, measured from the laser pulse; peak laser power; and average ion beam current) will be recorded event-by-event. The TDPSD data will be collected and histogrammed (real-time and/or off-line) using the VAX based data acquisition system at the ORNL Van de Graaff facility. Neutrals produced by photodetachment will be isolated from those produced by gas stripping by processing the 2D events in time coincidence with their known TOF (negligible accidental coincidence rate). This method of processing should also eliminate problems from laser light scattering and the TDPSD dark count. The dependence of cross sections on magnetic field strength will also be investigated to check the theoretical scaling relationship over the range of 0.02 - 1 T. Modeling calculations performed for the case of our pulsed laser indicate that data statistics similar to that shown in Fig. 2 should be obtainable in about 1 hour if the peak cross section were at least 1 x 10^{-19} at a peak pulsed laser power of 5 kW. If the shakedown experiments are successful, then the laser will be replaced by an IR laser so that negative-ion spectrocopy and cross sections for the production of electron and nuclear-spin aligned H beams in a magnetic field can be investigated.

IV. ACKNOWLEDGEMENT

This work was supported by the U.S. Army SDC, Huntsville, Al. and the U.S. Department of Energy, Division of Chemical Sciences under Contact No. DE-AC05-84OR21400 with Martin Marietta Energy Systems, Inc.

REFERENCES

1. W.A.M. Blumberg, W. M. Itano, and D. J. Larson, Phys. Rev. Lett. **40**, 1320 (1978).
2. W.A.M. Blumberg, W. M. Itano, and D. J. Larson, Phys. Rev. D**19**, 139 (1979).
3. O. H. Crawford, private communication.
4. A. I. Hershcovitch and E. A. Hinds in the Proc. Third Int. Symp. on Production and Neutralization of Negative Ions and Beams; K. Prelec, Editor, AIP Conf. Proc. No. 111 (1983), p. 763.
5. A. Hershcovitch, Nucl. Instr. Meth. in Phys. Res. **A243**, 271 (1986).
6. L. Wahlin, Nucl. Instr. Meth. in Phys. Res. **27**, 55 (1964).
7. J. D. Jackson, CLASSICAL ELECTRODYNAMICS, John Wiley and Sons, New York.
8. H. F. Krause, to be published.

DISCUSSION

Roberts: When you make the transformation to the beam frame so that the electric field goes to zero, the Lorentz force is invariant, it's zero in both frames. What happens to the magnetic field? It has to transform. The electric field has to show up as a magnetic field. Your magnetic field is not also going to zero, is it? Or is it just parallel to the beam?

Krause: Here, in general, are the equations. This is the projectile frame...the primes...and the unprimes are the laboratory. When you plug the v into there, of course, what will happen is the electric field will disappear. What will happen, really, is that the laboratory magnetic field, in the projectile frame, transforms as a slightly diminished field. So the electric field disappears, but the magnetic field is just slightly diminished. Incidentally, the magnetic field in the transformation is very homogeneous. That's an important consideration, because we have to have very high magnetic purity, otherwise that would be a broadening factor. Did that answer your question, Tom?

Roberts: I guess. It looks like you transferred to what would have been the drift velocity to make the E disappear. But I'm worried about your final results. I'm not so sure that the electric charge times the electric field, with v crossed into the magnetic field, and your final results are still zero. I think it has to be.

Krause: This is in the projectile frame...the electric field...the perpendicular field goes to zero, and...

Roberts: I'm worried about the magnetic field if it doesn't turn out to be parallel.

Krause: It's going to be parallel, it's just going to be diminished in size.

DETERMINATION OF NEUTRAL BEAM DIRECTION FROM RADIATION EMITTED BY PHOTODETACHED H⁻*

A. Hershcovitch, AGS Department,
Brookhaven National Laboratory, Associated Universities, Inc.
Upton, New York 11793

ABSTRACT

Non-destructive diagnostic techniques to determine beam direction of 200 MeV H° atoms are analyzed. These methods are based on excited hydrogen atoms in n=2 and n=3 levels due to photodetachment of H⁻ ions. With some development of hot cathodes, an e-beam driven ArF laser can produce $H^*(2s)$ atoms for laser resonance fluorescence by photoneutralizing H⁻ ions in a quantity comparable to that of a gas cell. Observation of fluorescence from spontaneous decay of $H^*(2p)$ or induced decay of $H^*(2s)$ can be readily used to indicated beam orientation with a 40 μrad accuracy. Measurements of minute Doppler shifts of this Lyman Alpha radiation by a spectrograph could in principle resolve beam direction to within 2.8 μrad. For schemes requiring n=3 hydrogen atoms, a Xe laser can produce $H^*(3s)$ or $H^*(3p)$ atoms in quantities larger than previously published.

INTRODUCTION

A number of methods, which are based on excited hydrogen atoms, are analyzed based on devices which are either available or could be developed in the near future. The basic principle is to neutralize H⁻ ions with a laser whose wavelength is such that the photodetachment process results in the formation of excited hydrogen atoms in either the n=2 level, or the n=3 level. These methods can be grouped into two categories: [1] direct optical measurements of Lyman Alpha radiation emitted by $H^*(2p)$, [2] production of $H^*(2s)$ or $H^*(3p)$ atoms for schemes already proposed, but hinging on the availability of these excited atoms.

Measuring the total fluorescence of Lyman Alpha radiation from $H^*(n=3)$ levels can be easily used to determine beam direction to within 40 μrad. Calculations of minute Doppler shifts of the Lyman Alpha radiation indicate that a resolution of $\Delta\lambda/\lambda$ of 10^{-6}, which could be measured by spectrograph, would sense angular changes of 2.8 microradians in beam orientation. In addition, spatial intensity and Doppler broadening measurements can yield information regarding density profile and velocity spread.

Sources of photodetachment produced $H^*(2s)$ and $H^*(3p)$ can greatly benefit the proposed laser fluorescence resonance technique

---

*Work done under the auspices of the U.S. Air Force Office of Scientific Research.

and the Doppler shift sensing technique respectively. There are many advantages to have excited hydrogen atoms as a result of photodetachment rather than collisional detachment. But the two most important advantages are the elimination of cascading, especially in the laser fluorescence resonance technique and the actual provision of $H^*(3p)$ atoms for the Doppler shift sensing technique, since no satisfactory source had been identified for it.

RELEVANT CROSS-SECTIONS

The process under consideration is

$$H^- + \gamma \rightarrow H^* + e^-$$

with the hydrogen atom being excited either into the n=2 or the n=3 levels. The photodetachment cross section has resonances. The n=2 resonance is also known as the shape resonance. As the following literature review indicates, the total cross sections are known and there is a reasonable agreement between theory and experiments. However, the branching ratio of each total cross section is not known. One can nevertheless make a good guess which should be better than 50%.

Partial Cross Sections in the Shape Resonance

The threshold for production of $H^*(n=2)$ lies at 10.959 eV. The most reliable calculations of its position are probably those of Broad and Reinhardt[1] and Taylor and Burke[2] as analyzed by Macek and Burke.[3] Broad and Reinhardt find that the shape resonance lies at 10.977 eV, 18 meV above the n=2 threshold in excellent agreement with Taylor and Burke. Thus, the photon energy required to excite the peak of the shape resonance is E_γ = 10.971 eV and the wavelength is λ = 1130.1 Å = 113.01 nm. The position of this resonance relative to the ^1P Feshbach resonance below n=2 has been confirmed experimentally by Bryant et al.[4] but the absolute position was not measured.

According to Broad and Heinhardt[1] the cross section at the peak of this resonance is $3.4a_0^2$ (9.5×10^{-17} cm^2) and the width is 15 meV. Taylor and Burke[2] find a width of 14 meV in good agreement but do not calculate a cross section. The experiments of Bryant et al.[4-5] do not determine absolute cross sections but when their data are normalized in the low energy continuum so as to agree with Broad and Reinhardt's value there, the measured cross section at the peak of the shape resonance is completely consistent with theory. Figure 1, taken from ref. 5, shows this comparison. Thus, the best values to take are probably the theoretical cross section, $\sigma = 9.5 \times 10^{-17}$ cm^2 and the measured width ΔE_γ = 23 MeV or $\Delta\gamma$ = 2.4 Å.

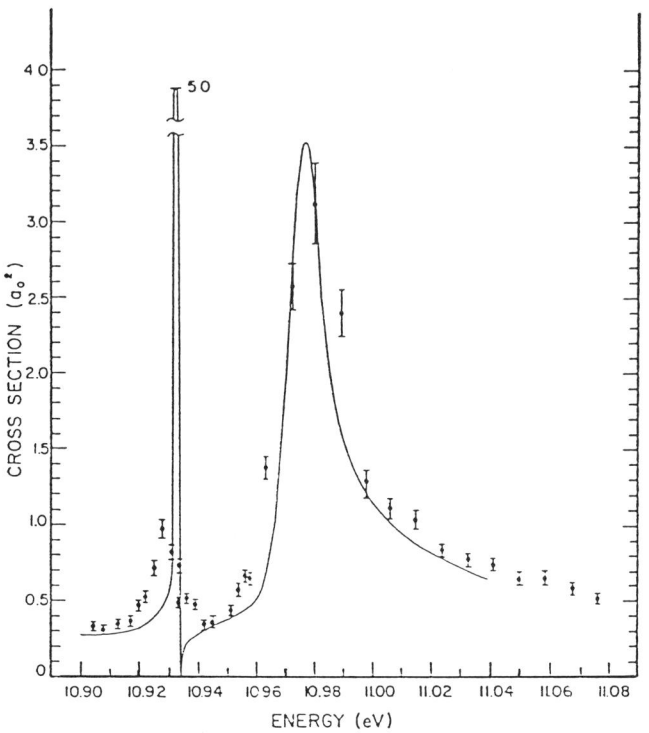

Fig. 1. Comparison of theory, ref. 1, and experiment, ref. 5. The measured cross section is normalized to the theory at 10.9 eV. The error bars are statistical only.

We now turn to the partial cross sections σ_{1s}, σ_{2s} and σ_{2p}. The only attempt to calculate these cross sections separately was made by Hyman, Jacobs and Burke.[6] Even though their theory was incomplete, Macek has argued[7] that the results should be fairly insensitive to the absent effects. Figure 2 shows the partial cross sections calculated by Hyman, Jacobs and Burke.[6] The sum of these partial cross sections is in almost exact agreement with the total cross section calculated by Macek[7] and hence gives essentially the correct cross section. Unfortunately it is not clear how to do the scaling correctly as the partial cross sections do not all vary with energy in the same way. Instead, consider the peak cross sections which have the ratio $\sigma_{2p}^p:\sigma_{12}^p:\sigma_{2s}^p = 4:3:1$. In the absence of any more detailed knowledge, it is not unreasonable to suppose that the true partial cross sections are related in roughly the same way. We therefore take the following values

$$\sigma_{2p}^p = 4.8 \times 10^{-17} \text{ cm}^2 \quad (1)$$

$$\sigma_{2s}^p = 1.2 \times 10^{-17} \text{ cm}^2 \quad (2)$$

$$\sigma_{1s} = 3.6 \times 10^{-17} \text{ cm}^2 \qquad (3)$$

These values are qualitatively consistent with Broad and Reinhardt's "very reasonable estimates" (Fig. 6 of ref. 1) which give $\sigma_{2s} + \sigma_{2p} = 6.4 \times 10^{-17}$ cm^2 and $\sigma_{1s} = 3.1 \times 10^{-17}$ cm^2.

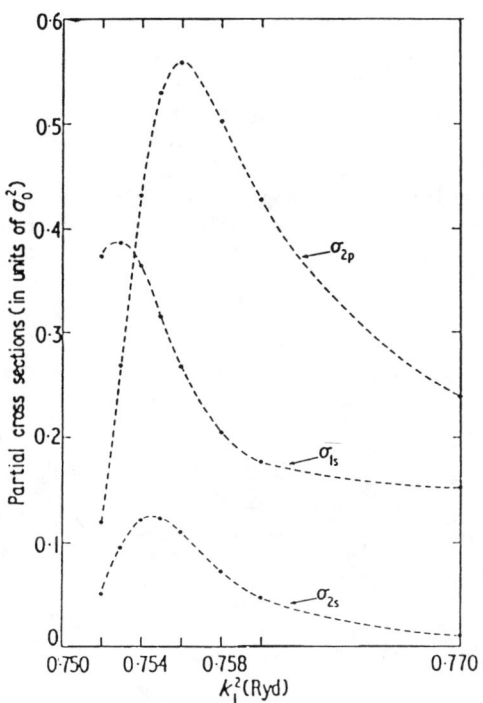

Fig. 2. Theoretical partial cross sections for formation of 1s, 2s, and 2p states of hydrogen by photodetachment of H$^-$. The abscissa is the photon energy in Rydbergs (1 Rydberg = 13.605 eV). The ordinate is the partial cross section in units of the Bohr radius squared ($a_0^2 = 2.8005 \times 10^{-17}$ cm^2).

The n=3 Resonance

Hamm et al.[8] have investigated experimentally the H$^-$ photodetachment cross section near the n=3 threshold and have shown agreement with theoretical calculations.[9] These studies were focussed on the structure (two dips) before the n=3 threshold. However, the n=3 Threshold was investigated as well. From the experimental results[8] the cross section for H$^-$ + γ → H*(n=3) +e$^-$ is $\sigma_{n=3} = 8.4 \times 10^{-18}$ cm^2 for a photon energy of 12.85 eV. The branching ratio of n=3 into 3p, 3s and 3d has never been calculated.[10] However, a good guess would be[10] that the 3p and 3s states are comparably excited with only less than 20% of the n=3 atoms in the 3d state and the rest are equally divided into the 3p and 3s states, i.e.

$$\sigma_{3p} = 3.36 \times 10^{-18} \text{ cm}^2 \tag{4}$$

EXPERIMENTAL ARRANGEMENT

A possible experimental set up is shown in Fig. 3. Beam parameters from the low divergence 200 MeV Neutral Beam Test Facility now under construction at BNL, are used to illustrate this method. In this device, the expected beam parameters: beam current 0.9 mA, pulse length 0.5 msec. The beam has an elliptical cross section with major and minor axis of 9.1 mm and 50 mm respectively.

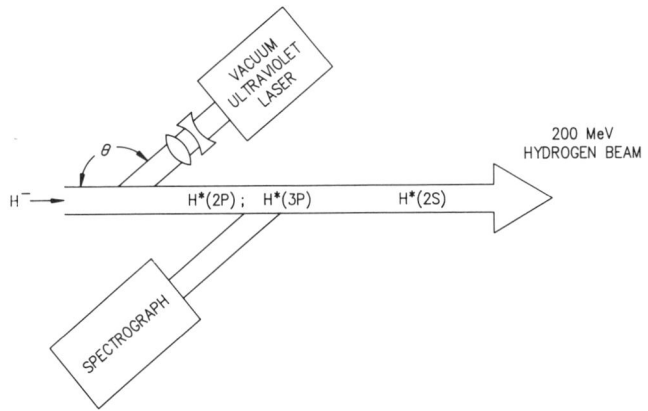

Fig. 3. Schematic of a possible experimental system.

In order to excite the shape resonance, a 113 nm laser is required for a cross beam experiment. At this wavelength only very short pulse moderate power lasers exist.[11] However, for fast H⁻ beams, other lasers can be used. If the wavelength of the laser is λ_0, we can use the Doppler shift to give the required 113 nm in the frame of reference of the ion beam by correctly choosing the angle between the two beams. Let the angle be θ, i.e., $\theta=0$ means the beams are parallel. Then the correct angle is given by

$$\cos\theta = \frac{1}{\beta} \left(1 - \frac{\lambda_0}{113\gamma}\right) \tag{5}$$

where $\beta = v/c$ and $\gamma = (1-\beta^2)^{-1/2}$ are the usual parameters of relatively. They are related to the ion beam energy by $\gamma = \varepsilon + 1$; $\beta = \left[1 - \frac{1}{(\varepsilon+1)^2}\right]^{1/2}$ where ε is the beam energy in MeV divided by 938.

In our case of a 200 MeV H⁻ beam give $\beta = 0.566$, $\gamma = 1.213$. The possible laser wavelengths for exciting the shape resonance in this beam lie between 214 nm ($\theta = 180$) and 60 nm ($\theta = 0$) as calculated from Eq. (5). In this wavelength range, the ArF _eximer_ laser operating at 193 nm, for which the correct angle for Doppler tuning is $\theta = 136.1°$ is best. Maxwell Laboratories, Inc., can make a 10 Joule e-beam driven ArF laser with a 5 µsec pulse duration, and probably stretch these numbers to as high as 100 J and 10 µsec.

To photodetach H⁻ ions, leaving the residual hydrogen atom in an n=3 states, the needed photon in the rest frame of the beam has a wavelength of 96.48 nm. Using Eq. (5) (replacing 113 by 96.5) to calculate the usable wavelength range to excite the n=3 resonance on a 200 MeV beam, one obtains a range of 185.25 nm ($\theta=180°$) to 51.34 nm ($\theta=0$). The best laser to use is a xenon laser at 175 nm. At this wavelength $\theta=151.06°$. Although power levels exceeding 10^8 W are available from these lasers, their pulse length cannot exceed 50 nsec due to the need of high e-beam currents which "load up" the foils thermally. For comparison, ArF lasers need only 5-15 Amps/cm². Thus, for 200 MeV beams, one can use Xe lasers with pulse lengths of up to 50 nsec. But, for 400 MeV beams, one can use ArF lasers with much longer pulse lengths.

APPLICATIONS

When an H⁻ beam undergoes photodetachment, its current reduces as

$$I = I_o \exp(-\sigma \Gamma \tau) \qquad (6)$$

where σ is the total photodetachment cross section, τ is the interaction time and Γ (photons/cm²-sec) is the photon flux. If the value of the exponent of Eq. (6) reaches 1, only exp(-1) or less than 37% of the ions are left intact and over 63% of the beam is stripped. For exponent values approaching 3 to 4, the beam can be considered as being fully stripped.

Lasers considered in the previous section have a beam width of 10 cm, and their orientation $\theta \sim 150°$. This means that a laser faces the H⁻ beam at 30° off the beam axis. Therefore, the interaction length, which is the beam length illuminated by the laser is equal to 10 cm/sin(180-θ) = 20 cm. Since the beam $\beta=0.566$, the interaction time over that length is $\tau \sim 1.2$ nsec. The total cross section σ for the shape resonance is about 10^{-16} cm². Therefore in order for $\sigma \Gamma \tau$ to exceed unity, the photon flux needed is $\Gamma \geqslant 1/\sigma\tau$, hence; $\Gamma \geqslant 8.3 \times 10^{24}$ photon/cm²-sec.

Each one of the photons from a Xenon or an Argon Fluoride laser carry energies of about 7 eV and 6.4 eV respectively. Therefore, a photon flux of 8.3×10^{24} photons/cm²-sec corresponds to 8.3 MW/cm² for a Xenon laser and 8.5 MW/cm² for an Argon Fluoride laser. In the preceding calculation the conversion from photons/ sec to Watt is based on 1 eV = 1.6 x 10^{-19}J and 1 J/sec = 1 Watt. The current distribution of the H⁻ beam is not known, but it will probably be Gaussian, i.e., high concentrated in the center. If the smaller (9.1 mm) of the transverse dimensions of the beam is irradiated by the laser, it is sufficient to illuminate 1 cm of beam height to expose practically all the H⁻ ions to the laser light. Thus, one of the laser beam dimensions would be focussed to 1 cm. Since the unfocussed beam dimension is 10 cm, the total area to be irradiated is 10 cm^2. Consequently, the laser power needed to photodetach the vast majority of H⁻ ions is under 100 MW (93 MW for Xe and 85 MW for ArF). Both of these lasers have demonstrated power levels which are even higher than 100 MW. Hence, to a first approximation, it can be assumed that the H⁻ beam can be fully neutralized for the duration of the laser pulse.

$H^*(2s)$ for Laser Resonance Fluorescence

In Laser Resonance Fluorescence (LRF), $H^*(2s)$ atoms are raised to the 3p level with a laser, and the resultant decay is observed. Two channels of decay are possible: either the H_α to the 2p state, or the much stronger Lyman Betha to ground. Many proposals to test this technique rely on the availability of $H^*(2s)$ atoms resulting from H⁻ neutralization in a gas cell. At best only 6% of the neutral hydrogen atoms are in the 2s, and only 50% of the H⁻ ions can be neutralized. Thus, under the most optimistic conditions only 3% of the H⁻ ions can be neutralized to the 2s level in a gas cell. In addition, many problems such as cascading and other particles exiting the neutralizer result in difficulties for fluorescence detection.

To evaluate the full merit of utilizing photodetachment produced $H^*(2s)$ in LRF, a quantitative comparison the the $H^*(2s)$ yield must also be done. Under the most optimistic condition using an Argon Fluoride laser in NBTF, the H⁻ beam can be almost completely photodetached using the shape resonance. From the partial cross sections discussions in section II, Eqs. (1) through (3), close to 13% of the neutral atoms emerging from the stripper are expected to be in the 2s state, i.e., about 13% of the original H⁻ beam could result in $H^*(2s)$. However, the H⁻ pulse length is 500 μsec, and the ArF laser pulse could be stretched to 10 μsec (from the already achieved 5 μsec), while a gas cell is active for the full duration of the beam pulse. Thus, H⁻ photodetachment by the ArF laser has a maximum duty factor of 0.02, during the H⁻ pulse. Taking this into account, in a single pulse a maximum of 0.26% of the H⁻ could be photodetached to $H^*(2s)$.

However, if e-beam drivers are developed with hot cathodes, ArF lasers with 100 μsec pulse length can be built. Thus, it is conceivable that, with some development, ArF laser can achieve a duty factor of 0.2 and photodetach 2.6% of the H⁻ to the $H^*(2s)$ state.

Estimates of the most optimistic $H^*(2s)$ yields from a gas cell and an ArF laser stripper show comparable yields per H^- pulse, if a hot cathode e-beam driver is developed.

Lyman Alpha Emission from a Photodetached H^- Beam

According to the partial cross sections discussed in section II, Eqs. (1)-(3), over 50% of the neutral atoms emerge in the 2p state. These atoms decay spontaneously with a rest frame lifetime of 1.56 μsec at a mean distance of 32 cm downstream from the laser beam. The large homogeneous width of the shape resonance ($\Delta\lambda/\lambda$ = 0.2%) is an advantage in this stripping scheme. It would take a large (~4 mrad) divergence in one of the beams or a 0.3% velocity spread (1.2 MeV energy spread) in the ion beam to create a comparable inhomogeneous width. Thus there is no particular virtue in using the so-called magic angle $\theta = \cos^{-1}(\beta)$ where the first order Doppler width vanishes.

On the other hand, the shape resonance is by no means insensitive to the angle between the two beams and thus the neutralizer itself provides information which could be used in a feedback loop to stabilize the particle beam orientation. In view of the large amount of fluorescence from the neutral beam (2p-1s decay), stability to 1% of the shape resonance linewidth (40 μrad) should be readily achieved. While this method of steering is not as accurate as an auxiliary direction monitor, based for example on the 2s-3p transition, it has the obvious virtue of simplicity.

The fluorescence from the beam carries information about velocity spread in the wavelength distribution and about density profile in the intensity distribution. Thus, the emitted Lyman Alpha radiation can be used to unfold these quantities, which are outside the scope of this work.

Measurements of Minute Doppler Shifts in Lyman Alpha Due to Small Angular Changes in Beam Orientation

The 2p-1s decay wavelength is 121.6 nm in the rest frame of the atoms. Thus in the laboratory frame the observed wavelength depends on the angle of emission relative to the beam velocity in the following way

$$\lambda = 121.6 \; \gamma(1 - \beta \cos \theta) \text{ nm} \qquad (7)$$

where again $\theta=0$ corresponds to the direction parallel to the particle beam. In our example of a 200 MeV beam, the decay radiation spans the large wavelength range 64-231 nm. There is also a relativistic modification of the intensity distribution; the low energy probability of decay at an angle θ into solid angle $d\Omega$ is multiplied at high beam energy by the relativistic factor R

$$R = \frac{1}{\gamma(1-\cos \theta)} \qquad (8)$$

The Doppler shift formula, which is equivalent but more general to Eq. (5) is

$$\nu = \nu_o \gamma(1 + \beta \cos\theta) \qquad (9)$$

For very small angular changes

$$\Delta\nu \simeq d\nu = -\nu_o\gamma\beta \sin\theta d\theta \simeq \nu_o\gamma\beta\sin\theta\Delta\theta \qquad (10)$$

using the first and last terms of Eq. (10), converting from $\Delta\nu$ to $\Delta\lambda$ and solving for $\Delta\lambda$, one obtains

$$\Delta\lambda = \frac{c}{\nu + \Delta\nu} - \frac{c}{\nu}$$

From which $\Delta\lambda/\lambda$ can be calculated to obtain for very small $\Delta\theta$ (neglecting corrections which are order of magnitudes lower),

$$\frac{\Delta\lambda}{\lambda} = \left| \frac{\beta\cos\theta}{1 + \beta\cos\theta} \Delta\theta \right| \qquad (11)$$

To measure minute Doppler shifts, detection devices must be oriented at the so-called "magic angle" determined when $\cos\theta = \beta$ in order to avoid $\Delta\beta$ broadening. For our 200 MeV beam, this occurs at $\theta = 55.5°$. At this angle (Eq. 7), the Lyman Alpha radiation has a wavelength of 1926.66Å. And, the relativistic correction factor R=0.65 (from Eq. (8)). For these values a Doppler shift measurement with $\Delta\lambda/\lambda = 10^{-6}$ would yield from Eq. (11) $\underline{\Delta\theta = 2.8~\mu rad}$. This kind of resolution (of $\Delta\lambda \simeq 0.002$Å) at this wavelength can be reached in a spectrograph. McPherson Instruments, examined the feasibility of fabricating a spectrograph to perform this type of measurement for the pertinent parameters of the Lyman Alpha radiation and the spectrograph orientation, i.e., 2.8×10^{10} $H^*(2p)$ per pulse, $\theta = 55.5°$, for the spectrograph at which angle $\lambda = 1926.66$Å and R = 0.65. The conclusion was that one of their spectrographs (Model 265) could be modified to perform such a measurement.

Generation of $H^*(3p)$ Atoms

From the discussion in section II, the cross section for H^- photodetachment at n=3 resonance is an order of magnitude lower than at the shape resonance. Therefore, power requirements for the Xenon laser are a factor of 10 higher. Nevertheless, these power levels of about 10^9 watt are either feasible or have been demonstrated with Xenon lasers for pulse duration of up to 50 nsec. The proposed Doppler shift sensing technique needs $H^*(3p)$ atoms to study Doppler shifts in 3p-2s decay photons. No satisfactory source of $H^*(3p)$ has been identified in that proposal.

A 10^9 Watt, 50 nsec pulse for a Xenon laser can fully neutralize the H^- beam yielding 5.6×10^{15} $H°(n=3)$/sec. For a 50 nsec pulse duration, one obtains 2.8×10^8 H*(n=3) atoms per pulse. Given the partial cross sections estimated in section II, i.e., about 40%, ends up as H*(3p) atoms, the H*(3p) yield could be as high as 1.125×10^8 H*(3p) atoms/pulse. This yield when compared to an estimated 3.4×10^6 H*(3p)/marco-pulse obtainable for LRF is a factor of 33 higher and it does not require the use of a potentially troublesome gas cell. For 400 MeV H⁻ beams or higher, lasers like ArF and KrF with much longer pulse duration could be used, yielding orders of magnitude more n=3 hydrogen atoms per pulse.

ACKNOWLEDGMENTS

The author is thankful to the Air Force Office of Scientific Research for supporting this work. Most of the cross section information, especially regarding partial cross sections was provided by Professor Ed. Hinds (Yale University). The feasibility study for the VUV spectrograph was done by McPherson Instruments Inc. Finally, the author would like to thank Dr. Gary McAllister and Mr. Steve Miscikowski of Maxwell Laboratories, Inc. for providing important information regarding VUV lasers.

REFERENCES

1. J.T. Broad and W.P. Reinhardt, Phys. Rev. A14, 2159 (1976).
2. A.J. Taylor and P.G. Burke, Proc. Phys. Soc. 92, 336 (1967).
3. J. Macek and P.G. Burke, Proc. Phys. Soc. 92, 351 (1967).
4. H.C. Bryant et al. in "Atomic Physics 7," Eds. D. Kleppner and F.M. Pipkin, Plenum Press, New York, 1980, p. 29.
5. H.C. Bryant et al., Phys. Rev. Lett. 38, 228 (1977).
6. H.A. Hyman, V.L. Jacobs and P.G. Burke, J. Phys. B5, 2282 (1972).
7. J. Macek, Proc. Phys. Soc. 92, 365 (1967).
8. M.E. Hamm et al., Phys. Rev. Lett. 43, 1715 (1979).
9. C.H. Greene, J. Phys. B. Lett. 13, L39 (1980).
10. C.H. Greene, (LSU), private communication 1986.
11. F. Tompkins and R. Mahon, Optics Letters 7, 304 (1982).

DISCUSSION

<u>Michels</u>: That work you referred to by Hyman, Jacobs, and Burke, is that a photodetachment cross-section, or is that electron excitation?
<u>Hershcovitch</u>: That was a photodetachment cross-section.
<u>Stewart</u>: The stripping process itself is going to introduce some divergence in the beam that I would think would be large, especially at these shorter wavelengths.
<u>Hershcovitch</u>: Divergence is introduced only in ground-state atoms. In 2p or 2s atoms, all the energy in excess of that required for stripping ends up in the excited atom.

SYSTEMS AND APPLICATIONS

JAPANESE NEGATIVE ION BEAM PROGRAM

T. Kuroda
Institute of Plasma Physics, Nagoya University
Nagoya 464, Japan

ABSTRACT

The negative-ion beam program and status in Japan are described. A program to develop the negative ion source with the capability of 24 A beam at 500 kV is prepared at JAERI. Fundamental research of negative ion production and its application are continued in universities. Status of their activities are briefly reviewed.

INTRODUCTION

Since the effectiveness of neutral beam injection heating has been demonstrated in ATC of Princeton Tokamak and in ORMAK of Oak Ridge National Laboratory, neutral beam injection heating has been used in many tokamak experiments and positive ion based neutral beam technologies have been highly progressed. In Japan, during past ten years, positive ion based neutral beam injectors have been developed and many neutral beam injection heating experiments have been carried out at Tokamak (JFT-2 at JAERI, JIPP T-II at IPP Nagoya) and Heliotron at Kyoto University and high power neutral beam has been also used for specific function in Mirror Machine (Gamma 10 at Tsukuba University). Recently, neutral beam injection system, 75 keV, 20 MW, 10 sec has been developed for JT-60[1] and high current ion source of 120 keV, 90 A, 1 sec[2] has been developed for R-Tokamak project which was proposed in IPP Nagoya. However, as for applying this technique to future large Tokamak Device, the required energy is so high that the neutralization efficiency become too low to realize an economical injector as well known. For these future plan, negative ion based neutral beam injector has been required for heating plasma and for current drive method. On the other hand RF heating has been progressed and has potentiality for heating method in future toroidal device. Negative ion based neutral beam technology can compete in future heating and current drive in large torus device with Radio Frequency technology.

In fusion research of Japan a preliminary experiment for high current negative ion source was started at IPP Nagoya University in 1978. Surface ion production has been investigated in a magnetron type ion source. Some new types of negative ion source have been proposed during last ten years and preliminary experiments of negative ion production have been carried out at some group of universities. With increasement of requirement of negative ion based NBI, fundamental studies of negative ion production and source have been active in university side.

On the other hand, the negative ion source development for neutral beam injector has been started at 1984 in JAERI after the neutral beam injection heating system development has been almost completed. On the basis of the preliminary results of the negative ion source, an experiment using negative ion based neutral beam injector has been proposed on next big project.

This paper describes the activities of negative ion source and negative ion bases NBI development and some proposal of the development in Japan.

NEGATIVE ION BASED NBI DEVELOPMENT PROGRAM IN JAERI

Negative ion source development at JAERI was started at 1984 and multiple cusp type negative ion source due to volume production method was chosen for following reason: 1. Simple construction and ease to extract long pulse beam. 2. Ease to scale up. 3. No need of technology for cesium handling. 4. Good beam characteristics.

For this multi cusp source, development has been concentrated in the increasement of output of H^- current density and electron suppression on extraction electrode[3]. The current density of less than 18 mA/cm^2 has attained at 1984. Performance of suppression of electron in extraction electrode has been also improved by using an electron suppression electrode. Beam power attained up to now is shown in Fig.1. Beam of 21 keV, 1.2 A, 0.1 sec has already attained now[4].

An FER Project has been proposed as next project in JAERI[5] and the design are discussed. Neutral beam injection of 50 MW at beam energy of 500 keV is required in the project. The requirement of NBI is shown in Table 1[6]. A negative ion based neutral beam injector has been conceptually designed to meet these requirement on the basis of experimental results on their multi cusp source. Figure 2 shows a conceptual design of negative ion based neutral beam injector for FER. A feature of the injector is to adopt a long gas neutralizer. As a results of the very long neutralizer, neutron and tritium back streaming to ion source can be reduced and also big element of beam line (cryo-pump, etc.) can be eliminated around the torus. The schedule for development of negative ion source of NBI on FER is shown in Fig.1. The research and development of negative ion beam in JAERI have the potential of producing neutral beam system which will be used in future project, FER program.

ACTIVITIES OF RESEARCH AND DEVELOPMENT OF NEGATIVE ION IN UNIVERSITIES OF JAPAN

Development of NBI in universities and neutral beam injection experiments have been carried out in three universities - Institute of Plasma Physics, Nagoya University, Kyoto University and Tsukuba University. Beam energy, beam current and duration of beam in most their injector are 30 keV to 40 keV, up to 60 A and 0.1 sec respectively. Several years ago, R-project was proposed as next big project in university side which required NBI heating of 15 MW at 120 keV. However this project was not started and instead of R-project, helical system has been investigated as a next big project in university. Although NBI heating will be adopted in the future

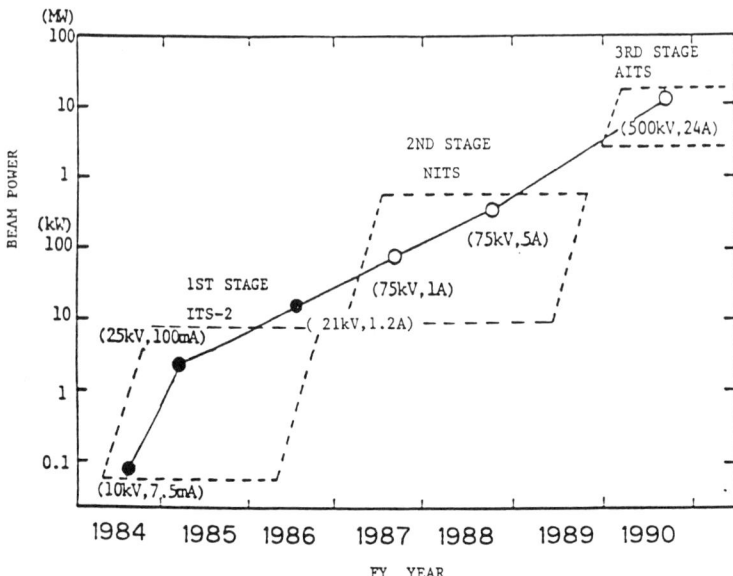

Fig.1. A long-range plan on development of negative ion source.

Table I. Performance of a 500 keV/20 MW injector.

Overall
    Neutral Beam Power   22.5 MW
    Power Density   56.3 MW/m^2 (average)
    Beam Energy   200 - 500 keV
    Pulse Length   30 sec
    Ion Species   D
    Power Efficiency   45 % (at 500 keV)
        Acceleration   96 %
        Neutralization   59 %
        Geometrical   84 %
        Re-ionization   95 %

Ion Source
    Number   1
    Type of D$^-$ Source   Volume Production
    Size   0.2 m x 2.4 m (grid)
    Current   100 A
    Current Density   50 mA/cm^2
    Divergence   5 mrad

Cryo-Pump
    Source Chamber   750 m^3/s
    Injection Chamber   200 m^3/s

Injection Port
    Size   0.4 m x 1.0 m x 9 m
    Pressure   4 x 10^{-3} Pa
    Gas Flow into Torus   <0.03 Pa m^3/s

Neutralizer
    Size   0.3 m x (2.4-1.4) m x 3 m
    Line Density   7.4 x 10^{15} molecules/cm^2
    Pressure   1.5 x 10^{-2} Pa (at entrance)
    Shield   20 cm iron

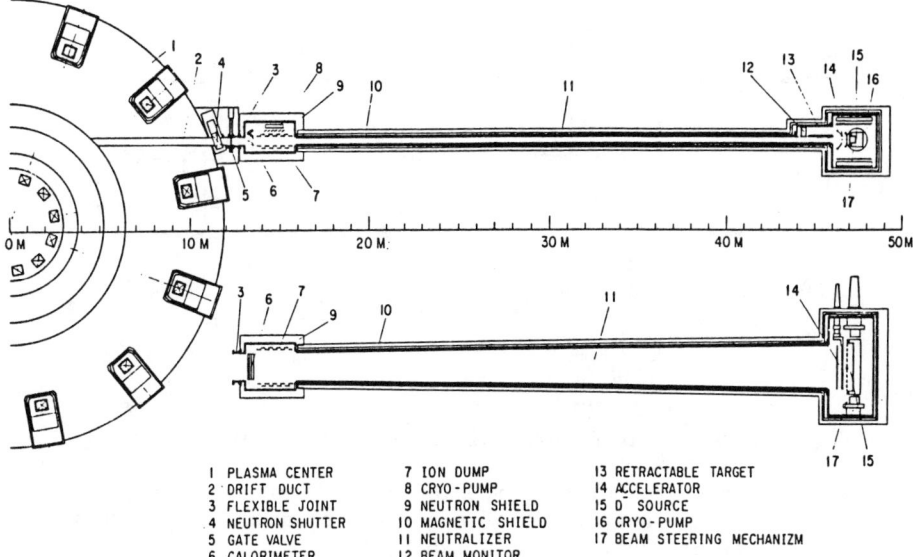

Fig.2. Conceptual design of a 500 keV, 20 MW negative-ion-based neutral beam line for FER.

Table II Status of negative ion research in university

1. New type ion source study
    1. Sheet plasma type negative ion source   ··· IPP Nagoya
    2. RF type (EBT source) negative ion source ··· Kyoto Univ.
    3. Permeation type H⁻ ion source   ··· Kyoto Univ.
2. Small experimental studies for H⁻ production
    1. Wall material effect on H⁻ negative production
                                                  ··· Yamaguchi Univ.
    2. Impurity effect on H⁻ negative production··· Kyoto Univ.
    3. Numerical studies of volume production negative ion source
                                                  ··· Yamaguchi Univ.

project, the needs of negative ion based neutral beam injector is very small. Therefore, the research and development of negative ion based NBI and high current negative ion sources are almost concentrated on the experiment of fundamental production process of negative ion and new type negative ion sources which have potentiality of an alternative for the present source. Most of studies in universities are not to scale-up the present promising source but to study the fundamental technology for NBI.

There are some small groups of negative ion research in universities. They have their own future plan to research and development of negative ion source related technology. In the following, the

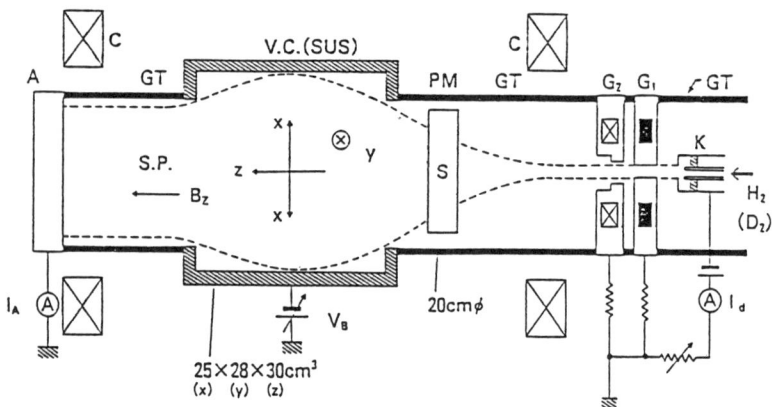

Fig.3 A schematic diagram of sheet plasma source.

activities of research of negative ion beam in each university are described. The present status of negative ion source research in each university are summarized in Table II.

1. Program in Institute of Plasma Physics, Nagoya University

Application of negative ion based neutral beam injection as well as positive ion based NBI of 120 keV 15 MW was planned in R-project and in a preliminary study of the project, high current negative ion source development has been carried out. A sheet plasma source which has wide area of high density plasma has been developed for high current negative ion source. Figure 3 shows a schematic diagram of a sheet plasma source[7]. So far the negative ion beam current density of up to 30 mmA/cm^2 is obtained at the beam energy of up to 5 keV in the sheet plasma source. Although R-project has not been started, the development of high current negative ion source are continuing to advance a sheet plasma source to actual negative ion source. For negative ion based NBI development, the fundamental study of elements of NBI, neutralization and transport of beams will be continued by NBI Group for beam energy of up to 500 keV which corresponds to the need of FER NBI. In future negative ion based NBI, higher voltage and high current interrupter is needed. In FER, switching system which interrupt 100 A at 500 keV is also needed. On the basis of successful results of GTO power switch in test stand of 120 keV, fundamental studies of interrupter[8] up to 500 kV will be done.

On the other hand, the alpha-particle measurement in reacting plasma due to He negative ion source has been started in preliminary study of R-project[9]. The diagnostics of confined alpha particles will be feasible by the beam probe method where double change exchange reactions with alpha particles are utilized. A prototype negative ion source of surface conversion type has been constructed and investigated the production of He$^-$. The alpha-particle diagnostic system by beam probe method is designed conceptually. The effort to development of the system are being continued.

2. Small experiments and research of negative ion in university

Experimental and numerical studies are carried out by several goups in university. New type plasma sources which is produced plasma by different method with multicusp source has been investigated. Permeation type H⁻ ion source[10] and EBT type negative[11] in source (Electron Heating in Bucket) has been studied in Kyoto University. The former is now in preliminary and fundamental phase. In the latter source which is operated with 1 kW of RF at 2.45 GH, it is reported that electron density and H⁻ yield are proportion to the RF power and becomes maximum at a certain hydrogen pressure. Extracted H⁻ current increases when positive bias (7-8 V) is applied to the extraction electrode. For the production and the attenuation of H⁻ ion, the effect of wall material and impurities on production of negative ion production has been investigated. It is reported by Yamaguchi University group that Al-wall generally produces the highest H⁻ ion current although the difference in H⁻ yield among Al, Cu and S,S lines depends apparently on the source pressure of hydrogen gas.[12] For impurities effect, it is reported that production of negative ions depends on the value of water vapor.[13] Numerical study on volume production of negative ions in hydrogen plasma by Yamaguchi University group[14] indicated that 1) not only n_e but also n_{fe} has some optimum value on H⁻ yield, 2) with increasing Te, H⁻ yield decreases monotonically, 3) H⁻ yield is enhanced by proper selection of the wall material and 4) there are optimum.

Summary

JAERI group is required to develop a high current negative ion source which is directly applicable to the neutral beam injectors for future device. A long-range plan on development of megawatt beam source is prepared. Universities take a part of fundamental research of physics and technology in negative ion beam production. Research and development for intense neutral beam will be continued at JAERI and universities in Japan.

References

1. Y. Ohara, Proc. Int. Ion Eng. Congress - ISIAT'83 & IPAT'83, Kyoto, 447 (1983).
2. Y. Oka, K. Sakurai, O. Kaneko and T. Kuroda, Proc. 13th Symp. on Fusion Technology, Varese, 565 (1984).
3. T. Shibata, H. Horiike, H. Inami, S. Matsuda, Y. Ohara, Y. Okumura and S. Tanaka, in Proc. of the IAEA Technical Committee Meeting on Negative Ion Beam Heating, Grenoble (1985).
4. Y. Okumura, H. Horiike, T. Inoue, T. Kurashima, S. Matsuda, Y. Ohara, and S. Tanaka, contributed paper of this symposium.
5. Annual Report of April 1, 1984 to March, 1985, JAERI-M, 85-205.

6. H. Horiike, Y. Ohara, Y. Okumura, T. Shibata and S. Tanaka, Japan Atomic Energy Research Institute Report, JAERI-M, 86-064 (1986) (in Japanese).
7. J. Uramoto, Institute of Plasma Physics, Nagoya University, Research Report IPPJ-731 (1985).
8. T. Kuroda, O. Kaneko, Y. Watanabe, N. Seki, O. Higa and F. Saito, Proc. of Power Electronics Conf., Vol.1, 808 (1983).
9. M. Sasao and K. N. Sato, IPPJ-695 (Institute of Plasma Physics Nagoya University, 1984).
10. R. Itatani, T. Namura, Y. Tomomura, A. Sugimura and I. Akrikata, Proc. 9th Symp. on ISIAT'85, Tokyo, 105 (1985).
11. R. Itatani, T. Namura, J. Kondo and H. Murayama, Proc. 10th Symp. on ISIAT'86, Tokyo, 135 (1986).
12. O. Fukimasa and S. Saeki, Proc. 10th Symp. on ISIAT'86, Tokyo, 187 (1986).
13. I. Arikata, T. Murakamai and R. Itatani, Proc. 10th Symp. on ISIAT'86, Tokyo, 193 (1986).
14. O. Fukumasa and S. Saeki, Proc. 10th Symp. on ISIAT'86, Tokyo, 181 (1986).

STRATEGIC DEFENSE INITIATIVE ORGANIZATION
NEUTRAL PARTICLE BEAM PROGRAM OVERVIEW

Brian R. Strickland

U.S. Army Strategic Defense Command
Huntsville, AL 35807-3801

ABSTRACT

The paper gives a brief overview of the Strategic Defense Initiative Organization (SDIO) Neutral Particle Beam (NPB) program giving the general goals and possible usages of a NPB system. The major program elements are given along with individual program objectives. The ion source and neutralizer technologies, which are most pertinent to the conference, are then outlined in more detail.

STRATEGIC DEFENSE INITIATIVE ORGANIZATION
NEUTRAL PARTICLE BEAM PROGRAM

The major responsibilities/goals of the SDIO's NPB program are shown below:

o Demonstrate the technical feasibility of the generation, high gradient acceleration, neutralization and accurate sensing of a high current, high brightness, negative ion beam in an integrated system at discrimination or weapons level parameters.

o Investigate and develop advanced components that will result in a more lethal, efficient, compact, and capable neutral particle beam system at discrimination or weapons level parameters.

o Investigate and assess other particle beam concepts that may have ballistic missile defense (BMD) utility.

o Perform associated support tasks such as evaluating particle beams for discrimination tasks, and intelligence data simulation and analysis efforts.

NPB systems can be used for ICBM defense in several ways: (1) As a space based platform useful for boost phase kill, midcourse kill, and midcourse discrimination, and (2) as a pop-up discriminator or RV killer. This is shown in figure 1.

© American Institute of Physics 1987

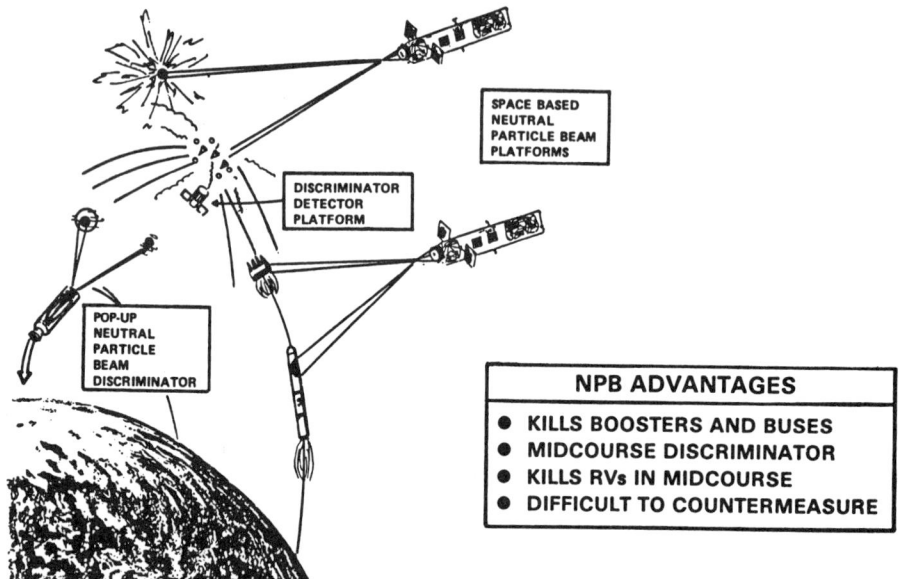

FIGURE 1: NPB Capabilities Essential to SDI

The U.S. Army Strategic Defense Command (USASDC) is responsible for developing the technology needed to demonstrate the feasibility of NPBs and to support the space demonstrations of the Air Force. In addition, the Army will be responsible for investigation and development of advanced or enhanced concepts. These enhanced concepts include laser photodetachment, plasma neutralizers, and other far term technologies which will not directly support space demonstrations. The near term technologies will have first priority over the far term technologies. It must be emphasized that while basic research is encouraged and needed, especially with regard to the enhanced technologies, this research must be focused to provide the technology needed for the near term and far term weapon or discrimination system.

The NPB program consists of the following concurrent major demonstrations or facilities: the accelerator test stand (ATS) at Los Alamos National Laboratory (LANL), the ground test accelerator (GTA) at LANL, the high resolution atomic beam (HIRAB) beam sensing facility at LANL, the neutral beam test facility (NBTF) at Brookhaven

National Laboratory (BNL), the radiation effects facility (REF) at
BNL, the Argonne National Laboratory (ANL) phase A and phase B 50 MeV
beam lines, the beams aboard rockets (BEAR), the continuous wave
deuterium demonstrator (CWDD) (site to be determined), and the
integrated space experiment (ISE). These programs, facilities, and
experiments performed at these facilities will lead to a system level
validation demonstration (SLVD) decision in the early 1990's. This
decision will determine whether development will begin on a system
level demonstrator. Figure 2 shows how several of these programs
interplay with each other to lead to the SLVD or prototype decision.

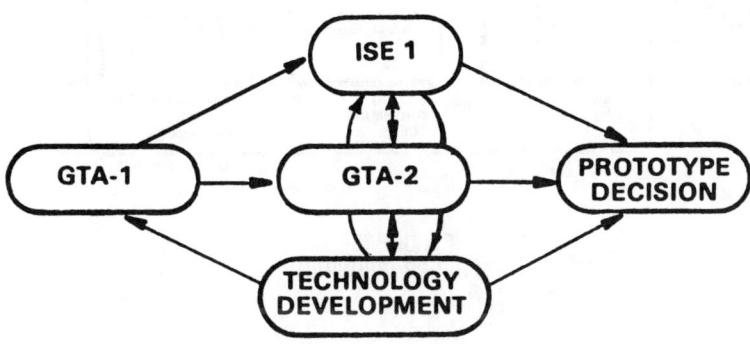

FIGURE 2
SDIO NPB PROGRAM

Figure 3 shows the total SDIO NPB program broken up by work
package area while figure 4 shows the objectives of the individual
USASDC work packages.

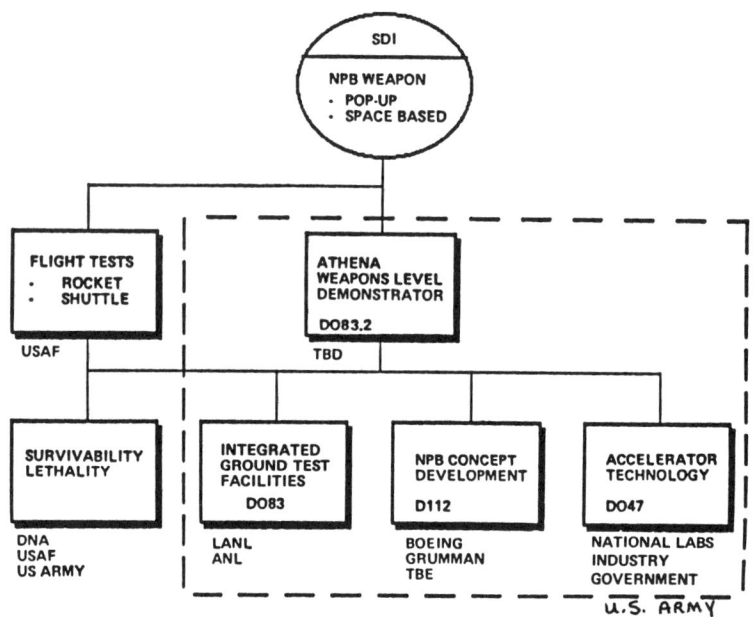

FIGURE 3
SDIO NPB PROGRAM

NPB PROGRAM WORK PACKAGES   OBJECTIVES

SBNPB CONCEPT DEFINITION	WPD112	- DEFINE PLATFORM REQUIREMENTS - DESIGN PLATFORM - DEVELOP TECHNOLOGY ROADMAP
ACCELERATOR TECHNOLOGY	WPD047	- PERFORM NPB COMPONENT DEVELOPMENT FOR ISE AND WEAPON/DISC SYSTEMS - VALIDATE LOW ENERGY INTEGRATED OPERATION - EXPLORE ALTERNATIVES AND ENHANCED TECHNOLOGIES
INTEGRATED GROUND EXPERIMENTS	WPD083	- FACILITY FOR TESTING COMPONENT TECHNOLOGIES - DEMONSTRATE PHYSICS OF FULLY INTEGRATED SYSTEM - LINK BETWEEN TECHNOLOGY AND INTEGRATED SPACE FLIGHTS
INTERACTIVE DISCRIMINATION SENSORS	WPDS501	- MEASURE NPB INDUCED TARGET EMISSIONS AND BACKGROUND - VALIDATE SIGNATURE CODES - DETERMINE HIT ASSESSMENT SIGNATURES AND SENSOR REQUIREMENTS - DEVELOP, DESIGN, FABRICATE, AND TEST NEUTRON SENSORS

FIGURE 4
USASDC NPB PROGRAM

NPB Component Technology

A typical prototype NPB weapon system is shown in figure 5. The major subsystems and their functions are given below:

(1) Ion Source - Generation of low emittance ions and acceleration to low energy for injection into the first stage of acceleration.

(2) Low Energy Beam Transport (LEBT) - Matching section between the ion source and first accelerator.

(3) Low Energy Accelerator - Typically a radio frequency quadruple (RFQ) accelerating the ions to approximately 2 MeV/nucleon.

(4) High Energy Accelerator - Typically a drift tube linac (DTL) structure. This includes the final stages of acceleration.

(5) Debuncher - Expands the ions in the $\vec{z}$ plane (direction of acceleration) prior to momentum compactor.

(6) Momentum Compacter - Decreases the momentum spread ($\Delta P/P$) in the $\vec{z}$ plane. This is necessary to keep the abberations in final optics at an acceptable level.

(7) Telescope (output optics) - This includes the beam expansion optics necessary to lower the output beam divergence ($\theta_d \propto \mathcal{E}/D$) and the final steering magnets.

(8) Neutralizer - Produces neutral particles from the accelerated negative ions.

(9) Beam Sensor - Senses direction of the particle beam.

(10) Acquisition, Pointing, and Tracking - Ties the beam sensing system into an overall target tracking system.

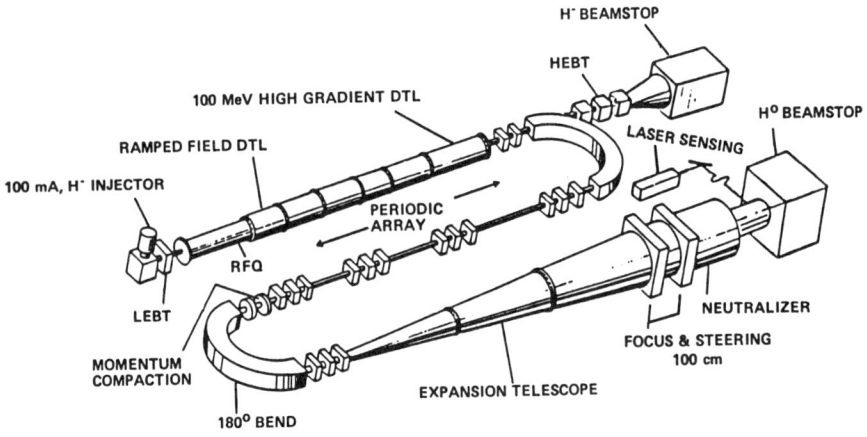

FIGURE 5
GROUND TEST ACCELERATOR (GTA)

The component technology research is being performed by the following organizations:

- Los Alamos
  GTA Facility
  Ion Source
  RFQ
  DTL
  Magnetic Optics
  Neutralizers
  Beam Sensing
  Discrimination

- Oak Ridge
  Ion Source
  Neutralizers

- Brookhaven
  Neutralizers
  Beam Sensing
  Lethality

- Lawrence Berkeley
  Ion Sources
  Magnetic Optics
  Plasma Neutralizer

- Draper
  APT

- HEDI
  Foils

- SNL
  Beam Jitter

- ANL
  50 MeV Facility

- Industry
  Cryogencis
  Funneling
  Neutralizers
  Beam Sensing
  RF Technology
  Support - CW

Neutralizer Technology

It is desired to develop a high efficiency, low emittance growth, large diameter neutralizer that is space qualifiable. Another way to say this is large diameter neutralization with minimum loss in beam brightness. Typically, SDIO is interested in diameters greater than 1 meter and better than 50% efficiency (neutralization efficiency combined with divergence factor). Two of the major items that determine the performance of a neutralizer are as follows: neutralization efficiency, defined as the neutral beam current divided by the negative ion current; and emittance growth, defined as the increase in beam emittance caused by the neutralizer. The root mean square (RMS) beam divergence caused by the beam emittance of a negative ion beam with velocity $v = \beta c$, normalized rms emittance ε_n, and rms beam diameter D (approximately ± 1 sigma for a Gaussian beam) is given by

$$\theta_\varepsilon = \frac{\varepsilon_n}{\pi \gamma \beta \; D/2}$$

This equation shows the direct relationship of beam emittance to beam divergence and the inverse relationship of beam diameter and beam divergence, therefore, showing that low emittance and large diameter beams (and therefore large diameter neutralizers) are needed to reduce the beam divergence to acceptable valves. However, there are limits on the neutralizer diameter due to the difficulty in making large diameter neutralizers and the increase in divergence caused by other components in the beam line as the beam diameter is increased (such as the output optics).

A material interaction neutralizer such as a foil or gas system will have a maximum divergence caused by the neutralizer (divergence growth therefore emittance growth) given by

$$\theta_n = \left(\frac{- m_e \; E_B}{m_I \; E_K} \right)^{1/2}$$

where

m_e = mass of the electron

m_I = mass of the ion

E_B = binding energy of the electron

E_K = Ion longitudinal kinetic energy

For H- at 250 MeV, θ_n = 1.28 microradians

For typical NPB system and typical target engagement parameters, the divergence growth of material interaction neutralizers coupled with their typical neutralization efficiencies (55%) yield on overall efficiency (neutral beam current on target divided by negative ion beam current) of approximately 20 to 30%. All of the above indicates the need for higher efficiency, lower divergence growth neutralizer techniques such as photodetachment or plasmas.

Other factors that contribute to the output beam divergence besides the divergence caused by the neutralizer, θ_N, are the divergence caused by beam emittance θ_ε, the divergence caused by momentum spread in the beam (chromatic aberration) θ_γ, the divergence caused by the beam jitter θ_J, and the divergence caused by spherical aberrations in the output optics θ_A. All of the factors add in rms fashion to give the total output beam divergence as

$$\theta_T^2 = \theta_N^2 + \theta_\varepsilon^2 + + \theta_\gamma^2 + \theta_J^2 + \theta_A^2$$

The various neutralizer concepts that are being considered at this time are discussed below. The basic list has not changed from last year.

1. <u>Gas Systems</u> - These include all systems that employ gas as a target material to achieve collisional electron detachment. These systems involve placing an optimum amount of gas at a certain pressure and temperature in the path of the negative ion beam to achieve detachment of the weakly bound electron. The maximum neutral particle efficiency that can be obtained by gas systems is 60% because of the production of positive ions by this method. Gas cells are static systems that do not take advantage of aerodynamic effects to help confine the gas. Gas jets are all gas concepts that do take advantage of aerodynamic effects to help confine the gas. Vapor curtains are gas jets that use relatively high mass effects.

2. <u>Film or Foils</u> - These include all systems that employ liquid films or solid foils as the target material to achieve collisional electron detachment. These systems are similar to the gas systems except that thinner neutralizers are possible (necessary) due to the increased density of the target material. These techniques will reduce the neutralizer length, therefore eliminating many neutralizer problems such as that caused by the Earth's magnetic field. However, the neutralizer lengths (foil thickness) for optimum neutralization are so short as to cause a problem with manufacture, durability, and survivability of these systems.

3. <u>Plasma Neutralizers</u> - These are concepts that utilize plasmas to convert energetic negative ions to neutrals by detachment of the weakly bound extra electron from the atom without removing any additional electrons. Plasmas are believed to give higher neutral yields than un-ionized gases, due to the lower production of positive ions, and possibly lower emittance growth.

4. <u>Photodetachment</u> - This is a concept that employs a laser or other photon emitter to impart enough energy to a negative ion to cause the weakly bound excess electron to detach. Due to the energy required to produce this detachment and the higher energy required to detach the more closely bound electron, less positive ions are produced by this technique. This technique offers the possibility of both enhanced neutralization efficiency and reduced angular divergence (lower emittance growth).

The neutralizer critical issues are given below:

o For Each Option:

1. Emittance Growth
2. Conversion Efficiency
3. Power Requirements
4. Package Requirements
5. Environmental Impact
6. System Impacts
7. Technology Assessment
8. Schedule
9. Fundings

The critical issues are coupled to each other (not independent). The amount of coupling depends on many factors, primarily the neutralizer concept. In emittance growth the primary interest is the transverse emittance growth, although there may be some longitudinal emittance growth which will lead to variations in longitudinal velocity. Conversion efficiency is the number of negative ions converted to neutral atoms. The actual number of usable particles, i.e., particles on target, is a combination of the emittance growth, conversion efficiency, and beam diameter as discussed above. Another critical issue is the type power, amount of power and power conditioning required. The environmental and system impacts must also be addressed. With regards to gas neutralizers, some of these are: the effect of gas discharge from the space platform on the platform vibration and stability, the effect of gas contamination on beam sensing, and the possible over stripping caused by gas contamination in the beam path. Other considerations are startup time, space charge effects, gas loss, gas flow, and plasma instabilities.

The technical developmental status of the neutralizer concept must be considered when choosing a concept for a particular mission such as the first space flight. The question must be asked "is major technology development necessary or is the technology basically almost there?"

Some of the neutralizer participants are given below:

BNL	Plug Nozzle
	Foils (with GA Tech)
	Plasma
LBL	Plasma
ORNL	Photodetachment
	Foils
LANL	Gas Cell
	Foils
	Photodetachment
HEDL	Foils

Ion Source Technology

Since the ion source is the first component in any NPB system its importance is obvious. The system emittance can never be reduced from whatever the ion source produces. All the other NPB components will only make the emittance worse. Some of the factors contributing to the importance of ion source research is given below:

o Initial component in beam production

o System emittance only as good as ion source emittance

o System only as reliable as ion source

o High brightness source essential to test other accelerator system components.

o Accelerator tests necessary in the next 3-6 years for near term system full scale engineering development (FSED) decision.

o Technology for ion sources not yet available.

The objective of the ion source research is to produce a CW (100% duty factor), low emittance, high current ion source capable of operating automatically and reliably for greater than 2 minutes. Low emittance and high current corresponds to high brightness commonly defined as the ion current divided by the two transverse emittances. The goal is to achieve as high a brightness as possible. The system impacts of the ion source component is tremendous. For example, if the ion source brightness is increased by a factor of two then the optics and neutralizer can be correspondingly reduced by a factor of 2. Some of the important considerations in ion source design if the current density, arc efficency, electron/ion contamination, gas neutralization in the LEBT and ease of operation.

A technical assessment of the risk associated with various aspects of an ion source is shown in Table 1.

REQUIREMENTS	SURFACE SOURCE	VOLUME SOURCE
HIGH CURRENT	LOW RISK	HIGH RISK
LOW EMITTANCE	LOW RISK	LOW RISK
HIGH DUTY FACTOR	HIGH RISK	LOW RISK
AUTOMATED OPERATION	HIGH RISK	MEDIUM RISK

TABLE 1

ION SOURCE TECHNICAL RISK

The ion sources under research presently can be broken down into two major classes - volume and surface. Each one has its particular advantages and disadvantages. For example, surface sources have high current density but require cesium for operation which makes CW and automatic operation difficult. Volume sources inherently have low emittance and non-cesium operation, but cannot match the surface sources in current density.

Some of the ion source participants are given below:

LANL	DUDNIKOV
	REFLEX VOLUME
	MODIFIED LBL CUSPED
LBL	CUSPED VOLUME
CULHAM	CUSPED VOLUME
ORNL	VITEX VOLUME
ORNL/ECOLE POLYTECHNIQUE	DIAGNOSTICS

CONCLUSIONS

NPB is essential to the SDIO mission because it can perform multiple missions, provides multiple kill mechanisms (electronics kill, warhead activation or dud, structural kill), and is difficult to countermeasure. The SIDO NPB programs support weapon system research by:

o Providing platform definition, design, and systems engineering

o Technology development and component validation

o Performance evaluation in a integrated weapons level NPB system in ground test facilities

o Lethality assessments

o Space demonstration via rocket and shuttle flights

MAGNETIC FUSION INTEREST IN NEGATIVE IONS

H. Stanley Staten
Office of Fusion Energy, U.S. Department of Energy
Washington, DC 20545

The problem of controlled thermonuclear fusion"... seems to have been created especially for the purpose of developing close cooperation between the scientists and engineers of various countries working at the problem according to a common plan..."
L. Artsimovich

I have been asked to talk about the magnetic fusion energy programs interest in negative ion based neutral beams. Let me answer the obvious question first by pointing out that there has always been a recognized desire for high power neutral beams of energy higher than that which can be efficiently produced with positive ion technology now presently in use. Therefore, let me turn the title of the talk around a bit and address the idea of "Negative Ion Interest in Magnetic Fusion". The point to be explored is whether there is a capability and interest in the negative ion community that can be brought to bear to propose development of negative ion based systems that the magnetic fusion energy researchers would like to incorporate into present and future experiments.

Since most of you are not directly associated with the magnetic fusion energy program, it is probably useful to first review the recent history of the magnetic fusion energy program. With the 1973 oil embargo there was a great desire to explore many and varied energy options. Energy development funds were readily available in the US and as one would expect, effort to develop magnetic fusion energy increased. The Tokamak Fusion Test Reactor (TFTR) was started. The goals for TFTR include demonstration of scientific breakeven. Later, the magnetic fusion energy engineering act was passed. This act mandated increased efforts on developing a fusion reactor, provided an aggressive schedule and directed that a plan to implement this schedule be prepared .

Subsequently, the prices of petroleum products stabilized and started to fall. There was even talk about an oil glut. The perception of urgent need for fusion to solve the energy crisis was gone and a different set of priorities associated with balancing the federal budget received attention. This of necessity led to decreases in the budget for fusion. Even with the pressures to reduce expenditures, fusion has maintained a credible effort and a respectable budget.

In February 1985 the Magnetic Fusion Program Plan was published. This plan has received wide acceptance inside and outside of government. It is a top level policy document that reflects the present perception of the need for energy development balanced with the desire to determine the potential for fusion as an energy source. A more detailed technical program analysis analysis has been undertaken. This has produced a draft technical program plan that should be read and commented upon by those who have an interest in its contents with respect to need for and utilization of negative ion based neutral beams.

The 1983 Versailles Economic Summit started a process of enhanced international cooperation in fusion. The 1986 Summit Working Group on Thermonuclear Fusion (FWG) acknowledged that a common medium-term goal for all fusions programs is an Engineering Test Reactor (ETR). The FWG reached a consensus that the joint design, construction and operation of this ETR is a reasonable objective for international collaboration. The FWG summary conclusions document noted that it is the view of the FWG that expanded technological cooperation can be successfully pursued with due regard to proprietary interests and respective national export control limitations.

As a result of this emphasis on common development I have been encouraged to seek increased cooperation in the development of negative ion neutral beams. Coming out of the above high level international meetings has been a process of joint planning that I believe can also be applied to negative ion neutral beam development. With the large expense of negative ion development and particularly the expense of building test facilities, I sincerely believe that the best way to proceed is to join efforts.

The joint planning process can be summarized by the following steps whereby we jointly:

- Determine goals
- Identify critical issues
- Prepare a plan to resolve critical issues and to reach our common goal
- Plan to share efforts

This process is applicable to negative ion neutral beam development.

Now I would like to look for a moment at the status of heating system development for large tokamak fusion test facilities. All the large tokamaks (JET, TFTR, JT-60, Doublet-III) include heating by neutral beams. The neutral beam systems are based on positive ion technology and operate in the 40 to 80 KeV per nucleon range. Even though these neutral beams are not sufficiently energetic to penetrate into the core of a high density plasma the exciting physics results now coming from TFTR and JET with respect to enhanced plasma confinement are based on use of the neutral beams.

High power radio frequency heating systems have also been installed on JET and JT-60. A High power radio frequency system is also scheduled to be installed on TFTR next Summer. These systems have the capability to provide heating power directly in the plasma center. Comparative heating data and experience is just now coming in from these experiments.

For the future large plasma experiments the preference is toward some form of radio frequency heating. In some cases neutral beams are included as a backup option. The US plans for the compact ignition device are based on Ion Cyclotron Resonant Heating (ICRH) while the TIBER design in addition to various radio frequency systems is considering use of neutral beams to drive the plasma current. It is premature to determine the ultimate heating method, yet the clear preference for radio frequency heating would cause one to question spending significant sums on development of negative ion based neutral beams. It is because of this apparent preference for radio frequency heating that I believe it is time for the negative ion community to step forward and present the best that they have to offer. I further believe that it is more likely to be successful if it is undertaken jointly.

The real question is then not what is the magnetic fusion interest in negative ions, but rather what does the negative ion community have to offer to magnetic fusion that can make the magnetic fusion effort better. Now is the time to make your proposals so that the competition with radio frequency heating can be completed in a way that provides the best possibility for neutral beams to be successful.

DISCUSSION

Schlachter: Is tangential access planned for CIT stand, or could neutral beam heating be done radially, where I believe access is planned?
Staten: Yes, they just increased the port size a little bit on CIT. They are in the process of doing some soul-searching about margins, and doing some redesign to relieve mechanical stress, whatever. It seems like it's an appropriate time to step in, and maybe you can't get everything you want out of the design, but the design is still fluid enough that you could probably get something to help more toward tangential access. My own reading is that it would probably have to be not normal to the plasma axis, but certainly not tangential; somewhere in between, like we do on a lot of the experiments today.
Stewart: Your viewgraph on the present plans mention heating, but not current drive. What are the primary and back-up alternatives for current drive?
Staten: There's a lot of things being talked about. CIT doesn't have current drive. On the engineering device, it's really anybody's ballgame at this point.
Peterson: You mentioned the fact that one of the encouraging aspects of the success on TFTR is based on the neutral beam heating. What you're saying is that its success has occurred because of the neutral beam heating; is that what you mean?
Staten: Yes.
Peterson: Have they tried rf heating on that at all?
Staten: RF will be installed about a year from now. ICRF will be.
Moses: Stan, I realize that DOE does a lot of wishful thinking, but I don't understand how one can assume FELs and ECRH heating as back-up for ICRH heating. Could you explain the logic behind that?
Staten: There are some FELs that are being operated today, and people are proposing an experiment to test an FEL into the Alcator-C if it's moved to Livermore.
Moses: Would you not agree that it's more realistic to consider negative ion beams than it is to ever consider an FEL heated plasma?
Staten: Possibly. I mean, it depends a lot on who is successful in developing their technologies. I come from the negative ion camp. I don't come from an FEL camp, so you know where my bias is. But, I'm not smart enough to tell you which is the most likely one to work.
Moses: I guess what I would have hoped is that this community would take an active role in making DOE aware that they feel that they have a better solution than the FEL.
Staten: That's reasonable.

NEUTRAL-BEAM CURRENT DRIVE IN TOKAMAKS

R. S. Devoto
Lawrence Livermore National Laboratory
University of California
Livermore, CA 94550

ABSTRACT

The theory of neutral-beam current drive in tokamaks is reviewed. Experiments are discussed where neutral beams have been used to drive current directly and also indirectly through neoclassical effects. Application of the theory to an experimental test reactor is described. It is shown that neutral beams formed from negative ions accelerated to 500-700 keV are needed for this device.

INTRODUCTION

A method of driving a steady plasma current in tokamaks has been a long-sought goal. Neutral beams, rf waves, and relativistic electrons, among others, have been proposed as possible current drivers. Many successful current-drive experiments have been carried out with slow waves in the lower hybrid (LH) range of frequencies (1-5 GHz). The measured current-drive efficiency, η, which is defined as the plasma current, I, divided by the absorbed power, P, has been found to be within a factor of two of the theoretical efficiency. The discrepancy can be explained and is chiefly due to the generation of back currents from parasitic waves launched in the opposite direction. Unfortunately, lower-hybrid waves of a parallel refractive index suitable for driving electron currents in an Experimental Test Reactor (ETR) cannot penetrate even halfway to the axis. It has been proposed that slow waves be supplemented by fast LH waves which can drive current in the core. However, some predictions show that the fast waves would be damped on fast alpha particles and thus be inefficient in driving current.

Electron cyclotron heating (ECH) has also been shown to generate plasma current in small experiments. The theoretical efficiency is slightly less than that for lower-hybrid waves. Until the recent invention of the free-electron laser there was no reasonably efficient source at a frequency appropriate for ECH drive in the higher magnetic field tokamaks. Experiments are currently underway in large tokamaks such as DIIID at GA Technologies.

A third major candidate for current drive, and the subject of this paper, is neutral beams. Two recent observations, both in the

TFTR tokamak at Princeton[1], serve to make neutral beams the most attractive option at present: the demonstration of current driven by neutral beams in a large tokamak, and the recognition that this driven current was serving as a seed current for the long-sought bootstrap current.[2]. The bootstrap current arises as a result of the radial gradient of pressure and temperature set up by the deposition of beam ions at the magnetic axis and can be larger than the original seed current. In this paper I discuss the theory behind neutral-beam current drive, the experimental results to date, and then the application to a ETR design-TIBER[3].

THEORETICAL PREDICTIONS

A theoretical treatment of neutral-beam current drive starts with the Fokker-Planck equation for the hot ions. In the full, non-simplified form, it includes unlike collisions with the background Maxwellian ions and electrons as well as like-particle collisons among the hot, beam-deposited ions. One approach is to solve this equation directly on several radial flux surfaces in the plasma finding the local deposition rate of hot ions from a simultaneous treatment of the beam absorption. The background plasma density and temperature can either be specified a priori as a function of radius or the radial transport equations can be solved simultaneously if the transport coefficients are known. These ideas are incorporated into a code called FPT[4] which was recently modified to compute the neutral-beam driven current, j_{nb}.

Because the electrons are much more mobile than the ions, they will speed up by drag on the hot ions and tend to cancel the ion current. If the charge of the beam ions, eZ_b, differs from that of the background ions, eZ_{eff}, the resulting difference in drag allows the electrons to only partially cancel the beam current. Furthermore, because a portion of the electrons are trapped in the toroidal well, the fractional cacellation depends also on the local inverse aspect ratio, $\epsilon \equiv r/R$. The net current can be written as

$$j_{net} = j_{nb} F_{nb}(Z_b, Z_{eff}, \epsilon), \qquad (1)$$

where F_{nb} is of the form[5,6]

$$F_{nb} = 1 - Z_b / Z_{eff} + f(Z_{eff}, \epsilon). \qquad (2)$$

At small ϵ, f is proportional to $\epsilon^{\frac{1}{2}}$ so the trapped-electron correction is small near the the magnetic axis. As a result, although j_{nb} may be peaked on axis, j_{net} may actually show a dip. An example of this is given in Fig. 1 which shows the current density for injection of 500 keV deuterium into the TIBER ETR[3] as computed with the FPT code.

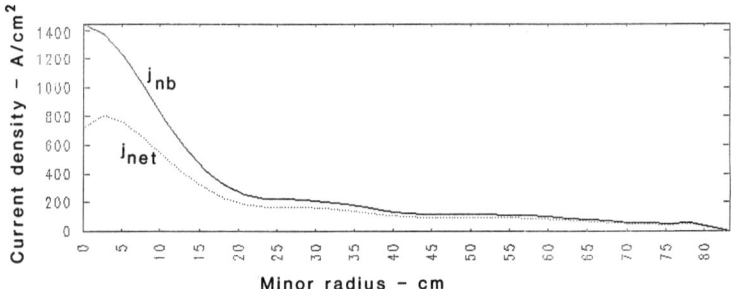

Figure 1. Current profile for TIBER showing the neutral-beam-driven curren, j_{nb}, and the net current after allowance for the induced electron current, j_{net}.

An additional source of current is the neoclassical current generated by radial gradients of pressure and temperature, the so-called bootstrap current.[2] In the banana regime in a $Z_{eff} = 1$ plasma the current obeys the relation

$$j_{bs} = \frac{\epsilon^{1/2}}{B_\theta} \{ -2.44 [\frac{\partial p_e}{\partial r} + \frac{\partial p_i}{\partial r} - 1.17 n_e \frac{\partial T_i}{\partial r}] + 1.75 n_e \frac{\partial T_e}{\partial r} \} \quad (3)$$

with B_θ the poloidal magnetic field which must be found from the local total current density using Ampere's law. This expression has been incorporated into the FPT code to yield an estimate of the neoclassical current for fixed pressure and temperature profiles. The profiles should come from a solution of the radial transport problem, but, because of the anomolous radial energy transport, there is as yet no accepted form for the transport coefficients. For TIBER, a model pressure profile is used to compute the MHD equilibrium, and, in accord with TFTR results, the density profile is taken as peaked near the center

$$n_e = n_e [1 - (r/a)^2]. \quad (4a)$$

Temperature profiles of both ions and electrons are assumed to have the same shape and from the p and n_e profiles a suitable approximate form is

$$T = T [1 - (r/a)^2]^{0.6}. \quad (4b)$$

These profiles have been used for the case of Fig. 1 to obtain the current profiles shown in Fig. 2. Since no neo-classical current flows on axis, the dip at the origin is slightly more pronounced. This particular case was computed for a neutral-beam source with a 5 mrad divergence located at 13 m from the midplane of the tokamak and focussed at the midplane in both the vertical and horizontal directions. The beam was aimed tangential to the major radius in this case. If a source with the same divergence is located further from the tokamak, then the beam must be aimed inside of the major radius to obtain an equivalent peaking.

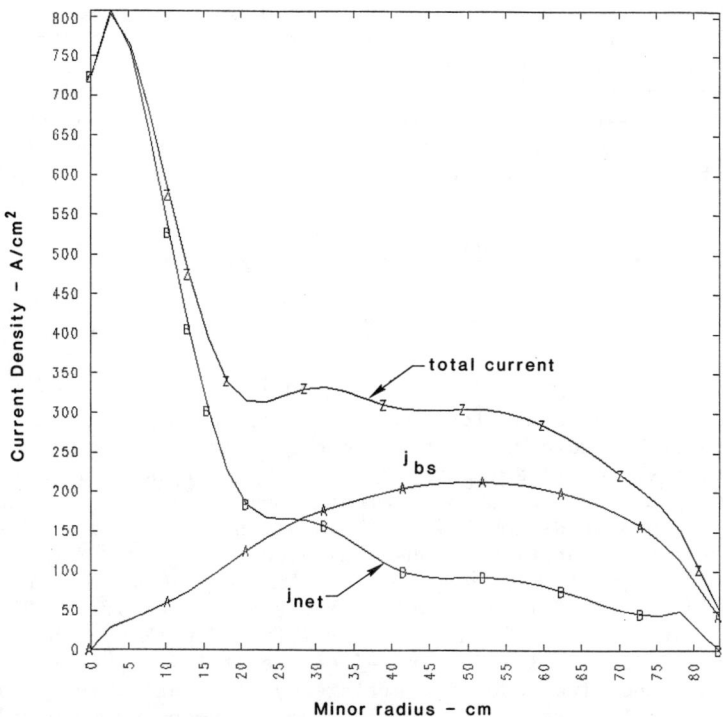

Figure 2. Profile of net beam current, j_{net}, and bootstrap current, j_{bs}, for typical density and temperature profiles in TIBER.

We can form a simple scaling formula for the fraction of current due to bootstrap by estimating the terms in Eq. (3) and noting that $B_\theta \approx I_p / a$ to get

$$\frac{I_{bs}}{I_p} \approx \left(\frac{a}{R}\right)^{\frac{1}{2}} \beta_p,$$

where $\beta_p \equiv 2\mu_o p / B_\theta^2$ is the poloidal β. Applying this to TIBER, which has an inverse aspect ratio a/R = 0.83 / 3.0 and $\beta_p \approx$ 1, we expect the fraction of current carried by bootstrap to be about 50 %. For the case of Fig. 2, the fraction is 3.52 MA / 5.73 MA = 61%. It should be noted that Eq. (3) is only valid for a Z_{eff} = 1 plasma; some changes can be expected when an expression including high-Z impurities is used.

In most cases, the hot-ion density is much less than the background density and self collisions can be neglected. Furthermore, the electron thermal speed is usually much greated than the beam speed. Several authors have made use of these simplifications to obtain a series solution for the hot-ion distribution function. A convenient formulation is the one presented by Gaffey.[7] Mikkelson and Singer[8] have integrated this distribution function to obtain an expression for the current-drive efficiency in a circular tokamak

$$\eta = \frac{4.25}{R_o \ln \Lambda} \frac{\int_0^a T_e v_{||} H J(x,y) F_{nb} 2\pi r dr}{n_e v_b a^2} \quad (5)$$

where R_o is in m, T_e in keV, and n in 10^{20} m^{-3}. $v_{||}$ is the local component of the beam velocity parallel to the B field, H is the fraction of the total beam power deposited at radius r, a is the minor radius, and $J(x \approx [E_b/10m_b T_e]^{\frac{1}{2}}, y \approx 0.8 Z_{eff}/m_b)$ is obtained from a quadrature which must be evaluated numerically. A plot of J(x,y), reproduced from Ref. 8, is given in Fig. 3. The beam energy for deuterium ions injected into a Z_{eff} = 2 plasma with T_e = 25 keV is given along the top axis. For TIBER, y ≈ 0.6 (using the full expressions for y given in Refs. 7,8), and we see that an increase in beam energy from 500 keV, the nominal value for TIBER, to 1 MeV, increases the efficiency by some 30 %. Although the increase in efficiency is desirable from the standpoint of a reduction in the size or number of neutral-beam injectors, it may not be advantageous for overall plasma performance. This point is discussed in the last section.

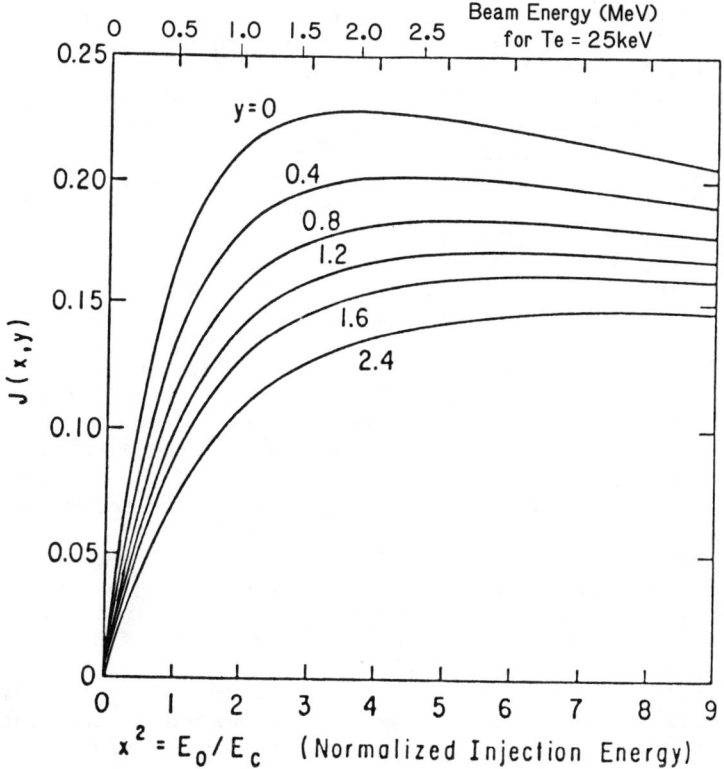

Figure 3. J(x,y) for Eq. (5).

Although we could use Eq. (5) together with a computation of H to determine η, we choose instead to use it as a scaling formula for the FPT code. We replace T_e and n_e in Eq. (5) by their volume-averaged values, take $v_{||} = v_b$ and $\epsilon = 0.1$ for evaluating F_{nb} in TIBER,

$$\eta = \frac{C \langle T_e \rangle J(\langle x \rangle, \langle y \rangle) F_{nb}}{\langle n_e \rangle R_0 \ln \Lambda} \qquad (6)$$

For plasmas with profiles like those in Eqs. (4a-b), beam energies from 0.5-2 MeV and average densities of 0.7-1 x 10^{20} m^{-3}, the constant C is about 3.

EXPERIMENTS

Current drive by neutral beams has been demonstrated in four experiments: the Culham superconducting Levitron[9,10], DITE[11,12],

JET[13] and TFTR[1]. In the Levitron, a hydrogen neutral beam generated from a positive-ion source with mean energy ≈ 8.5 keV and current ≈ 0.2 A was injected into a Z_{eff} ≈ 1 plasma which had T_e varying from 1 to 4.7 eV. Under these conditions, the electron thermal speed varies from about ½ to 1 x the beam speed. At the lower value, the electrons are moving too slowly to cancel the ion current and nearly the full ion current was observed. At the other limit, the measured current was found to be opposite to the ion current. In a theoretical paper which treated the electrons without assuming $v_e \gg v_b$, Cordey, et al.[10] were able to predict the entire dependance of the measured current on T_e, including the observed reversal of current direction at T_e = 4.7 eV.

The first application of neutral-beam current drive to a tokamak ocurred in DITE[11,12]. They injected approximately 1 MW of 24 keV neutral beam current into a plasma with peak density from 3.1 to 7.8 x 10^{19} m^{-3} and peak electron temperature from 0.67 to 1.1 keV. They did not have sufficient power to drive the entire plasma current with neutral beams, so they measured the reduction in loop voltage while the plasma current was held fixed with the ohmic transformer. They verified that co-injection lowered the voltage while counter-injection raised it. Since heating of the electrons by the beam ions could cause a change in the loop voltage, the drop of the loop voltage can not be taken as definitive evidence that beam-injected ions are carrying plasma current. They verified the existance of hot-ion current by computing the theoretical loop voltage from the experimental temperature and density profiles together with a Fokker-Planck computation of the fast-ion current and comparing with the measured loop voltage. The good agreement demonstrated the presence of neutral-beam driven current. The maximum current driven was 38 kA out of a total plasma current of 80 kA. A simple ratio of the current to total injected power for this case yields an apparent current-drive efficiency of 38 / 900 = 0.04. Note, however, that the electric field also accelerates ions so the true efficiency is probably less. The authors report that the effects of electron trapping, neoclassical current and plasma rotation are less than 10 % for this experiment and hence negligible.

Cordey[13] has also recently reported that the neutral beams in JET were probably driving some 0.6 MA of plasma current. For this experiment they find that the neoclassical current is negligible.

The recent TFTR experiments are the main reason for increased optimism that neutral beams can drive steady state currents in eventual reactors. The analysis of these experiments will probably continue over the next several months, but some significant early results are described by Zarnstorff[1]. If the edge density is low enough, as obtained with good outgassing of the walls, and, if the magnetic axis is shifted outward sufficiently (the so-called Shafranov shift), then the neutral beams penetrate to the plasma center and drive substantial plasma current (some 200 kA out of 0.9-1 MA). The experimental signal of these good conditions is the reversal of the surface loop voltage. Like DITE, the plasma current is held fixed by the transformer. The experiment is analyzed by predicting the loop voltage from theoretical expressions together with the measured density and temperature profiles. The measured voltage cannot be duplicated unless both beam-driven and neoclassical currents are included. Thus, the TFTR experiments have confirmed that bootstrap current does exist and can substantially enhance the beam-driven current in a large tokamak.

APPLICATION TO AN ETR-TIBER

The above ideas have been used to predict the performance of a small, superconducting ETR-TIBER[3]. Major parameters for this machine are major (minor) radius = 3 (0.83) m, ellipticity ≈ 2.1, triangularity ≈ 0.5, fusion power = 250-350 MW, average wall neutron load = 1.8-2.2 MW/m^2, plasma current = 8-12 MA, and current drive power = 40-70 MW. Plasma properties have been computed using the CIT physics guidelines[14] with the addition of the neutral-beam current drive as described above. Properties are listed in Table I for cases with and without half of the current supplied by neoclassical effects. We see that the incorporation of

Table I. Comparison of TIBER parameters with current supplied solely by NBI and with ½ of the current supplied by neoclassical effects.

	No Neoc.	w. Neoc.
Fusion power-MW	278	345
NBI power-MW @500 keV	72	53
Q	3.9	6.6
Plasma current-MA	8	8
Vol-ave. T_e-keV	20	17
Vol-ave. T_i-keV	28	18
Vol-ave. N_e-10^{20} m^{-3}	0.88	1.2
Z_{eff}	2.0	2.0
shine-through-%	1.6	0.3

neoclassical effects has a very favorable effect on the beam power and Q, but even without these effects, this ETR would still be an attractive machine for physics and engineering testing. The neutral-beam energy was chosen to be 500 keV for these studies because of the report[15] of a design at that energy with a reasonable electrical efficiency of 45 %. The shine-through for this energy is so low (Table I), that it should be possible to raise the energy somewhat and still be able to tolerate the increased shine-through power on the beam dump. Interestingly enough, the plasma parameters, as computed from this point model, do not change appreciably when the beam energy is raised up to 1 MeV, a possible limit for DC acceleration of ions. The cause is not difficult to find: as the beam energy is increased, the power generated by beam-background fusion reactions decreases. There is then less alpha-particle energy to heat the plasma and the temperatures will decrease somewhat. If the decrease in electron temperature is not too great, then the current-drive efficiency can increase because of the increased beam energy, as shown in Fig. 3. But the density will also increase if one operates at the beta limit, and the efficiency could actually decrease with increased beam energy. Both cases, increased or decreased efficiency have been observed. But, there is possibly a more important reason for higher energy: better penetration to the magnetic axis. Figures 1-2 showed results for a plasma with $<n_e>$ ≈ 8×10^{10} m^{-3}. For higher densities, such as in the second column of Table I, higher beam energies may be desirable for peaked current profiles.

ACKNOWLEDGEMENTS

Part of this work was carried out in collaboration with M. Fenstermacher and A. A. Mirin and will be reported in more detail elsewhere.

REFERENCES

1. M. Zarnstorff, "Enhanced Confinement Neutral Beam Heated Discharges in TFTR," Bull. APS 31, 1383 (1986).

2. R. J. Bickerton, J. W. Connor, J. B. Taylor, Nature Physical Science 229, 110 (1971).

3. "TIBER II-Tokamak Ignition/Burn Experimental Reactor," C. D. Henning and B. G. Logan, Scientific Editors, University of California, Lawrence Livermore National Laboratory Report, UCID-20863 (1986).

4. A. A. Mirin, J. Killeen, K. D. Marx, and M. E. Rensink, J. Compu. Physics 23, 23 (1977); J. Killeen, A. A. Mirin, and M. G. McCoy, Modern Plasma Physics (IAEA, Vienna, 1981) 395.

5. D. F. H. Start, J. G. Cordey and E. M. Jones, Plasma Phys. $\underline{22}$, 303 (1980).

6. D. F. H. Start and J. G. Cordey, Phys. Fluids $\underline{23}$, 1477 (1980).

7. J. D. Gaffey, J. Plasma Phys. $\underline{16}$, 149 (1979).

8. D. R. Mikkelsen and C. E. Singer, Nucl. Tech./Fusion $\underline{4}$, 237 (1983).

9. D. F. H. Start, P. R. Collins, E. M. Jones, et al., Phys. Rev. Letters $\underline{40}$, 1497 (1978).

10. J. G. Cordey, E. M. Jones, D. F. H. Start, et al., Nucl. Fusion $\underline{19}$, 249 (1979).

11. W. H. M. Clark, J. G. Cordey, M. Cox et al., Phys. Rev. Letters $\underline{45}$, 1101 (1980).

12. K. B. Axon, W. H. M. Clark, J. G. Cordey, et al., in <u>Plasma Physics and Controlled Fusion Research</u>, Vol. I (IAEA, Vienna, 1981) 413.

13. J. G. Cordey, "Beam Current Drive in JET: Experiments and Theory," presented at the IAEA INTOR-Related Specialists' Meeting on Non-Inductive Current Drive, Garching, BRD, 15-17 September, 1986.

14. J. Sheffield, "Physics Guidelines for the Compact Ignition Tokamak," presented at ANS Meeting, Reno, Nevada (June, 1986).

15. Y. Okumura, H. Horiike, H. Inami, et al., Proceedings, 11th Symposium of Fusion Engineering (IEEE, 1986), 113.

"Work performed under the auspices of the U.S. Department of Energy by the Lawrence Livermore National Laboratory under contract number W-7405-ENG-48."

DISCUSSION

Moses: The TIBER design is not an ignition device and all the other devices which have been proposed by the other countries are ignition devices. It would seem to me that we should be driving toward an ignition device rather than a driven one.

Devoto: I'm sorry, I don't really think that the other devices are ignition devices. They're all driven; they're all heated in some way or another.

Moses: But they go to ignition, where the alpha particles start contributing and maintain the temperature.

Devoto: It depends on how you define ignition, I guess, doesn't it? But they all are driven in some way or another. You've got to drive them up until they reach the ignition temperature, but then you immediately shut them off and drive them back again. And so, if you integrate over time, they certainly do not have $q = \infty$, do they? It depends on how you define ignition, I guess.

Lietzke: From your perspective, Steve, what is your prognosis for the availability of an ECRH source for current drive, as compared to neutral beams?

Devoto: The ECH current drive option is based upon the success which Livermore has had recently in the FEL program, and they've gotten quite high efficiencies; my recollection is 45 or 50% efficiency. That's energy conversion efficiency, that's not plug-to-plug efficiency. This is on the pulsed FELs. If you look at the whole system, in the pulsed FEL and extrapolate on out, the electrical efficiency should be down in the low 30% efficiency. And it does have good possibility for control of frequency. On the other hand, TRW has looked at a steady-state FEL, not a pulsed FEL. They don't have as much efficiency within the FEL itself – it's only at the order of 10% – but then they recirculate, direct convert, a lot of energy, and so they project 45% electrical efficiency. If you look at the JAERI neutral beam design, where they project 45% efficiency without a plasma neutralizer, just with a gas neutralizer, there certainly seems to be considerable chance that that's going to be even better, providing all of their other estimates are okay, if you switch. So it looks like, at least from an efficiency standpoint, the negative ion neutral beams are better. There's one other feature, too, which I didn't really emphasize. We're assuming, in the properties I showed for ECH, that we can also get bootstrap current there. Well, neutral beams, of course, are much different from ECH in that neutral beams deposit particles as well as energy, and they go where you want them to, generally. If the particle source on axis is really important, too, for bootstrap current, then you're going to have to get some very good high energy pellet injectors to get that particle source on axis. And it's not clear that you're going to be able to have it so peaked on axis as you probably can on neutral beams. And so that's why neutral beams look a heck of a lot better than ECH right now.

PROPOSED NEUTRAL-BEAM DIAGNOSTICS FOR FAST CONFINED ALPHA PARTICLES IN A BURNING PLASMA*

A. S. Schlachter and W. S. Cooper

Lawrence Berkeley Laboratory
University of California
Berkeley, CA 94720

ABSTRACT

Diagnostic methods for fast confined alpha particles are essential for a burning plasma experiment. Several methods which use energetic neutral beams have been proposed. We review these methods and discuss system considerations for their implementation.

INTRODUCTION

Alpha particles (He^{2+} ions) produced in the d-t fusion reaction

$$d + t \rightarrow \alpha(3.5 \text{ MeV}) + n(14.1 \text{ MeV}) \qquad (1)$$

are born at an energy of 3.5 MeV, or 880 keV/u. Most are trapped in the magnetic field confining the plasma and slow down by interaction with plasma ions and electrons, thereby heating the plasma. The plasma is called "ignited" when the alpha particles produced by the d-t fusion reaction provide sufficient heat to the plasma to sustain the reaction, i.e., a "burning" plasma.

A new generation of tokamaks is being planned in which a primary objective is to achieve ignition. An essential diagnostic is measurement of the spatial, temporal, and velocity distribution of the fast alpha particles as they slow down, to ascertain whether they slow down classically or by some other energy-loss mechanism, and to answer questions concerning alpha-particle confinement and efficiency of alpha-particle heating.

Alpha particles will have a distribution of velocities as they slow down. Most alpha particles will be confined by the magnetic field of the tokamak; some fraction will escape. Measurements can be made relatively easily of the escaping alpha particles; their distribution, however, might not be characteristic of the confined alpha particles. Neutrons from the d-t reaction (1) will certainly escape the plasma, and measurements can be made of their distribution. Alpha particles which are confined will become thermalized. This paper addresses methods of detecting fast confined alpha particles.

Development of a diagnostic method for fast confined alpha particles is difficult because of the high energy of the alpha

*Supported by U.S. DOE Contract Number DE-AC03-76SF00098.

LBL-22386

particles and because of the minimal access to and large background emission from a burning-plasma reactor. Parameters for the current in the United-States project, CIT (Compact Ignition Tokamak), include electron and ion densities of the order of 5×10^{14} cm^{-3}, electron and ion temperatures of 10 keV, alpha-particle density of 3×10^{13} cm^{-3}, and minor radius of less than 50 cm. CIT will be small and compact, with limited access for diagnostics. The burning plasma will generate a considerable flux of neutrons and photons, and activation will require remote-handling capability for equipment exposed to the plasma.

The topic of alpha-particle diagnostics and related atomic-physics issues has been recently reviewed by members of the Princeton Plasma Physics group (Post, Zweben, Grisham, and Fonck)[1-3] and by Schumacher,[4] while alpha-particle effects were the topic by a recent workshop.[5] The purpose of this paper is to review proposed methods for fast confined alpha-particle diagnostics in a burning-plasma reactor which are based upon energetic atomic beams, including discussion of accelerator requirements to implement a diagnostic. Much of what is presented here has been proposed and discussed in important articles by Post, Grisham, and co-workers at PPPL,[6,7] by partipants in the NYU workshop,[5] and in reports from Nagoya.[8,9]

A variety of methods has been proposed[5] for a diagnostic for fast confined alpha particles. These methods include laser techniques (small-angle infra-red Thomson scattering), microwave techniques (lower-hybrid-wave damping, ion-cyclotron emission), nuclear-reaction methods (gamma-ray emission), and charge-transfer methods. The charge-transfer methods use an energetic beam of atoms injected into the plasma as a target for charge transfer by the alpha particles. This method is described in the following paragraphs.

The major neutral-beam-based methods considered for a diagnostic for fast confined alpha particles are
a) single-electron capture by He^{2+} from an energetic neutral atom beam injected into the plasma to serve as a charge-transfer target (charge-exchange recombination spectroscopy), leading to emission of a photon from excited states of He$^+$:

$$He^{2+} + X \rightarrow He^{+*}(n\ell) + [X^+] ; \qquad (2)$$
$$h\nu \hookrightarrow He^+$$

(b) two-electron capture by He^{2+} from an energetic neutral atom beam injected into the plasma, creating a fast He atom which, being neutral, can escape the confining magnetic field:

$$He^{2+} + X \rightarrow He^0 + [X^{2+}]; \quad and \qquad (3)$$

(c) nuclear methods, in which the alpha particle interacts with target nuclei which are introduced into the plasma, resulting in emission of a gamma ray.[10] This latter method seems especially difficult, or requires an appreciable admixture of the target specie in the plasma, and will not be discussed in this paper. Non-beam-based methods for alpha-particle diagnostics have also been proposed, as well as diagnostics on the neutrons which are also emitted in the d-t reaction. Methods (a) and (b) have been discussed by Post et al.,[7] while an implementation of method (b) using a Li^0 beam has been discussed by Grisham et al.[6] and, using a He^0 beam, by Sasao and Sato.[8]

Methods (a) and (b) are shown schematically in Fig. 1 for the most likely candidates for the target beam: H^0 for reaction (2)

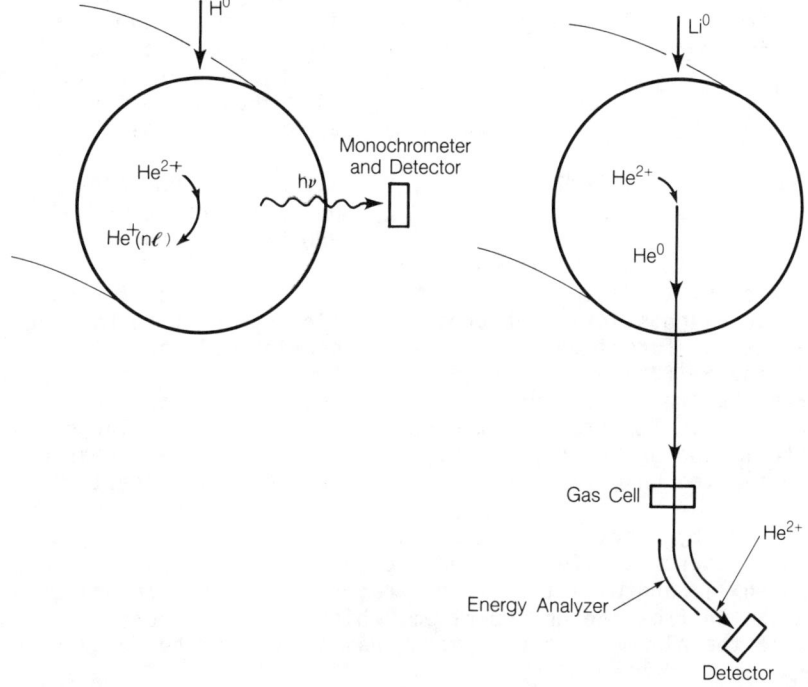

Fig. 1 Schematic diagram of neutral-beam methods for fast confined alpha-particle diagnostics, showing likely candidates for injected beams.

and He^0 or Li^0 for reaction (3); these choices are explained in more detail in the following sections. The corresponding cross sections for reactions (2) and (3) are shown in Fig. 2. As Post et al. have discussed,[7] the cross sections for the

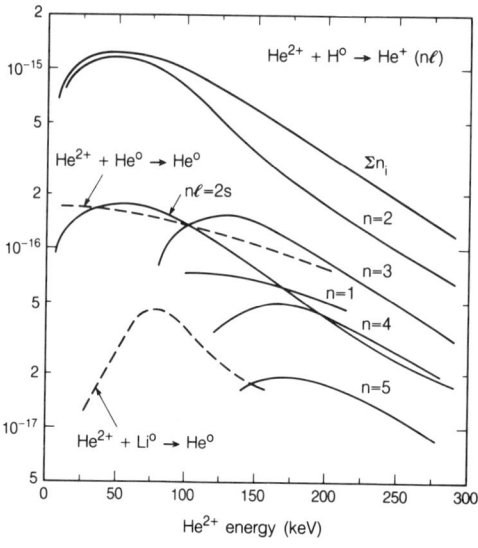

Fig. 2 Cross sections in units of cm^2 relevant to fast alpha-particle diagnostics. The cross sections for formation of He$^+$ (Σn_i) and He$^+$(2s) are measured, as are the cross sections for formation of He^0. The cross sections for formation of He$^+$(n=2,3,4,5,) are estimated only (see text).

reactions shown in Fig. 2 are small for high velocities, and thus the diagnostic method has velocity selectivity, i.e., fast alpha particles interact essentially only with injected target atoms of the same velocity (magnitude and direction). Varying the velocity of the injected target beam and varying its direction, if possible (given limited access), will allow measurement of the velocity distribution of the fast alpha particles.

SINGLE-ELECTRON CAPTURE TO He$^+$(nℓ) FOLLOWED BY PHOTON EMISSION

The method of single-electron capture by a fast confined alpha particle from a neutral atomic beam injected to serve as a charge-transfer target is a form of charge-exchange recombination spectroscopy.[11] Most of the He$^+$ formed by electron capture will be in a excited state, He$^+$(nℓ), where nℓ refers to the quantum numbers of the state (n>1), and will therefore decay radiatively with emission of a photon from the He$^+$ spectrum, which can be detected outside the plasma. This approach has been described by Post et al.[7]

The neutral beam serving as a charge-transfer target must have a velocity of the order of the alpha particles to be useful, because the cross section for electron capture is appreciable only at small relative energies. For example, Fig. 2 shows estimated cross sections for reaction 2 with X being an atomic hydrogen beam. This is a logical candidate, because intense H^0 beams are presently being developed at LBL and elsewhere. The cross sections for He^{2+} + H^0 → He$^+$(nℓ) were estimated by comparing calculations by R. E. Olson[12] with measured cross sections for electron capture into all states of He$^+$ and into the He$^+$(2s) state.[13] The accuracy of these estimated cross sections is, at best, a factor of 2, and could be considerably worse. Electron capture into the n=2 state predominates because

the reaction is accidentially resonant for collision with a
ground-state H^0 atom. The $He^+(n=2)$ state decays radiatively
with the emission of a photon at 304 Å, well into the vacuum
ultraviolet spectral region. The estimated cross section for
formation of $He^+(n=3)$ shows a broad peak at higher energy than
the cross section for formation of $He^+(n=2)$, and down in
magnitude by approximately a fractor of ten. The estimated cross
sections for formation of $He^+(n=4)$ and $He^+(n=5)$ are down by
approximately a factor of 2 and 4 from the $He^+(n=3)$ cross
section. The $He^+(n=3)$ level decays radiatively with the
emission of a photon at 1640 Å, which is in the vacuum
ultraviolet, and where refractive optics are useable; the
$He^+(n=4\rightarrow3)$ transition is in the visible (blue) spectral range
at 4686 Å. These transactions are shown in Fig. 3. Note that
the He^+ spectrum is the same as the H^0 spectrum, but with the
energies multiplied by a factor a 4; e.g., $He^+(4\rightarrow2)$ radiates
at the same energy as $H^0(2\rightarrow1)$, which is Lyman-alpha for H
atoms.

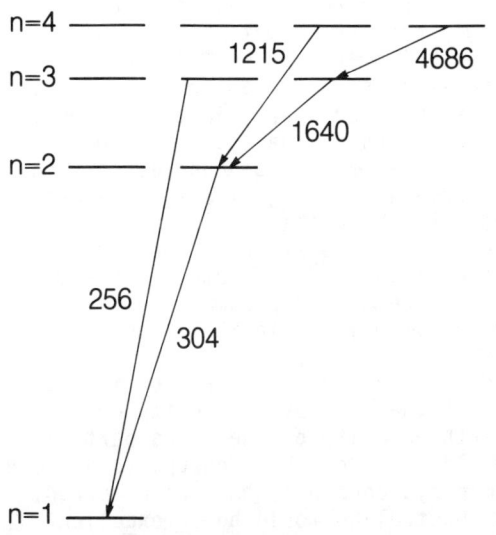

Fig. 3 Levels of $He^+(n\ell)$ shown schematically. The numbers are transition wavelengths in Angstrom units.

The photons radiated by $He^+(n\ell)$ will be Doppler-shifted because the ions are moving, and the photons can thus be distinguished from radiation from any slow He^+ ions which might be in the plasma or along the line of sight. The photons are, however, emitted into 4π solid angle, and only a small fraction will be detected, depending primarily upon constraints imposed by limited access. Mirror and refractive optics can be used for $He^+(n=3\rightarrow2)$ radiation at 1640 Å; a high-reflectivity multi-layer mirror will be used for He^+ $(n=2\rightarrow1)$ radiation at 304 Å. Estimated signal level for reasonable target-beam parameters in either case is estimated to be smaller than the bremsstrahlung background from the plasma, so that some signal-processing techniques will be needed.[15] The collecting mirror or lens will be exposed to an intense x-ray and neutron flux, so that radiation hardening will be a consideration in the design of a diagnostic system.

There are additional considerations. One is that a Li⁰ beam could be used as a charge-transfer target; this would give a greater yield of He⁺(n=3). However, intense Li⁰ beams are not presently under development, while intense H⁰ beams are being actively developed. Another consideration is that He⁺ (nℓ) must radiate before being collisionally ionized to produce a photon; this limits the usefulness of He⁺ states with large values on n. Finally, the intense magnetic field of the tokamak (10 Tesla) will cause changes in state lifetimes, state populations, and decay rates.

TWO-ELECTRON CAPTURE TO He⁰

A fast alpha particle which captures 2 electrons from a target atom will become a fast neutral He atom, which has some likehood of traversing the plasma and, being neutral, will escape the confining magnetic field, where it can be reionized and energy analyzed, or otherwise detected. This approach has been described by Grisham et al,[6] who proposed the use of a fast Li⁰ beam as the injected charge-transfer target. The cross section for 2-electron transfer[14] from Li⁰ to He⁺⁺ is shown in Fig. 2; it is appreciable only at small relative velocities, hence the injected Li⁰ beam must have an energy in the 5-6 MeV range to interact with alpha particles at their birth velocity.

Beam penetration through the plasma must be considered both for the Li⁰ beam injected to serve as a charge-transfer target and for the He⁰ beam produced when the alpha particles capture 2 electrons. The latter depends upon the state of the He⁰ atoms; any state other than the ground $1s^2$ ^1S state is likely to be reionized in traversing the remaining plasma. However, it is likely that most of the He⁰ beam will be in the ground state.[12]

The choice of a target beam for 2-electron transfer is dictated primarily by the requirement that the beam velocity be approximately equal to the birth velocity of the alpha particles, thus use of a heavy specie would require a high energy to achieve the necessary velocity. Atomic hydrogen only has one electron, and is thus not a candidate. Neutral He would be an excellent choice, because the 2-electron-transfer cross section,[16] He^{2+} + He^0 → He^0 + He^{2+} is larger than 10^{-16} cm², i.e., at least a factor of 3 larger than for a Li⁰ target beam. However, fast He⁰ in the ground state is not the predominant product of neutralization of energetic He⁻ in a gas target,[17] rather the $1s2s\,^3S$ metastable state is formed. Options could include neutralization and dissociation of a positive molecular ion, e.g., HeH⁺ or He$_2^+$, or of the negative ion He$_2^-$, which could result in fast ground-state He⁰, or decay of metastable He⁻ in a long drift space, as proposed by Sasso and Sato.[8] These ideas notwithstanding, Li⁰ is the likely candidate for a target beam. However, a fast beam of ground-state He⁰ would be very desirable.

The fast He⁰ beam that exits the plasma can be reionized and energy analyzed, to distinguish fast He⁰ from other ions and atoms emitted by the plasma. Although the cross section for 2-electron transfer is 1 to 2 orders of magnitude smaller than that for 1-electron transfer to He⁺(n=2), essentially all the He⁰ will exit the plasma (excepting those which are collisionally ionized) in the direction of the incident target beam, and can thus be detected.

SYSTEM CONSIDERATIONS

The successful development of atomic-beam-based alpha-particle diagnostic schemes will pose challenges in the area of ion source and accelerator development. Since only hydrogen or deuterium beams are presently under development, we shall confine our discussion to accelerators for these species. In addition, for reasons of economy of power and space utilization, we consider only systems capable of dc operation in a multiple-aperture configuration.

The confined alpha-particle diagnostic requires beam energies up to 1 MeV, thus H⁻ ions must be accelerated, as only negative ions can be converted to neutral atoms at these energies with high efficiencies. The beam energy must also variable, from a few hundred keV up to approximately 1 MeV; it would be desirable for beam current to be approximately constant, i.e., independent of the beam energy.

Although a conventional Pierce column has reached 350 mA of H⁺ at 600 keV dc operation,[18] another approach must be followed to allow variation of the beam energy with minimal change of current. The Pierce type of accelerator achieves control of the beam space charge by a suitable choice of axial gradient in the accelerating electric field, which, through Maxwell's equations, introduces a radial component of field that just balances the repulsive force due to the beam's space charge. This balance of forces is achieved if the beam current is proportional to the three-halves power of the accelerating voltage:

$$I = P \, V^{3/2} \tag{4}$$

where P is the perveance of the beam. This is the familiar Child-Langmuir formulation for space-charge limited flow. It is clear from this expression that a reduction of voltage in a straight-forward electrostatic accelerator will necessarily be accompanied by a substantial reduction in current.

The function of controlling the beam's space charge must be separated from that of acceleration in order to achieve variable beam energy at constant current. This can be done by devising a series of focussing elements such as electrostatic quadrupoles (ESQ)[19] or curved plates employing transverse-field focussing (TFF),[20] with a potential difference applied between successive sets of focussing elements, to accelerate a beam from a modest energy, of the order of 100 keV, to the final desired energy. Both concepts have been tested and shown to work.[21,22] A highly schematic drawing of an ESQ accelerator and beamline is shown in Fig. 4.

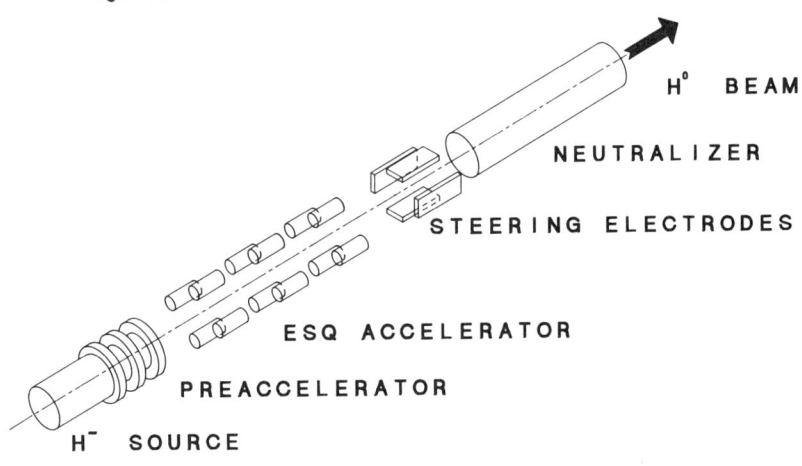

Fig. 4 Schematic diagram of an ESQ accelerator for producing a high-current H⁻ beam, and an H⁰ beam following neutralization.

This approach also promises a certain resistance to electrical breakdown, as the electrostatic fields required for focussing and acceleration can be kept modest, probably 50 kV/cm or less, and there are transverse fields everywhere to impede the motion of newly formed electrons and ions in the axial direction.

Beam intensities of up to 1 A will probably be required. Small volume-production sources of H⁻ ions have recently achieved high current densities, in excess of 250 mA/cm^2 for 5 msec pulses,[23] and it is not unreasonable to expect that, with continued development, current densities of 30-50 mA/cm^2 will be achieved in dc operation. It may be possible to make the sources so small that one could use one "sourcelet" per beamlet, with an ESQ accelerator, thus providing a relatively inexpensive approach to developing a system to meet the design requirements. A module of this type might accelerate 0.2 A of H⁻ to 1 MeV in dc operation, and deliver 0.1 A of H⁰ to the plasma as a diagnostic beam. Higher total currents would be achieved by stacking of modules. In principle, individual sourcelets could be switched on and off to provide signal coding of the beam as an aid in detection of a weak signal from the alpha particles, and individual beamlets could be steered electrostatically to move the beam to different positions within the target plasma.

An energetic H⁻ beam can be neutralized in a gas or vapor cell or jet with an efficiency of 50-60%. Higher neutralization effeciency could be achieved with a plasma (85%) or with a photo-neutralizer (which could, in principle, approach 100% neutral fraction).

A high-current ion source and accelerator for Li^- would be required to implement a diagnostic method using Li^0. Although Li^- production has been discussed,[24] we are not aware of any work in progress on a Li^- ion source and accelerator capable of the performance required for use of a Li^0 beam for an alpha-particle diagnostic. Neutralization of Li^- has, however, been measured.[25]

A high-current He^0 beam poses atomic-physics questions, as production of He^- in a neutralizer will not provide a beam of He^0 primarily in the ground state. Furthermore, He^- is generally more difficult to produce than H^- or Li^-, although high currents have been produced.[26] Alternative methods of producing high-energy ground-state He^0 should be sought. The method[8] of neutralization by spontaneous decay of metastable He^- would require a drift space of the order of 100 m. A speculative method would be acceleration of a molecular ion containing He, resulting in energetic He^0 after neutralization and dissociation; it is not known if this method would produce an appreciable yield of ground-state He atoms.

We are presently beginning to consider questions of beam penetration into the plasma for H^0, He^0, and Li^0, and penetration for He^0 (neutralized alpha particles) to escape the plasma. These considerations will include the enhancement of cross sections due to multistep collision processes.[27]

A substantial development program will be required to demonstrate that these concepts can be developed into a working alpha-particle diagnostic system.

REFERENCES

1. D. E. Post, S. J. Zweben, and L. R. Grisham, "Alpha-Particle Diagnostics," in proceeding of Morena School on Diagnostics, 1986.
2. S. J. Zweben, "Approaches to the Diagnostics of Alpha Particles in Tokamaks," Rev. Sci. Instrum. 57, 1723 (1986).
3. D. E. Post, L. R. Grisham and R. J. Fonck, "Several Atomic Physics Issues Connected with the Use of Neutral Beams in Fusion Experiments," Physica Scripta. T3, 135, 1983.
4. U. Schumacher, "Considerations for Alpha-Particle Diagnostics," Physica Scripta, 1986.
5. Alpha-Particle Effects Workshop, Courant Institute of Mathematical Sciences, New York University, December 5-6, 1985.
6. L. R. Grisham, D. E. Post, and D. R. Mikkelsen, "Multi-MeV Li^0 Beam as a Diagnostic for Fast Confined Alpha Particles," Nucl. Technol./Fusion 3, 121 (1983).

7. D. E. Post, D. R. Mikkelsen, R. A. Hulse, L. D. Stewart, and J. C. Weisheit, "Techniques for Measuring the Alpha-Particle Distribution in Magnetically Confined Plasmas," J. Fusion Energy, 1, 129 (1981).
8. M. Sasso and K. N. Sato, "Alpha-Particle Diagnostics with High-Energy Neutral Beams," Nagoya University, IPPJ-695, Sept. 1984.
9. M. Sasso, K. N. Sato, Y. Nakamura and M. Wakatani, "Active Diagnostics of Magnetically Confined Alpha Particles by Pellet Injection," Nagoya University, IPPJ-757, Jan. 1986.
10. L. R. Grisham, D. E. Post, and J. M. Dawson, Nucl. Technology/Fusion 4, 452 (1983); F. E. Cecil, S. J. Zweben, and S. S. Medley, Nucl. Instrum. Methods A245, 547 (1986); D. R. Slaughter, Rev. Sci. Instrum. 56, 1100 (1985).
11. R. J. Fonck, "Charge-Exchange Spectroscopy as a Plasma Diagnostic Tool," Rev. Sci. Instrum. 56, 885 (1985).
12. R. E. Olson, private communication.
13. M. B. Shah and H. B. Gilbody, "Electron Capture and He^+ (2s) Formation in Fast He^{2+} + H and He^+ + H Collisions," J. Phys. B 11, 121 (1978).
14. G. A. Murray, J. Stone, M. Mayo, and T. J. Morgan, "Single and Double Electron Transfer in He^{2+} + Li Collisions," Phys. Rev. A 25, 1805 (1982); R. W. McCullough, T. V. Goffe, M. B. Shah, M. Lennon, and H. B. Gilbody," Electron Capture by He^{2+} and He^+ Ions in Lithium Vapor," J. Phys. B 15, 111 (1982).
15. W. S. Cooper, "Use of the Kalman Filter in Signal Processing to Reduce Beam Requirements for Alpha-Particle Diagnostics," Rev. Sci. Instrum. 57, 1846 (1986); "Use of Optimal Estimation Theory, in Particular the Kalman Filter, in Data Analysis and Signal Processing," to be published in Rev. Sci. Instrum. (Nov. 1986).
16. K. H. Berkner, R. V. Pyle, J. W. Stearns, and J. C. Warren, "Single- and Double-Electron Capture by 7.2- to 181-keV $^3He^{++}$ Ions in He," Phys. Rev. 166, 44 (1968); M. B. Shah and H. B. Gilbody, "Formation of He^+(2s) Metastable Ions in Passage of 10-60 keV^3 He^{2+} Ions through Gases," J. Phys. B 7, 256 (1974).
17. A. S. Schlachter, "Metastable Contamination of Fast He^0 Beams Produced from He^-," Bull. Am. Phys. Soc. 25, 698 (1980).
18. O. B. Morgan, G. G. Kelley, and R. C. Davis, "Technology of Intense dc Ion Beams," Rev. Sci. Instrum. 38, 467 (1967).
19. W. R. Baker, D. A. Goldberg, C. F. Burrell, D. B. Hopkins, H. Kim, and L. J. Laslett, "A Multi-Aperture Electrostatic Quadrupole (MESQ) Accelerator for Neutral Beam Injection," Bull. Am. Phys. Soc. 27, 1142 (1982).
20. O. A. Anderson, "Transverse-Field Focusing Accelerator," in Production and Neutralization of Negative Ions and Beams, Brookhaven (1983) AIP Conf. Proc. No. 111.

21. G. M. Gamel, A. W. Maschke, and R. M. Mobley, "New High-Brightness Multiple Beam Linear Accelerator," Rev. Sci. Instrum. <u>52</u>, 971 (1981).
22. J. W. Kwan, G. D. Ackerman, O. A. Anderson, C. F. Chan, W. S. Cooper, L. Soroka, and W. F. Steele, "Transverse Field Focusing Beam Transport Experiment," these proceedings (1986); J. W. Kwan, G. D. Ackerman, O. A. Anderson, C. F. Chan, W. S. Cooper, P. Purgalis, L. Soroka, W. F. Steele, and A. Wandesforde, "Testing of H$^-$ Beam Transport in the TFF Matching/Pumping Section," Bull. Am. Phys. Soc. <u>31</u>, 1598 (1986).
23. K. W. Ehlers, K. N. Leung, R. V. Pyle, and W. B. Kunkel, "Characteristics of a Small Multicusp H$^-$ source," these proceedings; also, Bull. Am. Phys. Soc. <u>31</u>, 1506 (1986).
24. See discussions in AIP Conf. Proc. No. 111 and references in Ref. 6; J. R. Mowat, E. E. Fisch, A. S. Schlachter, J. W. Stearns, and Y. K. Bae, "Equilibrium Charge-State Fractions of Li$^-$, Li^0, and Li$^+$ in Mg, Sr, and Cs Vapors," Phys. Rev. A <u>31</u>, 2893 (1985).
25. L. R. Grisham, D. E. Post, B. M. Johnson, K. W. Jones, J. Barrette, T. H. Kruse, I. Tsemeya, and Wang Da-Hai, "Efficiences of Gas Neutralizers for Multi-MeV Beams of Light Negative Ions," Rev. Sci. Instrum. <u>53</u>, 281 (1982); A. I. Hershcovitch, B. M. Johnson, V. J. Kovarik, M. Meron, K. W. Jones, K. Prelec, and L. R. Grisham, "Neutralization of Multi-MeV Light Negative Ions by Plasma Neutralizers," Rev. Sci. Instrum. <u>55</u>, 1744 (1984).
26. E. B. Hooper, Jr., P. A. Pincosy, P. Poulsen, C. F. Burrell, L. R. Grisham, and D. E. Post, "High-Current Source of He$^-$ Ions," Rev. Sci. Instrum. <u>51</u>, 1066 (1980).
27. C. D. Boley, R. K. Janev, and D. E. Post, "Enhancement of the Neutral-Beam Stopping Cross Section in Fusion Plasmas due to Multistep Collision Processes," Phys. Rev. Lett. <u>52</u>, 534 (1984).

DISCUSSION

Allison: You didn't mention injecting the He^- into a magnetic field. Does that speed up the transition to the ground state after they've been auto ionized?

Schlachter: If they auto ionize, they're already in the ground state. He^- auto ionizes to the ground state.

Allison: Isn't that what you want to get, then?

Schlachter: Yes. The only reason I said that this isn't a trivial undertaking is that you need a drift space of 100 m. And I don't know about that beam optics, what the practicality is of having to drift 100 m - and that's one e-fold length. So I'd rather have a method that can get your He atom beam in the ground state in some more compact fashion. I'm not saying it's not a possible method.

Allison: Is that the conclusion, then, in a 10T field?

Schlachter: Maybe I'm not understanding the question. The He^- auto detaches to the ground state already. I don't know that the magnetic field effects that. I also don't know that the magnetic field effects the lifetime of He^-.

Allison: Well, the He^- will strip very rapidly in any sort of strong electric or magnetic field, and I wonder if that doesn't also speed up the transition to the ground state beyond this....Okay, the 10 µs is the natural lifetime. That's much shorter in a strong magnetic field.

Schlachter: You can't put negative ions into a strong magnetic field and have them go where they're supposed to go. You could put them into a solonoidal field, I suppose. I don't know what effect that has.

Graham: I was interested in your comment about needing ground-state He. You don't happen to know, offhand, what the binding energy of Li neutrals is? I wouldn't have thought it was all that much different from the binding energy of the excited He state, about 4 eV.

Schlachter: We have not calculated the penetration of a Li atom beam into a hot plasma. But your concern is well placed because, in fact, the first ionization potential of Li is considerably smaller than that of H or of He. It's a real concern, and we're working on beam penetration.

Peterson: Suppose you use a double capture from He - assume you could get ground-state He - and you would inject them at about the same energy that they have in the source. How would you distinguish the neutrals emerging from the beam in the same direction from what you shot in? I don't see that.

Schlachter: There's a couple of things you can do to tell what's going on. One is, you can inject He^3, instead of He^4, so that it looks different than alpha particles; but what comes out, the ones that you want coming out, are the neutralized alpha particles. So maybe that's the answer. You just use He^3. Also, you've got a lower energy accelerator that way.

CONCLUDING REMARKS

K.W. Ehlers

Well, we have done it again. We have spent a full week of intense, comprehensive, and detailed discussions of negative hydrogen ion generation, neutralization, and acceleration. As is usual for this Symposium, the group assembled here consists of the majority of the key workers in this field from the entire world and, perhaps because of this, the time sure seems to fly by.

I rather enjoyed Stan Staten's review, earlier today, of the fusion H^- program, so I decided that I would use this time to briefly review the H^- ion effort as I have seen it develop over the past 35 years. I'll do my best to keep it brief, as I know that most of you have flights to catch and places to be; Stan, for example, hopes to get home in time to participate in "trick-or-treat". In that tonight is Halloween, I'd best watch what I say, as I'd hate to be the one to "spook" this whole program.

Prior to about 1950, there was essentially no interest, in the physics community, in H^- ions or beams. The existence of negative ions was not in doubt; in fact, I first became acquainted with negative ions during the early part of World War II, when all our small cathode ray tubes in the early radars and oscilloscopes developed black, or dead spots, in regions where the electron beam traced repeatedly. It was soon recognized that the phosphor was being damaged by negative ions which were being accelerated along with the electrons to impact on the screen. These negative ions, which were being generated back at the cathode by surface production, were foiled by the development of the "bent" gun and a small magnet which separated the contaminant negative ions from the electron beams. Now today, our problem is the reverse, as we must use similar means to separate our wanted H^- ion beam from the contaminant electrons. I guess time changes everything.

Things changed in the early 1950's when L.W. Alvarez proposed, and demonstrated, a new system of ion acceleration. Negative hydrogen ions, extracted from an ion source at ground potential, were accelerated to the terminal of a positive high voltage supply. By passing through a very thin foil mounted in the terminal, two electrons were stripped from the incoming H^- ions, and the resulting H^+ ions could then be reaccelerated back to ground potential, giving a beam with twice the energy of the simple high voltage power supply. It didn't take long before those of us at Berkeley named this device the "Swindletron", as, indeed, it did seem that we were getting something for nothing. This effect was later developed by Rose and others, to provide a series of multi-MeV tandem accelerators.

The next use for H^- ions, as I recall, surfaced with the introduction of the Isochronous Cyclotron. In this device, unlike the original Lawrence Cyclotron, B increases with radius, and although one can obtain more beam energy in a given size pole tip, the beam separation per turn can be so close that one cannot insert the usual "septum" needed to extract the beam from the cyclotron.

A rather neat solution to this difficult problem of beam extraction was provided in 1962 by Rickey & Smythe when they tuned the University of Colorado isochronous cyclotron to accelerate H^-. By placing a thin foil at the outer orbit, the H^- was stripped to H^+ which, of course, was immediately self-extracted. An additional wrinkle was also recognized; namely, by stripping only to H^0, the beam would also exit the machine, and if one judiciously placed the foil at various radial positions, the cyclotron became a variable energy machine.

It was these developments that created my interest in H^- ions, as only ridiculously small currents were available. At UCLA, Wright and Richardson were actively heading a design effort for a proposed variable - energy, negative ion, meson factory type cyclotron. In about 1964, they sent me $20,000 to develop an H^- ion source for this machine, which would generate a usable beam of a few milliamperes CW of H^- ions. Even in those days, $20,000 didn't go very far, but it did allow a few "seat of the pants" ideas to be tried. The modifications worked exceedingly well, and beams as high as 5 mA dc were routinely obtained with reasonable source lifetime. The main modification was to relieve the arc column and, as such, we accidentally built in a "magnetic filter" which we all now recognize as essential to volume generation of H^- ions. The original logic for making this fortunate modification was wrong, but it took about 25 years for us to realize this fact.

Well, the AEC didn't fund the UCLA machine, so the design, along with Reg Richardson, went to Vancouver, BC. The machine was built and is called TRIUMF.

The Cyclotron Corporation in Berkeley, CA built the H^- source for TRIUMF and, in doing so, they became sufficiently interested to convert their line of cyclotrons, used for the commercial production of medical radioactive isotopes, to H^- ion acceleration.

In about 1963, in Russia, Dimov and Budker proposed a very interesting idea for the use of H^- ions to inject into large ringtype accelerators. This system, which is now used at Fermilab, starts with energetic H^- ions which are stripped to H^+ ions, so as to allow the particles to side-step their way onto the equilibrium orbit of the ring accelerator. This basic idea is sufficiently powerful that one can easily feel that all the next generation of large accelerators will demand H^- ion beams.

(Post Summary Remarks note: Chuck Schmidt of Fermilab has sent me a copy of Luis Alvarez's original paper in RSI $\underline{22}$, 705, 1951, which, indeed, discloses that he proposed the use of H^- ions for charge-exchange injection into a synchrotron, in 1951.)

Aside from the rather limited activity that I've discussed, there was very little other activity for the next decade, and publications involving H^- production were few and far between.

I recall a rather large ion source symposium held right in Brookhaven in 1971 and, though I could be in error, I don't recall a single person on H^- ions or beams. However, in 1974 we held an International Symposium on Ion Sources and Beams in Berkeley and perhaps as much as 20% of this meeting was devoted to H^- ions. This increase in activity was due to interest in the use of H^- ions to

generate high-energy neutral beams for fusion plasma heating. The
initial papers from Novosibirsk were presented at this conference,
and it was apparent that new workers in Europe, Russia, and the U.S.
were entering this field.

This increase in activity became obvious three years later
when, in 1977, the first meeting of this Symposium was held here at
Brookhaven. This Symposium, organized by BNL was the first devoted
entirely to H^- ions and beams. The Chairman of this meeting, Krsto
Prelec, had decided that the rather small list of attendees would be
limited solely to those who participated by presenting a paper. I
remember this meeting very well, as I was invited (for historical
reasons, perhaps), but I was not working in the field, so I appeared
with no paper to present. To maintain the purity of the Symposium
aims, Krsto decided to add a Conference Summary, and it would be my
duty to present it.

In the Symposia that followed, in 1980 when Theo Sluyters was
Chairman, and in 1983 when Krsto returned to the helm, this spot on
the program was preserved. Even though I was now in the program and
had papers to present, the summary remained my job - simple force of
habit, I guess. Now this year, 1986, it was a great opportunity for
the Chairman, Jim Alessi, to get some new blood. I could suggest
Joan Collins or perhaps Whoopi Goldberg but, in that I was to be
around, the job remains mine.

So, this brief summary brings us up to date: the conclusion of
our fourth meeting here at Brookhaven. As some of you know, at the
conclusion of our last meeting, I was very fearful that this fourth
meeting might not be held, due to lack of sponsorship. But Brookhaven came through, carried the ball again, and here we are. I was
absolutely elated to hear Dr. Lowenstein, in his opening remarks,
assure us that a fifth meeting will be held in 1989, here at Brookhaven. Indeed, this meeting would just not be this meeting if it
were not held at Brookhaven. You know, I sure wish that someone
with authority would present a large plaque to Brookhaven National
Lab for the effort and foresight in hosting this Symposium series.
In turn, they introduced a policy which has allowed full publication
of all presentations, even including questions and answers. The net
is that the Proceedings of these four Symposia contains the complete
record of the developmental effort (both theoretical and experimental) of H^- ions and beams, world-wide. I'm very certain that you
who publish in this field are fully aware of the numerous times you
quote these proceedings. I must say, in this regard, BNL could have
done us all a favor by selecting a shorter name. I shudder each
time I start the reference section of a paper, as it takes so long
to repeatedly write or type: "The Third International Symposium on
the Production and Neutralization of Negative Ions and Beams, da, da
da". But it's nice to have most of what you need in a few volumes,
and that's the price we pay! I think that's the main reason I was
so pleased with Dr. Lowenstein's assurance of a fifth of this
series, as it means that this record will be continued.

I feel very fortunate to have been a part of all of this - up
till now, at least. I doubt I'll make the meeting three years from
now. That's a long time away and, by then, I'll be pushing 70, and

I've been told that pushing 70 is all the activity one requires. But I'm only saying this to issue a warning; namely, any one of you who now plans to come to the next meeting without a paper, well, you could just inherit this job!

Now, speaking of Jim Alessi, and I was a few moments ago, Jim was our Chairman this year and he has certainly performed in the tradition of the Chairmen who preceded him. I'm sure you are all aware of the perturbation in your life-style which comes with the job - it's a lot of work. He has had good help: the girls at the desk, as well as Ron, who jumped to assist us with all our needs; and the folks who manned the roving microphones at the question-and-answer periods. Let's all show our appreciation for Jim and the crew here at Brookhaven with a rousing, standing, hand of applause.

Now, before I stop, let me offer a few words about the program's future, and I'll offer these remarks, as very likely I've either progressed or regressed to the position of an interested bystander. I think I feel better about the program's future now than I did three years ago.

I feel that the excellent results at PPPL with TFTR, heated with neutral beams, may well spur renewed activity and sponsorship of negative ion based neutral beam systems. With neutral beam systems we now have ion temperatures of approximately 20 keV, and an $n\tau$ that is approaching the Lawson Criteria, and you know there well could be simple reasons why one heating system is superior to another. To say the least, any time you do anything to a plasma, you must be careful not to add changes which are disruptive. An example of this is "plasma potential", and I've had a "seat of the pants" feeling for a long time that the eventual, successful reactor will have a plasma potential that is either the same as the walls, or perhaps a bit negative by a few volts. This is the best way to reduce wall effects, which dirty up the plasma. Needless to say, a limiter can not arc to the plasma if the plasma is negative. So, it could just be that neutral beam systems have side effects that are good rather than bad. I'm going to Baltimore next week to the APS Plasma Physics Division meeting to hear all about the new TFTR results.

Aside from TRIUMF, a large number of accelerators are now turning to H^- ion injection. We have followed these efforts at Fermilab, and now we have LANL, and KEK in Japan, and I'm sure this interest will spread if we can come up with good sources. I think there is now full agreement that volume produced H^- ions are cold, even colder than protons, and this means that ions produced via volume will certainly be in demand, and lots of work remains here to be completed.

In this meeting this week, we have had work sponsored by SDI, which means our list of sponsors has increased. So in all, things don't look all that bad.

So, now it's time to close these proceedings. I wish you all a safe trip home, or to your next destinations, and I wish you the best of luck with your H^- programs. It's been a great pleasure for me to have seen you all again.

APPENDICES

APPENDIX I: List of Participants

Major Richard Abbott
7615 Peacock Drive
Huntsville, AL 35801

Dr. James G. Alessi
Brookhaven National Laboratory
AGS Department, Bldg. 930
Upton, NY 11973

Dr. Paul Allison
Los Alamos National Laboratory
AT-2, Mail Stop H818
P.O. Box 1663
Los Alamos, NM 87545

Dr. L. Wilmer Anderson
Department of Physics
University of Wisconsin
Madison, WI 53706

Dr. Roger Azria
MRCG on Radiation Sciences
Faculte de Medecine
Universite de Sherbrooke
Sherbrooke, Quebec J1H 5N4
CANADA

Dr. Marthe Bacal
Laboratoire de Physique des
  Milieux Ionises
Ecole Polytechnique
91128 Palaiseau Cedex
FRANCE

Dr. William F. Bailey
Air Force Institute of
  Technology
Department of Engineering Physics
Building 640, Area B
Wright-Patterson, AFB, OH 45433

Capt. David P. Ball
Air Force Weapons Laboratory/AWYW
Kirtland Air Force Base
Albuquerque, NM 87117-6008

Dr. J.H.M. Bonnie
FOM-Institute for Atomic and
  Molecular Physics
Kruislaan 407
NL-1098 SJ Amsterdam
THE NETHERLANDS

Dr. Jean Bretagne
Laboratoire de Physique des
  Gaz et Plasmas
Universite de Paris-Sud.
91405 Orsay Cedex
FRANCE

Mr. John Brodowski
Brookhaven National Laboratory
AGS Department, Bldg. 911A
Upton, NY 11973

Dr. Jack Bruneteau
Laboratoire de Physique des
  Milieux Ionises
Ecole Polytechnique
91128 Palaiseau Cedex
FRANCE

Dr. Jacob Brynolf
Luthags Espl. 17A
Uppsala University
Box 534
S-75121 Uppsala
SWEDEN

Dr. Roy Champion
Department of Physics
College of William & Mary
Williamsburg, VA 23185

Dr. Wayne D. Cornelius
Los Alamos National Laboratory
AT-4, Mail Stop H821
P.O. Box 1663
Los Alamos, NM 87545

Dr. Oakley H. Crawford
Oak Ridge National Laboratory
Health & Safety Research Division
Bldg. 9201-2, Mail Stop 2
P.O. Box Y
Oak Ridge, TN 37831

Dr. W. Kelly Dagenhart
Oak Ridge National Laboratory
Bldg. 9201-2, Mail Stop 2
P.O. Box Y
Oak Ridge, TN 37831

Mr. Basil DeVito
Brookhaven National Laboratory
AGS Department, Bldg. 911B
Upton, NY 11973

Dr. R. Stephen Devoto
Lawrence Livermore National Laboratory
Mail Stop L-644
P.O. Box 5511
Livermore, CA 94550

Dr. James J. Donaghy
Physics Department
Washington & Lee University
Lexington, VA 24450

Dr. Henri J. Doucet
Laboratoire de Physique des
    Milieux Ionises
Ecole Polytechnique
91128 Palaiseau Cedex
FRANCE

Dr. Kenneth W. Ehlers
3129 Via Larga
Alamo, CA 94507

Dr. Joel W. Fink
Negion, Inc.
4023 East Avenue
Hayward, CA 94542

Dr. A. Theodore Forrester
University of California
Boelter Hall, Rm. 7731
Los Angeles, CA 90024

Dr. William G. Graham
Physics Department
University of Ulster
Coleraine BT52 1SA
NORTHERN IRELAND

Mr. Pierre Grand
Brookhaven National Laboratory
Nuclear Energy Department
Building 902C
Upton, NY 11973

Dr. Larry R. Grisham
Plasma Physics Laboratory
Princeton University
P.O. Box 451
Princeton, NJ 08544

Professor Dr. Willy Gruebler
Institut fur Mittelenergiephysik
Eidg. Techn. Hochschule
ETH-Honggerberg
CH-8093 Zurich
SWITZERLAND

Dr. Richard I. Hall
Universite Pierre et Marie Curie
4 Place Jussieu T12-E5
75252 Paris Cedex 05
FRANCE

Dr. Goran Hellblom
Libbyvagen 6
S-18365 Taby
SWEDEN

Dr. P.R. Wolfgang Henkes
Kernforschungszentrum Karlsruhe
Institut fur Kernverfahrenstechnik
Postfach 3640
D-7500 Karlsruhe 1
WEST GERMANY

Dr. Ady Hershcovitch
Brookhaven National Laboratory
AGS Department, Bldg. 911B
Upton, NY 11973

Dr. John R. Hiskes
Lawrence Livermore National Laboratory
Mail Stop L-630
P.O. Box 5511
Livermore, CA 94550

Dr. Andrew J. Holmes
UKAEA/Culham Laboratory
Applied Physics & Technology Division
Abingdon, Oxfordshire OX14-3DB
ENGLAND

Dr. Michael B. Hopkins
Physics Department
University of Ulster
Coleraine BT52 1SA
NORTHERN IRELAND

Dr. Brant M. Johnson
Brookhaven National Laboratory
Department of Applied Science
Building 815
Upton, NY 11973

Captain Rick G. Jones
Department of Engineering Physics
Air Force Institute of Technology
Building 640, Area B
Wright-Patterson AFB, OH 45433

Dr. Arnold M. Karo
Lawrence Livermore National
 Laboratory
Department of Chemistry &
 Materials Science
Mail Stop L-325
P.O. Box 808
Livermore, CA 94550

Mr. Vincent J. Kovarik
Brookhaven National Laboratory
AGS Department, Bldg. 911B
Upton, NY 11973

Dr. Ahovi Kponou
Brookhaven National Laboratory
AGS Department, Bldg. 911B
Upton, NY 11973

Dr. Herbert F. Krause
Oak Ridge National Laboratory
Mail Stop 377, Bldg. 5500
P.O. Box X
Oak Ridge, TN 37831-6377

Mr. Fred Kuehne
Grumman Aerospace
Mail Stop A01-26
Bethpage, NY 11714

Dr. Tsutomu Kuroda
Institute of Plasma Physics
Nagoya University
Nagoya 464
JAPAN

Dr. Joe Kwan
Lawrence Berkeley Laboratory
University of California
Mail Stop 16/108, Bldg. 4
Berkeley, CA 94720

Dr. Robert B. Leachman
Lockheed Missiles & Space Company
Orgn. 51-83, Bldg. 572
Sunnyvale, CA 94088

Dr. Ka-Ngo Leung
Lawrence Berkeley Laboratory
University of California
Building 4
Berkeley, CA 94720

Dr. Alan F. Lietzke
Lawrence Berkeley Laboratory
University of California
Mail Stop 16/108
Berkeley, CA 94720

Dr. George Maise
Brookhaven National Laboratory
Nuclear Energy Department
Building 902C
Upton, NY 11973

Dr. Peter Massmann
JET Joint Undertaking
Abingdon, OX14-3EA
ENGLAND

Dr. R. McAdams
UKAEA/Culham Laboratory
Abingdon, Oxfordshire OX14-3DB
ENGLAND

John B. McBride
Science Applications International
1200 Prospect Street
La Jolla, CA 92038

Dr. Malcolm W. McGeoch
AVCO Research Laboratory
 (Textron)
2385 Revere Beach Parkway
Everett, MA 02149

Captain Matthew W. McHarg
Air Force Weapons Laboratory/AWYW
Kirtland Air Force Base
Albuquerque, NM 87117-6008

Mr. R.B. McKenzie-Wilson
Brookhaven National Laboratory
Nuclear Energy Department
Building 902C
Upton, NY 11973

Dr. Mati Meron
Brookhaven National Laboratory
Department of Applied Science
Building 815
Upton, NY 11973

Dr. H. Harvey Michels
Physics Department
United Technologies Research
 Center
400 Main Street
East Hartford, CT 06108

Dr. J.B.A. Mitchell
Department of Physics and Centre
 for Chemical Physics
The University of Western Ontario
London, Ontario N6A 3K7
CANADA

Dr. Yoshiharu Mori
KEK, National Laboratory for
 High Energy Physics
Oho-machi, Tsukuba-gun
Ibaraki-ken 305
JAPAN

Dr. Kenneth G. Moses
JAYCOR
Plasma Technology Division
2908 Oregon Court, Bldg. I-7
Torrance, CA 90503

Dr. M.P.G. Nightingale
UKAEA/Culham Laboratory
Applied Physics & Technology Division
Abingdon, Oxfordshire OX14-3DB
ENGLAND

Dr. Alexander A. Panasenkov
Kurchatov Institute of Atomic Energy
Ulitza Kurchatova
123 182 Moscow
USSR

Dr. James R. Peterson
Chemical Physics Laboratory
Stanford Research Institute/
    International
Menlo Park, CA 94025

Dr. Krsto Prelec
Brookhaven National Laboratory
AGS Department, Bldg. 911B
Upton, NY 11973

Dr. Thomas G. Roberts
2815 Bentley Street
Huntsville, AL 35801

Dr. Kourosh Saadatmand
Grumman Aerospace
Bethpage, NY 11714

Dr. Alfred S. Schlachter
Lawrence Berkeley Laboratory
University of California
Mail Stop 4-230, Bldg. 5
Berkeley, CA 94720

Dr. Pierre A. Schmelzbach
Institut fur Mittelenergiephysik
Eidg. Techn. Hochschule
ETH Honggerberg
CH-8093 Zurich
SWITZERLAND

Dr. Charles Schmidt
Fermi National Accelerator
    Laboratory
Mail Stop 307
P.O. Box 500
Batavia, IL 60510

Dr. Paul W. Schmor
TRIUMF, UBC Campus
4004 Wesbrook Mall
Vancouver, BC V6T 2A3
CANADA

Dr. Milos Seidl
Department of Physics/
    Engineering Physics
Stevens Institute of Technology
Castle Point Station
Hoboken, NJ 07030

Dr. Nikolai N. Semashko
Kurchatov Institute of Atomic
    Energy
Ulitza Kurchatova
123 182 Moscow
USSR

Dr. Joseph Sherman
Los Alamos National Laboratory
AT-2, Mail Stop H818
P.O. Box 1663
Los Alamos, NM 87545

Dr. Murray Shubaly
Los Alamos National Laboratory
P.O. Box 1663
Los Alamos, NM 87545

Dr. Theo Sluyters
Brookhaven National Laboratory
AGS Department, Bldg. 911B
Upton, NY 11973

Major Bruce Smith
AFOSR/NP, Bldg. 410
Bolling Air Force Base
Washington, DC 20332-6448

Dr. H. Vernon Smith, Jr.
Los Alamos National Laboratory
AT-2, Mail Stop H818
P.O. Box 1663
Los Alamos, NM 87545

Dr. Kenneth Smith
Los Alamos National Laboratory
AT-2, Mail Stop H818
P.O. Box 1663
Los Alamos, NM 87545

Dr. Santosh K. Srivastava
Jet Propulsion Laboratory
California Institute of Technology
Mail Stop 183-601
4800 Oak Grove Drive
Pasadena, CA 91109

Mr. H. Stanley Staten
U.S. Department of Energy
Office of Fusion Energy
Mail Stop G-234, ER-531, GTN
Component Development Branch
Washington, DC 20545

Dr. J. Warren Stearns
Lawrence Berkeley Laboratory
University of California
Mail Stop 5-119, Bldg. 5
Berkeley, CA 94720

Dr. Ralph R. Stevens, Jr.
Los Alamos National Laboratory
AT-2, Mail Stop H818
P.O. Box 1663
Los Alamos, NM 87545

Dr. Larry Stewart
G.A. Technologies, Inc.
P.O. Box 85608
San Diego, CA 92138

Dr. Will L. Stirling
Oak Ridge National Laboratory
Bldg. 9201-2, Mail Stop 2
P.O. Box Y
Oak Ridge, TN 37831

Mr. Brian Strickland
U.S. Army, SDC
DASD-H-D
P.O. Box 1500
Huntsville, AL 35807

Dr. John R. Trow
JAYCOR, Plasma Technology Division
2908 Oregon Court, Bldg. I-7
Torrance, CA 90503

Dr. C.C. Tsai
Oak Ridge National Laboratory
Bldg. 9201-2, Mail Stop 2
P.O. Box Y
Oak Ridge, TN 37831

Dr. Jogindra M. Wadehra
Department of Physics
Wayne State University
Detroit, MI 48202

Dr. John H. Whealton
Oak Ridge National Laboratory
Bldg. 9201-2, Mail Stop 2
P.O. Box Y
Oak Ridge, TN 37831

Dr. Steven L. Wilson
Los Alamos National Laboratory
AT-2, Mail Stop H818
P.O. Box 1663
Los Alamos, NM 87545

Dr. Richard L. Witkover
Brookhaven National Laboratory
AGS Department, Bldg. 911B
Upton, NY 11973

Dr. Larry A. Wright
Mission Research Corporation
1720 Randolf Road, S.E.
Albuquerque, NM 87106

Dr. Rob L. York
Los Alamos National Laboratory
MP-DO, Mail Stop H823
Los Alamos, NM 87545

Dr. Dick Heling Yuan
TRIUMF, UBC Campus
4004 Wesbrook Mall
Vancouver, BC V6T 2A3
CANADA

APPENDIX 2

LIST OF AUTHORS

A

Ackerman, G. D., 507
Agagu, A., 643
Akerman, M. A., 404
Alessi, J. G., 419, 585
Allison, P. W., 181, 356, 465
Anderson, L. W., 593
Anderson, O. A., 507
Azria, R., 83

B

Baartman, R., 346
Bacal, M., 120, 246
Bailey, W. F., 16, 106
Ball, D. P., 356
Barber, G. C., 194
Becraft, W. R., 404
Bell, M. A., 404
Berry, L. A., 404
Bonnie, J. H. M., 133
Bretagne, J., 48
Bruneteau, J., 246

C

Capitelli, M., 48
Carr, W. E., 432
Champion, R. L., 536
Chan, C. F., 507
Cooper, W. S., 507, 727
Crawford, O. H., 663
Curtis, C. D., 425

D

Dagenhart, W. K., 194, 366, 404
DeBoni, T. M., 97
Derevyhankin, G. E., 395
Devoto, R. S., 717
DeVries, G. J., 356
Devynck, P., 246
Donaghy, J. J., 194, 366
Doucet, H. J., 453
Doverspike, L. D., 536
Dudnikov, V. G., 395

E

Eenshuistra, P. J., 133
Ehlers, K. W., 271, 282, 356, 739

F

Fink, J. H., 618
Firsov, P. S., 334
Forrest, M. J., 154
Fukumoto, S., 378, 605

G

Gardner, W. L., 404
Goranson, P. L., 404
Gorse, C., 48
Graham, W. G., 39, 145, 166
Green, T. S., 208, 298
Greer, J. T., 404
Grouard, J. P., 91
Gruebler, W., 573

H

Hagena, O. F., 384
Hall, R. I., 91
Haselton, H. H., 194, 404
Hauck, C. A., 231, 259
Hellblom, G., 219
Henkes, P. R. W., 384
Hershcovitch, A., 585, 682
Hillion, F., 246
Hinton, M. D., 298
Hiskes, J. R., 2, 34, 97, 231
Holmes, A. J. T., 208, 298
Hopkins, M. B., 145, 166
Hopman, H. J., 133
Horiike, H., 309
Hu, B.-L., 643

I

Ikegami, K., 378, 605
Inoue, T., 309

J

Jackson, L. T., 356
Johnson, B. M., 643
Jones, K. W., 643
Jones, R. G., 16, 106

K

Kaneko, O., 289
Karo, A. M., 97
Keller, R., 181
Kendall, K. R., 346

Kponou, A., 585
Krause, H. F., 673
Krylov, A. I., 334
Kunkel, W. B., 282
Kurashima, T., 309
Kuroda, T., 289, 694
Kuznetsov, V. V., 334
Kwan, J. W., 507

L

Law, M., 610
Lea, L. M., 298
Leung, K. N., 271, 282, 356, 631
Levy, C. D. P., 605, 610
Lewis, W. T., 356
Lietzke, A. F., 231, 259
Lopes, J. L., 432

M

Maor, D., 643
Matsuda, S., 309
McAdams, R., 298
McDonald, M., 346, 610
McGeoch, M. W., 567
McHarg, M. G., 356
Melnychuk, S. T., 432
Meron, M., 643
Michels, H. H., 555
Mitchell, J. B. A., 26, 39
Montgomery, J. A., Jr., 555
Montmagnon, J. L., 91
Mori, Y., 378, 605
Moses, K. G., 651
Mosscrop, D. R., 346

N

Newman, A. F., 298
Nightingale, M. P. S., 154, 208, 298
Niinikoski, T. O., 585

O

Ohara, Y., 309
Oka, Y., 289
Okumura, Y., 309
Olson, R. E., 631

P

Penent, F., 91
Peterson, J. R., 113
Pyle, R. V., 282

R

Raridon, R. J., 404
Rothe, K. E., 404
Ryan, P. M., 194, 366, 404

S

Sakurai, K., 289
Sanche, L., 83
Schechter, D. E., 194, 366, 404
Schlachter, A. S., 631, 727
Schlier, R. E., 567
Schmelzbach, P. A., 573
Schmidt, C. W., 425
Schmor, P. W., 346, 605, 610
Scott, D. M., 536
Seidl, M., 432
Semashko, N. N., 334
Singy, D., 573

Smith, H.V., Jr., 181
Soroka, L., 507
Srivastava, S. K., 69
Staten, H. S., 712
Stearns, J. W., 356, 631
Steele, W. F., 507
Stevens, R. R., Jr., 271
Stirling, W. L., 194, 366, 404
Strickland, B., 701
Swenson, D. R., 593

T

Takagi, A., 378, 605
Tanaka, S., 309
Tompa, G. S., 432
Trow, J. R., 651
Tsai, C. C., 194, 366
Tupa, D., 593

U

Uegaki, J., 610
Ueno, A., 605
Uramoto, J., 319

W

Wadehra, J. M., 59, 547
Wang, Y., 536
Whealton, J. H., 194, 366, 404, 482
Williams, M. D., 356

Y

York, R. L., 271
Yuan, D. H., 346

Z

Zhang, Z. W., 573

APPENDIX III: List of Sessions
FOURTH INTERNATIONAL SYMPOSIUM ON THE PRODUCTION AND NEUTRALIZATION OF NEGATIVE IONS AND BEAMS

Brookhaven National Laboratory
October 27-31, 1986

Monday, October 27

8:30   Opening Remarks
         D. Lowenstein, Chairman, AGS Department
         J. Alessi, Symposium Chairman

Session I - Fundamental Processes
Chairperson: M. Bacal

8:45   Review of Progress in the Theory of Volume Production
         J.R. Hiskes (LLNL) (30)

9:20   Electron Energy Distributions in Magnetic Multicusp Hydrogen Discharges
         Wm. F. Bailey and R.G. Jones (AFIT) (15)

9:40   The Role of H_3^+ Ions in H_2 ($v \geqslant 6$) Production in Negative Hydrogen Ion Sources
         J.B.A. Mitchell (Univ. of Western Ontario) and W.G. Graham (Univ. of Ulster) (15)

10:00   BREAK

10:30   Modeling of Electron Energy Distribution Functions and Vibrational Distributions in Volume H^- Ion Sources
         J. Bretagne (Univ. Paris), M. Capitelli and C. Gorse (Univ. Bari) (25)

11:00   Mutual Neutralization--Three Body Effects
         J.M. Wadehra (Wayne State Univ.) (15)

11:20   Present Status of Measured Dissociative Attachment Cross Sections
         S.K. Srivastava (JPL) (20)

11:45   Electron-Stimulated Desorption Negative Ions from Condensed Molecules
         R. Azria (Univ. Paris) and L. Sanche (Univ. Sherbrooke) (20)

12:10   Computer Simulations of Particle-Surface Dynamics
         A.M. Karo, J.R. Hiskes (LLNL), and T.M. DeBoni (Univ. of Texas) (25)

12:40   LUNCH

(Monday continued)

Session II - Source Diagnostics
Chairperson: A. Holmes

2:00   Diagnostic Techniques for Negative Ion Sources
        M. Bacal (Ecole Polytechnique) (30)

2:35   Exploration of a Hydrogen Discharge Using Resonant Multi-
        photon Ionization
        J.H.M. Bonnie, P.J. Eenshuistra, and H.J. Hopman
        (FOM-Amsterdam) (25)

3:05   Visible and V.U.V. Emission Measurements in a Tandem
        Multicusp Ion Source
        W.G. Graham and M.B. Hopkins (Univ. Ulster) (20)

3:30   BREAK

4:00   Spectroscopic Measurements of Atomic Species in Volume
        Sources
        M.P.S. Nightingale and A.J. Forrest (Culham) (15)

4:20   Spatially and Temporally Resolved EEDF Measurements in a
        Hydrogen Discharge
        M.B. Hopkins and W.G. Graham (Univ. Ulster) (20)

4:45   Spectroscopic Investigation of H^- and D^- Ion Source
        Plasmas
        H.V. Smith, Jr., P. Allison (LANL), and
        R. Keller (GSI-Darmstadt) (15)

5:05   Discharge Characteristics of a Plasma Generator for SITEX
        and VITEX Ion Sources
        C.C. Tsai, W.K. Dagenhart, W.L. Stirling, G.C. Barber,
        H.H. Haselton, P.M. Ryan, D.E. Schechter, J.H. Whealton
        (ORNL), and J.J. Donaghy (Washington & Lee Univ.) (20)

5:30-7:00 WINE AND CHEESE RECEPTION (courtesy of Associated
           Universities, Inc.)

Tuesday, October 28

Session III - H⁻ Sources
Chairperson: K. Ehlers

8:30   A Model for H⁻ Volume Production Ion Sources
       T.S. Green, A.J.T. Holmes, and M.P.S. Nightingale (Culham) (20)

8:55   A Numerical Model of the Mirror Electron Cyclotron Resonance (MECR) Source
       G. Hellblom (KTH-Stockholm) (20)

9:20   Analysis and Interpretation of a High Density Tandem Negative Ion Source
       J.R. Hiskes (LLNL), A.F. Lietzke and C. Hauck (LBL) (30)

9:55   BREAK

10:30   Volume Production of H⁻ Ions at Ecole Polytechnique. A Method for Extracting Volume Produced Negative Ions
       M. Bacal, J. Bruneteau, P. Devynck, and F. Hillion (Ecole Polytechnique) (30)

11:05   Extraction of Negative Ion in Reflex Type Sheet Plasma Negative Ion Source
       T. Kuroda, K. Sakurai, Y. Oka, and O. Kaneko (Nagoya Univ.) (25)

11:35   Production and Formation of Intense H⁻ Beams
       R. McAdams, A.J.T. Holmes, M.P.S. Nightingale, L.M. Lea, M.D. Hinton, A.F. Newman, and T.S. Green (Culham) (25)

12:05   Production of Negative Ions by Double Charge Exchange and Their Acceleration
       N.N. Semashko, V.V. Kuznetsov, A.I. Krylov, and P.S. Firsov (Kurchatov) (20)

12:30   LUNCH

(Tuesday continued)

Session IV - Posters
Chairperson: A. Hershcovitch

1:30 -
3:30   POSTER SESSION

1. Electron-Ion Collision Processes Relevant to H^- Ion Sources
   J.B.A. Mitchell (Univ. Western Ontario)

2. Electron Excitation of $H_2(v")$ Levels to Yield Vibrationally Excited H_2 Molecules H Molecules
   J.R. Hiskes (LLNL)

3. Energy and Angular Distribution of Electrons Detached in H^-- He Collisions (50 eV - 2 keV)
   F. Penent, J.P. Grouard, R.I. Hall, and J.L. Montmagnon (Univ. Pierre et Marie Curie)

4. Diffusion and Free Flow Through a Magnetic Filter
   R.G. Jones and W.F. Bailey (AFIT)

5. Comments on H^- Volume Production in Cs-Seeded Ion Sources
   J.R. Peterson (SRI)

6. H^- Ion Source Scaling Studies at LBL
   A.F. Lietzke and C.A. Hauck (LBL)

7. Operation of a Magnetically Filtered Multicusp Volume Source
   R.R. Stevens, Jr., R.L. York (LANL), K.N. Leung, and K.W. Ehlers (LBL)

8. Characteristics of a Small Multicusp H^- Source
   K.W. Ehlers, K.N. Leung, R.V. Pyle, and W.B. Kunkel (LBL)

9. A High Current Volume H^- Ion Source with Multi-Aperture Extractor
   Y. Okumura, H. Horiike, T. Inoue, T. Kurashima, S. Matsuda, Y. Ohara, and S. Tanaka (JAERI)
   (Presented by K.N. Leung, LBL)

10. Volume Produced H^-, D^- Ion Source for Proton Accelerator and Thermo-Nuclear Fusion Research by Sheet Plasma
    J. Uramoto (Nagoya)
    (Presented by A. Hershcovitch, BNL)

(Tuesday continued)

11. Surface-Plasma Source of H⁻ Ions
    G.E. Derevyankin and V.G. Dudnikov (Novosibirsk)
    (Presented by K. Prelec, BNL)

12. High Brightness Ion Source: The Penning Ringatron
    J.H. Whealton, W.L. Stirling, P.M. Ryan, M.A. Akerman, W.R. Becraft, W.L. Gardner, H.H. Haselton, K.E. Rothe, M.A. Bell, R.J. Raridon, D.E. Schechter, L.A. Berry, P.L. Goranson, J.T. Greer, and W.K. Dagenhart (ORNL)

13. A Circular Aperture Magnetron for Injection into an RFQ
    J.G. Alessi (BNL)

14. Operation of the Fermilab H⁻ Magnetron Source
    C.W. Schmidt and C.D. Curtis (FNAL)

15. Polarized H⁻ Source Development at BNL
    A. Kponou, J. Alessi, A. Hershcovitch (BNL), and T. Niinikoski (CERN)

16. Status of the TRIUMF Optically Pumped Polarized H⁻ Source
    M. Law, C.D.P. Levy, M. McDonald, P.W. Schmor, and J. Uegaki (TRIUMF)

17. Transverse-Field Focussing Beam Transport Experiment
    J.W. Kwan, G.D. Ackerman, O.A. Anderson, C.F. Chan, W.S. Cooper, L. Soroka, and W.F. Steele (LBL)

Session V - H⁻ Sources
Chairperson: W. Stirling

3:40 Extraction of H⁻ Ions from a DC Cusp Source
D.H. Yuan, R. Baartman, K.R. Kendall, M. McDonald, D.R. Mosscrop, and P.W. Schmor (TRIUMF) (15)

4:00 Operation of a Dudnikov Type Penning Source with LaB_6 Cathodes
K.N. Leung, G.J. DeVries, K.W. Ehlers, L.T. Jackson, J.W. Stearns, M.D. Williams (LBL), M.G. McHarg, D.P. Ball, W.T. Lewis (AFWL), and P.W. Allison (LANL) (25)

4:30 Accelerated Beam Experiments with the ORNL SITEX and VITEX H⁻/D⁻ Sources
W.K. Dagenhart, C.C. Tsai, W.L. Stirling, P.M. Ryan, D.E. Schechter, J.H. Whealton, (ORNL), and J.J. Donaghy (Washington & Lee Univ.) (25)

5:00 The Cusp H⁻ Ion Source at KEK
Y. Mori, A. Takagi, K. Ikegami, S. Fukumoto (KEK) (20)

5:25 A Surface Conversion Source of H⁻ with Hot Walls and Variable Converter Temperature
O.F. Hagena and P.R.W. Henkes (Karlsruhe) (20)

8:30 CONCERT - The Concord String Quartet
Berkner Hall

Wednesday, October 29

Session VI - H⁻ Production and Beam Formation
Chairperson: K. Prelec

8:30   Surface Production of Negative Hydrogen Ions by Hydrogen and Cesium Ion Bombardment
M. Seidl, W.E. Carr, J.L. Lopes, S.T. Melnychuk, and G.S. Tompa (Stevens Institute) (25)

9:00   DISCUSSION: Relative Importance of Reflection and Desorption in Surface H⁻ Sources (30)

9:35   Production of Intense H⁻ Ion Beams in High Power Pulsed Diodes
H.J. Doucet (Ecole Polytechnique) (25)

10:05   BREAK

10:35   Some Comments on Emittance of H⁻ Ion Beams
P. Allison (LANL) (30)

11:10   Review of Computer Modeling of Negative Ion Beam Formation
J.H. Whealton (ORNL) (30)

11:45   PANEL SESSION: Transport of Negative Ion Beams
Moderator: P. Allison

12:45   LUNCH

3:00   Leave for Statue of Liberty CRUISE and BANQUET
(return to BNL by 10:00 p.m.)

Thursday, October 30

Session VII – Production of Other Negative Ions
Chairperson: W. Gruebler

8:30 Fundamental Processes in Low Energy Collisions of Alkali Anions and Atoms
R.L. Champion, L.D. Doverspike, D.M. Scott, and Y. Wang (College of William and Mary) (30)

9:05 Dissociative Attachment to Lithium Dimers
J.M. Wadehra (Wayne State Univ.) (20)

9:30 Negative Ion Formation in Lithium Atom Collisions
H.H. Michels and J.A. Montgomery, Jr. (United Technologies) (20)

9:55 Density Scaling of an Optically Pumped Lithium Negative Ion Source
M.W. McGeoch and R.E. Schlier (AVCO) (20)

10:20 BREAK

10:45 Production of a Cooled, Polarized Atomic Hydrogen Beam
P.A. Schmelzbach, W. Gruebler, D. Singy, and Z.W. Zhang (ETH-Zurich) (25)

11:15 Optical Pumping of Na for Use in Polarized H^- Ion Sources
L.W. Anderson, D.R. Swenson, and D. Tupa (Univ. of Wisconsin) (25)

11:45 Recent Progress on the Optically Pumped Polarized H^- Ion Source at KEK
Y. Mori, A. Takagi, K. Ikegami, S. Fukumoto (KEK), A. Ueno (Kyushu Univ.), C.D.P. Levy and P.W. Schmor (TRIUMF) (20)

12:10 LUNCH

(Thursday continued)

Session VIII - Neutralization
Chairperson: T. Roberts

2:00   Neutralizer Options for High Energy H⁻ Beams
         J.H. Fink (Negion) (30)

2:35   A High-Charge-State Plasma Neutralizer for an Energetic H⁻ Beam
         A.S. Schlachter, K.N. Leung, J.W. Stearns (LBL), and R.E. Olson (Univ. Missouri) (20)

3:00   Measurement of Plasma Production and Neutralization in Gas Neutralizers
         D. Maor, M. Meron, B.M. Johnson, K.W. Jones, A. Agagu, and B. -L. Hu (BNL) (20)

3:25   BREAK

3:50   Characteristics of an RF Plasma Neutralizer
         J.R. Trow and K.G. Moses (JAYCOR) (20)

4:15   Oscillations in Photodetachment Cross Sections for Ions in Magnetic Fields
         O.H. Crawford (ORNL) (20)

4:40   A Novel High-Resolution Experimental Approach for Studying Laser Photodetachment in a Magnetic Field
         H.F. Krause (ORNL) (20)

5:05   Determination of Neutral Beam Direction from Radiation Emitted by Photodetached H⁻
         A. Hershcovitch (BNL) (20)

Friday, October 31

Session IX - Systems and Applications
Chairperson: L. Stewart

8:30   Japanese Negative Ion Beam Program
         T. Kuroda (Nagoya Univ.) (25)

9:00   Strategic Defense Initiative Organization Neutral Particle Beam Program Overview
         B. Strickland (30)

9:30   Magnetic Fusion Interest in Negative Ions
         H.S. Staten and D.H. Crandall (DOE) (20)

9:55   BREAK

10:25   Neutral-Beam Current Drive in Tokamaks
         R.S. Devoto (LLNL) (25)

10:55   Proposed Neutral-Beam Diagnostics for Fast Confined Alpha Particles in a Burning Plasma
         A.S. Schlachter and W.S. Cooper (LBL) (25)

11:25   Concluding Remarks
         K.W. Ehlers (LBL)

AIP Conference Proceedings

		L.C. Number	ISBN
No. 1	Feedback and Dynamic Control of Plasmas – 1970	70-141596	0-88318-100-2
No. 2	Particles and Fields – 1971 (Rochester)	71-184662	0-88318-101-0
No. 3	Thermal Expansion – 1971 (Corning)	72-76970	0-88318-102-9
No. 4	Superconductivity in d- and f-Band Metals (Rochester, 1971)	74-18879	0-88318-103-7
No. 5	Magnetism and Magnetic Materials – 1971 (2 parts) (Chicago)	59-2468	0-88318-104-5
No. 6	Particle Physics (Irvine, 1971)	72-81239	0-88318-105-3
No. 7	Exploring the History of Nuclear Physics – 1972	72-81883	0-88318-106-1
No. 8	Experimental Meson Spectroscopy –1972	72-88226	0-88318-107-X
No. 9	Cyclotrons – 1972 (Vancouver)	72-92798	0-88318-108-8
No. 10	Magnetism and Magnetic Materials – 1972	72-623469	0-88318-109-6
No. 11	Transport Phenomena – 1973 (Brown University Conference)	73-80682	0-88318-110-X
No. 12	Experiments on High Energy Particle Collisions – 1973 (Vanderbilt Conference)	73-81705	0-88318-111-8
No. 13	π-π Scattering – 1973 (Tallahassee Conference)	73-81704	0-88318-112-6
No. 14	Particles and Fields – 1973 (APS/DPF Berkeley)	73-91923	0-88318-113-4
No. 15	High Energy Collisions – 1973 (Stony Brook)	73-92324	0-88318-114-2
No. 16	Causality and Physical Theories (Wayne State University, 1973)	73-93420	0-88318-115-0
No. 17	Thermal Expansion – 1973 (Lake of the Ozarks)	73-94415	0-88318-116-9
No. 18	Magnetism and Magnetic Materials – 1973 (2 parts) (Boston)	59-2468	0-88318-117-7
No. 19	Physics and the Energy Problem – 1974 (APS Chicago)	73-94416	0-88318-118-5
No. 20	Tetrahedrally Bonded Amorphous Semiconductors (Yorktown Heights, 1974)	74-80145	0-88318-119-3
No. 21	Experimental Meson Spectroscopy – 1974 (Boston)	74-82628	0-88318-120-7
No. 22	Neutrinos – 1974 (Philadelphia)	74-82413	0-88318-121-5
No. 23	Particles and Fields – 1974 (APS/DPF Williamsburg)	74-27575	0-88318-122-3
No. 24	Magnetism and Magnetic Materials – 1974 (20th Annual Conference, San Francisco)	75-2647	0-88318-123-1
No. 25	Efficient Use of Energy (The APS Studies on the Technical Aspects of the More Efficient Use of Energy)	75-18227	0-88318-124-X

No. 26	High-Energy Physics and Nuclear Structure – 1975 (Santa Fe and Los Alamos)	75-26411	0-88318-125-8
No. 27	Topics in Statistical Mechanics and Biophysics: A Memorial to Julius L. Jackson (Wayne State University, 1975)	75-36309	0-88318-126-6
No. 28	Physics and Our World: A Symposium in Honor of Victor F. Weisskopf (M.I.T., 1974)	76-7207	0-88318-127-4
No. 29	Magnetism and Magnetic Materials – 1975 (21st Annual Conference, Philadelphia)	76-10931	0-88318-128-2
No. 30	Particle Searches and Discoveries – 1976 (Vanderbilt Conference)	76-19949	0-88318-129-0
No. 31	Structure and Excitations of Amorphous Solids (Williamsburg, VA, 1976)	76-22279	0-88318-130-4
No. 32	Materials Technology – 1976 (APS New York Meeting)	76-27967	0-88318-131-2
No. 33	Meson-Nuclear Physics – 1976 (Carnegie-Mellon Conference)	76-26811	0-88318-132-0
No. 34	Magnetism and Magnetic Materials – 1976 (Joint MMM-Intermag Conference, Pittsburgh)	76-47106	0-88318-133-9
No. 35	High Energy Physics with Polarized Beams and Targets (Argonne, 1976)	76-50181	0-88318-134-7
No. 36	Momentum Wave Functions – 1976 (Indiana University)	77-82145	0-88318-135-5
No. 37	Weak Interaction Physics – 1977 (Indiana University)	77-83344	0-88318-136-3
No. 38	Workshop on New Directions in Mossbauer Spectroscopy (Argonne, 1977)	77-90635	0-88318-137-1
No. 39	Physics Careers, Employment and Education (Penn State, 1977)	77-94053	0-88318-138-X
No. 40	Electrical Transport and Optical Properties of Inhomogeneous Media (Ohio State University, 1977)	78-54319	0-88318-139-8
No. 41	Nucleon-Nucleon Interactions – 1977 (Vancouver)	78-54249	0-88318-140-1
No. 42	Higher Energy Polarized Proton Beams (Ann Arbor, 1977)	78-55682	0-88318-141-X
No. 43	Particles and Fields – 1977 (APS/DPF, Argonne)	78-55683	0-88318-142-8
No. 44	Future Trends in Superconductive Electronics (Charlottesville, 1978)	77-9240	0-88318-143-6
No. 45	New Results in High Energy Physics – 1978 (Vanderbilt Conference)	78-67196	0-88318-144-4
No. 46	Topics in Nonlinear Dynamics (La Jolla Institute)	78-57870	0-88318-145-2
No. 47	Clustering Aspects of Nuclear Structure and Nuclear Reactions (Winnepeg, 1978)	78-64942	0-88318-146-0
No. 48	Current Trends in the Theory of Fields (Tallahassee, 1978)	78-72948	0-88318-147-9

No. 49	Cosmic Rays and Particle Physics – 1978 (Bartol Conference)	79-50489	0-88318-148-7
No. 50	Laser-Solid Interactions and Laser Processing – 1978 (Boston)	79-51564	0-88318-149-5
No. 51	High Energy Physics with Polarized Beams and Polarized Targets (Argonne, 1978)	79-64565	0-88318-150-9
No. 52	Long-Distance Neutrino Detection – 1978 (C.L. Cowan Memorial Symposium)	79-52078	0-88318-151-7
No. 53	Modulated Structures – 1979 (Kailua Kona, Hawaii)	79-53846	0-88318-152-5
No. 54	Meson-Nuclear Physics – 1979 (Houston)	79-53978	0-88318-153-3
No. 55	Quantum Chromodynamics (La Jolla, 1978)	79-54969	0-88318-154-1
No. 56	Particle Acceleration Mechanisms in Astrophysics (La Jolla, 1979)	79-55844	0-88318-155-X
No. 57	Nonlinear Dynamics and the Beam-Beam Interaction (Brookhaven, 1979)	79-57341	0-88318-156-8
No. 58	Inhomogeneous Superconductors – 1979 (Berkeley Springs, W.V.)	79-57620	0-88318-157-6
No. 59	Particles and Fields – 1979 (APS/DPF Montreal)	80-66631	0-88318-158-4
No. 60	History of the ZGS (Argonne, 1979)	80-67694	0-88318-159-2
No. 61	Aspects of the Kinetics and Dynamics of Surface Reactions (La Jolla Institute, 1979)	80-68004	0-88318-160-6
No. 62	High Energy e^+e^- Interactions (Vanderbilt, 1980)	80-53377	0-88318-161-4
No. 63	Supernovae Spectra (La Jolla, 1980)	80-70019	0-88318-162-2
No. 64	Laboratory EXAFS Facilities – 1980 (Univ. of Washington)	80-70579	0-88318-163-0
No. 65	Optics in Four Dimensions – 1980 (ICO, Ensenada)	80-70771	0-88318-164-9
No. 66	Physics in the Automotive Industry – 1980 (APS/AAPT Topical Conference)	80-70987	0-88318-165-7
No. 67	Experimental Meson Spectroscopy – 1980 (Sixth International Conference, Brookhaven)	80-71123	0-88318-166-5
No. 68	High Energy Physics – 1980 (XX International Conference, Madison)	81-65032	0-88318-167-3
No. 69	Polarization Phenomena in Nuclear Physics – 1980 (Fifth International Symposium, Santa Fe)	81-65107	0-88318-168-1
No. 70	Chemistry and Physics of Coal Utilization – 1980 (APS, Morgantown)	81-65106	0-88318-169-X
No. 71	Group Theory and its Applications in Physics – 1980 (Latin American School of Physics, Mexico City)	81-66132	0-88318-170-3
No. 72	Weak Interactions as a Probe of Unification (Virginia Polytechnic Institute – 1980)	81-67184	0-88318-171-1
No. 73	Tetrahedrally Bonded Amorphous Semiconductors (Carefree, Arizona, 1981)	81-67419	0-88318-172-X

No. 74	Perturbative Quantum Chromodynamics (Tallahassee, 1981)	81-70372	0-88318-173-8
No. 75	Low Energy X-Ray Diagnostics – 1981 (Monterey)	81-69841	0-88318-174-6
No. 76	Nonlinear Properties of Internal Waves (La Jolla Institute, 1981)	81-71062	0-88318-175-4
No. 77	Gamma Ray Transients and Related Astrophysical Phenomena (La Jolla Institute, 1981)	81-71543	0-88318-176-2
No. 78	Shock Waves in Condensed Matter – 1981 (Menlo Park)	82-70014	0-88318-177-0
No. 79	Pion Production and Absorption in Nuclei – 1981 (Indiana University Cyclotron Facility)	82-70678	0-88318-178-9
No. 80	Polarized Proton Ion Sources (Ann Arbor, 1981)	82-71025	0-88318-179-7
No. 81	Particles and Fields –1981: Testing the Standard Model (APS/DPF, Santa Cruz)	82-71156	0-88318-180-0
No. 82	Interpretation of Climate and Photochemical Models, Ozone and Temperature Measurements (La Jolla Institute, 1981)	82-71345	0-88318-181-9
No. 83	The Galactic Center (Cal. Inst. of Tech., 1982)	82-71635	0-88318-182-7
No. 84	Physics in the Steel Industry (APS/AISI, Lehigh University, 1981)	82-72033	0-88318-183-5
No. 85	Proton-Antiproton Collider Physics –1981 (Madison, Wisconsin)	82-72141	0-88318-184-3
No. 86	Momentum Wave Functions – 1982 (Adelaide, Australia)	82-72375	0-88318-185-1
No. 87	Physics of High Energy Particle Accelerators (Fermilab Summer School, 1981)	82-72421	0-88318-186-X
No. 88	Mathematical Methods in Hydrodynamics and Integrability in Dynamical Systems (La Jolla Institute, 1981)	82-72462	0-88318-187-8
No. 89	Neutron Scattering – 1981 (Argonne National Laboratory)	82-73094	0-88318-188-6
No. 90	Laser Techniques for Extreme Ultraviolt Spectroscopy (Boulder, 1982)	82-73205	0-88318-189-4
No. 91	Laser Acceleration of Particles (Los Alamos, 1982)	82-73361	0-88318-190-8
No. 92	The State of Particle Accelerators and High Energy Physics (Fermilab, 1981)	82-73861	0-88318-191-6
No. 93	Novel Results in Particle Physics (Vanderbilt, 1982)	82-73954	0-88318-192-4
No. 94	X-Ray and Atomic Inner-Shell Physics – 1982 (International Conference, U. of Oregon)	82-74075	0-88318-193-2
No. 95	High Energy Spin Physics – 1982 (Brookhaven National Laboratory)	83-70154	0-88318-194-0
No. 96	Science Underground (Los Alamos, 1982)	83-70377	0-88318-195-9

No. 97	The Interaction Between Medium Energy Nucleons in Nuclei – 1982 (Indiana University)	83-70649	0-88318-196-7
No. 98	Particles and Fields – 1982 (APS/DPF University of Maryland)	83-70807	0-88318-197-5
No. 99	Neutrino Mass and Gauge Structure of Weak Interactions (Telemark, 1982)	83-71072	0-88318-198-3
No. 100	Excimer Lasers – 1983 (OSA, Lake Tahoe, Nevada)	83-71437	0-88318-199-1
No. 101	Positron-Electron Pairs in Astrophysics (Goddard Space Flight Center, 1983)	83-71926	0-88318-200-9
No. 102	Intense Medium Energy Sources of Strangeness (UC-Sant Cruz, 1983)	83-72261	0-88318-201-7
No. 103	Quantum Fluids and Solids – 1983 (Sanibel Island, Florida)	83-72440	0-88318-202-5
No. 104	Physics, Technology and the Nuclear Arms Race (APS Baltimore –1983)	83-72533	0-88318-203-3
No. 105	Physics of High Energy Particle Accelerators (SLAC Summer School, 1982)	83-72986	0-88318-304-8
No. 106	Predictability of Fluid Motions (La Jolla Institute, 1983)	83-73641	0-88318-305-6
No. 107	Physics and Chemistry of Porous Media (Schlumberger-Doll Research, 1983)	83-73640	0-88318-306-4
No. 108	The Time Projection Chamber (TRIUMF, Vancouver, 1983)	83-83445	0-88318-307-2
No. 109	Random Walks and Their Applications in the Physical and Biological Sciences (NBS/La Jolla Institute, 1982)	84-70208	0-88318-308-0
No. 110	Hadron Substructure in Nuclear Physics (Indiana University, 1983)	84-70165	0-88318-309-9
No. 111	Production and Neutralization of Negative Ions and Beams (3rd Int'l Symposium, Brookhaven, 1983)	84-70379	0-88318-310-2
No. 112	Particles and Fields – 1983 (APS/DPF, Blacksburg, VA)	84-70378	0-88318-311-0
No. 113	Experimental Meson Spectroscopy – 1983 (Seventh International Conference, Brookhaven)	84-70910	0-88318-312-9
No. 114	Low Energy Tests of Conservation Laws in Particle Physics (Blacksburg, VA, 1983)	84-71157	0-88318-313-7
No. 115	High Energy Transients in Astrophysics (Santa Cruz, CA, 1983)	84-71205	0-88318-314-5
No. 116	Problems in Unification and Supergravity (La Jolla Institute, 1983)	84-71246	0-88318-315-3
No. 117	Polarized Proton Ion Sources (TRIUMF, Vancouver, 1983)	84-71235	0-88318-316-1

No. 118	Free Electron Generation of Extreme Ultraviolet Coherent Radiation (Brookhaven/OSA, 1983)	84-71539	0-88318-317-X
No. 119	Laser Techniques in the Extreme Ultraviolet (OSA, Boulder, Colorado, 1984)	84-72128	0-88318-318-8
No. 120	Optical Effects in Amorphous Semiconductors (Snowbird, Utah, 1984)	84-72419	0-88318-319-6
No. 121	High Energy e^+e^- Interactions (Vanderbilt, 1984)	84-72632	0-88318-320-X
No. 122	The Physics of VLSI (Xerox, Palo Alto, 1984)	84-72729	0-88318-321-8
No. 123	Intersections Between Particle and Nuclear Physics (Steamboat Springs, 1984)	84-72790	0-88318-322-6
No. 124	Neutron-Nucleus Collisions – A Probe of Nuclear Structure (Burr Oak State Park - 1984)	84-73216	0-88318-323-4
No. 125	Capture Gamma-Ray Spectroscopy and Related Topics – 1984 (Internat. Symposium, Knoxville)	84-73303	0-88318-324-2
No. 126	Solar Neutrinos and Neutrino Astronomy (Homestake, 1984)	84-63143	0-88318-325-0
No. 127	Physics of High Energy Particle Accelerators (BNL/SUNY Summer School, 1983)	85-70057	0-88318-326-9
No. 128	Nuclear Physics with Stored, Cooled Beams (McCormick's Creek State Park, Indiana, 1984)	85-71167	0-88318-327-7
No. 129	Radiofrequency Plasma Heating (Sixth Topical Conference, Callaway Gardens, GA, 1985)	85-48027	0-88318-328-5
No. 130	Laser Acceleration of Particles (Malibu, California, 1985)	85-48028	0-88318-329-3
No. 131	Workshop on Polarized ^{3}He Beams and Targets (Princeton, New Jersey, 1984)	85-48026	0-88318-330-7
No. 132	Hadron Spectroscopy–1985 (International Conference, Univ. of Maryland)	85-72537	0-88318-331-5
No. 133	Hadronic Probes and Nuclear Interactions (Arizona State University, 1985)	85-72638	0-88318-332-3
No. 134	The State of High Energy Physics (BNL/SUNY Summer School, 1983)	85-73170	0-88318-333-1
No. 135	Energy Sources: Conservation and Renewables (APS, Washington, DC, 1985)	85-73019	0-88318-334-X
No. 136	Atomic Theory Workshop on Relativistic and QED Effects in Heavy Atoms	85-73790	0-88318-335-8
No. 137	Polymer-Flow Interaction (La Jolla Institute, 1985)	85-73915	0-88318-336-6
No. 138	Frontiers in Electronic Materials and Processing (Houston, TX, 1985)	86-70108	0-88318-337-4
No. 139	High-Current, High-Brightness, and High-Duty Factor Ion Injectors (La Jolla Institute, 1985)	86-70245	0-88318-338-2

No. 140	Boron-Rich Solids (Albuquerque, NM, 1985)	86-70246	0-88318-339-0
No. 141	Gamma-Ray Bursts (Stanford, CA, 1984)	86-70761	0-88318-340-4
No. 142	Nuclear Structure at High Spin, Excitation, and Momentum Transfer (Indiana University, 1985)	86-70837	0-88318-341-2
No. 143	Mexican School of Particles and Fields (Oaxtepec, México, 1984)	86-81187	0-88318-342-0
No. 144	Magnetospheric Phenomena in Astrophysics (Los Alamos, 1984)	86-71149	0-88318-343-9
No. 145	Polarized Beams at SSC & Polarized Antiprotons (Ann Arbor, MI & Bodega Bay, CA, 1985)	86-71343	0-88318-344-7
No. 146	Advances in Laser Science–I (Dallas, TX, 1985)	86-71536	0-88318-345-5
No. 147	Short Wavelength Coherent Radiation: Generation and Applications (Monterey, CA, 1986)	86-71674	0-88318-346-3
No. 148	Space Colonization: Technology and The Liberal Arts (Geneva, NY, 1985)	86-71675	0-88318-347-1
No. 149	Physics and Chemistry of Protective Coatings (Universal City, CA, 1985)	86-72019	0-88318-348-X
No. 150	Intersections Between Particle and Nuclear Physics (Lake Louise, Canada, 1986)	86-72018	0-88318-349-8
No. 151	Neural Networks for Computing (Snowbird, UT, 1986)	86-72481	0-88318-351-X
No. 152	Heavy Ion Inertial Fusion (Washington, DC, 1986)	86-73185	0-88318-352-8
No. 153	Physics of Particle Accelerators (SLAC Summer School, 1985) (Fermilab Summer School, 1984)	87-70103	0-88318-353-6
No. 154	Physics and Chemistry of Porous Media—II (Ridge Field, CT, 1986)	83-73640	0-88318-354-4
No. 155	The Galactic Center: Proceedings of the Symposium Honoring C. H. Townes (Berkeley, CA, 1986)	86-73186	0-88318-355-2
No. 156	Advanced Accelerator Concepts (Madison, WI, 1986)	87-70635	0-88318-358-0
No. 157	Stability of Amorphous Silicon Alloy Materials and Devices (Palo Alto, CA, 1987)	87-70990	0-88318-359-9